GEOMORPHOLOGY AND GROUNDWATER

Geomorphology and Groundwater

Editor:

Quan Cui, Ph.D.

Delve Publishing

www.delvepublishing.com

Geomorphology and Groundwater

Editor: **Quan Cui, Ph.D.**

708 3rd Avenue, 6th Floor
New York, NY 10017
United States of America

www.delvepublishing.com

ISBN: 978-1-68095-826-3

Library of Congress Control Number: 2016951568

About the Editor

Quan Cui, Ph.D.

Quan Cui obtained her PhD from Beijing Normal University in 2015. She worked as environmental impact assessment engineer for years. Aiming at linking science to policy, her study interest spans from ecosystem modelling to water resources management. She published peer-reviewed articles about using machine learning models in inflow forecasting, CO2 management and watershed ecosystem health assessment.

Contents

List of Contributors

J.E. Chambers
British Geological Survey, Environmental Science Centre, Nottingham, NG12 5GG, UK

P.B. Wilkinson
British Geological Survey, Environmental Science Centre, Nottingham, NG12 5GG, UK

D. Wardrop
Lafarge Aggregates Ltd, Panshanger Park, Hertford, SG14 2NA, UK

A. Hameed
University of Leicester, Department of Geology, University Road, Leicester, LE1 7RH, UK

I. Hill
University of Leicester, Department of Geology, University Road, Leicester, LE1 7RH, UK

C. Jeffrey
University of Leicester, Department of Geology, University Road, Leicester, LE1 7RH, UK

M.H. Loke
Geotomo Software Sdn. Bhd., 115, Cangkat Minden Jalan 5, Minden Heights, 11700 Gelugor, Penang, Malaysia

P.I. eldrum
British Geological Survey, Environmental Science Centre, Nottingham, NG12 5GG, UK

O. Kuras
British Geological Survey, Environmental Science Centre, Nottingham, NG12 5GG, UK

M. Cave
British Geological Survey, Environmental Science Centre, Nottingham, NG12 5GG, UK

D.A. Gunn
British Geological Survey, Environmental Science Centre, Nottingham, NG12 5GG, UK

Brioch Hemmings
School of Earth Sciences, University of Bristol, Bristol, UK

Daren Gooddy
British Geological Survey, Wallingford, Oxfordshire, UK

Fiona Whitaker
School of Earth Sciences, University of Bristol, Bristol, UK

W. George Darling
British Geological Survey, Wallingford, Oxfordshire, UK

Alia Jasim
School of Earth Sciences, University of Bristol, Bristol, UK

Joachim Gottsmann
School of Earth Sciences, University of Bristol, Bristol, UK

John Lee Higson
Department of Earth and Environmental Sciences, University of St Andrews, United Kingdom

Michael Bliss Singer
Department of Earth and Environmental Sciences, University of St Andrews, United Kingdom
Earth Research Institute, University of California Santa Barbara, United States of America

S. Uhlemann
British Geological Survey, Natural Environment Research Council, Kingsley Dunham Centre, Keyworth, Nottingham NG12 5GG, United Kingdom
ETH Zurich, Institute of Geophysics, Sonneggstrasse 5, CH-8092, Zurich, Switzerland

A. Smith
Loughborough University, School of Civil and Building Engineering, Loughborough LE11 3TU, United Kingdom

J. Chambers
British Geological Survey, Natural Environment Research Council, Kingsley Dunham Centre, Keyworth, Nottingham NG12 5GG, United Kingdom

N. Dixon
Loughborough University, School of Civil and Building Engineering, Loughborough LE11 3TU, United Kingdom

T. Dijkstra
British Geological Survey, Natural Environment Research Council, Kingsley Dunham Centre, Keyworth, Nottingham NG12 5GG, United Kingdom

E. Haslam
British Geological Survey, Natural Environment Research Council, Kingsley Dunham Centre, Keyworth, Nottingham NG12 5GG, United Kingdom

P. Meldrum
British Geological Survey, Natural Environment Research Council, Kingsley Dunham Centre, Keyworth, Nottingham NG12 5GG, United Kingdom

A. Merritt
Plymouth University, School of Geography, Earth & Environmental Sciences, Drake Circus, Plymouth PL4 8AA, United Kingdom

D. Gunn
British Geological Survey, Natural Environment Research Council, Kingsley Dunham Centre, Keyworth, Nottingham NG12 5GG, United Kingdom

J. Mackay
British Geological Survey, Natural Environment Research Council, Kingsley Dunham Centre, Keyworth, Nottingham NG12 5GG, United Kingdom

Hironobu Kan
Graduate School of Education, Okayama University, Okayama, Japan
Graduate School of Natural Science and Technology, Okayama University, Okayama, Japan

Kensaku Urata
Osaka University of Economics and Law, Yao, Japan

Masayuki Nagao
Institute of Geology and Geoinformation, Advanced Industrial Science and Technology (AIST), Tsukuba, Japan

Nobuyuki Hori
Department of Geography, Nara University, Nara, Japan

Kazuhiko Fujita
Department of Geology, University of Ryukyus, Okinawa, Japan

Yusuke Yokoyama
Atmosphere and Ocean Research Institute (AORI), University of Tokyo, Kashiwa, Japan

Yosuke Nakashima
Ariake National College of Technology, Omuta, Japan

Tomoya Ohashi
Graduate School of Education, Okayama University, Okayama, Japan

Kazuhisa Goto
International Research Institute of Disaster Science (IRIDeS), Tohoku University, Sendai, Japan

Atsushi Suzuki
Institute of Geology and Geoinformation, Advanced Industrial Science and Technology (AIST), Tsukuba, Japan

David R. Bridgland
Department of Geography, Durham University, South Road, Durham DH1 3LE, UK

Rob Westaway
Faculty of Mathematics, Computing and Technology, The Open University, Abbots Hill, Gateshead NE8 3DF, UK
Newcastle Institute for Research on Sustainability (NIReS), Newcastle University, Newcastle-upon-Tyne NE1 7RU, UK

Mohammad Abou Romieh
National Earthquake Center, Rasheed Karameh Street, Al-Adawi, Damascus, Syria

Ian Candy
Department of Geography, Royal Holloway, University of London, Egham, Surrey TW20 0EX, UK

Mohamad Daoud
National Earthquake Center, Rasheed Karameh Street, Al-Adawi, Damascus, Syria

Tuncer Demir
Department of Geography, Harran University, 63300 Şanlıurfa, Turkey

Nikolaos Galiatsatos
Department of Geography, Durham University, South Road, Durham DH1 3LE, UK

Danielle C. Schreve
Department of Geography, Royal Holloway, University of London, Egham, Surrey TW20 0EX, UK

Ali Seyrek
Department of Soil Science, Harran University, 63300 Şanlıurfa, Turkey

Andrew D. Shaw
Department of Archaeology, University of Southampton, Highfield, Southampton SO17 1BJ, UK

Tom S. White
Department of Zoology, University of Cambridge, Downing Street, Cambridge CB2 3EJ, UK

John Whittaker
Department of Palaeontology, Natural History Museum, Cromwell Road, London SW7 5BD, UK

John Lewin
Institute of Geography and Earth Sciences, Aberystwyth University, Llandinam Building, Penglais, Aberystwyth, SY23 3DB, UK

Philip J. Ashworth
Division of Geography and Geology, School of Environment and Technology, University of Brighton, Lewes Road, Brighton, BN2 4GJ, UK

Lars Rosendahl Appelquist
UNEP Risoe Centre, Technical University of Denmark, UN City Copenhagen, Marmorvej 51, 2100 Copenhagen Ø, Denmark

Thomas Balstrøm
Department of Development and Planning, Aalborg University, A. C. Meyers Vænge 15, 2., 2450 Copenhagen SV, Denmark

Andres Payo
Oxford University Centre for the Environment, South Parks Road, Oxford OX1 3QY, UK

Jim W. Hall
Oxford University Centre for the Environment, South Parks Road, Oxford OX1 3QY, UK

Jon French
UCL Department of Geography, Coastal and Estuarine Research Unit, University College London, Gower St., London WC1E 6BT, UK

James Sutherland
HR Wallingford, Howbery Park, Wallingford, Oxfordshire OX10 8BA, UK

Barend van Maanen
University of Southampton, Faculty of Engineering and the Environment, Southampton SO17 1BJ, UK

Robert J. Nicholls
University of Southampton, Faculty of Engineering and the Environment, Southampton SO17 1BJ, UK

Dominic E. Reeve
Swansea University, College of Engineering, Singleton Park, SA2 8PP, UK

Edwin R.C. Baynes
School of GeoSciences, University of Edinburgh, Edinburgh EH8 9XP, UK

Mikaël Attal
School of GeoSciences, University of Edinburgh, Edinburgh EH8 9XP, UK

Andrew J. Dugmore
School of GeoSciences, University of Edinburgh, Edinburgh EH8 9XP, UK

Linda A. Kirstein
School of GeoSciences, University of Edinburgh, Edinburgh EH8 9XP, UK

Kathryn A. Whaler
School of GeoSciences, University of Edinburgh, Edinburgh EH8 9XP, UK

Jon French
Coastal and Estuarine Unit, UCL Department of Geography, University College London, Gower Street, London WC1E 6BT, UK

Andres Payo
Oxford University Centre for the Environment, South Parks Road, Oxford OX1 3QY, UK

Brad Murray
Division of Earth and Ocean Sciences, Nicholas School of the Environment, Center for Nonlinear and Complex Systems, Duke University, Box 90227, Durham, NC 27708, USA

Julian Orford
School of Geography, Archaeology and Palaeoecology, Queen's University Belfast, Belfast BT7 1NN, Northern Ireland, UK

Matt Eliot
Damara WA Pty Ltd, 5/19 Wotan St, Innaloo WA 6018, Australia

Peter Cowell
School of Geosciences, University of Sydney, NSW 2006, Australia

Dominic E. Reeve
College of Engineering, Swansea University, Wales, UK

Harshinie Karunarathna
College of Engineering, Swansea University, Wales, UK

Shunqi Pan
School of Engineering, Cardiff University, Wales, UK

Jose M. Horrillo-Caraballo
College of Engineering, Swansea University, Wales, UK

Grzegorz Różyński
Institute of Hydro-Engineering, Polish Academy of Sciences, Gdańsk, Poland

Roshanka Ranasinghe
UNESCO-IHE/Delft University of Technology, The Netherlands

Anna Malagò
European Commission, Joint Research Centre, Institute for Environment and Sustainability, Ispra, VA, Italy
University of Ferrara, Engineering Department, Ferrara, Italy

Dionissios Efstathiou
Technical University of Crete, 73100 Chania, Crete, Greece

Fayçal Bouraoui
European Commission, Joint Research Centre, Institute for Environment and Sustainability, Ispra, VA, Italy

Nikolaos P. Nikolaidis
Technical University of Crete, 73100 Chania, Crete, Greece

Marco Franchini
University of Ferrara, Engineering Department, Ferrara, Italy

Giovanni Bidoglio
European Commission, Joint Research Centre, Institute for Environment and Sustainability, Ispra, VA, Italy

Marinos Kritsotakis
Decentralized Administration of Crete, Directorate of Water, 71202 Heraklion, Greece

Peter E. Robins
Centre for Applied Marine Sciences, Bangor University, LL59 5AB, United Kingdom

Martin W. Skov
School of Ocean Sciences, Bangor University, LL59 5AB, United Kingdom

Matt J. Lewis
School of Ocean Sciences, Bangor University, LL59 5AB, United Kingdom

Luis Giménez
School of Ocean Sciences, Bangor University, LL59 5AB, United Kingdom

Shelagh K. Malham
Centre for Applied Marine Sciences, Bangor University, LL59 5AB, United Kingdom

Simon P. Neill
School of Ocean Sciences, Bangor University, LL59 5AB, United Kingdom

Timothy A. Whitton
Centre for Applied Marine Sciences, Bangor University, LL59 5AB, United Kingdom

Suzanna E. Jackson
School of Ocean Sciences, Bangor University, LL59 5AB, United Kingdom
School of Biological Sciences, Bangor University, LL57 2UW, United Kingdom

Colin F. Jago
Centre for Applied Marine Sciences, Bangor University, LL59 5AB, United Kingdom
School of Ocean Sciences, Bangor University, LL59 5AB, United Kingdom
School of Biological Sciences, Bangor University, LL57 2UW, United Kingdom

Günther Prasicek
Department of Geoinformatics – Z_GIS, University of Salzburg, Hellbrunnerstr. 34, 5020 Salzburg, Austria
Department of Geography and Geology, University of Salzburg, 5020 Salzburg, Austria

Jan-Christoph Otto
Department of Geography and Geology, University of Salzburg, 5020 Salzburg, Austria

David R. Montgomery
Department of Earth and Space Sciences, University of Washington, Seattle, WA 98195, USA

Lothar Schrott
Department of Geography, University of Bonn, 53115 Bonn, Germany

Christopher M. Darvill
Department of Geography, Durham University, South Road, Durham DH1 3LE, UK

Michael J. Bentley
Department of Geography, Durham University, South Road, Durham DH1 3LE, UK

Chris R. Stokes
Department of Geography, Durham University, South Road, Durham DH1 3LE, UK

L.E. Christianson
The Conservation Fund, 1098 Turner Road, Shepherdstown, WV 25443, USA

R.D. Harmel
Grassland Soil and Water Research Laboratory, USDA-ARS, 808 East Blackland Road, Temple, TX 76502, USA

Stephen J. Livingstone
Department of Geography, Sheffield University, Sheffield, UK

Robert D. Storrar
Department of Geography, Queen Mary University of London, UK

John K. Hillier
Department of Geography, Loughborough University, LE11 3TU, UK

Chris R. Stokes
Department of Geography, Durham University, South Road, Durham, UK

Chris D. Clark
Department of Geography, Sheffield University, Sheffield, UK

Lev Tarasov
Department of Physics and Physical Oceanography, Memorial University of Newfoundland and Labrador, Canada

Jeremy C. Ely
Department of Geography, University of Sheffield, Sheffield S10 2TN, UK

Chris D. Clark
Department of Geography, University of Sheffield, Sheffield S10 2TN, UK

Matteo Spagnolo
School of Geosciences, University of Aberdeen, Aberdeen AB24 3UF, UK

Chris R. Stokes
Department of Geography, Durham University, Durham DH1 3LE, UK

Sarah L. Greenwood
Department of Geological Sciences, Stockholm University, Stockholm SE-106 91, Sweden

Anna L
Department of Geography, Durham University, Durham DH1 3LE, UK

Hughes
Department of Earth Science, University of Bergen and Bjerknes Centre for Climate Research, Bergen N-5020, Norway

Paul Dunlop
School of Geography and Environmental Sciences, Ulster University, Coleraine BT52 1SA, UK

Dale Hess
Department of Earth and Environmental Sciences, The University of Rochester, Rochester, NY 14627, United States

Preface

Geomorphology and Groundwater is a state-of-the-art scientific field that focuses on the overlaps between geomorphology and hydrology. The study of groundwater helps to understand the local geomorphology, deposition of minerals, and even climate change records. The groundwater transportation carries the minerals, plays important role in deposition, sets up the distribution of them and affects water quality simultaneously. This lead to environmental issues and geographic phenomena. The understanding of groundwater processes contributes valuable information to a broad of geological topics. In this book, each article covers a specific geomorphology and groundwater topic, including studies on geotechnical methodologies for groundwater or beneath waters, weathering data from the angle of geomorphology, building models to understand cause and dynamic of landscapes or landforms and studies about hazard assessment and management.

Editor
Quan Cui, Ph.D.

CHAPTER 1

BEDROCK DETECTION BENEATH RIVER TERRACE DEPOSITS USING THREE-DIMENSIONAL ELECTRICAL RESISTIVITY TOMOGRAPHY

J.E. Chambers[a], P.B. Wilkinson[a], D. Wardrop[b], A. Hameed[c], I. Hill[c], C. Jeffrey[c], M.H. Loke[d], P.I. eldrum[a], O. Kuras[a], M. Cave[a], D.A. Gunn[a]

[a] British Geological Survey, Environmental Science Centre, Nottingham, NG12 5GG, UK

[b] Lafarge Aggregates Ltd, Panshanger Park, Hertford, SG14 2NA, UK

[c] University of Leicester, Department of Geology, University Road, Leicester, LE1 7RH, UK

[d] Geotomo Software Sdn. Bhd., 115, Cangkat Minden Jalan 5, Minden Heights, 11700 Gelugor, Penang, Malaysia

ABSTRACT

We describe the use of a fully volumetric geophysical imaging approach, three-dimensional electrical resistivity (3D ERT), for bedrock detection below mixed sand and gravel deposits typical of fluvial valley-fill terraces. We illustrate the method through an analysis of terrace deposits of the Great Ouse River (UK), where up to 4 m of sand and gravel have filled the valley bottom during the latest Pleistocene. We use an edge detector to identify the steepest gradient in first-derivative resistivity profiles, which yields an estimate of bedrock depth (verified by drilling) to a precision better than 0.2 m (average) and 0.4 m (standard deviation). Comparison of a range of drilling techniques at the site has revealed that borehole derived interface depths suffered from levels of uncertainty similar to those associated with the 3D ERT — indicating that the reliability of bedrock interface depths determined using these two approaches is comparable in this case. The 3D ERT method provides a high spatial resolution that enabled a previously unknown erosional bedrock structure, associated with the change from deeper first terrace to second terrace deposits, to be identified in the Great Ouse valley. The method provides a relatively quick method to quantify terrace fill volume over large sites to a greater degree of precision than currently available.

INTRODUCTION

River terrace deposits are a focus of considerable scientific, archaeological, and economic interest. Terrace architecture can provide important information regarding uplift, incision, and landscape evolution (e.g., Boreham et al., 2010 and Bridgland, 2010), with the formation of aggradational terraces in some settings correlating closely with climatic cycles (e.g., Bridgland, 2006). These deposits are a particularly rich source of archaeological artefacts preserving a record of Palaeolithic human activity (e.g., Wymer, 1988) and are also a major economic resource of groundwater (Gomme and Buss, 2006) and sand and gravel aggregates for construction (Smith and Collis, 2001).

River terrace deposits can be highly variable and difficult to characterise in terms of structure and lithology, particularly where the deposits of multiple or dissected terraces are present (Gibbard, 1982 and Peterson et al., 2011). Typical approaches to the characterisation of these deposits include geomorphological and geological mapping,

---

Citation: J.E. Chambers, P.B. Wilkinson, D. Wardrop, A. Hameed, I. Hill, C. Jeffrey, M.H. Loke, P.I. Meldrum, O. Kuras, M. Cave, D.A. Gunn, Bedrock Detection Beneath River Terrace Deposits using Three-dimensional Electrical Resistivity Tomography, doi:10.1016/j.geomorph.2012.03.034

remote sensing, and intrusive investigations (e.g., Suzuki et al., 2004 and Guccione, 2008). Perhaps the most detailed and commonly undertaken subsurface investigations of river terrace deposits are for mineral exploration, where drilling is the principal investigative tool (Merritt, 1992, Crimes et al., 1994 and Smith and Collis, 2001). However, because of the complexity of some deposits, even drilling using densely spaced boreholes can fail to adequately reveal the three-dimensional (3D) structure of a deposit in terms of thickness and composition (Wardrop, 1999).

To provide greater insights into subsurface heterogeneity, geophysical techniques such as seismic refraction, ground penetrating radar, and electrical methods are being increasingly applied (Hirsch et al., 2008 and Tye et al., 2011). Electrical resistivity tomography (ERT) is one such method that has been demonstrated to be an effective means of studying the architecture of these deposits for a range of applications, including the investigation of landscape evolution (Froese et al., 2005, Hickin et al., 2009 and Hsu et al., 2010), geological mapping (Tye et al., 2011), groundwater studies (Revil et al., 2005 and Hirsch et al., 2008), and mineral exploration (Baines et al., 2002 and Beresnev et al., 2002).

The principal benefits of ERT are that it provides high resolution images of the subsurface and is noninvasive. It is an effective means of characterising the subsurface because of the sensitivity of resistivity to variations in hydrogeological (e.g., saturation, pore fluid composition) and geological properties (e.g., mineral grain composition, porosity). In unconsolidated sediments, such as river terrace deposits, the major lithological control on resistivity is the type and proportion of clay minerals (Shevnin et al., 2007), with increasing clay content causing a decrease in resistivity.

Limitations of the technique include inaccuracies because of 3D structures to the side of the survey line or area and the indistinct appearance of boundaries resulting from the smoothness-constrained inversion techniques typically used for ERT imaging. Most previous ERT surveys of river terrace deposits have employed 2D, rather than 3D, imaging, because of its comparative rapidity and simplicity. However, for heterogeneous subsurface conditions, the two-dimensional (2D) assumption is violated because of the influence of 3D features in close proximity to the survey lines, which can cause significant inaccuracies in the resulting 2D resistivity models (Chambers et al., 2002 and Sjodahl et al., 2006). More accurate subsurface reconstruction can therefore be achieved by applying fully 3D ERT imaging approaches. However, the smoothness-constrained images can make it difficult to accurately determine the position of geological boundaries, such as the river terrace deposit–bedrock interface. To address this problem, Hsu et al. (2010) described an automated approach to bedrock edge detection, although their study was restricted to 2D ERT. They provided both synthetic and field based examples with borehole control, both of which showed good visual agreement between the ERT derived interfaces and the known interface locations.

Here we present a study in which fully volumetric 3D ERT imaging is used to investigate river terraces from the Great Ouse valley, Bedfordshire, UK. The principal advance described here is the development and validation of an approach to bedrock surface detection in a river terrace setting based on 3D rather than 2D imaging. We propose that a fully volumetric approach is particularly preferable for highly variable deposits that have a fundamentally 3D structure. The specific aims of this study are (i) to quantitatively assess an automated approach to bedrock surface detection below highly heterogeneous valley fill deposits from the 3D resistivity model and (ii) to consider the respective merits of 3D ERT and conventional intrusive approaches for river terrace deposit characterisation.

STUDY AREA

The study area is located within the valley of the Great Ouse, near the village of Willington, 4 km to the east of Bedford, UK (Fig. 1). The Great Ouse is an important component of The Wash fluvial network, preserving a record of late Quaternary uplift and climate variation and of human activity during the Palaeolithic, and as such is of international significance (e.g., Boreham et al., 2010). The geology comprises Quaternary alluvium and river terrace sand and gravel overlying Oxford Clay Formation bedrock of the middle Jurassic (Barron et al., 2010). In this area the Oxford Clay bedrock consists of the Peterborough member, which is a brownish grey, fissile mudstone, with an approximate thickness of 20 m. The Oxford Clay outcrops to both the southeast and northwest of the survey area, and has been exposed by extractive activities within the river valley (Fig. 1). The river terrace deposits are of the Ouse Valley Formation and are likely to have been deposited by braided rivers under periglacial conditions during different Quaternary cold stages (Rogerson et al., 1992, Green et al., 1996 and Bridgland, 2010). Three principal terrace deposits are observed in the area (Horton, 1970, Barron et al., 2010 and Boreham et al., 2010). The first, and lowest, terrace overlies the Felmersham member, which is ~ 3 m thick, with a surface between 0.6 and 2 m above the floodplain. The second terrace overlies the Stoke Goldington member and has a surface hereabouts between 2 and 7 m above the floodplain. The third terrace overlies the Biddenham member, which has a thickness of up to 7 m and a surface between 11 and 13 m above the floodplain. The sands and gravels of the three terraces display a similar composition, comprising planar-bedded, brownish yellow sand and gravel for which the gravel component mainly consists of flint and limestone. The present day floodplain is covered by a brown clay and silt alluvium, with a thickness of up to 4 m, which overlies the Ouse Valley Formation and in places may occupy channels cut in the Felmersham member by meandering rivers under temperate climate conditions (Barron et al., 2010). Extensive removal and reworking of the superficial deposits in this area have occurred from mineral extraction and, in particular, the quarrying of sand and gravel from the river terrace deposits. In many places the removal of sand and gravel has resulted in the exposure of the Oxford Clay Formation bedrock (Fig. 1).

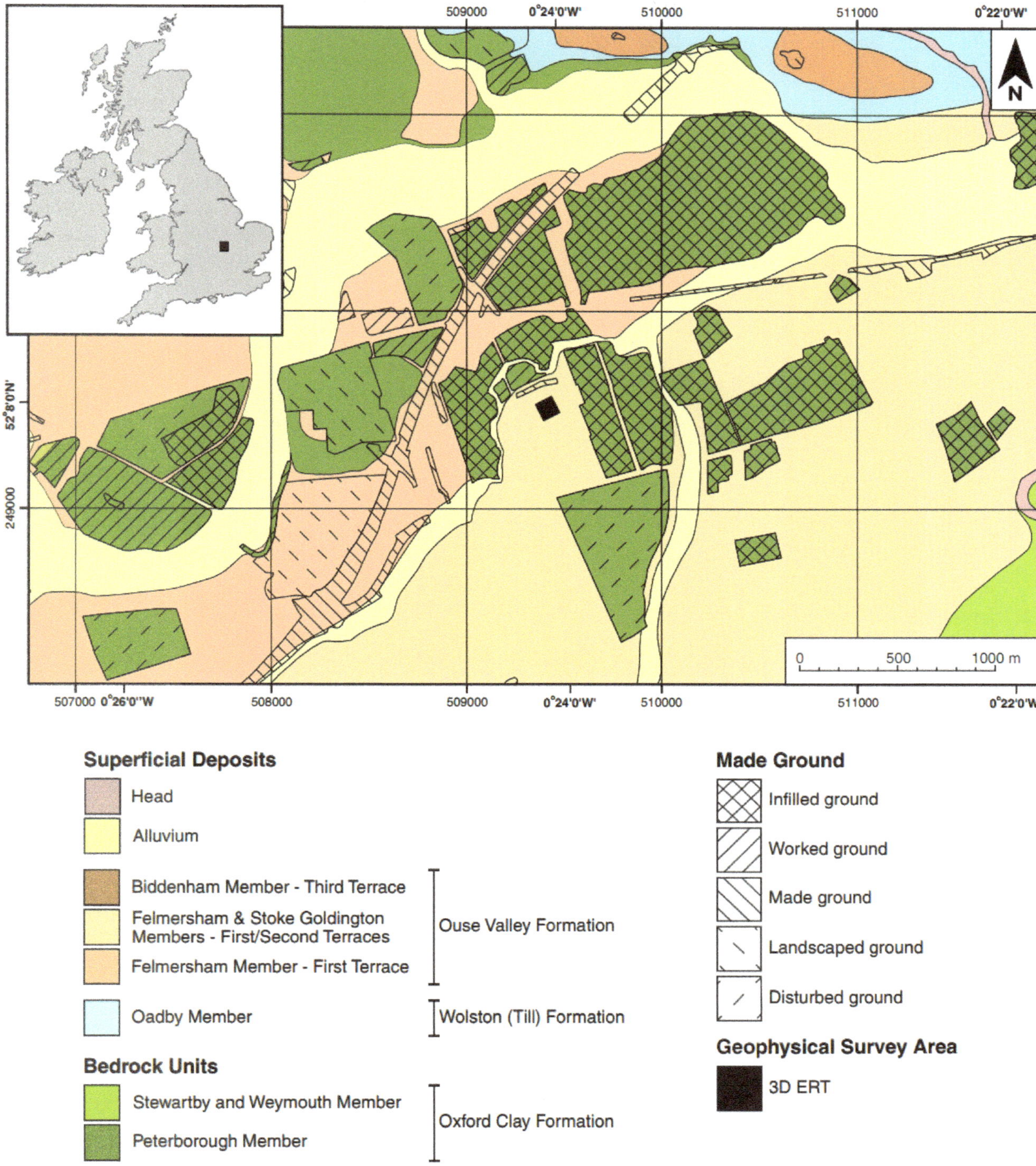

Figure 1: Geological map based on a recent geological resurvey of the area (Barron et al., 2010), showing the location of the study site and the distribution of artificially modified ground associated with extractive activities. Coordinate systems are given as longitude and latitude (bold) and British National Grid (normal). Inset map (top left) shows the location of the study site within the UK.

The study site is situated on terrace deposits of the undifferentiated Felmersham and Stoke Goldington members (Fig. 1), overlying Oxford Clay Formation bedrock. The terrace deposits at this site are the focus of a long-standing sand and gravel operation. At the time of this study, the topsoil (which was ~ 0.2 m thick), had been stripped and banked (Fig. 2) exposing alluvium at the surface. The alluvial materials observed across the survey area are probably modern overbank deposits, which are distinct from the thicker alluvium recorded on the geological map (Fig. 1). The area was selected because good subsurface data in the form of borehole logs was available with which to interpret and calibrate the geophysical results. Furthermore, mineral extraction activities immediately to the south of the study site and electromagnetic geophysical reconnaissance surveys (Hill et al., 2011) had revealed that the river terrace deposits in this area were extremely variable in terms of thickness and composition, thereby providing a complex target with which to test 3D ERT. The deposits were unsaturated because of dewatering associated with the mineral workings immediately to the south of the study site (Fig. 2).

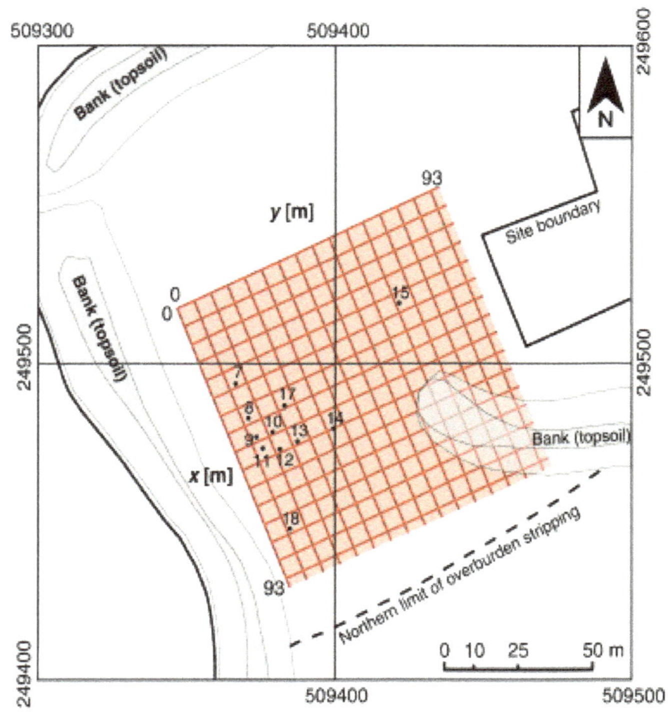

Figure 2: Three-dimensional ERT survey area (red shading), site boundary (black line), and line locations (red lines, 6-m separation), and borehole positions (black dots). Banked topsoil stockpiles (grey shading) crest heights are typically 3 m above ground level.

METHODOLOGY

Intrusive Investigations

Drilling at the site was carried out using a flight auger supplemented with holes drilled using other standard techniques, including shell and auger, reverse circulation, and sonic drilling. A total of 11 locations were drilled within the 3D ERT imaging area; five of the locations were drilled using only the flight auger; whilst the remaining six locations were drilled with a combination of two or more techniques. At each location bedrock was proven. For locations where multiple drilling techniques were applied, boreholes were drilled within ~ 1 m of one another. The drilling density achieved (i.e., about 11 holes per hectare) was considerably in excess of standard sand and gravel exploration drilling programmes that typically employ a 100-m drilling grid, which in complex situations can be reduced to 50 m. The drilling at the site was undertaken as a component of a separate project concerned with optimising sand and gravel deposit sampling strategies, which involved the geostatistical analysis of grading data and the comparison of different drilling technologies (Hill et al., 2011 and Jeffrey et al., 2011). Although the borehole locations were selected principally for the purpose of undertaking geostatistical analysis of grain size variations, they nevertheless provided a useful ground truth data set with which to assess the performance of 3D ERT for river terrace deposit characterisation and bedrock detection. Borehole locations are shown in Fig. 2, and summary information showing depth to bedrock determined by drilling is shown in Table 1.

Table 1: Drilling results by location (borehole ID), showing depth to bedrock determined using the flight auger (FA), shell and auger (SA), reverse circulation (RC), and sonic (SNC) drilling methods.

Borehole ID	Local grid		Ground lvl. [m AOD]	Depth to bedrock [m]					Range (m)
	x [m]	y [m]		FA	SA	RC	SNC	Average	
BH7	29.46	8.59	23.82	2.80				2.80	
BH8	41.19	8.15	23.96	3.00			3.40	3.20	0.40
BH9	49.28	8.44	23.95	4.00				4.00	
BH10	48.47	13.92	23.93	4.00				4.00	
BH11	51.93	8.82	23.87	4.00	3.50	4.15	4.50	4.04	1.00
BH12	54.39	14.00	23.88	4.00			4.40	4.20	0.40
BH13	54.51	20.44	23.87			3.30	2.90	3.10	0.40
BH14	55.59	33.23	23.83	3.20			3.00	3.10	0.20
BH15	27.82	69.19	23.73	2.60	3.00	2.90	3.00	2.88	0.40
BH17	42.25	20.95	23.87	4.00				4.00	
BH18	78.77	7.17	23.86	2.10				2.10	

Electrical Resistivity Tomography

The application of ERT can provide fully 3D volumetric models of subsurface resistivity distributions from which features of contrasting resistivity can be located and characterised. Methodologies for 3D data collection and modelling are well established in the literature (e.g., Chambers et al., 2007, Chambers et al., 2011 and Magnusson et al., 2010) and so only a brief summary is presented here.

Survey Design and Execution

The 3D ERT survey was carried out within an area of 93 m (x) by 93 m (y). Data were collected on a network of 32 orthogonal survey lines positioned at 6-m intervals, oriented in both x and y directions (Fig. 2). The dipole–dipole array with dipole sizes (a) of 3 and 6 m, and dipole separations (n) of 1a to 8a were used, and a full set of both normal and reciprocal measurements was collected. A line separation twice that of the along-line electrode separation was selected to avoid undersampling and to maximise survey coverage rate (Gharibi and Bentley, 2005). Likewise, the selected dipole sizes and separation were considered to be a reasonable compromise between vertical and lateral resolution and coverage rate. Orthogonal lines were employed to minimise bias in the resulting ERT model resulting from the use of a single line direction (Chambers et al., 2002). The dipole–dipole array was used because it is a well-tested array that can provide a relatively high level of resolution, it does not require a remote electrode, it can exploit the multichannel capabilities of modern ERT instruments, and crucially, it enables the efficient collection of reciprocal measurements (Dahlin and Zhou, 2004). For a normal four-electrode measurement of transfer resistance, the reciprocal is found by exchanging the current and potential dipoles, and in the absence of nonlinear effects should give the same result. Here, reciprocal error is defined as the percentage difference between the forward and reciprocal measurements. Reciprocal measurements are sensitive to both random and systematic sources of noise, and provide a particularly effective means of assessing data quality and determining robust data editing criteria (Dahlin and Zhou, 2004).

A real-time kinematic global positioning system (GPS) survey was undertaken to measure surface elevations across the area for incorporation into the resistivity inversion and forward modelling procedure. Although most of the survey area was very flat, the GPS survey was required to capture the topography of a 3-m-high bank of topsoil that encroached on the eastern corner of the ERT imaging area (Fig. 2).

Data Processing, Forward Modelling, and Inversion

The combined data set from the survey lines comprised 11,270 pairs of normal and reciprocal measurements. In general, data quality diminished with increasing geometric factors, which cause smaller measured potential differences. Data points with a reciprocal error of > 5% were removed, which in this case accounted for only 2% of the measured data, resulting in a filtered data set of 10,952 pairs. These were inverted using a 3D regularized least-squares optimisation method (Loke and Barker, 1996). The forward problem was solved using the finite-element method, in which node positions were adjusted to allow topography to be taken into account in the inversion process. In brief, the aim of the inversion process is to calculate a model that satisfies the observed data. A starting model is produced, which in this study was a homogeneous half-space, for which a response is calculated and compared to the measured data. The starting model is then modified in such a way as to reduce the differences between the model response and the measured data; these differences are quantified as a mean absolute misfit error value. This process continues iteratively until acceptable convergence between the calculated and measured data is achieved. In this case, a geologically realistic model was produced using L2-norm (smooth) model constraints because of the significant gradational lithological variations observed in the drift deposits and the undulating topography of bedrock (Loke et al., 2003). The final resistivity model consisted of 31 cells in the x-direction, 31 cells in the y-direction, and 11 layers in the z-direction, resulting in a total of 10,571 model cells.

Bedrock Detection

Amongst the most widely used approaches to edge detection are gradient techniques, which assume that interfaces are located where changes in image properties are at a maximum (e.g., Marr and Hildreth, 1980, Vafidis et al., 2005 and Sass, 2007). One of the only published examples of automated bedrock detection from ERT images is described by Hsu et al. (2010). They used a gradient method, which searches for values of zero in the Laplacian (second derivate) of the resistivity image in the horizontal and the vertical directions. Using this approach, they were able to accurately define the bedrock–sediment interface from a number of 2D ERT images. The principal drawback of the Laplacian technique was, according to their study, the prevalence of local zero lines that were difficult to differentiate from those associated with the larger magnitude gradients defining the primary bedrock interfaces.

Here we adopt a similar technique to Hsu et al. (2010). However, because of the added complexity of 3D image analysis compared to 2D, we have simplified their approach. We only consider variation in gradient in the vertical direction that although is less sensitive to very steeply dipping or vertical interfaces, is a reasonable approximation for the relatively layered structure of the river terrace deposits. We also only consider the gradient (first derivative) of the resistivity image, which tends to reduce the problem of the Laplacian method, which produces many more false interface (zero) lines. Although the first derivative eliminates false interfaces, it cannot discriminate between interfaces if multiple gradients are present. Consequently, we employ a two-stage heuristic approach for bedrock detection at the study site. First, if multiple gradients in the correct direction (i.e., decreasing resistivity with decreasing elevation) are present then the steepest gradient is chosen; this is because we anticipate that in most cases the steepest resistivity gradient in the subsurface will be between the relatively coarse-grained river terrace deposits and very clay rich Oxford Clay, rather than lithological boundaries within formations or between the alluvium and terrace deposits. Second, if the gradients are of a similar magnitude, we pick the deeper gradient, as the lower lithological interface in the ERT model is likely to be between the valley fill and bedrock surface.

Our implementation of the steepest gradient method involved extracting resistivity data,ρ, as a function of elevation, z, for each surface position (x, y). An interpolating curve was fitted through ρ(z) for each (x, y) point. In this case, a piecewise cubic hermite interpolating polynomial (PCHIP) was used. The coefficients of the polynomial are chosen so that the resistivity is continuous and smooth, its first derivative is continuous (although not necessarily smooth), and the interpolant is monotonic between data points (e.g., Fig. 3). This has the effect that the interpolant preserves the shape of the data (Fritsch and Carlson, 1980). Once the coefficients are determined, the first derivative can be calculated analytically. Then for interface detection, the depth corresponding to the steepest gradient on the interpolating curve that satisfied our heuristic was identified for each (x, y) point.

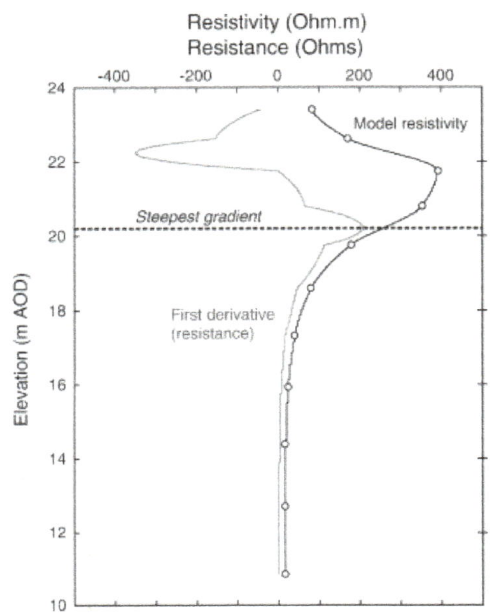

Figure 3: Example of a resistivity depth curve (black line) generated from PCHIP interpolation of resistivity data (circles) and the gradient (first derivative) of the resistivity (grey line). The maximum positive gradient is shown by the dashed black line.

RESULTS AND DISCUSSION

Direct Intrusive Sampling

The drilling results for the 11 locations (Fig. 2) in terms of the types of drilling techniques deployed, position, ground level, and depth to bedrock are shown in Table 1. The average depth to bedrock from each location, and hence river terrace and alluvium thickness, ranges from 2.1 to 4.2 m. Significant differences in deposit thickness were observed between the various drilling techniques for each location. The alluvium showed a consistent thickness of ~ 1 m across the survey area. Bedrock interface depths determined by multiple holes were not consistent (Table 1); the discrepancies ranged between 0.2 and 1 m, with an average of 0.46 m. The reasons for this apparent lack of agreement between drilling techniques are threefold: first, misidentification of interfaces because of contamination by material from the hole sides during stem withdrawal (a problem that is recognised in the interpretation of flight auger logging in particular); second, poor core recovery and slippage of core in the barrel during withdrawal (as observed to occur with, for example, sonic drilling); and third, true variation in bedrock surface elevation between clustered sampling points (i.e., ~ 1 m separation).

Three-dimensional Resistivity Model

Good convergence between the observed and model data was achieved, as indicated by the mean absolute misfit error of 2.4%. The resulting resistivity model has dimensions of 93 m (x) by 93 m (y) and extends to a depth of 14 m below ground level (z). Visualisations of the 3D ERT model are shown in Fig. 4 as a series of vertical and horizontal sections and volumetric images. The clay bedrock is defined as low resistivity material underlying more resistive and highly heterogeneous valley fill deposits. The banked topsoil in the eastern corner displays a similar resistivity range to that of the terrace deposits.

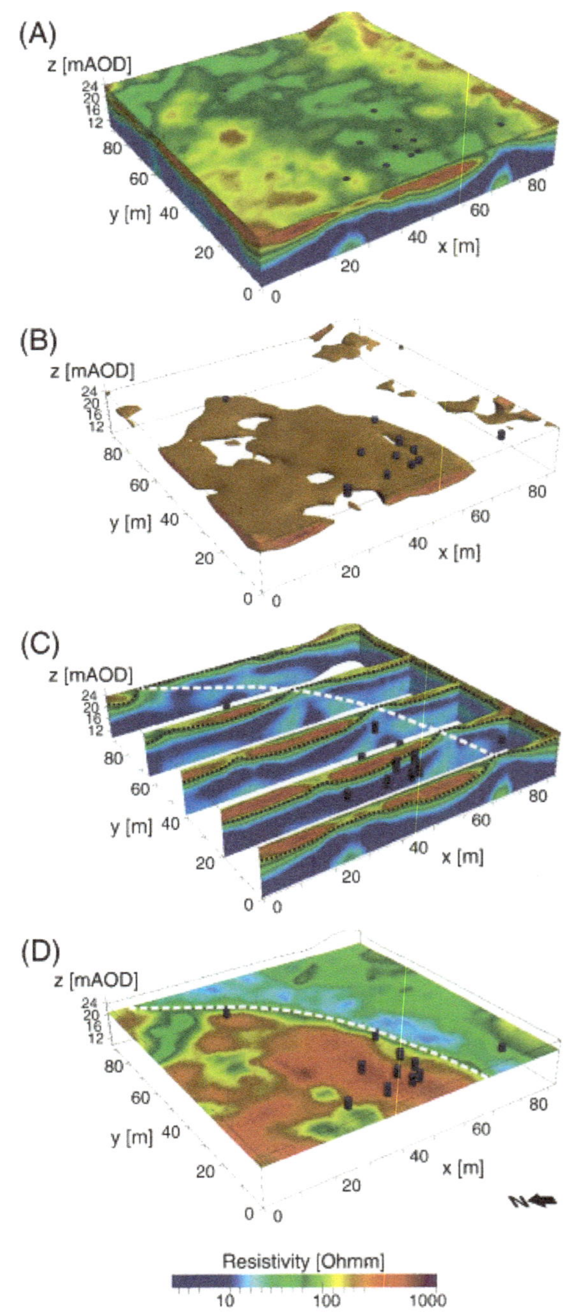

Figure 4: Three-dimensional ERT model displayed as (A) a solid volume, (B) a solid volume with opaque volume defining resistivities

above 200 Ωm, (C) vertical sections, (D) a horizontal section at 20 m AOD. Vertical extent of mineral and overburden, determined from drilling, shown as grey cylinders. The southeastern edge of the incised channel structure is indicated as a dashed white line.

The distribution of inverted resistivities is shown in Fig. 5, plotted as a probability density function (PDF). The PDF was estimated using a kernel smoothing algorithm (Sheather and Jones, 1991), which sets up a normal distribution at each of the measured values in the data set and adds these together to produce smoothed PDF. Using the standard deviation (SD) and relative proportions of points from an initial approximation as starting points, an optimisation routine (Rowan, 1990) that modifies the input parameters to minimise the root mean square error between the estimated PDF and the actual PDF was used to determine mean and standard deviations for each of the predicted resistivity populations. Three resistivity populations with means of 15, 60, and 125 Ωm, respectively, were estimated using this approach. The well-defined low resistivity peak (peak 1) corresponds to the Oxford Clay bedrock, whilst the higher resistivity and less distinct peaks are consistent with separate populations within the deposits of varying composition. For unsaturated valley fill deposits present at this site, the high resistivity population (peak 3) is likely to be associated with relatively clean coarse sand and gravel, whilst the lower resistivities (peak 2) are consistent with the more clay-rich alluvium.

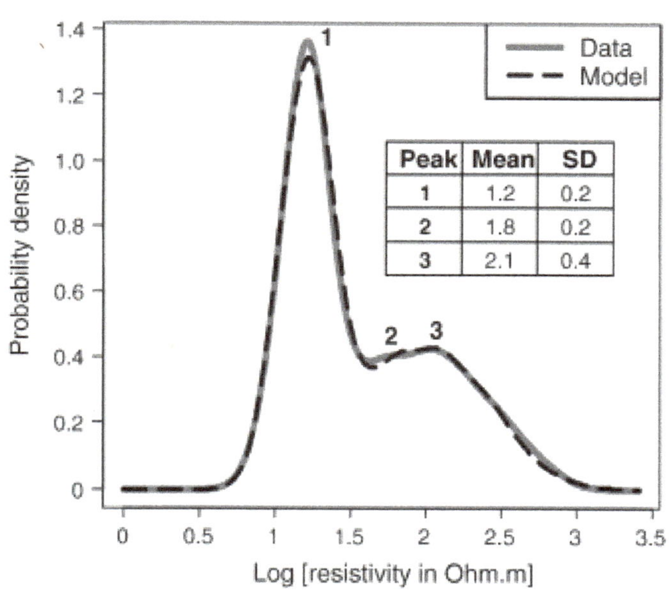

Figure 5: Probability distribution plot of the Willington 3D ERT data (solid line), and optimised probability distribution model (dashed line) for three normal distributions with peaks at log resistivities of 1.21, 1.75, and 2.09 Ωm (i.e., resistivities of 16, 56, and 123 Ωm).

The geological sequence at the site – comprising a thin layer of alluvium at the surface, river terrace sand and gravel, and Oxford Clay bedrock – is apparent in the 3D ERT image (Fig. 4). The alluvium is seen as a thin layer of relatively low resistivity (< 100 Ωm) material (e.g., Fig. 4 and Fig. 6), which indicates a higher clay content than the underlying sand and gravel. The alluvium appears to vary in composition across the area, with the northwestern corner and southern edge showing a higher resistivity, due perhaps to a lower clay content. The underlying terrace deposits are generally more resistive than both the alluvium and the Oxford Clay bedrock. They display a broad range of resistivities with a spatial distribution that is consistent with deposition as part of a braided river system, with silt and clay-rich channel fill and coarser bar deposits. The Oxford Clay bedrock is associated with a relatively homogeneous resistivity distribution. A number of slightly higher resistivity zones are seen within the bedrock, with the two strongest features at y = 0 m and x = 25 and 75 m, respectively. It is probable that these are artefacts of the inversion process rather than real bedrock features for three principal reasons. First, they are not consistent with known geological structure. Second, they are in a part of the model that has low model resolution (Wilkinson et al., 2012); in this case the model resolution reduces by more than an order of magnitude between 4 m below ground level and the base. Third, because they are at the base of the model they are influenced by measurements with higher geometric factors, which have poorer signal-to-noise characteristics.

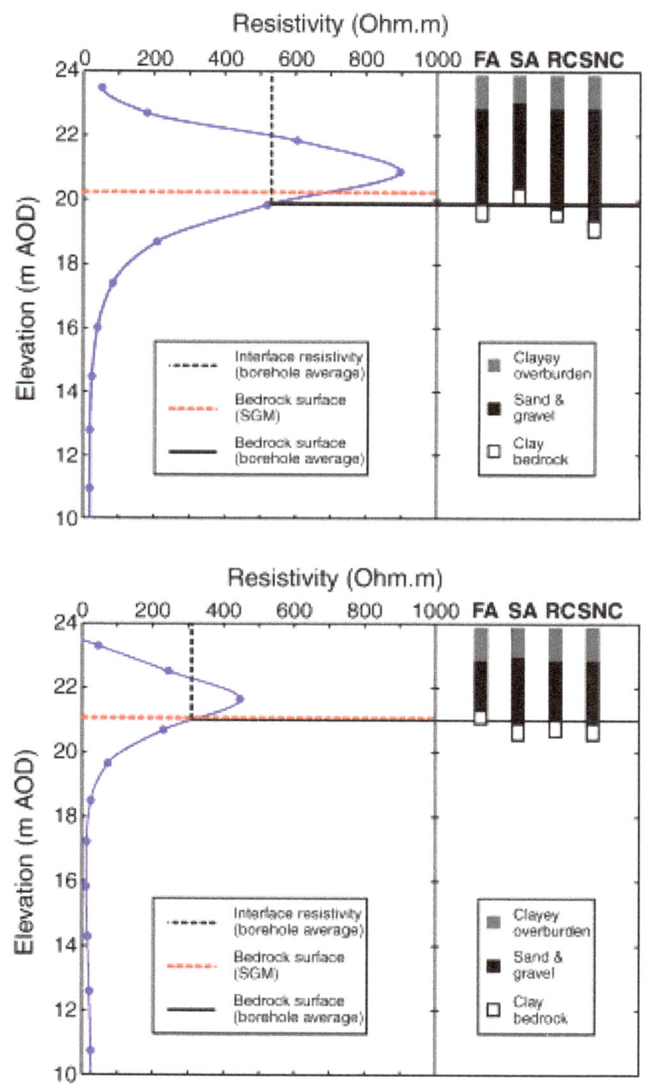

Figure 6: Resistivity data (circles) and interpolating curves (blue line) as a function of elevation, given as mAOD at surface positions corresponding to BH11 (top) and BH15 (bottom). The elevations associated with the steepest gradient method (SGM) and the intersections between the borehole-derived elevations and the resistivity depth curves (interface resistivities) are indicated. Drilling results for four different techniques at this location are shown: flight auger (FA); shell and auger (SA); reverse circulation (RC); sonic (SNC).

The primary structure is an arch-shaped feature (Fig. 4), running approximately SW to NE, which defines thicker terrace deposits and deeper bedrock to the NW. The transition from thicker to thinner deposits is likely to represent that transition from first to second terrace. Three lines of evidence corroborate this interpretation. First, it is close to the anticipated transition between the first and second terrace (Barron et al., 2010; A.J.M. Barron, British Geological Survey, personal communication, 2011). Second, the thickness and height change between the first and second terraces recorded in the area (Horton, 1970, Barron et al., 2010 and Boreham et al., 2010) is consistent with the structure observed in the ERT model. Third, the orientation of the erosional structure identified in the ERT model is subparallel to the long axis of the Great Ouse.

Steepest Gradient Method Bedrock Surface Detection

The bedrock surface extracted from the 3D ERT model using the steepest gradient (first derivative) method extends between 20 and 24 m above Ordnance Datum (AOD) (Fig. 7). The broad structure identified in the 3D ERT model, interpreted as the transition from first to second terrace, is clearly visible in the steepest gradient bedrock surface as a sharp upward step toward the eastern corner of the image. In addition, the steepest-gradient-derived surface contains a scattering of false high elevation points where our heuristic approach failed to capture the full complexity of resistivity variations in the model. These points appear as isolated spikes, or bull's-eyes, and are concentrated in the northwestern corner, below the higher resistivity alluvium, and in the southeastern corner, below the topsoil bank.

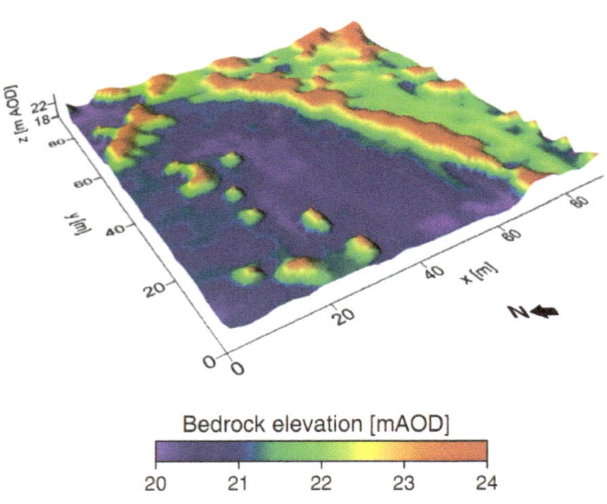

Figure 7: Bedrock surface determined using the steepest gradient (first derivative) method, showing the erosional structure associated with the transition from the first to the second terrace of the Great Ouse.

Examples of interpolated resistivity depth curves from the 3D ERT model, showing the location of the steepest gradient and 'known interface' resistivities, are given for borehole locations 11 and 15 (Fig. 6). The known interface resistivity is the value associated with the borehole-defined depth; an alternative to the steepest gradient approach is to use the known interface resistivity to define an isoresistivity surface, which is assumed to coincide with the bedrock surface (see discussion on the use of isoresistivity surfaces below). Summary data for each of the borehole locations is given in Table 2. Statistical analysis has been carried out using the Bland and Altman (1986) method, which provides a means of comparing two different methods of measurement (i.e., ERT and boreholes) where the true value of the measured parameter is unknown. It is used to calculate the bias and the agreement, or standard deviation, between the two methods. This approach has indicated a reasonable agreement between the boreholes and steepest-gradient-derived method as indicated by an SD of 0.38 m (Fig. 8A). A slight bias of 0.19 m caused by two outlying data points (BH8 and BH13) has been observed between the boreholes and steepest gradient method, with the ERT-derived bedrock elevations slightly higher than those recorded in the boreholes. Likewise, the Pearson correlation coefficient for the steepest gradient and borehole-derived bedrock elevations is 0.83, with a p-value of 0.001 (Fig. 8B), indicating good agreement between the two approaches and a high degree of statistical significance. Based on the steepest gradient method, a volume of 12,250 m^3 (SD 3240 m^3) has been calculated (using the trapezoidal rule) for the valley fill sediment (terrace sand and gravel, and alluvium) within the 3D ERT survey area.

Table 2: Bedrock elevations derived from drilling and from the steepest gradient method, and interface resistivities

Borehole ID	Bedrock surface elevation [m AOD]		Difference in elevation [m]	Interface resistivity [Ohm.m][b]
	SGM	Borehole[a]		
BH7	21.22	21.02	− 0.20	42
BH8	20.25	20.76	0.51	434
BH9	20.19	19.95	− 0.24	497
BH10	20.23	19.93	− 0.30	339
BH11	20.23	19.83	− 0.40	520
BH12	20.22	19.68	− 0.54	189
BH13	20.21	20.77	0.56	138
BH14	21.26	20.73	− 0.53	165
BH15	21.06	20.86	− 0.20	280
BH17	20.27	19.87	− 0.40	370
BH18	22.07	21.76	− 0.31	95

[a]Average.

[b]At level determined from boreholes.

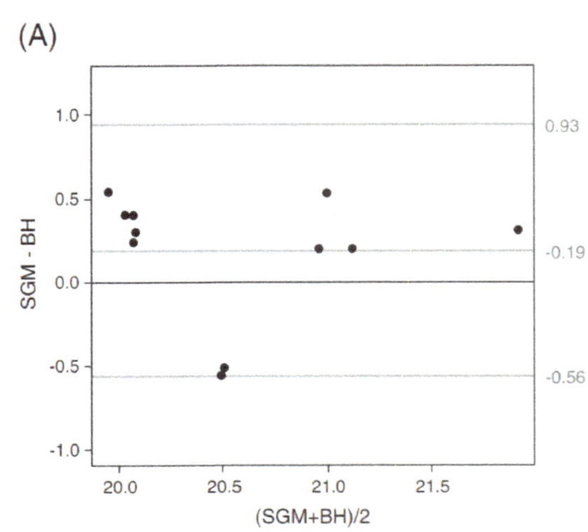

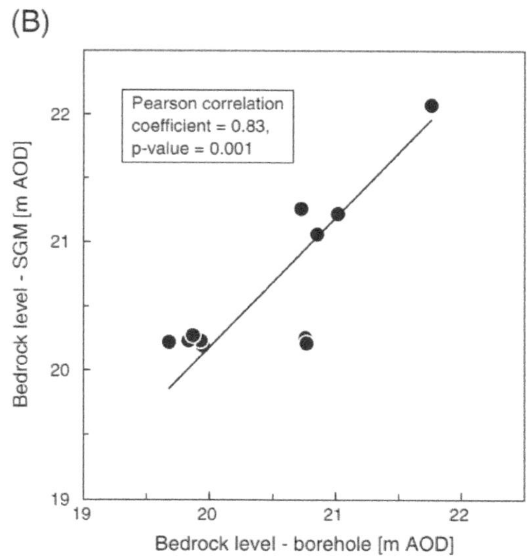

Figure 8: (A) Bland Altman plot of steepest gradient method and borehole-derived (BH) elevations, showing the 95% confidence limit between − 0.56 and 0.93 m. (B) Cross plot of steepest gradient method and borehole-derived bedrock elevations, showing Pearson correlation coefficient.

These results also confirm the findings of Hsu et al. (2010) that isoresistivity lines are not necessarily a good indicator of bedrock surface geometry. For isoresistivity lines to successfully define the bedrock surface, the interface must be characterised by a consistent value of resistivity. By comparing the results of the 11 drilling locations with the ERT model, it is clear that the range of interface resistivity values is considerable (Table 2), varying between 42 and 520 Ωm. The large range of interface resistivities is a function of the complexity of the deposit, with the valley fill deposits displaying a large resistivity range and significant heterogeneity. This is further illustrated with reference toFig. 6, where the interface resistivity for BH 11 is 520 Ωm, whilst for BH15 it is 280 Ωm. The reason for the difference between these two locations is that at BH11 the terrace deposits were significantly more resistive than at BH15, resulting in a large difference in interface resistivity values.

Comparison of 3D ERT and Borehole Results

Drilling and ERT produce very strongly contrasting types of information. Boreholes provide very detailed, very high resolution (centimetre to decimetre scale) information for vertical profiles at discrete locations but provide very poor lateral resolution, even for dense drilling grids or profiles considered here, because of separations that are typically on the scale of at least tens of metres between holes. Moreover, drilling can provide direct samples of subsurface materials. Conversely, 3D ERT provides high resolution (metre scale) spatially continuous volumetric subsurface models but provides indirect information on material properties. Interestingly, the uncertainty associated with bedrock surface elevation for both drilling and ERT was of a similar magnitude (i.e., tens of centimetres), with an average discrepancy between drilling techniques of 0.46 m (Section 4.1) and a standard deviation of 0.38 m for the difference between steepest gradient and average borehole-derived bedrock elevations (Section 4.3).

In this geological setting, the spatial information provided by ERT was essential for resolving the structure of the bedrock surface, due the complexity of the deposit, in terms of thickness variations and sediment heterogeneity. The relative success of ERT was a function of the spatial resolution (in the x-, y- and z-directions) of the technique, which was closer to the scale of deposit heterogeneity than the borehole data, which had sufficient resolution only in the z-direction. However, intrusive investigations and sampling will always be necessary for this type of investigation, whether it be for mineralogical assessment and dating for geological, geomorphological, or archaeological studies; hydrogeological testing for groundwater resource assessment; or particle size distribution determination for mineral exploration. Crucially, intrusive sampling is also essential for the calibration and validation of geophysical images. These two approaches are therefore complementary. The combined use of 3D ERT and boreholes has the potential to reduce the number of boreholes required, and the ERT images could also assist in the more effective targeting of boreholes.

Boreholes were also important for deposit characterization in this case, as they were able to differentiate between river terrace and alluvium. The 3D ERT model did reveal a thin, relatively conductive layer across much of the surface of the model, but in places alluvium was indistinguishable from the underlying sand and gravel due to insufficient resistivity contrasts (e.g., Fig. 4 and Fig. 5). For this reason the steepest gradient method was not applied to identify the interface between the alluvium and the sand and gravel.

CONCLUSIONS

Automated bedrock detection from 3D ERT imaging at a site in the Great Ouse Valley, UK, using the steepest gradient (first derivative) method was shown to correlate well with borehole-derived bedrock elevations. Comparison of the borehole and steepest gradient methods has enabled the performance of 3D ERT for bedrock detection to be quantitatively assessed and uncertainty associated with sediment volume calculations to be determined. Whilst the steepest gradient method was shown to provide a good quality bedrock elevation model, isoresistivity lines were shown to provide a very poor indication of bedrock rock surface depth and geometry in this situation. Interestingly, a comparison of a range of drilling techniques deployed at the site has indicated a level of uncertainty for borehole derived interface depths similar to that associated with 3D ERT steepest gradient edge detection — indicating that intrusive sampling cannot always be regarded as providing inherently more reliable information than geophysical investigations.

Subsurface geological variations (including the distribution of major formations, and lithological heterogeneity, and river terrace deposit thicknesses) were captured within the 3D ERT model. Crucially, a major erosional feature on the bedrock surface was identified as the boundary between first and second terrace deposits of the Great Ouse valley.

Three-dimensional ERT image analysis using the steepest gradient method has been shown to be an effective bedrock detection method in this locality, owing in part to the strong contrast in resistivity between the bedrock and river terrace deposits. It is therefore reasonable to presuppose that ERT would be similarly successful in other river terrace settings with strong resistivity contrasts between valley fill and bedrock materials. In particular, in areas of clay or mudstone bedrock, a good resistivity contrast could be expected with river terrace sand and gravel because of the large difference in the proportion of clay between the two material types.

The appropriateness of 3D ERT for any given setting will also be dependent on a number of other factors, including the required spatial coverage and level of resolution. The practical limit of survey coverage using 3D ERT is probably in the order of a few tens of hectares for individual surveys and, as such, is not equivalent to surface mapping approaches using remote sensing or towed ground-based systems that permit very rapid large-scale data collection. Therefore, in the context of river terrace deposit investigations, 3D ERT is best suited to targeted site-specific surveys associated with complex deposits displaying significant lateral variations where detailed information on subsurface structure is required.

ACKNOWLEDGEMENTS

The research was funded by Defra through the MIST Programme (grant MA/7/G/1/007) and in-kind contributions from project partners. This paper is published with the permission of the Executive Director of the British Geological Survey (NERC). The authors wish to thank A.J.M. Barron for his advice on the local geology.

REFERENCES

1. Baines, D., Smith, D.G., Froese, D.G., Bauman, P., Nimeck, G., 2002. Electrical resistivity ground imaging (ERGI): a new tool for mapping the lithology and geometry of channel-belts and valley-fills. Sedimentology 49, 441–449.
2. Barron, A.J.M., Sumbler, M.G., Morigi, A.N., Reeves, H.J., Benham, A.J., Entwisle, D.C., Gale, I.N., 2010. Geology of the Bedford District — A Brief Explanation of the Geological Map. 1:50 000 Sheet 203
3. Bedford (England and Wales), Nottingham. Beresnev, I.A., Hruby, C.E., Davis, C.A., 2002. The use of multi-electrode resistivity imaging in gravel prospecting. Journal of Applied Geophysics 49, 245–254.
4. Bland, J.M., Altman, D.G., 1986. Statistical methods for assessing agreement between two methods of clinical measurement. Lancet i, 307–310. Boreham, S., White, T.S., Bridgland, D.R., Howard, A.J., White, M.J., 2010. The Quaternary history of the Wash fluvial network, UK. Proceedings of the Geologists' Association 121, 393–409.
5. Bridgland, D.R., 2006. The Middle and Upper Pleistocene sequence in the lower Thames: a record of Milankovitch climatic fluctuation and early human occupation of southern Britain — Henry Stopes Memorial Lecture 2004. Proceedings of the Geologists' Association 117, 281–305.
6. Bridgland, D.R., 2010. The record from British Quaternary river systems within the context of global fluvial archives. Journal of Quaternary Science 25, 433–446.
7. Chambers, J.E., Ogilvy, R.D., Kuras, O., Cripps, J.C., Meldrum, P.I., 2002. 3D electrical imaging of known targets at a controlled environmental test site. Environmental Geology 41, 690–704.
8. Chambers, J.E., Wilkinson, P.B., Weller, A.L., Meldrum, P.I., Gilvy, R.D., Caunt, S., 2007. Mineshaft imaging using surface and crosshole 3D electrical resistivity tomography: a case history from the East Pennine Coalfield, UK. Journal of Applied Geophysics 62, 324–337.
9. Chambers, J.E., Wilkinson, P.B., Kuras, O., Ford, J.R., Gunn, D.A., Meldrum, P.I., Pennington, C.V.L., Weller, A.L., Hobbs, P.R.N., Ogilvy, R.D., 2011. Three-dimensional geophysical anatomy of an active landslide in Lias Group mudrocks, Cleveland basin, UK. Geomorphology 125, 472–484.
10. Crimes, T.P., Chester, D.K., Hunt, N.C., Lucas, G.R., Mussett, A.E., Thomas, G.S.P., Thompson, A., 1994. Techniques used in aggregate resource analyses of 4 areas in the UK. The Quarterly Journal of Engineering Geology 27, 165–192.
11. Dahlin, T., Zhou, B., 2004. A numerical comparison of 2D resistivity imaging with 10 electrode arrays. Geophysical Prospecting 52, 379–398.
12. Fritsch, F.N., Carlson, R.E., 1980. Monotone piecewise cubic interpolation. SIAM Journal on Numerical Analysis 17, 238–246.
13. Froese, D.G., Smith, D.G., Clement, D.T., 2005. Characterizing large river history with shallow geophysics: middle Yukon River, Yukon Territory and Alaska. Geomorphology 67, 391–406.
14. Gharibi, M., Bentley, L.R., 2005. Resolution of 3-D electrical resistivity images from inversions of 2-D orthogonal lines. Journal of Environmental and Engineering Geophysics 10, 339–349.
15. Gibbard, P.L., 1982. Terrace stratigraphy and drainage history of the plateau gravels of north Surrey, south Berkshire, and north Hampshire, England. Proceedings of the Geologists' Association 93, 369–384.
16. Gomme, J., Buss, S., 2006. Groundwater quality review: Thames Valley Gravels, Thames Region. Environment Agency Report 6441R8, Reading, UK. .
17. Green, C.P., Coope, G.R., Jones, R.L., Keen, D.H., Bowen, D.Q., Currant, A.P., Holyoak, D.T., Ivanovich, M., Robinson, J.E., Rogerson, R.J., Young, R.C., 1996. Pleistocene deposits at Stoke Goldington, in the valley of the Great Ouse, UK. Journal of Quaternary Science 11, 59–87.
18. Guccione, M.J., 2008. Impact of the alluvial style on the geoarcheology of stream valleys. Geomorphology 101, 378–401.
19. Hickin, A.S., Kerr, B., Barchyn, T.E., Paulen, R.C., 2009. Using ground-penetrating radar and capacitively coupled resistivity to investigate 3-D fluvial architecture and grain-size distribution of a gravel floodplain in northeast British Columbia, Canada. Journal of Sedimentary Research 79, 457–477.
20. Hill, I., Jeffrey, C.A., Hameed, A., 2011. Geophysical quality assessment of sand and gravel deposits. Proceedings of Near Surface 2011, European Association of Geoscientists and Engineerings, University of Leicester, UK.

21. Hirsch, M., Bentley, L.R., Dietrich, P., 2008. A comparison of electrical resistivity, ground penetrating radar and seismic refraction results at a river terrace site. Journal of Environmental and Engineering Geophysics 13, 325–333.

22. Horton, A., 1970. The drift sequence and sub-glacial topography in parts of the Ouse and Nene Basins. Report of the Institute of Geological Sciences 70/9.

23. HMSO, London. Hsu, H.L., Yanites, B.J., Chen, C.C., Chen, Y.G., 2010. Bedrock detection using 2D electrical resistivity imaging along the Peikang River, central Taiwan. Geomorphology 114, 406–414.

24. Jeffrey, C., Hill, I., Hameed, A., 2011. Deposit knowledge for efficient production. Defra, Mineral Industry Sustainable Technology Fund, Commissioned Report MA/7/G/ 002. University of Leicester, UK.

25. Loke, M.H., Barker, R.D., 1996. Practical techniques for 3D resistivity surveys and data inversion. Geophysical Prospecting 44, 499–523.

26. Loke, M.H., Acworth, I., Dahlin, T., 2003. A comparison of smooth and blocky inversion methods in 2D electrical imaging surveys. Exploration Geophysics 34, 182–187.

27. Magnusson, M.K., Fernlund, J.M.R., Dahlin, T., 2010. Geoelectrical imaging in the interpretation of geological conditions affecting quarry operations. Bulletin of Engineering Geology and the Environment 69, 465–486.

28. Marr, D., Hildreth, E., 1980. Theory of edge-detection. Proceedings of the Royal Society of London, Series B: Biological Sciences 207, 187–217. Merritt, J.W., 1992. A critical-review of methods used in the appraisal of onshore sand and gravel resources in Britain. Engineering Geology 32, 1–9.

29. Peterson, C.D., Minor, R., Peterson, G.L., Gates, E.B., 2011. Pre and post-Missoula flood geomorphology of the pre-Holocene ancestral Columbia River valley in the Portland forearc basin, Oregon and Washington, USA. Geomorphology 129, 276–293.

30. Revil, A., Cary, L., Fan, Q., Finizola, A., Trolard, F., 2005. Self-potential signals associated with preferential ground water flow pathways in a buried paleo-channel. Geophysical Research Letters 32, L07401.

31. Rogerson, R.J., Keen, D.H., Coope, G.R., Robinson, E., Dickson, J.H., Dickson, C.A., 1992. The fauna, flora and palaeoenvironmental significance of deposits beneath the low terrace of the River Great Ouse at Radwell, Bedfordshire, England. Proceedings of the Geologists' Association 103, 1–13.

32. Rowan, T., 1990. Functional stability analysis of numerical algorithms. Ph.D. thesis, Department of Computer Sciences, University of Texas at Austin.

33. Sass, O., 2007. Bedrock detection and talus thickness assessment in the European Alps using geophysical methods. Journal of Applied Geophysics 62, 254–269.

34. Sheather, S.J., Jones, M.C., 1991. A reliable data-based bandwidth selection method for kernel density estimation. Journal of the Royal Statistical Society B 53, 683–690.

35. Shevnin, V., Mousatov, A., Ryjov, A., Delgado-Rodriquez, O., 2007. Estimation of clay content in soil based on resistivity modelling and laboratory measurements. Geophysical Prospecting 55, 265–275.

36. Sjodahl, P., Dahlin, T., Zhou, B., 2006. 2.5D resistivity modeling of embankment dams to assess influence from geometry and material properties. Geophysics 71, G107–G114.

37. Smith, M.R., Collis, L., 2001. Aggregates: Sand, Gravel and Crushed Rock Aggregates for Construction Purposes. The Geological Society, London. Suzuki, R., Hiyama, T., Asanuma, J., Ohata, T., 2004. Land surface identification near Yakutsk in eastern Siberia using video images taken from a hedgehopping aircraft. International Journal of Remote Sensing 25, 4015–4028.

38. Tye, A.M., Kessler, H., Ambrose, K., Williams, J.D.O., Tragheim, D., Scheib, A., Raines, M., Kuras, O., 2011. Using integrated near-surface geophysical surveys to aid mapping and interpretation of geology in an alluvial landscape within a 3D soil-geology framework. Near Surface Geophysics 9, 15–31.

39. Vafidis, A., Economou, N., Ganiatsos, Y., Manakou, M., Poulioudis, G., Sourlas, G., Vrontaki, E., Sarris, A., Guy, M., Kalpaxis, T., 2005. Integrated geophysical studies at ancient Itanos (Greece). Journal of Archaeological Science 32, 1023–1036. Wardrop, D.R., 1999. A study on the accuracy of sand and gravel reserve estimates. The Quarterly Journal of Engineering Geology 32, 81–86.

40. Wilkinson, P.B., Loke, M.H., Meldrum, P.I., Chambers, J.E., Kuras, O., Gunn, D.A., Ogilvy, R.D., 2012. Practical aspects of applied optimised survey design for electrical resistivity tomography. Geophysical Journal International 189, 428–440.

41. Wymer, J., 1988. Paleolithic archaeology and the British Quaternary sequence. Quaternary Science Reviews 7, 79–98.

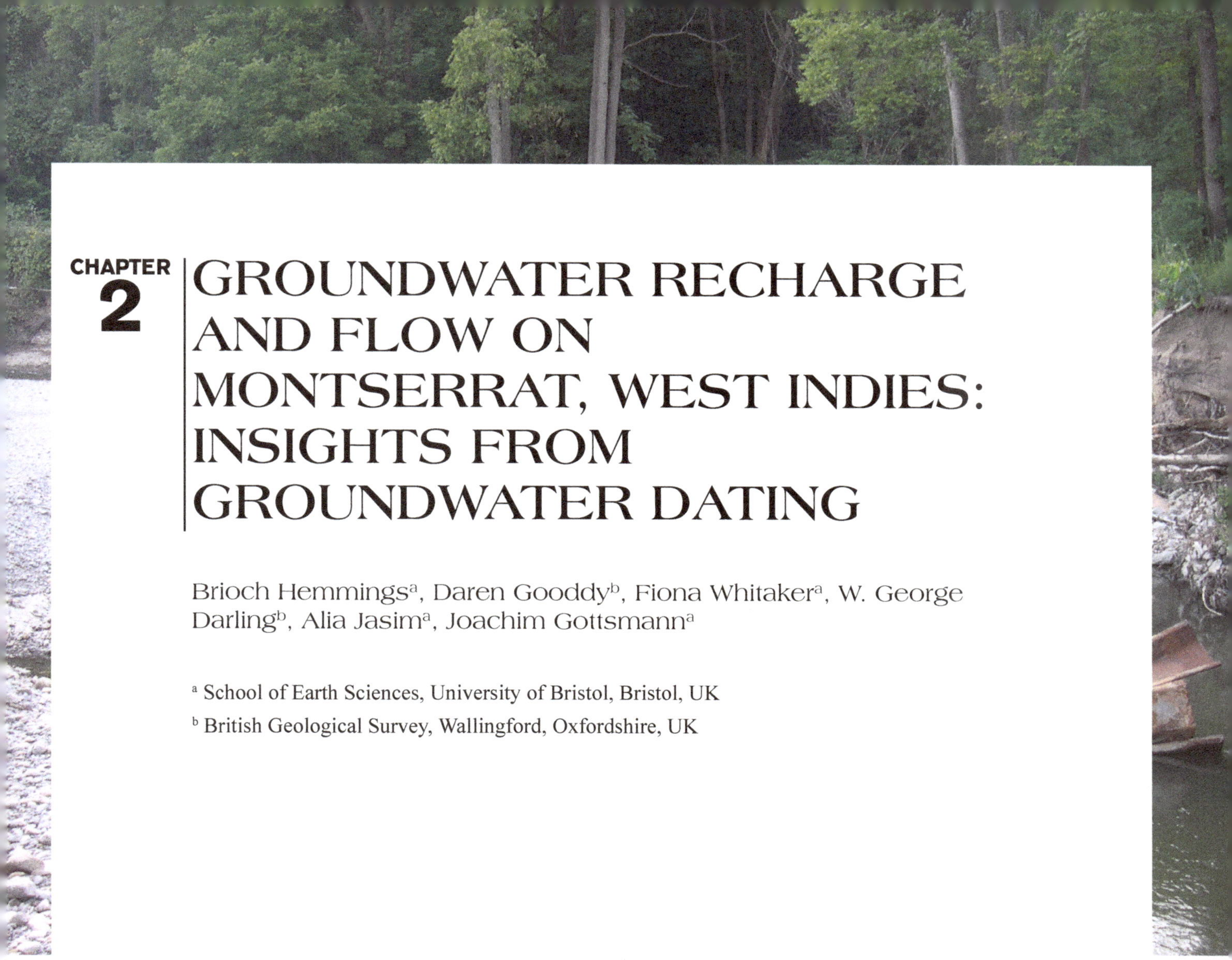

CHAPTER 2

GROUNDWATER RECHARGE AND FLOW ON MONTSERRAT, WEST INDIES: INSIGHTS FROM GROUNDWATER DATING

Brioch Hemmings[a], Daren Gooddy[b], Fiona Whitaker[a], W. George Darling[b], Alia Jasim[a], Joachim Gottsmann[a]

[a] School of Earth Sciences, University of Bristol, Bristol, UK

[b] British Geological Survey, Wallingford, Oxfordshire, UK

ABSTRACT

Analysis of δ^2H and $\delta^{18}O$ isotopes, and chlorofluorocarbon (CFC) anthropogenic tracers in Montserrat groundwater provides insights into the age and provenance of the spring waters.

New hydrological insights

δ^2H and $\delta^{18}O$ analysis indicates uniform recharge elevations for groundwaters on Montserrat. CFC-11 and CFC-12 analysis reveals age differences between isotopically similar, high elevation springs and low elevation aquifer waters. Low CFC concentrations within a confined low elevation aquifer suggest water ages of ~45 years. High CFC concentrations in the northern and western springs are explained by rapid infiltration of cool (high CFC concentration) rainfall into saturated compartments, with flow through the vadose zone to the phreatic zone dominated by compartment flow. Lower CFC concentrations in a number of aligned warmer springs suggest a contribution from older, warmer waters from depth. Temperatures and CFC concentrations indicate older component supply rates of up to 8 L/s to the highest yielding spring on Centre Hills, with contributions of up to 75% in the warmest spring waters.

INTRODUCTION

Despite high rainfall on the tropical, volcanic island of Montserrat, Lesser Antilles (Fig. 1a), surface water is limited and the deeply incised drainage channels (known locally as ghauts) are largely ephemeral. The only permanent streams are sourced from springs which emerge at elevations ranging from 200 and 400 m (amsl) (Hemmings et al., 2014). The freshwater demands of the entire island are met by potable supply from six productive springs on the flanks of the extinct volcanic complex of Centre Hills (CH) (Fig. 1b). In 2013 the combined discharge from the six supply springs was in excess of 50 L/s (Montserrat Utilities Ltd. (MUL), unpublished data). Supply at this rate easily meets the current demands of the population, (~19 L/s). However, consumptions rates are expected to rise as population and agriculture recover, during a period of relative quiescence, after >15 years of destructive and disruptive volcanic activity at Soufrière Hills Volcano. An expected increase in future freshwater demands, together with historical fluctuations in spring discharge (Hemmings et al., 2014), demonstrates the fragility of the freshwater resource on Montserrat.

Citation: Brioch Hemmings, Daren Gooddy, Fiona Whitaker, W. George Darling, Alia Jasim, Joachim Gottsmann, Groundwater recharge and flow on Montserrat, West Indies: Insights from groundwater dating, doi:10.1016/j.ejrh.2015.08.003

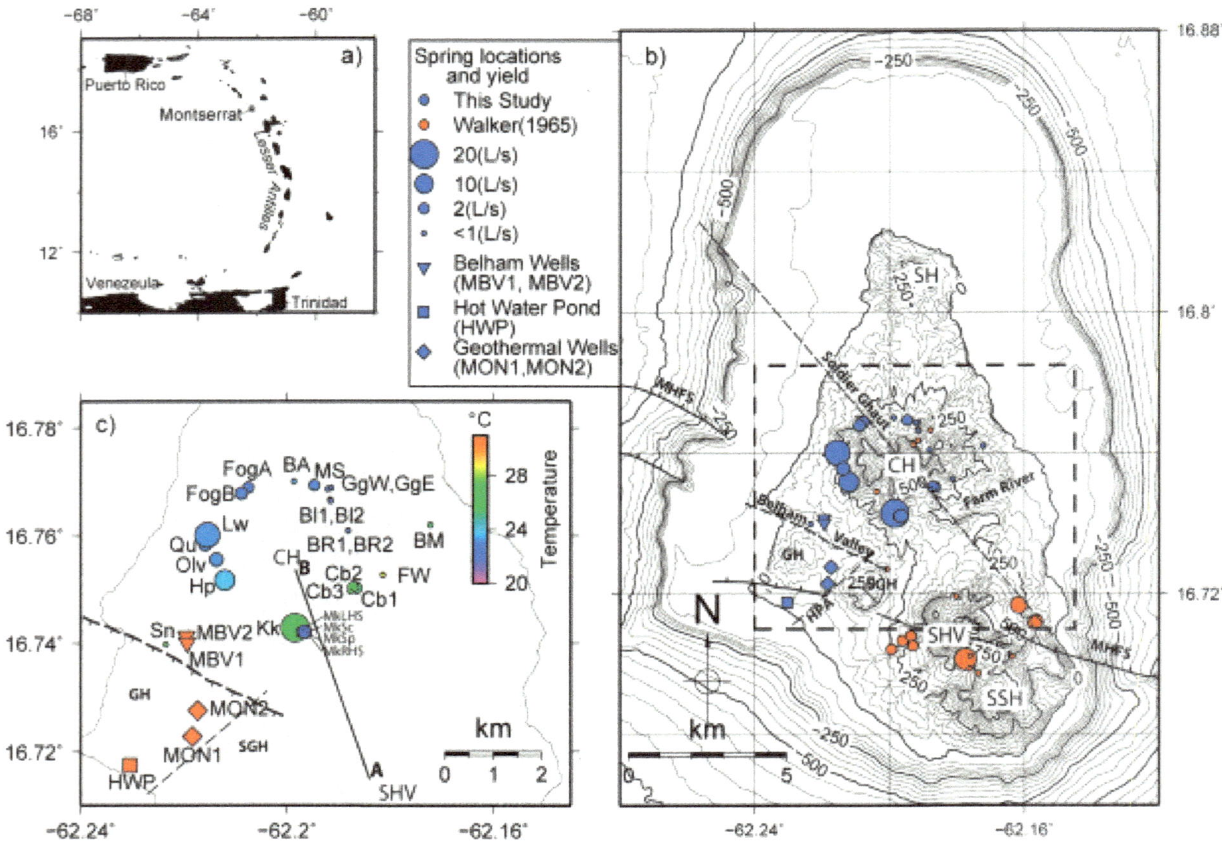

Figure 1: (a) Location of Montserrat within the Lesser Antilles arc. (b) Spring locations and discharge with major volcanic complexes (SH, Silver Hills; CH, Centre Hills; SHV, Soufrière Hills Volcano; SSH, South Soufrière Hills.) Also marked are GH, Garibaldi Hill; SGH, St George's Hill and the major active (solid lines) and inferred (dashed lines) fault systems: MHFS, Montserrat-Havers Fault System and Belham Valley fault, from Feuillet et al. (2010); Soldier Ghaut fault, inferred by Hautmann et al. (2010); HPA, the NE trending fault inferred by Ryan et al. (2013) from earthquake hypo-centres. (c) Temperatures of spring waters and Belham Valley artesian well waters (inverted triangles) Hemmings et al. (2014). Also marked is Hot Water Pond (HWP, square) at 56 °C and the locations of the two geothermal boreholes (MON1 and MON2, diamonds). Spring codes relate to springs in Table 1. Circle radius is proportional to spring discharge. The line A–B is the surface trace of the schematic cross-section presented in Fig. 8.

Fragile water resources are familiar to many island populations, particularly on volcanic islands where high surface permeabilities are typical and groundwater flow is critical to supply. Understanding the complex groundwater systems that exist in these settings is essential for effective and sustainable resource management. The results of this study highlight and illuminate the complexity that exists in volcanic island hydrology, particularly volcanic arc islands. The methods we describe will be useful for study of groundwater systems in other volcanic island settings. In a recent study on the basaltic ocean island of Hawaii, Kelly and Glenn (2015) combined anthropogenic tracers (Chlorofluorocarbons, CFCs) and isotope analysis to explore the residence times and recharge areas of groundwater in coastal drinking water aquifers. Here, we analyse CFCs and stable isotopes δ^2H and δ^{18}O, in Montserrat groundwater, aiming to provide insights into the age and provenance of the island's vital spring waters.

CFCs as Groundwater Tracers

Chlorofluorocarbon (CFC) analysis, particularly CFC-11 and CFC-12, have proven to be a valuable tool for dating groundwater younger than 60 years (Stuart et al., 2010). It also has the potential to provide insights into the extent of mixing between groundwater components of different age (Gooddy et al., 2006). The use of CFCs for dating of groundwater is based on known concentrations in the atmosphere over the last 60 years, together with observations that they are globally well-mixed, and the assumption that their solubility in water follows Henry's Law at the temperature of recharge (Darling et al., 2012). CFC concentrations in groundwater are assumed to be proportional to the partial pressure, pi, of the gas in the atmosphere at the time of recharge (Dunkle et al., 1993 and Plummer et al., 2006):

$$C_i = K_{H_i} p_i, \tag{1}$$

where Ci is the concentration of a particular CFC compound in water, and K_{Hi} is the Henry's Law coefficient for that CFC compound. KH is a function of both temperature and salinity and has the form (Warner and Weiss, 1985),

$$\ln K_H = a_1 + a_2\left(\frac{100}{T}\right) + a_3 \ln\left(\frac{T}{100}\right) + S\left[b_1 + b_2\left(\frac{T}{100}\right) + b_3\left(\frac{T}{100}\right)^2\right] \tag{2}$$

where T is temperature in degrees kelvin and S is salinity in parts per thousand by weight. a_1, a_2, a_3, b_1, $b_2$and b_3 are least squares fitting parameters, determined by and detailed in Warner and Weiss (1985).

Eq. (2) demonstrates the temperature and salinity dependance of Henry›s Law coefficient, KH. For groundwater dating application the salinity correction is often omitted

(Plummer et al., 2006). Partial pressure, pi, of CFC gases in the atmosphere is also dependent on temperature, due to the variation of water vapour pressure with temperature. pi is also a function of atmospheric pressure which is usually approximated using an estimate for recharge elevation (Plummer et al., 2006). Fig. 2 illustrates the effect that recharge temperature has on CFC-11 and CFC-12 concentrations in water recharged since 1930. CFC concentrations are higher when recharge occurs at lower temperatures. If recharge temperature, and to a lesser extent the recharge elevation, is well known, analysis of CFC concentration can be used to date groundwater. Fig. 2highlights that well constrained recharge temperature is particularly important for dating groundwater younger than 20 years.

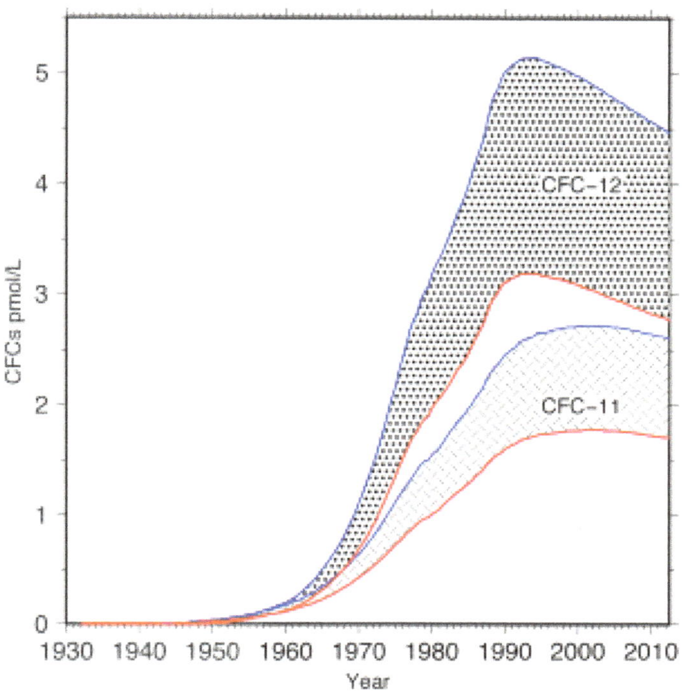

Figure 2: Groundwater concentration ranges for CFC-11 (hatched) and CFC-12 (dotted) at recharge temperatures between 12 °C (blue lines) and 22 °C (red lines), due to changes in atmospheric concentration over time. CFC concentrations in the atmosphere have declined as a result of the Montreal Protocol on Substances that Deplete the Ozone Layer. Based on data from water.usgs.gov/lab/software/air_curve. (For interpretation of the references to color in this figure legend, the reader is referred to the web version of the article.)

Combined CFC and Sulphur hexafluoride (SF_6), analysis of groundwater on Maltese islands of Malta and Gozo provided groundwater ages and saturated zone travel-time estimates of between 15 and 60 years (Stuart et al., 2010). On the volcanic island of Jeju, Korea, Koh et al. (2007) combined CFC-12, CFC-11, SF_6 and H^3 analysis to distinguish groundwater ages in basaltic aquifers. They established that groundwaters followed a binary mixing model between young (~10 years) and old (~50 years) end members.

Koh et al. (2007) also highlight issues with contamination of groundwater samples leading to potential tracer excess in volcanic settings. In Jeju they suggest that granitic basement rocks provide a terrigenic source of SF_6 that bias groundwater dating estimates towards younger ages, by up to 20 years. Elevated concentrations of SF_6 are potentially due to chemical interaction of groundwaters with rocks containing fluorite or sulphide minerals (Harnisch and Eisenhauer, 1998, Darling et al., 2012 and Roques et al., 2014). Possibly as a consequence of such contamination, SF_6 analysis on groundwater samples from Montserrat did not produce coherent or reliable results. Therefore, only results from CFC analysis are present here.

CFC concentrations in groundwater are also susceptible to contamination. Potential contamination sources include local anthropogenic variations in atmospheric levels (Ho et al., 1998), and environmental contamination, particularly from leaching from landfill (Höhener et al., 2003), and mixing with waste water (Kralik et al., 2014). It has also been speculated that volcanoes can produce chlorofluorocarbons (Gribble, 1994), thus disrupting both CFC and SF_6 analysis on samples from active volcanic environments.Frische et al. (2006) measured CFCs and SF_6 above detection limits in volcanic gases from Nicaraguan subduction volcanoes. However, they found that levels were in relative proportions characteristic of the ambient atmosphere, concluding that the levels were due to varied amounts of air entrainment into the volcanic edifice, with no significant additional CFC or SF_6 production by the volcanic activity.

Hydrological and Hydrogeological Background

In addition to the six tapped springs on Centre Hills, Montserrat, there are a further 19 springs which are untapped or abandoned, due to low yields or limited access. The springs are generally situated on the edges of steep and deeply incised drainage valleys or ghauts. Long-term spring yield monitoring (MUL, unpublished data), combined with one-off flow gauging during field visits in 2010–2013, provides a total spring discharge estimate of ~78 L/s. Individual springs discharge water at up to 19 L/s. Such high discharge rate at relatively high elevation (300 m amsl for Killiekrankie spring, the highest order spring), on an island with only ephemeral river flow and no other surface water, requires the presence of reasonably laterally continuous low permeability units. InHemmings et al. (2014) we proposed two potential conceptual hydrological models describing Montserrat springs. Type 1 (Fig. 3a), based on models developed on volcanic islands such as the Canaries (Cabrera and Custodio, 2004 and Custodio, 2007), describes a low permeability intrusive and (hydro)thermally altered core, overlain by fractured and jointed extrusive andesitic lava breccia and talus (autoclastic) and flanked by volcaniclastic and reworked material. Relatively high permeability autoclastic, volcaniclastic and reworked deposits host aquifers which, when exposed by erosion, form springs, normally located on the slopes of incised ghauts. In Type 2 scenario (Fig. 3b), low permeability fine ash and/or weathered palaeosol layers perch aquifers at high elevation. Again, springs occur where the aquifers are exposed by erosion.

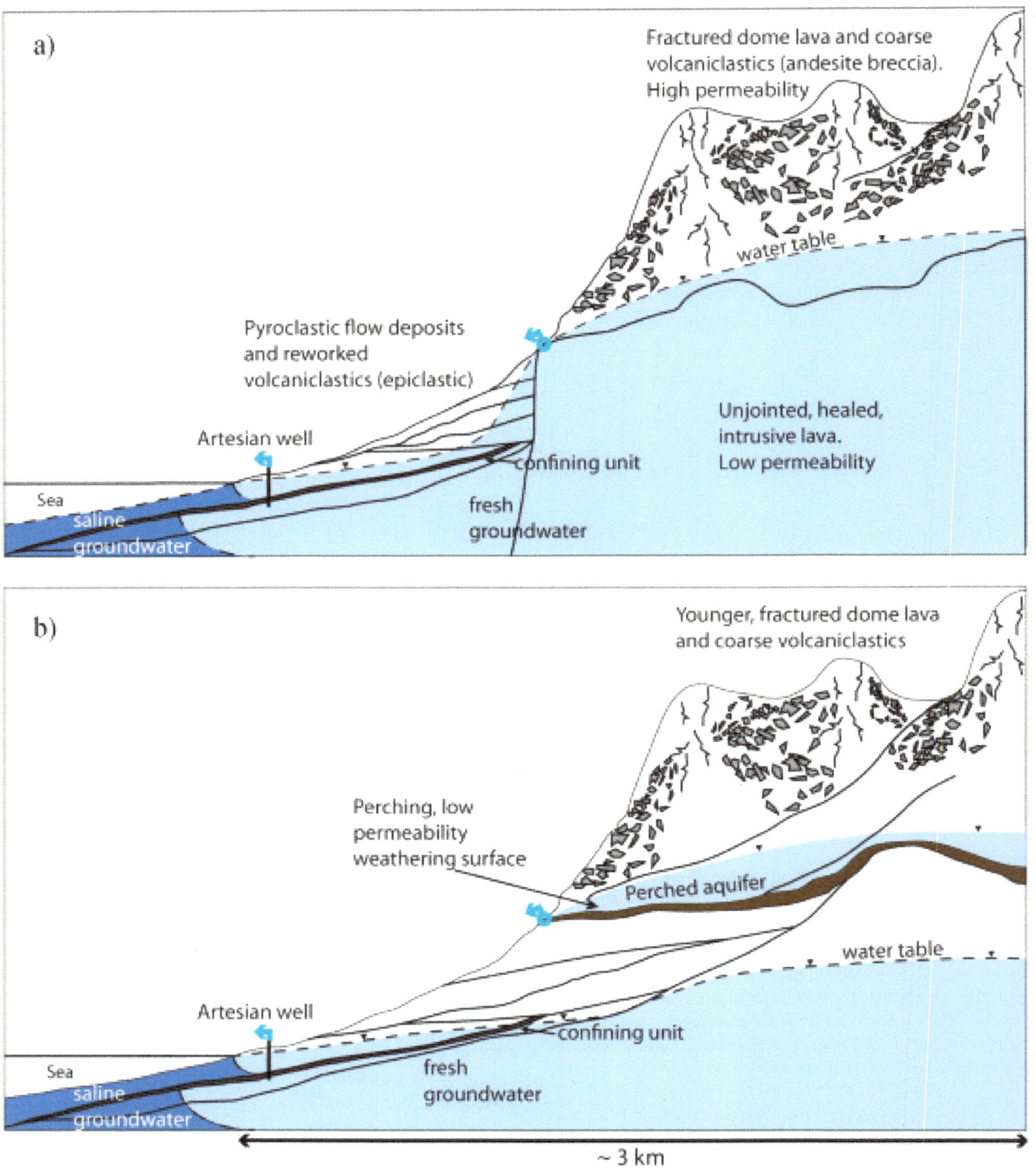

Figure 3: Type 1 conceptual model (a) and the Type 2 conceptual model (b) of the hydrology on Montserrat, afterHemmings et al. (2014)

The temperature of Centre Hills spring waters are temporally stable but vary spatially, from 22 °C to 29 °C (Hemmings et al., 2014, Fig. 1c). Warmer springs, with temperatures greater than 25 °C, are located in a line extending north-east from the highest yielding Killiekrankie spring (Fig. 1c). Specific electrical conductivity (SEC_{25}), referenced to 25 °C, is also elevated in the warmer springs (Table 1). To explain the systematic temperature variation, Hemmings et al. (2014) suggest a contribution of warmer waters from an aquifer at depth, possibly supplied through faults to the high elevation spring aquifers. Previous studies have not identified significant chemical differences between the Centre Hills spring waters (Chiodini et al., 1996, Davies and Peart, 2003 and Jones et al., 2010). Jones et al. (2010) attributed high SO_2/Cl in a few untrapped Centre Hills springs in 2008 to a contribution from run-off influenced by volcanically derived SO_4.

Table 1: Groundwater CFC concentrations, and $\delta^{18}O$ and $\delta^{2}H$ results from springs and wells surveyed in February 2013. Also displayed are temperature and specific electrical conductivity (SEC_{25}, referenced to 25 °C) data, initially presented in Hemmings et al. (2014)

Spring code	Spring name	Elev. (m amsl)	Date	Temp. (°C)	SEC_{25}(μS/cm)	CFC-12 (pmol/L)	CFC-11 (pmol/L)	$\delta^{18}O$ (‰)	$\delta^{2}H$ (‰)
BA	Blackwood Allen	315	10/02/13	23.6	458	1.89	3.16	−2.95	−11.3
Bl1	Bottomless Ghaut 1	332	04/02/13	22.1	554	2.63	4.85	−3.3	−13.7
Bl2	Bottomless Ghaut 2	356	04/02/13	22.1	255	2.23	5.27	−3.23	−12.3
BM	Bessy Mack	190	11/02/13	25.4	1288	1.98	2.53	−3.3	−13.6
BR1	Big River 1	398	04/02/13	22.7	331	2.17	4.52	−2.99	−10.5
BR2	Big River 2	398	04/02/13	22.1	353	2.37	5.24	−2.9	−9.9
Cb1	Corbett 1	306	02/02/13	25.2	467	1.90	4.01	−3.09	−11.3
Cb2	Corbett 2	325	02/02/13	26.1	423	1.34	2.32	−3.11	−11.8
Cb3	Corbett 3	327	11/02/13	26.9	419	1.55	2.42	−3.12	−11.3
FogA	Fogarty A	309	18/02/13	22.8	287	2.10	4.62	−3.01	−10.0
FogB	Fogarty B	306	18/02/13	22.9	283	2.24	3.87	−3.06	−10.9
FW	Fairy Walk	248	11/02/13	28.8	555	1.14	2.05	−3.48	−14.9
Hp	Hope	306	08/02/13	23.7	306	2.28	4.45	−2.98	−10.2
Kk	Killiekrankie	297	31/01/13	25.9	308	1.99	3.25	−3.05	−10.0
Kk	Killiekrankie	297	14/02/13	25.7	309	1.71	2.94		
Lw	Lawyers	188	30/01/13	23.2	281	2.04	8.27	−3.00	−9.6
Lw	Lawyers	188	10/02/13	23.2	279	2.49	6.02		
MS	Mongo/ Underwood	345	09/02/13	23.0	588	2.11	5.55	−3.13	−12.1
Olv	Olveston	240	01/02/13	23.2	282	2.12	4.51	−2.91	−10.7
Qu	Quashie	231	30/01/13	23.2	285	1.88	4.16	−2.95	−10.0
Qu	Quashie	231	10/02/13	23.1	285	2.11	3.50		
Sn	Sunny	15	06/02/13	27.4	1703	1.04	2.15	−2.89	−11.0
HWP1	Hot Water Pond seaward	1.0	07/02/13	40.4	49,100	0.79	1.41	2.47	−0.9
HWP2	Hot Water Pond up−valley	1.5	07/02/13	56.1	38,600	0.88	0.56	1.09	−6.5
MBV1	Belham Well 1	34	06/02/13	31.0	663	0.38	0.56	−2.92	−10.9
MBV2	Belham Well 2	29	06/02/13	31.0	630	0.56	0.97	−2.95	−11.2

Additional and notable hydrological features include two artesian wells within the Belham Valley: MBV1 and the flowing artesian MBV2 (triangles on Fig. 1b and c). These wells tap a confined aquifer, 15–38 m below mean sea level (bmsl), beneath one of the islands larger, though still ephemeral, drainage channels. Drilling well logs from 2004 (HydroSource, 2004) describe a thin (1 m) clay cap, beneath a thicker (12 m) lahar deposit, overlying the Belham Valley aquifer which is confined in reworked alluvial and channel deposits. These units are beneath a further ~35 m lahar deposits, alternating with volcanic breccia deposits (between 1.5 and 12 m thick), with occasional, thin (~1 m) lapilli fall and channel units. The remaining ~10 m to the 2004 drilling surface is composed of block, pumice and ash flow deposits. Since the onset of volcanic activity in 1995 the Belham valley has been inundated and infilled by remobilised pyroclastic flow and lahar deposits (Barclay et al., 2007), raising the channel surface by 10–15 m (Donnelly, 2007). This process has continued since the completion of the wells in 2004, raising the ground surface around the well housings by 0.5–1 m. The most recent deposits are composed of unconsolidated volcanic sand, with boulders and cobbles, and occasional clay and silt interbeds (Donnelly, 2007). Waters produced by the Belham wells are at 31 °C with SEC_{25} of 663 and 630 μS/cm, respectively; the wells are maintained for back-up supply. In other respects, the Belham well waters and the springs are chemically similar (Jones et al., 2010). Jones et al. (2010) suggest that the well waters were sourced from a deeper, longer residence time aquifer than the Centre Hills springs, based on detection of a slight enrichment in Na, Mg, Si, Cl, K, Ca, Fe Rb and Sr.Jones et al. (2010) also suggest that an enrichment in $\delta^{13}C$ may represent some hydrothermal contribution to the Belham aquifer.

A definite manifestation of the hydrothermal system exists at the coast, just north of the destroyed former capital Plymouth; the now buried Hot Water Pond (HWP). Previous studies identified seeps suppling HWP waters close to 90 °C (Chiodini et al., 1996).Chiodini et al. (1996) characterised these waters as "neutral Na-Cl type", enriched in Ca, K, Li, B NH_4, and SiO_2 and depleted in Mg and SO_4 relative to sea water. The hydrothermal system on Montserrat has been the subject of some recent interest regarding geothermal energy development (Younger, 2010, Ryan et al., 2013 and Poux and Brophy, 2012). HWP is considered to be a surface expression of hot upflow of the Soufrière Hills Volcano (SHV) geothermal system, concentrated at the intersection of the NW-WNW trending zone of weakness identified by Wadge and Isaacs (1988) – the Belham Valley fault system (Feuillet et al., 2010) – and a NE trending fault inferred by earthquake hypo centres (Ryan et al., 2013; HPA in Fig. 1). Recent seismological studies support the presence of such a NE-SW open fracture system in this location (Baird et al., 2015). Geothermal boreholes MON1 and MON2 (Figure 1), drilled in 2013, target this feature with reported temperatures of 230 °C and 265 °C at 2300 and 2800 m depth, respectively (Ryan et al., 2013 and Ryan and Shalev, 2014).

With the analysis of CFC-11 and CFC-12 concentrations and δ^2H and $\delta^{18}O$ of Montserrat groundwater samples, this study adds to temperature and SEC_{25} data published inHemmings et al. (2014). This tracer analysis allows us to distinguish different groundwater components in otherwise chemically similar waters.

METHODS

Twenty-six groundwater samples from Montserrat were collected for analysis of CFC and SF6 for age dating, during a February 2013 field campaign when spring temperature and specific electrical conductivity (SEC_{25}) were also recorded (see Hemmings et al., 2014for SEC_{25} and temperature observations). Groundwater samples were also collected for analysis of $\delta^{18}O$ and δ^2H isotopes. Twenty-two samples are from 19 springs on and around Centre Hills; two samples are from wells that tap a confined aquifer in the Belham Valley (MBV); two further samples are from shallow (50 cm) pits dug to intercept the now buried coastal Hot Water Pond (HWP), where geothermal groundwater is expected to mix with seawater (Hemmings et al., 2014).

For reliable CFC analysis careful sample collection is essential, to avoid contamination by atmospheric air (Plummer et al., 2006). Groundwater from springs and HWP were collected through PVC tubing using a battery powered impeller pump. To minimise contact with the air, samples were collected as close to the spring emanation point as possible and upstream of any water treatment (chlorination). A motor speed controller was devised to allow pumping at reduced flow rates in the low flow springs and also to prevent turbulence. Three times the tubing volume was pumped before sampling at each spring site. Sample collection followed the USGS single bottle method (Plummer et al., 2006). Glass, 125 mL sample bottles with foil-lined caps were rinsed three times in the spring waters, then filled inside a 10 L bucket. The sample bottles were filled, from the base, until the bucket overflowed. Critically, the bottles were capped underwater to prevent any air contamination. An extra seal of tape was then wrapped around the caps to prevent air contamination during transit.

For sampling of the Belham Valley well waters, the procedure was modified. The MBV1 sample was taken during a pumping test; the well was pumped continuously for 30 min at ~30 L/s before the sample was collected from the medium density polyethylene (MDPE) outflow pipe. Due to the very high discharge rate it was not possible to fill the sample bottle from the base. However, care was still taken to over fill the sample bottle and seal underwater, as above, to limit contact with the air. MBV2 is and artesian well, flowing at ~3.5 L/s, and is piped through an MDPE pipe to a discharge point ~10 m from the wellhead. The discharging pipe is too wide to fill the sample bottle from the base; once again care was taken to over-fill the bottle in the bucket to minimise air contact and exclude air from the sample.

Isotope samples were collected concurrently in 10 mL glass vials, and capped immediately to prevent evaporation. Waters were prepared for stable isotope analysis on an Isoprime 100 mass spectrometer using standard methods (CO2 equilibration for$\delta^{18}O$, on-line chromium reduction for δ^2H). Values are expressed as permil (‰) with respect to VSMOW2 (Vienna Standard Mean Ocean Water2). Analytical precisions are ±0.1‰for $\delta^{18}O$ and ±1‰for δ^2H.

Bubble formation due to dissolved air in CFC samples is normal and can be corrected for during analysis (Plummer et al., 2006). However, as no significant bubbles formed in the samples collected from Monserrat, such correction was not necessary. The CFC samples were analysed at the BGS GeoEnvironmental Laboratories using gas chromatography with cryogenic preconditioning (Oster et al., 1996 and Gooddy et al., 2006). The anticipated analytical precision of the technique is on the order of ±5% (Gooddy et al., 2006). Duplicates samples were collected for three springs: Lawyers, Quashie and Killiekrankie. Duplicate samples were taken between 11 and 14 days apart and show percentage differences of between ±10 and ±20%. There is large discrepancy in CFC-11 concentration for the two samples from Lawyers. CFC-11 concentration in the first sample from Lawyers is significantly elevated above modern CFC-11 levels; it is therefore considered contaminated and removed from further analysis. Although the results from the duplicate samples may genuinely represent different waters at different sampling times, the other field parameters (temperature, SEC_{25} and discharge) are the same at both sampling times. For the purposes of this study, differences between duplicate samples are assumed to represent potential sampling error.

RESULTS

The results of the CFC and $\delta^{18}O$ and $\delta^{2}H$ analysis are presented in Table 1, together with additional field parameters (temperature and SEC_{25}). $\delta^{18}O$ and $\delta^{2}H$ values for springs and the Belham Valley aquifer sampled in the two Belham Wells (MBV1 and MBV2) are tightly distributed arounds means of −3.08‰$\delta^{18}O$ and −11.3‰$\delta^{2}H$, with standard deviations a little outside measurement precision (0.16‰and 1.5‰, respectively). The springs lie offset from the meteoric water line, consistent with the offset of analysed rainfall and through fall samples, and also with previous analysis of spring samples collected in June 2008 (Jones et al., 2010; Fig. 4a). Fairy Walk (FW), the warmest of the springs (28.8 °C) is marginally depleted in ^{18}O and ^{2}H, relative to the other springs and the well waters (Fig. 4b), with −3.48‰$\delta^{18}O$ and −14.9‰$\delta^{2}H$. All the spring and well waters are noticeably depleted relative to means of −1.21‰$\delta^{18}O$ and −7.36‰$\delta^{2}H$ for rainfall and throughfall samples collected during the same campaign. HWP samples are significantly enriched in both ^{18}O and ^{2}H relative to the springs (Fig. 4a, Table 1).

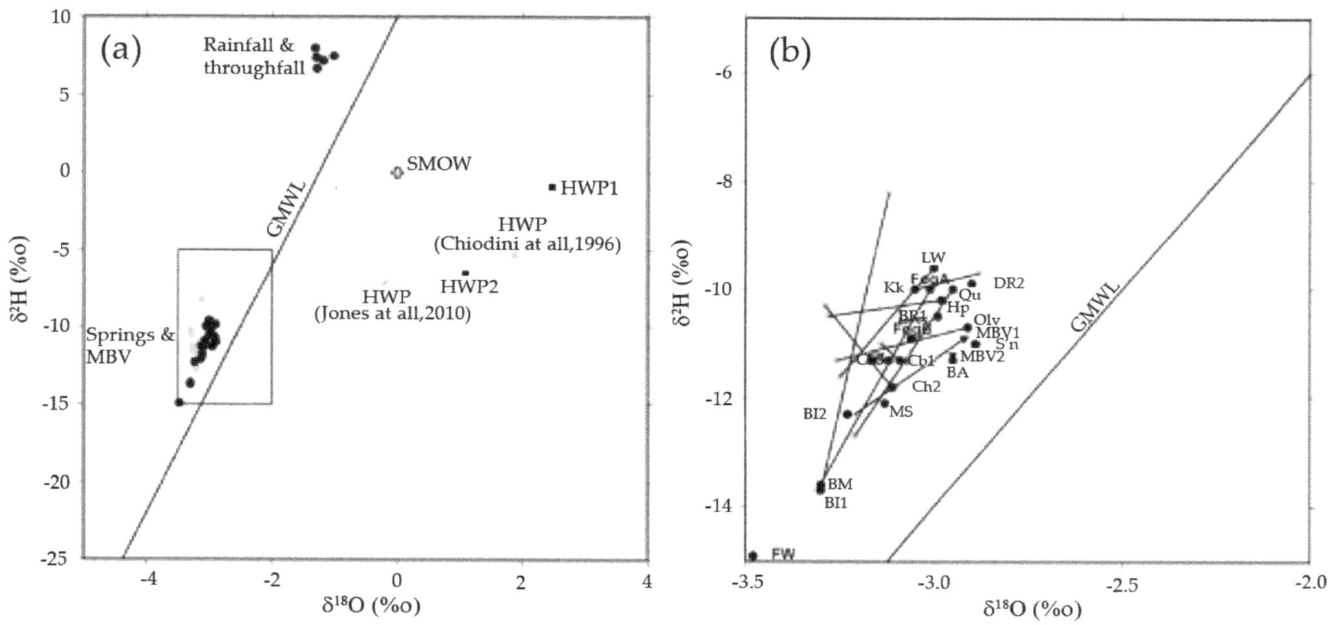

Figure 4: (a) $\delta^{18}O$ vs $\delta^{2}H$ for all Montserrat water samples. Isotopic values of springs, rainfall and throughfall (black dots), collected during the same campaign, lie offset from the global meteoric water line (GMWL, black line). Values for springs and Belham Valley well waters (MBV, inverted triangles) are consistent with observations of Jones et al. (2010) (grey dots and inverted triangles). Waters sampled from Hot Water Pond (HWP, black squares) are significantly enriched relative to the springs. HWP data from Chiodini et al. (1996) and Jones et al. (2010) are also plotted as grey squares in (a). The cross marks Standard Mean Ocean Water (SMOW). The box in (a) demarks the limits for (b). (b) is focussed on the data range covered by the springs and Belham Valley well waters. Symbol format is the same as (a).

The springs and well waters all have similar isotopic signatures. However, the results from CFC analysis reveal differences, previously unrecognised. Groundwaters from the confined aquifer beneath the Belham Valley (MBV1 and MBV2) have the lowest CFC concentrations, generally indicative of older water.

There is also a clear relationship between the CFC concentrations and the temperature of the springs (Fig. 5). This relationship cannot be accounted for by simple temperature dependence of CFC solubility. CFC solubility is inversely related to temperature. However, Henry's Law predicts CFC concentrations to be just 0.55 pmol/L lower for CFC-12, and 1 pmol/L lower for CFC-11, in water equilibrated with the atmosphere at 30 °C, compared to equilibration at 20 °C. The CFC data suggests that the warmer springs on the south side of Centre Hills produce older water than the cooler springs on the west and north, with the oldest waters sampled in the artesian Belham Valley aquifer.

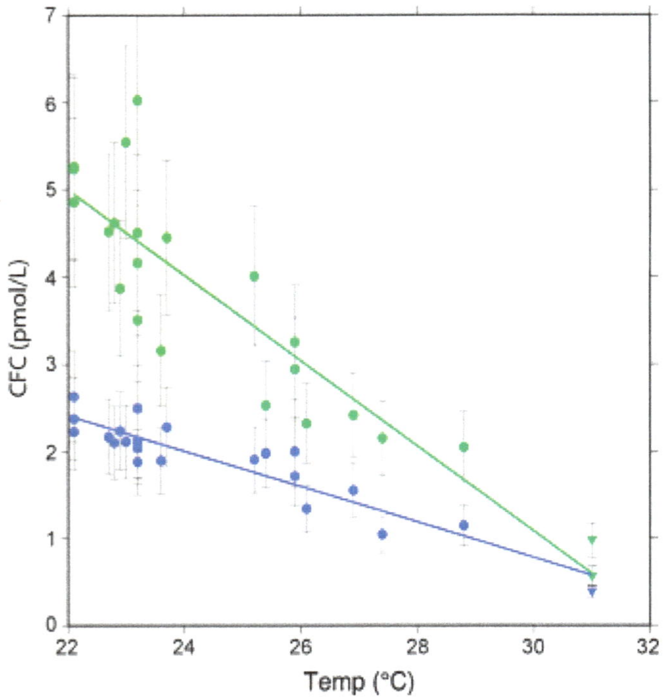

Figure 5: Relationship between spring CFC concentration and temperature in Montserrat groundwater. CFC-11 data (green) follows the linear regression trend $y = -0.49x + 15.75$, with $R^2 = 0.82$ ($n = 23$). CFC-12 concentrations (blue) follow linear regression trend $y = -0.21x + 6.93$, with $R^2 = 0.89$ ($n = 23$). Triangles represent data from Belham Valley wells (MBV1, MBV2). Error bars represent an conservative error estimate of ±20%. (For interpretation of the references to color in this figure legend, the reader is referred to the web version of the article.)

HWP samples also exhibit low CFC concentrations. However, these values are consistent with modern groundwater at HWP field temperatures, suggesting that sampled waters were free to exchange with atmospheric CFC, albeit at elevated temperature.

DISCUSSION

Hot Water Pond

The enrichment of stable isotopes in Hot Water Pond samples, relative to the springs is expected; a positive $\delta^{18}O$ offset from the global meteoric water line is consistent with the observations of Chiodini et al. (1996) and Jones et al. (2010), although the absolute values vary (Fig. 4a). Chiodini et al. (1996), who sampled near 90 °C thermal spring water, prior to the onset of the 1995 volcanic eruption, suggest that the offset may be the result of water–rock interaction within a high enthalpy geothermal system, or mixing with arc-type magmatic water. However, Jones et al. (2010) conclude that HWP waters, including those sampled by Chiodini et al. (1996), lie on a mixing line between seawater, with a slight ^{18}O shift due to interaction with hot rock at depth, and Belham Valley aquifer-type waters. They propose that a lack of pronounced enrichment in volcanic volatiles in HWP samples suggests negligible magmatic fluid contribution. The variation in HWP δ^2H and $\delta^{18}O$ values between previous studies (Chiodini et al., 1996 and Jones et al., 2010) and those presented here may represent temporal evolution or behavioural changes of the thermal spring. However, consistent sampling of the waters at this site is difficult due to dynamic geomorphology associated with its coastal location and recent deposition of volcanic material (Hemmings et al., 2014). Relative enrichment of the seaward sample (HWP1), with higher SEC_{25} and presumably higher seawater content compared to the hotter HWP2, suggest favour for the model of Jones et al. (2010): water on a mixing line between ^{18}O rich seawater and Belham Valley aquifer-type waters. The seawater end-member in this model requires $\delta^{18}O$ enrichment beyond the measured levels for seawater in this region ($\delta^{18}O \approx 1.0$ ‰, Schmidt et al., 1999 and Sanjuan et al., 2005), supporting the hypothesis for interaction with hot-rocks at depth. However, the presence of additional geothermal or magmatic fluid components cannot be ruled-out based on this data alone; detection of such a fluid component requires a more comprehensive chemical analysis.

Springs and Belham Valley Wells

δ^2H and $\delta^{18}O$

The isotopic composition of groundwater can be strongly affected by recharge elevation.$\delta^{18}O$ and δ^2H isotope depletion on the order of −0.15 and −1.0 ‰per 100 m, respectively, have been recorded in a number of island settings (see compilation in Scholl et al., 1996).Scholl et al. (1996) use $\delta^{18}O$ composition variations between −6 and −2 ‰to estimate recharge elevation for well and spring waters on Hawaii. The relative consistency in isotopic composition of the springs and the Belham Valley aquifer waters, presented here, suggest that these waters are derived from recharge at consistent elevations, presumably above the elevation of the springs (400 m amsl). The spring and well waters are depleted in δ^2H and $\delta^{18}O$ compared to rainfall and throughfall samples, collected from elevations below 250 m, during the same sampling campaign (Fig. 4a). This discrepancy is not unexpected; the isotopic composition of groundwater generally reflects the average isotopic composition of recharge. Although the relative depletion of spring waters may result from the groundwater being derived from recharge at higher elevations, this would imply an unusually high lapse rate. It is more likely that the discrepancies between isotopic signatures of groundwater and rainwater are due to temporal variations in the isotopic balance of rainfall and recharge (Clark and Fritz, 1997). Schol et al. (2009) identified different isotopic signatures for different rainfall types in Puerto Rico. Wet season rainfall, delivered largely by tropical easterly waves and low pressure systems, including tropical storms, was isotopically depleted compared to the predominantly orographic showers occurring in the dry season.

Spring and well waters, rainfall and throughfall samples plot offset from the global meteoric water line (Fig. 4a). A tentative local meteoroic water line based on the relatively depleted springs and well waters, and the rainfall sample is defined by a gradient of 9 and an intercept around 16 ‰. This is similar to the local meteoric water line defined for Puerto Rican GNIP stations above 300 m elevation (slope of 8.5, intercept around 16 ‰,IAEA/WMO, 2015 and Schol et al., 2009). It is also comparable to local meteoric water line suggested for the island of Martinique (slope of 8 and intercept of 16 ‰, Sanjuan et al., 2005).

δ^2H and $\delta^{18}O$ values from spring and Belham Valley well samples, collected in February 2013, are consistent with the values obtained by Jones et al. (2010) in June 2008 (Fig. 4b). Jones et al. (2010) attributed differences between waters sampled in June 2008 and August 2003 (Davies and Peart, 2003) to secular or seasonal variations. Rainfall data from Monserrat (provided by Montserrat Utilities Ltd., MUL) reveal that the depth of precipitation was slightly reduced in the year and six months prior to the August 2003 sampling by Davies and Peart (2003) compared to the respective time periods prior to the 2008 sampling by Jones et al. (2010), which itself was drier than prior to the 2013 sampling in this study. However, the quarter-year, and two-month period immediately preceding the 2003 sampling was significantly wetter than the respective periods prior to both the 2008 and 2013 sampling (Fig. 6). From the isotope surveys of Davies and Peart (2003) and Jones et al. (2010) it was not clear if the isotopic variations in groundwater are seasonal or related to longer term changes. The addition of the new δ^2H and $\delta^{18}O$ data presented here suggest that consideration also needs to be given to shorter timescale variations in the isotopic balance of groundwater, on the order of 2–3 months.

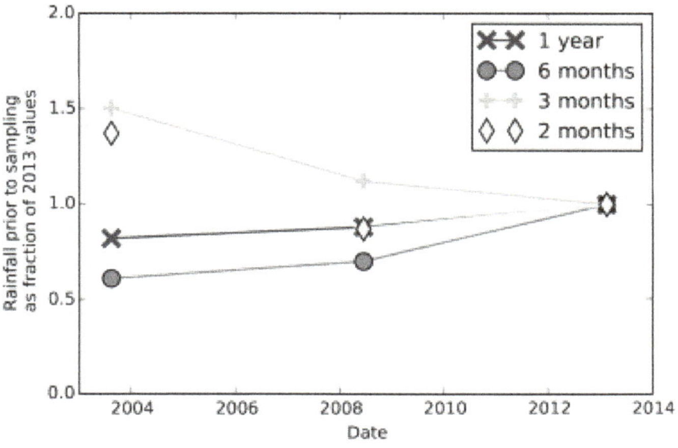

Figure 6: Rainfall for the one year, six, three and two months prior to 2003 and 2008 surveys of Davies and Peart (2003) and Jones et al. (2010) as a fraction of the rainfall over the same period prior to the 2013 sampling.

The relative depletion in isotopic values of the spring waters, compared to the dry season point samples of rainfall and throughfall presented here, may represent a recharge bias towards wet season recharge. However, confirmation of this would require further regular isotope sampling of rainfall and spring discharges.

CFC-11 and CFC-12

As discussed in (Hemmings et al., 2014), recharge at elevation may occur at temperatures below the apparent ambient annual average air temperature at sea level (~26 °C). A maximum recharge temperature of 22 °C can be estimated from the temperature of the springs in the northwest of Centre Hills. This is in-line with estimated air temperatures, due to adiabatic cooling, at ~600 m (amsl) (Blume, 1974). Assuming a recharge temperature of 22 °C the majority of the samples contain CFC concentrations higher than the current atmospheric CFC level (17 samples for CFC-12, 16 samples for CFC-11). Sixteen samples for CFC-12 and 14 samples for CFC-11 are still higher than would be expected for peak atmospheric concentrations in the early 1990 and the early 2000, respectively. Three major scenarios can produce CFC concentrations that are apparently higher than modern atmospheric concentrations:

1. Sample contamination from local CFC source. Contamination often results in an excess of one CFC relative to the other (Höhener et al., 2003 and Darling et al., 2012). An example of a potentially contaminated sample is the first sample from Lawyers spring, with a markedly elevated CFC-11 concentration (Fig. 7). However, the contamination source is unclear. There are no settlements and very little human activity around or above the springs. The other samples do not appear to demonstrate significant contamination.

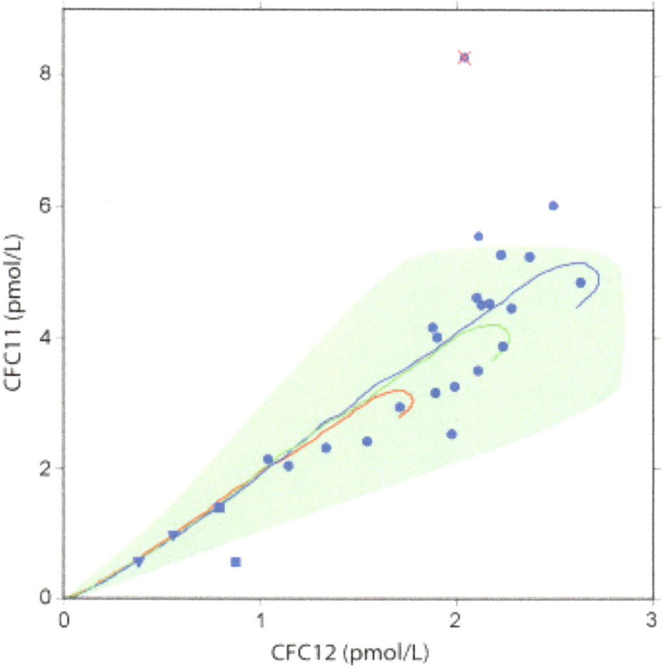

Figure 7: Cross-plot of CFC-12 vs CFC-11. Blue dots are results from Montserrat springs, triangles are from Belham Valley wells, and squares from Hot Water Pond. The red crossed dot is the suspected contaminated Lawyers sample. The red line represents the historical equilibrium curve for groundwater CFC concentration at recharge temperature of 22 °C blue at 12 °C and green at 16 °C. The shaded green area marks the range CFC concentrations that are within error (±20%) of the equilibrium curve at 16 °C. (For interpretation of the references to color in this figure legend, the reader is referred to the web version of the article.)

2. Significant excess air within groundwater. Excess air (Heaton and Vogel, 1981) is caused by the trapping of air within pore spaces below the water table due to fluctuations in water level. Excess air concentrations in most aquifer environments are <5 cm³/L (Standard Pressure and Temperature, STP) but may be significantly higher in fractured hard-rock aquifers (Wilson and McNeill, 1997 and Manning and Caine, 2007). The relative concentration changes in dissolved gases due to excess air are in inverse proportion to their solubility. The CFCs, which are relatively soluble, would for example require an excess air contribution of ~10 cm³/L (STP) to raise concentrations by 10% (Darling et al., 2012).
3. Overestimation of recharge temperatures. The solubility of CFC in water is inversely proportional to

temperature. For example the CFC concentration in water in equilibrium with the atmosphere at 10 °C is 4.98 pmol/L for CFC-11 and 2.85 pmol/L for CFC-12; at 20 °C the concentrations are 3.02 and 1.83 pmol/L, respectively. Clearly, overestimating the recharge temperature can lead to systematic misinterpretation of CFC concentrations.

The cross-plot of CFC-11 versus CFC-12 (Fig. 7) demonstrates that both CFC-11 and CFC-12 concentrations are "over modern" for a number of samples, for recharge temperatures of 22 °C, assuming no excess air correction is required. If lower recharge temperatures are assumed, the CFC concentrations of these "over modern" samples are closer to current or peak atmospheric concentrations. For example, a recharge temperature of 12 °C could explain the CFC-12 concentrations in all the samples, although a few of the CFC-11 values are still high, but within error. Allowing for errors in the analysis of ±20% a recharge temperature of 16 °C can fit the CFC data. Sixteen degrees celsius is significantly lower than the apparent ambient air temperature, even with elevation adjustment.

A potential hypothesis for recharge occurring at such low temperatures is rapid infiltration to a saturated zone before rainwater reaches thermal equilibrium. With a thick vadose zone, as is expected on Montserrat, such a saturated zone may be represented by shallow saturated compartments. Cascade and saturated channel flow between sequential compartments could potentially buffer the recharging water from CFC re-equilibration with the unsaturated zone, preserving the low temperature concentrations in the waters that recharge the regional aquifer, and eventually emanate from the springs. Although this mechanism is not tested here, in terms of the relative equilibration and diffusion times for temperature and CFCs in infiltrating rainwater, the notion of saturated compartmental flow in volcanic settings has been suggested before (Ecker, 1976).

By comparing CFC-11, CFC-12 and SF6 concentrations it can be possible to distinguish between piston flow mechanism and mixing of different aged groundwaters (Gooddy et al., 2006). In the absence of SF6 data and with the uncertainties in recharge temperature and excess air described above, it is not possible to ascribe a definitive flow mechanism or age to the Montserrat spring waters. Assuming piston flow, a recharge temperature of 16 °C, and recharge elevation above the elevation of the springs, for example, 500 m amsl, the CFC concentrations in the oldest waters (Belham Valley, MBV1 and MBV2) provide recharge date estimates of 1970 and 1967, respectively, assuming no old CFC-free water is admixed. Uncertainties in recharge temperatures shifts these estimates by a maximum of ± 2 years, for temperature ranges between 12 °C and 22 °C. Recharge elevation uncertainty has a negligible effect on the estimated groundwater age.

The working hypothesis presented in Hemmings et al. (2014) suggested a contribution of deeper warmer water to explain the anomalously high temperatures in the southern Centre Hills springs. The CFC analysis presented here supports this hypothesis; the warm spring waters represent mixtures of deeper older groundwater and the young meteoric water that supplies the western and northern springs. The SW-NE alignment of the warmer and older southern Centre Hills springs is suggestive of upflow from depth along a permeable fault and fracture network. Such a network may be an extension of, or related to, the NE trending fault alignment associated with HWP and the geothermal targets (HPA in Fig. 1), as described by Ryan et al. (2013).

If one assumes linear binary mixing between end-members, Belham well-type waters (~45 ya) and Bottomless Ghaut waters (modern), the relative contributions of young (cold) and old (warm) waters suggested by temperature, and CFC-11 and CFC-12 are in reasonable agreement. The estimated contribution of older, warmer water at the high discharge Killiekrankie spring is between 40 and 45%. At Fairy Walk, the warmest Centre Hills spring, up to 65–75% may be from the deeper, older and warmer water source. The relative depletion in δ^2H and $\delta^{18}O$ at Fairy Walk compared to the other groundwater samples may indicate some steam contribution to this spring. However, a complete geochemical survey would be required to definitively elucidate any signature of a steam component.

The assumption here is that the older, warmer end-member is represented by the Belham Valley well-type water. Although the CFC data reveals that the oldest sampled groundwaters on Montserrat were from the confined Belham Valley aquifer, we cannot rule-out the possibility that these waters are themselves intermediate. If the warmer groundwater end-member contributing to the southern Centre Hills springs is even older and warmer then the Belham Valley well waters, a lower volume contribution would be needed to produce the temperature and CFC results highlighted above.

An updated conceptual model, based on Fig. 3a, incorporating the additional constraints provided by combined temperature, SEC_{25}, CFC and isotope data is illustrated in Fig. 8. This model describes flow of recharging meteoric water through saturated fractures and breccia compartments within the high permeability vadose zone in the central portion of the Centre Hills complex. These cool, young waters feed the springs at elevations between 200 and 400 m amsl, on the flanks of Centre Hills. Aquifers associated with the southern warmer springs receive additional contributions of older waters from deeper warmer aquifers, supplied via high permeability fault or fracture networks.

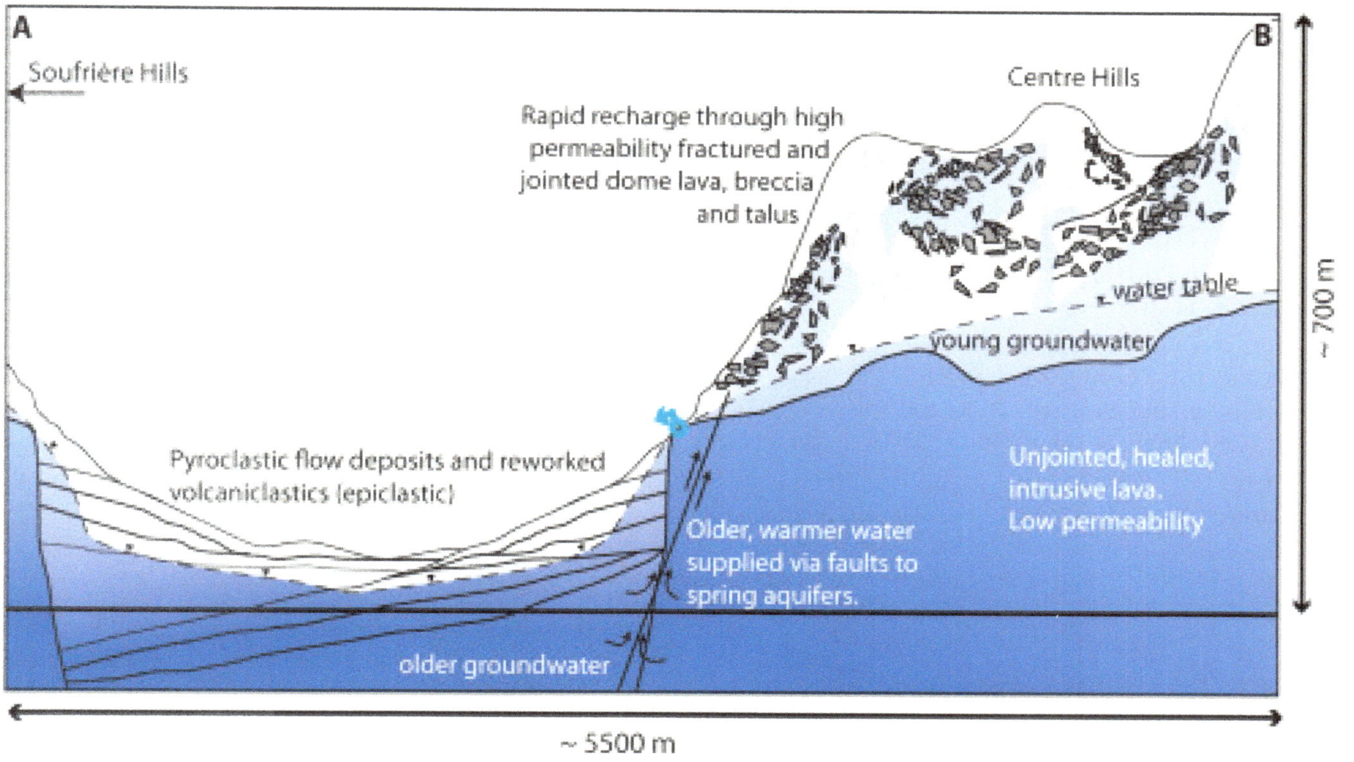

Figure 8: An update of Type 1 conceptual model (Fig. 3a). A schematic cross-section between **A** and **B**, with a southern Centre Hills spring supplied by both young cool meteoric waters (light blue), and older warmer waters (dark blue) from depth. White regions represent the unsaturated material. Sea level is marked by a thick horizontal black line. (For interpretation of the references to color in this figure legend, the reader is referred to the web version of the article.).

CONCLUSIONS

The consistency of δ^2H and $\delta^{18}O$ values for Monsterrat's Centre Hills springs and the Belham Valley confined aquifer waters indicates groundwaters are derived from similar recharge elevations. However, CFC-11 and CFC-12 analysis reveals age differences in otherwise similar groundwaters. Low CFC concentrations in warm (31 °C) waters from within the confined Belham Valley well aquifer are suggestive of water ages of ~45 years, if no old, CFC-free water is present. While CFC concentrations springs are indicative of predominantly young, modern groundwater. Lower CFC concentrations in the warmer springs, identified in Hemmings et al. (2014), suggest a contribution from older, warmer waters, probably supplied via faults and fractures from deeper aquifers. Temperatures and CFC concentrations indicate potential contributions of up to 75% for the warmest spring on Centre Hills and supply rates of up to 8 L/s to the highest yielding spring.

The CFC and δ^2H and $\delta^{18}O$ data also provide insights into the groundwater recharge regime on the island. High CFC concentrations in the northern and western springs can be explained by rapid infiltration of cool (high CFC concentration) rainfall into saturated compartments, with transport through the vadose zone to the phreatic zone dominated by compartment flow. Transport within saturated compartments reduces CFC equilibration with vadose zone air and allows relatively rapid movement through the unsaturated zone, providing young water to the springs. Groundwaters are depleted inδ^2H and $\delta^{18}O$ relative to rainfall and throughfall samples. High frequency isotopic analysis of groundwaters and rainfall is necessary to explore potential spatial and temporal variations of groundwater recharge. Such a study would reveal more detail of the residence time of the young spring waters, as well as the likely response to changes in recharge patterns.

ACKNOWLEDGEMENTS

Sample collection on Montserrat was assisted by Reuel Lee (MUL) and Eric Tomme. This work is funded by the NERC BUFI programme (studentship no. S186), The Royal Society, and EC-FP7 (#282759) "VUELCO". Daren Gooddy and George Darling publish with the permission of the Executive Director, British Geological Survey (NERC). The authors would also like to thank three anonymous reviewers whose comments and suggestions have helped to improve the manuscript

REFERENCES

1. Baird et al., 2015A.F. Baird, J.M. Kendall, R.S.J. Sparks, B. BaptieTranstensional deformation of Montserrat revealed by shear wave splittingEarth Planet. Sci. Lett., 425 (2015), pp. 179–186 http://dx.doi.org/10.1016/j.epsl.2015.06.006
2. Barclay et al., 2007J. Barclay, J. Alexander, J. Susnik Rainfall-induced lahars in the Belham Valley, Montserrat, West IndiesJ. Geol. Soc. London, 164 (2007), pp. 815–827 http://dx.doi.org/10.1144/0016-76492006-078
3. Blume, 1974 H. BlumeThe Caribbean IslandsLongman Group Ltd., London (1974)Cabrera and Custodio, 2004
4. M. Cabrera, E. CustodioGroundwater flow in a volcanic-sedimentary coastal aquifer: Telde area, Gran Canaria, Canary Islands, Spain

5. Hydrogeol. J., 12 (2004), pp. 305–320 http://dx.doi.org/10.1007/s10040-003-0316-y
6. Chiodini et al., 1996G. Chiodini, R. Cioni, A. Frullani, M. Guidi, L. Marini, F. Prati, B. Raco
7. Fluid geochemistry of Montserrat Island, West Indies
8. Bull. Volcanol., 58 (1996), pp. 380–392 http://dx.doi.org/10.1007/s004450050146
9. Clark and Fritz, 1997I.D. Clark, P. FritzEnvironmental Isotopes in HydrogeologyLewis (1997)
10. Custodio, 2007E. CustodioGroundwater in volcanic hard rocksJ. Krásný, J.M. Sharp (Eds.), Groundw. Fract. Rocks IAH Sel. Pap. Ser., Taylor and Francis (2007), pp. 95–108 http://dx.doi.org/10.1201/9780203945650.ch5 (chapter 5)
11. Darling et al., 2012W.G. Darling, D. Gooddy, A. MacDonald, B. MorrisThe practicalities of using CFCs and SF6 for groundwater dating and tracing Appl. Geochem., 27 (2012), pp. 1688–1697 http://dx.doi.org/10.1016/j.apgeochem.2012.02.005
12. Davies and Peart, 2003J. Davies, R. PeartA Review of the Groundwater Resources of Central and Northern Montserrat. Technical Report CR/03/257C British Geological Survey (2003)
13. Donnelly, 2007 L. DonnellyEngineering geology of landslides on the volcanic island of Montserrat, West Indies
14. Q. J. Eng. Geol. Hydrogeol., 40 (2007), pp. 267–292 http://dx.doi.org/10.1144/1470-9236/06-039
15. Dunkle et al., 1993 S.A. Dunkle, L.N. Plummer, E. Busenberg, P.J. Phillips, J.M. Denver, P.A. Hamilton, R.L. Michel, T.B. CoplenChlorofluorocarbons (CCl3F and CCl2F2) as dating tools and hydrologic tracers in shallow groundwater of the Delmarva Peninsula, Atlantic Coastal Plain, United States Water Resour. Res., 29 (1993), pp. 3837–3860 http://dx.doi.org/10.1029/93WR02073
16. Ecker, 1976A. EckerGroundwater behaviour in Tenerife, volcanic island (Canary Islands, Spain)J. Hydrol., 28 (1976), pp. 73–86 http://dx.doi.org/10.1016/0022-1694(76)90053-6
17. N. Feuillet, F. Leclerc, P. Tapponnier, F. Beauducel, G. Boudon, A. Le Friant, C. Deplus, J.F. Lebrun, A. Nercessian, J.M. Saurel, V. Clément
18. Active faulting induced by slip partitioning in Montserrat and link with volcanic activity: New insights from the 2009 GWADASEIS marine cruise data
19. Geophys. Res. Lett. (2010), p. 37 http://dx.doi.org/10.1029/2010GL042556
20. Frische et al., 2006M. Frische, K. Garofalo, T.H. Hansteen, R. Borchers, J. HarnischThe origin of stable halogenated compounds in volcanic gasesEnviron. Sci. Pollut. Res., 13 (2006), pp. 406–413 http://dx.doi.org/10.1065/espr2006.01.291Gooddy et al., 2006
21. D. Gooddy, W.G. Darling, C. Abesser, D. LapworthUsing chlorofluorocarbons (CFCs) and sulphur hexafluoride (SF6) to characterise groundwater movement and residence time in a lowland Chalk catchmentJ. Hydrol., 330 (2006), pp. 44–52 http://dx.doi.org/10.1016/j.jhydrol.2006.04.011
22. Article | PDF (230 K) | View Record in Scopus | Citing articles (62)Gribble, 1994G.W. GribbleThe natural production of chlorinated compoundsEnviron. Sci. Technol., 28 (1994), pp. 310A–319A http://dx.doi.org/10.1021/es00056a001
23. Harnisch and Eisenhauer, 1998J. Harnisch, A. EisenhauerNatural CF4 and SF6 on Earth Geophys. Res. Lett., 25 1998), pp. 2401–2404 http://dx.doi.org/10.1029/98GL01779
24. Hautmann et al., 2010S. Hautmann, J. Gottsmann, A.G. Camacho, N. Fournier, I.S. Sacks, R.S.J. SparksMass variations in response to magmatic stress changes at Soufrière Hills Volcano, Montserrat (W.I.): insights from 4-D gravity data Earth Planet. Sci. Lett., 290 (2010), pp. 83–89 http://dx.doi.org/10.1016/j.epsl.2009.12.004
25. Heaton and Vogel, 1981T. Heaton, J. Vogel Excess air in groundwater J. Hydrol., 50 (1981), pp. 201–216 http://dx.doi.org/10.1016/0022-1694(81)90070-6
26. Hemmings et al., 2014 B. Hemmings, F. Whitaker, J. Gottsmann, A. Hughes Hydrogeology of Montserrat, review and new insights J. Hydrol. Reg. Stud., 3 (2014), pp. 1–30 http://dx.doi.org/10.1016/j.ejrh.2014.08.008
27. Ho et al., 1998D.T. Ho, P. Schlosser, W.M. Smethie, H.J. Simpson Variability in atmospheric chlorofluorocarbons (CCl3F and CCl2F2) near a large urban area: Implications for groundwater dating
28. Environ. Sci. Technol., 32 (1998), pp. 2377–2382 http://dx.doi.org/10.1021/es980021h
29. Höhener et al., 2003P. Höhener, D. Werner, C. Balsiger, G. PasterisWorldwide Occurrence and Fate of Chlorofluorocarbons in GroundwaterCrit. Rev. Environ. Sci. Technol., 33 (2003), pp. 1–29 http://dx.doi.org/10.1080/10643380390814433
30. HydroSource, 2004HydroSource Groundwater Exploration and Development of High Yield Water Supply Wells for the Montserrat Water Authority. Technical ReportHydrosource Assoc. Inc., Ashland, New Hampshire (2004)
31. IAEA/WMO, 2015IAEA/WMO Global Network of Isotopes in Precipitation(2015) http://www.iaea.org/water
32. Jones et al., 2010 M.T. Jones, D.J. Hembury, M.R. Palmer, B. Tonge, W.G. Darling, S.C. LoughlinThe weathering and element fluxes from active volcanoes to the oceans: a Montserrat case study Bull. Volcanol., 73 (2010), pp. 207–222 http://dx.doi.org/10.1007/s00445-010-0397-0
33. Kelly and Glenn, 2015 J.L. Kelly, C.R. Glenn Chlorofluorocarbon apparent ages of groundwaters from west Hawaii, USA J. Hydrol., 527 (2015), pp. 355–366 http://dx.doi.org/10.1016/j.jhydrol.2015.04.069
34. Koh et al., 2007 D. Koh, L. Plummer, E. Busenberg, Y. Kim Evidence for terrigenic SF6 in groundwater from basaltic aquifers, Jeju Island, Korea: Implications for groundwater dating J. Hydrol., 339 (2007), pp. 93–104 http://dx.doi.org/10.1016/j.jhydrol.2007.03.011
35. Kralik et al., 2014M. Kralik, F. Humer, J. Fank, T. Harum, G. Klammler, D. Gooddy, J. Sültenfuß, C.

Gerber, R. PurtschertUsing 18O/2H, 3H/3He, 85Kr and CFCs to determine mean residence times and water origin in the Grazer and Leibnitzer Feld groundwater bodies (Austria)Appl. Geochem. (2014) http://dx.doi.org/10.1016/j.apgeochem.2014.04.001

36. Manning and Caine, 2007A.H. Manning, J.S. Caine Groundwater noble gas, age, and temperature signatures in an Alpine watershed: valuable tools in conceptual model development Water Resour. Res. (2007), p. 43 http://dx.doi.org/10.1029/2006WR005349
37. Oster et al., 1996 H. Oster, C. Sonntag, K.O. Münnich Groundwater age dating with chlorofluorocarbons
38. Water Resour. Res., 32 (1996), pp. 2989–3001 http://dx.doi.org/10.1029/96WR01775
39. Plummer et al., 2006 L. Plummer, E. Busenberg, P. Cook Use of Chlorofluorocarbons in Hydrology
40. International Atomic Energy Agency, Vienna (2006)
41. Poux and Brophy, 2012B. Poux, P. Brophy Geothermal exploration on the Island of Montserrat, Caribbean
42. GRC Trans., 36 (2012), pp. 737–744
43. Roques et al., 2014C. Roques, L. Aquilina, O. Bour, J.c. Maréchal, B. DewandelGroundwater sources and geochemical processes in a crystalline fault aquifer J. Hydrol., 519 (2014), pp. 3110–3128 http://dx.doi.org/10.1016/j.jhydrol.2014.10.052
44. Ryan and Shalev, 2014 G. Ryan, E. Shalev Seismic velocity/temperature correlations and a possible new geothermometer: insights from exploration of a high-temperature geothermal system on Montserrat, West Indies
45. Energies, 7 (2014), pp. 6689–6720 http://dx.doi.org/10.3390/en7106689
46. Ryan et al., 2013 G.a. Ryan, J.R. Peacock, E. Shalev, J. Rugis Montserrat geothermal system: a 3D conceptual model
47. Geophys. Res. Lett., 40 (2013), pp. 2038–2043 http://dx.doi.org/10.1002/grl.50489
48. Sanjuan et al., 2005B. Sanjuan, R. Millot, M. Brach, J.C. Foucher, J.Y. Roig, J.M. Baltassat
49. Geothermal exploration in the Mount Pelée Volcano-Morne Rouge and Diamant Areas (Martinique, West French Indies): geochemical data
50. Proc. World Geotherm. Congr., Antalaya, Turkey (2005), p. 12
51. Schmidt et al., 1999G. Schmidt, G. Bigg, E. Rohling Global Seawater Oxygen-18 Database – v1.21
52. (1999) http://data.giss.nasa.gov/o18data/
53. Schol et al., 2009 M.A. Schol, J.B. Shanley, J.P. Zegarra, T.B. Coplen The stable isotope amount effect: new insights from NEXRAD echo tops, Luquillo Mountains, Puerto Rico Water Resour. Res. (2009), p. 45 http://dx.doi.org/10.1029/2008WR007515
54. Scholl et al., 1996 M.a. Scholl, S.E. Ingebritsen, C.J. Janik, J.P. Kauahikaua Use of precipitation and groundwater isotopes to interpret regional hydrology on a tropical Volcanic Island: Kilauea Volcano Area, Hawaii
55. Water Resour. Res., 32 (1996), p. 3525 http://dx.doi.org/10.1029/95WR02837
56. Stuart et al., 2010 M. Stuart, L. Maurice, T. Heaton, M. Sapiano, M. Micallef Sultana, D. Gooddy, P. Chilton
57. Groundwater residence time and movement in the Maltese islands: a geochemical approach Appl. Geochem., 25 (2010), pp. 609–620 http://dx.doi.org/10.1016/j.apgeochem.2009.12.010
58. Wadge and Isaacs, 1988G. Wadge, M.C. Isaacs Mapping the volcanic hazards from Soufriere Hills Volcano, Montserrat, West Indies using an image processor J. Geol. Soc. London, 145 (1988), pp. 541–551 http://dx.doi.org/10.1144/gsjgs.145.4.0541
59. Warner and Weiss, 1985 M. Warner, R. Weiss Solubilities of chlorofluorocarbons 11 and 12 in water and seawater
60. Deep Sea Res. A: Oceanogr. Res. Pap., 32 (1985), pp. 1485–1497 http://dx.doi.org/10.1016/0198-0149(85)90099-8
61. Wilson and McNeill, 1997 G. Wilson, G. McNeillNoble gas recharge temperatures and the excess air component
62. Appl. Geochem., 12 (1997), pp. 747–762 http://dx.doi.org/10.1016/S0883-2927(97)00035-8
63. Younger, 2010 P. YoungerReconnaissance assessment of the prospects for development of high-enthalpy geothermal energy resources, Montserrat Q. J. Eng. Geol. Hydrogeol., 43 (2010), pp. 11–22 http://dx.doi.org/10.1144/1470-9236/08-083

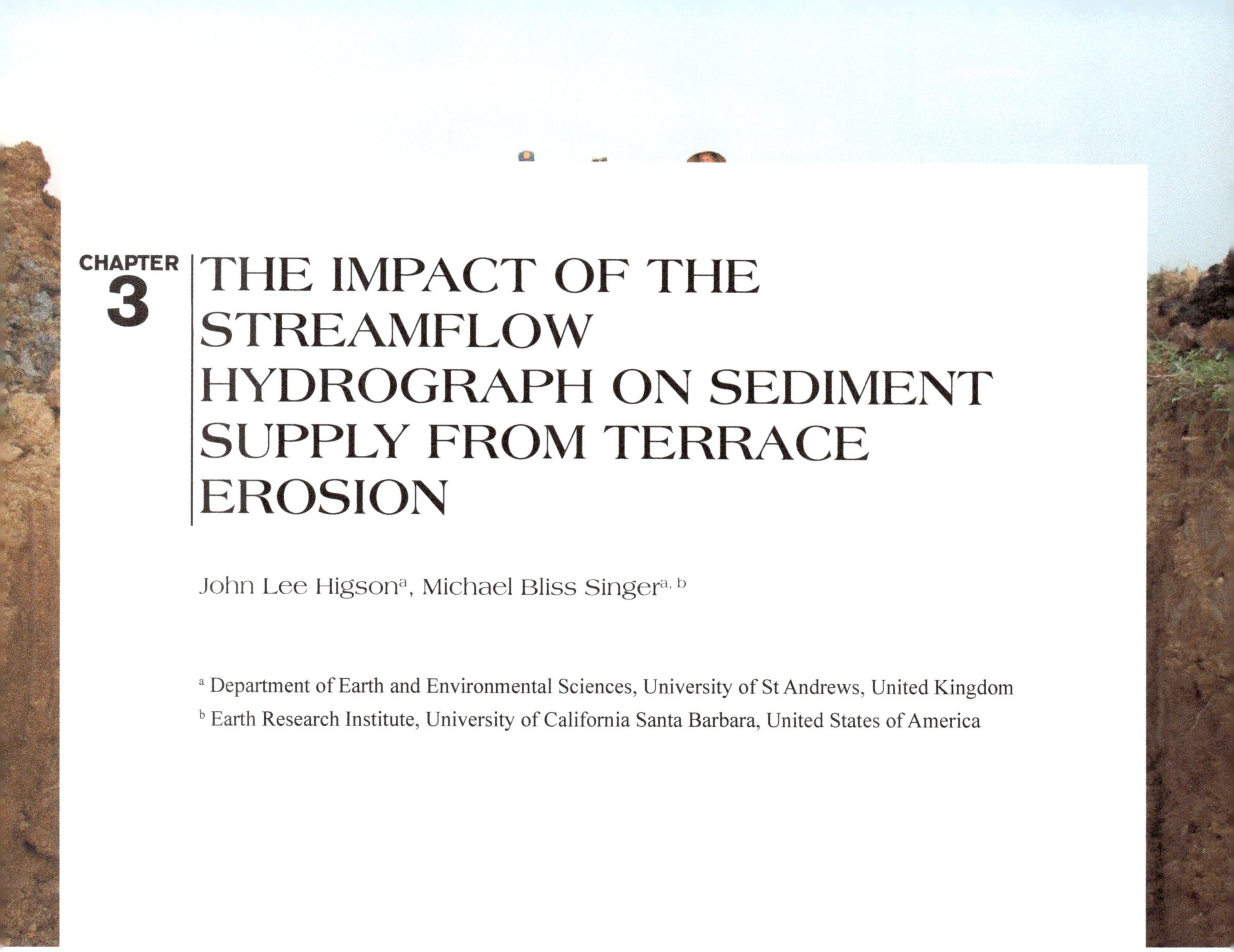

CHAPTER 3

THE IMPACT OF THE STREAMFLOW HYDROGRAPH ON SEDIMENT SUPPLY FROM TERRACE EROSION

John Lee Higson[a], Michael Bliss Singer[a, b]

[a] Department of Earth and Environmental Sciences, University of St Andrews, United Kingdom

[b] Earth Research Institute, University of California Santa Barbara, United States of America

ABSTRACT

Sediment supply from banks and terraces has important implications for grain-size distributions in alluvial rivers (and by extension for aquatic habitat), as well as for the delivery of floodplain-stored nutrients and contaminants to the aquatic environment. The interactions between streamflow hydrographs and lateral channel boundary failure control the sediment supply from banks and terraces. However, the relationships between variable flow and discrete sediment supply from catastrophic erosion of lateral boundaries and subsequent mass sediment flux in rivers are not well characterised by existing methods and models that focus only on one of several relevant interrelated processes. In order to improve predictive capability of catastrophic sediment supply from lateral boundaries, we adopt a new approach to modelling the process interactions between stream hydrology, erosion of banks/terraces, and the corresponding discrete supply of sediment to channels. We develop a modelling framework for terrace — channel coupling that combines existing theories of flow through porous media, bank stability, and fractional sediment flux. We demonstrate the utility of this modelling approach by assessing hydrologically driven erosion, evolution of grain size in the channel, and fine sediment flux from a study site along the Yuba River in California over individual flood hydrographs and over decadal historical flow series. We quantify the supply of sediment eroded from a contaminated nineteenth century fan terrace of hydraulic gold mining tailings intersecting the Yuba, and find that a threshold for erosion exists at a stage in the channel in excess of 8 m producing episodic sediment concentrations in excess of ~300 mg L^{-1}. The modelling produced erosion and fine sediment pulses from each of three major floods in the past several decades until the flow drops below 500 $m^3\ s^{-1}$ and a bed armor layer forms, while no sediment was generated from the terrace during smaller floods. We further assess the impact on terrace erosion of various river management scenarios with distinct hydrograph shapes and find increased erosion potential when the terrace contains antecedent moisture or the flood time series is run over an extended duration. Sensitivity analysis demonstrated that elevated antecedent moisture within the lateral boundary and increased hydrograph rising time each reduce bank stability and thus increase volumes of failed material. We also show that fluctuations in the hydrograph, typically associated with hydroelectricity generation, result in a more stable terrace than those of a longer duration

Citation: Higson, J. L., & Singer, M. B. (2015). The impact of the streamflow hydrograph on sediment supply from terrace erosion. Geomorphology, 248, 475–488. http://doi.org/http://dx.doi.org/10.1016/j.geomorph.2015.07.037

because there is less time for hyporheic stream water to infiltrate the lateral boundary. This study demonstrates that changes in hydrograph shape as a consequence of climatic forcing or anthropogenic dam releases may have considerable impacts upon sediment delivery and associated contaminants from banks and terraces to the downstream environment.

INTRODUCTION

The erosion of banks and terraces supplies sediment to channels and thus has important implications for grain-size distributions and morphology in alluvial rivers and by extension for aquatic habitat, as well as for the delivery of floodplain-stored nutrients and contaminants to the aquatic environment. Once sediment enters the channel from banks and terraces, a proportion of it is transported by competent flows, while the remainder becomes incorporated into the channel bed material, where it may affect channel capacity and thus flood risk (Slater et al., 2015). Thus, the problem of sediment supply to channels has important implications for the partitioning of fractional longitudinal sediment flux, the evolution of channel grain-size distributions (GSDs), and stream habitat and for morphological development along river reaches and over temporal periods ranging from multiple storms to millennia. Sediment supplied from lateral boundaries may comprise a large proportion of the total sediment load to rivers (Odgaard, 1987 and Simon and Darby, 2002). In the relatively low energy systems of the UK, Walling et al. (1999) showed that in the long-term up to 37% of the suspended sediment yield originated from lateral boundary erosion. In addition, sediment budget research on the Amazon River has demonstrated contributions of bank erosion that are 1.3 times larger than the total sediment flux at the river mouth (Óbidos) (Dunne et al., 1998). Contribution of sediment from lateral boundaries during high flow stage should be particularly large in fluvial systems with unconsolidated high banks or terraces along the channel margins (e.g.,Singer et al., 2013) and in basins where rainstorms produce major floods substantially larger than mean flows.

Localised sediment supply to channels from bank and terrace erosion (hereafter calledlateral boundary erosion) is a key source of sediment to lowland channel systems, in which tributaries may play a more limited role in terms of their influence on channel sediment dynamics (Rice, 1998) within lowland valley floors (Singer and Dunne, 2001,Singer, 2008a, Singer, 2008b and Singer, 2010). Lateral boundary erosion may be generated by mass wasting of banks and terraces as a result of failure induced by pore pressures once the bank toe has been removed (Simon et al., 2000). This has been observed following floods as extensive failed bank material along channel margins near riverbanks. We have documented many such failures within the Sacramento Valley of California, suggesting that lateral boundary failure did not occur during the rise to peak of the flood (otherwise bank material would have been carried away), but at the peak or on the falling limb of the hydrograph as a catastrophic mass wasting process in response to floodplain water table fluctuations (Rinaldi et al., 2004 and Luppi et al., 2009). The implication is that differences in hydrograph shape (whether caused by climate change or anthropogenic modification of hydrology) have the potential to affect erosion of channel boundaries and therefore impact the supply of sediment to the channel from these sources. Thus, streamflow may be the primary driver of lateral boundary erosion through its influence on near-bank stresses and their support of lateral hyporheic water tables that rise and fall within stream banks.

Failure induced by pore pressures in lateral boundaries has not typically been considered to be the dominant mechanism affecting the supply of sediment from banks and terraces and many floodplain-channel morphology modelling efforts have subsumed the relevant erosion processes into lateral boundary erodibility based on characterisations of the flow field (e.g., Johannesson and Parker, 1989 and Lauer and Parker, 2008). Such models typically explore the velocity field interacting within the geometry of river bends to identify loci and rates of lateral erosion. However, several other recent research efforts have placed great emphasis on links between boundary water status, bank stability, boundary failure, and in-channel sediment transport (Amiri-Tokaldany et al., 2003, Amiri-Tokaldany et al., 2007, Rinaldi et al., 2004 and Rinaldi et al., 2008). Various reviews of research on the mechanisms of river width adjustment and approaches to its modeling were undertaken in previous decades (Thorne, 1982,American Society of Civil Engineers, 1998a and American Society of Civil Engineers, 1998b), so we will not endeavour to repeat this summary here. What is important here is that it is challenging to systematically predict lateral boundary erosion and its contribution to sediment loads over short timescales (individual or sequences of storms) because bank/terrace erosion occurs catastrophically along channel margins during or following floods. Lateral boundaries are subjected to infiltrating rain into floodplain materials and, more significantly, hyporheic water tables as a function of rising river stage that affect material stability. As water content in lateral boundaries increases, frictional resistance between grains is reduced and the entire deposit becomes more prone to failure (Osman and Thorne, 1988 and Casagli et al., 1999).

The processes by which water content in lateral channel boundaries affects timing and volumes of bank/terrace failure are the subject of extensive research applying the theory of infinite slope stability (Terzaghi, 1942 and Selby, 1993) to riverbanks (Simon et al., 1991 and Darby and Thorne, 1996). The infinite slope stability method produces a factor of safety based on linear (Mohr–Coulomb) relationships between shear strength and the normal stress on the failure surface, which take into account material properties, slope geometry, and water content in the matrix. Later implementations have considered the local hydrology explicitly, including confining pressure of river stage, infiltration, and the hydrologic role of vegetation (Simon and Collison, 2001 and Darby et al., 2007). Coupled with a groundwater model that computes water table rise and fall within the bank/terrace, these components provide a good framework for predicting lateral boundary failure in response to hydrologic changes in the near-channel zone. These components can then be coupled to characterisations of the fate of the failed lateral boundary material (Amiri-Tokaldany et al., 2007). An important open question is how variations in the driving flow (discharge hydrographs) interact with river banks/terraces over various timescales from individual floods to decades. If developed, this understanding could be used to better assess the impact of hydrology on sediment supply from lateral channel boundary erosion.

Evidence is increasing that hydrologic variability has important implications for channel bed material transport and sorting (Hassan et al., 2006, Humphries et al., 2012, Mao, 2012, Alvarez and Schmeeckle, 2013, Singer et al., 2013 and Singer and Michaelides, 2014). Similarly, the shape of the discharge hydrograph and antecedent moisture conditions within lateral channel boundaries are likely to impact failure probabilities for particular bank materials, with implications for the fate and transport of sediment once failure of banks occurs (Rinaldi et al., 2004 and Luppi et al., 2009). In the case of fine-grained floodplain materials, a significant proportion of the failed sediment would likely become part of the river›s washload (nonsettling component of suspended load) and thus be transported through the reach and deposited on downstream floodplains (Kilham et al., 2012). Legacy terraces from past climatic forcing and/or anthropogenic activities (Bull, 1997 and Singer et al., 2013) are typically composed of noncohesive materials and are thus more prone to failure than relatively cohesive riverbanks constructed by modern lowland rivers carrying higher percentages of silt and clay. These lateral boundary units often contain a wide mix of grain sizes that interact with the riverbed in various ways. While some of the failed sediment certainly contributes to fine-grained suspended loads (Carson et al., 1973 and Bull, 1997), relatively coarse sediment typically becomes incorporated into the bed. Therefore, a partitioning of sediment loads occurs upon failure, such that the bed inherits a censored signature of bank erosion as has been found for hillslope sediment supply to channels (Michaelides and Singer, 2014).

In this paper, we couple two preexisting modelling frameworks: 1) water status in lateral boundaries is based on the channel discharge hydrograph; and 2) infinite slope stability analysis of terrace failure and subsequent grain size selective sediment transport is used to assess the impact of terrace erosion on bed material grain size and fine-grained sediment flux. We investigate whether consistent thresholds for terrace erosion exist and how various configurations of in-channel flow might be expected to affect the contribution of sediment from lateral boundaries. We then model the impact of individual storm hydrographs, as well as a decadal sequence of daily flows at a site along the Yuba River in California, which is impacted by contaminated historical river tailings terraces produced by nineteenth century hydraulic gold mining in the Sierra Nevada foothills (Singer et al., 2013). We further investigate in a simplistic manner how climate changes or dam operations may affect lateral boundary erosion.

FIELD SETTING AND FIELD DATA COLLECTION

Our field site is within the Yuba Fan, located in the northwestern Sierra Nevada foothills of California (Fig. 1) and has been intensively studied as a result of the legacy of mining debris driving the geomorphology of the area. Nineteenth century hydraulic gold mining resulted in the rapid delivery of mining tailings to channels and delivery of this sediment much farther downstream. The corresponding aggradation of piedmont valleys such as the nearby Bear River basin (James, 1989) led to sustained storage of much of the original deposit lateral to the main channel within terraces and banks (Singer et al., 2013). Twentieth century flood control measures in the form of levees have fixed these legacy sediments into the active Yuba River corridor (James and Singer, 2008).

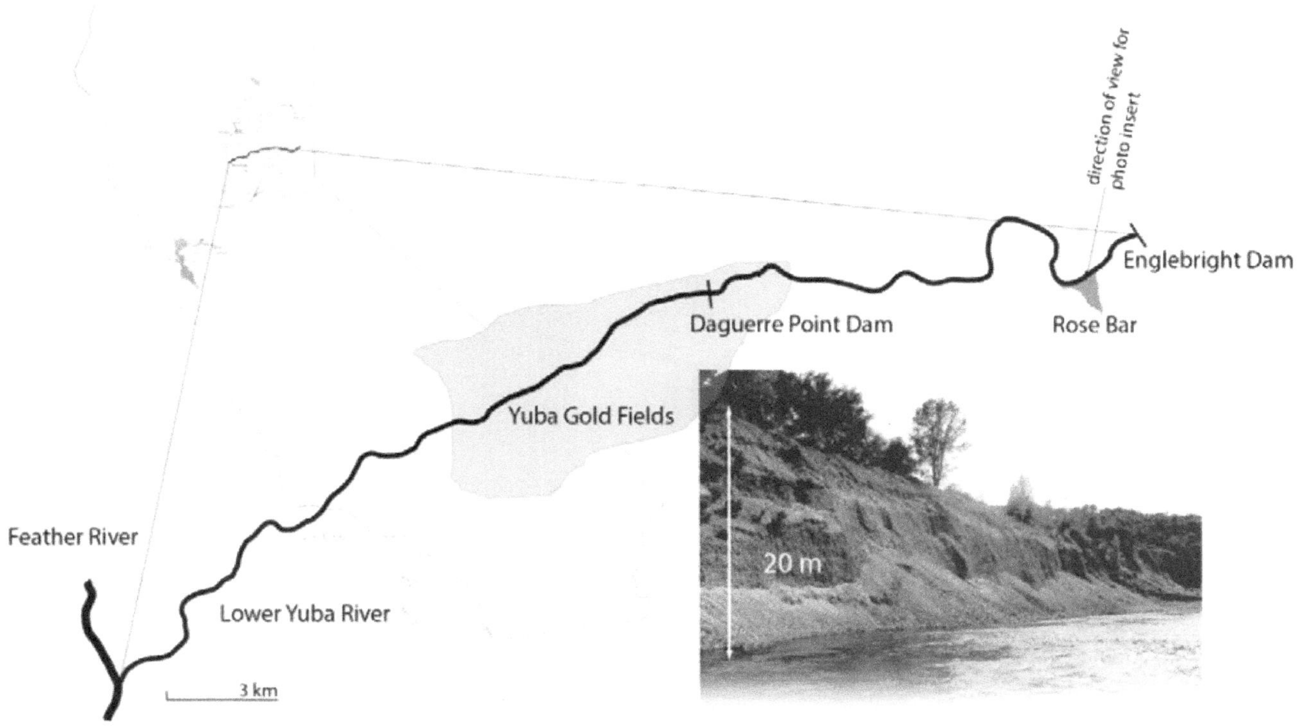

Figure 1: Location map of Lower Yuba River, California, with inset photograph of Rose Bar terrace near Timbuctoo Bend.

More than 4 × 10^6 kg of mercury (Hg), used for gold separation, is estimated to have been lost during the mining process (Aplers et al., 2005). Documented concentrations of total Hg in legacy terraces along the Yuba River vary over two orders of magnitude above background values (Singer et al., 2013). Exceedingly high Hg concentrations have also been documented within lowland Sacramento Valley sediments (Hornberger et al., 1999 and Springborn et al., 2011) and even in lowland biota and food chains (Greenfield et al., 2005, Eagles-Smith et al., 2009 and Greenfield and Jahn, 2010). Much of this Hg appears to emerge from legacy hydraulic mining sediment along the Yuba and other Sierran streams (Bouse et al., 2010, Gehrke et al., 2011 and Donovan et al., 2013). It is transported from the toxic upland terraces to the lowlands during major floods (Kilham et al., 2012 and Singer et al., 2013), leading to planform changes detectable in aerial photographs that bracket the timing of these floods (Ghoshal et al., 2010) as well as from recent DEM analysis (Carley et al., 2012). Because vertical incision has slowed considerably owing to the exposure of coarse bed sediments that create an armored surface layer in the Yuba and other reaches within the Sacramento Valley (Singer, 2008b), lateral erosion has emerged as the dominant processes of hydraulic sediment entrainment along the Yuba (Singer et al., 2013). Such episodic flooding of the Yuba River has been shown to force erosion of lateral boundaries and, consequently, valleywide reworking and redistribution of unvegetated legacy sediments into lowland environments (Singer et al., 2008, Singer et al., 2013, Singer and Aalto, 2009, Kilham et al., 2012 and Springborn et al., 2011) (Fig. 2).

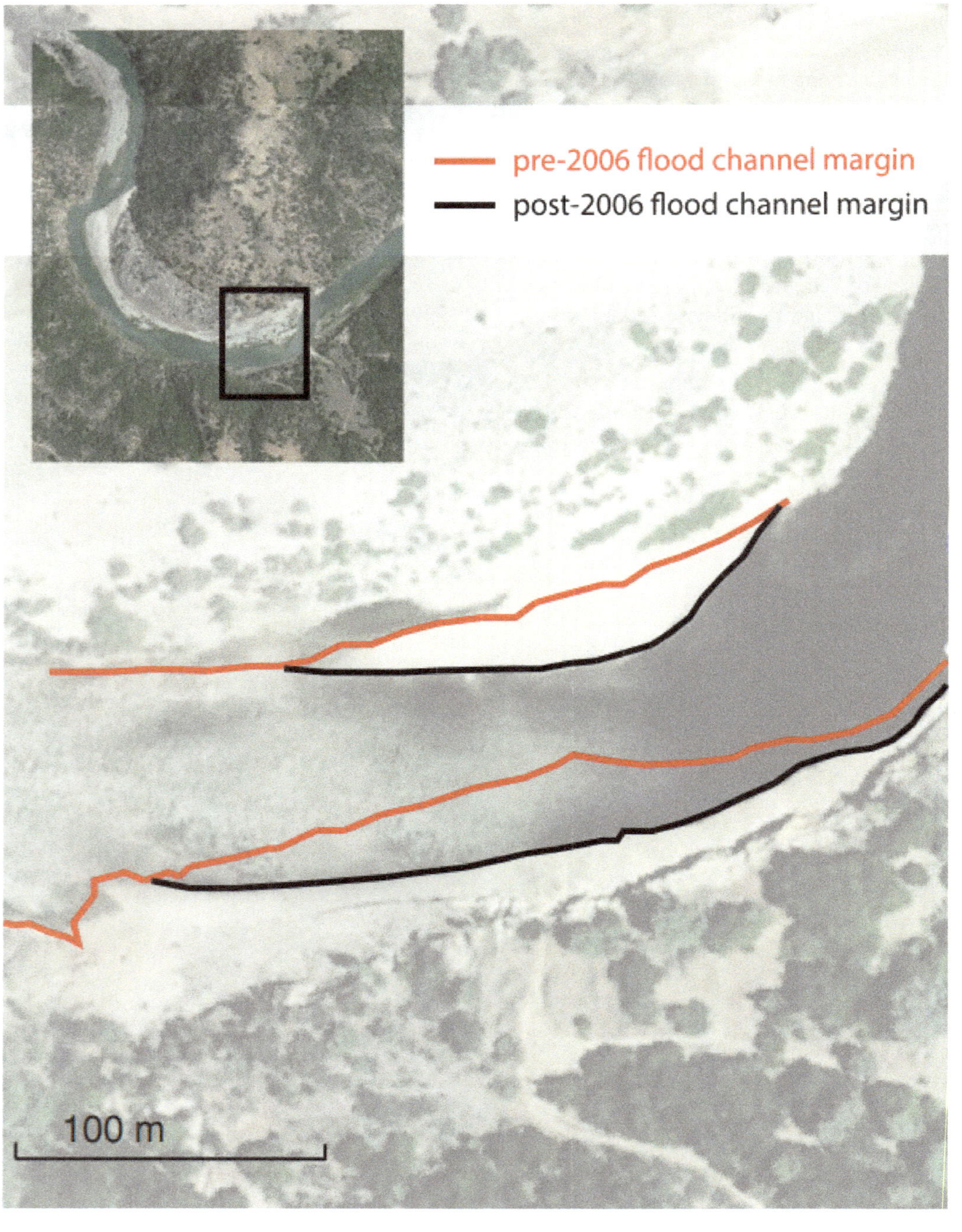

Figure 2: Channel migration at Rose Bar along the Yuba River from ~ June 2005 (pre-2006 flood channel margin) to ~ June 2007 (post-2006 flood channel margin), based on Google Historical Imagery.

In order to characterise terrace failure and entrainment into sediment loads, we have selected a study area called Rose Bar along a stretch of the Yuba River near Timbuctoo Bend (27 km upstream of the confluence with the Feather River), which includes the largest legacy tailings terrace along the Yuba downstream of Englebright Dam (Fig. 1), and appears to be recruiting sediment over decadal timescales. Sediment accumulation rates in Englebright Reservoir based on subaqueous coring (Snyder et al., 2006), are low relative to estimated rates of sediment flux into the Central Valley (Singer et al., 2013), suggesting the dominance of downstream, along-channel bank/terrace sediment on decadal sediment budgets for the Yuba that may contribute up to 45% of the total net flux from the Yuba into the downstream Feather River (Carley et al., 2012). The region is subjected to a Mediterranean climate with hot, dry summers and cool, wet winters and ~ 80% of the annual rainfall occurs between November and March (Snyder et al., 2006). The annual precipitation along the lower Yuba River is 500 mm and mean daily discharge is 100 $m^3 s^{-1}$. In spite of various upstream dams controlling some of the flood flow (Singer, 2007), variability in streamflow is great during floods and the hydrographs remain very peaked with rising and falling times that may be increasing and thus leading to longer periods of terrace inundation (Singer et al., 2013) and subsequent downstream transport of fine sediment during large floods (Kilham et al., 2012).

Our field measurements show that the Rose Bar terrace surface stands ~20 m above the surface of the river bed and the terrace slope angle is 75°. At this site, we collected representative GSDs of dry terrace sediment by bulk dry sieving in the field (full φ size classes from 0.063 to 256 mm), satisfying the criteria that the largest particle in the distribution comprised no more than 1% of the total sample mass, which is a slight relaxation from the recommendations of Church et al. (1987), also employed by Singer, 2008a and Singer, 2008b. Prior to the bulk collection, smaller samples were sieved from three elevations in the terrace to assess the variability of the GSD through the stratigraphy of the bank. We also characterised the bed material GSD through Wolman pebble counts (Wolman, 1954) on the bar surface and by bulk sieving of the subsurface material (after removing the surface layer). Bar surface and subsurface GSDs were averaged based on mass (Bunte and Abt, 2001) to generate an initial representative GSD for the riverbed, which was then permitted to adjust through model iterations. Lateral boundary GSDs are required as input to models of water table fluctuation and bank stability because they determine porosity and permeability and thus control hydraulic conductivity, while channel GSD is required for initial conditions for bed material transport modelling (see below). In order to gain a detailed cross-channel profile, we utilised 0.6-m horizontal resolution bathymetry collected by the U.S. Army Corps of Engineers in 1999 and determined cross-sectional bank geometry in the field (including bank and toe angles and lengths) using an Impulse 200 Laser Rangefinder with a resolution of 0.01 m.

MODELLING SEDIMENT SUPPLY FROM LATERAL BOUNDARIES

Various models have been developed in order to represent the interaction between lateral boundary erosion and river channels. These models largely fit into two categories: channel centerline migration models and bank failure models. Channel centerline models, originally conceived by Ikeda et al. (1981) and later developed by others (e.g.,Johannesson and Parker, 1989 and Howard, 1992), assume that sediment deposition occurs on the inside of river bends, building point bars and deflecting the high velocity core of the flow toward the outer bank. Erosion occurs based on depth-averaged near-bank velocity and the cross-sectionally averaged velocity multiplied by an empirical bank erodibility coefficient to maintain equilibrium channel geometry. Such modelling frameworks have been adapted recently for more complex problems, including the flux and residence time of contaminated fine-grained floodplains (Lauer and Parker, 2008) and the influence of lingering slump blocks on migration rates (Parker et al., 2011). Generally, channel centerline migration models have been used to predict long-term evolution (years to millennia) in the planform of meander channels (e.g., Howard, 1992 and Sun et al., 1996), but they are challenged to characterise the direct impact of event-based hydrology on bank erosion.

Mass failure models (lateral boundary failure models) simulate erosion on a discontinuous timescale, which is appropriate where large but infrequent events dominate lateral erosion (Thorne, 1982). A visual assessment of the Lower Yuba River suggests that planar (as opposed to rotational and/or cantilever) failure is the primary mechanism for mass failure on the reach. Planar mass failure occurs when gravitational forces exceed the internal resisting forces within the bank/terrace (Fredlund et al., 1978). They typically focus on (i) lateral boundary geometry and the influence of tension cracks (Osman and Thorne, 1988) or (ii) the effects of positive and negative pore-water pressures (Casagli et al., 1999, Rinaldi and Casagli, 1999 and Simon et al., 2000) in the saturated and unsaturated zones within the lateral boundary materials. Our emphasis here is to assess the influence of the saturated and unsaturated conditions within the lateral boundary (driven by fluctuations in the water table). We have thus adopted a multipronged modelling approach to investigate the hydrologic controls on mass failure and to address the contribution of failure material to channel sediment loads. The components are (Fig. 3): (i) channel hydraulics (average water depth, hydraulic radius, flow depth); (ii) water table elevation (WTE) in the lateral channel boundary; (iii) lateral boundary stability; and (iv) bed material flux and corresponding riverbed GSD evolution. We describe these components in detail below.

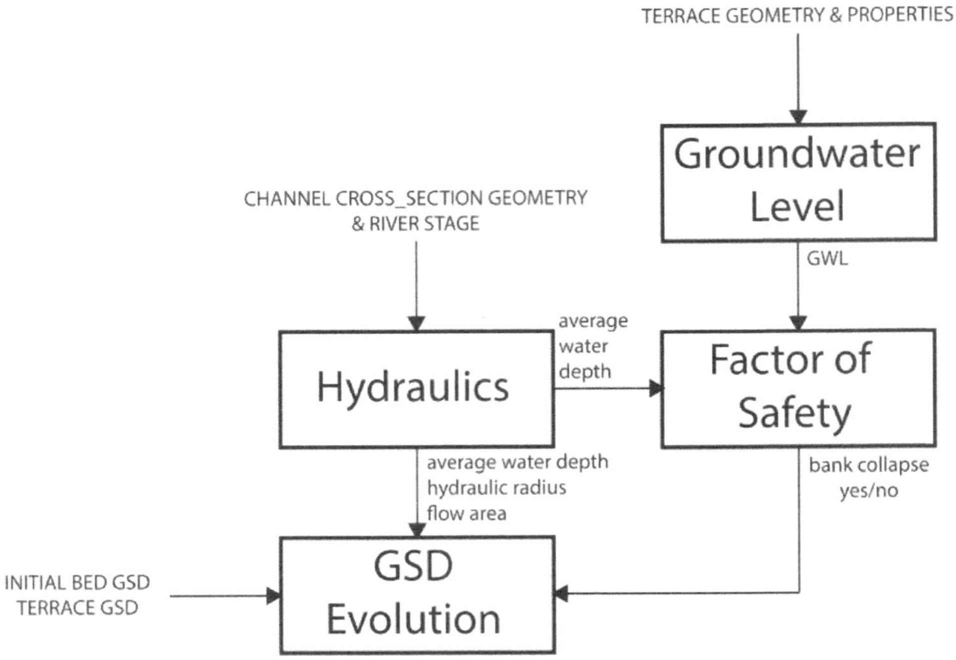

Figure 3: Flowchart of the model components, inputs, and workflow.

Channel Hydraulics

The magnitude of water flux into lateral boundary materials along a river is largely a function of local channel discharge, which integrates runoff generation processes in the contributing upstream drainage basin. Such local streamflow is straightforwardly characterised by historical flood records, which contain information on the various hydrograph characteristics (e.g., Singer, 2007) that may affect such near-channel erosion processes. We used the flow data and stage-discharge relationship from the nearest gauging station that was rated for flow (USGS site 11418000 — 3 km upstream). No tributaries enter the Yuba between this gauging station and Rose Bar, so we assume streamflow is approximately uniform and that the cross section at Rose Bar is similar in shape and configuration to the gauging site.

Water Table Elevation in the Lateral Boundary

The elevation of the water table within the lateral boundary strongly controls bank stability through its effect on pore-pressures and the unit weight of bank or terrace sediment (Simon and Collison, 2001). We employ a Darcian water table model (i.e., flow through the porous media is determined by a rise or fall in pressure gradient) based on the Dupuit–Forchheimer assumption that the groundwater moves horizontally and hydrostatically. We conceive of the Rose Bar terrace as a single sediment layer with uniform porosity and permeability determined from the measured terrace GSD. While this assumption may be limiting in cases where vertical variability exists in the lateral boundary, it is appropriate for many channel terraces that exhibit more uniform GSDs because of their anthropogenic history as tailings fans left over after mining. The Rose Bar terrace is one such tailings fan, so it is largely homogenous in grain-size distribution and by extension in porosity and permeability. We noted only a 4% increase in the gravel-to-sand ratio over roughly 10 m of the vertical terrace profile. In order to maintain a water table gradient, we divide the bank into a consecutive series of vertical columns and calculate the water table elevation in each accounting for lateral flow between neighbouring columns. We use local river stage (see above) as a boundary condition for the model and assume that water may only enter or exit the model domain from the channel (as lateral hyporheic flow). We therefore also assume that no recharge of the water table occurs from infiltrating precipitation at the top of the terrace and no vertical flow out from the base of the terrace. We make these assumptions because the latter two terms in the water budget of the lateral boundary are expected to be very small compared with the lateral hyporheic flow for this site. The Dupuit–Forchheimer equation can be written as

$$\frac{\partial h}{\partial t} = \frac{K}{\varepsilon} \cdot \frac{\partial}{\partial x}\left(h \cdot \frac{\partial h}{\partial x}\right) \tag{1}$$

where h = water table elevation (m), t = time (s), K = permeability (m/d), ε = porosity (decimal fraction of pore space), and x = horizontal distance (width) into the bank (m).

In order to maintain model stability, the water table elevation (WTE) time step was set at 15-minute intervals; the input values for the river stage were therefore interpolated from the USGS daily series. Column widths were delineated at 20-cm spacings (Fig. 4b), in order to retain sufficient detail to represent later boundary erosion, while encompassing the size of the largest clast (160 mm along the A-axis) found in the terrace.

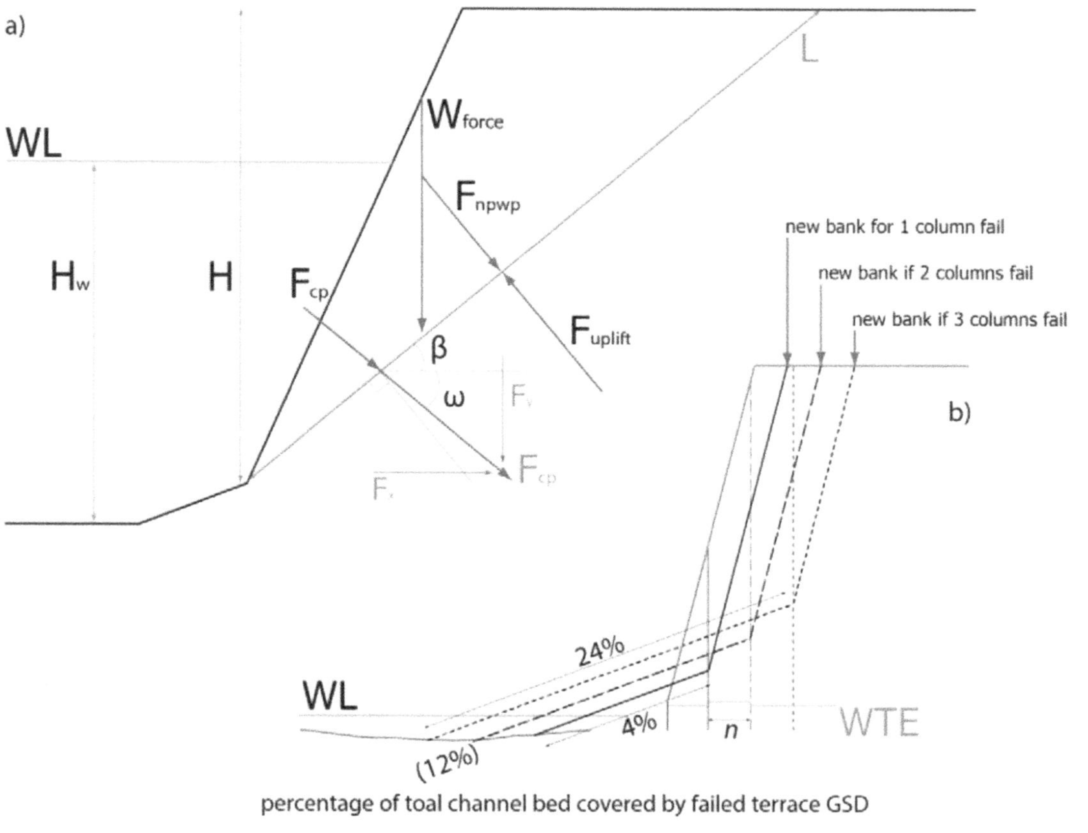

Figure 4: (a) Force diagram to resolve the factor of safety (FoS) (Eq. (2)) and minimum percentage representation of channel bed cover following terrace failure. WL = Water level (m); L = length of failure plane (m); H_w = depth of water in river (m); H = elevation between the emergence of the failure plane and the top of the bank (m); W_{force} = weight of unit width of the failure block (N m^{-1}); F_{cp} = hydrostatic confining pressure (N m^{-1}); F_{npwp} = negative porewater pressure (N m^{-1}); F_{uplift} = positive porewater pressure (N m^{-1}); β = angle of failure plane (degrees); ω = angle of resultant force through failure plane (degrees). (b) Diagram to show percentage bed cover by failed terrace material, 4% for a single column fail, 12% for a 2 column fail, and 24% for a 3 column fail; WTE = water table elevation (m); n = column width (20 cm).

Bank STABILITY

Based on the water table fluctuations discussed above, we modelled erosion of the lateral boundary of our cross-section via infinite slope stability adapted for riverbanks (Amiri-Tokaldany et al., 2003; Simon et al., 2003), which is applicable only for planar failures that apparently prevail at this site. We extracted a channel–terrace cross section by combining bathymetry with subaerial survey via laser rangefinder to obtain a representative transect of the Rose Bar terrace (Fig. 1). Our survey at this site indicated little variation in angle or shape of the cut bank. Bank stability, measured by Factor of Safety (FoS), is determined as the ratio between the physical resisting forces (F_r) to the forces driving bank erosion (F_d) (Fig. 4a):

$$FoS = \frac{F_r}{F_d}. \quad (2)$$

The resultant driving force (Fig. 4a) is equation(3)

$$F_d = F_w \sin\beta - F_{cp} \sin\delta \quad (3)$$

where F_w = weight of unit width of the failure block (N m^{-1}), β = angle of failure plane (degrees), F_{cp} = hydrostatic confining pressure (N m^{-1}), and δ = angle between the direction of the resultant hydrostatic confining pressure and at normal to the failure plane (degrees). The resultant resisting force (Fig. 4a) is

$$F_r = c'L + F_{suc} + (F_w \cos\beta + F_{cp} \cos\delta - F_{hu}) \tan\phi' \quad (4)$$

where c′ = effective cohesion averaged across each individual riverbank soil layer (kPa), L = length of failure plane (m), F_{suc} = suction force caused by negative pore-water pressure (N m^{-1}), F_{hu} = hydrostatic uplift force caused by positive pore-water pressure (N m^{-1}), and φ′ = angle of internal friction (degrees). The model calculates F_w as a summed weight of the saturated and unsaturated volumes:

$$F_w = (A_{unsat}\gamma_{sed}) + (A_{sat}\gamma_{sedsat})$$

where A_{unsat} = projected area by unsaturated layer (m^2), γ_{sed} = specific weight of sediment layer (N m^{-3}), A_{sat} = projected area by saturated layer (m^2), and γ_{sedsat} = specific weight of saturated soil layer (N m^{-3}). In order to determine the depth in the bank at which the sediment changes from saturated to unsaturated upon drainage (and the reverse during seepage into the bank), the WTE results are combined with the following equation for capillary rise between the pore spaces in the sediment (Amiri-Tokaldany et al., 2003):

$$h_{cr} = \frac{4\sigma \cos\theta}{\gamma_w^d} \quad (6)$$

where h_{cr} = height of capillary rise (m), σ = magnitude of the surface tension (N m^{-1}),θ = angle of contact between liquid and solid (degrees), γ_w = specific weight of water (N m^{-3}), and d = average diameter of the pore-spaces between the sediment (m) (Bowles, 1979):

$$d \cong \frac{1}{5} D_{10}. \quad (7)$$

In cases where no tension cracks are present, such as in our field area (noted after an extensive search along the terrace top), β is simply determined as the average of the angle of internal friction and the bank angle (Table 1) (Osman and Thorne, 1988).

Table 1: Field-measured and literature-based parameter values

Data from field		Data from literature		
Parameter	Value	Parameter	Value	Source
GSD (terrace)	0.01 (0.063 mm)	ε	30%	Selby (1993)
Fractional	0.01 (0.125 mm)			
Representation	0.04 (0.25 mm)	K	0.001 m/s	Selby (1993)
	0.15 (0.5 mm)			
	0.09 (1 mm)	φ′	38°	Selby (1993)
	0.07 (2 mm)			
	0.08 (4 mm)	c′	100 kPa	Selby (1993)
	0.12 (8 mm)			
	0.17 (16 mm)	Q	1986, 1997,	USGS
	0.19 (32 mm)		2006 WY	
	0.06 (64 mm)			
GSD (bed)	0 (0.063–1 mm)			
Fractional	0.05 (2 mm)			
Representation	0.14 (4 mm)			
	0.12 (8 mm)			
	0.15 (16 mm)			
	0.31 (32 mm)			
	0.22 (64 mm)			
H	16.7 m			
Bank angle	75°			
Toe angle	33°			

The F_{cp} on a riverbank is defined by compressive forces acting upon the bank face (Amiri-Tokaldany et al., 2003) as

$$F_{cp} = \sqrt{\left(F_x^2 + F_y^2\right)} \quad (8)$$

in which

$$F_x = \frac{1}{2} H_w^2 \gamma_w \quad (9)$$

$$F_y = \gamma_w V_w \quad (10)$$

where H_w = depth of water in river (m), V_w = volume of water per unit width on the sloping bank face (m^3 m^{-1}), and the subscripts x and y indicate the horizontal and vertical directions of the force (F), respectively.

Following Amiri-Tokaldany et al. (2003), δ is then determined by

$$\delta = 90 - (\beta + \omega) \quad (11)$$

in which

equation(12)

$$\omega = \tan^{-1} \frac{F_y}{F_x} \quad (12)$$

and ω = angle of resultant force through failure plane (degrees). The length of the failure surface is determined by

$$L = H_{fp} \sin\beta \quad (13)$$

where H_{fp} = elevation between the emergence of the failure plane and the top of the bank (m) (Fig. 4a).

We used GSDs measured in the field (Table 1) from which we can derive representative values for c′ and φ′ (Selby, 1993 and Simon et al., 2000). We introduce a cohesion value (Table 1) to account for the chemical cementation that was present within the lateral boundary. These are needed to determine the rate of increasing soil strength with increasing matric suction, i.e., pressure induced by soils in order to try and equalise moisture content (Simon and Collison, 2001 and Li et al., 2005) in the absence of hydrological field measurements.

We determined the shallow phreatic (water table) boundary elevation using WTE from Eqs. (1) and (6) with selected values of porosity and permeability (Table 1) and subsequently computed the weight acting upon the failure plane Eq. (5). Substituting this quantity and values from Eqs. (8), (11), and (13) into Eqs. (3) and (4), we obtain a value for the FoS Eq. (2). If Eq. (2) produces a value < 1, the bank is considered to fail at a single point in the bank. Our model then loops the FoS model through each column width into the terrace, specified in the WTE model, until the lateral boundary reaches a point of stability (any value > 1). We then compute a two-dimensional area of failure material from the lateral boundary along the cross section.

Sediment Flux and Bed Grain-Size Distribution Evolution

Once we determined the lateral boundary failure area (above) in the modelling framework, we account for the textural contribution of the failed sediment to the channel by updating the local river channel GSD. We determine how far into the channel the sediment supplied from the lateral boundary failure extends according to a simple procedure that captures the essence of the problem without detailed parameterisation. We assume that minor terrace failures cover the portion of the channel bed occupied by the initial bank/terrace toe (Fig. 4b). This assumption is made on the basis that no major flood events have occurred following the failure of the lateral boundary (a single-column failure) during a recent (2006) large flood event. Therefore, the present day terrace toe must be a minimum representation of the bed cover (4%) produced by the lateral boundary failure during that event. Deeper terrace failures (multiple columns) are assumed to be proportional to the volume of failed material, while maintaining the same toe angle (33° for the Rose Bar terrace) and are calculated geometrically with respect to the cross-sectional area of the column(s). These length scales of sediment supply from the lateral boundary are then used to update channel GSDs, information which feeds back into calculations of average channel sediment flux. We use these cover percentages of failed material to update the overall channel GSD by mixing the terrace GSD with the pre-existing channel GSD (from the previous time step) at specified percentages ratios of terrace:channel contribution (i.e., single-column fail = 4:96%; 2-column fail = 12:88%; and a 3-column fail = 24:76%).

Once we have updated the GSD following lateral boundary failure, we determine the partitioning of bedload and suspended load by computing fractional bed material sediment flux for each full phi grain size class at the same time steps as the WTE modelling. This allows us to characterise the grain size evolution of the riverbed, as well as the partitioning of sediment loads between coarse and fine fractions when the terrace fails. Note that we are not adjusting riverbed topography based on the assumption that this is much slower to change in response to terrace failure than is the channel GSD because of the relative magnitude of these failures to the channel width (~ 75 m). The fine sediment pulse associated with terrace failure is of particular importance for the Yuba study area given the high concentrations of Hg within these legacy sediments (Singer et al., 2013), the strong affinity of Hg and other transition metals for fine sediments (Gibbs, 1977), and the great risks to lowland ecosystems of its downstream delivery (Singer et al., 2013).

We calculate fractional transport rates for all grain-size classes via the Singer–Dunne equation (Singer and Dunne, 2004):

$$qs_n = \alpha \frac{\rho_s U^2 (\tau^* - \tau_c^*) \sqrt{\tau_n^*} \sqrt{\left(\frac{\rho_s}{\rho} - 1\right) g d_n^3}}{2gHS} f_n \quad (14)$$

where qs_n is the unit bed material transport rate (kg s^{-1}) per unit width of size class n,α is a dimensionless grain-size dependent parameter based on particle hiding and sorting, ρ_s is the density of sediment (assumed to be 2650 kg m^{-3}), U is the streamwise velocity (m s^{-1}), computed via an empirical fit of the Darcy–Weisbach formula based on grain size (Knighton, 1998), τ^* is dimensionless shear stress for D_{50},τ_c^* is the dimensionless critical shear stress for D_{50} (assumed to be 0.045), τ_n^* is the dimensionless shear stress computed for a particle in size class n, ρ is the density of water (assumed to be 1000 kg m^{-3} for 10 °C water), g is the gravitational acceleration

(9.81 m s^{-2}), d_n is the characteristic grain size (m) of the size class for which the computation is being made, f_n is the fraction of bed material in that grain-size class, His the flow depth (m), and S is the water surface slope (m m^{-1}).

$$\log_{10}\alpha = 1.62 * \sigma_\varphi + 0.58 * \left(\frac{D_i}{D_{50}}\right)^{-0.5} - 7.85 \quad (15)$$

where σφ is the sorting coefficient of bed material, Di is the characteristic grain size of each grain-size class, D_{50} is the characteristic (median) grain size of bed material. Sediment flux is computed as a cross-sectional average using average values of flow hydraulics. We compute fine sediment fluxes generated by terrace failure as suspended sediment concentrations by dividing daily flux of material in grain-size classes between 0.063 and 1 mm by the daily channel discharge, assuming this material is rapidly well mixed in the flow.

During periods where the lateral boundary does not fail, Eq. (14) simply iterates to compute fractional sediment flux based on the evolving GSD at every time step. Upon bank failure, the GSD is updated as detailed above and Eq. (14).

Model Operation and Goals

We use the modelling framework described above to investigate the impact of hydrograph shape on the magnitude and timing of terrace failure and associated sediment pulses at Rose Bar. We assess (i) individual event hydrographs from recent large floods (the 1986, 1997, and 2006 flood events were shown by Singer et al. (2013) to be the seminal events in the last several decades to produce geomorphic change); (ii) a decadal flow series that incorporates these three events but accounts for historical variability in flows in between these floods; (iii) characteristic hydrograph shapes modified from that of one historical flood (2006), but which retain the same overall flood volume; and (iv) simulated anthropogenic modification to the discharge hydrograph (e.g., power peaking).

RESULTS AND DISCUSSION

Hydrograph Shape

Our model shows a tight link between the WTE and the FoS of the lateral boundary when driven using channel stage data for the 1986 (Fig. 5a), 1997 (Fig. 5b) and 2006 (Fig. 5c) flood events. Across these major flood events of the last several decades, we identify in our model a threshold WTE range of 8–9.25 m above the datum that, once exceeded, induces terrace failure at the field site. This threshold range is supported by the absence of failure during the second, lower peak of the 1986 flood (Fig. 5a) and failure at the peak stage during the 2006 flood event (Fig. 5c), bounding a critical range of WTE. The 1997 flood event (Fig. 5b) was the only storm of high enough magnitude to produce a 2-column failure (i.e., 40 cm into the terrace); the other two events induced only single-column failures. The 1997 flood was previously shown to deliver a very high sediment load to into the Feather River (Kilham et al., 2012 and Singer et al., 2013). This suggests that storms of varying magnitudes and hydrograph shapes could dramatically impact the volume of sediment supplied from lateral boundaries merely as a function of changes in the WTE. The progressive increase in the WTE with each peak in the channel stage and the delayed decline in the WTE within the lateral boundary columns is most clearly visible through the 2006 water year (Fig. 5c). Following a rapid decline in channel stage, the WTE in the lateral boundary is left elevated several metres above the channel water surface. Over time, this retention of water in the bank weakens boundary material properties and enables collapse, suggesting the importance of antecedent moisture to terrace failure.

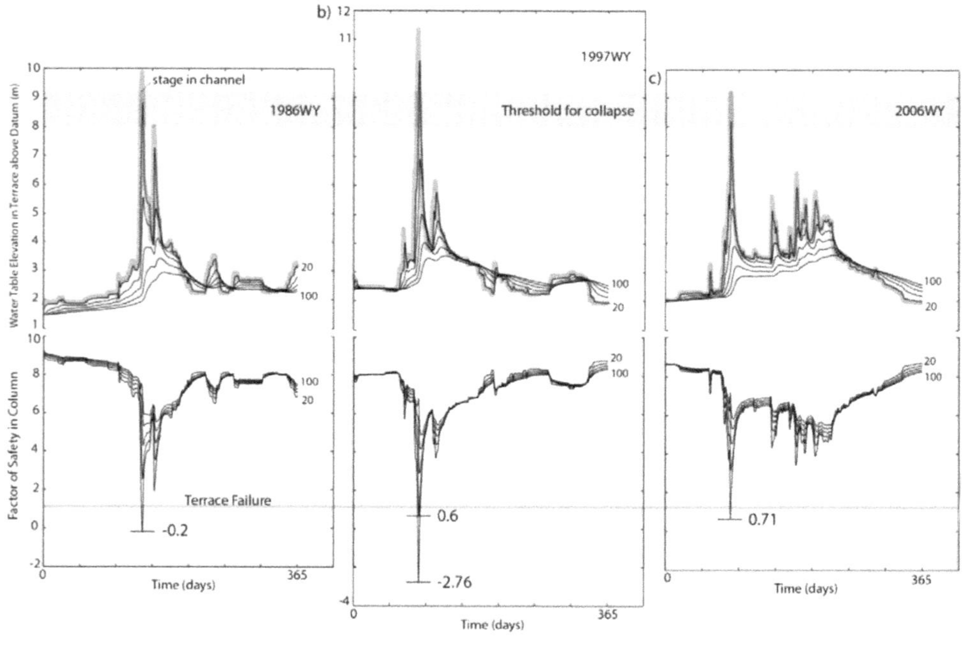

Figure 5: Timeseries of the WTE (m) and the FoS of the Rose Bar cross-section during three water years containing major floods: (a) 1986, (b) 1997, and (c) 2006. Isolines in the WTE and FoS plots are in increments of 20-cm width into the terrace. Lowest values in FoS are identified. Horizontal shaded areas across the WTE plots are our proposed range of stage threshold for inducing terrace collapse (8–9.25 m above datum).

Because discharge-driven WTE exerts a strong control on the stability of sediments within the lateral boundary (Fig. 5), we hypothesise that different hydrograph shapes should affect terrace failure through their influence on WTE and its duration. Singer et al. (2013) and Singer (2007) documented increased rising and falling limbs for the largest floods in recent decades along the Yuba River. Furthermore, regional climate changes are projected to produce larger and longer floods in the Sacramento River basin well into the future (Dettinger, 2011), as well as shifts in precipitation regimes toward more rainfall and less snow (Knowles et al., 2006).

One characteristic of a storm hydrograph that could be relevant to terrace stability is the baseline discharge above which a flood hydrograph is imprinted. This stage in the channel when the flood begins to affect the antecedent moisture (saturation) within the lateral boundary. Elevated water table levels can result from increased rainfall and runoff throughout the hydrologic year, or for near-channel terrace materials, through a series of successive flood events capable of saturating lateral channel boundaries. However, it is unclear how such antecedent moisture may affect failure probabilities within lateral channel boundaries, information that is important for anticipating the knock-on effects of sequences of floods.

To assess this, we adjusted the starting WTE in our model by altering the shape of the driving stream hydrograph while maintaining the same peak discharge. We conducted several modelling experiments that progressively increased the initial discharge (Fig. 6a–c) and therefore stage and terrace WTE. We find that increasing the initial channel stage and by extension, antecedent moisture in the lateral boundary rapidly decreases the FoS such that failure occurs at a much greater distance into the terrace (deeper failure generating more sediment supply). Specifically, a doubling of the initial (baseline) channel discharge produces 40-cm (40%) deeper failure, and a tripling of this discharge produces more than three times the depth of failure. In both cases, the failure also occurred one to several days earlier (Fig. 6a–c). In order to verify how antecedent moisture might affect terrace stability and to make an estimate on the probability of failure over a longer time-period, we also ran a simulation of 23 years that includes all the flood events shown in Fig. 5. The probability of failure on any day within this timeseries is 0.0009, indicating these failure events are quite rare. In fact, we expect this failure probability to be far lower for the entire hydrologic series over the last six decades. We note that while no change in the depth of failed material occurred (i.e., distance of failure into the terrace) for any of the recent major events, running the entire timeseries reduced (made less stable) the lowest FoS values for each event (Fig. 7) compared with those for modelled individual water years (Fig. 5). For example, the FoS for the 1986, 1997, and 2006 events at the 20-cm depth into the lateral boundary were 40%, 1%, and 7% lower in the 23-year series (Fig. 7). At the 40-cm distance into the boundary during the 1997 failure event, we note an 18% decrease in the FoS. These are the results of antecedent moisture in the river terrace prior to each flood event.

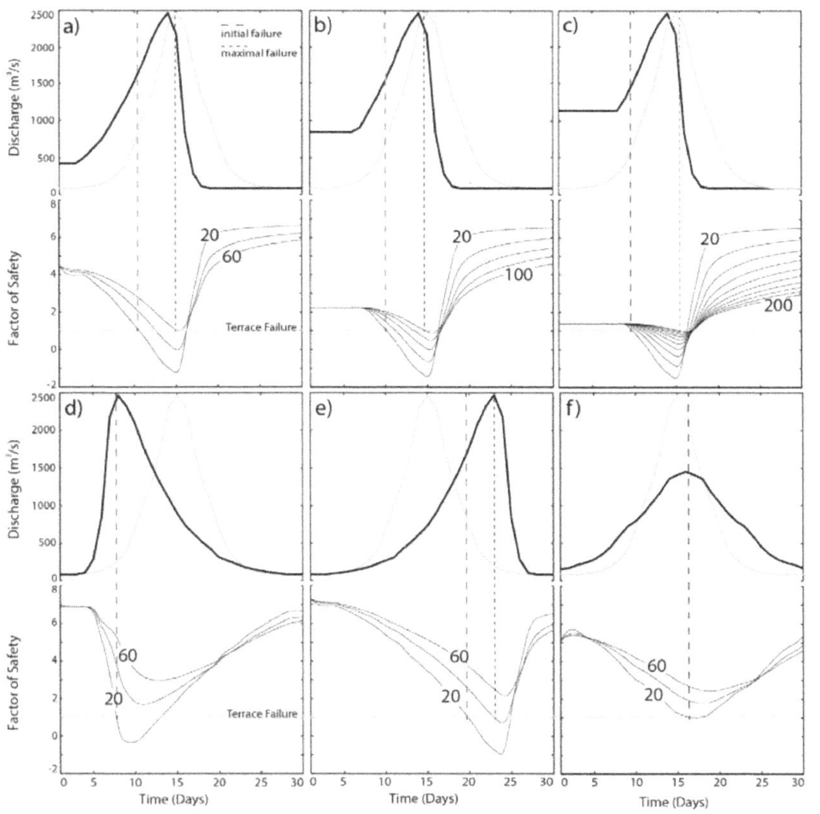

Figure 6: Effects of hydrograph shape on the FoS in the Rose Bar cross-section during a characteristic flood event. Modelled input flood hydrographs are shown above their corresponding FoS plot. Grey line depicts the characteristic flood hydrograph from which alterations were made while retaining the same flood peak and volume of water. Isolines in the FoS plots are in increments of 20-cm width into the terrace. (a–c) Three distinct antecedent water table conditions with the same peak discharge simulating flooding episodes in close proximity to one another; (d) accelerated rising limb to simulate flash flooding; (e) accelerated falling limb; and (f) lowered but extended peak discharge simulating dam-controlled flood release.

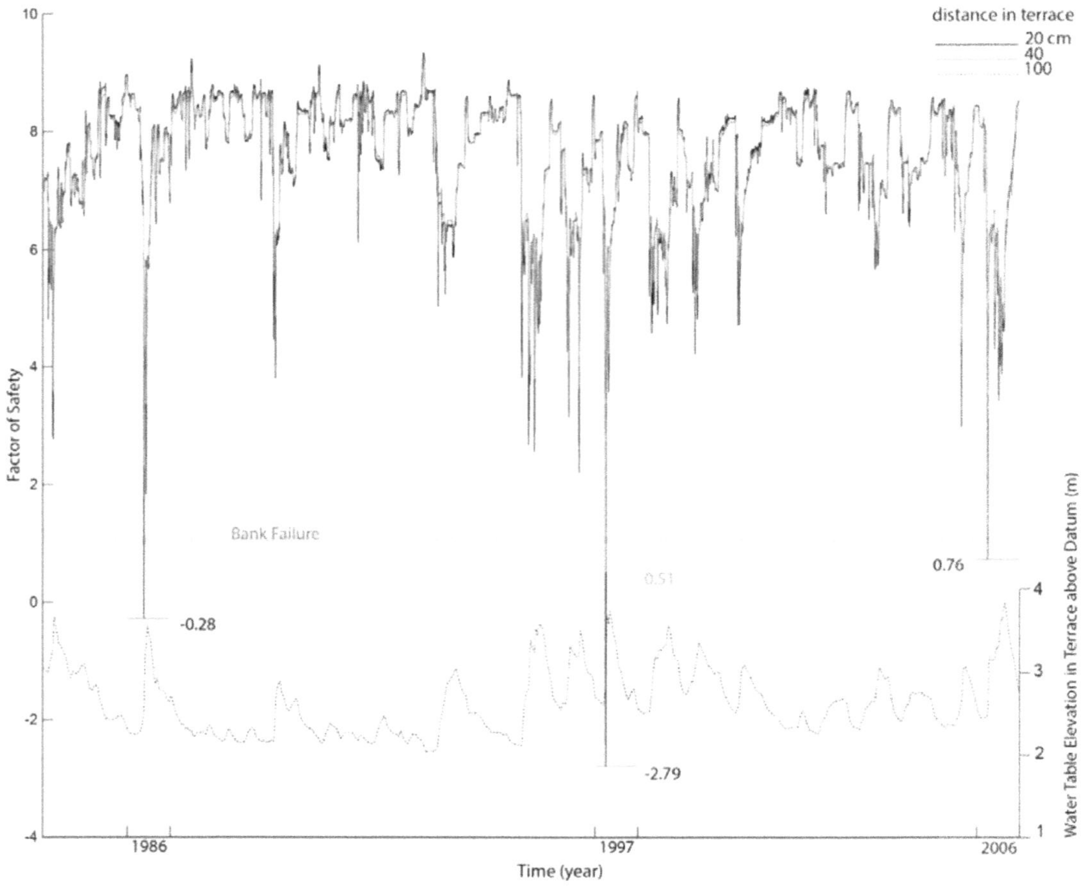

Figure 7: Timeseries of the FoS of the Rose Bar cross-section during a 23-year (1984–2007) model run. Lowest values in FoS are identified. The dashed line at the bottom of the plot displays modelled WTE at 1 m into the terrace.

Our extended timeseries modelling further highlighted that at increasing distances into the lateral boundary, the near-bank impact of hyporheic flow dissipates because the high flow in the channel is modulated as a smoothed and moderate water table elevation. For example, the WTE at 1 m into the lateral boundary for all three major flood peaks is nearly obscured and exhibits only small fluctuations with the varying hydrograph (Fig. 7). These results support the importance of antecedent moisture conditions within the near-channel zone when predicting lateral boundary stability. They suggest that a flood event that occurs when the hyporheic WTE is elevated (e.g., resulting from a previous flood or extended high rainfall), the resulting collapse would be more severe, i.e., failure at a greater distance into the terrace, and would occur earlier (c.f., Fig. 6a–c). In our modelling, an increase in the antecedent discharge from 15% to 50% of the peak discharge (a trebling of the baseline discharge from 400 to 1200 $m^3 s^{-1}$) shifts the initial lateral boundary failure one day earlier (Fig. 6a and c). While the timing of the initial failure remains close to day 10 within the modelled series, if the input timeseries instead begins when the discharge begins to rise above the antecedent conditions, failure occurs three days sooner as a result of doubling the antecedent discharge (Fig. 6a–b) and five days sooner when trebling the antecedent discharge (Fig. 6a–c).

Failure on the falling limb of the hydrograph has been noted elsewhere (Casagli et al., 1999, Rinaldi and Casagli, 1999 and Rinaldi et al., 2004). Rinaldi and Darby (2007) suggested that collapsed lateral boundary material produced on the falling limb is likely to remain at the toe for an extended period, compared to that produced by failure on the rising limb or at the flood peak. This is consistent with our field observations of sediment accumulation on the toe during low flow that was likely produced during the last major erosion event of 2006 (Fig. 1). Thus, hydrograph shape apparently impacts the timing of lateral boundary failure, but the length of rising/falling limbs with respect to an unchanging peak discharge should also influence the magnitude of terrace collapse. We find that extension (lengthening) of the rising limb produces a substantial reduction inFoS (i.e., increased probability of failure) compared with a contraction or steepening of the rising limb (c.f., Fig. 6d; e). Increasing the time to peak (from 5 days to 20 days) allows for greater hyporheic flow of water into the lateral boundary thus elevating the terrace WTE near the channel boundary, and shifting failure about a week earlier. This extended permeation of flow reduced the FoS from − 0.5 to − 0.9 for the 20-cm column but also induced failure at the 40-cm-width increment (Fig. 6e). The effect on the timing of failure is clear when comparing the initial failure (at the 20 cm width) between the two simulated flows. The extended rising limb (Fig. 6e) induces failure on the rise to peak at only 66% of the discharge required to induce failure with the steeper rising limb and shifts the timing of this failure 11 days later. Both simulations, however, produced initial failure on the rising limb of the hydrograph, consistent with Nardi et al. (2012), likely as a result of loss in apparent cohesion and the removal of sediment at the toe.

A common flood control measure is to cut off the discharge peak (e.g., by storing high flows in upstream reservoirs), thereby dampening the event hydrograph, although various strategies depend on the flood control capacity in the drainage basin (Singer, 2007). Cutting off the peak yields a notable increase in our modelled value of FoS (Fig. 6f), indicating that clear links between anthropogenic modifications to flow regimes and lateral boundary stability exist.

To further investigate such links, we conducted simulations of four reduced peak flow regimes (stepped percentage reduction from the original peak discharge for the 1986 flood), while maintaining the same 70-day flow volume. We find that controlling the flood peak is an effective way to maintain lateral boundary stability. The 40% reduction in peak flow dramatically reduced the depth of failure, and a 50% reduction imparted stability to the terrace (Fig. 8). However, this simulation required a 50% increase in the duration of flow at peak discharge, during which FoS hovers around the value of terrace failure. We can see in this example that while the FoS in the lateral boundary, following initial failure on day 9 in the event, is always above the threshold for failure (a value of 1), it oscillates above and below a value of 1.3, which is considered to be conditionally stable (Wilson et al., 2007 and Parker et al., 2008). However, to reduce the chance of terrace failure during a flood similar to that of 1986, the peak discharge would have to be reduced by 50%.

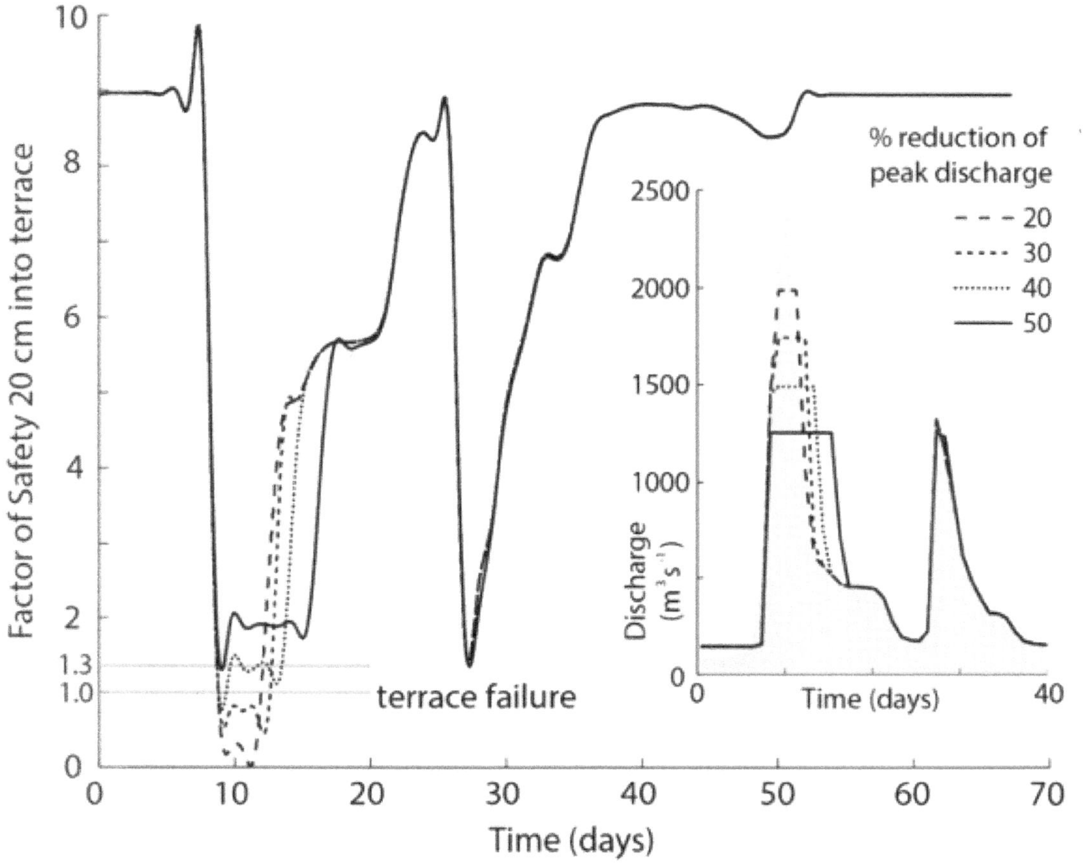

Figure 8: 70-day timeseries of FoS under simulated reduced peak discharges. Inset shows original hydrograph of 1986 flood in grey and modifications to it by the dashed lines. These line styles are maintained in the FoSplot.

Conversely, it may be desirable in some fluvial systems with altered streamflows (e.g.,Richter et al., 1996) to modify flow releases (e.g., Richter and Richter, 2000) in order to simulate flood events capable of scouring riverbeds and even eroding lateral boundaries to entrain fresh bed material substrate suitable for aquatic habitat (Kondolf and Wilcock, 1996, Poff et al., 1997 and Singer and Dunne, 2006). Our modelling framework would be useful for that purpose (e.g., by increasing flood peak discharge until a failure is predicted). The range of conditionally stable values discussed above might be useful for entraining new bed material from banks and terraces as substrate for aquatic habitat (Kondolf and Wolman, 1993), without producing disruptive geomorphic change that reshapes the channel (Wolman and Gerson, 1978). A 40% reduction in the peak discharge for the modelled event might just achieve this goal over the flood period.

Our final assessment of hydrograph shape is simulation of a hydropower flow regime (power peaking), conditions under which lateral boundaries are subjected to sharp rises and falls in discharge pulses (rather than the single peaks assessed in Fig. 6). We conducted two modelling experiments—one to assess the effect of a prolonged individual power peak on the FoS in the lateral boundary and the other to investigate a longer series of high discharge pulses (Fig. 9). Again, we used the 1986 flood as a base hydrograph and maintained the same total discharge volume while altering for power peaking in both scenarios. We find that an extended high flow release has a marked negative impact upon the stability of the lateral boundary. When the peak discharge is maintained over a period of three days, rather than comprised by a series of three individual peaks, the FoS declines from 1.9 to 0.5 causing failure. Shorter duration flow releases progressively reduce the FoS, but it remains above the failure threshold (Fig. 9a and b), consistent with the findings of Darby et al. (2000). This suggests that low flow power peaking that occurs in the summer months resulting from high electricity demand would not be expected to produce much bank/terrace failure in a channel like the Yuba. However, based on the impact to terrace failure of hydrograph shape in higher flows, close consideration should be paid to designing appropriate dam releases during flood event to meet management objectives with respect to the erosion of banks and terraces. Furthermore, in fluvial systems where banks are created by the modern river (rather than terraces created by legacy processes), power peaking may be an important consideration for downstream bank failure.

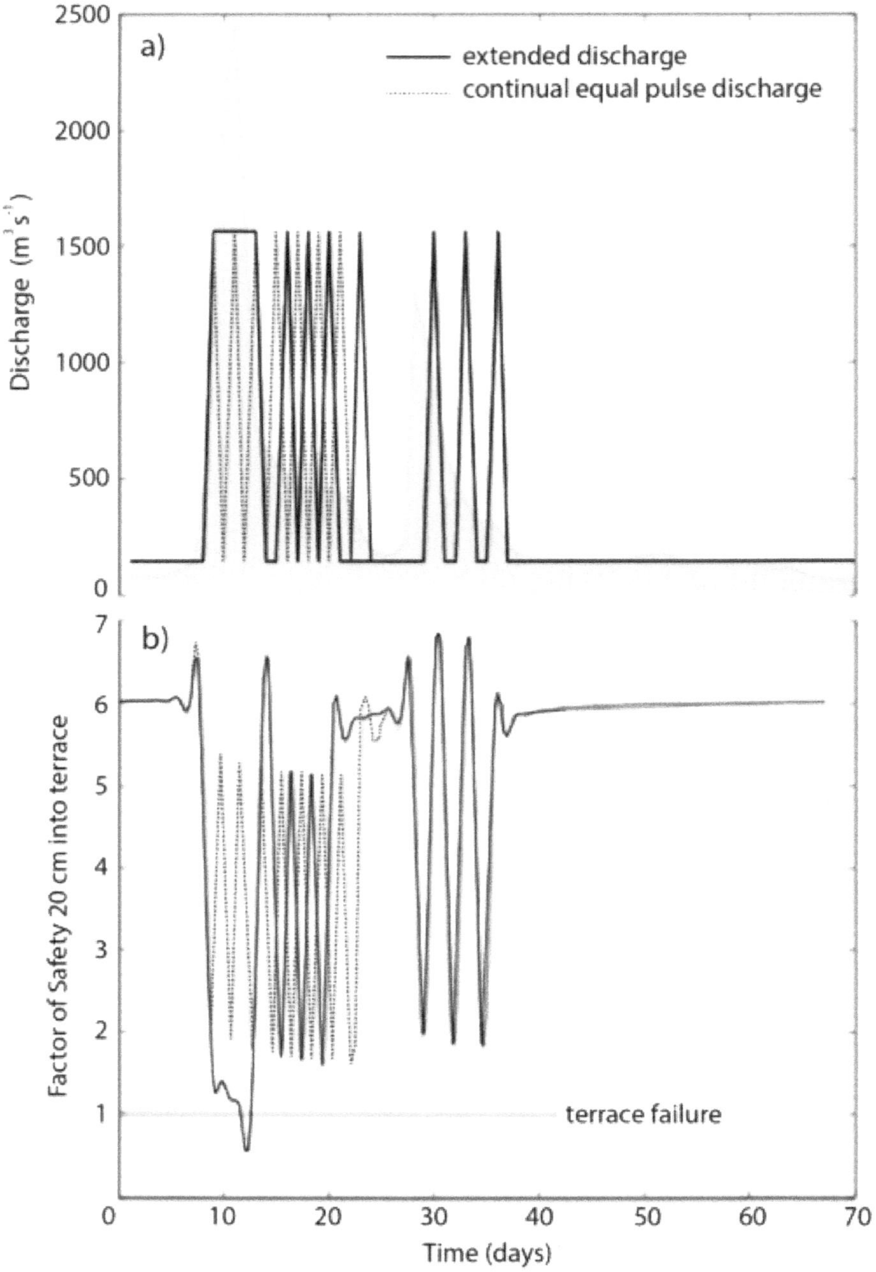

Figure 9: (a) 70-day timeseries of simulated hydropower flow regimes (power peaking). The background hydrograph is the unaltered USGS discharge data (from below Englebright Dam) for the 1986 flood event, the solid black line has an initial prolonged release of flow, and the grey line indicates a series of shorter pulse flows while maintaining the same volume of water as the unaltered discharge series. (b) The corresponding modelled FoS results for the flow simulations.

GSD Evolution and fine Sediment Fluxes

Our modelling framework also allows us to investigate links between the timing of lateral boundary failure and the downstream flux of fine sediment within the channel and to quantify the evolution of the riverbed GSD associated with terrace contributions. Following lateral boundary collapse, an injection of fine sediment (< 1 mm particle diameter) is delivered to the riverbed, but it is removed over a period of 25 days (Fig. 10a). As soon as an armor layer is reestablished (in this instance, during flows < 500 $m^3 s^{-1}$) in the model, the bed material grain-size distribution becomes stable (Fig. 10a) (as observed by Parker, 1990, Hassan et al., 2006 and Mao, 2012). However, in this case, the stable armor layer becomes slightly finer as a result of the terrace failure (Fig. 10a).

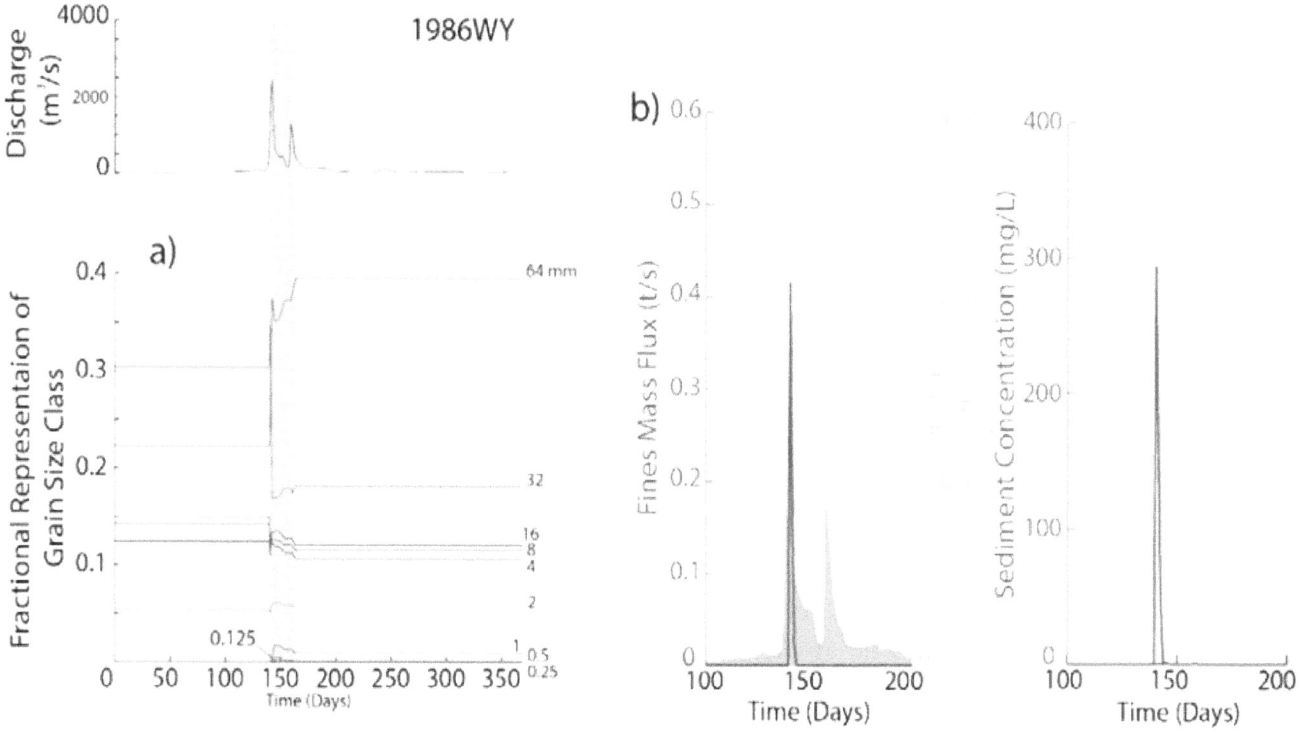

Figure 10: a) Timeseries of the bed material GSD evolution in the Rose Bar cross-section during the 1986 water year. Grain-size classes are represented by individual lines. Annual discharge hydrograph is shown above the GSD evolution plot, and shaded in grey is the period of time through which the GSD on the river bed was evolving. (b) Modelled fine sediment flux (< 1 mm) and sediment concentrations for the 1986 flood are represented by the dark lines in their respective plots and a discharge hydrograph is shown shaded in grey.

In terms of sediment pulses produced by terrace failure, our modelling suggests that during the 1986 water year, fine sediment flux at the Rose Bar cross section occurred only following bank collapse (Fig. 10b), producing mean sediment concentration of ~ 300 mg L^{-1}. Recent research (Kilham et al., 2012 and Singer et al., 2013) reported surface sediment concentrations > 100 mg L^{-1} persisting in the Yuba River up to 30 days following the peak discharge of the 1986 flood. While these concentrations cannot be directly compared, our higher concentration value is likely associated with the localised impact of terrace failure on sediment concentration at the peak of the event and does not account for the incorporation and dilution of other boundary failures along the course of the lower Yuba during the same event. Assessment of sediment concentrations from satellite imagery likely reflects the mixing of many discontinuous upstream failures, yielding a lower value of sediment concentration, especially well after the event peak (Kilham et al., 2012). In the Yuba, transport of the gravel surface layer that might trap underlying fine sediment is low and rare because of the presence of bed armoring that maintains bed stability for most flows. Thus, downstream transport of fine sediment is largely derived from lateral boundary failures such as those at Rose Bar. And it is precisely these intense episodes of fine sediment flux derived from failed terraces that are of concern to downstream biota because of the association of this sediment with Hg (Singer et al., 2013). These factors surely play into decadal, reach-scale sediment budgets for this and similar fluvial systems with a legacy of mining.

Sensitivity Analysis

Given the strong links between modelled WTE and FoS results (Fig. 5), we performed a sensitivity analysis to assess the effects of terrace materials (porosity and permeability, εand K) within the WTE model on FoS (Fig. 11a–c), using the 1986 flood event. FromSelby (1993) we can determine that altering the sand:gravel mix (GSD) within the terrace has a greater impact on K than on ε. Selby (1993) presented ε values for sand and gravel within an overlapping range; however, K values for sand are an order of magnitude lower than those presented for gravel. Varying ε between 25% and 40% (based on gravel and sand GSDs in Selby (1993)), we find no apparent impact on the timing at which the minimum FoS value occurs. The calculated minimum values of FoS for porosities of 25% and 40% are − 0.51 and − 0.17, respectively, resulting in a 66% increase in bank stability associated with more available pore space in the terrace matrix. We tested the effects of permeability (K) over a range from 0.01 to 0.0001 m s^{-1} and it is evident that the stability of the terrace is increased by lowering K (Fig. 11a and b). At 20 cm into the lateral boundary, K = 0.01 m s^{-1} results in a FoS of − 1.38, while K = 0.0001 increases FoS to 2.10 (more stable), an ~ 250% increase in lateral boundary stability. At 40 cm into the lateral boundary, the stability increases over 1000% (with FoS values from − 0.46 to 5.00) for the same assessed range in K values. Therefore, the material properties affecting the flow of water in the lateral boundary clearly have an important influence on the FoS, and the model is markedly more sensitive to permeability. Considering these factors, field measurements of permeability (e.g., via permeameter) would certainly reduce uncertainty in terrace/bank stability models.

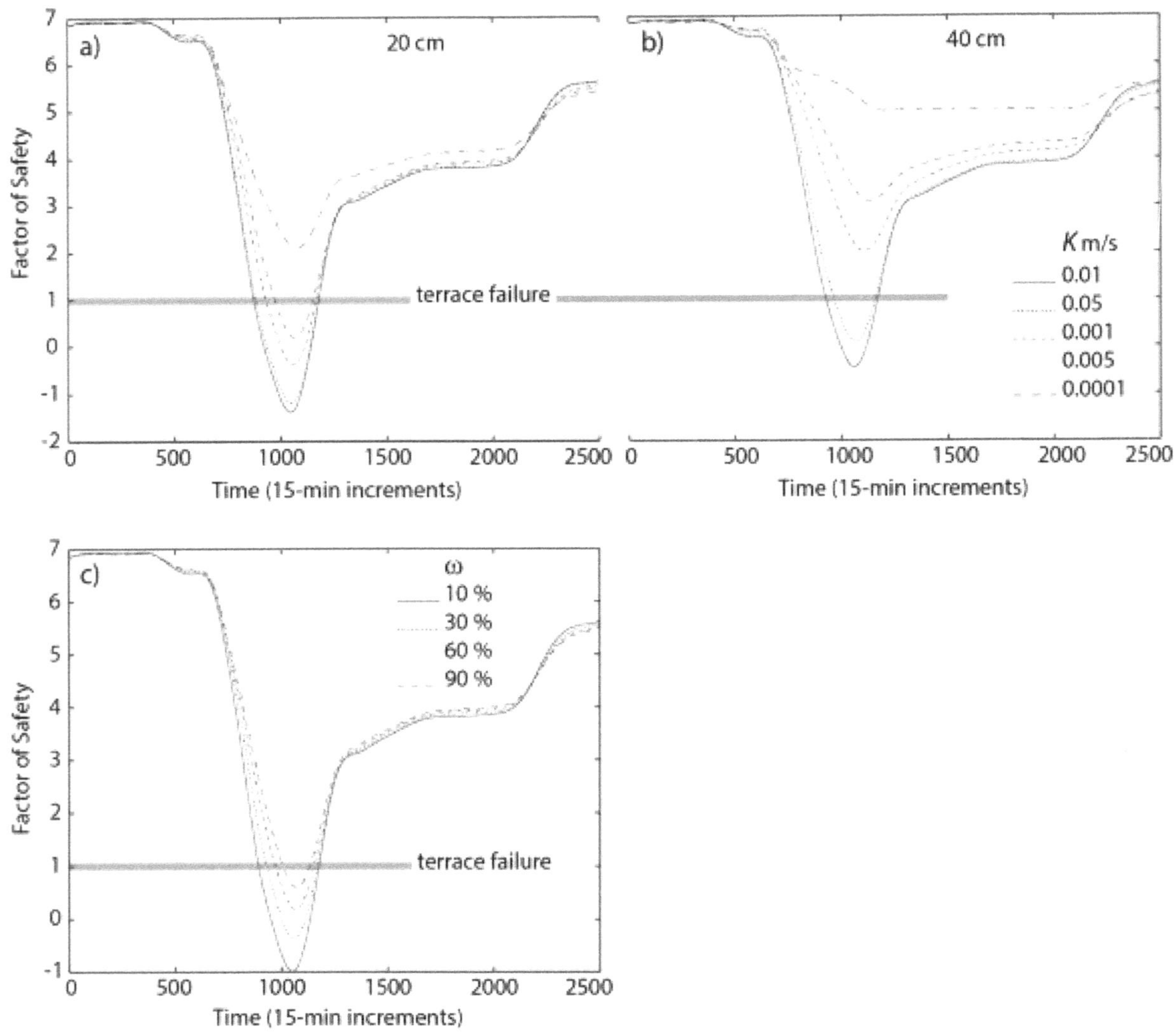

Figure 11: FoS model sensitivity analysis using the USGS 1986 flood event (70 days) discharge data varying permeability (a) at 20 cm and (b) at 40 cm into the terrace; (c) varying porosity at 20 cm into the terrace.

CONCLUSION

Episodic reworking of terrace material along the Lower Yuba River is responsible for mass delivery of fine sediments (in particular), which contain most of the Hg contamination (Springborn et al., 2011 and Singer et al., 2013), to the downstream environment. Here, we identified the mechanism for this flux as terrace failure during large flood events (Fig. 10), and we identified how it might vary according to hydrograph shape and antecedent moisture conditions (Fig. 6). In addition, we assessed the impact of various anthropogenically forced flow regimes (Fig. 8 and Fig. 10) on terrace failure.

Lateral boundary stability in the Rose Bar has been shown to be strongly linked to the magnitude of the peak discharge and its influence on the WTE within the lateral boundary (Fig. 5), the duration of time taken to reach that peak (Fig. 6d and e), and the antecedent conditions prior to the peak (Fig. 6a–c). Each of these conditions leads to a sustained high river stage acting on the face of the lateral boundary facilitating greater lateral infiltration, resulting in increased weight of the terrace materials acting upon the failure plane and ultimately generating mass failure. The longer the residence time of a high stage in the channel, the greater the time would be required for the WTE to respond to drawdown in stage/discharge visibility (Fig. 5c). We conclude that changes in the shape of the flood hydrograph, which may be forced by climate or by anthropogenic controls for damming, have the potential to drive increased/decreased mass wasting of lateral boundaries of rivers.

ACKNOWLEDGEMENTS

This work was funded in part by the Natural Environment Research Council in the form of a PhD studentship to Higson and by a U.S. National Science Foundation grant to Singer (EAR-1226741). We thank S. Mudd and J-P. Reneau for assistance in model development, and A. Simon for helpful discussions. We would also like to acknowledge Mikael Attal, Vincent Rinterknecht, and several anonymous reviewers for helpful input/advice as well as the journal editor for detailed editing.

REFERENCES

1. Alvarez and Schmeeckle, 2013 L.V. Alvarez, M.W. Schmeeckle Erosion of river sandbars by diurnal stage fluctuations in the Colorado River in the Marble and Grand Canyons: full-scale laboratory experiments
2. River Res. Appl., 29 (7) (2013), pp. 839–854
3. American Society of Civil Engineers, 1998a American Society of Civil Engineers River width adjustment. I.

Process and mechanisms J. Hydrol. Eng., 124 (9) (1998), pp. 881–902

4. American Society of Civil Engineers, 1998b American Society of Civil Engineers River width adjustment. II. Modeling J. Hydrol. Eng., 124 (9) (1998), pp. 903–917
5. Amiri-Tokaldany et al., 2003 E. Amiri-Tokaldany, S.E. Darby, P. Tosswell Bank stability analysis for predicting reach scale land loss and sediment yield J. Am. Water Resour. Assoc., 39 (2003), pp. 897–909
6. Amiri-Tokaldany et al., 2007 E. Amiri-Tokaldany, S.E. Darby, P. Tosswell Coupling bank stability and bed deformation models to predict equilibrium bed topography in river bends J. Hyrologic Eng., 133 (10) (2007), pp. 1167–1170
7. Aplers et al., 2005 C.N. Aplers, M.P. Hunerlach, J.T. May, R.L. Hothem Mercury contamination from historical gold mining in California USGS Fact Sheet 2005-3014 (1.1) (2005)
8. Bouse et al., 2010 R.M. Bouse, C.C. Fuller, S. Luoma, M.I. Hornberger, B.E. Jaffe, R.E. Smith Mercury-contaminated hydraulic mining debris in San Francisco Bay: San Francisco Estuary Watershed Sci. J., 8 (1) (2010), p. 28
9. Bowles, 1979 J.E. BowlesPhysical and Geotechnical Properties of Soils McGraw-Hill Inc., United Kingdom (1979), p. 478
10. Bull, 1997 L.J. Bull Magnitude and variation in the contribution of bank erosion to the suspended sediment load of the River Severn, UK Earth Surf. Process. Landf., 22 (1997), pp. 1109–1123
11. Bunte and Abt, 2001 K. Bunte, S.R. Abt Sampling surface and subsurface particle-size distributions in wadable gravel- and cobble-bed streams for analysis in sediment transport, hydraulics, and streambed monitoring US Dept. of Agriculture, Forest Service, Rocky Mountain Research Station, General Technical Report (2001)
12. Carley et al., 2012 J.K. Carley, G.B. Pasternack, J.R. Wyrick, J.R. Barker, P.M. Bratovich, D.A. Massa, G.D. Reedy, T.R. Johnson Significant decadal channel change 58–67 years post-dam accounting for uncertainty in topographic change detection between contour maps and point cloud modelsGeomorphology, 179 (2012), pp. 71–88
13. Carson et al., 1973 M.A. Carson, C.H. Taylor, B.J. Grey Sediment production in a small Appalachian watershed during spring runoff, Eaton Basin 1970–1972Can. J. Earth Sci., 10 (1973), pp. 1707–1734
14. Casagli et al., 1999 N. Casagli, M. Rinaldi, A. Gargini, A. Curini Pore water pressure and streambank stability: results from a monitoring site on the Sieve River, Italy Earth Surf. Process. Landf., 24 (1999), pp. 1095–1114
15. Church et al., 1987
16. M. Church, D. Mclean, J.F. Wolcott River bed gravels: sampling and analysis C.R. Thorne, J.C. Bathurst, R.W. Hey (Eds.), Sediment Transport in Gravel Bed Rivers, Wiley, Chichester (1987), pp. 43–79
17. Darby and Thorne, 1996 S.E. Darby, C.R. Thorne Numerical simulation of widening and bed deformation of straight sand-bed rivers. I: model development J. Hydrol. Eng., 122 (1996), pp. 184–193
18. Darby et al., 2000 S.E. Darby, D. Gressler, C.R. Thorne Computer program for stability analysis of steep, cohesive riverbanks Earth Surf. Process. Landf., 25 (2000), pp. 175–190
19. Darby et al., 2007 S.E. Darby, M. Rinaldi, S. Dapporto Coupled simulations of fluvial erosion and mass wasting for cohesive river banks J. Geophys. Res. Earth Surf., 112 (2007)
20. Dettinger, 2011 M. Dettinger Climate change, atmospheric rivers, and floods in California - A multimodel analysis of storm frequency and magnitude changes J. Am. Water Resour. Assoc., 47 (3) (2011), pp. 514–523
21. Donovan et al., 2013 P.M. Donovan, J.D. Blum, D. Yee, G.E. Gehrke, M.B. Singer An isotopic record of mercury in San Francisco Bay sediment Chem. Geol., 349–350 (2013), pp. 87–98
22. Dunne et al., 1998 T. Dunne, L. Mertes, R. Meade, J. Richey Exchanges of sediment between the flood plain and channel of the Amazon River in Brazil Geol. Soc. Am. Bull., 110 (4) (1998), pp. 450–467
23. Eagles-Smith et al., 2009 C.A. Eagles-Smith, J.T. Ackerman, S.E.W. De La Cruz, J.Y. Takekawa Mercury bioaccumulation and risk to three waterbird foraging guilds is influenced by foraging ecology and breeding stage
24. Environ. Pollut., 157 (2009)
25. Fredlund et al., 1978 D.G. Fredlund, N.R. Morgenstern, R.A. Widger Shear-strength of unsaturated soils Can. Geotech. J., 15 (1978), pp. 313–321
26. Gehrke et al., 2011 G.E. Gehrke, J.D. Blum, D.G. Slotton, B.K. Greenfield Mercury isotopes link mercury in San Francisco Bay forage fish to surface sediments Environ. Sci. Technol., 45 (4) (2011), pp. 1264–1270
27. Gibbs, 1977 R.J. Gibbs Transport phases of transition metals in the Amazon and Yukon Rivers Geol. Soc. Am. Bull., 88 (6) (1977), pp. 829–843
28. Ghoshal et al., 2010 S. Ghoshal, L.A. James, M.B. Singer, R. Aalto Channel and floodplain change analysis over a 100-year period: Lower Yuba River, California Remote Sens., 2 (7) (2010), pp. 1797–1825
29. Greenfield and Jahn, 2010 B.K. Greenfield, A. Jahn Mercury in San Francisco Bay forage fish Environ. Pollut., 158 (2010)
30. Greenfield et al., 2005 B.K. Greenfield, J.A. Davis, R. Fairey, C. Roberts, D. Crane, G. Ichikaway Seasonal, interannual, and long-term variation in sport fish contamination, San Francisco Bay Sci. Total Environ., 336 (2005)
31. Hassan et al., 2006 M.A. Hassan, R. Egozi, G. Parker Experiments on the effect of hydrograph characteristics on vertical grain sorting in gravel bed rivers Water Resour. Res., 42 (2006)
32. Hornberger et al., 1999 M.I. Hornberger, S.N. Luoma, A. Van Geen, C. Fuller, R. Anima Historical trends of

metals in the sediments of San Francisco Bay, California Mar. Chem., 64 (1999)

33. Howard, 1992 A.D. Howard Modelling channel migration and flooplain sedimentation in meandering streams P.A. Carling (Ed.), Lowland Floodplain Rivers: Geomorphological Perspectives, John Wiley and Sons, New York (1992), pp. 1–39

34. Humphries et al., 2012 R. Humphries, J.G. Venditti, L.S. Sklar, J.K. Wooster Experimental evidence for the effect of hydrographs on sediment pulse dynamics in gravel-bedded rivers Water Resour. Res., 48 (2012)

35. Ikeda et al., 1981 S. Ikeda, G. Parker, K. Sawai Bend theory of river menaders: 1. Linear development J. Fluid Mech., 112 (1981), pp. 363–377

36. James, 1989 L.A. James Sustained storage and transport of hydraulic gold mining sediment in the Bear River, California Ann. Assoc. Am. Geogr., 79 (1989), pp. 570–592

37. James and Singer, 2008 L.A. James, M.B. Singer Development of the Lower Sacramento Valley flood-control system: an historic perspective Nat. Hazards Rev., 9 (3) (2008), pp. 125–135

38. Johannesson and Parker, 1989 H. Johannesson, G. Parker Linear theory of river meanders S. Ikeda, G. Parker (Eds.), River Meandering, American Geophys. Union, Washington D.C. (1989), pp. 181–214

39. Kilham et al., 2012 N.E. Kilham, D. Roberts, M.B. Singer Remote sensing of suspended sediment concentration during flood conditions on the Feather River, California: a modeling approach Water Resour. Res., 48 (1) (2012), p. W01521

40. Knowles et al., 2006 N. Knowles, M.D. Dettinger, D.R. Cayan Trends in snowfall versus rainfall in the Western United States J. Clim., 19 (18) (2006), pp. 4545–4559

41. Knighton, 1998 D. Knighton Fluvial Forms and Processes: A New Perspective Hodder Arnold Publication, London (1998), p. 383

42. Kondolf and Wilcock, 1996 G.M. Kondolf, P.R. Wilcock The flushing flow problem: Defining and evaluating objectives Water Resour. Res., 32 (1996), pp. 2589–2599

43. Kondolf and Wolman, 1993 G.M. Kondolf, M.G. Wolman The sizes of salmonid spawning gravels Water Resour. Res., 29 (7) (1993), pp. 2275–2285

44. Lauer and Parker, 2008 J.W. Lauer, G. Parker Modeling framework for sediment deposition, storage, and evacuation in the floodplain of a meandering river: application to the Clark Fork River, Montana Water Resour. Res., 44 (2008) Li et al., 2005 A.G. Li, Z.Q. Yue, L.G. Tham, C.F. Lee, K.T. Law Field-monitored variations of soil moisture and matric suction in a saprolite slope Can. Geotech. J., 42 (2005), pp. 13–26

45. Luppi et al., 2009 L. Luppi, M. Rinaldi, L.B. Teruggi, S.E. Darby, L. Nardi Monitoring and numerical modelling of riverbank erosion processes: a case study along the Cecina River, Central Italy Earth Surf. Process. Landf., 34 (4) (2009), pp. 530–546

46. Mao, 2012 L. Mao The effect of hydrographs on bed load transport and bed sediment spatial arrangement J. Geophys. Res. Earth Surf., 117 (2012) Michaelides and Singer, 2014 K. Michaelides, M.B. Singer Impact of coarse sediment supply from hillslopes to the channel in runoff-dominated, dryland fluvial systems J. Geophys. Res. Earth Surf., 119 (6) (2014) (2013JF002959)Nardi et al., 2012

47. L. Nardi, M. Rinaldi, L. Solari An experimental investigation on mass failures occurring in a riverbank composed of sandy gravelGeomorphology, 163–164 (2012), pp. 36–69

48. Odgaard, 1987 A.J. Odgaard Streambank erosion along two rivers in Iowa Water Resour. Res., 23 (7) (1987), pp. 1225–1236

49. Osman and Thorne, 1988 A.M. Osman, C.R. Thorne Riverbank stability analysis: 1. Theory J. Hydraul. Eng. ASCE, 114 (1988), pp. 134–150

50. Parker, 1990 G. Parker Fluvial armor J. Hydraul. Res., 28 (4) (1990), pp. 529–544

51. Parker et al., 2008 C. Parker, A. Simon, C.R. Thorne The effects of variability in bank material properties on riverbank stability: Goodwin Creek, Mississippi Geomorphology, 101 (2008)

52. Parker et al., 2011 G. Parker, Y. Shimizu, G.V. Wilkerson, E.C. Eke, J.D. Abad, J.W. Lauer, C. Paola, W.E. Dietrich, V.R. Voller A new framework for modeling the migration of meandering rivers Earth Surf. Process. Landf., 36 (2011)Poff et al., 1997

53. N.L. Poff, J.D. Allan, M.B. Bain, J.R. Karr, K.L. Prestegaard, B.D. Richter, R.E. Sparks, J.C. Stromberg The natural flow regime Bioscience, 47 (11) (1997), pp. 769–784\

54. Rice, 1998 S. Rice Which tributaries disrupt downstream fining along gravel-bed rivers? Geomorphology, 22 (1998), pp. 39–56

55. Richter and Richter, 2000 B.D. Richter, H.E. Richter Prescribing flood regimes to sustain riparian ecosystems along meander rivers Conserv. Biol., 14 (5) (2000), pp. 1467–1478

56. Richter et al., 1996 B.D. Richter, J.V. Baumgatner, J. Powell, D.P. BraunA method for assessing hydrologic alteration within ecosystemsConserv. Biol., 10 (4) (1996), pp. 1163–1174

57. Rinaldi and Casagli, 1999 M. Rinaldi, N. CasagliStability of streambanks formed in partially saturated soils and effects of negative pore water pressures: the Sieve River, Italy Geomorphology, 26 (1999), pp. 253–277

58. Rinaldi et al., 2004 M. Rinaldi, N. Casagli, S. Dapporto, A. Gagini Monitoring and modelling of pore water pressure changes and riverbank stability during flow eventsEarth Surf. Process. Landf., 29 (2004), pp. 237–254

59. Rinaldi and Darby, 2007 M. Rinaldi, S.E. DarbyModelling river-bank erosion processes and mass failure mechanisms: progress towards fully coupled simulations H. Habersack, H. Piegay, M. Rinaldi (Eds.), Gravel-Bed Rivers VI: from Process Understanding to River Restoration, Elsevier, Amsterdam, The Netherlands (2007), pp. 213–239

60. Rinaldi et al., 2008M. Rinaldi, B. Mengoni, L. Luppi, S.E. Darby, E. Mosselman Numerical simulation of hydrodynamics and bank erosion in a river bendWater esour. Res., 44 (9) (2008)Selby, 1993.J. SelbyHillslope Materials and ProcessesOxford University Press, Oxford, United Kingdom (1993) (480 pp.)

61. Simon and Collison, 2001 A. Simon, A.J.C. Collison Pore-water pressure effects on the detachment of cohesive streambeds: seepage forces and matric suction Earth Surf. Process. Landf., 26 (2001)

62. Simon and Darby, 2002 A. Simon, S.E. Darby Effectiveness of grade-control structures in reducing erosion along incised river channels: the case of Hotophia Creek, Mississippi Geomorphology, 42 (3–4) (2002), pp. 229–254

63. Simon et al., 2003A. Simon, A. Collison, A. Layzell Incorporating bank-toe erosion by hydraulic shear into the ARS bank-stability model: Missouri River, Eastern Montana World Water & Environ. Res. Cong (2003), pp. 1–11

64. Simon et al., 1991A. Simon, W.J. Wolfe, A. Molinas Mass wasting algorithms in alluvial channel model

65. Proceedings 5th Federal Interagency Sedimentation Conference, Las Vegas, 2 (1991) (8-22-28-29)

66. Simon et al., 2000 A. Simon, A. Curini, S.E. Darby, E.J. Langendoen Bank and near-bank processes in an incised channel Geomorphology, 35 (2000), pp. 193–217

67. Singer and Dunne, 2006 M.B. Singer, T. Dunne Modeling the influence of river rehabilitation scenarios on bed material sediment flux in a large river over decadal timescales Water Resour. Res., 42 (12) (2006), p. 14 http://dx.doi.org/10.1029/2006wr004894 (W12415,)

68. Singer, 2007M.B. SingerThe influence of major dams on hydrology through the drainage network of the Sacramento

69. Singer, M.B., Aalto, R., James, L.A., 2008. Status of the lower Sacramento Valley flood-control system within the context of its natural geomorphic setting. Nat. Hazards Rev. 9 (3), 104–115.

70. Singer, M.B., 2008a. A new sampler for extracting bed material sediment from sand and gravel beds in navigable rivers. Earth Surf. Process. Landf. 33 (14), 2277–2284.

71. Singer, M.B., 2008b. Downstream patterns of bed-material grain size in a large, lowland alluvial river subject to low sediment supply. Water Resour. Res. 44, W12202. http://dx.doi.org/10.1029/2008WR007183.

72. Singer, M.B., 2010. Transient response in longitudinal grain size to reduced gravel supply in a large river. Geophys. Res. Lett. 37 (18)L18403. http://dx.doi.org/10.1029/ 2010GL044381.

73. Singer, M.B., Aalto, R., 2009. Floodplain development in an engineered setting. Earth Surf. Process. Landf. 34, 291–304.

74. Singer, M.B., Dunne, T., 2001. Identifying eroding and depositional reaches of valley by analysis of suspended-sediment transport in the Sacramento River, California. Water Resour. Res. 37 (12), 3371–3381.

75. Singer, M.B., Dunne, T., 2004. Modeling decadal bed material sediment flux based on stochastic hydrology — art. no. W033032. Water Resour. Res. 40, 14.

76. Singer, M.B., Aalto, R., James, L.A., Kilham, N.E., Higson, J.L., Ghoshal, S., 2013. Enduring legacy of a toxic fan via episodic redistribution of California gold mining debris. Proc. Natl. Acad. Sci. 110 (46), 18436–18441.

77. Singer, M.B., Michaelides, K., 2014. How is topographic simplicity maintained in ephemeral dryland channels? Geology 42 (12), 1091–1094. Slater, L.J., Singer, M.B., Kirchner, J.W., 2015. Hydrologic versus geomorphic drivers of trends in flood hazard. Geophys. Res. Lett. 42 (2), 370–376.

78. Snyder, N.P., Wright, S.A., Alpers, C.N., Flint, L.E., Holmes, C.W., Rubin, D.M., 2006. Reconstructing depositional processes and history from reservoir stratigraphy: Englebright Lake, Yuba River, northern California. J. Geophys. Res. Earth Surf. 111.

79. Springborn, M., Singer, M.B., Dunne, T., 2011. Sediment-adsorbed total mercury flux through Yolo Bypass, the primary floodway and wetland in the Sacramento Valley, California. Sci. Total Environ. 412–413, 203–213.

80. Sun, T., Meakin, P., Jossang, T., Schwarz, K., 1996. A simulation model for meandering rivers. Water Resour. Res. 32, 2937–2954.

81. Terzaghi, K., 1942. Theoretical Soil Mechanics. Wiley, New York 978-0-471-85305-3.

82. Thorne, C.R., 1982. Processes and mechanisms of river bank erosion. In: Hey, R.D., Bathurst, J.C., Thorne, C.R. (Eds.), Gravel-bed Rivers. J. Wiley & Sons, London, pp. 227–271.

83. Walling, D.E., Owens, P.N., Leeks, G.J.L., 1999. Fingerprinting suspended sediment sources in the catchment of the River Ouse, Yorkshire, UK. Hydrol. Process. 13, 955–975.

84. Wilson, G.V., Perketi, R.K., Fox, G.A., Dabney, S.M., Shields, F.D., Cullum, R.F., 2007. Soil properties controlling seepage erosion contributions to streambank failure. Earth Surf. Process. Landf. 32, 447–459.

85. Wolman, M.G., 1954. Method of sampling coarse river bed material. Trans. AGU 35, 951–956.

86. Wolman, M.G., Gerson, R., 1978. Relative time scales of time and effectiveness of climate in watershed geomorphology. Earth Surf. Process. Landf. 3, 189–208.

CHAPTER 4

ASSESSMENT OF GROUND-BASED MONITORING TECHNIQUES APPLIED TO LANDSLIDE INVESTIGATIONS

S. Uhlemann[a, b] A. Smith[c], J. Chambers[a], N. Dixon[c], T. Dijkstra[a], E. Haslam[a], P. Meldrum[a], A. Merritt[d], D. Gunn[a], J. Mackay[a]

[a] British Geological Survey, Natural Environment Research Council, Kingsley Dunham Centre, Keyworth, Nottingham NG12 5GG, United Kingdom

[b] ETH Zurich, Institute of Geophysics, Sonneggstrasse 5, CH-8092, Zurich, Switzerland

[c] Loughborough University, School of Civil and Building Engineering, Loughborough LE11 3TU, United Kingdom

[d] Plymouth University, School of Geography, Earth & Environmental Sciences, Drake Circus, Plymouth PL4 8AA, United Kingdom

ABSTRACT

A landslide complex in the Whitby Mudstone Formation at Hollin Hill, North Yorkshire, UK is periodically re-activated in response to rainfall-induced pore-water pressure fluctuations. This paper compares long-term measurements (i.e., 2009–2014) obtained from a combination of monitoring techniques that have been employed together for the first time on an active landslide. The results highlight the relative performance of the different techniques, and can provide guidance for researchers and practitioners for selecting and installing appropriate monitoring techniques to assess unstable slopes. Particular attention is given to the spatial and temporal resolutions offered by the different approaches that include: Real Time Kinematic-GPS (RTK-GPS) monitoring of a ground surface marker array, conventional inclinometers, Shape Acceleration Arrays (SAA), tilt meters, active waveguides with Acoustic Emission (AE) monitoring, and piezometers. High spatial resolution information has allowed locating areas of stability and instability across a large slope. This has enabled identification of areas where further monitoring efforts should be focused. High temporal resolution information allowed the capture of 'S'-shaped slope displacement-time behaviour (i.e. phases of slope acceleration, deceleration and stability) in response to elevations in pore-water pressures. This study shows that a well-balanced suite of monitoring techniques that provides high temporal and spatial resolutions on both measurement and slope scale is necessary to fully understand failure and movement mechanisms of slopes. In the case of the Hollin Hill landslide it enabled detailed interpretation of the geomorphological processes governing landslide activity. It highlights the benefit of regularly surveying a network of GPS markers to determine areas for installation of movement monitoring techniques that offer higher resolution both temporally and spatially. The small sensitivity of tilt meter measurements to translational movements limited the ability to record characteristic 'S'-shaped landslide movements at Hollin Hill, which were identified using SAA and AE measurements. This high sensitivity to landslide movements indicates the applicability of SAA and AE monitoring to be used in early warning systems, through detecting and quantifying accelerations of slope movement.

Citation: S. Uhlemanna, A. Smith, J. Chambersa, N. Dixon, T. Dijkstra, E. Haslam, P. Meldrum, A. Merritt, D. Gunn, J. Mackaya, Assessment of ground-based monitoring techniques applied to landslide investigations, doi:10.1016/j.geomorph.2015.10.027

INTRODUCTION

Landslides form one of the major natural hazards causing loss of life and damaging of infrastructure worldwide. In the seven year period between 2004 and 2010, 2620 fatal, non-seismically triggered landslides were recorded worldwide, causing 32,322 fatalities (Petley, 2012). Although the landslide frequency in the UK is comparably low (e.g. some 754 reported events between 2006 and August 2015), the economic impact is high. Failure of engineered earthworks (embankments and cuttings) or adjacent natural slopes causes interruptions to transportation and utilities networks — a process that is affected by ongoing climate change and ageing of slope materials (Dijkstra and Dixon, 2010,Foster et al., 2011, Dijkstra et al., 2014, Glendinning et al., 2015 and Pennington et al., 2015).

Since one of the primary triggering mechanism for landslides is intense or prolonged precipitation (Highland and Bobrowsky, 2008), the frequency and severity of landslides are expected to fluctuate with changes in precipitation patterns (spatial and temporal variations of duration and intensity). In a context of climate change, precipitation cannot be regarded as a steady state input and it is essential to develop robust models of adequate complexity that allow evaluation of possible future changes in slope instability due to forecasted changes in precipitation (Dijkstra and Dixon, 2010). Mid- to high-latitude regions are likely to face an increase in precipitation of up to 20%, including increased flash floods due to more frequently appearing high-intensity rainfall events (Defra, 2012 and Füssel et al., 2012). This increase in precipitation is likely to occur during the winter season, while summers will become drier (Defra, 2012). Wetter winters and drier summers will lead to an increased and deeper weathering of the topsoil, due to larger amplitudes in the wetting and drying cycles. This is likely to reduce the strength of the material and cause more frequent shallow slope instabilities. Understanding triggering mechanisms and failure potentials, to improve landslide forecasting, is therefore a major focus of research internationally.

Monitoring of kinematic, hydrological, and climatic parameters plays a significant role in supporting the development of slope stability models (e.g. Buchli et al., 2013 and Springman et al., 2013), since without understanding movement patterns and responses to climate events, forecasting is not possible (Angeli et al., 2000). This requires not just monitoring of actual movements, but also environmental factors including rainfall, temperature, soil moisture content and relative (air) humidity, as well as geotechnical parameters, such as, pore water pressure. This enables the correlation of movement events with their triggering mechanisms and helps to inform the underlying causalities in the process–response models.

Previous studies have related deformation measurements by GPS/GNSS, inclinometer, extensometer or tilt meter readings to rainfall events (e.g. Malet et al., 2002, Corsini et al., 2005, García et al., 2010 and Brückl et al., 2013) to study the deformation behaviour of rainfall triggered landslides. Additionally, Malet et al., 2002 and Corsini et al., 2005, andBrückl et al. (2013) compare measurements of two or three of the mentioned conventional deformation monitoring techniques, showing good correlation between surface (e.g. GPS) and subsurface (e.g. inclinometer) deformations in terms of movement occurrences.

In this paper we introduce and compare conventional techniques, such as GPS, inclinometer and tilt meter, and recently emerging deformation monitoring techniques, such as acoustic emission (AE) monitoring using active waveguides (AEWG), and Shape Acceleration Array (SAA). To our knowledge, this is the first time that these monitoring techniques have been combined on an active landslide, providing long-term (2009–2014) measurements. This paper highlights the relative performance of these techniques focusing on different movement periods, and it provides detailed, integrated interpretations of movement, environmental, and geotechnical data of the Hollin Hill landslide. The paper reports how these long-term monitoring results have enabled a step-change in the understanding of slope dynamics, building upon previously published work.

The study site, where a Lias mudstone formation is failing, is typical for many inland landslides in lowland settings. Thus, the conclusions drawn from this study can provide guidance for researchers and practitioners for selecting and installing appropriate monitoring techniques to assess unstable slopes.

Slope Monitoring Instruments and Techniques

There is a clear need to monitor landslides and marginally stable slopes to provide early warning of instability. This will allow for timely evacuation of vulnerable people, as well as timely repair and maintenance of critical infrastructure. The cost of remediation subsequent to landslide failure is several times higher than the cost of corrective measures and repairs if conducted prior to collapse (Glendinning et al., 2009); this highlights the importance of early warning through monitoring. Monitoring also provides: (1) the information necessary for slope stability analysis and remediation design, (2) knowledge of stability to and through construction, and subsequent to remediation, as well as (3) understanding of the condition (both serviceability and ultimate limit states) of adjacent infrastructure that has the potential to be impacted upon by slope instability (Dunnicliff, 1988 and Machan and Beckstrand, 2012).

There are many different techniques and types of instrumentation typically used in slope monitoring, and numerous emerging technologies. No single technique or instrument can provide complete information about a landslide and therefore various combinations are usually employed. The primary parameters of interest are deformation and pore-water pressure. Information on these is necessary to assess the rate and magnitude of movement, as well as changes to effective stress and hence stability. Performance of monitoring techniques and instruments is usually assessed in terms of accuracy and precision, spatial and temporal resolutions, sensitivity, and reliability (Dixon et al., 2015). However, on most projects the predominant factor driving the choice of instrumentation and techniques is their cost. For this reason the majority of slope monitoring programmes comprise installation of inclinometer casings and standpipes, which are usually read at discrete and infrequent intervals of the order of months. Inclinometers allow the depth to any shear surface(s) to be identified and standpipes provide information on the ground water conditions; information that is necessary for stability assessment and remediation design. However, this mode of monitoring provides relatively low spatial and temporal resolutions, which is usually insufficient to provide early warning of instability.

Detailed in Table 1 are the monitoring instruments and techniques employed in this study, together with an indication of their spatial and temporal resolutions. The resolution of the methods is a function of the nature of the installation and the extent of the sensor network. Note that the classification shown in Table 1 specifically refers to the installation at the Hollin Hill study site. This combination of monitoring approaches was selected in order to: (1) provide relatively high spatial resolution of ground surface movements, (2) determine the depth to shear surfaces, (3) monitor subsurface deformation at localised areas with high temporal resolution, and (4) to monitor pore-water pressures at shear surface depths with high temporal resolution. This significant level of information provides the basis for a thorough assessment of each of the approaches in order to make recommendations for other landslide investigations.

Table 1: Monitoring devices installed at the Hollin Hill landslide field laboratory. For each system a brief description as well as an indication of its spatial and temporal resolutions is given. Spatial resolution is given on a local, i.e. measurement, and global, i.e. slope, scale

Instrument/technique	Description	Spatial resolution		Temporal resolution
		Local	Global	
Surface deformation monitoring				
Global positioning system (GPS)	• Repeated survey of permanently installed markers using RTK/GNSS GPS with centimetric accuracy providing time-series of 3D topographical data. Allows for measurement of surface movements and deformations • Permanent GNSS stations allow for high temporal resolution	Moderate	Moderate	Low
Tilt meter	• Installed near the ground surface providing an indication of landslide activity and an indirect measurement of deformation • Most suitable for monitoring of landslides with a large rotational component of movement; translational movements are unlikely to be detected	High	Low	High
Subsurface deformation monitoring				
Conventional inclinometer	• Installed in a borehole penetrating stable stratum below any shear surface (or potential shear surface). The annulus around the casing is backfilled with material having similar properties as the surrounding soil (e.g. bentonite-grout) • A torpedo probe measures the inclination of the casing with depth, providing a horizontal deformation vs. depth profile • Time-series data can be used to determine magnitude and rate of displacements, and depth to shear surfaces (Mikkelsen, 2003, Simeoni and Mongiovì, 2007 and Wan and Standing, 2014) • If an in-place probe or probe-string is installed and deformation is logged continuously then high temporal resolution can be acquired	High	Low	Low

Shape Acceleration Array (SAA)	• Comprises a string of MEMS (micro-electrical–mechanical systems) sensors installed at regular intervals along the depth of a borehole, which penetrates any shear surface or potential shear surface	High	Low	High
	• Each sensor monitors displacement in x-, y- and z-directions with high accuracy and temporal resolution (Abdoun et al., 2012)			
Active waveguide and Slope ALARMS sensor (i.e. AE monitoring)	• Installed in a borehole penetrating any shear surface or potential shear surface. Comprises a steel waveguide (i.e. to transport AE signals generated at the shear surface to the ground surface with low attenuation) and granular backfill (i.e. to generate high energy AE as the slope deforms)	High	Low	High
	• AE is generated when the granular backfill is deformed by deformation of the host slope			
	• AE is measured at the ground surface with a Slope ALARMS system (Dixon and Spriggs, 2011)			
	• Recorded are AE rates or ring-down counts (RDC) per unit time, which are the number of times the signal amplitude crosses a pre-defined voltage threshold per pre-defined time period			
	• AE rates (e.g. RDC/h) are proportional to slope displacement rates, enabling detection of accelerations and decelerations of slope movement (Dixon et al., 2003, Dixon et al., 2014, Dixon et al., 2015, Smith et al., 2014 and Smith and Dixon, 2015)			
Ground water monitoring				
Piezometer	• Piezometer tip is installed to depth(s) of interest (e.g. a shear surface) inside a borehole to monitor pore-water pressures of a certain depth range	Moderate	Low	Low to High
	• The annulus around the tip is backfilled with either sand or grout (dependent on the permeability characteristics of the host slope material)			
	• The remainder of the hole is backfilled with low permeability grout to form a closed/sealed system (Wan and Standing, 2014)			
	• Water levels in the piezometer can either be logged manually (low temporal resolution) or automatically (high temporal resolution) and can be used to calculate pore-water pressures within the screened interval of the piezometer tip			

STUDY SITE

The landslide observatory is set on a south-facing hillslope, Hollin Hill, with a mean slope angle of 12°, located south of the village of Terrington, North Yorkshire, UK (54°06′38″ N, 0°57′30″W; Fig. 1). It is set in the Howardian Hills, an undulating landscape running approximately NW–SE shaped in four bedrock formations of Lower to Middle Jurassic age and covered by superficial deposits of variable thickness (Fig. 1b). Hollin Hill is a flat-topped hill capped by calcareous sandstone and ferruginous limestone belonging to the Dogger Formation (DF). There is a sharp boundary to the underlying Whitby Mudstone Formation (WMF), the failing formation at this site (Gunn et al., 2013). The WMF, showing a thickness of about 25 m (Chambers et al., 2011), is composed of grey to dark grey mudstone and siltstone with scattered bands of calcareous and sideritic concretions (Chambers et al., 2011). It overlies the Staithes Sandstone Formation (SSF). The boundary between these two formations is represented by an upward transition from sandstone or siltstone to shaley mudstone. This change in depositional facies is marginally episodic and, although a fining upward trend persists, the boundary between the SSF and WMF is manifested by sequences of ferruginous, micaceous siltstone, with fine-grained sandstone and thin mudstone partings. The SSF is heavily bioturbated and shows local occurrences of siderite and pyrite (Gaunt et al., 1980). The formation underlying the SSF is the Redcar Mudstone Formation (RMF) that comprises grey, silty, calcareous, and sideritic mudstone with thin shelly limestones (Powell, 1984 and Powell, 2010). The DF represents a potential aquifer above the WMF. Hydrogeological characterisation of the WMF and SSF is very complex as a result of variations in particle size in sedimentary sequences that, in turn, may be disturbed by bioturbation and inter-connected joint systems.

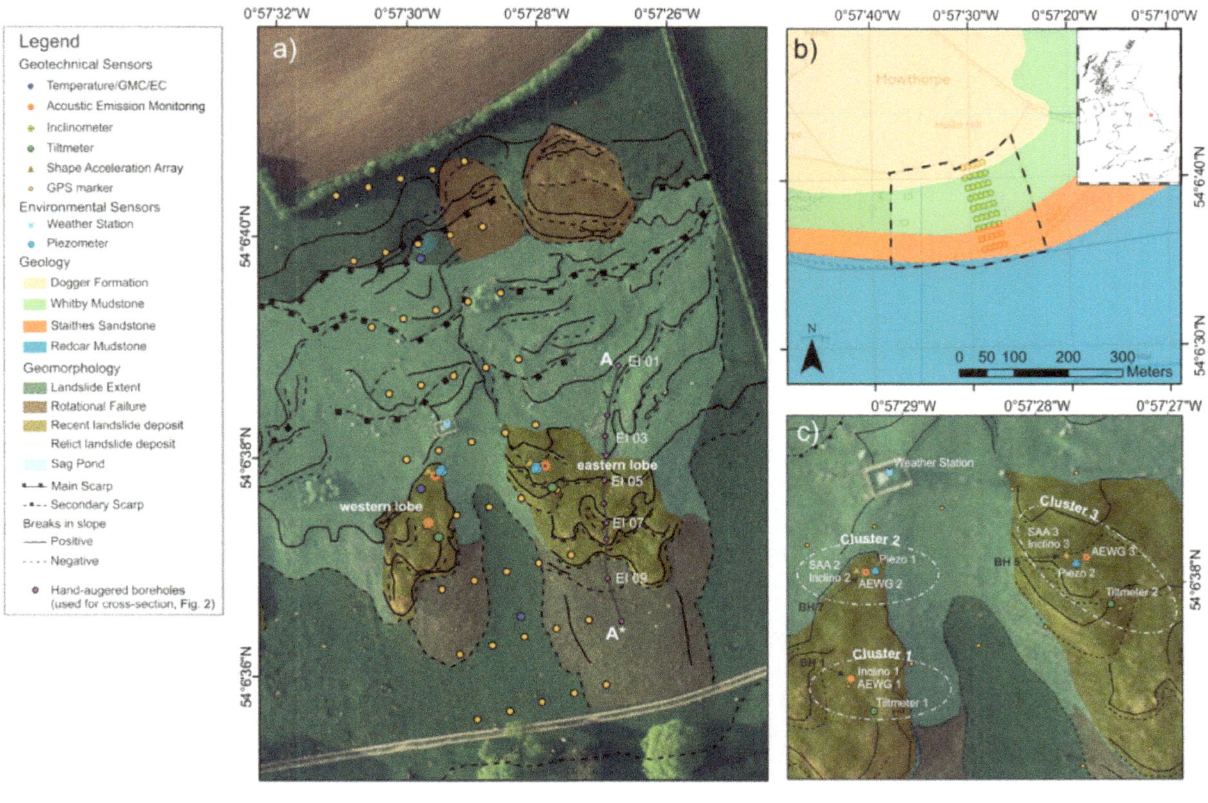

Figure 1: Overview maps of the study site. a) Geomorphological map of the landslide indicating different landslide bodies and features (after Merritt et al., 2013). Also shown are the locations of the monitoring equipment that are deployed at the site and discussed in this paper. b) Geological setting of the study site; inlet shows location within the UK. c) A section of Fig. 1a at a larger scale showing the locations of sensor clusters comprising inclinometer, AEWG, SAA, piezometer, and tiltmeter deployed on the two lobes, as well as the location of the weather station. Note that cluster 1 comprised AEWG logging during February 2010 to December 2010; the locations of the SAAs at clusters 2 and 3 are adjacent to the locations of inclinometers 2 and 3.

On top of much of the lower slopes in the SSF, a thin (up to 2 m thick) fine sand deposit is found, preserving the old slope and head deposits. It is likely that this sequence represents a periglacial slope with associated coversand (e.g. Catt, 1977, Ballantyne and Harris, 1994 and Bateman, 1995). The persistence of a thin head deposit on top of the SSF suggests that the old (periglacial) slope profile has been largely preserved by the subsequent cover of the aeolian sands.

The Hollin Hill landslide is in many ways representative of inland landslides in stiff clays and is acting as a field laboratory to provide, amongst others, a test site for a range of monitoring techniques, including the ALARMS (e.g., Dixon et al., 2010, Dixon et al., 2014 and Smith et al., 2014) and ALERT systems (e.g. Kuras et al., 2009). So far, the geological and geomorphological characterisation was mainly informed by geophysical data, borehole logs and inclinometer readings during periods of very-slow movement (~ 0.5 m $year^{-1}$; Chambers et al., 2011 and Merritt et al., 2013). This paper utilises previously unavailable data to focus on periods of landslide reactivation with movements of up to 3.5 m $year^{-1}$, leading to a better informed interpretation of landslide processes and characterisation of the site.

Merritt et al. (2013) provide a preliminary geomorphological characterisation of the slope together with an initial assessment of a ground model. An updated and more detailed ground model is presented here based on the interpretation of borehole and geotechnical data. Characterisation of the morphogenesis of this slope is supported by a section constructed on the basis of a series of shallow (up to 4 m depth), hand-augered boreholes (Fig. 2).

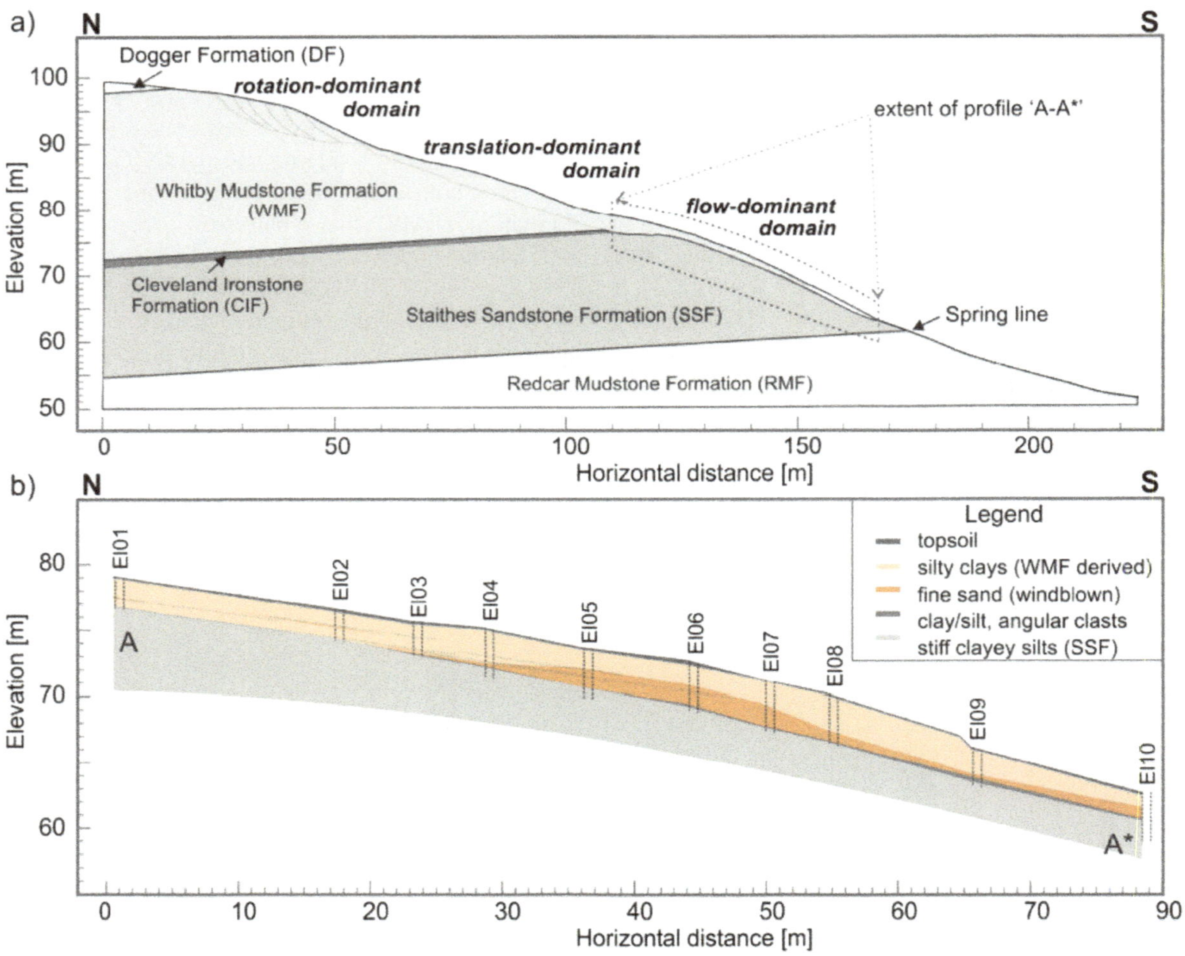

Figure 2: Representative section of the stratigraphy and slope instability domains for Hollin Hill. a) Diagram modified after Gunn et al. (2013). b) Profile A–A* through the eastern lobe informed by hand augered holes down to depth of 4 m (profile location indicated in Fig. 1).

In summary, Hollin Hill comprises a largely linear slope that is affected by different types of slope instability that can be sub-divided into three domains (with fuzzy boundaries; Fig. 2).

- In the central section translational landsliding in the WMF dominates. Multiple translational components can be distinguished and discrete shear zones have been observed in boreholes in this central part of the slope complex.
- With progressive displacement of the translational slides, disintegration of the slide mass occurs, leading to slope deformation that is increasingly characteristic of flow-like behaviour further down the slope.
- The slopes above the central section display recent, rotational deformation. Progressive translational deformation in the central domain leads to unloading and oversteepening of the toe of the upper slopes, resulting in landsliding that is dominated by sliding along curved slip surfaces.

Note that the instrumentation clusters described in this study are installed in the translation/flow-dominant zone. The translation/flow-dominant lobes located in this zone are simply referred to as lobes hereafter.

Instability in the WMF involved sequences of thin translational slide or flow events. Multiple slip surfaces and stretched out, thin bands of aeolian sands provide some evidence that phases of deformation have occurred subsequent to the aeolian deposition.

The steepest slope angles are found where the slope is developed in the SSF. As the translational-flow slide deformation in the WMF reaches this steeper section, it would be logical to expect the development of further runout, but instead a series of ridges are observed. Progressive deformation of the translational-flow slides appears to be impeded by base draining as the landslide moves over the fine sands further down the slope.

A substantial thickness of fine sand is found between boreholes EI04 and EI08 (hand-augered boreholes, Fig. 2). This local accumulation may reflect uneven aeolian deposition, but it is also plausible that landslide activity in the WMF incorporated some sand in the slope deformation process, driving this surface veneer onto the sand further down the slope and thus creating a local accumulation. This latter scenario is supported by the inclinometer data of Fig. 3. It shows borehole logs and the results of cumulative inclinometer measurements conducted over a period of 2.5 to 3 years. It is evident that the lowest active shear surfaces occur in the fine sandy deposits at a depth of approximately 1.5 m at the western lobe (BH 1 and 7). The profiles also suggest the existence of further slip surfaces or flow-like deformation in more shallow deposits.

• Three clusters of active waveguides monitoring acoustic emission (2009–ongoing)
• Three inclinometers (2008–2012)
• Two shape acceleration arrays (2013–ongoing)
• Two tilt meters (2008–ongoing)
• GPS tracking of 45 marker points (2008–ongoing)

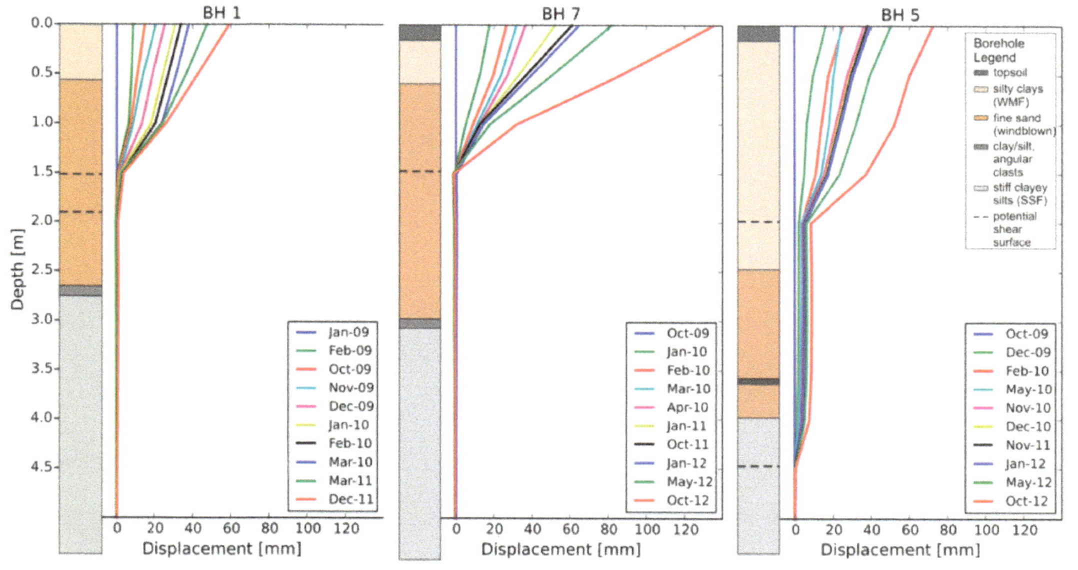

Figure 3: Borehole logs and corresponding inclinometer data highlighting slip surfaces. Boreholes 1 and 7 (BH1 and BH7) are located on the western lobe (Inclinometer 1 and adjacent to SAA2 in Fig. 1c), Borehole 5 (BH5) on the eastern lobe (adjacent to AEWG 3 in Fig. 1c).

The profile representing the eastern lobe (BH 5) is more complex. This borehole is set in an accumulation ridge where ongoing deformation in the upper WMF is inhibited by toe draining through the underlying fine sands. This is clearly illustrated by the relatively thick silty clays overlying the fine sands. In these silty clay landside deposits shear surfaces are observed at about 2.0 m depth, i.e. about 0.5 m above the boundary with the fine sands, and these observations are supported by the inclinometer data. At approximately 4 m depth there appears to be a second shear surface resulting in very slight deformation of up to 10 mm over a period of 3 years.

The aeolian sands within and below the landslide deposits have a complex relationship with the deformation behaviour of this slope. When incorporated in the slide body, it is possible that these sand lenses act as interceptors for diffuse inflow of water, and are potentially capable of generating artesian pore pressures at critical depths.

INSTRUMENTATION INSTALLATION

As already mentioned and shown, this landslide is equipped with a range of monitoring systems to serve as a research landslide observatory to improve our understanding of triggering mechanisms for first time failure and landslide reactivation. The following monitoring systems are deployed on the slope and were actively monitoring as indicated:

- Three clusters of active waveguides monitoring acoustic emission (2009–ongoing)
- Three inclinometers (2008–2012)
- Two shape acceleration arrays (2013–ongoing)
- Two tilt meters (2008–ongoing)
- GPS tracking of 45 marker points (2008–ongoing)
- Weather station (2008–ongoing)
- Two piezometers (2009–ongoing)

Additionally, soil moisture content, bulk conductivity and temperature are monitored at three different locations, covering the backscarp, active lobes, and an area outside of the landslide, with sensors deployed at different depths below ground level (bgl) ranging from 0.1 to 6.5 m. The location of the different monitoring systems can be found in Fig. 1a, c. This range of different monitoring systems, each characterized by different accuracy, sensitivity, and spatial and temporal resolutions provides the basis for a thorough assessment of techniques used for slope monitoring.

In the following a brief description is provided for each monitoring system.

Active waveguides (AEWG)

The active waveguides were installed in 130 mm diameter boreholes to depths of 5.7 m bgl; each comprises 6 m long, hollow steel tubing of 50 mm diameter and 3 mm thickness. The annulus around the steel tubing was backfilled with angular 5 to 10 mm gravel compacted in nominally 0.25 m high lifts. The top 0.3 m of the boreholes were backfilled with a bentonite grout plug to seal against the ingress of surface water (Fig. 4). The gravel backfill, which is acoustically noisy, generates high-amplitude AE signals. This overcomes the limitation of detecting low-amplitude AE signals caused by slope movement of the fine-grained, acoustically quiet, host material (Dixon et al., 2003). Transducers were coupled to each of the waveguides at the ground surface and were connected to Slope ALARMS (Dixon and Spriggs, 2011) sensors which were powered by batteries and recharged by solar panels; logging was at 30-min intervals. These AE monitoring systems, which inform only about the behaviour of the installation borehole, provide globally a low spatial resolution, but deliver information on rates of subsurface deformation (i.e. accelerations and decelerations) with high temporal resolution.

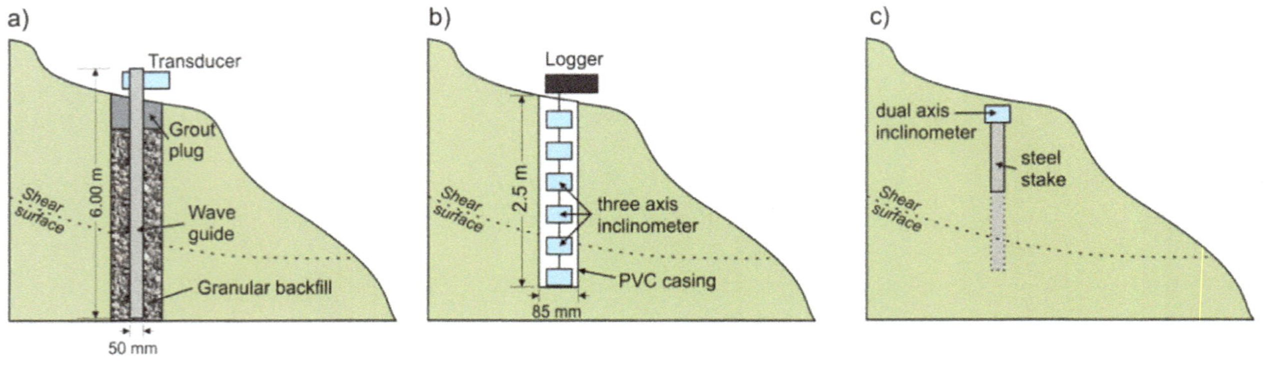

Figure 4: Schematic illustration of permanently installed subsurface movement monitoring techniques; a) AEWG monitoring, b) SAA, and c) tilt meter. Note, employing a longer stake for the tilt meter and fixing it below the shear surface (dashed installation) will induce a larger proportion of rotational movement..

Inclinometers

Inclinometer casings were installed at the three cluster locations and approximately 1 m east from the active waveguides with keyways orientated along the slope dip- and strike-directions. The inclinometers penetrated to depths of 6 to 7 m bgl. They ceased to be usable after 3 years of monitoring in November 2012 due to significant shear surface displacement inducing localised bends within the casing sufficient to no longer allow the torpedo probe to pass.

Inclinometers provide relatively low global spatial resolution as a survey of the casing only informs of deformation in the soil surrounding the borehole. If casings are surveyed manually using a torpedo probe, relatively low temporal resolution with intervals of the order of months can be obtained (NB — higher temporal resolutions can be achieved by using in-place probes).

Shape Acceleration Array (SAA)

SAAs were installed approximately 1 m west of AEWG2 in May 2013 and AEWG3 in December 2013 to provide continuous subsurface deformation measurements; their installation was instigated subsequent to the inclinometer casings no longer being usable. They were installed to depths of 2.5 m bgl. Each of the MEMS sensors that comprise the SAA log deformation at 1-hour intervals. The SAA operates in a similar manner to an in-place inclinometer in that it provides deformation vs. depth measurements with high temporal resolution (Fig. 4).

Tilt Meter

Tilt meters were installed to monitor small changes in surface inclination, related to rotational downslope movement of mass that is evident at several trees on site. Each tilt meter consists of a dual axis inclinometer sensor with an accuracy of 0.1° mounted on a 0.5 m long coated iron stake (Fig. 4). The two tilt meters were connected to a data logger recording tilt angles hourly.

Tilt meters provide point measurements only that are representative for a small area within the landslide (and thus low global spatial resolution) and monitor tilt as opposed to actual deformation; however, they log data with high temporal resolution enabling correlation of movement with rainfall-induced instability.

Tracking OF GPS Markers

GPS markers were distributed regularly as five parallel lines of nine markers each, and made of wooden pegs. Inline marker separation was about 20 m, while line spacing was about 10 m. Their initial positions were surveyed using a real-time kinematic GPS equipment, providing centimetric accuracy. Every 1–2 months these markers were re-surveyed using the same equipment, building a time-series of landslide movements at those locations.

This approach provides a moderate spatial resolution, both, on local (i.e. marker) and global (i.e. slope) scales. It allows for identification and mapping of areas of activity across the slope, forming the basis for deciding where more sensitive instrumentation is to be installed (NB — depending on the available resources fine grids and permanent GNSS logging stations can be employed, delivering high spatial and temporal resolutions).

Weather Station

The weather station was an all-in-one system measuring barometric pressure, humidity, precipitation, temperature, and wind speed and direction, with an hourly frequency. To minimize the effects upon the system from vegetation and other installations at the site, especially on the measurement of wind speed and direction, the weather station was placed on top of a pole at a height of 3 m. The data were recorded and stored using the same data logger as for recording the tilt meter readings.

Piezometers

Piezometers were installed in boreholes hand augered to 2.85 and 2.8 m on the western and eastern lobes, respectively. These depths were chosen in order to place the active zone of the piezometers in the vicinity of the depth to predicted shear surfaces, which were determined using CPT measurements (Gunn et al., 2013). We installed a 19 mm uPVC pipe, fitted with a 0.9 m slotted, porous piezometer tip in the borehole. The hole was backfilled with clean sand to 1.95 and 1.75 m below ground for the western and eastern piezometers, respectively, forming an active zone that allows for monitoring of the pore water pressure in the region of the slip surface located at about 1.6 and 1.9 m depth, respectively. The remainder was backfilled using bentonite granules to ensure sealing. The piezometric head measured in these standpipes is referred to the top level of the sand pocket forming the active zone, thus positive heads present conditions of positive pore water pressure. Since the active zones of the piezometers are in close proximity to the shear surface (< 0.35 m), the measured pore water pressures are indicative for the conditions at the shear surface. The water levels in these piezometers were logged with a 30 min interval. Note that the eastern piezometer experienced severe bending and became unusable in January 2013. It was reinstalled in December 2013, although slightly deeper (now 0.5 m below the shear surface).

It is important to highlight that the monitoring clusters are distributed over the two lobes, rather than installed adjacent to each other. Thus, times of activation may vary depending on their location on the lobe.

INTERPRETATION OF SLOPE DEFORMATION

Delineation of Regions of Slope Instability

Table 1 shows that the different monitoring techniques deployed have different spatial and temporal resolutions. To indicate their ability to differentiate movements at the system and over a typical slope scale, their spatial resolution has to be divided into two scales, local and global. While the tilt meter, AEWG, and SAA are sensitive to very small movements and thus have high local spatial resolution, their global spatial resolution is rather poor, since their recordings are determined by movements in their vicinity only. However, they can record movements with a high temporal resolution, allowing for a correlation of movement events with climatic circumstances, such as specific rainfall events. GPS marker and inclinometer measurements on the other hand have low temporal resolution, but can easily be used to determine areas of instability (GPS markers) and shear surfaces (inclinometer) on a slope scale. Fig. 5 highlights this ability of GPS marker measurements to determine regions of slope instability and active movements. These measurements can be used to (1) show the direction and amplitude of surface deformations at monitoring points (Fig. 5a) and (2) to provide a map of the 5-year average landslide movement rates covering the entire slope (Fig. 5b) by interpolation of the monitoring points (Uhlemann et al., 2015). Between 2009 and 2014 the lobes experience significant surface deformations, with the eastern lobe showing significantly higher activity than the western, while the remaining areas show little or no deformation.

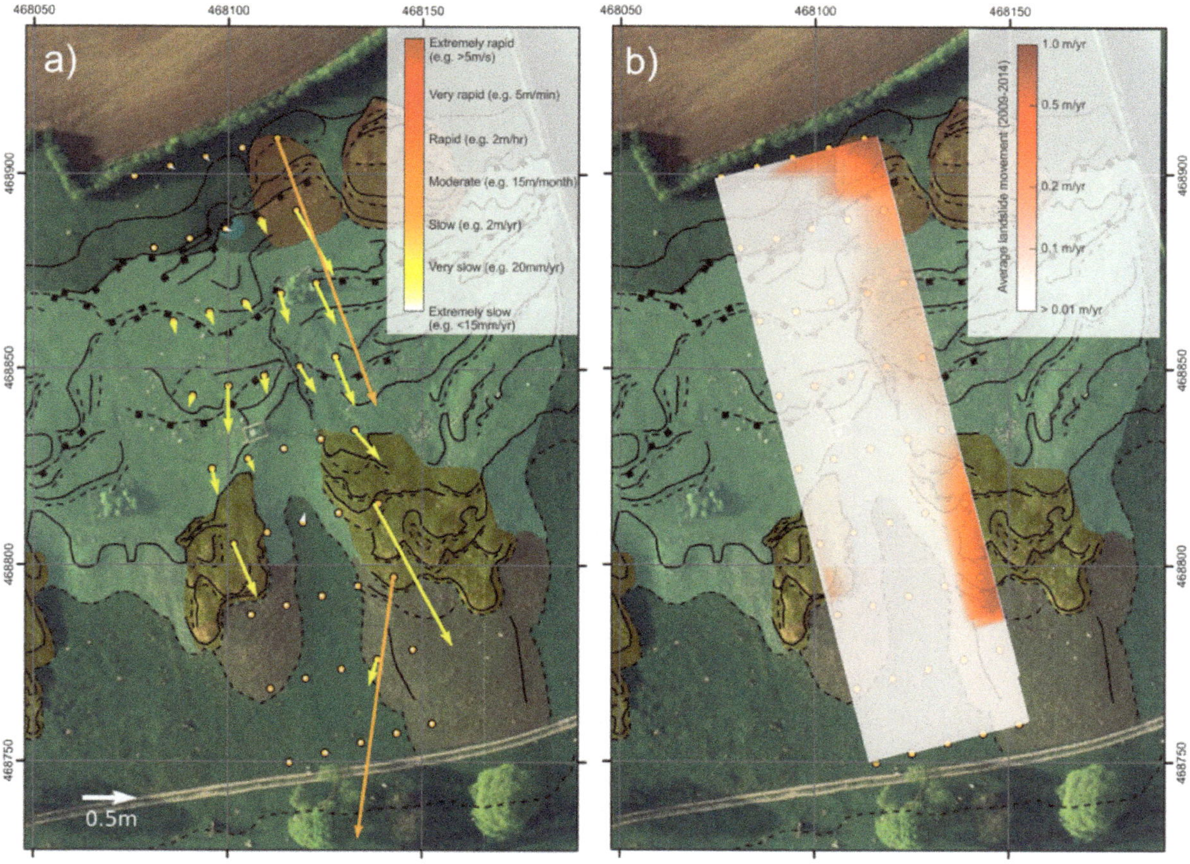

Figure 5: Displacements determined from GPS marker surveys between 2009 and 2014 superimposed on the geomorphological map after Merritt et al. (2013). a) Displacement vectors (i.e. direction and amplitude); arrows show absolute surface deformation over the 5-year monitoring period. The colour of the arrows corresponds to the standard classification for landslide movement rates after Cruden and Varnes (1996). b) Map of average landslide movement rate (after Uhlemann et al., 2015) highlighting the activity of the eastern lobe and the connection between rotational failure of the backscarp and translational movements of the lobe.

Fig. 5 shows that the rotational movements in the backscarp are directly connected with translational movements of the lobes. Such investigation is only possible with a distributed network of deformation measurements offering reasonable spatial resolution on both scales. Hence, studies of movement patterns are only possible either with distributed subsurface monitoring devices or limited to surface deformation monitoring. In its current application the data suffer from low temporal resolution due to the need to survey positions manually. Therefore, the timings of periods of activity can only be approximately determined. The approach introduced by Buchli et al. (2012) would overcome this limitation by employing wireless GPS sensor networks that could log deformations nearly continuously. Recent studies on tracking electrode movements by employing electrical resistivity monitoring data (Wilkinson et al., 2010 and Wilkinson et al., 2015) introduce another technique to obtain surface movement information on a global scale with high temporal and spatial resolutions.

Characteristics of the 5-year Monitoring Period

Fig. 6 shows rainfall, water level, cumulative tilt, and GPS marker movements for the monitoring period from the beginning of 2009 to 2014 (five years). Note that the GPS movement data shown in Fig. 6 correspond to markers adjacent to Cluster 2 on the western lobe and close to the southernmost boundary of the eastern lobe, respectively. Rainfall data are shown in terms of absolute rainfall (black line) and effective rainfall (green bars, potential evotranspiration calculated using Hydrus1D; Šimůnek et al., 2008). Cumulating the effective rainfall over the summer (red bars) and the winter (blue bars) shows a clear seasonality in the first two years, with dry summers and wet winters. It also indicates a negative water balance (i.e. potential evotranspiration larger than the amount of rainfall), which got expressed as an 18-months drought between 2010 and 2012. This seasonality is interrupted in the years 2011 and 2012 (Fig. 6). This can be attributed to a cooler and wetter than average summer in 2011 resulting in enhanced effective rainfall (absolute rainfall 10–20% higher than 30 year average; Belcher et al., 2014) followed by a drier than average winter. The following summer of 2012 was characterised by prolonged and intense rainfall of up to 65 mm day^{-1}. It has been recorded as the wettest summer since 1912 (Belcher et al., 2014).

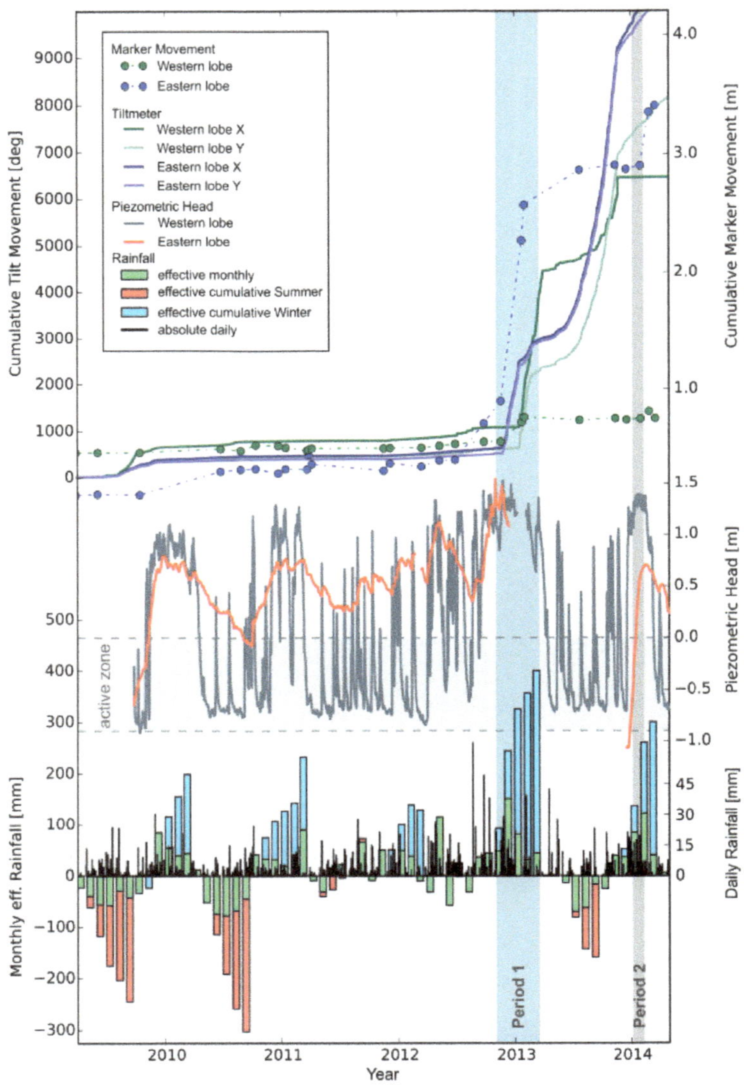

Figure 6: Data overview showing rainfall, water level, cumulative tilt movements (i.e. absolute change in tilt meter reading), and cumulative GPS marker movements. Note the interruption in seasonality in rainfall pattern and the correlation between peaks in piezometric heads and reactivation of the landslide movements. Note that the reason for the drop of the head of the eastern piezometer below the active zone is its reinstallation to a slightly deeper level.

The seasonality of the years 2009 to 2011 can also be seen in the piezometric heads. During the winters of 2009–10 and 2010–11 the effective rainfall percolates downwards to recharge the water table resulting in elevation of piezometric heads. Conversely, in the summer, recharge diminishes and groundwater flow and discharge causes the piezometric heads to fall. The difference between summer and winter piezometric heads is about 1.8 and 1.0 m for the western and eastern lobes, respectively. Not only is the difference between summer and winter heads larger on the western lobe, it also shows more fluctuation and faster response times to rainfall events in comparison to the eastern lobe, which shows a much smoother response. This indicates that the two lobes are set within different hydrological environments. These can be characterised by (1) stronger bypass flow on the western lobe due to a higher degree of fissuring as indicated by recent infiltration experiments, and (2) the existence of sand lenses within the WMF of the western lobe providing preferential pathways for moisture drainage due to the higher permeability of those sand lenses. These two effects are less pronounced on the eastern lobe.

The surface movement indicators shown in Fig. 6 (i.e. cumulative GPS marker and tilt movement) indicate three periods of significant movement within the five year monitoring period. The first movements were recorded in October 2009, with a total of approximately 0.25 m. This movement period was brief, lasting for no longer than 2 months. After this, an 18-month period of persistent drier than average conditions caused the slope to remain in a stable condition with no movements until July 2012 when movement commenced due to the wet summer. First indication of movement during this period (Fig. 6) is given by the GPS measurements located at the front of the eastern lobe. The tilt meter of the eastern lobe shows the movements commencing in November 2012, during a period when significant movements were also recorded on the GPS markers of the eastern lobe (1.4 m between November 2012 and January 2013). This offset in the start of movement recordings may be caused by the different sensor locations (the tilt meter is located about 20 m upslope from the marker) or by the insensitivity of the tilt meter to record pure translational movements. Movements on the western lobe commenced about 2 months after the eastern lobe, in January 2013, and lasted until April 2013. Although the tilt meter of the western lobe shows a response comparable to that on the eastern lobe, the GPS marker measurements reveal a smaller degree of movement of only about 0.25 m between November 2012 and January 2013. This period of landslide reactivation and acceleration lasted until February 2013, after which the landslide entered a period of deceleration due to a reduction of rainfall, which allowed dissipation of pore-water pressures in the slope. GPS marker measurements show that between February and October 2013 the eastern lobe moved about 0.3 m, while the western lobe reached a stable state. In the same period, the tilt meter recorded movements on both lobes. This may be explained by (1) localised surface disaggregation at or near the tilt meter, or (2) small magnitude movement that was not captured using RTK-GPS measurements (i.e. sub-mm scale). Another acceleration of movements occurred in the winter of 2013–2014, where the eastern lobe moved another 0.3 m in just one month.

The dashed horizontal lines in the piezometric head section of Fig. 6 indicate the active zone of the piezometer; the upper dashed line shows the location of the upper level of the sand pocket, which was used as reference for the piezometric head. Thus, positive heads indicate positive pore water pressure close to the slip surface. Extended periods of positive pore water pressure correlate clearly with reactivation and acceleration of the landslide on the western lobe (i.e., movement between December 2009–March 2010, and November 2012–March 2013). The piezometer on the eastern lobe shows positive pore water pressures throughout the years 2011 and 2012, with an average piezometric head of + 0.6 m above the slip surface. A global peak showing a piezometric head of + 1.5 m was recorded in November 2012 when the fastest movements commenced on this lobe. The positive piezometric head level sufficient to induce instability and movement at this location is therefore between 0.6 and 1.5 m. These movements led to a severe bending of the piezometer pipe, causing the loss of the logger. Another piezometer was installed in November 2013. This new piezometer showed positive pore water pressures immediately after its installation, correlating well with another acceleration of this lobe.

Two periods of winters 2012–13 and 2013–14 are considered in detail below. These two periods represent times at which the landslide reactivated or movement accelerated, respectively, as indicated by the GPS marker measurements. To understand the actual kinematic behaviour of the deformations, higher temporal resolution than available for the GPS marker data is required. This is provided by the continuous or near-continuous recordings of the AEWG, SAA, and tilt meter.

Period 1 — winter 2012/2013

Fig. 7 shows period 1, which presents a time at which the landslide reactivated subsequent to more than 2 years of stability. This period, set between November 2012 and April 2013, follows a very wet summer, with rainfall exceeding 50% more than the 30-year average (Belcher et al., 2014). This intense rainfall caused the highest piezometric heads measured at the site during the 5-year monitoring period (Fig. 6). Although the high piezometric heads and thus positive pore water pressures are likely to be one of the main reasons for the reactivation and acceleration of the landslide, their highest peak has not been the ultimate trigger for the reactivation of the lobe movements. For this period the reasons for reactivation were: (1) intense rainfall leading to increased total stress due to increased moisture content; (2) loading from a rotational failure up-slope; and (3) reduction of material strength caused by elevations in pore water pressure. Fig. 7 shows that rainfall of up to 30 mm day^{-1} on 25th November 2012 caused the stable slope to move. These movements have been recorded by the AEWG on the western lobe and the tilt meter on the eastern lobe. The two techniques recorded the onset of instabilities at the same time, despite being located on different lobes. At this time, no AEWG was operational on the eastern lobe. Inclinometer readings (Fig. 3) show very slow subsurface deformation from the beginning of the monitoring period in 2009 (< 50 mm year^{-1}) with a significant acceleration between May and October 2012 (when they eventually ceased being useable). The tilt meter and GPS markers are installed at different locations to the AEWG and inclinometer. Hence, the difference between their recordings may be an indication that the lobes move as separate blocks. Tilt meter recordings from the western and eastern lobes (and also GPS marker measurements, but with limited temporal resolution) reveal that major movements on the western lobe commenced on 16th January 2013, 51 days after movement began on the eastern lobe. At the time of slope movement initiation no intense rainfall has been recorded. Snow build-up during the several days of low temperature prior to this reactivation event resulted in 5 to 10 cm of snow cover. This snow melted as temperatures increased on 16th January 2013, which provided additional moisture to a nearly saturated subsurface (Uhlemann et al., 2014) leading to elevated pore water pressures. While smaller rainfall events (27th and 29th January with 4 and 9 mm day^{-1}, respectively) succeeding the snowmelt period caused further movement acceleration, a heavy rainfall event on 5th February 2013 (39 mm day^{-1}) did not increase the rate of movement. This could be due to already fully saturated conditions of the lobe and thus surface runoff rather than infiltration of moisture. This heavy rainfall event was followed by a dry period leading to drainage of moisture, thus a reduction of pore water pressures, and movement decelerated. All movement indicators exhibit characteristic 'S'-shaped slope displacement-time behaviour (e.g. Allison and Brunsden, 1990, Petley et al., 2005 and Smith et al., 2014); periods of acceleration followed by deceleration and stability, due to oscillations in pore water pressures in the vicinity of the shear surface. The GPS marker information did exhibit this behaviour; however, it had low temporal resolution and numerous 'S'-shaped events were not captured. The tilt meter and AEWG both captured all of the phases of reactivated movement and with a sufficient temporal resolution. This demonstrates the significance of temporal resolution in capturing landslide deformation behaviour. The time series measurements show that these periods of reactivated movement were detected at different times on the different instruments; this is because the instruments are installed at different locations on the lobes. For this five-month period of winter 2012–2013 the GPS marker measurements showed surface movements on the eastern lobe of up to 1.7 m.

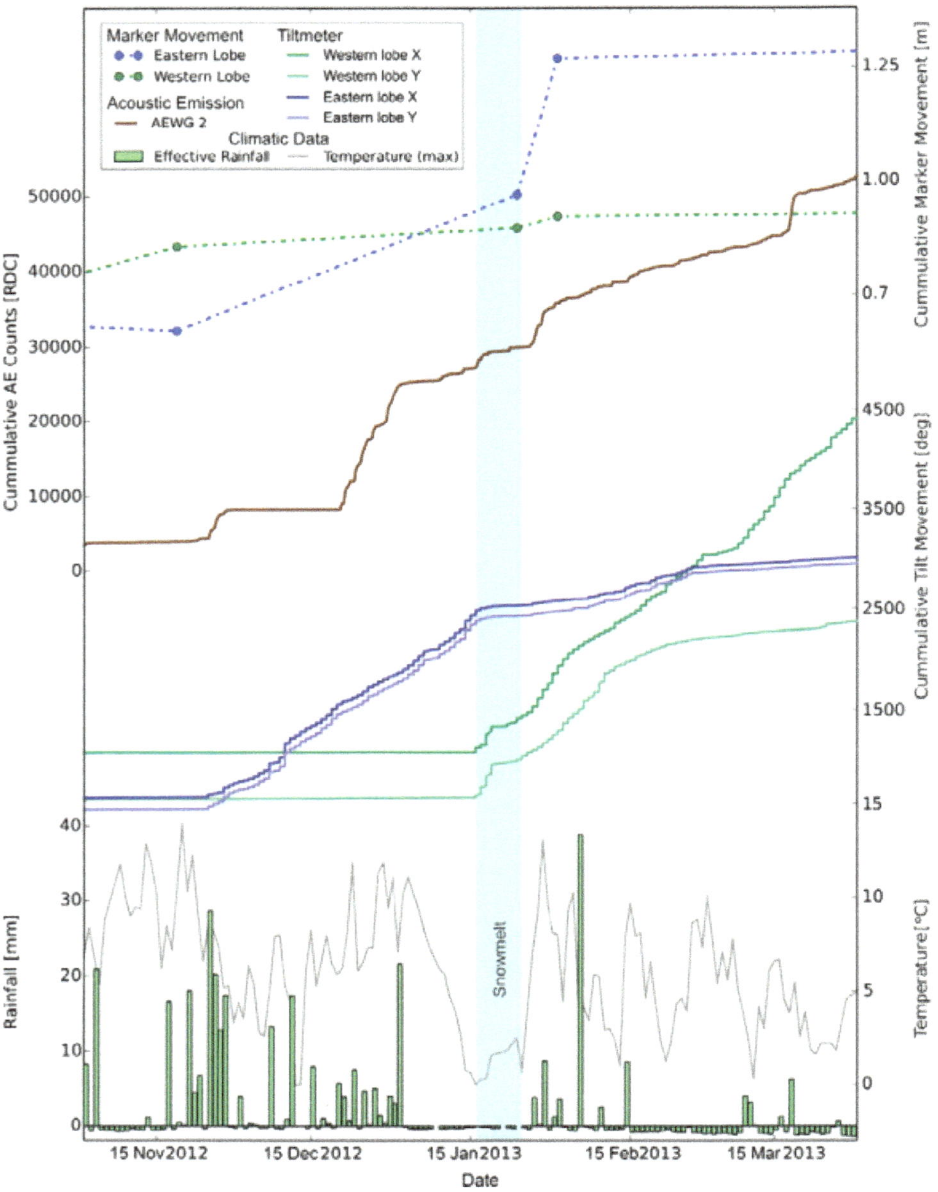

Figure 7: Comparison of movement measurements for period 1. Shown are GPS marker, cumulative AE (RDC), and tilt meter measurements. Also shown are effective rainfall and temperature.

Period 2 — winter 2013/2014

The second period of landslide reactivation investigated in this paper occurred between 9th January 2014 and 4th February 2014. Period 2 has a different prehistory compared to Period 1. While Period 1 followed one of the wettest summers of recent records, Period 2 follows an average summer that led to drainage of the moisture accumulated in the slope during the winter before. Note that the weeks prior to this period were characterized by strong rainfall events, resulting in approximately 220 mm cumulative 90-day antecedent rainfall. For this period, also SAA data for the two lobes is available.

Fig. 8 shows continuous cumulative RDC- and deformation-time series measurements from active waveguides (AEWG2 and AEWG3), SAAs (SAA2 and SAA3), GPS marker, and tilt meter measurements recorded from the two lobes. The shape of the cumulative time series at both lobes are characteristic of 'S'-shaped slope movements (e.g. Allison and Brunsden, 1990, Petley et al., 2005 and Smith et al., 2014). The series of slide movement events are preceded by periods of rainfall that induced transient elevations in pore water pressures along the shallow shear surface and induced movement. The slide mass accelerated towards a peak velocity and subsequently decelerated as pore-water pressures dissipated. Shear resistance was mobilised internally within the slide mass and through remoulding of the landslide toe (Dixon et al., 2014 and Dixon et al., 2015). This process was repeated subsequent to each rainfall trigger as shown in Fig. 9. Fig. 9shows the AE rates, which are the first derivative of cumulative AE and are proportional to slope velocity (Smith et al., 2014 and Smith and Dixon, 2015), and SAA measured velocity at Clusters 2 and 3 for the same period of movements as in Fig. 8. Although movement appears to trigger on both lobes simultaneously, the eastern lobe clearly experienced greater magnitudes of movement as shown by GPS measurements.

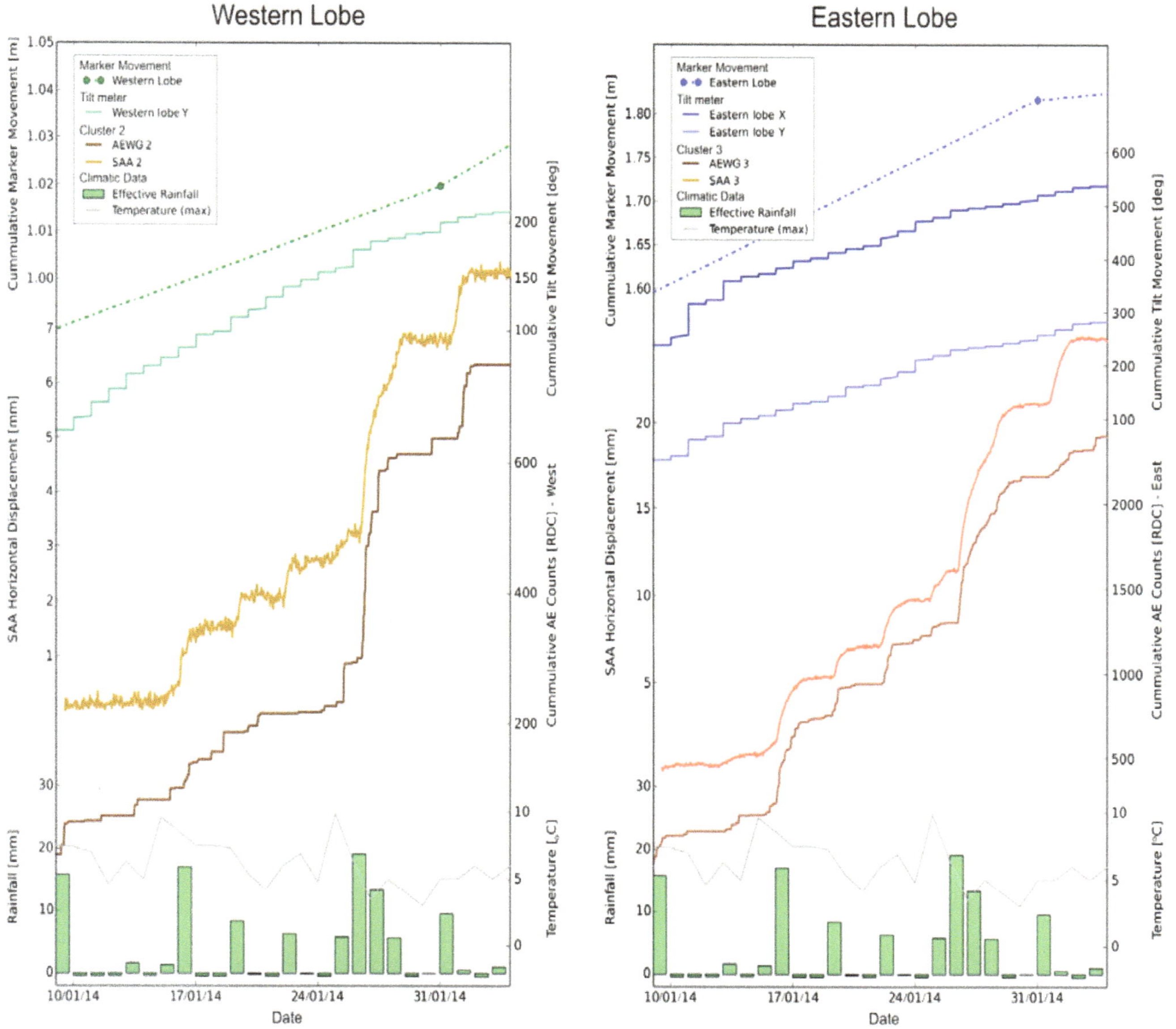

Figure 8: Comparison of movement measurements for period 2. Shown are GPS marker, cumulative AE (RDC), SAA horizontal displacement measurements, and tilt meter measurements, together with effective rainfall and temperature. Note that axis for the movement measurements differ for the western and eastern lobes.

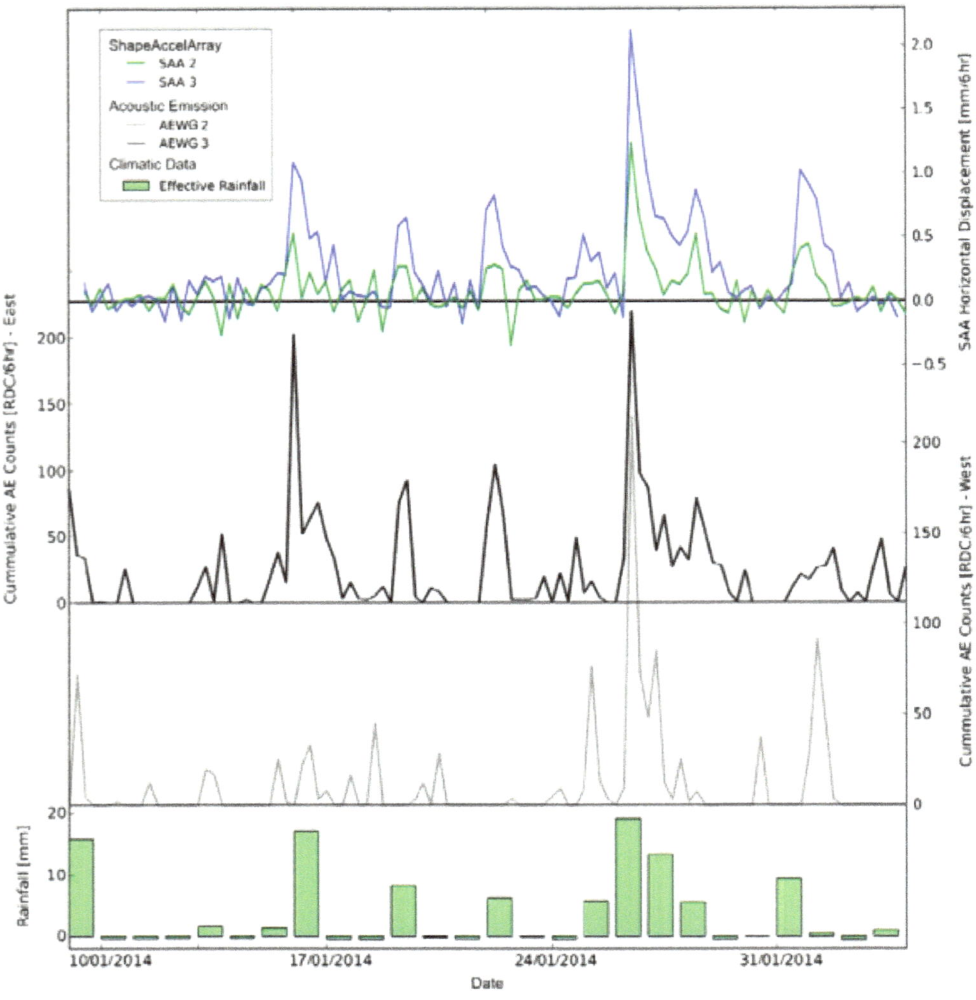

Figure 9: First derivative of data presented in Fig. 8 showing AE rates and SAA measured velocity. Note that periods of acceleration correlate well with rainfall events. A subset of the SAA and AE measurements from the eastern lobe are also analysed in Smith et al. (2014).

Monitoring GPS markers did not provide sufficient temporal resolution to capture the 'S'-shaped phases of reactivated slope movement. The tilt meter did not capture the reactivation phases because the movements had a predominantly translational component. It is only possible to capture this slope deformation behaviour using deformation monitoring techniques that provide continuous information with high temporal resolution (e.g. AEWG and SAA), and can capture the appropriate behaviour (i.e. tilt meters are inappropriate for measuring translational movements). For a system to be capable of providing an early warning of slope instability, it must provide this level of information, to detect accelerations of movement as early as possible and communicate them in real-time using appropriate communication systems.

The information gained from the evaluation of the movement monitoring techniques in conjunction with the accompanying rainfall and pore water pressure data led to an improved understanding of the movement pattern forming the Hollin Hill landslide in recent years. Fig. 10 shows a simplified representation of the geomorphological features and example cyclic processes underpinning the movements. Rainfall induced instabilities in the WMF of the back scarp and subsequent downslope movement of material (1 in Fig. 10) lead to an accumulation of slipped material at the WMF–SSF boundary (2). These processes, as they are rainfall dependent, show varying movement rates. Once a threshold of accumulated material and pore water pressure has been exceeded, a breakthrough occurs, transporting slipped WMF further downslope with high movement rates (3), which decrease as pore water pressures dissipate. Consequently, this material moves in a flow-like behaviour on top of the aeolian sands and the SSF, where changes in the pore water pressure levels determine the movement rates. Pore water pressures are elevated by rainfall-induced moisture input and reduced by toe-drainage through the aeolian sands and SSF. To maintain this flow-dominated process that forms the lobes of slipped WMF on top of the SSF material, input from upslope is required. The flow material is supplied episodically, following a cyclic pattern of rainfall induced deformation in the WMF, and accumulated at the WMF–SSF boundary until a stress/pore water pressure threshold is exceeded that triggers a breakthrough of a flow lobe. This cyclic behaviour is characteristic of the slopes movement during winter months when it is at its wettest and the pore water pressures in the slope are at their greatest magnitude. The slope remains predominantly stable for the remainder of the year, unless rainfall with high intensity and duration occurs.

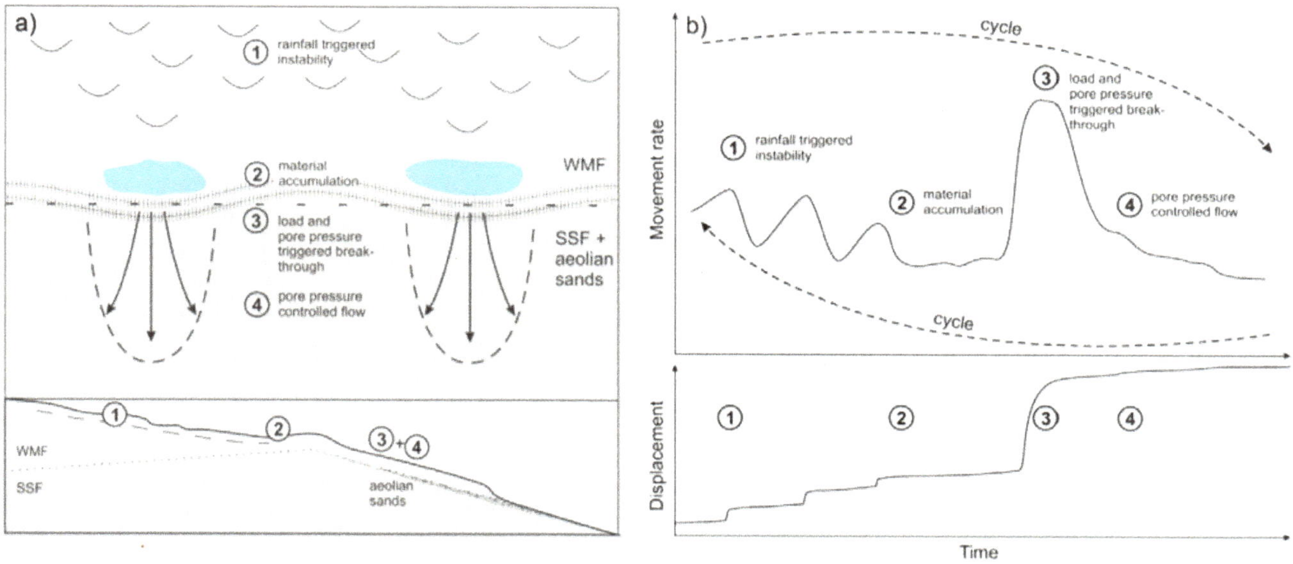

Figure 10: Conceptual diagram of processes defining the landslide behaviour at Hollin Hill. a) Simplified plan view and cross section of Hollin Hill showing the geomorphological processes forming the landslide. b) Characteristic, cyclic movement pattern of the Hollin Hill landslide.

DISCUSSION OF TECHNIQUE APPLICABILITY AND CONCLUSIONS

This paper has detailed the application of various monitoring instruments and techniques to a reactivated landslide that moves in response to rainfall-induced pore water pressure elevations. The case study has served as a comparison of the various monitoring approaches and particular attention has been given to the spatial and temporal resolution offered by the approaches.

Monitoring an array of markers installed across the slope using GPS during site visits has allowed the quantification of surface movement vectors at each marker location and therefore the identification of areas of stability/instability. This high spatial resolution information is particularly important where a large slope area is of suspected instability. The information obtained from monitoring the GPS markers can be utilised in identifying areas across the slope where further monitoring effort should be focused.

The centimetric accuracy offered by the GPS surveys is sufficient in identifying areas of relative stability across the slope and periods of movement; however, for most monitoring applications, but especially when a slope threatens infrastructure, greater accuracy in time and space is necessary. While GPS measurements are sensitive to surface deformation, acquisition of subsurface deformation information is often critical. The conventional inclinometer allows the depth to shear surface(s) to be identified. Movements over a localised depth, such as the shear surface, can be determined with an accuracy of the order of ± 0.2 mm (Mikkelsen, 2003). The inclinometer casings at Hollin Hill became unusable (i.e. would no longer allow the torpedo probe to pass the shear surface) after localised shear surface deformations of approximately 60 mm, as evidenced by the incremental readings (note Fig. 3 shows cumulative readings). Consideration should therefore be given to the life-span of such instrumentation, keeping in mind the anticipated magnitudes of movement.

Monitoring markers using GPS and inclinometer casings using torpedo probes during site visits offer low temporal resolution. This mode of monitoring informs the user whether movement occurred since the last reading but does not provide information on how and when the slope moved during this interval. It cannot provide real-time information for use in early warning. This can be overcome by using in-place probes and probe-strings installed within boreholes, providing high temporal resolution information as deformation is logged near-continuously at user-defined intervals. This study at Hollin Hill utilised SAA and AEWG to perform this function. The SAA logged deformation at 1-hour intervals using a series of MEMS installed along the depth of the borehole with accuracy of the order of ± 1.5 mm per 30 m (Abdoun et al., 2012). Additionally, AEWG monitoring provided continuous information on rates of subsurface movement, accelerations and decelerations in particular, sampled with a high temporal resolution of 30 min intervals. This high temporal resolution information provided by SAA and AEWG monitoring, was paramount in understanding the deformation behaviour of the slope in response to rainfall-induced instability. It allowed the identification of the 'S'-shaped deformation time cycles. This was not captured using the other deformation monitoring approaches. SAA and AEWG have therefore demonstrated the potential to provide early warning of movement by detecting accelerations of movement in real-time if connected to an appropriate communication system. Note that the selection of appropriate measurement intervals for these approaches is very important. An understanding of the way in which the slope will behave should inform this decision. For example, the slope at Hollin Hill is reactivated and moves in response to rainfall-induced pore water pressure elevations; these episodes of movement occur over a period of 2–3 days (Fig. 8). A monitoring frequency of the order of hours is therefore sufficient to capture this behaviour. However, when monitoring a slope that has the potential to undergo progressive first-time failure, i.e., collapse as a result of brittle strength loss (Chandler, 1984), which may experience rapid accelerations of movement over minutes or hours, a monitoring frequency of the order of minutes would be necessary to capture this behaviour and provide early warning.

Tilt meters also provide continuous information with high temporal resolution. However, as evidenced by the study at Hollin Hill, they do not successfully detect movements with predominantly translational components. This is because the instrument monitors rotation at the near-surface. Anchoring the base of the tilt meter below the slip surface offers an approach to capturing translational movements. Movement would induce bending within the tilt meter conduit, which would be measured as rotation by the tilt meter (dashed installation in Fig. 4c). Consideration should therefore be given to the use of tilt meters for slopes that are expected to have a significant rotational component of movement. For slopes with predominantly translational deformation their installation should be adapted accordingly.

Rainfall-induced elevations in pore water pressure are the main trigger for slope instability in the UK and the majority of the world. Thus, monitoring both rainfall and pore water pressures continuously with high temporal resolution provides significant opportunities in identifying critical conditions for instability. Resistance to movement is dictated primarily by the stability along shear surface(s) and therefore knowledge of pore water pressures at shear surface depth(s) is critical in understanding the stability of the slope. Piezometer tips were installed at predicted shear surface depths, which were inferred from CPT measurements at the same time as the installation of inclinometer casings. This resulted in the active zone of the piezometers being installed below the depth of the shear surfaces, as identified from subsequent inclinometer measurements. More accurate measurements of pore water pressures at the shear surface would be required for stability assessment. Therefore, it would be advisable to install piezometers at shear surfaces after their depths have been determined from inclinometer measurements. The piezometers were logged with high temporal resolution and this, in conjunction with continuous rainfall data, allowed the identification of periods of potential instability (e.g. Fig. 6 and Section 4.2). This level of information can be taken to develop rainfall intensity–duration relationships (e.g. Caine, 1980) and critical pore water pressures to allow prediction/forecasting of instability for the slope.

Stability assessment of slopes requires information that inclinometers and standpipes can provide such as shear surface depths and ground water levels. However, in order to inform ongoing risk assessments and numerical modelling of slope stability, and to provide early warning of instability, greater knowledge of the slopes behaviour must be acquired. This can be achieved by using combinations of instruments and techniques. This combination should cover instrumentation that provides high temporal and spatial resolutions, both at the measurement and slope scale. While high spatial resolution at the slope scale usually omits high spatial resolution at the measurement scale and high temporal resolution, it provides the necessary information for targeted installation of sensors that can subsequently provide the necessary high resolution data at those scales. For high temporal resolution data that inform early warning systems, a telemetric link between the field site and a database is required. Methods to deliver those data in real-time to web-platforms informing not only engineers and scientists but the wider public about potential hazards are becoming more frequently implemented in recent years (Reid et al., 2012 and Huang et al., 2015), but still require more research on how to manage the available data and present it in an informative fashion.

Although it is a significant constraint on the majority of projects, the cost of monitoring and of each instrument and technique has not been discussed in this paper. However, with rapidly developing technologies and greater awareness of the value that monitoring can bring to projects, such instruments and techniques are expected to become more widely used in the future. This case study provides an example of various techniques and instruments applied to a landslide over a 5-year period. It has demonstrated the level of information attainable from the combination of instruments and techniques, in order to capture both spatial and temporal behaviour. It showed that only a well-balanced choice of monitoring devices covering all scales of spatial and temporal resolutions delivers information that can be used to understand triggering mechanisms and deformation behaviour of a slope in great detail.

ACKNOWLEDGEMENTS

We thank the editor, Takashi Oguchi, Federica Bardi, and two anonymous reviewers for their constructive comments and suggestions that helped to improve this paper. We would like to extend our sincerest gratitude to Steve and Josie Gibson (the Hollin Hill landowners) for their support and cooperation in the research. The Engineering and Physical Sciences Research Council (EPSRC), UK, funded much of the Slope ALARMS research and development. This paper is published with the permission of the Executive Director of the British Geological Survey (NERC).

REFERENCES

1. Abdoun, T., Bennett, V., Desrosiers, T., Simm, J., Barendse, M., 2012. Asset management and safety assessment of levees and earthen dams through comprehensive realtime field monitoring. Geotech. Geol. Eng. 31, 833–843. http://dx.doi.org/10.1007/s10706-012-9569-3.
2. Allison, R.J., Brunsden, D., 1990. Some mudslide movement patterns. Earth Surf. Process. Landf. 15, 297–311. http://dx.doi.org/10.1002/esp.3290150402
3. Angeli, M.-G., Pasuto, A., Silvano, S., 2000. A critical review of landslide monitoring experiences. Eng. Geol. 55, 133–147. http://dx.doi.org/10.1016/S0013-7952(99)00122-2
4. Ballantyne, C.K., Harris, C., 1994. The Periglaciation of Great Britain. Cambridge University Press, Cambridge.
5. Bateman, M.D., 1995. Thermoluminescence dating of the British coversand deposits. Quat. Sci. Rev. 14, 791–798. http://dx.doi.org/10.1016/0277-3791(95)00053-4.
6. Belcher, S., Slingo, J., McCarthy, R., Burton, C., Betts, R., Brown, S., Clark, R., Kahana, R., Kendon, E., Knight, J., McCarthy, M., McNeall, D., Sexton, D., Stott, P., 2014. Too Hot Too Cold, Too Wet, Too Dry : Drivers and Impacts of Seasonal Weather in the UK. MetOffice, Exeter

7. Brückl, E., Brunner, F.K., Lang, E., Mertl, S., Müller, M., Stary, U., 2013. The Gradenbach Observatory-monitoring deep-seated gravitational slope deformation by geodetic, hydrological, and seismological methods. Landslides 10, 815–829. http://dx.doi.org/ 10.1007/ s10346-013-0417-1.

8. Buchli, B., Sutton, F., Beutel, J., 2012. GPS-equipped wireless sensor network node for high-accuracy positioning applications. In: GianPietro, P., Heinzelman, W. (Eds.), Wireless Sensor Networks. Springer, Berlin Heidelberg, pp. 179–195 http://dx.doi. org/10.1007/978-3-642-28169-3_12.

9. Buchli, T., Merz, K., Zhou, X., Kinzelbach, W., Springman, S.M., 2013. Characterization and monitoring of the Furggwanghorn Rock Glacier, Turtmann Valley, Switzerland: results from 2010 to 2012. Vadose Zone J. 12, 1–15. http://dx.doi.org/10.2136/ vzj2012.0067.

10. Caine, N., 1980. The rainfall intensity: duration control of shallow landslides and debris flows. Geogr. Ann. Ser. A, Phys. Geogr. 62, 23–27. http://dx.doi. org/10.2307/520449.

11. Catt, J.A., 1977. Loess and coversands. In: Shotton, F.W. (Ed.), British Quaternary Studies — Recent Advances. Clarendon Press, Oxford, pp. 221–230.

12. Chambers, J.E., Wilkinson, P.B., Kuras, O., Ford, J.R., Gunn, D.A., Meldrum, P.I., Pennington, C.V.L., Weller, A.L., Hobbs, P.R.N., Ogilvy, R.D., 2011. Three-dimensional geophysical anatomy of an active landslide in Lias Group mudrocks, Cleveland Basin, UK. Geomorphology 125, 472–484. http://dx.doi. org/10.1016/j.geomorph.2010.09.017

13. Chandler, R., 1984. Recent European experience of landslides in over-consolidated clays and soft rocks. 4th International Symposium on Landslide. Toronto, pp. 61–80.

14. Corsini, A., Pasuto, A., Soldati, M., Zannoni, A., 2005. Field monitoring of the Corvara landslide (Dolomites, Italy) and its relevance for hazard assessment. Geomorphology 66, 149–165. http://dx.doi. org/10.1016/j.geomorph.2004.09.012.

15. Cruden, D.M., Varnes, D.J., 1996. Landslide types and processes. In: Turn, A.K., Schuster, R.L. (Eds.) Landslides Investig. Mitigation. Spec. Rep. 247, 36–75.

16. Defra, 2012. UK Climate Change Risk Assessment: Evidence Report (No. GA0204). Defra, London.

17. Dijkstra, T.A., Dixon, N., 2010. Climate change and slope stability in the UK: challenges and approaches. Q. J. Eng. Geol. Hydrogeol. 43, 371–385. http://dx.doi. org/10.1144/1470- 9236/09-036.

18. Dijkstra, T., Dixon, N., Crosby, C., Frost, M., Gunn, D., Fleming, P., Wilks, J., 2014. Forecasting infrastructure resilience to climate change. Proc. Inst. Civ. Eng. 167, 269–280.

19. Dixon, N., Spriggs, M., 2011. Apparatus and method for monitoring soil slope displacement rate. UK Patent 2467419A.

20. Dixon, N., Spriggs, M.P., Smith, A., Meldrum, P., Haslam, E., 2014. Quantification of reactivated landslide behaviour using acoustic emission monitoring. Landslides http://dx.doi.org/10.1007/s10346-014-0491-z. Dixon, N., Hill, R., Kavanagh, J., 2003. Acoustic emission monitoring of slope instability: development of an active waveguide system. Proc. ICE Geotech. Eng. http://dx.doi. org/10.1680/geng.2003.156.2.83.

21. Dixon, N., Spriggs, M., Meldrum, P., Ogilvy, R., 2010. Development of a low cost acoustic emission early warning system for slope instability. In: Williams, A.L. (Ed.), Geologically Active: Proceedings of the 11th IAEG Congress. CRC Press, Auckland, New Zealand, pp. 1803–1810.

22. Dixon, N., Smith, A., Spriggs, M., Ridley, A., Meldrum, P., Haslam, E., 2015. Stability monitoring of a rail slope using acoustic emission. Proc. Inst. Civ. Eng. — Geotech. Engl. J. 168, 373–384. http://dx.doi.org/10.1680/ jgeen.14.00152.

23. Dunnicliff, J., 1988. Geotechnical Instrumentation for Monitoring Field Performance. John Wiley & Sons, New York. Foster, C., Pennington, C.V.L., Culshaw, M.G., Lawrie, K., 2011. The national landslide database of Great Britain: development, evolution and applications. Environ. Earth Sci. 66, 941–953. http://dx.doi. org/10.1007/s12665-011-1304-5.

24. Füssel, H.-M., Jol, A., Hildén, M., 2012. Climate change, impacts and vulnerability in Europe 2012 — An Indicator-based Report (No. 12/2012). EEA — European Environment Agency, Luxembourg http://dx.doi. org/10.2800/66071.

25. García, a., Hördt, A., Fabian, M., 2010. Landslide monitoring with high resolution tilt measurements at the Dollendorfer Hardt landslide, Germany. Geomorphology 120, 16–25. http://dx.doi.org/10.1016/j. geomorph.2009.09.011.

26. Gaunt, G.D., Ivimey-Cook, H.C., Penn, I.E., Cox, B.M., 1980. Mesozoic Rocks Proved by IGS Boreholes in the Humber and Acklam Areas. Institute of Geological Studies, Nottingham.

27. Glendinning, S., Hall, J., Manning, L., 2009. Asset-management strategies for infrastructure embankments. Proc. ICE Eng. Sustain. http://dx.doi.org/10.1680/ensu. 2009.162.2.111.

28. Glendinning, S., Helm, P.R., Rouainia, M., Stirling, R.A., Asquith, J.D., Hughes, P.R.N., Toll, D.G., Clarke, D., Powrie, W., Smethurst, J., Hughes, D., Harley, R., Karim, R., Dixon, N., Crosby, C., Chambers, J., Dijkstra, T.A., Gunn, D., Briggs, K., Muddle, D., 2015. Research-informed Design, Management and Maintenance of Infrastructure Slopes: Development of a Multi-Scalar Approach. IOP Conference Series, Earth and Environmental Science (EES).

29. Gunn, D.A., Chambers, J.E., Hobbs, P.R.N., Ford, J.R., Wilkinson, P.B., Jenkins, G.O., Merritt, A., 2013. Rapid observations to guide the design of systems for long-term monitoring of a complex landslide in the Upper Lias clays of North Yorkshire, UK. Q. J. Eng. Geol. Hydrogeol. 46, 323–336. http://dx.doi.org/10.1144/ qjegh2011-028.

30. Highland, L., Bobrowsky, P., 2008. The Landslide Handbook: A Guide to Understanding Landslides. U.S. Geolo. ed, Reston, Virginia.
31. Huang, H., Yi, W., Yi, Q., Lu, S., 2015. Real-time monitoring and data analysis of an active landslide in the Three Gorges Reservoir. In: Lollino, G., Giordan, D., Crosta, G.B., Corominas, J., Azzam, R., Wasowski, J., Sciarra, N. (Eds.), Engineering Geology for Society and Territory — Volume 2 SE — 253. Springer International Publishing, pp. 1433–1436 http://dx.doi.org/10.1007/978-3-319-09057-3_253.
32. Kuras, O., Pritchard, J.D., Meldrum, P.I., Chambers, J.E., Wilkinson, P.B., Ogilvy, R.D., Wealthall, G.P., 2009. Monitoring hydraulic processes with automated time-lapse electrical resistivity tomography (ALERT). C. R. Geosci. 341, 868–885. http://dx.doi. org/10.1016/j.crte.2009.07.010.
33. Machan, G., Beckstrand, D., 2012. Practical considerations for landslide instrumentation. In: Eberhardt, E., Froese, C., Turner, K., Leroueil, S. (Eds.), Landslides and Engineered Slopes: Protecting Society through Improved Understanding, pp. 1229–1234.
34. Malet, J.-P., Maquaire, O., Calais, E., 2002. The use of Global Positioning System techniques for the continuous monitoring of landslides: application to the Super-Sauze earthflow (Alpes-de-Haute-Provence, France). Geomorphology 43, 33–54. http://dx.doi.org/10. 1016/S0169-555X(01)00098-8.
35. Merritt, A.J., Chambers, J.E., Murphy, W., Wilkinson, P.B., West, L.J., Gunn, D.A., Meldrum, P.I., Kirkham, M., Dixon, N., 2013. 3D ground model development for an active landslide in Lias mudrocks using geophysical, remote sensing and geotechnical methods. Landslides 11, 537–550. http://dx.doi.org/10.1007/s10346-013-0409-1.
36. Mikkelsen, P., 2003. Advances in inclinometer data analysis. In: Myrvoll, F. (Ed.), 6th International Symposium on Field Measurements in Geomechanics. Oslo, Norway, pp. 1–13.
37. Pennington, C., Freeborough, K., Dashwood, C., Dijkstra, T., Lawrie, K., 2015. The national landslide database of Great Britain: acquisition, communication and the role of social media. Geomorphology 249, 44–51. http://dx.doi.org/10.1016/j.geomorph.2015.03.013.
38. Petley, D., 2012. Global patterns of loss of life from landslides. Geology 40, 927–930. http://dx.doi.org/10.1130/G33217.1.
39. Petley, D.N., Mantovani, F., Bulmer, M.H., Zannoni, A., 2005. The use of surface monitoring data for the interpretation of landslide movement patterns. Geomorphology 66, 133–147. http://dx.doi.org/10.1016/j.geomorph.2004.09.011.
40. Powell, J.H., 1984. Lithostratigraphical nomenclature of the Lias Group in the Yorkshire Basin. Proc. Yorks. Geol. Soc. 45, 51–57. http://dx.doi.org/10.1144/pygs.45.1-2.51.
41. Powell, J.H., 2010. Jurassic sedimentation in the Cleveland Basin: a review. Proc. Yorks. Geol. Soc. 58, 21–72. http://dx.doi.org/10.1144/pygs.58.1.278.
42. Reid, M.E., Lahusen, R.G., Baum, R.L., Kean, J.W., Schulz, W.H., Highland, L.M., 2012. Realtime monitoring of landslides. U.S. Geological Survey Fact Sheet 2012–3008 (4 pp.).
43. Simeoni, L., Mongiovì, L., 2007. Inclinometer monitoring of the Castelrotto landslide in Italy. J. Geotech. Geoenviron. Eng. 133, 653–666. http://dx.doi.org/10.1061/ (ASCE)1090-0241(2007)133:6(653).
44. Šimůnek, J., van Genuchten, M.T., Šejna, M., 2008. Development and applications of the HYDRUS and STANMOD software packages and related codes. Vadose Zone J. 7, 587–600. http://dx.doi.org/10.2136/vzj2007.0077.
45. Smith, A., Dixon, N., 2015. Quantification of landslide velocity from active waveguide– generated acoustic emission. Can. Geotech. J. 52, 413–425. http://dx.doi.org/10. 1139/cgj-2014-0226.
46. Smith, A., Dixon, N., Meldrum, P., Haslam, E., Chambers, J., 2014. Acoustic emission monitoring of a soil slope: comparisons with continuous deformation measurements. Geotech. Lett. 4, 255–261. http://dx.doi.org/10.1680/geolett.14.00053.
47. Springman, S.M., Kienzler, P., Friedel, S., 2013. A long-term field study for the investigation of rainfall-induced landslides. Géotechnique 63, 1177–1193. http://dx.doi.org/ 10.1680/geot.11.P.142.
48. Uhlemann, S., Chambers, J., Merritt, A., Wilkinson, P., Meldrum, P., Gunn, D., Maurer, H., Dixon, N., 2014. Integrated interpretation of geophysical, geotechnical, and environmental monitoring data to define precursors for landslide activation. AGU Fall Meeting. San Francisco, USA.
49. Uhlemann, S., Wilkinson, P., Chambers, J., Maurer, H., Merritt, A., Gunn, D., Meldrum, P., 2015. Interpolation of landslide movements to improve the accuracy of 4D geoelectrical monitoring. J. Appl. Geophys. 121, 93–105. http://dx.doi.org/10.1016/j. jappgeo.2015.07.003.
50. Wan, M.S.P., Standing, J.R., 2014. Lessons learnt from installation of field instrumentation. Proc. ICE Geotech. Eng. 167, 491–506.
51. Wilkinson, P.B., Chambers, J.E., Meldrum, P.I., Gunn, D.A., Ogilvy, R.D., Kuras, O., 2010. Predicting the movements of permanently installed electrodes on an active landslide using time-lapse geoelectrical resistivity data only. Geophys. J. Int. 183, 543–556. Wilkinson, P.B.,
52. Uhlemann, S., Chambers, J.E., Meldrum, P.I., Loke, M.H., 2015. Development and testing of displacement inversion to track electrode movements on 3-D electrical resistivity tomography monitoring grids. Geophys. J. Int. 200, 1566–1581. http://dx.doi.org/10.1093/gji/ggu483.

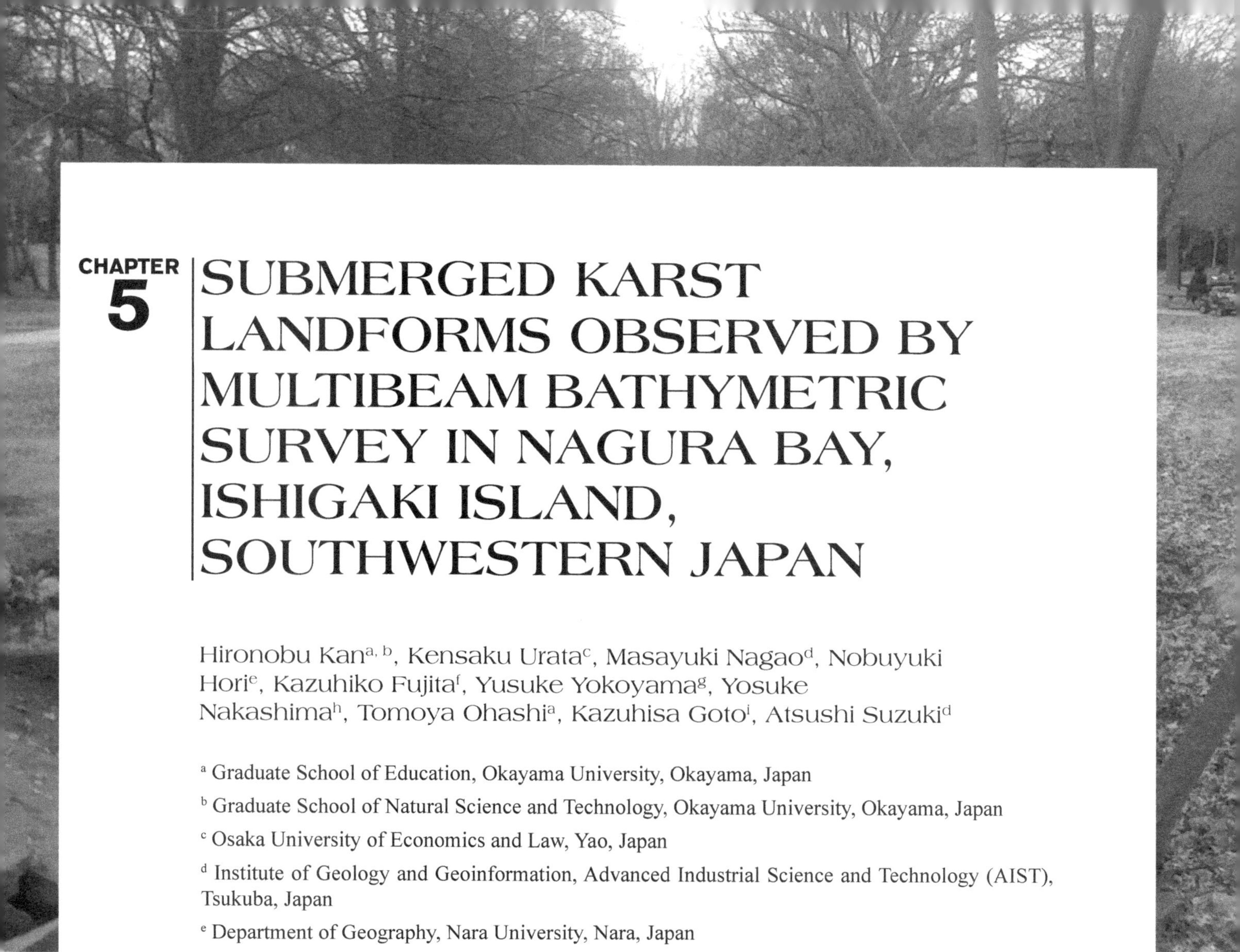

CHAPTER 5

SUBMERGED KARST LANDFORMS OBSERVED BY MULTIBEAM BATHYMETRIC SURVEY IN NAGURA BAY, ISHIGAKI ISLAND, SOUTHWESTERN JAPAN

Hironobu Kan[a, b], Kensaku Urata[c], Masayuki Nagao[d], Nobuyuki Hori[e], Kazuhiko Fujita[f], Yusuke Yokoyama[g], Yosuke Nakashima[h], Tomoya Ohashi[a], Kazuhisa Goto[i], Atsushi Suzuki[d]

[a] Graduate School of Education, Okayama University, Okayama, Japan

[b] Graduate School of Natural Science and Technology, Okayama University, Okayama, Japan

[c] Osaka University of Economics and Law, Yao, Japan

[d] Institute of Geology and Geoinformation, Advanced Industrial Science and Technology (AIST), Tsukuba, Japan

[e] Department of Geography, Nara University, Nara, Japan

[f] Department of Geology, University of Ryukyus, Okinawa, Japan

[g] Atmosphere and Ocean Research Institute (AORI), University of Tokyo, Kashiwa, Japan

[h] Ariake National College of Technology, Omuta, Japan

[i] International Research Institute of Disaster Science (IRIDeS), Tohoku University, Sendai, Japan

ABSTRACT

Submerged tropical karst features were discovered in Nagura Bay on Ishigaki Island in the southern Ryukyu Islands, Japan. The coastal seafloor at depths shallower than ~ 130 m has been subjected to repeated and alternating subaerial erosion and sedimentation during periods of Quaternary sea-level lowstands. We conducted a broadband multibeam survey in the central area of Nagura Bay (1.85 × 2.7 km) and visualized the high-resolution bathymetric results over a depth range of 1.6–58.5 m. Various types of humid tropical karst landforms were found to coexist within the bay, including fluviokarst, doline karst, cockpit karst, polygonal karst, uvalas, and mega-dolines. Although these submerged karst landforms are covered by thick postglacial reef and reef sediments, their shapes and sizes are distinct from those associated with coral reef geomorphology. The submerged landscape of Nagura Bay likely formed during multiple glacial and interglacial periods. According to our bathymetric results and the aerial photographs of the coastal area, this submerged karst landscape appears to have developed throughout Nagura Bay (i.e., over an area of approximately 6 × 5 km) and represents the largest submerged karst in Japan.

Citation: Hironobu Kan, Kensaku Urata, Masayuki Nagao, Nobuyuki Hori, Kazuhiko Fujita, Yusuke Yokoyama, Yosuke Nakashima, Tomoya Ohashi, Kazuhisa Goto, Atsushi Suzuki, Submerged karst landforms observed by multibeam bathymetric survey in Nagura Bay, Ishigaki Island, southwestern Japan, doi:10.1016/j.geomorph.2014.07.032

INTRODUCTION

Evidence suggests that coastal regions shallower than ~ 130 m have been subjected to repeated alternations between subaerial and marine conditions during the frequent glacio-eustatic sea-level changes that occurred throughout the Quaternary, with extensive environmental change occurring between glacial and interglacial periods (e.g.,Yokoyama et al., 2000, Camoin et al., 2004 and Fujita et al., 2010). The geomorphology of shallow coastal regions has been modulated by repeated subaerial and submarine processes during glacio-eustatic sea-level change. However, in contrast to the vast knowledge that has been accumulated regarding terrestrial landforms, few previous studies have dealt with shallow seafloor landforms, which represent former terrestrial landscapes modified by present marine processes, from a geomorphological perspective. In this regard, karst geomorphology is no exception, although Taviani (1984)provided some discussion regarding karst-like features on the ocean bottom from various perspectives.

Submerged karst areas are typically recognized by the subaerial exposure of karst landforms in coastal regions. For example, partly submerged tower karsts have been recognized in Ha Long Bay in Vietnam (Waltham and Hamilton-Smith, 2004), Phang Nga Bay in Thailand (Kiernan, 1994), Dweyra Bay and the Blue Grotto in Malta (Paskoff, 2005), and the Rock Islands in Palau (Dieter, 1991). Similarly, submerged karst landforms have been found in Marseille (Collina-Girard, 1996), the Adriatic Coast of Croatia (Surić, 2002), and near Lesbos Island and along the Argolid Peninsula in Greece (Scheffers et al., 2012). In Japan, submerged dolines have been reported only from Minami Island (one of the Ogasawara Islands of southeastern Japan), where circular depressions in shallow coastal areas were mapped based on aerial photographs (Hori, 1996). Nevertheless, many previous studies have reported the existence of submerged caves (e.g., Kitamura et al., 2007, De Waele et al., 2009, Surić et al., 2010, Taviani et al., 2012 and Mylroie and Mylroie, 2013). These facts suggest the presence of submerged karst features in many coastal areas globally. Moreover, in tropical coral reef areas, antecedent karst underlying postglacial coral reef complexes has been investigated previously and interpreted to control present reef geomorphology (e.g., Purdy, 1974,Purdy, 1998, Rasmussen and Neumann, 1988, Burke, 1993 and Macintyre et al., 2000).

Recent developments in multibeam bathymetric survey technology have enabled the visualization of submarine landforms in three dimensions. For example, this technique has been used to investigate landforms associated with coral reefs (e.g., Webster et al., 2006, Beaman et al., 2008 and Abby et al., 2011), bedform structures (e.g., Barnard et al., 2013 and Li et al., 2014), tectonic geomorphology (e.g., Tibor et al., 2010 and Barrie et al., 2013), glacial features (e.g., Howe et al., 2012), and submerged river landforms (e.g., Harris et al., 2013a). A pioneering study involving the observation of submerged karst by multibeam bathymetric survey was presented by Taviani et al. (2012). That study focused on the Adriatic shelf margin in Italy where eleven submerged dolines with diameters of 50–155 m were found, confirming the presence of thin postglacial deposits covering pre-existing landforms. Similar submerged dolines have been found to the north of the Maltese Islands in the Mediterranean Sea (Micallef et al., 2013). Here, we present the first study to describe a submerged humid tropical karst landscape in Nagura Bay, Ishigaki Island, southern Ryukyu Islands, Japan (Fig. 1A). In particular, we describe the landforms of this region based on a high-resolution bathymetric map with a grid size of 1 m, which we created with reference to observations obtained using our multibeam bathymetric echo-sounding survey combined with SCUBA diving observations.

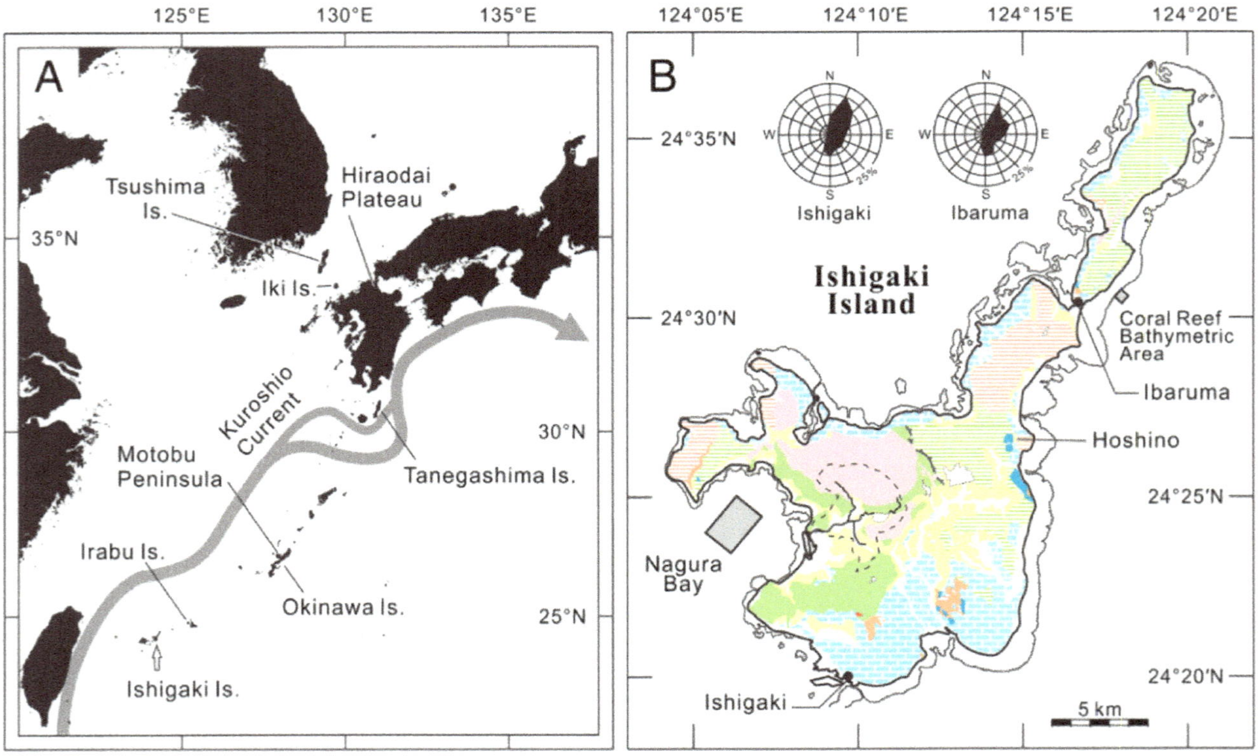

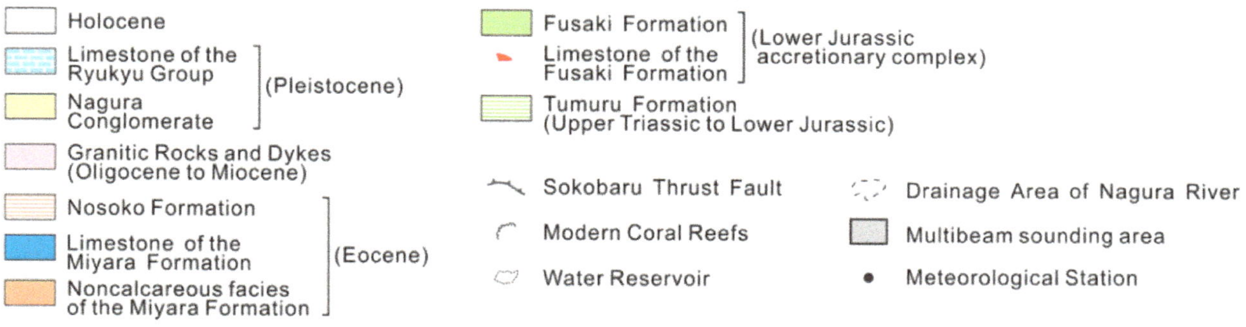

Figure 1: (A) Location and (B) lithology of Ishigaki Island, showing surveyed area. The flow axis of the Kuroshio Current in (A) illustrates conditions in November 2013, as presented by the Japan Meteorological Agency (2013). Wind roses in (B) have been summarized over the period 1991–2000, as presented by the Ishigakijima (Ishigaki Island) Local Meteorological Observatory for Ishigaki (southwest) and Ibaruma (northeast). Each of the concentric circles in the wind roses represents 5% frequency. The surface lithology in (B) was produced with reference to Nakae et al. (2009).

STUDY AREA

Geographical Setting

Ishigaki Island is one of the Ryukyu Islands and is located in the northwestern Pacific at latitudes and longitudes of 24°19.8–34.8′N, and 124°13.2–20.5′E, respectively (Fig. 1A). For Ishigaki Island, the average annual mean (1981–2010), monthly maximum, and monthly minimum temperatures are 24.3 °C, 29.5 °C, and 18.6 °C, respectively; the annual rainfall is 2106.8 mm and the average monthly maximum and minimum sea surface temperatures (SSTs) are 29 °C and 23 °C, respectively (Japan Meteorological Agency, 2013). The major western boundary current in the north Pacific, the Kuroshio Current, originates from the Western Pacific Warm Pool (WPWP) and flows through the Ryukyu Islands. Fringing reefs are developed around Ishigaki Island and a 30 × 20 km barrier reef (Sekisei Reef) is developed to the west of the island, lying between Ishigaki and Iriomote Islands. The prevailing wind directions in the vicinity of Ishigaki Island are NNE–NE from September to May and S–SSW during June to August (Japan Meteorological Agency, 2013) under the influence of the East Asian monsoon. An average of 7.6 typhoons passed within 300 km of the Ryukyu Islands annually from 1981 to 2010 (Japan Meteorological Agency, 2013), and it has been shown that the monsoon reversal and frequent typhoons in the area were responsible for the development of the high-energy fringing reef environment that surrounds most of the islands in the region (Kan, 2011).

Nagura Bay (which spans 6 km and 5 km in the north–south and east–west directions, respectively) is located on the western coast of Ishigaki Island. Wind roses (Fig. 1B) presented by the Ishigakijima Local Meteorological Observatory (2013) demonstrate that the frequency of the westerly wind is ~ 1%, suggesting that relatively calm sea conditions are prevalent in Nagura Bay. The environment of Nagura Bay is influenced by the Nagura River and its catchment (Fig. 1B). This river has a mainstream length of 4.6 km and a catchment area of 16.1 km^2 (Ikeda et al., 2009), with a semi-closed river mouth formed by the development of a sand spit and a river mouth wetland that is registered according to Ramsar Convention (Convention on Wetlands of International Importance Especially as Waterfowl Habitat). A mangrove swamp became established at the river mouth ~ 2 ka, resulting in the accumulation of a thin layer (~ 30 cm) of mangrove peat (Fujimoto et al., 1995 and Miyagi et al., 2003). Additionally, the present environment of Nagura Bay is thought to be influenced by the influx of a terrestrial red soil and nutrient load induced by land-use change to produce sugarcane and pineapple plantations in the drainage area (Ikeda et al., 2009). Distinct submerged groundwater discharge has not been reported around Nagura Bay.

Geological Setting

The geological basement of Ishigaki Island is formed primarily by the Tumuru and Fusaki Formations (Foster, 1965). The Tumuru Formation is an Upper Triassic to Lower Jurassic metamorphic complex (high-pressure low-temperature type) that is composed mainly of pelitic and mafic schists and is distributed throughout the eastern part of the island and on the Yarabu Peninsula at the northern end of Nagura Bay (Fig. 1B). The Fusaki Formation is a Lower Jurassic accretion complex composed primarily of chert, mudstone, sandstone, limestone, and basalt, and is distributed primarily in the southwestern part of the island. The limestone in this formation has yielded Permian fusulinids (Isozaki and Nishimura, 1989). The Tumuru Formation has been thrust upon the Fusaki Formation by the Sokobaru thrust fault and both formations are strongly deformed.

The Middle to Upper Eocene Miyara Formation is a shallow marine deposit, composed of limestones and noncalcareous facies including conglomerates, sandstones, and mudstones, which rests unconformably on the basement Fusaki and Tumuru Formations. The Nosoko Formation is composed of andesitic lavas and tuff and overlies the Miyara Formation conformably. These Eocene rocks have experienced negligible deformation since their deposition (Ujiié, 1996) and it is within the limestone of the Miyara Formation that cone karsts have developed in the eastern part of Ishigaki Island.

After deposition of the Miyara and Nosoko Formations, the study area underwent a period of intense igneous activity. Accordingly, the Oligocene to Miocene geological record (~ 29.9–21.0 Ma) in this region is characterized by granitic plutonic rocks and dacitic to rhyolitic intrusions (Ujiié, 1996). Subsequently, the Middle Miocene sequence of the Yaeyama Group was deposited in a shallow marine to terrestrial environment. This sequence is composed primarily of conglomerates, sandstones, and siltstones and is distributed throughout the southern Ryukyu Islands. However, these strata are not exposed anywhere on Ishigaki Island; rather, they can be found on Kohama, Iriomote, and Yonaguni Islands, which lie to the west of Ishigaki Island. Most prominently, a limestone layer with a thickness of ~ 30 m can be observed in the Yaeyama Formation off Miyako Island, 100 km east of Ishigaki Island (Tsuburaya and Sato, 1985).

After the second phase of extension of the Okinawa Trough commenced ~ 2 Ma (Sibuet et al., 1987), the Ryukyu Islands experienced remarkable development of coral reefs that accumulated to form the Pleistocene limestone of the Ryukyu Group. This limestone is composed of shallow water (0–50 m depth) coral framestone and deep water (50–150 m depth) rhodolith and *Cycloclypeus–Operculina* rudstone covering the insular shelf (Nakamori, 1986, Nakamori et al., 1995 and Iryu et al., 1998). Around Ishigaki Island, a relatively thin coral reef limestone formed after 0.41 Ma (Iryu and Matsuda, 2010), when it grew on a Pleistocene conglomerate known as the Nagura Gravel (Foster, 1965). $^{230}Th/^{234}U$ ages and altitudes of in situ *Porites* corals from the upper part of the Pleistocene coral reef limestone have provided ages of 118.5 ± 2.0 ka at 1.3 m at Ibaruma (in the northern part of Ishigaki Island) and 119.9 ± 2.0 ka at 15 m at Arakawa (in the southern part of Ishigaki Island), permitting their assignment to Marine Isotope Stage 5 (MIS5). Therefore, this limestone is best assigned to the last interglacial period (Marine Isotope Stage 5e; hereafter, MIS5e). Based on these ages and altitudes, uplift rates after MIS5e have been estimated to be − 0.01 to 0.03 m ka^{-1} and 0.1 to 0.4 m ka^{-1} in the northern and southern parts, respectively, of Ishigaki Island (Yamada et al., 2007). Taking this northward tilting of Ishigaki Island into account, an intermediate uplift rate is presumed for Nagura Bay, which is located in the middle part of the island.

METHODOLOGY

Multibeam Survey

A broadband multibeam echosounder (Sonic 2022, R2 Sonic, LLC) and its accessory system were used for this research. The Sonic 2022 has a variable ultrasonic frequency of 200–400 kHz and a 60 kHz signal bandwidth throughout its frequency range. It has 256 ultrasonic beams and selectable swath coverage of 10–160°. The typical ultrasonic beam widths parallel and orthogonal to the direction of travel are within one degree of each other when an ultrasonic frequency of 400 kHz is selected. For the present study, the accessory system of the Sonic 2022 consisted of a VS111 GPS compass system with A20 and A30 antennas (Hemisphere Inc.) combined with a dynamic motion sensor (DMS-10, Teledyne TSS Ltd.), a sea surface sound velocity sensor (miniSVS, Valeport Ltd.), a sound velocity profiler (MicroSVP, AML Oceanographic Ltd.), a sonar interface module (SIM), a junction box, and an operation PC with hydrographic survey software (HYPACK2010). The theoretical vertical resolution for the entire frequency range of the Sonic 2022 is 1.25 cm; however, the actual vertical resolution has been estimated to be 5–10 cm, because the vertical resolution depends on the measurement accuracy of both the accessory system and the echosounder (Nagao et al., 2011). The HYPACK2010 software was used for both hydrographic survey and data processing. Because complex undulations of submarine topography are known to exist in the surveyed area, we did not use the automatic noise filters provided in HYPACK2010. Rather, we identified noise by determining geomorphological patterns and continuity and removed the noise from every beam profile manually. The JGD2000/Japan Plane Rectangular CS XVI coordinate system was adopted for Ishigaki Island. IVS3D Fledermaus (Interactive Visualization Systems Inc.) was used for three-dimensional visualization with a grid size of 1 m and as an aid for description.

Multibeam bathymetric surveys were conducted in a rectangular area spanning 1.85 × 2.7 km in the central part of Nagura Bay (Fig. 1B). The surveyed area was defined by a rectangle with four corners: 24°25′13.4"N, 124°06′32.4″E (northwest); 24°24′36.6″N, 124°07′20.1″E (northeast); 24°23′30.0″N, 124°06′17.9″E (southeast); and 24°24′08.3″N, 124°05′29.3″E (southwest). The coastal areas within the bay were too shallow to conduct any surveys. Because the observed geomorphology of Nagura Bay was distinct from other coral reef areas around Ishigaki Island, an additional bathymetric survey was conducted for coral reef morphology of reef edge and reef slope areas of northeastern Ishigaki Island to enable comparison of the bathymetric results for Nagura Bay with typical coral reef geomorphology. In general, the ultrasonic frequency of 400 kHz was selected for bathymetric survey, however, we adopted 200 kHz in the southwestern (i.e., oceanward) third of the Nagura Bay area owing to the occurrence of occasional weak reflections produced by partially distributed soft bottom sediments and to increase swath width. Overlap of at least ~ 20% (typically ~ 50%) was implemented throughout the bathymetric survey to ensure 100% coverage of the surveyed area.

SCUBA Observations

The ground truthing was achieved by SCUBA diving observations at seven sites down to a depth of 40 m in November 2011. Surface features (sediments, rocks, coral assemblages) were also observed by SCUBA diving. Underwater photographs were taken using a Nikon D300 with a frame-filling fisheye lens (AF DX Fisheye Nikkor, 10.5 mm, f/2.8G) mounted in a Nauticam housing.

Sampling and Analysis of Geological Materials

During the SCUBA diving surveys, we collected three bedrock carbonate samples using a hammer and chisel from the overhanging walls at the edge of the submarine plateau to define the main lithologies affected by karst processes. Two sediment samples were also collected for a preliminary description of sediment types draping the bottom of the submarine canyon in the central part of the study area.

Optical microscopy of bulk samples was conducted using a binocular stereoscopic microscope. X-ray diffraction (XRD) analyses to determine the mineralogy of rock and sediment samples were conducted using a Rigaku MiniFlex II X-ray diffractometer. Quantitative analyses of minerals were undertaken using powder diffraction analysis software (Rigaku PDXL) employing the reference intensity ratio (RIR) method with reference to the mineralogical database of the International Center for the Diffraction Data (ICDD).

In November 2013, we started drilling down to 60 m depth at 24°24′32.5″N, 124°06′42.3″E. We discuss here some preliminary information, because this is, to date, the only source of borehole information for the entire Nagura Bay area.

RESULTS

Seafloor Landforms

The bathymetric results obtained for the central part of Nagura Bay (spanning 1.85 × 2.7 km) are presented in Fig. 2. The depth of the surveyed area is within the range 1.6–58.5 m. In general, the seafloor landforms of Nagura Bay are characterized by large, frequent undulations (~ 30 m) with numerous depressions (represented by closed contours in the figure). The undulations decrease in the seaward (i.e., southwestward) direction, such that a flat seafloor without closed depressions appears below 40 m.

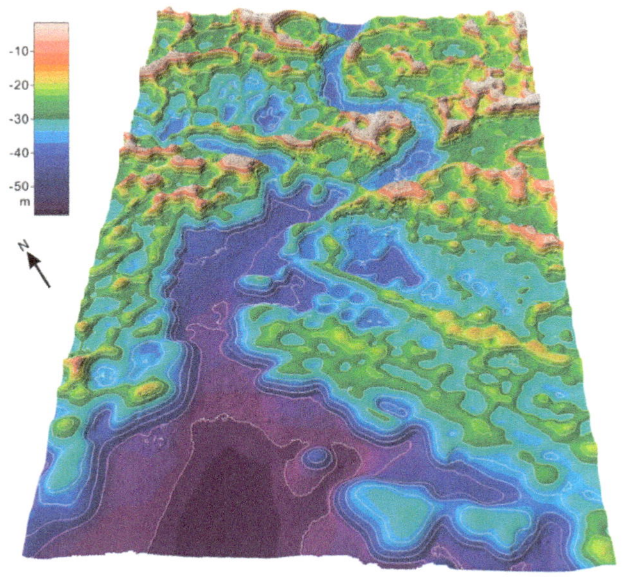

Figure 2: Multibeam bathymetry of Nagura Bay. A high-resolution model of our bathymetric results, visualized with a horizontal grid size of 1 m. Color scales show depths. Contours indicate 5 m isobaths. Vertical exaggeration is times 3.

The terrestrial karst landscapes break up the integrated and scalable patterning typical of fluvial systems by encouraging the development of small centripetal drainage basins and closed depressions exhibiting irregular patterning (Jennings, 1985). The seafloor morphology of Nagura Bay resembles that associated with terrestrial karst landforms and may be attributed to a submerged karst that developed during a sea level lowstand. Although there may be some uncertainty attached to applying terminology related to subaerial geomorphology to the seafloor landforms, as suggested by Taviani (1984), we tentatively apply the terminology of terrestrial karst landforms to describe the submarine landforms of Nagura Bay because no terminology exists to describe either the landforms found in the coral reef or the geomorphology of the seafloor in this context. In particular, we observed five types of karst landform in the surveyed area (Fig. 3) and suggest that these features may represent the styles and stages of karst development.

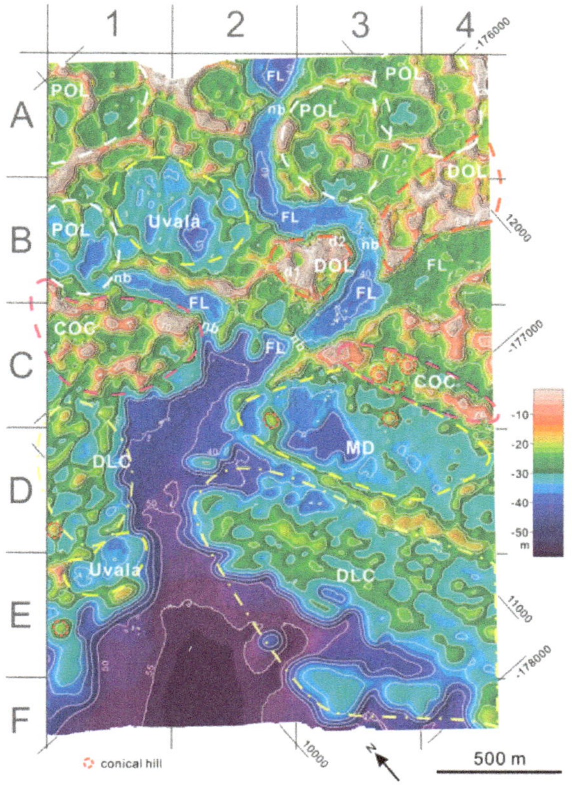

Figure 3: Geomorphological interpretation of Nagura Bay. A plan view with 5 m contours, with locations represented in the JGD2000/ Japan Plane Rectangular CS XVI coordinate system. Grid lines are drawn at 500 m intervals to identify different geomorphological sites. Color scales show depths. FL: fluviokarst, nb: natural bridge, DOL: doline karst, COC: cockpit karst, POL: polygonal karst, MD: mega-doline, DLC: dissected low-relief cockpit karst. d1, d2: individual doline cited in the text.

Fluviokarst

A meandering valley was observed to traverse the study area from northeast to southwest. This valley is interpreted as the trace of a paleo-river (i.e., forming a fluviokarst landscape). The fluviokarst landscape here is characterized by crescent-shaped closed depressions with widths of 150–200 m, with a maximum depth of 41 m occurring in the deepest parts of the valley (i.e., in regions A2, B2, B3, and C3 in Fig. 3). In the upper and lower reaches of the stream, the depression has lengths of ~ 800 m (A2–B3) and ~ 500 m (B3–C3), respectively. In this region, the steepest gradients are found on the undercut slopes of the meandering valley. The transversal ridges separating the upstream and downstream regions of the fluviokarst depressions may represent the remnants of natural bridges, where the upstream and downstream parts may have been connected previously by a lower corridor.

A shallower valley is observed beside the main valley in the B3–B4 regions. This valley has a maximum depth of ~ 30 m and a width of ~ 200 m. An isolated mound with diameter of 50 m remains in the central part of the valley (i.e., at B4). Furthermore, another fluviokarst region with length and width of 500 m and 150 m, respectively, is developed at B1–B2, with natural bridges dividing its western and southeastern reaches. However, such valleys with natural bridges do not appear to exist at depths greater than 40 m in Nagura Bay, where a flat seafloor extends offshore from C2 to E2.

Doline Karst

Two submarine plateaus (i.e., eastern and western plateau) with concordant ridge elevations that are separated by a fluviokarst depression are observed in shallow water between B2 and B4 (i.e., water depths of less than 8 m). These plateaus have a width of 200–300 m. Enclosed depressions on the plateau, referred to here as dolines, exhibit bowl-shaped basin forms with flattish floors.

The top depth of the western plateau located in B2 and B3 is in the range 5–7 m. Two dolines are developed on this plateau: one has diameter and depth of 100 m and 18 m, respectively (d1 in Fig. 3), whereas the other has diameter and depth of 30 m and 8 m, respectively (d2). Additionally, we found the northern and western edges of this plateau to be fringed by numerous dolines with diameters of 80–100 m.

The upper surface of the eastern plateau is less extensive than that of the western plateau. The dolines of the eastern plateau typically exhibit diameters and depths of ~ 100 m and ~ 15 m, respectively, although a small doline (50 m diameter, 10 m depth; d3) was found in the central part of the plateau. Similarly, the dolines found in Nagura Bay were mostly ~ 100 m in diameter, although smaller dolines with diameters of ~30–50 m have also been established on the upper surfaces of the submarine plateau in this region.

Cockpit Karst

The cockpit karst is described as the "egg-box" style of polygonal karst, which is characterized by steep-sided enclosed lobate depressions surrounded by residual hills, typically developed in Cockpit Country in Jamaica (Day and Chenoweth, 2004). Two cockpit karst areas are recognized in Nagura Bay (C1–C2 and C3–C4), adjacent to the central fluviokarst region. These areas exhibit typical cockpit karst forms, with "egg-box" depressions and intervening conical hills. In the eastern areas (C3–C4), conical hills (although some of the tops are flat, with the relief appearing to be mesa-like) and enclosed lobate depressions are aligned with a ridge extending in the NNW–SSE direction; four distinct conical hills (with diameters and heights of 60–100 m and 10–15 m, respectively) are observed in this area. Conversely, in the western area (C1–C2), typical "egg-box" depressions with diameters of 70–100 m are apparent. The relief formed by undulations between the tops of the conical hills and the bottoms of the intervening depressions is typically 20–25 m, although the tops of the conical hills in the western area (C1–C2) are typically shallower than those in the eastern area (C3–C4).

Polygonal Karst with Central Hill

Submarine landforms resembling polygonal karst with a cellular network are recognized in the surveyed area (Fig. 3). Although the above-mentioned cockpit karst is a type of polygonal karst, this additional category of polygonal karst is characterized by the relative lack of conical hills. Five areas within the study area exhibit polygonal karst features, as follows. (1) Five narrow saddles extend radially in different directions from the central residual hill in A1. These saddles form topographic divides between adjoining depressions, producing a pentagonal pattern in plan view. The top of the central hill lies at a depth of 6.2 m, whereas the saddles of the radial narrow ridges are at depths of 13.5–23.1 m. The adjoining depressions exhibit a maximum depth of 31.85 m and typically have diameters of ~ 100–200 m, as with the dolines in Nagura Bay. Small conical hills with heights of 3–4 m remain in some depressions. Although the northern margin of this polygonal karst lies outside the bathymetric survey area, we believe that the diameter of this formation may be ~ 500 m based on the distance between the central hill and the peripheral ridges of the depressions. (2) A polygonal karst with a diameter of ~ 500 m is observed in the B1 area. The depth of the top of the central hill is 19.4 m and the depression exhibits a maximum depth of 39.3 m. Two distinct radial narrow saddles extend from the central hill; however, all other saddles are discontinuous. (3) The polygonal karst located in the A3 area exhibits a diameter of ~ 500 m, corresponding to other polygonal karsts in Nagura Bay. The central part of this polygonal karst consists of a series of complex hills, the highest of which exhibits a square shape in plan view (100 × 80 m). The top and bottom depths of this hill are ~ 5 m and ~ 27 m, respectively, whereas the depth of the top of the central bean-shaped ridge (120 × 60 m in plan view) is 8.9 m. Three other conical hills (with diameters and heights of 40–50 m and 10–15 m, respectively) constitute the remainder of the central hill complex. The depressions between these conical hills are ~ 80 m in diameter. This central hill complex resembles the cockpit karst in the C1 area, although larger depressions (~ 100–200 m diameter) are recognized around the central hill complex. The maximum depth of the depressions in this area is 31.7 m, and the southwestern depression is connected to the adjoining fluviokarst. (4) A similar polygonal karst can be seen in the A4 area, to the east of the A3 polygonal karst.

Uvalas and Mega-dolines

The large compound closed depressions are evident within the study area (Fig. 3). (1) A large depression of 650 × 400 m has formed around the B1–B2 area and is subdivided internally into smaller second-order depressions with diameters of ~ 50–300 m. This arrangement resembles an uvala (i.e., compound doline), which is typically formed by the enlargement and merging of individual dolines (Ćalić, 2011). The largest and deepest (maximum depth: 40.9 m) depression is located in the southern part of this uvala. A small cone with a height and diameter of ~ 50 m and ~ 3 m, respectively, remains in the central part of the uvala (i.e., at the eastern end of the B1 area) and second-order depressions are located around the central cone. Based on the distribution of the second-order depressions, we assume that the original form of this uvala was a polygonal karst. Another uvala with dimensions of 360 × 180 m has formed around the D1–E1 area. (2) A large depression (tentatively referred to as a mega-doline) with a length and width of 1000 m and 500 m, respectively, is seen between the C2 and D4 areas. The long axis of this depression is oriented NNW–SSE, corresponding to the long axis of the above-mentioned uvala. However, in contrast to the uvala, second-order depressions are indistinct in this mega-doline. Six isolated conical hills remain in this area, forming topography similar to that expressed by the terrestrial karst landform known as "hum" in the Dinaric classical karst. These solitary conical hills appear to be the remnants of the conical hills typical of polygonal karst.

Low-relief Cockpit Karst

Low-relief cockpit karsts are distributed in two seaward areas of the study area: the D2–D4 and C1–E1 areas (Fig. 3). The tops of the conical hills in these regions lie below 20 m depth, and the slopes of these hills are gentler than those of the cockpit karst found around areas C1 and C3–C4. This low-relief cockpit karst constitutes the outer margin of the submerged karstic landforms in Nagura Bay.

SCUBA Observations

SCUBA diving surveys were conducted for geomorphological and sedimentological observations in three areas: the central submarine plateau, the western polygonal karst, and the edges of the eastern submarine plateau.

The surface of the central submarine plateau (Fig. 4A) at a water depth of ~ 5–6 m was found to be covered by living corals that consist primarily of large free-living fungiids (Fig. 5A) and branching corals. The interior slope of the d1 doline (~ 100 m in diameter) was found to be covered by living foliaceous corals and free-living fungiids and branching corals and their rubble at the midslope to baseslope (Fig. 5B). The bottom of the doline lies at a water depth of 24 m; here, the presence of a living coral that widens toward its top (Fig. 5C) suggests that the bottom of the doline has experienced waveless conditions for an extended period of time. Reef sediments consisting primarily of branching coral rubble were found to have formed talus creep deposits along the northern edge of the submarine plateau (Fig. 5D). Furthermore, some monospecific living coral communities consisting of branching or foliaceous corals (Fig. 5E) were found in the midslope region, and a grayish white carbonate mud with rubble has accumulated on the bottom of the valley. We also discovered a large living branching coral community dominated by*Acropora* (Fig. 5F) extending over a natural bridge lying to the south of the central submarine plateau (C3 in Fig. 3; f in Fig. 4A).

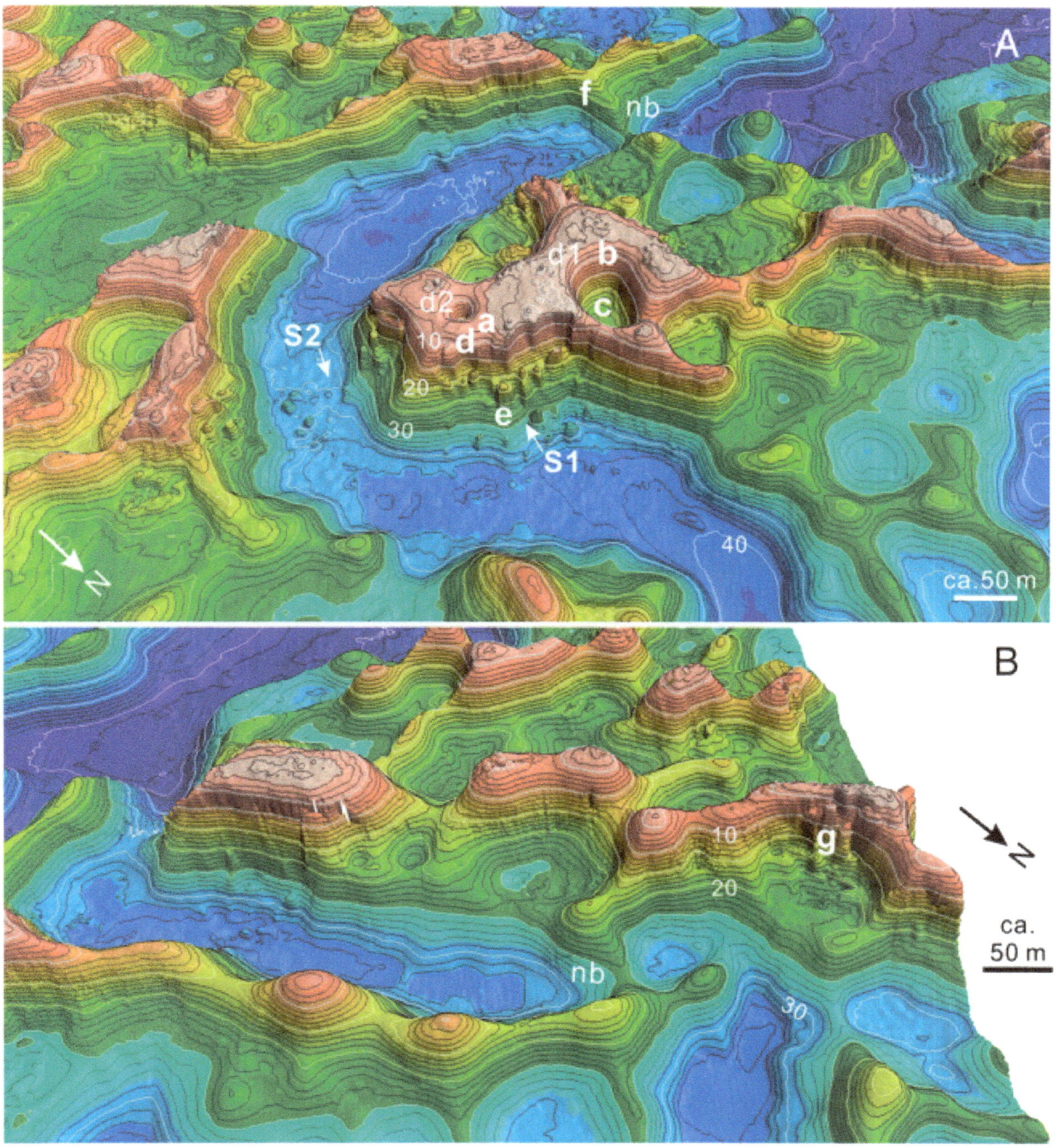

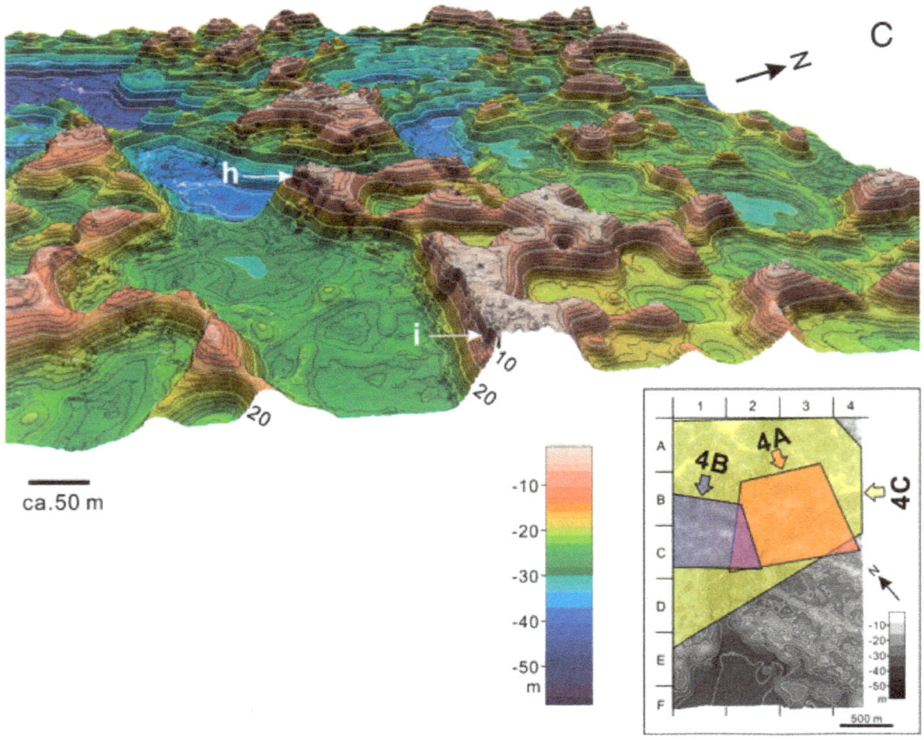

Figure 4: Three-dimensional views of submerged karst in Nagura Bay showing observation and sampling sites. Contours indicate 5 m (white) and 1 m (black) isobaths. Color scales show depths. Scale bars represent approximately 50 m. Panels show (A) doline karst and fluviokarst viewed from the northeast, (B) polygonal karst (right) and fluviokarst (left) viewed from the northeast, and (C) fluviokarst viewed from the southeast. d1 and d2: dolines corresponding to Fig. 3, nb: natural bridge, a–i: photographic sites illustrated in Fig. 5. Sampling sites of rocks and surface sediments (as shown in Table 1) are indicated by h/i and S1/S2, respectively. The map shown in the bottom right panel shows the areas of Fig. 4A (red), 4B (blue), and 4C (yellow) overlaid on a plan view corresponding to Fig. 3. Arrows indicate the direction of three-dimensional views. Vertical exaggeration is times 3.

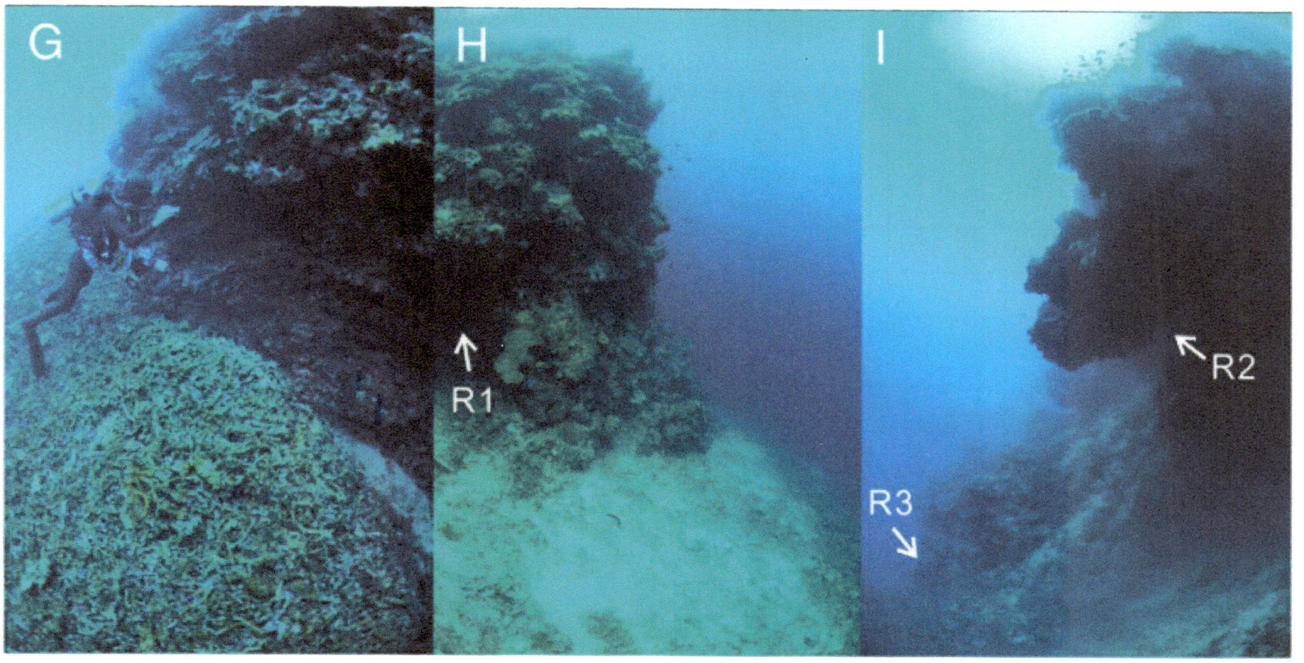

Figure 5: SCUBA diving observations of geomorphology and seafloor conditions in Nagura Bay. Photographic sites are indicated in Fig. 4. A: Top surface of the submarine plateau covered by living solitary and branching corals (depth: 5 m). B: Foliaceous *Montipora* in the middle of an interior slope of doline d1 (depth: 15 m). C: Living coral head at the bottom of doline d1 (depth: 23 m). D: Edge of the submarine plateau where branching coral rubble forms talus creeps that extend toward the fluviokarst valley (depth: 10 m). E: Monospecific communities of foliaceous coral (depth: 22 m). F: Large thicket of branching *Acropora* sp. (depth: 23 m). G: Accumulation of branching coral rubble with height of ~ 1.5 m at the foot of an overhanging wall (depth: 15 m). H: Upper steep slope and lower slope covered by reef talus along the slope of the fluviokarst valley (depth: 12 m). I: An overhanging wall at the top of the fluviokarst valley (depth: 10 m). All photographs were taken by H. Kan in November 2011.

The interior slope of the polygonal karst in the B1 area (Fig. 4B) is particularly steep at water depths shallower than 24 m. Along the upper steep slope, branching coral rubble has accumulated to a thickness of ~ 1.5 m at the foot of an overhanging wall (Fig. 5G). Living large coral communities here include laminar *Montipora* sp., branching *Acropora*, and foliaceous corals distributed along the gentle slope at 24–28.7 m.

At the edge of the eastern submarine plateau (Fig. 4C), the upper slopes are formed by very steep or overhanging walls that extend to ~ 12 m and ~ 15 m depth at sites "h" (Fig. 5H) and "i" (Fig. 5I), respectively. Below these walls, modern reef sediments form talus that covers the slope surface. No corrosional features were observed during our SCUBA surveys.

Surface Lithology and Sedimentology

Two bottom sediment samples (S1 and S2 in Table 1) were collected from two sites within the main valley (S1 and S2 in Fig. 4A) at depths of 30.8 m (S1) and 35.2 m (S2). The sediments in these samples were found to consist of grayish white carbonate mud with rubble of *Halimeda* and fragments of coral–algal framestone. Three rock samples (R1–R3 in Table 1) were collected from the overhanging walls (Fig. 6) at the edge of submarine plateau (sites "h" and "i" in Fig. 4C). The depths of the sampling sites were 12.7 m, 8.3 m, and 15.8 m for R1, R2, and R3, respectively. In terms of lithofacies, we classed these samples as coral–algal boundstones (R1 and R3) and a carbonate grainstone (R2) (Fig. 6). Owing to the lack of evidence of subaerial exposure, we believe such carbonates to have been formed recently. The detected mineralogies of these samples consist of aragonite and high-magnesium (high-Mg) calcite minerals (Table 1).

Table 1: Mineralogical composition of sediment and rock samples collected from Nagura Bay.

Sample number	Site inFig. 4	Depth (m)	Mineralogical composition (%)			Lithofacies
			Aragonite	High-Mg calcite	Others	
S1		30.8	38.8	61.2	0.0	Sediment
S2		35.2	41.6	58.4	0.0	Sediment
R1	h	12.7	77.6	22.4	0.0	Coral–algal boundstone
R2	i	8.3	39.7	60.3	0.0	Grainstone
R3	i	15.8	89.7	10.3	0.0	Coral–algal boundstone

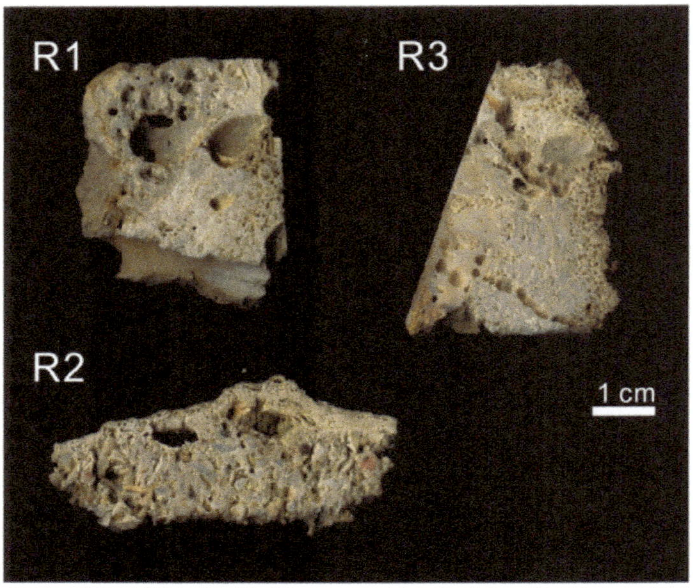

Figure 6: Rock samples collected from the upper layer of submerged karst in Nagura Bay. (R1) coral–algal boundstone collected from site "m" at 12.7 m depth. (R2) carbonate grainstone collected from site "n" at 8.3 m depth. (R3) coral–algal boundstone collected from site "n" at 15.8 m depth. Sampling sites (as shown in Table 1) are indicated in Figs. 4C and 5H/I.

In general, the rock and sediment samples described here indicate that the upper lithofacies of Nagura Bay consist of recent coral reef and reef sediments that have accreted or accumulated after the postglacial sea level rise. The maximum thickness of postglacial reef around Ishigaki Island has been reported to be 22 m for the fringing reef along the northeastern coast of the island (Hongo and Kayanne, 2009) and ~ 20 m for the barrier reef developed 2–10 km southwest of Ishigaki Island (Kan and Kawana, 2006). Because the top of the submarine plateau lies at ~ 5 m water depth, the recent coral–algal boundstone collected from 15.8 m water depth suggests that more than 10 m thickness of reef has been accreted in this postglacial period. Smaller karst features at the meter scale (i.e., less than 10 m), such as solution runnels, were not observed in Nagura Bay, likely because they were buried by the postglacial reef complex.

DISCUSSION

Difference from Coral Reef Geomorphology

Because the surface lithology of Nagura Bay is formed by the modern reef structure, it is important to discuss how the geomorphology of Nagura Bay is distinct from that of coral reefs. In general, the coral reef geomorphology of Ishigaki Island includes a reef crest, shallow lagoon, and reef slope. This sea level reef exhibits a zonal geomorphology in which the reef crest (with a width of ~ 200–300) has developed 500–1000 m offshore, oriented along the coastline of the island (e.g., Goto et al., 2010). The shallow lagoon (~ 3 m deep) has formed between the reef crest and shore. These zonal reef features around Ishigaki Island were built up ~ 4 ka (Yamano et al., 2001, Kan and Kawana, 2006 and Hongo and Kayanne, 2009); that is, they were generally formed when the reef reached sea level after the middle Holocene and matured during the late Holocene in the Ryukyu Islands (Kan, 2011). In particular, the spurs and grooves making up the smaller reef features at present were developed under stable sea level during the late Holocene owing to reef flat accretion at the reef edge (Kan et al., 1997 and Kan, 2011). Such distinct zonal reef features at such large scales (i.e., hundreds of meters in the horizontal direction) have not been reported previously for offshore reefs.

Fig. 7 illustrates a 500 × 500 m area of the submarine landscape in Nagura Bay (Fig. 7A; around the C1 area in Fig. 3) alongside the coral reef morphology of the reef edge and reef slope areas along the northeastern coast of Ishigaki Island (Fig. 7B) at the same scale. In the latter, the depth is in the range 2–40 m, and distinct spurs and grooves have developed with long axes parallel to wave orthogonals. This reef topography can be considered typical of the Ryukyu Islands. Spur widths vary with water depth: widths are 20–40 m, 10–20 m, and ~ 10 m at the reef edge, below 10 m water depth, and at depths of 25–30 m, respectively. Conversely, spur lengths are relatively constant with depth and are predominantly 50–70 m. The relative heights of spurs with respect to the intervening grooves are ~ 5–10 m at all depths down to the 30 m isobath. This pattern of reef geomorphology corresponds to the marine conditions that prevailed during the middle to late Holocene; in particular, the arrangement of spurs and grooves corresponds to wave orthogonals.

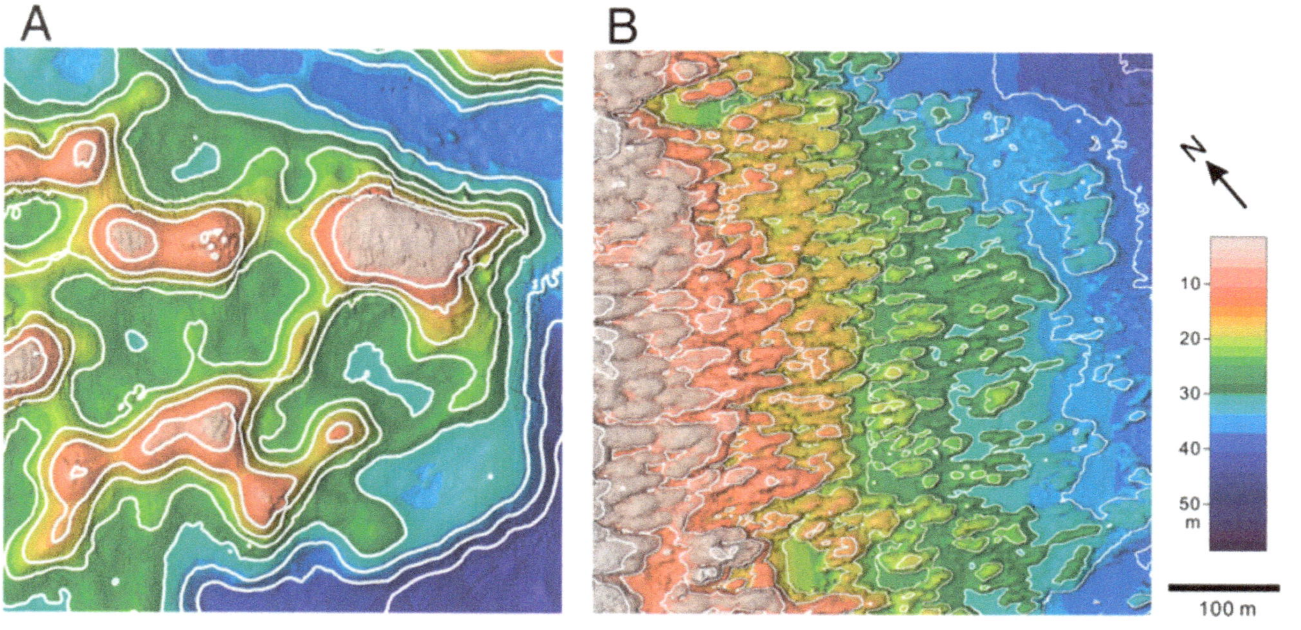

Figure 7: Geomorphology of submerged karst and coral reef at Ishigaki Island. (A) Submerged karst in Nagura Bay. (B) Coral reef around northeastern Ishigaki Island (see Fig. 1B). Bathymetric results are in plan view and cover areas of 500 × 500 m, visualized for a horizontal grid size of 1 m. Color scales show depths. Contours indicate 5 m isobaths.

In contrast to the reef geomorphology described above, the bathymetric map of Nagura Bay (Fig. 7A) is characterized by large geomorphological units with gently curved isobaths and large undulations. The dominant horizontal scale of the geomorphology of Nagura Bay is greater than 100 m, similar to the ~ 100 m diameters of dolines but much smaller than the ~ 500 m diameters observed for polygonal karsts. Based on our results, the submarine landscape of Nagura Bay is entirely distinct from the geomorphology of the present coral reef in terms of both spatial scale and pattern. Thus, although the submerged landforms in Nagura Bay are covered by the recent reef complex (rocks and sediments), the general outlines of landforms at the scale of hundreds of meters have been preserved.

Present-day Marine Environment in Nagura Bay

Nagura Bay is located on the western coast of Ishigaki Island, where the wave intensity is lower than elsewhere around the island. Moreover, the distance of 5 km from the offshore area to the bayhead produces a gradual decrease of wave intensity toward the shore, and the bottom of the drowned doline experiences a semi-enclosed and waveless environment due in part to the protected nature of the bay (e.g., Fig. 5C). The large undulations are another remarkable feature of the submarine landscape of Nagura Bay: the submarine plateaus at ~ 5 m water depth lie adjacent to valleys at ~ 40 m water depth. Such a configuration is relatively unique in coral reef geomorphology, even in comparison to the morphology of the reef system around the rest of Ishigaki Island. Owing to these large undulations and the terrestrial influence, underwater irradiance tends to be lower in Nagura Bay than in other coral reef areas around Ishigaki Island. Based on these gradients in physical properties (i.e., wave properties and irradiance), we assume the existence of various environments and associated biota in Nagura Bay.

It has been reported that the living coral coverage of only ~ 10% in Nagura Bay is composed primarily of massive (hemispherical) coral *Porites* sp. (Nature Conservation Bureau, Ministry of the Environment, 2002). This coverage ratio is the lowest observed throughout the fringing reefs around Ishigaki Island. However, we observed large coral communities during our preliminary SCUBA observations (e.g., Fig. 5E, F). In particular, abundant corals with branching and solitary growth forms were found living on top of the central submarine plateau. Furthermore, plentiful living corals with branching, foliaceous, and laminar growth forms were also observed on the intermediate gentle slopes at ~ 20–30 m depth in regions of steep slopes beside fluviokarst valleys or polygonal karst slopes, with coral rubble covering the steep slopes as talus and muddy sediments accumulated on the bottoms of valleys and depressions. In general, reduced irradiance has been shown to give rise to extensive mesophotic coral communities in shallower environments, typically at depths of 30–150 m (Lesser et al., 2009 and Slattery et al., 2011), and such conditions are thought to be important in forming refuges for shallow coral communities (Bak et al.,

2005, Bridge et al., 2013 and Harris et al., 2013b). Thus, the abundant living corals found along the deeper gentle slopes in Nagura Bay may play an important role in supporting the regional coral reef ecosystem.

Paleoenvironment in the Glacial Period Around Ishigaki Island

The development of karst landforms is closely related to regional climate variables such as temperature and precipitation (e.g., Daoxian, 2013). However the karst morphology in coral reef areas may reflect past climatic conditions, particularly those that prevailed during sea-level lowstands (e.g., Purkis et al., 2010). Cockpit karst typically develops in humid tropical or subtropical climates and is morphologically distinct from most temperate dolines as suggested by Lehmann (1936, in Ford and Williams., 2007). The present climate of Ishigaki Island provides conditions favorable for the development of cockpit karsts, as indicated by Zhu et al. (2013).

The climate of the Ryukyu Islands is influenced by the warm Kuroshio Current of the major western boundary current and the East Asian monsoon. In contrast to present conditions, where the Kuroshio Current flows through the Ryukyu Islands, the results of deep sea drilling have suggested that this current was weakened during the last glacial period such that it did not enter the East China Sea fully (e.g., Diekmann et al., 2008). Reef growth during the last glacial period has been constrained by drilling of the island shelf off Irabu Island in the southern Ryukyu Islands (Sasaki et al., 2006). This reef grew 22–15 ka at the present depth of 126.2–123.2 m and was drowned by the subsequent postglacial sea level rise. Massive faviid corals were the dominant species in the drowned reef, which resembled the higher latitude coral reefs reported for the Iki and Tsushima Islands in Japan at 34°N (Yamano et al., 2012). The oxygen isotope and Sr/Ca records of a faviid coral of 16.17 ka suggest that the paleo-SSTs were ~ 5 °C lower than present-day SSTs, whereas a stronger winter monsoon and weaker summer monsoon have been inferred based on sea surface salinity (SSS), which was at least ~ 0.2‰ higher than present-day SSS (Mishima et al., 2009). Furthermore, an ocean drilling core obtained from the Ryukyu Trench slope off the south of Ishigaki Island (International Marine Past Global Changes Study (IMAGES), core MD01-2398, located at 23°59.51′N, 124°24.76′E) also indicates MIS-2 SSTs that were ~ 2 °C and ~ 5 °C lower in summer and winter, respectively, than present-day SSTs (Ujiié and Ujiié, 2006).

Present-day winter SSTs of 23 °C and 19 °C have been reported for Ishigaki and Tanegashima Islands, respectively; the latter is located at 30°N, in the northern Ryukyu Islands (Japan Meteorological Agency, 2013). Coral facies and paleo-SSTs indicate that the environment of the southern Ryukyu Islands during the last glacial period resembled present-day conditions at latitudes of 30–34°N. Thus, the submerged karst in Nagura Bay would have developed under a climate that was slightly cooler and drier than the present day but still humid tropical or subtropical.

Corrosion rates of 87.3–102.6 mm ka^{-1} have been reported for tropical and subtropical karst areas in China (Zhu et al., 2013), and similar rates have been observed or calculated (using theoretical denudation equations) for other regions globally (Gunn, 2013). Assuming that dissolution does not always contribute only to direct lowering of the land surface but is also involved in the development of epikarst (Gunn, 2013), and considering the timescales involved in limestone corrosion, the development of the karst landforms in Nagura Bay likely proceeded throughout multiple glacial and interglacial periods. Then, the karst landforms in Nagura Bay would have been submerged for the final time during a period of postglacial sea level rise, with no subsequent rapid subsidence or uplift occurring in Nagura Bay since MIS-5e (Yamada et al., 2007).

Geomorphological Features of Submerged Karst in NAGURA Bay

Ishigaki Island can be considered a composite island based on the island type classification of the Carbonate Island Karst Model (Mylroie, 2013). For this island type, both carbonate and non-carbonate rocks are exposed at the surface and both allogenic and autogenic catchments are present. Allogenic streams have been shown to contribute to the enlargement of cave passages (Gunn, 2013). The present-day Nagura River flows through the non-carbonate rocks in its catchment (Fig. 1B). However, during the last glacial period, allogenic water may have entered the carbonate rocks in Nagura Bay and provided dissolutional potential for the formation of the transverse valleys of the fluviokarsts. Thus, the Nagura River may have been formed as a limestone gorge in the lower reaches of the river during the last glacial period.

Gunn (2004) suggested that fluviokarst can be considered as one end member in a continuous sequence of surface landform assemblages for carbonate rocks, where the other end member is polygonal karst, the entire surface of which is typically pitted by dolines with little or no trace of any previous valley form. These two end members seem to coexist in the submerged karst of Nagura Bay. Moreover, it is particularly remarkable that various landform assemblages spanning the sequence from dolines to polygonal karst and uvalas are present in Nagura Bay. The dolines in Nagura Bay exhibit either bowl-shaped forms or flat floors with vertical walls and have sizes that correspond to those of general karst dolines, which are typically tens to hundreds of meters in diameter and exhibit depths of a few meters to tens of meters (Williams, 2004). Their sizes are also similar to submerged dolines (observed by aerial photographs) in the Ogasawara Islands, southeastern Japan (Hori, 1996).

The cockpit landscape in Nagura Bay, which exhibits dominant depression/cone diameters of ~ 100–150 m, resembles the humid tropical type example of the "egg-box" morphology developed in polygonal karst at Cockpit Country in Jamaica (Day and Chenoweth, 2004), where the diameter of depressions and cones is approximately 200 m. In the Ryukyu Islands, a typical cockpit karst is developed in the Nakijin Limestone of the Triassic accretionary wedge of the Motobu Peninsula, Okinawa Island (Arakawa and Miura, 1990). The dominant diameters of cockpit depressions and conical hills are in the ranges 100–200 m and 50–100 m, respectively, in the Yamazato area (Fig. 8A, C). The porosity of the Nakijin Limestone is ~ 0%, whereas a porosity of 15.2–48.4% was observed for the Pleistocene limestone of the Ryukyu Group (Maekado, 1989). Matsukura et al. (2007) reported a corrosion rate of 205 mm ka^{-1} for a Pleistocene reef terrace. Because the porous limestone is more soluble, a slower corrosion rate is assumed for the Nakijin Limestone; this may have allowed long-term karstification to occur, thus developing the cockpit karst. Similarly, at Ishigaki Island, conical hills with horizontal scales of ~ 100–500 m (Fig. 8B, D) developed in the Eocene limestone (Miyara Formation) in the Hoshino area (Fig. 1B).

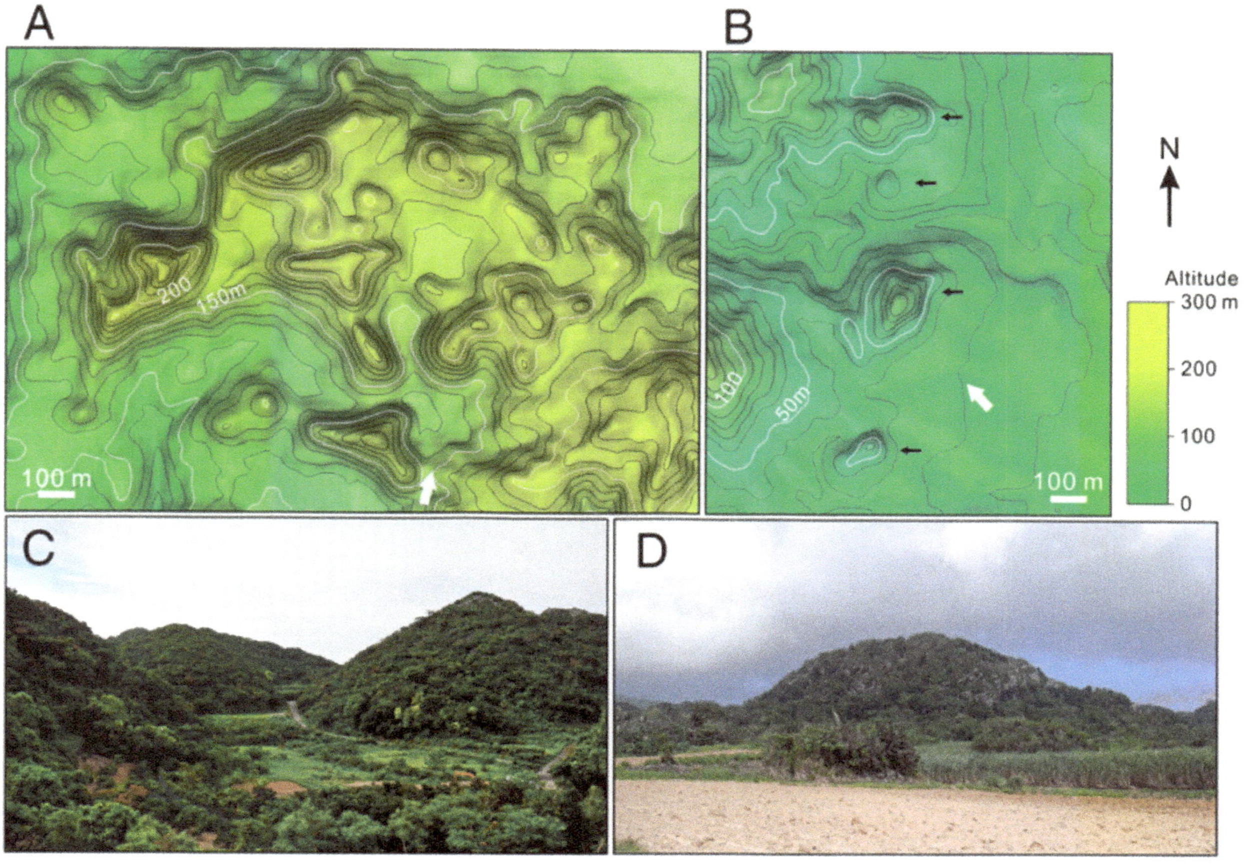

Figure 8: Cone karsts in the Ryukyu Islands. (A, C) Yamazato area of the Motobu Peninsula, Okinawa Island. (B, D) The Hoshino area of Ishigaki Island. The maps were prepared using a 10 m mesh DEM published by the Geospatial Information Authority of Japan (http://www.gsi.go.jp) and visualized using IVS3D Fledermaus. Color scales show altitudes. Black arrows in (B) indicate limestone hills. Note that the entirety of (A) consists of limestone. White arrows in (A) and (B) indicate the direction of photography for (C) and (D), respectively. Photographs were taken by H. Kan.

The pentagonal polygonal karsts observed in Nagura Bay have many features in common with the polygonal karst in Papua New Guinea described by Williams (1971). In particular, all of the available space in these karst landscapes is occupied by depressions, rather like an egg box, and the topographic divides separating adjoining solution depressions exhibit polygonal patterns in plan view. Once a cellular polygonal karst has been established, its geometry typically undergoes only minor modification through time (Williams, 1972). In Nagura Bay, an uvala is located next to the dissected polygonal landform located in the B1 area (Fig. 3). These landforms may represent the corrosional sequence of polygonal karst. Meanwhile, the low-relief cockpit karsts found primarily in the most seaward part of Nagura Bay may indicate that the "egg-box" style of polygonal karst did not always develop to form pentagonal polygonal karst.

An aerial photograph (Fig. 9) illustrates that the submarine plateaus observed in our bathymetric result extend to the shallow coastal area of Nagura Bay, beyond the area covered by our bathymetric survey. Many closed depressions can also be seen in this photograph. For example, the shallow sea geomorphology of the river mouth is also formed by closed depressions. Thus, the entirety of Nagura Bay appears to exhibit characteristics consistent with those of submerged karst. Nagura Bay extends for 6 km and 5 km in the north–south and east–west directions, respectively. This is approximately the same size as Minami Daito Island in the northwestern Pacific and is larger than the limestone area of the Hiraodai Plateau, a typical karst area on Kyushu Island, Japan. Thus, Nagura Bay appears to be the largest submerged karst in Japan.

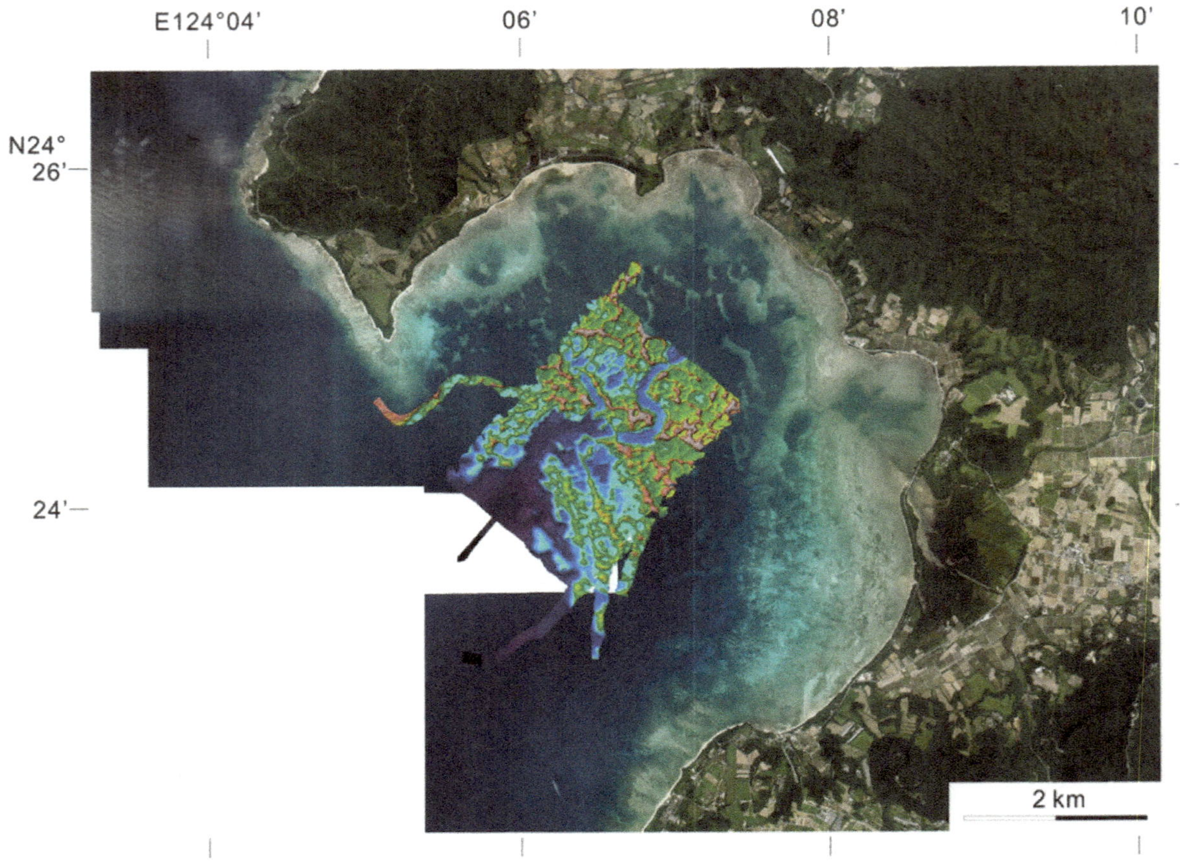

Figure 9: Coastal geomorphology with reference to the bathymetric results showing submerged karst in Nagura Bay. The aerial photograph was provided by Pasco Co. Ltd.

Perspectives on Geological Structure

Because the submerged karst in Nagura Bay was covered by reef structures, the host rock of the karst could not be identified during our SCUBA diving surveys. Thus, knowledge of this karstified rock remains lacking. The karst processes may affect not only limestone bedrock but also soluble evaporites. However, evaporites are not distributed in the Ryukyu Islands. Based on consideration of the geology of Ishigaki Island, the following strata are possible host rocks: (1) the Pleistocene limestone of the Ryukyu Group, (2) the Eocene limestone of the Miyara Formation, and (3) the limestone of the Permian Fusaki Formation.

The Pleistocene limestone of the Ryukyu Group (Ohama Formation) is distributed broadly on Ishigaki Island (Fig. 1B). It is a porous reef limestone composed of corals, foraminifers, and coralline algae, among other constituents. Cockpit karsts are typically developed in pure and hard carbonate strata without major impermeable layers (Zhu et al., 2013), and it is assumed that the landforms in Nagura Bay developed in solid limestone with low permeability, with dolines developed along joints. However, large karst features have not developed in the Pleistocene limestone areas of the Ryukyu Islands, where only collapsed dolines or dish-shaped shallow dolines have formed. Moreover, the cone karsts developed in areas consisting of the Pleistocene limestone of the Ryukyu Group typically appear as solitary cones (Hirose et al., 2013). We suggest that the porous Pleistocene limestone with high permeability would not have allowed the formation of the large karst features developed in Nagura Bay.

The limestone of the Permian Fusaki Formation is an accretionary wedge with a relatively limited surface distribution covering an area of ~ 0.05 km^2 (Fig. 1B). Thus, the submerged karst in Nagura Bay likely developed in the Eocene limestone of the Miyara Formation, which accumulated on the continental shelf and has not been deformed. This limestone is distributed widely throughout the terrestrial parts of Ishigaki Island: conical hills are developed in the Hoshino district on the eastern coast of the island, whereas karren are developed along the northwestern coast of Nagura Bay. Thus, we tentatively assume that the Miyara Formation limestone is the host rock of the submerged karst landscape in Nagura Bay.

Despite the presumed presence of a large limestone area underlying Nagura Bay, older (i.e., Tertiary or older) limestones are seldom found distributed on land near the bay (Fig. 1B). We interpret this as follows. In contrast to terrestrial areas, where karst corrosion has occurred continuously owing to subaerial exposure, karst in the tropical island shelf has experienced repeated accumulation of carbonate sediments due to submergence during interglacial periods. Corrosion that occurred during glacial periods presumably acted to remove the most recent deposits. However, the older limestones likely remained in the submerged karst on the island shelf and were less affected by corrosion.

The preliminary results of our latest ongoing drilling survey, which penetrated to 60 m drilling depth in the central submarine plateau between dolines d1 and d2 (Fig. 3), indicate that a considerable thickness (~ 34 m) of postglacial reef complex overlies the Pleistocene limestone. Although detailed analysis of this core has yet to be completed, these preliminary results show that this is the thickest record of postglacial reef obtained in the Ryukyu Islands to date.

Moreover, preliminary observations of Pleistocene stratigraphy in our drilling core indicate the presence of a layer of rigid in situ reef boundstone with a thickness of ~ 2 m, underlain by loose sediments consisting of Pleistocene limestone debris and soil down to a drilling depth of 60 m. We suspect that the Pleistocene limestone may be underlain by a dense rigid Tertiary limestone that constituted the original form of the submerged karst landforms. However, as we have not yet reached this older rigid limestone, our knowledge of the entire geological structure of the submerged karst in Nagura Bay remains incomplete.

In contrast to the relatively thin (~ 3 m) postglacial deposits covering submerged karst landforms reported in the Adriatic Sea (Taviani et al., 2012), the thick postglacial reef complex of Nagura Bay has accumulated on pre-existing landforms. These pre-existing prototypical landforms are assumed to be old and to lie at much deeper levels within the strata of the region. Subsequently, the undulations of landforms in this region may have been amplified by active reef growth, which typically occurs in convex topography during multiple interglacial periods, whereas the depressions have been buried by allochthonous reef sediments. Thus, the development of submerged karst landforms in tropical coral reef areas may be different in both style and history to those developed in temperate shelves.

CONCLUSIONS

We have discovered and visualized an extensive submerged karst landscape in Nagura Bay, Ishigaki Island, southwestern Japan, using multibeam bathymetric survey. Based on our bathymetric map with a grid size of 1 m, various types of humid tropical karst features coexist in the bay, including fluviokarst, doline karst, cockpit karst, polygonal karst, uvalas, and mega-dolines. The dominant spatial scale of the geomorphology observed in Nagura Bay is hundreds of meters, which is an order of magnitude larger than the present coral reef geomorphology at a similar depth.

The present marine conditions of Nagura Bay are characterized by low energy (calm sea) and low irradiance owing to the terrestrial influence. Such conditions have been emphasized by the presence of large undulating landforms, which cause decreases in wave intensity and irradiance with depth. These characteristics have acted to establish unique conditions compared to other coral reef areas in the Ryukyu Islands.

However, because the karst landforms are covered by a thick postglacial reef complex, the host rock remains unknown. We assume that Nagura Bay was a large karst basin in which older limestone remained submerged, thus preventing corrosion and the accumulation of reef sediments during periods of submersion, whereas the limestone outcropping on land was corroded during multiple interglacial and glacial periods. Based on our bathymetric result together with aerial photographs of the coastal area, we conclude that the submerged karst landscape has likely developed throughout the whole of Nagura Bay, covering an area of ~ 6 × 5 km. Accordingly, this area hosts the largest submerged karst in Japan.

ACKNOWLEDGMENT

This study was supported by the Japan Society for the Promotion of Science (JSPS) through a Grant-in-Aid for Scientific Research (A) (Nos. 22240084 and 25242026; chief investigator: H. Kan). The broadband multibeam echosounder (Sonic 2022) was introduced to HK›s laboratory using the above-mentioned grant #22240084. We would also like to express our appreciation for Shin Takada and Kouichi Nakano of the Toyo Corporation for technical support with R2 Sonic 2022. Field surveys were supported by the following people: Tateru Yarabu and Minoru Uehara, who steered the research boat; Ken›ichi Togashi, who assisted with the SCUBA diving surveys; and Tsuguo Chimura of the Chuo Kaihatsu Co. Ltd., who assisted with the drilling survey. Akira Maekado kindly provided information regarding limestone corrosion in the Ryukyu Islands. The constructive comments of Davide Baioni and two anonymous reviewers and the editor Jo De Waele helped to improve this contribution.

REFERENCES

1. Abby, E.,Webster, J.M.,Braga, J.C.,Sugihara, K.,Wallace, C.,Iryu, Y.,Potts, D.,Done, T.,Camoin, G., Seard, C., 2011. Variation in deglacial coralgal assemblages and their paleoenvironmental significance: IODP Expedition 310, "Tahiti Sea Level". Glob. Planet. Chang. 76, 1–15.
2. Arakawa, T.,Miura, H., 1990. Karst landforms in Japan. In: Hori, N., Toguchi, K., Hasegawa, H., Nakai, T. (Eds.), Tropics in Japan: Natural History of Coral IslandsSeries of Coral Islands in Japan. 1. Kokon-shoin Ltd, Tokyo, pp. 215–229 (in Japanese).
3. Bak, R.P.M.,Nieuwland, G.,Meesters, E.H., 2005. Coral reef crisis in deep and shallow reefs: 30 years of constancy and change in reef of Curacao and Bonaire. Coral Reefs 24, 475–479.
4. Barnard, P.L.,Erikson, L.H.,Elias, E.P.L.,Dartnell, P., 2013. Sediment transport patterns in the San Francisco Bay Coastal System from cross-validation of bedform asymmetry and modeled residual flux. Mar. Geol. 345, 72–95.
5. Barrie, J.V., Conway, K.W.,Harris, P.T., 2013. The Queen Charlotte Fault, British Columbia: seafloor anatomy of a transform fault and its influence on sediment processes. GeoMar. Lett. 33, 311–318.
6. Beaman, R.J.,Webster, J.D.,Wust, R.A.J., 2008. New evidence for drowned shelf edge reefs in the Great Barrier Reef, Australia. Mar. Geol. 247, 17–34.
7. Bridge, T.C.L.,Hughes, T.P.,Guinotte, J.M.,Bongaerts, P., 2013. Call to protect all coral reefs. Nat. Clim. Chang. 3, 528–530.
8. Burke, R.B., 1993. How have Holocene sea level rise and antecedent topography influenced Belize Barrier Reef development? Proc. Colloquium on Global Aspects of Coral Reefs. Univ. Miami, Miami, Florida, pp. 14–20.
9. Ćalić, J., 2011. Karstic uvala revisited: toward a redefinition of the term. Geomorphology 134, 32–42.
10. Camoin, G.F., Montaggioni, L.F., Braithwaite, C.J.R., 2004. Late glacial to post glacial sea levels in the Western Indian Ocean. Mar. Geol. 206, 119–146.

11. Collina-Girard, J., 1996. Préhistoire et karst littoral: la grotte Cosquer et les Calanques Marseillaises (Bouches-du-Rhône, France). Karstologia 27, 27–40 (re-published in English); Collina-Girard, J., 2004. Prehistory and coastal karst area: Cosquer Cave and the "Calamques" of Marseille. Speleogenesis and Evolution of Karst Aquifers. 2(2), pp. 1–13.
12. Daoxian, Y., 2013. Variation of karst geomorphology over geoclimatic gradients. In: Shroder, J.F., Frumkin, A. (Eds.), Treatise on Geomorphology 6. Academic Press, San Diego, CA, pp. 319–326.
13. Day, M., Chenoweth, S., 2004. Cockpit Country cone karst, Jamaica. In: Gunn, J. (Ed.), Encyclopedia of Caves and Karst Science. Fitzroy Dearborn, New York, pp. 233–235.
14. De Waele, J., Brook, G.A., Oertel, A., 2009. Monk seal (Monachus monachus) bones in Bel Torrente Cave (central-east Sardinia) and their paleogeographical significance. J. Cave Karst Stud. 71, 16–23.
15. Diekmann, B.,Hofmann, J.,Henrich, R.,Fütterer, D.K.,Röhl, U.,Wei, K.-Y., 2008. Detrital sediment supply in the southern Okinawa Trough and its relation to sea-level and Kuroshio dynamics during the late Quaternary. Mar. Geol. 255, 83–95.
16. Dieter, K., 1991. Main trends of Palau Islands› coastal evolution, identified by air and ground truthing. GeoJournal 24, 77–85.
17. Ford, D.,Williams, P., 2007. Karst Hydrogeology and Geomorphology. John Wiley & Sons Ltd., Chichester, (562p).
18. Foster, H.L., 1965. Geology of Ishigaki-shima Ryukyu-retto. U.S. Geological Survey, Professional Paper. 339A, pp. 1–119.
19. Fujimoto, K.,Yamanouchi, H.,Mezaki, S.,Hasegawa, H.,Maekado, A., 1995. Development of lowland and mangrove forest in Nagura River, Ishigaki Island. Proc. 47th General Meeting Association of Japanese Geographers, pp. 384–385 (in Japanese).
20. Fujita, K.,Omori, A.,Yokoyama, Y.,Sakai, S.,Iryu, Y., 2010. Sea-level rise during Termination II inferred from large benthic foraminifers: IODP Expedition 310, Tahiti Sea Level. Mar. Geol. 271, 149–155.
21. Goto, K.,Miyagi, K.,Kawamata, H.,Imamura, F., 2010. Discrimination of boulders deposited by tsunamis and storm waves at Ishigaki Island, Japan. Mar. Geol. 269, 34–45.
22. Gunn, J., 2004. Fluviokarst. In: Gunn, J. (Ed.), Encyclopedia of Caves and Karst Science. Fitzroy Dearborn, New York, pp. 363–364.
23. Gunn, J., 2013. Denudation and erosion rates in karst. In: Shroder, J.F., Frumkin, A. (Eds.), Treatise on Geomorphology 6. Academic Press, San Diego, CA, pp. 72–81.
24. Harris, M.S., Sautter, L.R., Johnson, K.L.,Luciano, K.E., Sedberry, G.R.,Wright, E.E., Siuda, A.N.S., 2013a. Continental shelf landscapes of the southern United States since the last interglacial. Geomorphology 203, 6–24.
25. Harris, P.T., Bridge, T.C.L., Beaman, R.J.,Webster, J.M., Nichol, S.L., Brooke, B.P., 2013b. Submerged banks in the Great Barrier Reef, Australia, greatly increase available coral reef habitat. ICES J. Mar. Sci. 70, 284–293.
26. Hirose, T., Sangawa, T.,Aoki, H.,Matsukura, Y.,Maekado, A., 2013. A study of cone karsts developed on Quaternary Ryukyu Limestone in Okinawa. Okinawa J. Geogr. Stud. 12, 25–32 (in Japanese).
27. Hongo, C., Kayanne, H., 2009. Holocene coral reef development under windward and leeward locations at Ishigaki Island, Ryukyu Islands, Japan. Sediment. Geol. 214, 62–73.
28. Hori, N., 1996. Submerged karst topography around Minami-Shima and Minami-Saki Chichijima, Ogasawara Islands. In: Nakamura, K., Ujiié, H., Ikehara, S., Tagawa, H., Hori, N. (Eds.), Nature in Japan. 8. Iwanami Shoten, Tokyo, pp. 86–87 (in Japanese).
29. Howe, J.A., Dove, D., Bradwell, T.,Gafeira, J., 2012. Submarine geomorphology and glacial history of the Sea of the Hebrides, UK. Mar. Geol. 315–318, 61–76.
30. Ikeda, S., Osawa, K., Akamatsu, Y., 2009. Sediment and nutrients transport in watershed and their impact on coastal environment. Proc. Jpn. Acad. Ser. B 85, 374–390.
31. Iryu, Y., Matsuda, H., 2010. Neogene and Quaternary, Ryukyu Islands. In: Geological Society of Japan (Ed.), Regional geology of Japan. 8. Asakura Shoten, Tokyo, pp. 149–154 (in Japanese).
32. Iryu, Y., Nakamori, T., Yamada, T., 1998. Pleistocene reef complex deposits in the Central Ryukyus, south-western Japan. In: Camoin, G.F., Davies, P.J. (Eds.), Reefs and Carbonate Platforms in the Pacific and Indian Oceans. Special Publication of the International Association of Sedimentologists. 25, pp. 197–213.
33. Ishigakijima Local Meteorological Observatory, 2013. http://www.jma-net.go.jp/ishigaki/ index.html (accessed on November 2013).
34. Isozaki, Y.,Nishimura, Y., 1989. Fusaki Formation, Jurassic subduction–accretion complex on Ishigaki Island, southern Ryukyus and its geologic implication to Late Mesozoic convergent margin of East Asia. Mem. Geol. Soc. Jpn. 33, 259–275 (in Japanese with English abstract).
35. Japan Meteorological Agency, 2013. http://www.jma.go.jp (accessed on November 2013).
36. Jennings, J.N., 1985. Karst Geomorphology. Basil Blackwell, Oxford, UK, (293 pp).
37. Kan, H., 2011. Ryukyu Islands. In: Hopley, D. (Ed.), Encyclopedia of Modern Coral Reefs: Structure, Form and Process, Encyclopedia of Earth Science Series. Springer, Dordrecht, pp. 940–945
38. Kan, H.,Kawana, T., 2006. 'Catch-up' of a high-latitude barrier reef by back-reef growth during post-glacial sea-level rise, Southern Ryukyus, Japan. Proc. 10th International Coral Reef Symposium, Okinawa, Japan, pp. 494–503.

39. Kan, H., Hori, N., Ichikawa, K., 1997. Formation of a coral reef-front spur. Coral Reefs 16, 3–4.
40. Kiernan, K., 1994. The coastal tower karst of Phang Nga Bay. International Symposium on the Exploitation and Protection of Karst and Cave Scenic Tourist Resources: Study of Karst and Cave Scenic Tourist Resources. University of Tasmania ePrints, pp. 9–21.
41. Kitamura, A., Yamamoto, N., Kase, T., Ohashi, S., Hiramoto, M., Fukusawa, H.,Watanabe, T., Irino, T., Kojitani, H., Shimamura, M., Kawakami, I., 2007. Potential of submarinecave sediments and oxygen isotope composition of cavernicolous micro-bivalve as a late Holocene paleoenvironmental record. Glob. Planet. Chang. 55, 301–316
42. Lehmann, H., 1936. Morphologische studies auf Java. Series 3, No. 9, GeographischeAbhandlungen, Stuttgart, Germany, (114 pp).
43. Lesser, M.P.,Slattery, M.,Leichter, J.J., 2009. Ecology of mesophotic coral reefs. J. Exp. Mar. Biol. Ecol. 375, 1–8.
44. Li, M.Z., Shaw, J., Todd, B.J., Kostylev, V.E.,Wu, Y., 2014. Sediment transport and development of banner banks and sandwaves in an extreme tidal system: Upper Bay of Fundy, Canada. Cont. Shelf Res. 83, 86–107.
45. Macintyre, I.G., Precht, W.F., Aronson, R.B., 2000. Origin of the Pelican Cay Ponds, Belize. Atoll Res. Bull. 466, 1–11.
46. Maekado, A., 2000. Some physical and mechanical properties of the limestones in Okinawa Island. Bulletin of the Okinawa Geographical Society 2, 29–33.
47. Matsukura, Y.,Maekado, A.,Aoki, H.,Kogure, T.,Kitano, Y., 2007. Surface lowering rates of uplifted limestone terraces estimated from the height of pedestals on a subtropical island of Japan. Earth Surf. Process. Landf. 32, 1110–1115.
48. Micallef, A., Foglini, F., Le Bas, T., Angeletti, L.,Maselli, V., Pasuto, A., Taviani, M., 2013. The submerged paleolandscape of the Maltese Islands: morphology, evolution and relation to Quaternary environmental change. Mar. Geol. 335, 129–147.
49. Mishima, M.,Kawahata, H.,Suzuki, A.,Inoue, M.,Okai, T.,Omura, A., 2009. Reconstruction of the East China Sea palaeoenvironment at 16 ka by comparison of fossil and modern Faviidae corals from the Ryukyus, southwestern Japan. J. Quat. Sci. 24, 928–936.
50. Miyagi, T.,Ajiki, K., Fujimoto, K., 2003. Mangrove Habitat Dynamics and Human Relation. Kokon-Shoin Ltd., Tokyo, (193 pp) (in Japanese).
51. Mylroie, J.E., 2013. Coastal karst development in carbonate rocks. In: Lace, M.J., Mylroie, J.E. (Eds.), Coastal Karst Landforms, Coastal Research Library. 5. Springer, Dordrecht, pp. 77–109.
52. Mylroie, J.E.,Mylroie, J.R., 2013. Flank margin caves in carbonate islands and the effects of sea level. In: Shroder, J.F., Frumkin, A. (Eds.), Treatise on Geomorphology 6. Academic Press, San Diego, CA, pp. 351–362.
53. Nagao, M.,Kan, H.,Nakano, K.,Takada, S.,Ogasawara, H.,Nakamura, T.,Ohashi, T.,Suzuki, A., 2011. An attempt to find small artificial objects in the shallow sea bottom using a broadband multibeam echosounder. Proceedings of International Sessions in Conference on Coastal Engineering JSCE. 2, pp. 51–55.
54. Nakae, S., Nagamori, H., Miyazaki, K., Komazawa, M., 2009. Geological Map of Japan 1: 200,000, Ishigaki Jima. Geological Survey of Japan, AIST, Tsukuba (in Japanese with English abstract).
55. Nakamori, T., 1986. Community structures of Recent and Pleistocene hermatypic corals in the Ryukyu Islands, Japan. The Science Reports of the Tohoku University, 2nd Series (Geology) 56, 71–133.
56. Nakamori, T., Iryu, Y.,Yamada, T., 1995. Development of coral reefs of the Ryukyu Islands (southwest Japan, East China Sea) during Pleistocene sea-level change. Sediment. Geol. 99, 215–231.
57. Nature Conservation Bureau, Ministry of the Environment, 2002. Annual report of coral reef monitoring around Ishigaki Island 2002 Japan (in Japanese).
58. Paskoff, R.P., 2005. Karst coasts. In: Schwartz, M.L. (Ed.), Encyclopedia of Coastal Science. Springer, Dordrecht, pp. 581–586.
59. Purdy, E.G., 1974. Karst-determined facies patterns in British Honduras: Holocene carbonate sedimentation model. AAPG Bull. 58, 825–855.
60. Purdy, E.G., 1998. Structural termination of the southern end of the Belize Barrier Reef. Coral Reefs 17, 231–234.
61. Purkis, S.J., Rowlands, G.P., Riegl, B.M., Renaud, P.G., 2010. The paradox of tropical karst morphology in the coral reefs of the arid Middle East. Geology 38, 227–230.
62. Rasmussen, K.A., Neumann, A.C., 1988. Holocene overprints of Pleistocene paleokarst: bight of Abaco, Bahamas. In: James, N.P., Choquette, P.W. (Eds.), Paleokarst. SpringerVerlag, New York, pp. 132–148.
63. Sasaki, K.,Omura, A.,Miwa, T.,Tsuji, Y.,Matsuda, H.,Nakamori, T.,Iryu, Y.,Yamada, T.,Sato, Y., Nakagawa, H., 2006. 230Th/234U and 14C dating of a lowstand coral reef beneath the insular shelf off Irabu Island, Ryukyus, southwestern Japan. Island Arc 15, 455–467
64. Scheffers, A.M.,Scheffers, S.R.,Kelletat, D.H., 2012. The Coastlines of the World With Google Earth. Springer, Dordrecht, (293p).
65. Sibuet, J.-C.,Letouzey, J.,Barbier, F.,Charvet, J.,Foucher, J.-P.,Hilde, T.W.C.,Kimura, M.,Ling-Yun, C.,Marsset, B.,Muller, C.,Stéphan, J.-F., 1987. Back arc extension in the Okinawa Trough. J. Geophys. Res. 92, 14041–14063.
66. Slattery, M.,Lesser, M.P.,Brazeau, D.,Stokes, M.D.,Leichter, J.J., 2011. Connectivity and stability of mesophotic coral reefs. J. Exp. Mar. Biol. Ecol. 408, 32–41.
67. Surić, M., 2002. Submarine karst of Croatia — evidence of former lower sea levels. Acta Carsologica 31 (3), 89–98.
68. Surić, M.,Lončarić, R.,Lončar, N., 2010. Submerged caves of Croatia: distribution, classification and origin. Environ. Earth Sci. 61, 1473–1480.

69. Taviani, M., 1984. Submarine 'sinkholes': a review. Proc. First Multidiciplinary Conference on Sinkholes, Orlando, Florida, USA, pp. 117–121.

70. Taviani, M.,Angeletti, L.,Campiani, E.,Ceregato, A.,Foglini, F.,Maselli, V.,Morsilli, M.,Parise, M.,Trincardi, F., 2012. Drowned karst landscape offshore the Apulian margin (Southern Adriatic Sea, Italy). J. Cave Karst Stud. 74, 197–21

71. Tibor, G.,Niemi, T.M.,Ben-Avraham, Z.,Al-Zoubi, A.,Sade, R.A.,Hall, J.K.,Hartman, G.,Akawi, E., Abueladas, A., Al-Ruzouq, R., 2010. Active tectonic morphology and submarine deformation of the northern Gulf of Eliat/Aqaba from analyses of multibeam data. Geo-Mar. Lett. 30, 561–573

72. Tsuburaya, H.,Sato, T., 1985. Borehole of MITI Miyakojima-Oki. J. Jpn. Assoc. Pet. Technol. 50, 25–33 (in Japanese)

73. Ujiié, H., 1996. Formation of the foundation of Ryukyu Island Arc. In: Nakamura, K., Ujiié, H., Ikehara, S., Tagawa, H., Hori, N. (Eds.), Nature in Japan. 8. Iwanami Shoten, Tokyo, pp. 93–111 (in Japanese).

74. Ujiié, Y., Ujiié, H., 2006. Dynamic changes of the surface and intermediate waters in the Ryukyu Arc region during the past ~250,000 years: based on planktonic and benthic foraminiferal analyses of two IMAGES cores. Fossils 79, 43–59.

75. Waltham, T.,Hamilton-Smith, E., 2004. Ha Long Bay, Vietnam. In: Gunn, J. (Ed.), Encyclopedia of Caves and Karst Science. Fitzroy Dearborn, New York, pp. 413–414.

76. Webster, J.M., Clague, D.A., Braga, J.C., Spalding, H., Renema, W., Kelley, C., Applegate, B., Smith, J.R., Paull, C.K.,Moore, J.G., Potts, D., 2006. Drowned coralline algal dominated deposits off Lanai, Hawaii; carbonate accretion and vertical tectonics over the last 30 ka. Mar. Geol. 225, 223–246

77. Williams, P.W., 1971. Illustrating morphometric analysis of karst with examples from New Guinea. Z. Geomorphol. 15, 40–61.

78. Williams, P.W., 1972. Morphometric analysis of polygonal karst in New Guinea. Geol. Soc. Am. Bull. 83, 761–796.

79. Williams, P., 2004. Dolines. In: Gunn, J. (Ed.), Encyclopedia of Caves and Karst Science. Fitzroy Dearborn, New York, pp. 304–310.

80. Yamada, S.,Matsuda, H.,Omura, A., 2007. Uranium-series dates of fossil hermatypic corals in Ishigaki-jima Island, Okinawa Prefecture, and their significance. J. Geol. Soc. Jpn. 113, 19–22 (in Japanese with English abstract).

81. Yamano, H.,Kayanne, H.,Yonekura, N., 2001. Anatomy of a modern coral reef flat: a recorder of storms and uplift in the late Holocene. J. Sediment. Res. 71, 295–304.

82. Yamano, H., Sugihara, K.,Watanabe, T., Shimamura, M., Hyeong, K., 2012. Coral reefs at 34°N, Japan: exploring the end of environmental gradients. Geology 40, 835–838.

83. Yokoyama, Y., Lambeck, K. De, Deckker, P., Johnstone, P., Fifield, I.K., 2000. Timing of the Last Glacial Maximum from observed sea-level minima. Nature 406, 713–716.

84. Zhu, X., Zhu, D., Zhang, Y.,Lynch, E.M., 2013. Tower karst and cone karst. In: Shroder, J.F., Frumkin, A. (Eds.), Treatise on Geomorphology 6. Academic Press, San Diego, CA, pp. 327–340.

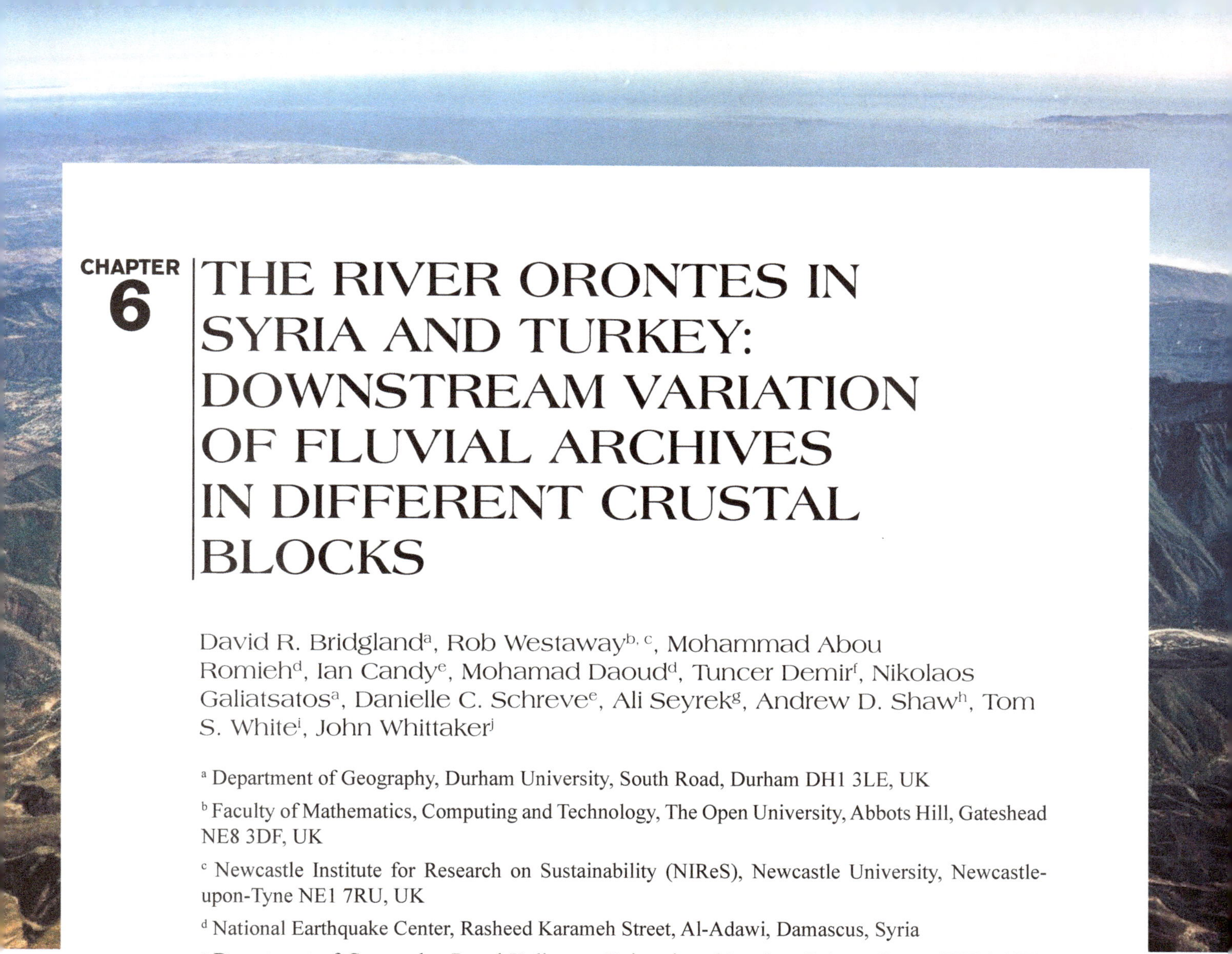

CHAPTER 6

THE RIVER ORONTES IN SYRIA AND TURKEY: DOWNSTREAM VARIATION OF FLUVIAL ARCHIVES IN DIFFERENT CRUSTAL BLOCKS

David R. Bridgland[a], Rob Westaway[b, c], Mohammad Abou Romieh[d], Ian Candy[e], Mohamad Daoud[d], Tuncer Demir[f], Nikolaos Galiatsatos[a], Danielle C. Schreve[e], Ali Seyrek[g], Andrew D. Shaw[h], Tom S. White[i], John Whittaker[j]

[a] Department of Geography, Durham University, South Road, Durham DH1 3LE, UK

[b] Faculty of Mathematics, Computing and Technology, The Open University, Abbots Hill, Gateshead NE8 3DF, UK

[c] Newcastle Institute for Research on Sustainability (NIReS), Newcastle University, Newcastle-upon-Tyne NE1 7RU, UK

[d] National Earthquake Center, Rasheed Karameh Street, Al-Adawi, Damascus, Syria

[e] Department of Geography, Royal Holloway, University of London, Egham, Surrey TW20 0EX, UK

[f] Department of Geography, Harran University, 63300 Şanlıurfa, Turkey

[g] Department of Soil Science, Harran University, 63300 Şanlıurfa, Turkey

[h] Department of Archaeology, University of Southampton, Highfield, Southampton SO17 1BJ, UK

[i] Department of Zoology, University of Cambridge, Downing Street, Cambridge CB2 3EJ, UK

[j] Department of Palaeontology, Natural History Museum, Cromwell Road, London SW7 5BD, UK

ABSTRACT

The geomorphology and Quaternary history of the River Orontes in western Syria and south-central Turkey have been studied using a combination of methods: field survey, differential GPS, satellite imagery, analysis of sediments to determine provenance, flow direction and fluvial environment, incorporation of evidence from fossils for both palaeoenvironments and biostratigraphy, uranium-series dating of calcrete cement, reconciliation of Palaeolithic archaeological contents, and uplift modelling based on terrace height distribution. The results underline the contrasting nature of different reaches of the Orontes, in part reflecting different crustal blocks, with different histories of landscape evolution. Upstream from Homs the Orontes has a system of calcreted terraces that form a staircase extending to ~200 m above the river. New U-series dating provides an age constraint within the lower part of the sequence that suggests underestimation of terrace ages in previous reviews. This upper valley is separated from another terraced reach, in the Middle Orontes, by a gorge cut through the Late Miocene–Early Pliocene Homs Basalt. The Middle Orontes terraces have long been recognized

Citation: David R. Bridglanda, Rob Westawayb, Mohammad Abou Romiehd, Ian Candye, Mohamad Daoudd, Tuncer Demirf, Nikolaos Galiatsatosa, Danielle C. Schrevee, Ali Seyrekg, Andrew D. Shawh, Tom S. Whitei, John Whittakerj, The River Orontes in Syria and Turkey: Downstream variation of fluvial archives in different crustal blocks, doi:10.1016/j.geomorph.2012.01.011

as a source of mammalian fossils and Palaeolithic artefacts, particularly from Latamneh, near the downstream end of the reach. This terraced section of the valley ends at a fault scarp, marking the edge of the subsiding Ghab Basin (a segment of the Dead Sea Fault Zone), which has been filled to a depth of ~ 1 km by dominantly lacustrine sediments of Pliocene–Quaternary age. Review of the fauna from Latamneh suggests that its age is 1.2–0.9 Ma, significantly older than previously supposed, and commensurate with less uplift in this reach than both the Upper and Lower Orontes. Two localities near the downstream end of the Ghab have provided molluscan and ostracod assemblages that record somewhat saline environments, perhaps caused by desiccation within the former lacustrine basin, although they include fluvial elements. The Ghab is separated from another subsiding and formerly lacustrine depocentre, the Amik Basin of Hatay Province, Turkey, by a second gorge, implicit of uplift, this time cut through Palaeogene limestone. The NE–SW oriented lowermost reach of the Orontes is again terraced, with a third and most dramatic gorge through the northern edge of the Ziyaret Dağı mountains, which are known to have experienced rapid uplift, probably again enhanced by movement on an active fault. Indeed, a conclusion of the research, in which these various reaches are compared, is that the crust in the Hatay region is significantly more dynamic than that further upstream, where uplift has been less rapid and less continuous.

INTRODUCTION

The Orontes ('Asi' in Arabic) is the principal river draining to the Levant coastline of the Mediterranean Sea. From its source in the Bekaa Valley of Lebanon, on the flank of the Lebanon mountain range, it flows northwards across western Syria through the cities of Homs and Hama and into Hatay Province, southern Turkey, before turning sharply south-westward to reach the sea ~ 30 km downstream of Antakya (Fig. 1). In north-west Syria the Orontes forms the axial drainage of the Ghab Basin, a linear valley marking the Dead Sea Fault Zone (DSFZ), the boundary between the African plate (to the west) and the Arabian plate (to the east), along which left-lateral relative plate motion is accommodated (Fig. 1). Upstream of the Ghab Basin, the terrace sequence of the Middle Orontes has been well documented, largely on account of attention from archaeologists interested in its Palaeolithic contents (e.g. Burkhalter, 1933, Modderman, 1964, Clark, 1966a, Clark, 1966b, Clark, 1966c, Clark, 1967, Clark, 1968, Van Liere, 1966, Besançon et al., 1978a,Besançon et al., 1978b, Besançon and Sanlaville, 1993a, Dodonov et al., 1993 and Bartl and al-Maqdissi,, 2005). Indeed, the work described here stemmed initially from an archaeological survey of the Homs region (Philip et al., 2005), which included ~ 700 km^{2}of the upper catchment of the Orontes. An extensive sequence of river terraces, previously unrecognized as such, was recorded in the upper Orontes valley as a result of this initiative (Bridgland et al., 2003 and Bridgland and Westaway, 2008a). In seeking to obtain a full understanding of the context for this newly discovered long-timescale fluvial record, research on the Orontes was extended downstream and has now been undertaken along the length of the valley between Homs and the Mediterranean, revealing marked contrasts between different reaches. Separating the upper and middle catchments, both of which have terraces, is a gorge, named after the town of Rastan, ~ 25 km north of Homs. Two further gorges occur: one between the Ghab Basin and the border with Hatay Province and the other downstream of Antakya. Between them is another former lacustrine basin (Fig. 1), occupied until the mid 20th Century by Lake Amik.

This paper relates how multi-disciplinary research has allowed the complex and contrasting records from different reaches of this unusual river course to be reconstructed and reconciled one with another. Methods have included field survey, recording and analyses of fluvial sediments, remote surveying of fluvial landforms (both depositional and erosional) and the use of Geographical Information Systems (GIS) techniques to obtain height data (of considerable value in areas remote from known-height markers such as bench marks). Also valuable has been the study of the fossil and artefact contents of the Orontes deposits, which have provided information on palaeo-environments and possible ages. In addition, the incision recorded by river terraces, which is interpreted as a response to uplift, can be modelled mathematically against time, using computer programs designed for the purpose (Westaway, 2002, Westaway, 2004a and Westaway et al., 2002). These take account of forcing mechanisms that affect the rate of uplift, which are considered to be driven by climatic fluctuation and linked to surface processes, with the aim of providing an age framework for the interpretation of terrace sequences (Bridgland and Westaway, 2008a, Bridgland and Westaway, 2008b and Westaway et al., 2009a). This approach has been applied previously to the upper Orontes terrace sequence (Bridgland et al., 2003 and Bridgland and Westaway, 2008a; see below).

The aim of the paper is to establish age-related stratigraphical frameworks for those reaches with accessible sedimentary evidence, in the form of river terraces, by pooling the available evidence and applying the most appropriate of the above-mentioned techniques. This will allow correlation and comparison between these reaches and also demonstrate the contrast with reaches in which terraces have not been developed. The importance of these findings is that they can be related to different histories of valley evolution in different reaches, corresponding with separate crustal blocks, for which the causes can be discussed in terms of crustal deformation.

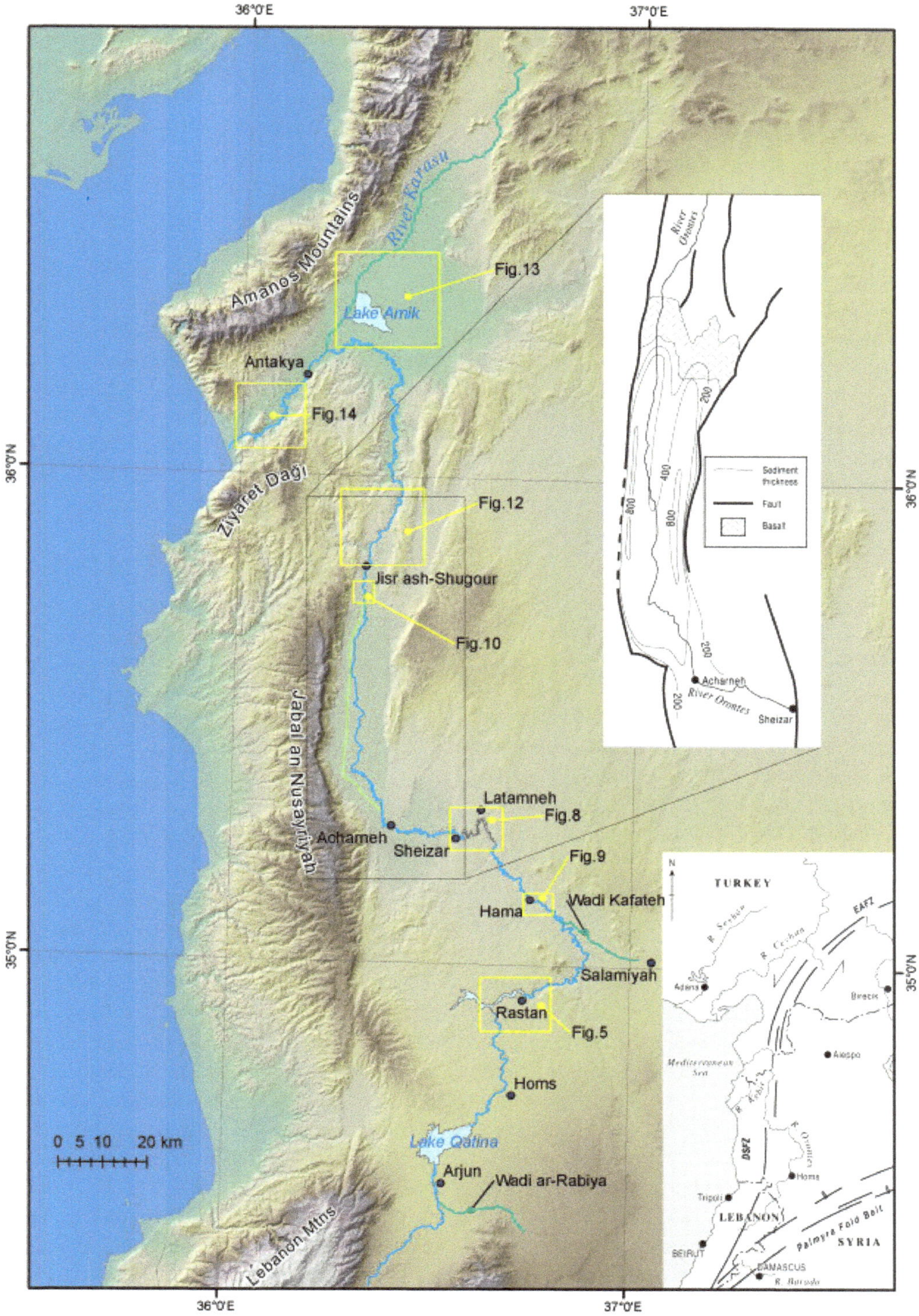

Figure 1: Course of the Orontes in relation to topography (main image, a DEM derived from SRTM) and structural setting (lower inset). Locations are shown of places described in the paper and of other figures. Abbreviations: DSFZ = Dead Sea Fault Zone; EAFZ = East Anatolian Fault Zone. The upper inset shows the fault control and sediment thickness of the Ghab Basin.

SURVEY METHODS

Orontes river terraces have been mapped previously between Rastan and the Ghab Basin, based on surveys by geologists and archaeologists in the 1970s and 1980s (Besançon et al., 1978a, Besançon et al., 1978b, Besançon and Sanlaville, 1993a and Dodonov et al., 1993). The resultant maps, although detailed, have been found to simplify the complexity of the terrace sequence and to be heavily dependent on geomorphology, with little attention paid to underlying fluvial sediment bodies. Nonetheless, these pre-existing maps represent an excellent resource and, with 'ground-truthing', are superior to anything yet produced in most other reaches, where the more recent surveys conducted by the authors are, by comparison, rudimentary. The only other reach of the Orontes that has hitherto been documented in comparable detail is that between Antakya and the coast, described by Erol (1963). Given the constraints of research visits of limited extent, the recent surveys have relied on a combination of different GIS resources, with field surveys designed to determine the terrace sequence at key points along the course. The locations of these were often determined by accessibility and permissions, although published sources of Palaeolithic and palaeontological evidence were targeted for investigation and re-survey, and searches for new data of these types were undertaken wherever possible.

New and supplementary mapping, including topographic information, has made use of differential global positioning system (dGPS) equipment, specifically Leica System 300, operated in static survey mode, with reference to temporary base stations on high points and to known heights such as bench marks. Optimal results were obtained when the roving station occupied each survey point for at least 100 two-second recording epochs and in cases where such points were not more than 50 km from the base station. Under favourable conditions, however, the technique has been shown to work satisfactorily with the base station and roving stations up to 100 km apart (cf. Demir et al., 2009 and Demir et al., 2012). At certain locations, with limited sky visibility, it was necessary to survey to a point in the open and calculate the distance and height difference, the latter making use of an Abney level. In the earlier surveys, generally those in the Upper Orontes, the dGPS data were processed using Leica SKI v2.3 software. Later survey data have been processed using Leica GeoOffice (version 4.1, 5.0 or 5.1), which incorporates an improved algorithm for eliminating phase ambiguities in its differential GPS solutions (in part because it uses a different approach for treating propagation delays of GPS signals through the ionosphere) and generally provides better data resolution. Where original dGPS raw output had been retained it proved possible to reprocess older data with this improved software but, unfortunately, since the provision of improved software had not been foreseen, much of the Upper Orontes survey output had been retained only in processed form.

GIS resources have included satellite imagery, with emphasis on CORONA high-resolution photographic images from 1960–1972, before the large-scale expansion of agriculture (Galiatsatos et al., 2008), and shuttle radar topographic mission (SRTM) altimetry. Since the 1995 declassification of CORONA (KH-4), it has been thoroughly studied and used in many applications, mostly related to change detection, and photo-interpretation (e.g. Galiatsatos, 2004, Galiatsatos, 2009, Sohn et al., 2004 and Dashora et al., 2007). The archaeological community has shown a keen interest in CORONA, particularly in areas where it is difficult to obtain detailed historical photography, such as the Near East. For the Homs Survey area a digital elevation model (DEM) is now available, obtained from CORONA and ground-truthed by dGPS. The research reported here has also used satellite imagery from the Fragile Crescent Project, a large-scale archaeological investigation of the Middle East (Galiatsatos et al., 2009). In addition to CORONA, the project took advantage of a GAMBIT image that was acquired on 25 April 1966 in the area of Hama. GAMBIT (or KH-7) was, like CORONA, a spy satellite program; it flew 38 missions from July 1963 to June 1967 and was declassified in 2002. It has a spatial resolution of 0.6–1.2 m but, as the relevant documentation remains classified, little is known about the camera system.

Of four versions of SRTM data, Version 1 is the 'raw' data, Version 2, used in the present study, is the result of editing for water bodies and the removal of spikes and wells, whereas Versions 3 and 4 were created and distributed by the CGIAR Consortium for Spatial Information and include improvements in the filling of voids. The SRTM data used (cf. JPL, 1998, JPL, 2005 and JPL (Jet Propulsion Laboratory), 1998) have a resolution of 3 arc seconds of latitude, or roughly 90 m, with a global vertical accuracy of better than 10 m, and horizontal accuracy of ~ 10 m, depending on the relief of the ground (Rodriguez et al., 2006). However, various applications from Hungary (Kay et al., 2005), Portugal (Gonçalves and Fernandes, 2005) and Turkey (Jacobsen, 2005, Westaway et al., 2006, Westaway et al., 2009b and Demir et al., 2012) have demonstrated a vertical accuracy of better than 5 m in a variety of terrain. Data of this type provide a particularly valuable source of height information for parts of the world where large-scale topographic maps are not readily available; they can be rendered as DEMs in a range of formats using standard GIS techniques. Throughout this study, satellite imagery and imagery generated from SRTM data are displayed using Universal Transverse Mercator (UTM) co-ordinates expressed using the WGS-84 datum; the same co-ordinate system is therefore used for reporting co-ordinates of field sites.

Supplementary Geological Techniques

A technique of assistance in terrace surveys is clast-lithological analysis of gravels. This has been particularly valuable for identifying Orontes deposits and distinguishing them from the products of local (tributary) rivers. Clast analysis of cemented gravels, which are common in certain reaches, was, of necessity, conducted in the field. A cardboard sieve plate was used to estimate clast size in order to count only pebbles between the desired sizes of 16 and 32 mm (a recommended standard size range for such analyses:Bridgland, 1986). This works well enough where gravel components, such as the limestones and flints/cherts that characterize the Orontes in Syria (Table 1), are readily distinguished in outcrop, so analyses can be carried out with the aid of a hand lens, a small knife to determine hardness, 10% HCl for confirmation of calcareous lithologies and a suitable pen to mark clasts as they are counted. In reaches where the Orontes gravels were less consolidated it was possible to collect gravel samples and identify loose clasts, although the difficulty and expense of transporting heavy samples meant that analyses were still conducted during fieldwork; with bagged samples this could take place during evenings or even during journeys between locations, optimizing field time for mapping and making records of sections. Downstream of Hama the Orontes gravels become completely dominated by flint (Table 1), there being a rich source of brown flint in the Upper Cretaceous chalk and chalky limestone, through which the river has incised (Clark, 1967), although flint has also been reported in Palaeogene limestones in the region (Ponikarov, 1986). This flint has provided the raw material for the Palaeolithic industries that are well represented in the gravels of this reach, and which first drew the attention of researchers to the Orontes system (e.g., Clark, 1966a, Clark, 1966b, Clark, 1967, Muhesen, 1985, Copeland and Hours, 1993, Shaw, 2008 and Shaw, in press). The relative monotony of the Middle Orontes gravels, however, means that their analysis is of less value here. Further downstream, in Hatay, the gravels consist largely of crystalline rocks from the local area, including the Hatay ophiolite (latest Cretaceous), supplemented by further travelled material from the Amanos Mountains to the north of Antakya, derived by way of the River Karasu, a right-bank Orontes tributary (Fig. 1;Table 1).

Table 1: Gravel clast lithology counts from deposits attributed to the River Orontes and related systems (in approximate downstream sequence)

Reach	Site	Coordinates	Chert and flint	Limestone	Voids	Quartzite	Basalt	Coarse basaltic	Ophiolite	Others	Total count
Upper Orontes	Al Hauz[a]	BU 73478 28799	53.5	36.0	10.5						275
	Arjun Quarry[b]	BU 74580 25832	33.6	66.4							256
	Al Hussainiyeh	BU 90236 27562	37.9	62.1							253
	Dmaynah[a]	BU 87074 28915	65.9	33.7	0.4						258
	Um al-Sakhr[a]	BU 86728 31399	28.1	56.6	15.2						256
	Tir Mala[a]	BU 90149 53013	60.4	38.5	1.1						273
Wadi ar-Rabiya	Motorway[c]	BU 93392 14769	11.6	85.2						3.2	250
	SE of Al Qusayr[d]	BU 80187 19083	12.9	87.1							255
Middle Orontes	Rastan[e]	BU 93451 66988	40.7	45.8						13.4	253
	Jnan[f]	CU 02201 85047	29.0	71.0							276
	Ain Kassarine[g]	CU 01128 88403	61.5	37.8			0.4			0.4	275
	Latamneh south[h]	BV 83384 09911	27.7	70.8			1.2			0.4	253
	Latamneh Lower[i]	BV 83022 10451	99.2	0.8							251
	Latamneh Upper[j]	BV 83021 10459	99.6			0.4					276
Ghab	Karkour sluice[k]	BV 57282 59073	96.6	3.4							268
Darkush	Gorge[l]	BV 63445 77174	0.7	99.3							149
Amik	Alaattin Köyü[m]	BA 52477 15506	9.8	73.9		0.7	10.1	3.1	0.4	1.8	287
Lower	Antakya bypass[n]	BA 41382 06943	0.4			1.1	29.8	51.6	14.4	2.9	285
Orontes	Şahin Tepesi[o]	BV 38313 99525	1.7	7.0		1.0	42.4	40.4	2.6	5.0	302

[a]Voids probably represent dissolved limestone clasts.

[b]See Fig. 4B.

[c]Others are calcareous sandstone (2.4%) and marl (0.8%).

[d]Chert includes one worked flake.

[e]Gravel also contains boulders of basalt, highly weathered; voids might represent smaller basalt clasts (weathered away) as well as dissolved limestones.

[f]Basalt clasts are conspicuous at larger sizes.

[g]Others = 1 x calcareous sandstone.

[h]Sample was collected from a disused small quarry between Latamneh village and the River Orontes, at a lower level than and ~ 600 m southeast of the extant Latamneh quarry; the flint includes two worked flakes and four banded cherts of Upper Orontes type; others = 1 weathered bone fragment.

[i]Sample was collected from the lower gravel in the extant' Latamneh quarry; flint includes three worked flakes.

[j]Sample was collected from the upper gravel in the extant Latamneh quarry; flint includes two worked flakes.

[k]Flint includes four cherts of Upper Orontes type.

[l]Undersized sample from valley-floor gravel beneath loam, collected at Hammam ash Sheykh Isa, between Jisr esh-Shugur and Darkush.

[m]Ophiolite includes a single clast of amphibolite; others includes two clasts of weathered schist and a single pottery fragment (?Roman or similar). The schist probably originated in the Amanos Mountains to the north, having been worked into the Amik Basin by the River Karasu. Other constituents of this gravel, such as limestone, quartziite and basalt, may have shared this provenance.

[n]Others are kaolinized crystalline rocks (2.5%) and a single vein quartz (0.4%).

[o]Sample collected from a flat overlooking the left bank of the Orontes apparently from a tributary deposit]; others are weathered crystalline, perhaps related to ophiolite; ophiolites include chert (0.7%), weathered foliated rock (0.7%), schistose with serpentine (1.0%) and olivine-rich ultramafic (0.4%).

UPPER ORONTES: LEBANON BORDER TO THE RASTAN GORGE

The open, low-relief landscape of the Upper Orontes (Fig. 2A), upstream of the Rastan Gorge, is floored by Neogene lacustrine marl of inferred 'Pontian' (latest Miocene) age (Dubertret and Vautrin, 1938). This marl is interbedded with the Late Miocene–Early Pliocene Homs basalt, for which recent re-dating using the Ar–Ar technique indicates an age range of ~ 6–4 Ma (Searle et al., 2010 and Westaway, 2011), superseding earlier whole-rock K–Ar dating that gave generally older (~ 8–5 Ma) numerical ages (Mouty et al., 1992, Sharkov et al., 1994 and Butler and Spencer, 1999). Boreholes east of Homs confirm that marl occurs both above and below the Homs Basalt (Ponikarov et al., 1963a;Fig. 3), suggesting that lacustrine conditions (possibly caused by the damming of proto-Orontes drainage by the earliest basalt eruptions) persisted into the Pliocene. It is also evident, from its interbedding with the marl and from pillow lava formation, seen at a number of localities to the north of Homs, that the Homs Basalt erupted into the 'Pontian' lake.

Figure 2: The Upper Orontes valley, field photographs: A — Typical exposure of cemented gravel in the land surface beside a rural road (at BU 75492 26442), showing the subdued relief of the Upper Orontes terrace staircase (looking northwards); this deposit is ~ 520 m a.s.l. and is interpreted as forming part of the Tir Mala terrace (Fig. 4; Table 2). B — Section at Arjun, showing lower cemented gravel filling a channel cut in (much softer) Pliocene marl bedrock, overlain by calcified fine-grained fluvial sediments, with post-cementation solution structures. The calcreted channel gravel illustrated here was sampled for U-series dating (see online Appendix 2).C — Cemented gravels of the lowest (al-Hauz) terrace of the Orontes, exposed in the right bank of the river at the type locality, ~ 25 km south-west of Homs.

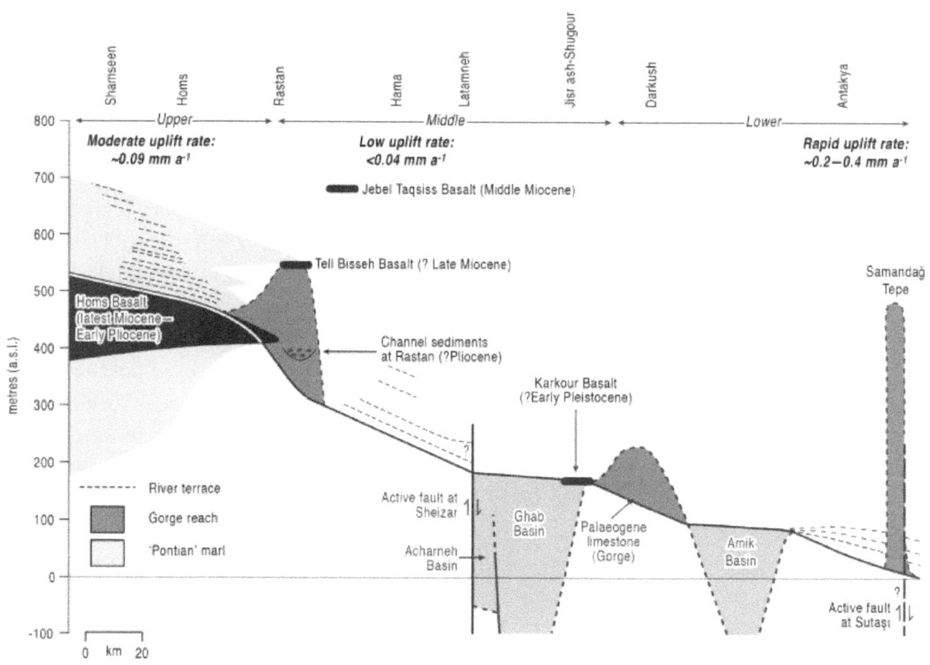

Figure 3: Schematic long profile of the Orontes valley floor and its terraces in relation to gorge reaches, subsiding fluvio-lacustrine basins and basalt flows. Summaries of pos-Early Pleistocene histories of the terraced reaches are also shown (for explanation, see text).

The Quaternary record of the Upper Orontes received little attention prior to the Homs Survey, although Van Liere (1961) alluded to Neogene–Pleistocene conglomeratic sediments that he attributed to braided fans; these can now be interpreted as (cemented) Quaternary river terrace deposits (Bridgland et al., 2003 and Bridgland and Westaway, 2008a; Fig. 2). The eastern valley side of the Orontes in this region slopes gently from > 750 m a.s.l. (above sea level) to river level at 480–510 m, over a distance of ~ 15 km, within which is preserved an extensive sequence of Late Cenozoic terraces (Fig. 4). In the marl areas these are represented by localized but conspicuous chert/flint-rich conglomerates, densely cemented by carbonate, of the type noted by Van Liere (1961); occasionally these record the three-dimensional form of fluvial channels (Fig. 2B), which suggests that they represent 'channel calcretes' (cf. Wright and Tucker, 1991, Nash and Smith, 1998, Nash and Smith, 2003 and McLaren, 2004). The conglomerates are rarely more than ~ 1 m thick, which has allowed farmers to displace them from field surfaces to boundary lines or informal clearance cairns. In situ conglomerates are seen in occasional quarry sections and road cuttings, or in surface outcrop amongst farmland where they have proved too difficult to remove; they are also seen frequently in the sides or beds of unpaved tracks, such outcrops having been the basis of much of the terrace mapping in the Upper Orontes (Bridgland et al., 2003). In exposure the conglomerates are seen to be interbedded with finer-grained alluvial sediments (Fig. 2B) in sequences showing fluvial bedding structures. The recognition of these conglomerates as Orontes terrace gravels is confirmed by the occurrence of the most extensively preserved conglomerate forming a low-level right-bank terrace (the al-Hauz Terrace), which can be traced for several kilometres beside the river upstream of Lake Qatina (Bridgland et al., 2003;Fig. 1 and Fig. 2C, Fig. 3 and Fig. 4). Sections in the cemented Orontes gravels show that their uppermost levels, especially where they are immediately below the land surface, have generally been decalcified, with prominent 'pipe' structures reaching depths of a metre or more (Fig. 2B; online supplement, Fig. A.1.1).

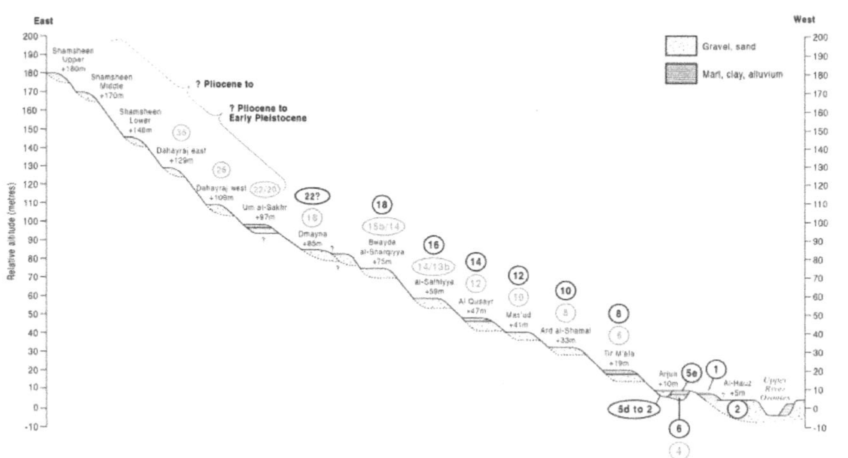

Figure 4: Idealized transverse section through the terrace sequence upstream from Homs (after Bridgland et al., 2003, with modifications). MIS attributions suggested by Bridgland et al. (2003) are shown, together with correlations of the older terraces now suggested (see text).

A single example of channel calcrete has been analysed in some detail, in part to determine the potential for dating the calcite cement using the uranium-series method (e.g. Ivanovich et al., 1992; see below). Samples of the cemented channel-fill gravel at Arjun (Fig. 2B) were subjected to petrological and geochemical analyses, which revealed evidence for two phases of calcite cement precipitation. The first, fine grained and brown in colour, was restricted to occasional and sometimes fragmentary pebble coatings, only intermittently present and lacking any preferred orientation. The second, in contrast, was pinkish grey and filled interstices within a clast-supported sandy, silty matrix, thus forming the bulk of the calcrete cement (see online Appendix 2). The fabric of the conglomerates forming the higher (and therefore older) terraces has been considerably modified by repeated decalcification and re-cementation; clasts (presumably calcareous ones) have been weathered out to leave cavities, sometimes partly filled with re-precipitated calcium carbonate. This means that the fluvial bedding structures are best preserved in the younger conglomerates, such as those forming the Arjun and al-Hauz terraces (Figs. 2C and 4).

Bridgland et al. (2003) reported Orontes terrace deposits at up to 130 m above the modern river (Table 2). They constructed an age model for this sequence, assuming climatically generated terrace formation in approximate synchrony with 100 ka (Milankovitch) climatic fluctuation (cf. Bridgland, 2000 and Bridgland and Westaway, 2008a) and using a correlation, based upon height above the modern river, with the sequence in the Middle Orontes, for which there is vertebrate biostratigraphical evidence (see below). Further consideration of the palaeontological evidence, however, indicates that the age model used in 2003 was a significant underestimate (see below). Subsequent attempted U-series dating of the Arjun channel calcrete (see above) has provided an adjusted age pinning point for the Upper Orontes terrace staircase, albeit rather low in the sequence. Different U-series age estimates were obtained for the brown pebble coating s and the pinkish-grey interstitial cement: > 350,000 years for the former and 157,000 ± 31,000 years for the latter (see online Appendix 2). The pebble coatings are interpreted as reworked older cement that was already present on clasts derived from older gravels within the sequence. It is considered, therefore, that the age of the interstitial cement is more representative of the channel gravel. It should be noted that channel calcretes form in active (semi-arid) fluvial environments, by infiltration of calcareous water into recently deposited underlying sediment, so there is every reason to believe that the cement was precipitated during the same geological episode as the gravel (cf. Nash and McLaren, 2003 and McLaren, 2004).

Table 2: Upper Orontes terrace stratotypes and other key localities, numbered sequentially. To obtain accurate positioning, co-ordinates of sites were measured in the field using a portable GPS receiver, and are expressed as 8- or 10-digit grid references using the Universal Transverse Mercator (UTM) system. Height information has been obtained from dGPS and/or SRTM topographic imagery

Terrace	Type locality (UTM coordinates)	Above river	Height a.s.l.	MIS 2003[1]	Revised MIS
al-Hauz	Right bank of the Orontes (BU 7345 2783)	5 m	506 m	2	2
Arjun	Quarry (BU 7458 2573)	10 m	512 m	5 d–2	6–2
Tir M'ala	Bluff exposure (BU 9023 5296)	19 m	471 m	6	8
Ard al-Shamal	Surface exposure (BU 7762 2598)	33 m	534 m	8	10
Mas'ud	Surface exposure (BU 7877 2570)	41 m	542 m	10	12
Al Qusayr	Surface exposure (BU 7918 2288)	47 m	551 m	12	14
al-Salhi-yya	Surface exposure (BU 8209 2876)	59 m	552 m	14 or 13b	16
Bwayda al-Sharqi-yya	Surface exposure (BU 8482 3052)	75 m	567 m	16 or 15b	18
Dmayna	Bluff exposure (BU 8512 2897)	85 m	580 m	18	?22
Um al-Sakhr	Quarry (BU 8673 3142)	97 m	584 m	22 or 20	–
Dahayraj West	Surface exposure (BU 8537 2249)	109 m	613 m	26	–
Dahayraj East	Surface exposure (BU 8846 2614)	129 m	625 m	36	–
Shamseen Lower	Surface exposure (BU 91260 28896)	~ 140 m	637 m	–	–
Shamseen Middle	Surface exposure (BU 92897 22448)	~ 170 m	689 m	–	–
Shamseen Upper	Surface exposure (BU 93026 21473)	~ 180 m	698 m	–	–

Notes: 1 — MIS correlations proposed by Bridgland et al. (2003). Information from more recent research suggests that ages have been underestimated (see revised MIS column).

The basal channel gravel at Arjun is thus attributed to Marine oxygen Isotope Stage (MIS) 6, rather than the MIS 4 age favoured by Bridgland et al. (2003). A comparison of the earlier 2003 and revised age models is provided in Fig. 4. An important implication of the revision is that uplift has been significantly less rapid than previously supposed: 85 m since MIS 22, instead of 97 m (see Fig. 4).

Fieldwork in 2002–3, reported here for the first time, has revealed that the Upper Orontes terrace staircase extends higher than was reported by Bridgland et al. (2003), the highest gravel remnants being found in cuttings along the Homs–Damascus motorway as it crosses the interfluve between the Orontes and its prominent tributary, the Wadi ar-Rabiya (Fig. 1). The oldest gravels occur in the vicinity of Shamseen (at BU 925240), the highest being south of the village (BU 2930 3821), ~ 700 m a.s.l. and ~ 180 m above the level of the modern Orontes (Fig. 4). Mapping of in situ fluvial conglomerates has revealed at least fifteen Upper Orontes terraces (Fig. 3 and Fig. 4), although the wider vertical gaps between the higher terrace remnants suggest that others await discovery or have been removed by erosion.

The Upper Orontes gravels are characterized by two main rudaceous components: (1) flint and/or chert, highly variable in character, and (2) limestone. The siliceous rocks are believed to have been derived both from the valley sides (from Cretaceous and Palaeogene flint-bearing strata) and from upstream in the Orontes catchment, whereas the limestone probably represents Cretaceous–Miocene occurrences in the wider region (e.g. Ponikarov et al., 1963a). The limestone component varies from approximately one third of the (16–32 mm) total to nearly three-quarters (counting voids as limestone clasts removed by solution: see above and Table 1). Thus the siliceous lithologies never fall below 25% of the total count in those Orontes gravels analysed (Table 1). Substantial gravels have also been produced by the Wadi ar-Rabiya (Fig. 1); calcreted gravels, presumably from the last climate cycle, are well exposed in a wadi-floor quarry (BU 85311 19993). These wadi gravels, however, are invariably dominated by limestone clasts, although up to 14% flint/chert occurs in them (Table 1), presumably reworked from the older Orontes terraces, across which the tributary has flowed. Thus, clast analyses can be used as a means of distinguishing the deposits of the main river, with their larger siliceous component.

THE RASTAN GORGE

The gorge at Rastan reflects the lateral constriction of the river in its passage through the relatively resistant Homs Basalt. The puzzling disposition of the gorge close to the eastern margin of the basalt outcrop (Fig. 5) suggests that the course of the Orontes here has been superimposed from a valley originally developed in less resistant overlying marl. Indeed, the basalt is known from borehole data (Ponikarov et al., 1963a) to have an eastward inclination, which suggests that its exhumation from beneath overlying marl will have caused the cessation of the westward migration of the river exemplified by the distribution of terraces further upstream. Geological mapping (Ponikarov et al., 1963a) indicates that a flow unit of Homs Basalt reached the Rastan area from the main outcrop west of the Orontes valley. Its main outcrop, on the north side of the valley, reaches a location ~ 2 km north-east of Rastan (~ BU 940700), where its upper surface is ~ 420 m a.s.l., dying out just east of the point where the gorge is crossed by the Damascus–Aleppo motorway viaduct. A smaller basalt outcrop south of the river, also depicted on the 1963 geological map, reaches ~ 1 km west of Rastan (~ BU 920680), where it is ~ 400 m a.s.l. This outcrop is both overlain and underlain by the 'Pontian' marl; however, the overlying marl is locally no more than a few tens of metres thick, although its above-basalt thickness increases southwards to ~ 100 m in the Homs area (Fig. 3), according to borehole data compiled by Ponikarov et al. (1963a). The larger outcrop on the north side of the valley is underlain by the marl, but any overlying marl has been lost to erosion. The Orontes is locally ~ 320 m a.s.l.; the ~ 100 m depth of the Rastan Gorge below the top of the basalt and overlying marl, determined from GIS data (Fig. 5), thus provides a measure for fluvial incision in the ~ 4–5 million years since the basalt eruption and cessation of marl deposition. Fluvial deposits reported by Bridgland et al. (2003) in the southern approach cutting to the Rastan motorway viaduct (online supplement, Fig. A.1.2) are at roughly the same height as the basalt south of the river and thus (rather than the early Middle Pleistocene age previously suggested) probably indicate the Early Pliocene level of the Orontes.

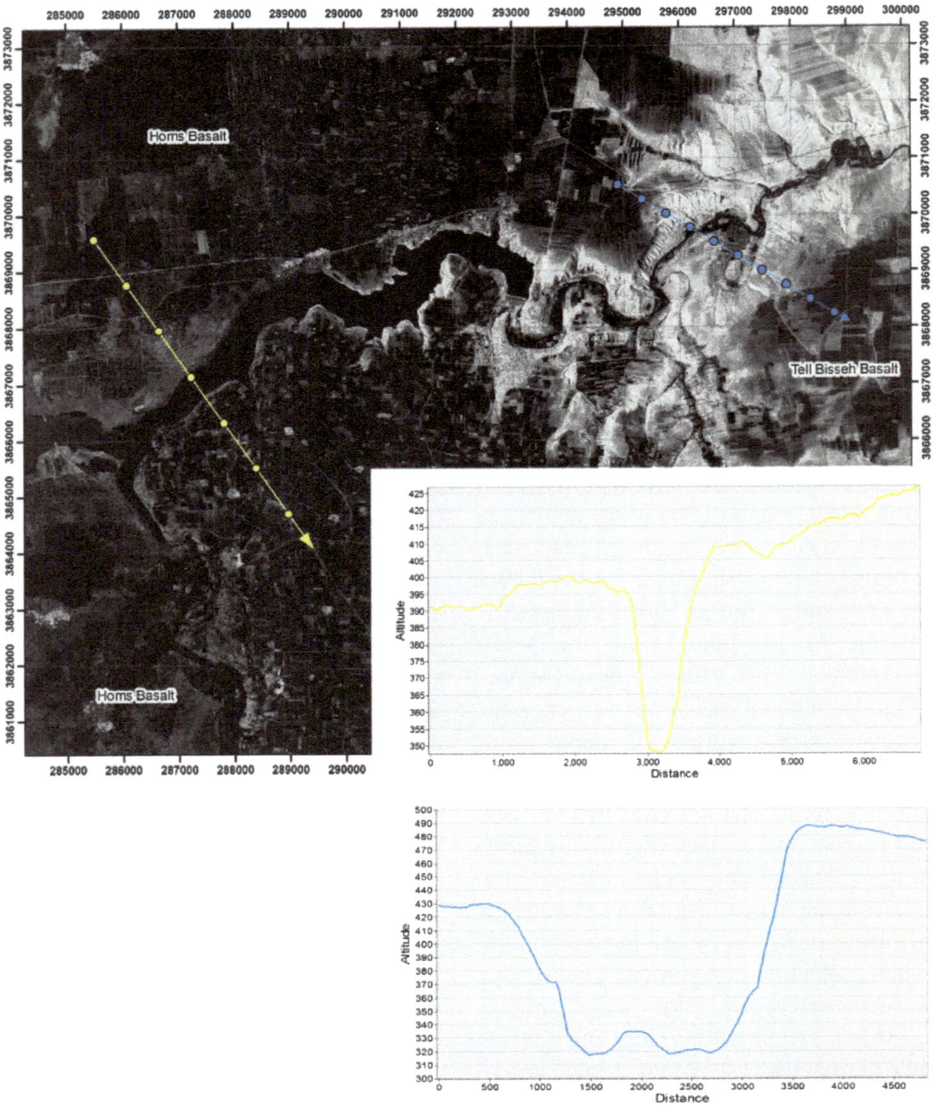

Figure 5: The Rastan Gorge. This is a CORONA image from the 1960s, with derived transverse profiles (colour coded). Positions of the two basalts, of different age, are indicated (see text). The westernmost (upper) cross section is located on the Homs Basalt on both sides of the river, whereas the easternmost (lower) cross section shows the higher Tell Bisseh Plateau on the right bank of the river.

On the southern side of the river, ~ 4–12 km downstream of Rastan, an older basalt has been exhumed from beneath the marl to form the capping of the Tell Bisseh Plateau (Fig. 3 and Fig. 5), up to 550 m a.s.l.; it is mapped as Upper Miocene (Ponikarov et al., 1963a), although it has not been dated directly. Furthermore, higher-level basalts further downstream, in the Middle Orontes, have yielded Middle Miocene ages (Sharkov et al., 1994; see below).

MIDDLE ORONTES: RASTAN TO THE GHAB BASIN

For much of its length, the Middle Orontes forms a deep valley up to 400 m below a succession of flat-topped hills capped with basalt mapped as Upper Miocene (comparable with that capping the Tell Bisseh Plateau, noted above: Ponikarov, 1986); the highest of these, Jebel Abou Dardeh and Jebel Taqsiss, reach 682 and 685 m a.s.l., according to Besançon and Sanlaville (1993a). The river loops east of Jebel Taqsiss, north of Rastan, and then turns northwards, flowing to the west of other mesas in the area north and east of Hama (see Fig. 6A and B). Sharkov et al. (1994) obtained whole-rock K–Ar dates of 17.3 ± 0.6 and 12.8 ± 0.6 Ma for basalt samples from Jebel Taqsiss, implying eruption in the Middle rather than the Late Miocene, as well as 12.0 ± 0.5, 10.8 ± 0.3, and 7.8 ± 0.3 Ma for basalt samples from the area north-east of Hama. It is now apparent, however, that K–Ar dates of this whole-rock type frequently result in numerical ages that are significantly older than the true age of the volcanism, as a result of inherited argon in phenocrysts (Kelley, 2002); this problem, alluded to above in relation to the Homs Basalt (Westaway, 2011), raises doubts about the accuracy of the above ages. Nonetheless, it is apparent, from its relation with the Pontian lacustrine marl, that the Tell Bisseh Plateau Basalt is older than the latest Miocene–Early Pliocene Homs Basalt. It is also clear that the basalts capping Jebel Taqsiss and the mesas north and east of Hama (also mapped as Upper Miocene) belong to the older group, although determining whether they represent a single eruptive phase must await future dating.Bridgland et al. (2003) were thus incorrect in their correlation of the Jebel Taqsiss and Homs basalts, with the inference that the whole ~ 400 m of entrenchment of the Middle Orontes had developed since the Early Pleistocene. On the contrary, it is now clear that the Middle Orontes valley is of much greater antiquity, apparently dating back at least to the Middle Miocene, with the implication that fluvial incision has been correspondingly slower than was previously supposed. The basalt-capped mesas of the Middle Orontes valley are formed in Palaeocene–Eocene nummulitic limestones, lying to the north of the latest Miocene lacustrine basin (Van Liere, 1961).

Figure 6: Terraces of the Middle Orontes: A — View looking downstream from CU 02087 83210, a location on the inside of a meander loop on the left (west) side of the Orontes, ~ 10 km south-east of Hama. Landforms on this side of the valley hereabouts were mapped by the French as part of their numbered 'glacis' and terrace system but clearly developed in bedrock, rather than being a depositional features (see text). On the right skyline is a basalt-capped mesa (a volcano), which is ~ 10 km distant, with its summit at CU 034927. B — Surface exposure in cemented gravel at CU 09511 78980, ~ 410 m a.s.l. (> 100 m above modern river level). The view is northward, looking across the valley of an Orontes tributary, the Wadi Kafat, towards the basalt-capped plateau in the distance, from a point ~ 3 km east of the modern river. C — Limestone fan gravels at the Khattab 2 locality (see also online supplement, Fig. A.1.4; see text for explanation). The gravels are exposed in the bluffs beneath the building on the skyline.

The fluvial sediments of the Middle Orontes have been studied since the early 1930s (Burkhalter, 1933), although a major catalyst for research in this area was the discovery of an extensive mammalian fossil assemblage in gravel quarries ~ 1.5 km south of the village of Latamneh (Van Liere, 1960, Hooijer, 1961 and Hooijer, 1965; Fig. 1; see below). This led to a series of excavations in which unabraded Lower Palaeolithic artefacts were recovered from fine-grained fluvial deposits overlying the gravels containing the vertebrate fauna (e.g., Modderman, 1964, Clark, 1967 and Clark, 1968). Latamneh is the single major source of fossils in the Middle Orontes and provided the vertebrate biostratigraphical evidence used by Bridgland et al. (2003), by extrapolation upstream, as an age indicator for their Upper Orontes sequence (see above). In addition, a tooth of the ancestral mammoth *Mammuthus meridionalis* was reported from a gravel quarry at Sharia, east of Hama (Van Liere and Hooijer, 1961), in an area subsequently built over during the expansion of the city. This contrasts with the teeth of the more evolved mammoth *Mammuthus trogontherii* from Latamneh (e.g. Van Liere, 1960, Hooijer, 1961 and Hooijer, 1965), ~ 25 km downstream of Hama (Fig. 1).

During the late 1970s and early ‹80s a team from the French Centre National de la Réchèrche Scientifique (CNRS) instigated a survey of Pleistocene deposits in the Middle Orontes valley in order to place the discoveries from Latamneh within a local chronostratigraphical sequence (Besançon et al., 1978a, Besançon et al., 1978b,Besançon and Sanlaville, 1993a and Copeland and Hours, 1993). This led to the discovery of Lower and Middle Palaeolithic artefacts at a number of localities and the identification of a sequence, above the valley-floor alluvium, of up to five terraces, as follows (from Besançon and Sanlaville, 1993a):

Qf0	Floodplain and Holocene alluvium of the valley bottom
QfI	Lowest Pleistocene terrace; ~10 m above river
Qf II	Second Pleistocene terrace; ~25 m above river
QfIII	Third Pleistocene terrace; includes Latamneh; 30–60 m above river
QfIV	Fourth Pleistocene terrace; ~80 m above river
QfV	Highest Pleistocene terrace

The CNRS team mapped these five terraces along the Orontes from the Rastan Gorge to the southern end of the Ghab Basin, with preservation distributed on both sides of the valley, although QfV was identified as an erosion surface that was devoid of fluvial sediments and QfIV was mapped as a 'glacis' with occasional (poorly documented) traces of fluvial conglomerate. Fieldwork during 2007 and 2009 has shown the CNRS feature mapping to be generally sound but, whereas the right-bank terraces are generally formed from fluvial gravels and floodplain silts, those on the left bank (including those below QfIV) are often 'glacis' type features formed in bedrock or slope deposits and incorporating valley-side fans and colluvial slope aprons (Fig. 6.A). It was also found that the terrace sequence in this reach extends higher above the river on the eastern side of the valley than had been realised previously, with a series of cemented gravels capping hills south–east of Hama and west of Salamiyeh (Fig. 1 and Fig. 6B and 7). The highest level at which ancient Orontes sediments have been observed hereabouts is alongside its right-bank tributary, the Wadi Kafateh, which joins the main river some 8 km upstream of Hama (online supplement, Fig. A.1.3). These cemented gravels, which crop out widely hereabouts, can be confirmed as Orontes deposits from their flint content; in every respect, including their disposition within the landscape, they resemble the high-level conglomerates of the reach above Homs. There are several facets underlain by gravel, the highest reaching 410 m a.s.l. (Fig. 6B), with lower-level 'flats' down to a prominent level at ~ 370 m a.s.l., well developed around CU 08825 81661. The sequence thus ranges between 80 and 120 m above the modern level of the river (Fig. 7). Whether these are erosional facets or whether they mark 'cut-and-fill' events cannot be determined.

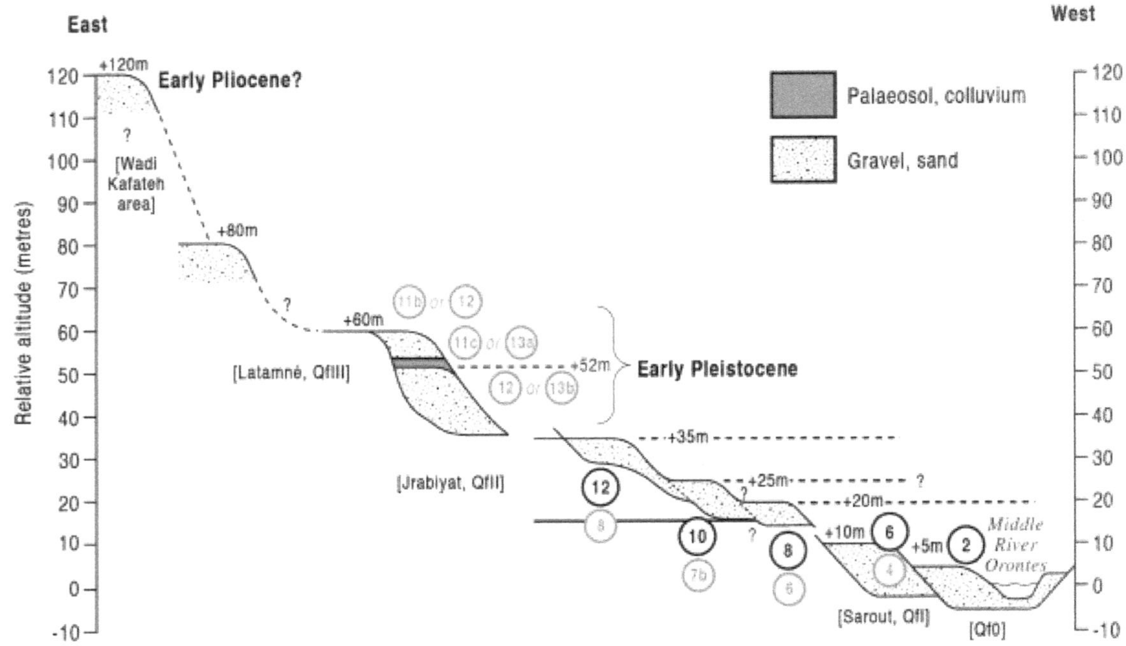

Figure 7: Idealized transverse section through the terrace sequence of the Hama–Latamneh area, showing the MIS correlation suggested by Bridgland et al. (2003), now thought to be underestimated, and the greater ages suggested in this paper (see text). Uses data from Besançon and Sanlaville (1993a) and Dodonov et al. (1993); artwork modified from Bridgland et al. (2003).

The height of these cemented gravels, in comparison with the fossiliferous deposits at Latamneh (see below), helps confirm the great antiquity of Orontes drainage in north-western Syria and their disposition implies westward migration of the river during the Late Cenozoic, helping to explain the absence of fluvial terrace deposits on the left bank of the river. On the basis of height above the river, bearing in mind the disposition of the fluvial deposits at Rastan (see above), an Early Pliocene age is tentatively estimated for these high-level deposits of the Middle Orontes.

In contrast to these new discoveries, the single locality at which the French workers recorded substantial gravels beneath their QfIV terrace, Khattab 2 (regarded by them as the type locality of the Khattabian Palaeolithic Industry: Copeland and Hours, 1993; location: BU 88795 96553), was visited in 2007 and found to expose a cemented limestone gravel of presumed local, perhaps alluvial-fan origin. It comprises mainly rounded limestone pebbles, although with large angular flints that have clearly not been subjected to significant fluvial transport. Thus, despite occurring in the flanks of a steeply incised reach of the Orontes, this deposit must be rejected at a product of that river, since all Orontes gravels in the Middle reach are strongly dominated by subangular siliceous clasts. Indeed, the location of the cemented gravel adjoins a confluence with a tributary wadi (cf. Besançon and Sanlaville, 1993a, their Fig. 7A), which is the possible source of the material. This interpretation is supported by the poorly stratified nature of the deposit, more akin to fan gravel than a fluvial channel facies (Fig. 6C; online supplement, Fig. A.1.4). Besançon and Sanlaville (1993a) regarded the Khattab gravel as older than the QfIII deposits encountered at Latamneh and elsewhere, largely because the former is well-cemented, whereas the QfIII deposits of the Middle Orontes are not; its height, no more than ~ 30 m above the modern river, is not a basis for considering it as ancient. The cemented nature of the deposit, however, can probably be attributed to its limestone clast composition, since evidence from other reaches of the Orontes shows that in calcareous groundwater areas gravels can be cemented rapidly, as in the lowest terraces of the valley upstream from Homs (see above).

Copeland and Hours (1993) applied the name Khattabian to a series of artefact assemblages lacking handaxes that they considered older than those from handaxe-bearing QfIII terrace deposits such as at Latamneh. This material was reportedly from pockets of fluvial conglomerate on the QfIV terrace glacis, some of which were at much greater heights than the supposed type locality at Khattab. Recent re-examination of the artefacts from the type locality at Khattab (Shaw, 2008 and Shaw, in press) failed to identify any pieces unequivocally of human manufacture, nor indeed were any definite artefacts identified amongst the small collections from the five other 'Khattabian' findspots in the Orontes (Abu Obeida, Mahardeh 2, Ard Habibeh, el-Farcheh 1 and Khor el-Aassi). Not only is the age attribution of the Khattab 2 type locality suspect, therefore, but the status of a separate and earlier non-handaxe industry is also open to question, particularly since handaxes are known from the site at Ubeidiya, further south in the Levant, in deposits dating from the mid Early Pleistocene, ~ 1.4 Ma (Tchernov,

1987 and Tchernov, 1999). The only one of these 'Khattabian' sites that is both at a relatively high level and, from the brief descriptions by Besançon and Sanlaville (1993a)and Copeland and Hours (1993), clearly in deposits of the Orontes, is el-Farcheh (~ CU 045 850); at ~ 340 m a.s.l. it is ~ 55 m above the river. Being significantly lower, this is evidently younger than the gravels described above in the vicinity of the Wadi Kafateh; the deposits at el-Farcheh can be tentatively ascribed to the latest Pliocene–earliest Pleistocene.

The aforementioned Latamneh locality in fact constitutes a number of separate sites (Fig. 8) that were worked for gravel during the latter half of the 20th and into the present century, details of which have been reviewed recently by Shaw, 2008 and Shaw, in press. Remote sensing data from different dates have now been used to determine the disposition of the sediments. In summary of these various findings, it has been concluded that the large extant quarry (online supplement, Fig. A.1.5) studied by the present authors is an expanded version of what was called Latamneh 2 by Copeland and Hours (1993); it is, however, a later working than was available in the 1960s, when most of the collections were made (Clark, 1966a, Clark, 1966b, Van Liere, 1966 and de Heinzelin, 1968) at localities 1–1.5 km to the south and east (Fig. 8). Much of the recorded mammalian material came from Latamneh Quarry 1 (Hooijer, 1961 and Van Liere, 1966) and from two sondages (A and B) to the south. There were also finds from the working that Van Liere (1966) described as Quarry 2 (not the extant quarry) and from excavations (the 'Living Floor excavations') to the east of that quarry (Hooijer, 1965, Clark, 1966a,Clark, 1966b, Clark, 1967 and Clark, 1968; see Fig. 8C), although they represent a small proportion of the total. The thick gravel sequence exposed in the extant quarry, mapped as QfIII (see above), is disposed between ~ 260 m and ~ 280 m a.s.l., the upper level being ~ 55 m above the level of the modern River Orontes, with a further ~ 10 m of silt above the quarried levels. There is nothing to contradict the view that all the various sites have exposed the same set of mammal-bearing fluvial deposits, these being the 'lower gravels' within the thick aggradational sequence here (cf. Shaw, in press).

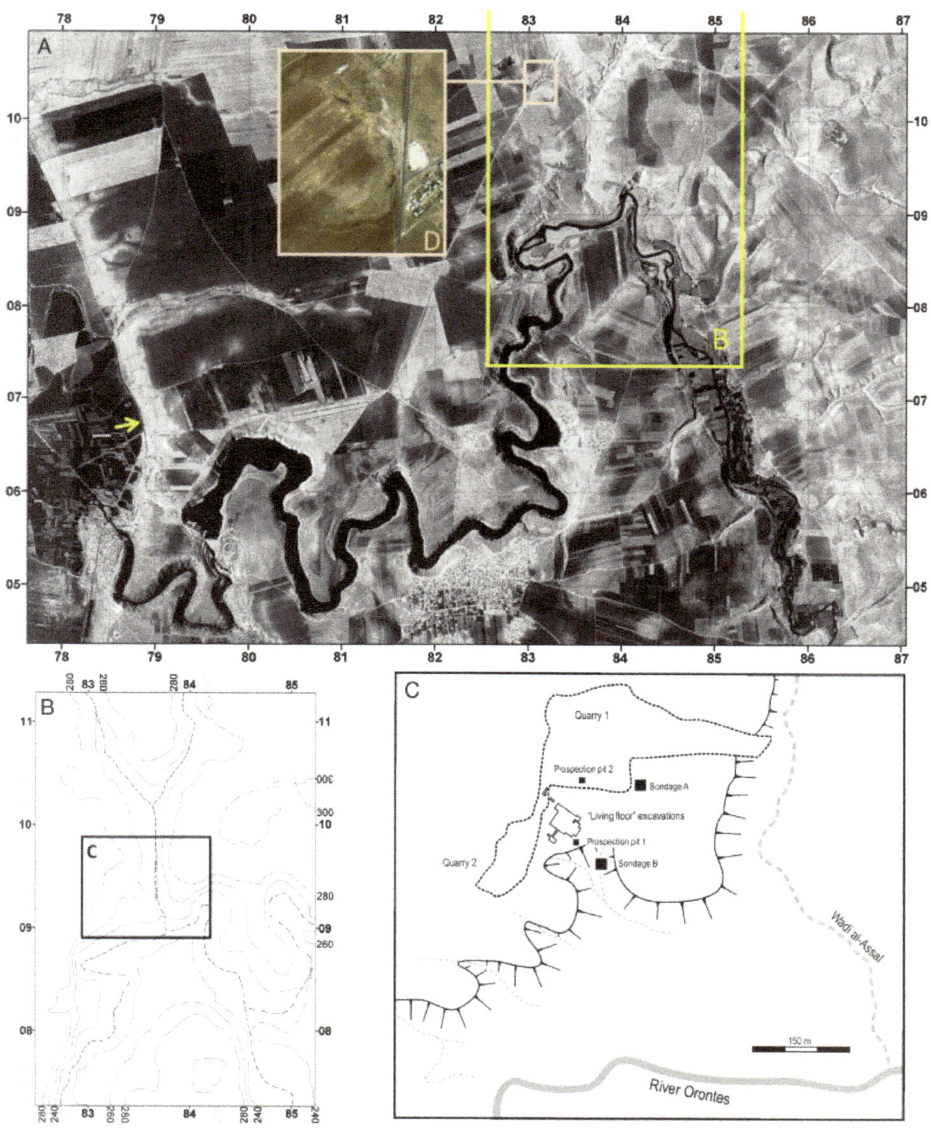

Figure 8: The important archaeological and fossiliferous locality at Latamneh: A — CORONA image from the 1960s of the area between Latamneh and the Sheizar fault scarp, the latter being indicated with an arrow (top left); the southern part of Latamneh village is at the northern edge of this image, within the area of B. B — Contour map (5 m interval) derived from Shuttle Radar Topographic Mission (SRTM). C — Location of the earlier palaeontological and archaeological localities at Latamneh (from Shaw, in press).D — Inset showing a modern Google Earth view of the extant quarry at Latamneh, to the north of the earlier localities depicted in C.

Biostratigraphical Evidence from the Middle Orontes

The mammalian remains from Latamneh include *Crocuta crocuta*, *Hippopotamus* cf.*behemoth*, *Camelus* sp., *Giraffa camelopardalis*, *Praemegaceros verticornis*, *Bos primigenius*, *Bison priscus*, Bovidae 'de type antilope', gen. et sp. indet., cf. *Pontoceros*(?), *Equus* cf. *altidens*, *Stephanorhinus hemitoechus*, *Mammuthus trogontherii* and*Stegodon* cf. *trigonocephalus* (Guérin and Faure, 1988 and Guérin et al., 1993). During the 2007 field season further vertebrate fossils were obtained from exposures in the large extant Latamneh quarry (see online supplement, Figure A.1.6), as follows:

1. Indeterminate bone fragment, white–pale yellow preservation, heavily weathered
2. Distal diaphysis of right humerus of fallow deer *Dama* sp. (probably *mesopotamica*)
3. Molar fragment (very crushed) of cf. *Stegodon* sp., encrusted with gravel
4. Left upper molar fragment of *Equus* sp.

These new discoveries contribute little to the list reported by Guérin et al. (1993); *Dama*cf. *mesopotamica* was recorded previously, along with *Camelus* cf. *dromedarius* and a molar tentatively attributed to *Gazella soemmering*, from the 'Living floor' excavations (Hooijer, 1965; Fig. 8C).

The Latamneh assemblage is of Lower–Middle Pleistocene affinity and combines mammoth and giant deer species that are unknown in Europe after the Elsterian (excluding the dwarf form of *M. trogontherii* that characterizes MIS 7 faunas: Lister and Sher, 2001) with a rhinoceros (*S. hemitoechus*) that first appears in Europe immediately after that glacial, in the Holsteinian. On that basis Bridgland et al. (2003) suggested correlation with MIS 13 or 11 (perhaps a MIS 12 refugial population would be equally likely), considering this to be an important pinning point for Terrace QfIII within the Orontes sequence and for their uplift–incision modelling. There are, however, reasons to doubt that interpretation, which owed much to the occurrence of *S. hemitoechus*. In a recent review of Early and Middle Pleistocene faunas and archaeological evidence in the wider region, Bar-Yosef and Belmaker (2010) have grouped Latamneh, despite the occurrence of *S.* cf. *hemitoechus*, with sites thought to represent the interval 1.2–1.0 Ma. In this they were influenced by the small-mammal assemblage, in which Mein and Besançon (1993) reported the arvicolid *Lagurodon arankae*, which is not thought to have survived into the Middle Pleistocene. Mein and Besançon (1993) noted that all the small-mammal genera at Latamneh are represented at Ubeidiya, but by different species or more primitive morphotypes. The most abundant species at Latamneh is a gerbil,*Meriones maghrebianus*, which is defined on the basis of North African material. The species recorded as *Arvicola jordanica* (now attributed to a different genus: *Tibericola jordanica* Haas 1966) is also present at Ubeidiya (Koenigswald et al., 1992) and Gesher Benot Yaaqov (Goren-Inbar et al., 2000), which, respectively, pre-date and post-date the supposed age of Latamneh (Bar-Yosef and Belmaker, 2010). *L. arankae* is indeed the most significant element, biostratigraphically; it first appeared in late Early Pleistocene Biharian faunas in eastern Europe but had disappeared by the early Middle Pleistocene (Markova, 2007), one of its last appearances being in the Karapürçek Formation in Turkey (Ünay et al., 1995, Ünay et al., 2001 and Demir et al., 2004), which is dated to ~ 0.9 Ma.

The location of the earlier discovery of a mandible, with teeth, of *M. meridionalis* at Sharia, near Hama (see above), also coincides with deposits mapped as QfIII by the CNRS workers, although in the original description Van Liere and Hooijer (1961)regarded it as older. This would seem to relate, at least in part, to the fact that Van Liere (1966) underestimated the height of the deposits at Latamneh. Artefacts were also discovered but the assemblage is too small for the absence of evidence for handaxe making to be meaningful. This leaves a conundrum in that ancestral mammoths of different types, and presumed to be at different evolutionary stages, are recorded in deposits that have been attributed to the same Orontes terrace. In an attempt to resolve this problem, the photographs of mammoth teeth provided for Sharia by Van Liere and Hooijer (1961) and for Latamneh by Hooijer (1965) were re-examined. This confirmed that the single R m3 tooth from Latamneh figured by Hooijer (1965) is consistent with an identification of *Mammuthus trogontherii*. Sixteen plates are apparent on the image but there has been a limited amount of wear at the front of the tooth so this should be regarded as a minimum count. Early Pleistocene *M. meridionalis* is characterized by low hypsodonty and a mean of 10–14 plates in the third molar, whereas *M. trogontherii* is more hypsodont, with between 16 and 22 plates (Lister and Sher, 2001). The published images of the Hama tooth allow confirmation of its identification as *M. meridionalis*, implying a greater age than the Latamneh fauna.

Consultation of SRTM imagery (Fig. 9) has confirmed that the deposits at Sharia (≤ 10 m thick) are aggraded to ~ 40 m above the modern river, as originally suggested by Van Liere and Hooijer (1961), placing them within the height range of the sequence at Latamneh and making the attribution of both sites to QfIII seem entirely reasonable. Possible differences in uplift history between the two localities cannot be ruled out however, which could explain the apparently older mammalian remains at Sharia. Indeed, Van Liere (1966) suggested that an older set of deposits, equivalent to those at Sharia, occurred at Latamneh and that the fossiliferous and handaxe-bearing sediments had subsequently been incised into them, although no later descriptions have confirmed a sequence of this sort.

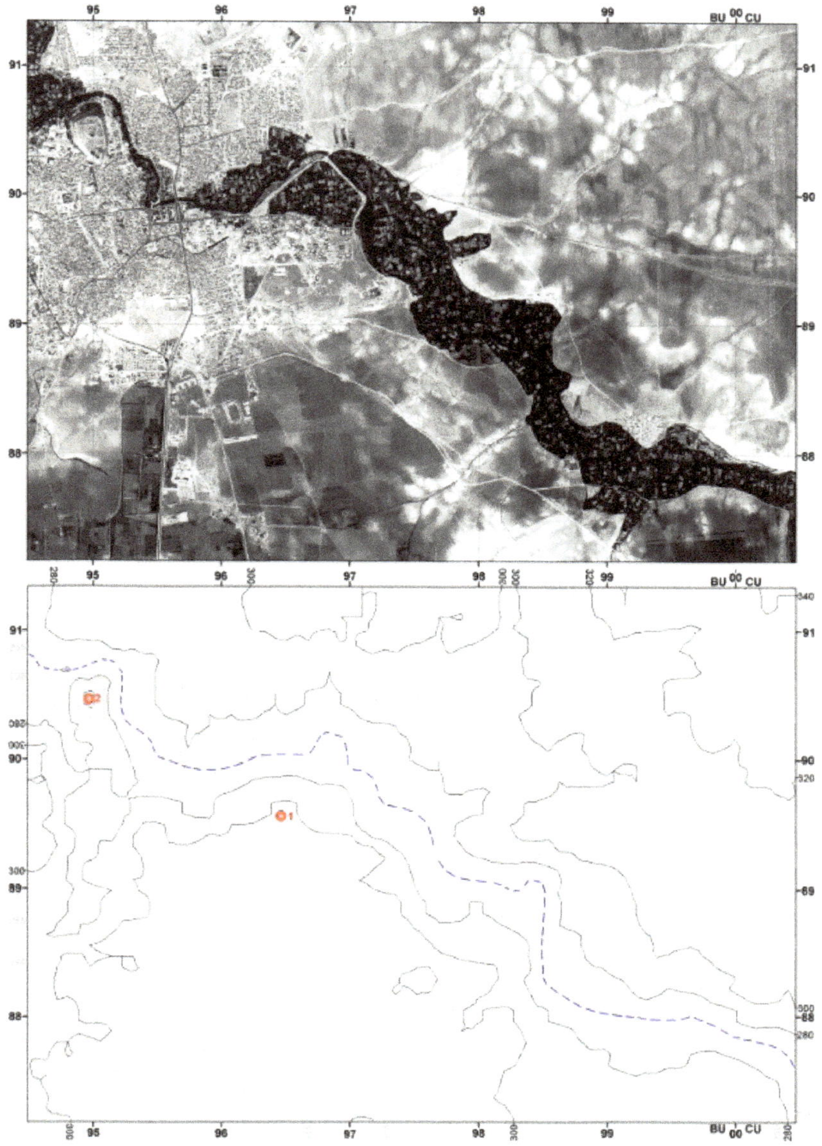

Figure 9: Paired images showing the *M. meridionalis* locality at Sharia, Hama: upper — GAMBIT image, taken in 1967; lower SRTM-derived contour image, with selected 5 m interval contours labelled and the Sharia locality indicated, centre left (as Point 1) in what was in 1967 the south-eastern edge of the Hama conurbation (UTM coordinates: BU 965 896). Point 2 is the ancient citadel in the centre of Hama. Coordinates are marked for 1 km grid squares.

The various biostratigraphical evidence thus requires the reattribution of the QfIII terrace, as represented at Latamneh and Sharia, to an age in the region of 1.2–0.9 Ma. This has a significant implication for the age model established for the Orontes by Bridgland et al. (2003; see above), which is confirmed as underestimated. Revised suggestions for the ages of terraces in the Hama–Latamneh area are indicated in Fig. 7. The ~ 40 m height of the Sharia deposit suggests a significantly younger age than the aforementioned el-Farcheh site; the somewhat greater height of the deposits at Latamneh, notwithstanding the younger age, is attributed to a component of localized uplift in the vicinity of the adjacent Sheizar Fault (see below and Fig. 8A). Given this revision, it is no longer tenable to consider the Middle Orontes terraces as representative of formation in response to 100 ka Milankovitch climatic forcing (cf. Bridgland and Westaway, 2008a).

THE GHAB BASIN

About 5 km downstream of Latamneh (along the valley axis) the Orontes leaves the incised part of its middle reach and enters the Acharneh Basin (located to the east of the southern part of the larger Ghab Basin: Fig. 1, inset), passing through a well-marked west-facing scarp slope close to the town of Sheizar (Fig. 8A). Further downstream the valley widens out within the Acharneh Basin to form a large flat plain and all but the lowest Pleistocene terraces disappear. These lowest terraces are represented on the maps of Besançon and Sanlaville, 1993a and Besançon and Sanlaville, 1993b by widespread glacis, mostly forming apparent fans that extend westwards from the scarp; a few of the highest fragments of these are mapped as QfIII, although QfI and QfII dominate and extend along the course of the river downstream to Acharneh, beyond which they too cease to be represented (Fig. 10). Sections beneath the glacis surfaces in the latter area reveal them to be formed on lacustrine sediment of the Acharneh Basin (Devyatkin et al., 1997), probably Neogene in age, that have been slightly uplifted. Downstream of Acharneh the river resumes its northward course, flowing above the stacked infill of the Ghab Basin (Fig. 3); this has been widely interpreted as an actively developing pull-apart basin (Devyatkin et al., 1997, Brew et al., 2001, Westaway, 2004b and Westaway, 2010) the formation of which was probably broadly synchronous with the comparable Hula Basin further south, the initiation of the latter being reliably dated to ~ 4 Ma from its associated volcanism (Heimann and Steinitz, 1989 and Heimann et al., 2009). These structures thus reflect the development of the present geometry of the DSFZ, which came into being around that time, at ~ 3.7–3.6 Ma (Westaway, 2004b,Westaway, 2010 and Seyrek et al., 2007). The fact that fluvial aggradational terraces occur only upstream of the Sheizar scarp implies that the latter has formed in response to Pleistocene dip-slip (down-to-the-west) movement of a fault at this location (Fig. 3 and Fig. 8A). This interpretation was suggested by de Heinzelin (1966) but was overlooked by later workers; indeed, de Heinzelin envisaged that the succession at Latamneh (see above) was deposited before this fault became active, at a time when much of the present relief of the area did not exist. Movement on this fault can be presumed to account for the lack of fluvial incision in the area downstream and the greater-than-expected height (give their biostratigraphical age) of the deposits at Latamneh, ~ 5 km upstream.

Figure 10: Fossiliferous sites in the Ghab Basin: 1960s CORONA image showing the location of (from north to south) the Karkour railway bridge and Karkour sluice sites at the northern end of the Ghab; note that the railway was under construction, with the bridge not yet in place. Insets show the location of sample points from present-day Google Earth. Coordinates are marked for 1 km grid squares, although the image shows a 0.5 km grid.

Much work was carried out in the Ghab by Russian scientists during the Soviet era, as reviewed by Domas (1994) and Devyatkin et al. (1997), who described a mixed lacustrine and fluvial Pliocene infill that, from geophysical evidence, attains a maximum thickness of 0.8–1.0 km. According to Devyatkin et al. (1997), the maximum sediment thickness occurs in the central–southern part of the basin, thinning to < 200 km in the north (Fig. 1, inset). They also reported a lacustrine fill of up to ~ 300 m in the Acharneh Basin. A longitudinal section through the northernmost part of the Ghab, based on borehole evidence (Besançon and Sanlaville, 1993b), shows the majority of the infill, proved to a maximum thickness of 40 m (but unbottomed), to be Pliocene shelly clay, with interbedded sands, volcanic ash and, at the northern end of the basin, basaltic lava (Fig. 11B). This lava, erupted from poorly preserved cones to the north and east of the downstream limit of the Ghab, has been attributed to the Pliocene (Ponikarov et al., 1963b, Ponikarov, 1986 and Domas, 1994), although it has yielded K–Ar dates as young as 1.1 ± 0.2 and 1.3 ± 0.9 Ma (e.g. Devyatkin et al., 1997).

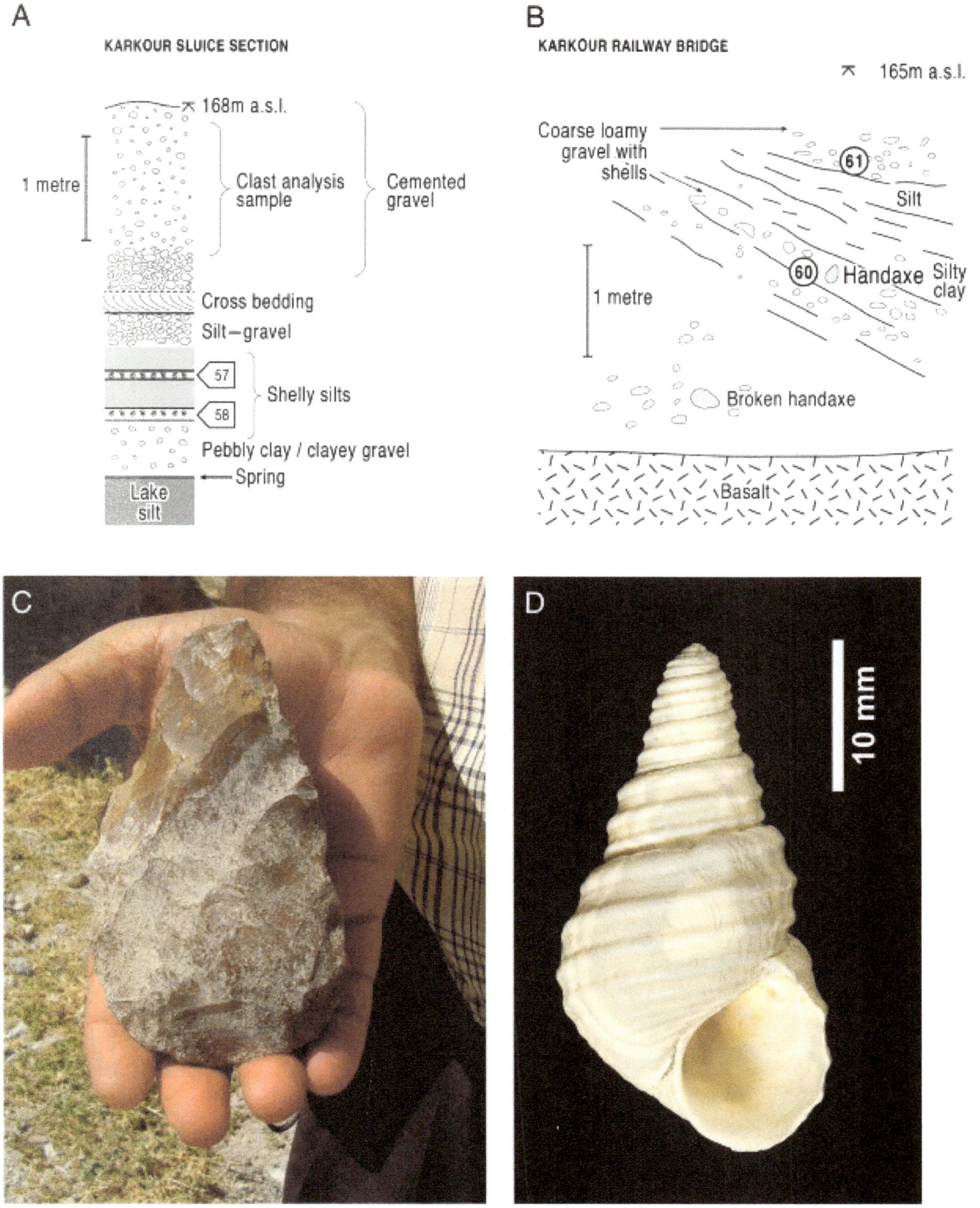

Figure 11: Karkour, northern Ghab Basin: A — Section log from the Karkour sluice locality; B — Section log from the Karkour railway bridge locality; C — hand axe from the Karkour railway bridge section; D — *Apameaus apameae* from the Karkour railway bridge section. This species is named after Apamaea, a Roman site in the Ghab.

It is unclear whether the Ghab has been continuously occupied by a lake throughout the Pleistocene or whether there were periods when the Orontes flowed across a dry lake floor, as at present (following anthropogenic drainage). As it approaches the northern end of the Ghab, the modern Orontes channel becomes increasingly deeply incised, partly, perhaps, as a result of the artificial drainage during the 1950s of what was, in historical times, a wetland (Besançon and Sanlaville, 1993b). In part, however, this is likely to be a consequence of the river cutting into the basalt barrier as it passes from the subsiding basin interior into an area that has clearly been uplifting. There is evidence for uplift of the basalt barrier during the Quaternary, in the form of inset low-level river terrace gravels that border the course of the Orontes along the northernmost 15 km of the Ghab (Domas, 1994). These deposits are well exposed at Karkour, near the northern end of the basin, where Quaternary gravel was recorded by Van Liere (1961) and Besançon and Sanlaville (1993b, their Fig. 19). Two gravel exposures were recorded during fieldwork in 2007, both in the incised banks of the Orontes. First, at Karkour sluice (BV 57282 59073; ~ 168 m a.s.l.), 1–2 m of cemented medium–fine gravel was exposed, capping an unconsolidated sequence of shelly silts and fine sands above a lower clayey gravel–pebbly clay and culminating in fine gravel with an argillaceous matrix; the basal cemented gravel reveals cross bedding indicative of northward palaeoflow (Fig. 11A; online supplement, Fig. A.1.7). Analysis of the cemented gravel, carried out in the field (as in the Upper Orontes), showed that it comprises > 95% flint/chert (Table 1), the only other constituent being limestone, which has presumably been introduced from the sides of the basin. Some 2.5 km further north, at the Karkour railway bridge (BV 57306 61493; 165 m a.s.l.), a second gravel exposure was observed. The deposit here, which overlies the basalt that gives rise to the knick point at the northern end of the Ghab, is much coarser than that further south, containing large weathered basalt clasts. It is ~ 3 m thick, has a variable argillaceous matrix and contains shells, including large gastropods (see below; online supplement, Fig. A.1.8). It also yielded a well-preserved handaxe (Fig. 11C) and a butt fragment from a larger handaxe. The flint/chert reaching Karkour has evidently been transported from the Middle Orontes upstream of Sheizar, avoiding deposition as part of the thick stacked succession in the Ghab Basin. Analysis of samples from the two Karkour sites has shown that the sediments contain ostracods as well as molluscs.

Palaeontology of the Ghab Sediments

Mollusca were analysed from both Karkour exposures (Table 3). Samples 57 and 58, from silty deposits beneath the cemented gravel at Karkour sluice, yielded assemblages dominated by *Dreissena bourguignati* (~ 95% of the total). Samples 60 and 61, from Karkour railway bridge, also yielded well preserved shells, the most distinctive being the large viviparid gastropod *Apameaus apameae* (formerly *Viviparus apameae*; Fig. 11D; cf.Sivan et al., 2006). Previous authors have described other fossiliferous exposures in this general area (cf. Schütt, 1988 and Devyatkin et al., 1997; see online Appendix 3).

Table 3: Molluscan faunas from exposures at Karkour (for locations, see Fig. 10; for details see text)

	Sample			
	57	58	60	61
Theodoxus jordani (Sowerby)	+	+	+	+
Theodoxus orontis Schütt				+
Apameaus apameae (Blanckenhorn)			+	+
Melanopsis blanckenhorni Schütt	+	+	+	+
Melanopsis unicincta Blanckenhorn				+
Melanopsis bicincta Blanckenhorn				+
Melanopsis cylindrata Blanckenhorn				+
Lymnaea (?Radix) sp.	+			
Gyraulus piscinarium Bourguignat			+	+
Planorbis carinatus				+
Planorbis sp.		+	+	+
Valvata saulcyi Bourguignat	+	+	+	+
Semisalsa longiscata (Bourguignat)	+	+		
Bithynia applanata Blanckenhorn	+	+		+
Syrofontana fossilis Schütt	+			+
Falsipyrgula rabensis (Blanckenhorn)	+	+		+
Falsipyrgula ghabensis Schütt	+	+		+
Potomida kinzelbachi Schütt			+	+
Unio terminalis Bourguignat			+	+
Dreissena bourguignati Locard	+	+	+	+

First described by Blankenhorn (1897) based on shells from the Orontes, *A. apameae* is best known from the Jordan valley, where it has been used as an index fossil to define the 'upper freshwater series' or '*Viviparus* Beds' of the Benot Ya'aqov Formation at Gesher Benot Ya'aqov (Picard, 1963, Tchernov, 1973, Goren-Inbar and Belitzky, 1989 and Bar-Yosef and Belmaker, 2010), noted above as being somewhat more recent than the Latamneh deposits. As at Karkour, *A. apameae* is found in direct association with handaxes at Gesher Benot Ya'aqov (Goren-Inbar and Belitzky, 1989 and Goren-Inbar et al., 1992). The species became extinct in the Jordan Valley at ~ 240 ka, on the basis of U-series dating (Kafri et al., 1983, Moshkovitz and Magaritz, 1987 and Heller, 2007). Its first appearance there followed the eruption of the Yarda Basalt, in the vicinity of the Sea of Galilee, and of the Hazbani Basalt, which flowed from southern Lebanon into the Hula Basin in the northernmost Jordan valley. Although dating of these basalt eruptions has been attempted using the K–Ar method, this has not resulted in reliable age-determinations (e.g. Schattner and Weinberger, 2008). Given the data currently available, probably the best guide to the age of the Hazbani Basalt and of the stacked succession in the Hula Basin is from the biostratigraphical and oxygen isotope calibration by Moshkovitz and Magaritz (1987), based on borehole evidence. They recognized glacial-to-interglacial termination IX (the MIS 20–19 transition) at a depth of ~ 160 m in the Hula Basin fill, which implies that *A. apameae*, which occurred between depths of 67 and 168 m in the studied borehole, had appeared by MIS 21. If it is assumed that the occurrence of this gastropod in the Ghab was synchronous with its presence in the Jordan, then it can be concluded that the sediments at Karkour railway bridge are Middle Pleistocene.

Samples 57 and 61 were also examined for ostracods, Sample 57 yielding the most abundant and diverse fauna of seven species (Table 4), amongst which the most common was *Cyprideis torosa*. The assemblages appear to include reworked and indigenous components, indicated by differential preservation; the reworked material might be as old as Pliocene (see online Appendix 3). Pliocene–Pleistocene ostracod faunas from the Levant region have been poorly documented, making this an important record, particularly since these crustaceans are extremely valuable palaeolimnological indicators (Holmes, 1996 and Battarbee, 2000). For example, *C. torosa* generally occurs in brackish water, developing noded valves in salinities below ~ 6‰, but can also tolerate hypersaline conditions in lakes and water bodies prone to desiccation. The indigenous component from the Karkour samples, presumed to be Pleistocene, comprises both brackish (*C. torosa* and the loxoconchids) and freshwater elements (*Candona neglecta*,*Ilyocypris* spp. and an unnamed *Heterocypris*).

Table 4: Ostracod faunas from faunas from exposures at Karkour (for locations, see Fig. 10; for details see text).

	Sample	
	57	61
Cyprideis torosa (Jones)	+	+
Ilyocypris cf. *inermis* Kaufmann	+	
Ilyocypris sp. (spinose form)	+	+
Loxoconcha spp.	+	
Candona neglecta Sars	+	
Candona sp. (juveniles)		+
Heterocypris salina (Brady)		+
Herpetocypris sp.	+	

The analysis of these two faunal groups thus supports a Pleistocene age for the sediments at both Karkour sites, with reworked ostracods from earlier, possibly Pliocene sediments within the Ghab Basin. It is worth highlighting the close similarities between the palaeontological and archaeological material from the Ghab deposits at Karkour and the Benot Ya'aqov Formation in the Jordan Valley; these records, coupled with the suggested contemporaneity of the Ghab and Hula basins, permit a tentative broad correlation to be suggested. This and the above-mentioned K–Ar dates for the basalt at the northern end of the Ghab imply an age range from the latest Early Pleistocene to Middle Pleistocene for the sedimentary succession at Karkour, rather than the Pliocene age previously suggested (cf. Ponikarov et al., 1963b), making it therefore somewhat younger than the deposits at Latamneh.

THE ORONTES GORGE NORTH OF JISR ASH-SHUGOUR

The town of Jisr ash-Shugour is situated at the downstream end of the Ghab Basin, north of which the Orontes enters a gorge, ~ 100 m deep (although within a wider, deeper feature up to 300 m deep) and ~ 25 km long (along its sinuosity; see Fig. 12; online supplement, Fig. A.1.9), that it has cut through Palaeogene limestone. At the northern end of the gorge is the town of Darkush, only 3 km upstream of the border with Hatay Province (Turkey).

Figure 12: The gorge between Jisr ash-Shugour and Darkush: 1960s CORONA image of the gorge (for location, seeFig. 1); dotted lines show locations of derived cross sections (see online supplement, Fig. A.1.9). Coordinates are marked for 1 km grid squares. Insets show photographic images of the gorge: A — looking SSW (upstream) from BV 63205 76708, about 5 km east of Qanaya. The mountains visible in the far distance are part of the Jebel Nusayriyah range along the western side of the Ghab Basin.; B — Looking northwards, downstream along the valley, from BV 63234 76771. The gravel sample locality at Hammam ash Shaykh Isa is ~ 400 m to the north of the viewpoint, accessed from the nearer part of the valley-floor platform in the middle distance. Both photographs show features that might represent early river levels, although other explanations would be possible and ground-truthing has not been practicable.

Where this gorge could be accessed it was, unsurprisingly, found to contain no fluvial terraces, only a basal gravel ~ 0.3 m thick, beneath ~ 0.2 m of sand with gravel seams and capped by ~ 1.0 m of silty overbank deposits. This valley-floor sequence was observed and sampled at Hammam ash Sheykh Isa, where the gravel (126 m a.s.l.) proved to comprise mainly limestone (> 99% of a somewhat undersized sample:Table 1). This demonstrates that the bedload of the Orontes has been recharged with limestone from the local Palaeogene outcrop, this having greatly diluted the flint/chert clasts that accounted for > 95% of the gravel at Karkour.

The Amik Basin, Hatay Province

The meandering Orontes channel north-west of Darkush forms the border between Syria and Hatay for a (straight-line) distance of 23 km. To the north the river enters a second subsiding lacustrine basin, formerly occupied by Lake Amik. In this case the lake has disappeared during the last half century, having (along with its associated wetlands) occupied 53.3 km^2 in 1972 (Fig. 1 and Fig. 13) but been eliminated completely by 1987 as a result of drainage for agriculture (Kiliç et al., 2006 and Çalışkan, 2008; for notes on its earlier history, see Wilkinson, 1997 and Yener et al., 2000). It was also known as the Lake of Antioch, the ancient city of that name (modern Antakya) being situated on the south-western extremity of the Amik Basin, the flat-topped infill of the latter giving rise to the Antioch/Amik Plain. The lake was drained by an artificial channel (the Balıkgölü Canal) into the Orontes, which flows along the southern edge of the basin without reaching the former site of the lake (Fig. 13).

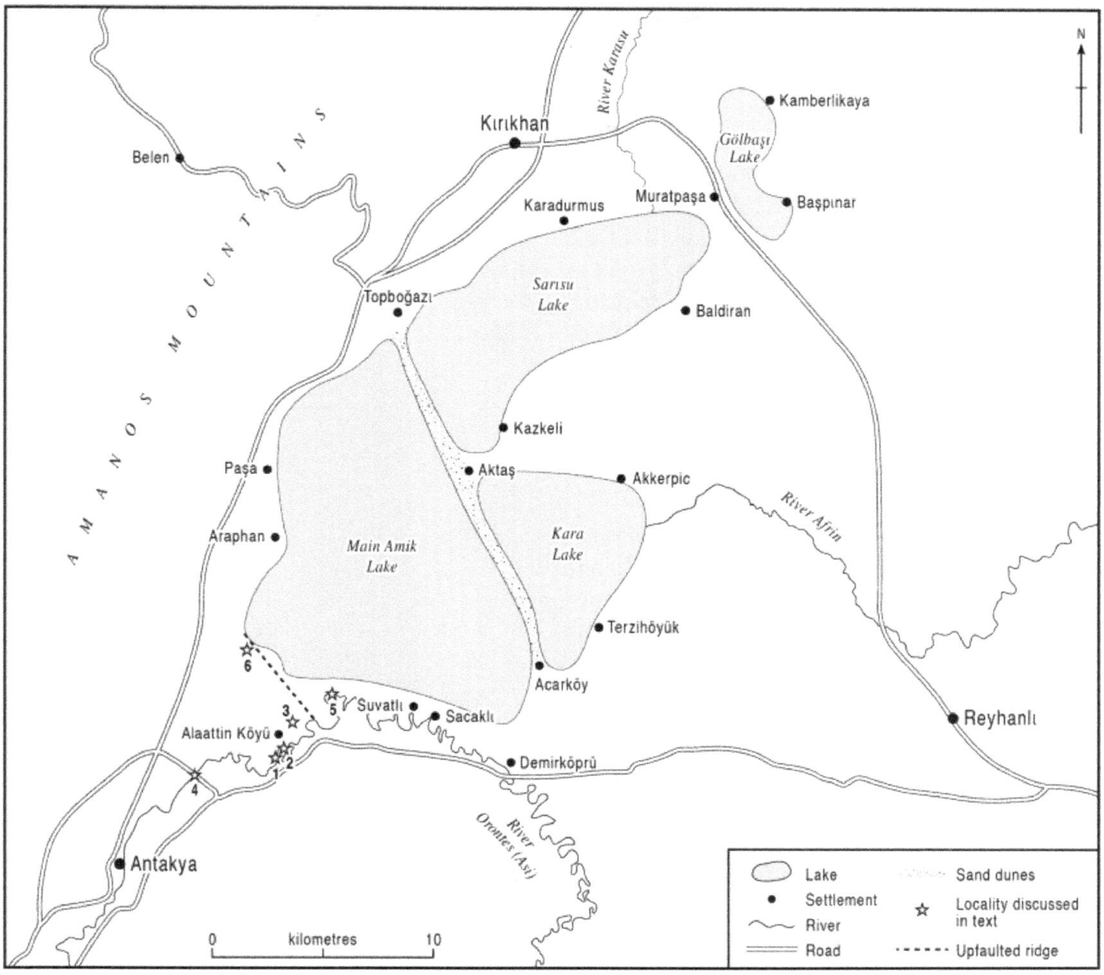

Figure 13: The Amik basin showing the extent of the former Lake Amik (modified from Çalışkan, 2008). For location, see Fig. 1. Field localities are numbered as follows (see text): 1, the Alaattin Köyü clast analysis locality (see Table 1); 2, river-bank Section 1 km south of Alaattin Köyü (BA 52477 15506); 3, quarry section in upfaulted basin sediments at Alaattin Köyü (BA 52626 16772); 4 and 5, downstream and upstream limits of measurement fluvial entrenchment beneath the Amik Plain (at BA 48448 15139 and BA 54539 18533, respectively); 6, spot-height in the upfaulted sediment observed at 3, at 159 m a.s.l. (~ BA 508 200).

Extending NNE (upstream) from the Amik Basin is the valley of the River Karasu, which follows the alignment of the DSFZ as it trends in that direction, passing into the East Anatolian Fault Zone (e.g. Westaway, 2004b and Westaway et al., 2006; Fig. 1). The nature of faulting in this region was for many years a subject of debate, in relation to controversy concerning the geometry of the DSFZ in western Syria. Yurtmen et al. (2002)showed that the evidence was consistent with active faulting along the Karasu valley continuing southward into the DSFZ, with no requirement for any other large-scale faulting extending offshore to the south-west, as others had previously inferred. In this view, the Amik Basin can be interpreted as marking a leftward step in the faulting, from the Amanos Fault, which runs along the western margin of the Karasu valley, to the DSFZ segment further south, which Westaway (2004b) called the Qanaya–Babatorun Fault. Subsequent analyses (e.g. Seyrek et al., 2007, Seyrek et al., 2008 and Westaway et al., 2008) have confirmed this general interpretation. The Amik Basin can therefore be interpreted as another pull-apart basin within the DSFZ, comparable with the Ghab. Although they have yet to be mapped in detail, the existence of active faults in the region is corroborated by records of numerous large historical earthquakes (e.g. Akyuz et al., 2006). The Karasu valley is the site of extensive eruptions of Quaternary basalt; offsets of dated basalt flows, where they cascade into the main valley along tributary gorges, have provided the principal method of measuring slip rates on faults in this region (e.g.Yurtmen et al., 2002 and Seyrek et al., 2007).

It is difficult to study the Orontes to the south-east of the Amik Basin because of the political sensitivity of the Syria–Hatay border, with which it coincides in this area. Once downstream of the border, the river channel can readily be observed as it meanders extravagantly across the Amik Plain. Simple measurements at intervals demonstrate that the bed of the river becomes progressively incised below the surface of the plain as the outlet from the basin, near Antakya, is approached. The Orontes channel was measured as declining from ~ 3.5 m below the Antioch Plain at BA 54539 18533, ~ 2.5 km upstream of Alaattin, to ~ 9.5 m on the northern outskirts of Antakya (BA 48448 15139), over a (straight-line) distance of ~ 7 km (see Fig. 13, points 4 and 5).

Thin cobble-gravel, interbedded with floodplain silt, cropping out in the river bank (north side) ~ 2 m above the Orontes, ~ 1 km south of Alaattin Köyü (~ 86 m a.s.l.; Fig. 13, point 2; the source of a clast analysis sample: Table 1), was found to contain pottery, demonstrating a mid–late Holocene age, as well as freshwater mussel shells. Flint/chert from the Middle Orontes has increased in frequency in the river›s bedload here (~ 10%), in comparison with the Darkush Gorge sample locality ~ 50 km upstream, despite the much greater distance from its source outcrops than from the Palaeogene limestone (which has fallen to < 75%: Table 1). This change can be attributed to the relative hardness and resistance to abrasion of the siliceous clasts, which persist into the gravels of the Lower Orontes, whereas the limestone disappears (see below; Table 1). The Amik Basin sample also contains a fresh input of crystalline material, notably local basalt (> 10%), as well as coarser mafic rocks and traces of quartzitic, schistose and amphibolitic rocks. The coarse mafic rocks are likely to be from the latest Cretaceous Hatay ophiolite, whereas the other constituents are probably derived from the Precambrian/Palaeozoic succession exposed in the Amanos Mountains, west of the Karasu, by way of its right-bank tributaries. All such material is exotic to the Orontes valley upstream of this point.

Also exposed in the incised channel sides of the Orontes in the Amik Basin are horizontally laminated fine sands and silts, interpreted as fluvio-lacustrine sediments of Neogene–Quaternary age. They were observed in a river-bank section, ~ 1 km south of Alaattin (Fig. 13, point 1). Comparable deposits, although rather more indurated and gently tilted, were also observed in quarry sections in low hills standing above the Amik plain at Alaattin Köyü (Fig. 13, point 3). These would appear to represent upfaulted blocks of Neogene basin fill; at ~ 125 m a.s.l., the deposits have been uplifted ~ 40 m above the general level of the plain. The same upfaulted ridge reaches 159 m a.s.l. to the north-west of Alaattin Köyü (Fig. 13, point 6). The absence of any other high ground suggests that this area has been a subsiding sedimentary depocentre for most of recent geological history and certainly since the present geometry of the DSFZ in this region came into being, in the Mid Pliocene (~ 3.6–3.7 Ma: e.g. Seyrek et al., 2007 and Westaway et al., 2008). A seismic reflection profile published by Perinçek and Çemen (1990) indicates that the sedimentary fill in the Amik Basin reaches a maximum thickness of ~ 1 km.

THE LOWER ORONTES, DOWNSTREAM OF ANTAKYA

Flowing south-westwards from Antakya, the Orontes again enters a high-relief area in which river terrace deposits are preserved sporadically along its incised course (Erol, 1963), before entering the most spectacular of its three gorges, more than 400 m deep, cut into resistant latest Cretaceous ophiolitic rocks (Fig. 1 and Fig. 14). This incised valley makes a dramatic contrast with the low relief in the area of the Amik Plain. Between Aknehir and Sutaşı (BA 30927 97820), for a distance of ~ 8 km, the Orontes follows a particularly deeply entrenched gorge between the Ziyaret Dağı mountain range to the south–east and the isolated Samandağ Tepe hill (summit 479 m a.s.l.) to the north–west. The abruptness of its downstream end (Fig. 14), where the Orontes flows out of a major escarpment that was recognized by Erol (1963), raises the possibility that it is caused by localized slip on an active dip-slip fault (Fig. 3 and Fig. 14). A fault in this location, near the village of Sutaşı, was indeed indicated by Tolun and Erentöz (1962). Recent interpretations of the regional kinematics (e.g. Westaway, 2004b, Gomez et al., 2006,Seyrek et al., 2008 and Abou Romieh et al., 2009) require a component of crustal shortening in the crustal blocks alongside the active left-lateral faults forming the northern DSFZ. It is thus probable that any dip-slip fault in the Antakya area, away from the main left-lateral faulting, would be a reverse fault (cf. Fig. 3) rather than a normal fault. Heights and tentative ages for the marine terraces on the Mediterranean coastline were reported by Erol (1963) and were used by Seyrek et al. (2008) to infer a typical uplift rate of ~ 0.2 mm a^{-1} during the latter part of the Middle Pleistocene and the Late Pleistocene. The rate of localized uplift in the supposedly fault-bounded gorge reach may well be significantly higher than this already-high estimate of the regional uplift rate (cf. Demir et al., 2012-this issue).

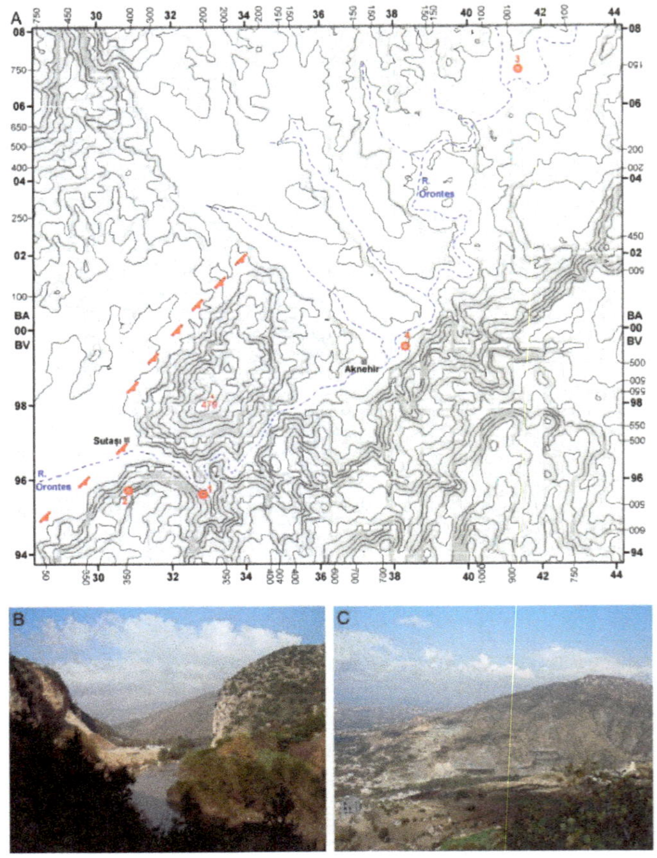

Figure 14: The Lower Orontes Gorge, Hatay Province, Turkey: A — Synthetic contour map generated from SRTM data showing the Orontes valley downstream of Antakya (for location, see Fig. 1). Faint contours are at 10 m intervals, with dark contours at 50 m intervals, the latter omitted where the relief is too steep. Note the contrast between the broad valley upstream of Aknehir and the gorge reach between there and Sutaşı. The Sutaşı fault is marked, with ticks on the hanging wall. The summit of Samandağ Tepe is marked by its ~ 479 m spot height. Numbered localities denote the following: 1, viewpoint for Fig. 14B; 2, viewpoint forFig. 14C; 3, Bostancık Köyü (Antakya western bypass) clast analysis locality; 4, the Şahin Tepesi clast analysis locality.B — Looking north-west (downstream) from BV 32785 95604 along the entrenched gorge reach shown in A. C — Looking NNE from BV 30827 95697 across the exit of the Orontes gorge at the village of Sutaşı, where the river emerges from what is interpreted as a locally uplifted fault block, the margin of which also creates the abrupt escarpment below the viewpoint.

In this gorge reach the Orontes falls 50 m in 16 km to the sea south of Samandağı. Former dock-side masonry of ancient Seleucia Pieria (the former port of Antioch) can be observed today within agricultural land (at YF 63841 00704), demonstrating relative sea-level fall in this area in the past two millennia. Relative sea-level decline has also been documented from the levels on the quayside masonry of borings by marine molluscs, at up to 0.75 m above modern sea level, and has been linked to historical earthquakes, notably one in AD 551 (Pirazzoli et al., 1991).

Study of gravel exposures in this Lower Orontes reach reveals a further input of the crystalline rocks first seen in the Amik Basin sample. For example, a sample collected near Bostancık Köyü (Fig. 14), from an Orontes terrace gravel exposed on the western side of the Antakya western bypass (76 m a.s.l.; ~ 25 m above the river), contained ~ 30% basalt and > 50% coarser basaltic lithologies, both including badly weathered examples (Table 1). Highly weathered ophiolitic rocks were also encountered (14.4%), as well as quartzose lithologies and kaolinized rocks (Table 1). Only a single flint/chert clast was recorded in this sample, which was devoid of limestone. In contrast, a sample collected from a bluff on the left bank of the Orontes at Şahin Tepesi, ~ 25 m above the river (therefore probably representing the same terrace), yielded 1.7% flint/chert and 7% limestone, the latter assumed to be locally derived (from the Cretaceous marine succession onto which the Hatay ophiolite has been emplaced; e.g. Tolun and Erentöz, 1962) rather than transported from Syria. It was comparable with the previous sample in that it was dominated by basalt and coarser basaltic rocks, together with ophiolite and a trace of quartzitic lithologies (Table 1). Erol (1963) included the terrace deposits here as part of his Orontes Terrace II, which is the lowest of the terraces depicted in Fig. 3 (his lowest terrace being too close to the river to show at this scale). He correlated this fluvial terrace with his Marine Terrace II, which Seyrek et al. (2008) concluded to be of last interglacial (MIS 5e) age. A direct correlation between fluvial and marine terraces, andErol›s (1963) resulting age assignment of this fluvial terrace to the 'Riss–Würm Interglacial', is unlikely to be tenable, given that current wisdom generally attributes river terrace gravels to cold-climate episodes (e.g. Bridgland, 2000, Bridgland and Westaway, 2008b and Bridgland et al., 2008). The reported heights of Erol›s (1963) five Orontes terraces in the reach between Bostancık Köyü and the upstream end of the Samandağ Tepe gorge, 5–7, 15–25, 40–50, 70–80, and 90–100 m above the modern river, are indicative of a regular pattern, suggesting terrace formation in response to 100 ka climatic cycles, as in other parts of the world (cf. Bridgland, 2000 and Bridgland and Westaway, 2008a). The reported terraces can perhaps be correlated with MIS 12, 10, 8, 6 and 2. Uplift rates in this reach of the river during this span of time would thus approach 0.2 mm a^{-1}, far higher than in localities further upstream. Like in the Ceyhan valley through the Amanos Mountains further north (Seyrek et al., 2008), and in the lower reaches of the Nahr el-Kebir near Latakia in north–west Syria (Bridgland et al., 2008), the absence of any earlier record in this reach of the Orontes can be attributed to the rapid uplift; the resulting high relief and the associated rates of slope processes have led to low probabilities for preservation of older river terrace deposits.

DISCUSSION: POSSIBLE AGE AND RELATION TO CLIMATIC FLUCTUATION

Unlike other nearby rivers in Turkey and Syria (Sharkov et al., 1998, Bridgland et al., 2007, Demir et al., 2007, Demir et al., 2012, Seyrek et al., 2008 and Westaway et al., 2009b), there are no Pleistocene lava flows interbedded within the Pleistocene terrace sequence of the Orontes to provide marker levels for dating, although the Homs Basalt provides a maximum age for the start of incision by the upper river below the level of the Late Miocene–Early Pliocene lake bed onto which this lava erupted. It has been thought, therefore, that the best indication of age within the sequence comes from the mammalian assemblage from the Middle Orontes (see above). Indeed, Bridgland et al. (2003)previously used the Latamneh terrace deposits as a pinning point for correlation between the Middle and Upper Orontes terrace sequences, having also modelled the latter (see also Bridgland and Westaway, 2008a) using the technique applied widely to terrace systems elsewhere (Westaway, 2004a and Westaway et al., 2002; see above). They attributed the formation of comparable terrace sequences in both the middle and upper reaches of the Orontes to cyclic climatic forcing of fluvial sedimentation and erosion with progressive incision in response to regional uplift. The similarity of these records to those in rivers elsewhere, notwithstanding proximity to plate boundaries (as in the case of the Orontes) or otherwise (as in rivers in NW Europe with similar uplift histories) was also noted (Bridgland et al., 2003 and Bridgland and Westaway, 2008a). The consistent pattern of uplift, seen widely and calibrated in systems with optimal dating control, records acceleration at around 3 Ma, followed in many cases by renewed acceleration around 2 Ma, then by decrease during the Early Pleistocene and a further acceleration at around the 'Mid-Pleistocene Revolution', when the 100 ka climate cycles began (Bridgland and Westaway, 2008b and Westaway et al., 2009a). Each of these phases of uplift is well developed in fluvial sequences within the Arabian Platform, notably that of the River Euphrates (e.g. Demir et al., 2007, Demir et al., 2008 and Westaway, 2010). This pattern of persistent uplift is seen, however, only in areas of post-Early Proterozoic (i.e. non-cratonic) crust, older crust being colder and more stable and showing either intermittent uplift and subsidence (Early Proterozoic crust) or, in the case of Archaean cratons, little vertical movement at all during the Late Cenozoic (Westaway et al., 2003,Westaway et al., 2009a and Westaway, 2012). Subsiding areas of sediment accumulation are further exceptions to the standard pattern; not necessarily fault bounded, their subsidence is presumed to be a response to sedimentary isostasy. Examples of substantial subsiding regions of this type are the Lower Rhine, beneath the Netherlands (Brunnacker et al., 1982 and Ruegg, 1994), and the Great Hungarian Plain (Gábris and Nádor, 2007 and Kasse et al., 2010). Smaller areas of subsidence are also recognized within the Rhine, such as the Neuwied Basin, where a substantial stacked sequence of fluvial deposits floors a down-faulted block (Meyer and Stets, 2002). The Ghab and Amik Basins compare closely with the last-mentioned European example, as they also occupy small subsiding fault-bounded blocks.

It is apparent from the disposition of fluvial gravel terraces in both the Upper and Middle Orontes that the river in both these reaches has migrated westwards during the Pleistocene, although throughout that time it flowed between them through the fixed and entrenched Rastan Gorge. The close coincidence of the modern Orontes course and the eastern margin of the Homs Basalt outcrop can be explained as the result of this westward migration being prevented once the river encountered the resistant basalt, with forms an eastward-inclined interbed within the 'Pontian' lacustrine marl.

The previous use of the Latamneh deposits as a dating level within the Orontes sequence must now be open to question, since, as noted above, the mammalian assemblage from that locality includes elements that date back to the late Early Pleistocene and would seem to imply an age in the region of 1.2–0.9 Ma. This revision (Fig. 7) effectively doubles the supposed age of the QfIII terrace at Latamneh, with the clear implication that uplift there has been at half the rate previously supposed. Account must also be taken of the contradictory indications from the occurrences of different mammoth species within deposits mapped as Terrace QfIII at Latamneh (*M. trogontherii*) and Sharia, Hama (*M. meridionalis*), the latter being potentially an older indicator than even the arvicolid *L. arankae* at Latamneh. It is also now evident that the Latamneh locality is within a few kilometres of an active fault, at Sheizar, that has experienced significant movement during the Pleistocene. It is thus possible that the deposits there might provide a less-than-reliable indication of the regional uplift rate in the Middle Orontes valley, which could explain the indication that older deposits occur at a lower height relative to the modern river in the Hama area. The deposits in the latter area are ~ 30 km upstream of the faulting and can be assumed to provide a genuine indication of the regional uplift. It is nonetheless clear that the deposits of Middle Orontes Terrace QfIII are significantly older than was suggested in previous publications (cf. Sanlaville, 1988 and Bridgland et al., 2003).

The reinterpretation of the Middle Orontes record raises the possibility that it is comparable with that from other parts of the Arabian Platform, further east in Syria and to the north-east in Turkey, where it has been discerned from the terrace sequence of the River Euphrates. Evidence from the Euphrates points to relatively slow uplift and to a brief period of subsidence in the late Early Pleistocene, during which the valley was partly backfilled with fluvial sediment (Demir et al., 2007, Demir et al., 2008 and Demir et al., 2012). The resultant thick aggradation, culminating at about MIS 22, has yielded numerous handaxes, although the Euphrates sediments are generally devoid of fossils. It is tempting, therefore, to suggest a correlation with the deposits at Latamneh, which also indicate thick sediment accumulation, have an age (from biostratigraphy) close to the Early–Middle Pleistocene boundary and contain numerous handaxes. The comparable localities in the Euphrates, indicative of subsidence in the late Early Pleistocene, include sites such as Ain Abu Jemaa, near Deir ez-Zor, Syria (Demir et al., 2007), and Birecik and Karababa, in south–eastern Turkey (Demir et al., 2008 and Demir et al., 2012). A physical mechanism for the reversals in vertical crustal motion that are implicit in this interpretation is suggested by Westaway (2012-this issue). The subsequent increase in regional uplift rates that caused a switch from aggradation back to incision in these rivers can be attributed to the Mid-Pleistocene Revolution, with the more severe cold stages that occurred within the subsequent 100 ka climate cycles leading to enhanced climatic forcing and a positive-feedback enhancement of erosional isostasy. At Latamneh this effect might well have been accentuated by the component of vertical slip on the Sheizar Fault (cf. de Heinzelin, 1966). Indeed, it is conceivable that increased rates of vertical motion in opposite senses (i.e. faster uplift in the Middle Orontes and faster subsidence in the southern Ghab Basin) following the Mid-Pleistocene Revolution affected the state of stress in the region, resulting in initiation or reactivation of slip on the Sheizar Fault (cf. Westaway, 2006).

There is scant evidence that late Early Pleistocene subsidence occurred further upstream in the Orontes. Recalibration of the ages of the Upper Orontes terraces (cf. Fig. 4) generally suggests that the Bridgland et al. (2003) age model underestimates their ages by a single Milankovitch (100 ka) cycle, given that the number and spacing of the terraces remain suggestive of formation in approximate synchrony with these glacial–interglacial cycles. The more rapid uplift, in comparison with the Middle Orontes, that is implicit in this interpretation is in keeping with the evident post-Pliocene uplift of the 'Pontian' lacustrine basin (cf. Fig. 3).

The dating of river gorges is, in general, difficult and typically relies upon projection of terraces from upstream and downstream (cf. Fig. 3). The Rastan Gorge is, in fact, an exception, since its incision can also be constrained by the ages of the Homs and Tell Bisseh basalts. In particular, the well-constrained age of the Homs Basalt provides an upper limit on the ages of the oldest terraces, comparable in height above the river (Fig. 3), of ~ 4–5 Ma. It is much more difficult to be clear about the age of the Darkush Gorge, since it is isolated from well-dated terrace sequences, falling between the subsiding Ghab and Amik basins.

In contrast to the Upper and Middle Orontes catchment, much higher rates of vertical crustal motion are indicated in coastal parts of the study region. Setting aside any local effects of active faulting, as already noted, the disposition of the Orontes and marine terraces downstream of Antakya indicates an uplift rate of ~ 0.2 mm a^{-1}, roughly an order-of-magnitude faster than estimates for the Middle Orontes, where the inferred ~ 120 m of incision in the Hama reach since ~ 4 Ma (Early Pliocene) equates to a rate of ~ 0.03 mm a^{-1}, although if the deposit at Sharia, ~ 40 m above the modern river, is as young as ~ 1 Ma it indicates an accelerated rate of ~ 0.04 mm a^{-1} during the last million years. This estimate of the regional uplift rate can be compared with an independent estimate of ~ 0.4 mm a^{-1} further north in the Amanos mountain range, based on the terraces of the River Ceyhan, which are capped by datable basalt flows (Seyrek et al., 2008). The uplift rate is also ~ 0.4 mm a^{-1} further south, in the Latakia area of north-west Syria, on the basis of marine and fluvial terraces in the region of the Nahr el-Kebir estuary (Bridgland et al., 2008), the latter river draining a catchment that lies entirely seaward from the course of the Orontes (Fig. 1, inset). Seyrek et al. (2008) reported on a modelling study that attempted to establish the cause of the high uplift rates in these coastal regions, which lie close enough to the DSFZ for it to be possible that they are affected by the distributed crustal shortening (as well as localized faulting) that is required given the orientation and slip sense of this fault zone. However, they found that this process is unlikely to be the main cause of the rapid regional uplift observed, concluding instead that this is primarily the isostatic response to erosion. They envisaged a complex sequence of events, with positive feedback effects, following the elevation of the coastal mountain ranges as a result of plate motions (once the modern geometry of the plate boundary zone was established at ~ 4 Ma). Once some initial topography had developed, orographic precipitation would have been initiated in this coastal region, as a result of westerly winds from the Mediterranean. The resulting rainfall in turn triggered erosion, and, given the physical properties of the underlying crust (cf. Bridgland and Westaway, 2008a and Bridgland and Westaway, 2008b), the resultant unloading drove the observed uplift isostatically. Conversely, sediment loads, triggered by the same combination of processes, may well have driven the observed subsidence in depocentres such as the Ghab and Amik basins. It thus appears likely that both plate motions and climate have contributed to the pattern of vertical crustal motions in this coastal region, whereas parts of the arid interior hinterland that are well away from any active faults, such as the Hama area, have experienced much slower uplift, reflecting the lower rates of erosion.

The difference between the ~ 0.2 mm a^{-1} uplift estimated for the lower Orontes downstream of Antakya and the ~ 0.4 mm a^{-1} rates in the other coastal regions may result from a significant part of the deformation in the former area being accommodated by local reverse faults, such as that at Sutaşı (see above; Fig. 3 and Fig. 14), whereas in the regions traversed by the Ceyhan and Nahr el-Kebir it is accommodated entirely by distributed crustal deformation. It is thus possible that the spatial average of the uplift rates in the terraced reach of the Lower Orontes downstream of Antakya and in the gorge reach in the hanging wall of the Sutaşı Fault equates to ~ 0.4 mm a^{-1}. However, there is no way of estimating directly the local uplift rate in this gorge reach to facilitate such a comparison.

In the Upper Orontes the uplift rate appears to be significantly higher than in the Middle Orontes; the age model in Fig. 4 indicates a time-averaged rate since the Mid-Pleistocene Revolution of ~ 0.09 mm a^{-1}, more than double the upper bound of ~ 0.04 mm a^{-1} for the Middle Orontes over a similar interval, based on the supposed age of the Sharia deposits (see above). This difference may possibly reflect the relative erodability of the 'Pontian' marl substrate in the Homs area (i.e., faster erosion is resulting in a faster isostatic uplift response). Alternatively, much of the study region south of Homs adjoins the NNE end of the Anti-Lebanon mountain range and the western end of the Palmyra Fold Belt (Fig. 1). It is thus possible that localities in this region are affected by components of localized deformation (possibly arising from slip on blind reverse faults beneath anticlines; cf. Demir et al., 2012-this issue). Abou Romieh et al. (2009) indeed noted localized deformation of Euphrates terraces where the Palmyra Fold Belt crosses this river valley in NE Syria, and inferred significant rates of deformation in more westerly parts of this deforming zone. Demir et al. (2012-this issue) have shown that the heights of Euphrates terraces in SE Turkey vary laterally because rates of localized deformation on active faults and folds can be significant compared with regional uplift rates.

CONCLUSIONS

The work on the Orontes, reported here, testifies to the value of a pragmatic approach, using multiple techniques and taking account of data from all relevant sources, in this case field and GIS survey of terrace morphology and sediments (assisted by dGPS), the use of fossil and artefact content as indicators of age for particular terrace formations, of clast analysis to identify the deposits of this river (as opposed to tributaries) and mathematical modelling, to provide a broad impression of likely ages, calibrated by pinning points and other constraints.

The variable Quaternary record from different reaches of the Orontes underlines the roles of crustal properties and climate in controlling landscape evolution within particular regions. Although less well dated than sequences in nearby catchments in Syria and Turkey, it is possible, using the starting point of the Homs Basalt eruption and the limited biostratigraphical and geochronological constraint (the last from the U-series dating reported here for the first time), to erect tentative age models for the sequences in the key reaches. This can be compared with those applicable to neighbouring catchments, in particular the slowly uplifting interior of the Arabian Platform (transected by the Euphrates), to which the Middle Orontes can be likened, and the rapidly uplifting coastal area that extends both north and south of, as well as including, the lowermost Orontes (cf. Fig. 3). The repeated changes along the course of the river between uplifting crustal blocks (with either terraces or gorges, according to the relative resistance of the bedrock) and subsiding ones is the major contribution of active Quaternary crustal deformation; otherwise the disposition of terraces in those reaches where they have formed is comparable with regions of post-Archaean crust elsewhere in the world (Bridgland and Westaway, 2008a, Bridgland and Westaway, 2008b and Westaway et al., 2009a). The contrasting record from the different reaches of this single river thus provides valuable insight into the contrasting types of records from rivers elsewhere that are wholly within individual crustal blocks.

ACKNOWLEDGEMENTS

This research includes work that was part of the project 'Settlement and Landscape Development in the Homs Region, Syria', a joint project of the University of Durham and the Directorate General of Antiquities and Museums (DGAM), Syria, and funded by the Council for British Research in the Levant (CBRL). We wish to thank the 'Fragile Crescent Project' (funded by the Arts and Humanities Research Council, UK) for providing us with satellite imagery data. The research also includes work undertaken using CBRL funding awarded to DRB, latterly supported through collaboration with the National Earthquake Center, Damascus, which has supplied both scientific expertise and logistical assistance. Work in Turkey has been funded in part by HÜBAK, the research fund of Harran University (grants to TD and ADS).

REFERENCES

1. Abou Romieh, M., Westaway, R., Daoud, M., Radwan, Y., Yassminh, R., Khalil, A., al-Ashkar, A., Loughlin, S., Arrell, K., Bridgland, D., 2009. Active crustal shortening in NE Syria revealed by deformed terraces of the River Euphrates. Terra Nova 27, 427–437.
2. Akyuz, H.S., Altunel, E., Karabacak, V., Yalçiner, C.C., 2006. Historical earthquake activity of the northern part of the Dead Sea Fault Zone, southern Turkey. Tectonophysics 426, 281–293.
3. Bartl, K., al-Maqdissi,, M., 2005. Achäologische Oberflächenuntersuchugen im Gebiet zwischen ar-Rastan und Qal'at Sayzar. Orte und Zeiten, 25 Jahre archäologische Forschung in Syrien, 1980–2005. Deutches Archäologische Institut, OrientAbteilung, Auβenstelle Damaskus, Damascus, pp. 136–141.

4. Bar-Yosef, O., Belmaker, M., 2010. Early and Middle Pleistocene Faunal and hominin dispersals through Southwestern Asia. Quaternary Science Reviews 30, 1318–1337.

5. Battarbee, R.W., 2000. Palaeolimnological approaches to climate change, with special regard to the biological record. Quaternary Science Reviews 19, 107–124.

6. Besançon, J., Sanlaville, P., 1993a. La vallée de l'Oronte entre Rastane et Aacharné. In: Sanlaville, P., Besançon, J., Copeland, L., Muhesen, S. (Eds.), Le Paléolithique de la vallée moyenne de l'Oronte (Syrie): Peuplement et environnement: British Archaeological Reports, International Series, vol. 587, pp. 13–39.

7. Besançon, J., Sanlaville, P., 1993b. Remarques sur la géomorphologie du Ghab (Syrie). In: Sanlaville, P., Besançon, J., Copeland, L., Muhesen, S. (Eds.), Le Paléolithique de la vallée moyenne de l'Oronte (Syrie): Peuplement et environnement: British Archaeological Reports, International Series, vol. 587, pp. 41–54.

8. Besançon, J., Copeland, L., Hours, F., Sanlaville, P., 1978a. Morphologie et préhistoire de la vallée de l'Oronte entre Rastane et le Ghab. Comptes Rendus de l'Académie des Sciences de Paris, 287, pp. 857–860.

9. Besançon, J., Copeland, L., Hours, F., Sanlaville, P., 1978b. The Palaeolithic Sequence in Quaternary Formations of the Orontes River Valley, northern Syria: A Preliminary Report, 15. Bulletin of the Institute of Archaeology, London, pp. 149–170.

10. Blankenhorn, M.L.P., 1897. Zur Kenntnis des Süsswasser Ablagerungen und Mollusken Syriens. Palaeontographica 44, 1–144.

11. Brew, G., Lupa, J., Barazangi, M., Sawaf, T., Al-Imam, A., Zaza, T., 2001. Structure and tectonic development of the Ghab basin and the Dead Sea fault system, Syria. Journal of the Geological Society of London 158, 665–674. Bridgland, D.R., 1986.

12. Clast lithological Analysis. Technical Guide, 3. Quaternary Research Association, Cambridge. 207 pp. Bridgland, D.R., 2000. River terrace systems in north-west Europe: an archive of environmental change, uplift and early human occupation. Quaternary Science Reviews 19, 1293–1303.

13. Bridgland, D.R., Westaway, R., 2008a. Climatically controlled river terrace staircases: a worldwide Quaternary phenomenon. Geomorphology 98, 285–315.

14. Bridgland, D.R., Westaway, R., 2008b. Preservation patterns of Late Cenozoic fluvial deposits and their implications: results from IGCP 449. Quaternary International 189, 5–38.

15. Bridgland, D.R., Philip, G., Westaway, R., White, M., 2003. A long Quaternary terrace sequence in the Orontes River valley, Syria: a record of uplift and of human occupation. Current Science (New Delhi) 84, 1080–1089.

16. Bridgland, D.R., Demir, T., Seyrek, A., Pringle, M., Westaway, R., Beck, A.R., Rowbotham, G., Yurtmen, S., 2007. Dating Quaternary volcanism and incision by the River Tigris at Diyarbakır, SE Turkey. Journal of Quaternary Science 22, 387–393.

17. Bridgland, D.R., Westaway, R., Daoud, M., Yassminh, R., Abou Romieh, M., 2008. River Terraces of the Nahr el Kebir, NW Syria, and their Palaeolithic record. CBRL Bulletin 3, 36–41. Brunnacker, K., Löscher, M., Tillmans, W., Urban, B., 1982. Correlation of the Quaternary terrace sequence in the lower Rhine valley and northern Alpine foothills of central Europe. Quaternary Research 18, 152–173.

18. Burkhalter, L., 1933. Note sur les stations préhistoriques du Gouvernement de Lattaquié. Bulletin de la Société Préhistorique Française 30 582-558. Butler, R.W.H., Spencer, S., 1999. Landscape evolution and the preservation of tectonic landforms along the northern Yammouneh Fault, Lebanon. In: Smith, B.J., Whalley, W.B., Warke, P.A. (Eds.), Uplift, Erosion, and Stability: Perspectives on Long-term Landscape Development: Geological Society of London Special Publication, 162, pp. 143–156.

19. Çalışkan, V., 2008. Human-Induced Wetland Degradation: A case study of Lake Amik (Southern Turkey), 27. Balwois, Ohrid, Republic of Macedonia, pp. 1–10.

20. Clark, J.D., 1966a. The Middle Acheulian occupation site at Latamne, northern Syria. Annales Archéologiques Arabes Syriennes 16, 31–74.

21. Clark, J.D., 1966b. Further excavations (1965) at the Middle Acheulian occupation site at Latamne, northern Syria: general results, definitions and interpretations. Annales Archéologiques Arabes Syriennes 16, 75–113.

22. Clark, J.D., 1966c. Acheulian occupation sites in the Middle East and Africa: a study in cultural variability. American Anthropologist 68, 202–229.

23. Clark, J.D., 1967. The Middle Acheulian occupation site at Latamne, northern Syria. Quaternaria 9, 1–68.

24. Clark, J.D., 1968. The Middle Acheulian occupation site at Latamne, northern Syria: further excavation. Quaternaria 10, 1–76.

25. Copeland, L., Hours, F., 1993. The Middle Orontes paleolithic flint industries. In: Sanlaville, P., Besançon, J., Copeland, L., Muhesen, S. (Eds.), Le Paléolithique de la vallée moyenne de l'Oronte (Syrie): Peuplement et environnement: British Archaeological Reports, International Series, 587, pp. 63–144.

26. Dashora, A., Lohani, B., Malik, J.N., 2007. A repository of Earth resource information — CORONA satellite programme. Current Science 92, 926–932.

27. de Heinzelin, J., 1966. Geological observations near Latamne. Annales Archeologiques Arabes Syriennes 16, 115–120.

28. de Heinzelin, J., 1968. Geological observations near Latamne. Quaternaria 10, 3–8.

29. Demir, T., Yeşilnacar, İ., Westaway, R., 2004. River terrace sequences in Turkey: sources of evidence for lateral variations in regional uplift. Proceedings of the Geologists' Association 115, 289–311.

30. Demir, T., Westaway, R., Bridgland, D., Pringle, M., Yurtmen, S., Beck, A., Rowbotham, G., 2007. Ar–Ar dating of Late Cenozoic basaltic volcanism in northern

Syria: implications for the history of incision by the River Euphrates and uplift of the northern Arabian Platform. Tectonics 26. doi:10.1029/2006TC001959.

31. Demir, T., Seyrek, A., Westaway, R., Bridgland, D., Beck, A., 2008. Late Cenozoic surface uplift revealed by incision by the River Euphrates at Birecik, southeast Turkey. Quaternary International 186, 132–163.

32. Demir, T., Seyrek, A., Guillou, H., Scaillet, S., Westaway, R., Bridgland, D., 2009. Preservation by basalt of a staircase of latest Pliocene terraces of the River Murat in eastern Turkey: evidence for rapid uplift of the eastern Anatolian Plateau. Global and Planetary Change 68, 254–269.

33. Demir, T., Seyrek, A., Westaway, R., Guillou, H., Scaillet, S., Beck, A., Bridgland, D.R., 2012. Late Cenozoic regional uplift and localised crustal deformation within the northern Arabian Platform in southeast Turkey: Investigation of the Euphrates terrace staircase using multidisciplinary techniques. Geomorphology 165–166, 7–24 (this issue).

34. Devyatkin, E.V., Dodonov, A.E., Sharkov, E.V., Zykin, V.S., Simakova, A.N., Khatib, K., Nseir, H., 1997. The El-Ghab rift depression in Syria: its structure, stratigraphy and history of development. Stratigraphy and Geological Correlation 5, 362–374.

35. Dodonov, A.E., Deviatkin, E.V., Ranov, V.A., Khatib, K., Nseir, H., 1993. The Latamne formation in the Orontes river valley. In: Sanlaville, P., Besançon, J., Copeland, L., Muhesen, S. (Eds.), Le Paléolithique de la vallée moyenne de l'Oronte (Syrie): Peuplement et environnement: British Archaeological Reports, International Series, 587, pp. 189–194.

36. Domas, J., 1994. The Late Cenozoic of the Al Ghab Rift, NW Syria. Sbornik Geologickych Ved: Antropozoikum, 21, pp. 57–73.

37. Dubertret, L., Vautrin, H., 1938. Sur l'existence du Pontien lacustre en Syrie et sur sa signification tectonique. Comptes Rendus Hebdomadaires des Séances de l'Academie des Sciences de Paris, 206, pp. 69–71.

38. Erol, O., 1963. Asi Nehri deltasının jeomorfolojisi ve dördüncü zaman denizakarsu şekilleri (Die Geomorphologie des Orontes-deltas und der anschliessenden pleistozänen Strand- und Flussterrassen, Provinz Hatay, Türkei). İstanbul Universitesi Dil ve Tarih-Coğrafya Fakültesi Yayınları [Publications of the Faculty of Language and History-Geography of İstanbul University], 148, pp. 1–110. [in Turkish and German].

39. Gábris, G., Nádor, A., 2007. Long-term fluvial archives in Hungary: response of the Danube and Tisza rivers to tectonic movements and climatic changes during the Quaternary: a review and new synthesis. Quaternary Science Reviews 26, 2758–2782.

40. Galiatsatos, N., 2004. Assessment of the CORONA series of satellite imagery for landscape archaeology: a case study from Orontes valley, Syria. PhD thesis, Durham University, United Kingdom, 373 pp.

41. Galiatsatos, N., 2009. The shift from film to digital product: focus on CORONA imagery. Photogrammetrie - Fernerkundung - Geoinformation 3, 251–260.

42. Galiatsatos, N., Donoghue, D.N.M., Philip, G., 2008. High resolution elevation data derived from stereoscopic CORONA imagery with minimal ground control: an approach using IKONOS and SRTM data. Photogrammetric Engineering and Remote Sensing 74, 1093–1106.

43. Galiatsatos, N., Wilkinson, T.J., Donoghue, D.N.M., Philip, G., 2009. The Fragile Crescent Project (FCP): Analysis of Settlement Landscapes Using Satellite Imagery. CAA 2009: Making history interactive, Williamsburg, Virginia, USA. .

44. Gomez, F., Khawlie, M., Tabet, C., Darkal, A.N., Khair, K., Barazangi, M., 2006. Late Cenozoic uplift along the northern Dead Sea transform in Syria and Lebanon. Earth and Planetary Science Letters 241, 913–931.

45. Gonçalves, J., Fernandes, J.C., 2005. Assessment of SRTM-3 DEM in Portugal with topographic map data. Proceedings of the EARSeL Workshop 3D-Remote Sensing, 10–11 June, Porto, Portugal (European Association of Remote Sensing Laboratories). unpaginated CD-ROM.

46. Goren-Inbar, N., Belitzky, S., 1989. Structural position of the Pleistocene Gesher Benot-Ya'aqov site in the Dead Sea Rift Zone. Quaternary Research 31, 371–376.

47. Goren-Inbar, N., Belitzky, S., Verosub, K., Werker, E., Kislev, M., Heimann, A., Cari, I., Rosenfeld, A., 1992. New discoveries at the Middle Pleistocene Acheulian site of Gesher Benot Ya'aqov, Israel. Quaternary Research 38, 117–128.

48. Goren-Inbar, N., Feibel, C.S., Verosub, K.L., Melamed, Y., Kislev, M.E., Tchernov, E., Saragusti, I., 2000. Pleistocene milestones on the out-of-Africa corridor at Gesher Benot Ya'aqov, Israel. Science 289, 944–948.

49. Guérin, C., Faure, M., 1988. Biostratigraphie comparée des grands mammifères du Pléistocène en Europe Occidentales et au Moyen Orient. Paléorient 14 (2), 50–56.

50. Guérin, C., Eisenmann, V., Faure, M., 1993. Les grands mammifères du gisement Pléistocène Moyen de Latamné (vallée de l'Oronte, Syrie). In: Sanlaville, P., Besançon, J., Copeland, L., Muhesen, S. (Eds.), Le Paléolithique de la vallée moyenne de l'Oronte (Syrie): Peuplement et environnement: British Archaeological Reports, International Series, 587, pp. 169–178.

51. Heimann, A., Steinitz, G., 1989. 40Ar/39Ar total gas ages of basalts from Notera 3 well, Hula Valley, Dead Sea Rift: stratigraphic and tectonic implications. Israel Journal of Earth Science 38, 173–184.

52. Heimann, A., Zilberman, E., Amit, R., Frieslander, U., 2009. Northward migration of the southern diagonal fault of the Hula pull-apart basin, Dead Sea Transform, northern Israel. Tectonophysics 476, 496–511.

53. Heller, J., 2007. A historic biogeography of the aquatic fauna of the Levant. Biological Journal of the Linnean Society 92, 625–639.

54. Holmes, J.A., 1996. Trace-element and stable-isotope geochemistry of non-marine ostracod shells in Quaternary palaeoenvironmental reconstruction. Journal of Paleolimnology 15, 223–235.

55. Hooijer, D.A., 1961. Middle Pleistocene mammals from Latamne, Orontes Valley, Syria. Annales Archéologiques de Syrie 11, 117–132.

56. Hooijer, D.A., 1965. Additional notes on the Pleistocene mammalian fauna of the Orontes valley. Annales Archéologiques de Syrie 15, 101–104.

57. Ivanovich, M., Latham, A.G., Ku, T.L., 1992. Uranium-series disequilibrium applications in geochronology, In: Ivanovich, M., Harmon, R.S. (Eds.), Uranium-Series Disequilibrium: Applications to Earth, Marine and Environmental Science, Second Edition. Oxford University Press, Oxford, pp. 62–94.

58. Jacobsen, K., 2005. Analyses of SRTM elevation models. Proceedings of the EARSeL Workshop 3D-Remote Sensing, 10–11

59. June, Porto, Portugal (European Association of Remote Sensing Laboratories). unpaginated CD-ROM. JPL, 1998. Seeing Earth's Surface in 3-D: Shuttle Radar Topographic Mission. Jet Propulsion Laboratory, California Institute of Technology, Pasadena, California. http:// www2.jpl.nasa.gov/srtm/factsheet_pub.pdf. JPL report 400–714 7/98. Available online.

60. JPL, 2005. Shuttle Radar Topographic Mission: The Mission to Map the World. Jet Propulsion Laboratory, California Institute of Technology, Pasadena, California. http://www2.jpl.nasa.gov/srtm/index.html. Available online:.

61. JPL (Jet Propulsion Laboratory), 1998. Shuttle Radar Topographic Mission: Mapping the World in Three Dimensions. Jet Propulsion Laboratory, California Institute of Technology, Pasadena, California. http:// www2.jpl.nasa.gov/srtm/factsheet_tech.pdf. JPL report 400–713 7/98. Available online.

62. Kafri, U., Kaufman, A., Magaritz, M., 1983. The rate of Pleistocene subsidence and sedimentation in the Hula Basin as compared with those of other time spans in other Israeli tectonic regions. Earth and Planetary Science Letters 65, 126–132.

63. Kasse, C., Bohncke, S.J.P., Vandenberghe, J., Gábris, G., 2010. Fluvial style changes during the last glacial–interglacial transition in the middle Tisza valley (Hungary). Proceedings of the Geologists' Association 121, 180–194.

64. Kay, S., Spruyt, P., Zielinski, R., Winkler, P., Mihály, S., Ivan, G., 2005. Quality checking of DEM derived from satellite data (SPOT and SRTM). Proceedings of the EARSeL Workshop 3D-Remote Sensing, 10–11 June, Porto, Portugal (European Association of Remote Sensing Laboratories). unpaginated CD-ROM. Kelley, S.P., 2002.

65. K–Ar and Ar–Ar Dating. Reviews in Mineralogy and Geochemistry 47, 785–818.

66. Kiliç, S., Evrendilek, F., Berberoğlu, B., Demirkesen, A.C., 2006. Environmental monitoring of land-use and land-cover changes in Amik Plain, Turkey. Environmental Monitoring and Assessment 114, 157–168.

67. Lister, A.M., Sher, A.V., 2001. The origin and evolution of the woolly mammoth. Science 294, 1094–1097.

68. Markova, A., 2007. Pleistocene chronostratigraphic subdivisions and stratigraphic boundaries in the mammal record. Quaternary International 160, 100–111.

69. McLaren, S., 2004. Characteristics, evolution and distribution of Quaternary channel calcretes, southern Jordan. Earth Surface Processes and Landforms 29, 1487–1507.

70. Mein, P., Besançon, J., 1993. Micromammifères du Pléistocène Moyen de Letamné. In: Sanlaville, P., Besançon, J., Copeland, L., Muhesen, S. (Eds.), Le Paléolithique de la vallée moyenne de l'Oronte (Syrie): Peuplement et environnement: British Archaeological Reports, International Series, 587, pp. 179–182.

71. Meyer, W., Stets, J., 2002. Pleistocene to recent tectonics in the Rhenish Massif (Germany). Netherlands Journal of Geosciences/Geologie en Mijnbouw 81, 217–221.

72. Modderman, P.J.R., 1964. On a survey of Palaeolithic sites near Hama. Annales Archéologiques Arabes Syriennes 14, 51–66.

73. Moshkovitz, S., Magaritz, M., 1987. Stratigraphy and isotope records of Middle and Late Pleistocene mollusks from a continuous corehole in the Hula Basin, Northern Jordan Valley, Israel. Quaternary Research 28, 226–237.

74. Mouty, M., Delaloye, M., Fontignie, D., Piskin, O., Wagner, J.-J., 1992. The volcanic activity in Syria and Lebanon between Jurassic and Actual. Schweizerische Mineralogische und Petrographische Mitteilungen 72, 91–105.

75. Muhesen, S., 1985. L'Acheuléen Récent Évolué de Syrie. British Archaeological Reports, International Series, 248.

76. Nash, D.J., Smith, R.F., 1998. Multiple calcrete profiles in the Tabernas Basin, southeast Spain: their origins and geomorphic implications. Earth Surface Processes and Landforms 23, 1009–1029.

77. Nash, D.J., McLaren, S.J., 2003. Kalahari valley calcretes: their nature, origin and environmental significance. Quaternary International 111, 3–22.

78. Nash, D.J., Smith, R.F., 2003. Properties and development of channel calcretes in a mountain catchment, Tabernas Basin, southeast Spain. Geomorphology 50, 227–250.

79. Perinçek, D., Çemen, Í., 1990. The structural relationship between the East Anatolian and Dead Sea fault zones in southeastern Turkey. Tectonophysics 172, 331–340.

80. Philip, G., Abdulkarim, M., Newson, P., Beck, A., Bridgland, D., Bshesh, M., Shaw, A., Westaway, R., Wilkinson, K., 2005. Settlement and landscape development in the Homs Region, Syria: report on work undertaken during 2001–2003. Levant 37, 21–42.

81. Picard, L., 1963. The Quaternary in Northern Jordan Valley. Proceedings of Israel Academy of Science and Humanities, 1, pp. 1–34.

82. Pirazzoli, P.A., Laborel, J., Saliège, J.F., Erol, O., Kayan, İ., Person, A., 1991. Holocene raised shorelines on the

Hatay coasts (Turkey): palaeoecological and tectonic implications. Marine Geology 96, 295–311.

83. Ponikarov, V. (Ed.), 1986. Geological Map of Syria (1: 1,000,000; 2nd Edition). Establishment of Geology and Mineral Resources (Ministry of Petroleum and Mineral Resources), Syrian Arab Republic.

84. Ponikarov, V., Kozlov, V., Artemov, A., Kalis, A., 1963a. Geological map of Syria, 1:200,000 scale, sheet I-36-XVIII, I-37-XIII, Homs-Trablus. Technoexport, Moscow, and Ministry of Industry, Damascus.

85. Ponikarov, V., Shatsky, V., Kazmin, V., Mikhailov, I., Aistov, L., Kulakov, V., Shatskaya, M., Shirokov, V., 1963b. Geological map of Syria, 1:200,000 scale, sheet I-36-XXIV, I-37-XIX, Hama-Latheqieh. Technoexport, Moscow, and Ministry of Industry, Damascus.

86. Rodriguez, E., Morris, C.S., Belz, J.E., 2006. A global assessment of the SRTM performance. Photogrammetric Engineering and Remote Sensing 72, 249–260.

87. Ruegg, G.H.J., 1994. Alluvial architecture of the Quaternary Rhine–Meuse river system in the Netherlands. Netherlands Journal of Geosciences/ Geologie en Mijnbouw 72, 321–330.

88. Sanlaville, P., 1988. Synthèse sur le paléoenvironnement. Paléorient 14, 57–60.

89. Schattner, U., Weinberger, R., 2008. A mid-Pleistocene deformation transition in the Hula Basin, northern Israel: implications for the tectonic evolution of the Dead Sea Fault. Geochemistry, Geophysics, Geosystems 9, Q07009. doi:10.1029/ 2007GC001937, 18 pp.

90. Schütt, H., 1988. Ergänzungen zur Kenntnis der Molluskenfauna oberpliozäner Süsswasser konglomerate Syriens. Archiv für Molluskenkunde 118, 129–143.

91. Searle, M.P., Sun-Lin, Chung, Ching-Hua, Lo, 2010. Geological offsets and age constraints along the northern Dead Sea fault, Syria. Journal of the Geological Society of London 167, 1001–1008.

92. Seyrek, A., Demir, T., Pringle, M., Yurtmen, S., Westaway, R., Beck, A., Rowbotham, G., 2007. Kinematics of the Amanos Fault, southern Turkey, from Ar–Ar dating of offset Pleistocene basalt flows: transpression between the African and Arabian plates. In: Cunningham, W.D., Mann, P. (Eds.), Tectonics of Strike-slip Restraining and Releasing Bends: Geological Society, London, Special Publication, 290, pp. 255–284.

93. Seyrek, A., Demir, T., Pringle, M., Yurtmen, S., Westaway, R., Bridgland, D., Beck, A., Rowbotham, G., 2008. Late Cenozoic uplift of the Amanos Mountains and incision of the Middle Ceyhan river gorge, southern Turkey; Ar–Ar dating of the Dűziçi basalt. Geomorphology 97, 321–355.

94. Sharkov, E.V., Chernyshev, I.V., Devyatkin, E.V., Dodonov, A.E., Ivanenko, V.V., Karpenko, M.I., Leonov, Y.G., Novikov, V.M., Hanna, S., Khatib, K., 1994. Geochronology of Late Cenozoic basalts in western Syria. Petrology 2, 385–394 [Russian original: Petrologiya 2, 439–448]..

95. Sharkov, E.V., Chernyshev, I.V., Devyatkin, E.V., Dodonov, A.E., Ivanenko, V.V., Karpenko, M.I., Lebedev, V.A., Novikov, V.M., Hanna, S., Khatib, K., 1998. Novye dannye po geochronologii Pozdnekainozoiskich platobazaltov severo-vostochnoi periferii Krasnomorskoi riftovoi oblasti (severnaya Siriya) [New data on the geochronology of the Late Cenozoic plateau basalt of the northeast margin of the Red Sea rift region (northern Syria)], 358. Doklady Akademii Nauk, pp. 96–99.

96. Shaw, A.D., 2008. The Earlier Palaeolithic of Syria: Settlement History, Technology and Landscape-use in the Orontes and Euphrates Valleys. Unpublished Ph.D Thesis, University of Durham.

97. Shaw, A.D. (in press). The Earlier Palaeolithic of Syria: Reinvestigating the Orontes and Euphrates Valleys. Southampton Monographs in Archaeology. Archaeopress, Oxford.

98. Sivan, N., Heller, J., Van Damme, D., 2006. Fossil Viviparidae (Mollusca: Gastropoda) of the Levant. Journal of Conchology 39, 207–219.

99. Sohn, H.G., Kim, G., Yom, J., 2004. Mathematical modelling of historical reconnaissance CORONA KH-4B imagery. The Photogrammetric Record 19, 51–66.

100. Tchernov, E., 1973. On the Pleistocene Molluscs of the Jordan Valley. Israel Academy of Science and Humanities, Jerusalem. Tchernov, E., 1987. The age of the 'Ubeidiya Formation, an Early Pleistocene hominid site in the Jordan Valley, Israel. Israel Journal of Earth Sciences 36, 3–30.

101. Tchernov, E., 1999. The earliest hominids in the southern Levant. Proceedings of the International Conference of Human Palaeontology, Orce, Spain, 1995. Museo de Prehistoria y Paleontologia, Orce, Spain, pp. 389–406.

102. Tolun, N., Erentöz, C. 1962. Hatay sheet of the Geological Map of Turkey, 1:500,000 scale. General Directorate of Mineral Research and Exploration, Ankara.

103. Ünay, E., Göktaş, F., Hakyemez, H.Y., Avşar, M., Şan, Ö., 1995. Büyük Menderes Grabeni'nin kuzey kenarındaki çökellerin Arvicolidae (Rodentia, Mammalia) faunasına dayalı olarak yaşlandırılması [Age determination of the sediments of the northern margin of the Büyük Menderes graben based on their Arvicolidae (Rodentia, Mammalia) fauna]. Bulletin of the Geological Society of Turkey 38, 75–80 (in Turkish with English abstract).

104. Ünay, E., Emre, Ö., Erkal, T., Keçer, M., 2001. The rodent fauna from the Adapazarı pullapart basin (NW Anatolia): its bearings on the age of the North Anatolian fault. Geodinamica Acta 14, 169–175.

105. Van Liere, W.J., 1960. Un gisement Paléolithique dans un niveau Pleistocène de l'Oronte à Latamné (Syrie). Annales Archéologiques de Syrie 10, 165–174.

106. Van Liere, W.J., 1961. Observations on the Quaternary in Syria. Berichten Rijksd. Ovdheidk. Bodemonderzoek 11, 7–69. Van Liere,W.J., 1966. The Pleistocene and Stone Age of the Orontes. Annales Archéologiques Arabes Syriennes 16, 7–30.

107. Van Liere, W.J., Hooijer, D.A., 1961. A paleo-Orontes level with Archidiskodon meridionalis (Nesti) at Hama. Annales Archéologiques de Syrie 11, 165–172.

108. von Koenigswald, W., Fejfar, O., Tchernov, E., 1992. Revision einiger alt- und mittelpleistozaner arvicoliden (Rodentia, Mammalia) aus dem östlichen Mittelmeergebiet ('Ubeidiya, Jerusalem und Kalymnos–Xi). Neues Jahrbuch für Geologie und Paläontologie, Abhandlungen 184, 1–23.
109. Westaway, R., 2002. The Quaternary evolution of the Gulf of Corinth, central Greece: coupling between surface processes and flow in the lower continental crust. Tectonophysics 348, 269–318.
110. Westaway, R., 2004a. Pliocene and Quaternary surface uplift revealed by sediments of the Loire-Allier river system, France. Quaternaire 15, 103–115.
111. Westaway, R., 2004b. Kinematic consistency between the Dead Sea Fault Zone and the Neogene and Quaternary left-lateral faulting in SE Turkey. Tectonophysics 391, 203–237.
112. Westaway, R., 2006. Investigation of coupling between surface processes and induced flow in the lower continental crust as a cause of intraplate seismicity. Earth Surface Processes and Landforms 31, 1480–1509.
113. Westaway, R., 2010. The relationship between initial (end-Pliocene) hominin dispersal and landscape evolution in the Levant; an alternative view. Quaternary Science Reviews 29, 1491–1500.
114. Westaway, R., 2011. Discussion of 'Geological offsets and age constraints along the northern Dead Sea fault, Syria' by Michael P. Searle, Sun-Lin Chung, and ChingHua Lo (Journal of the Geological Society, London, 167, 1001–1008, 2010). Journal of the Geological Society of London 168, 621–623.
115. Westaway, R., 2012. A numerical modelling technique that can account for alternations of uplift and subsidence revealed by Late Cenozoic fluvial sequences. Geomorphology 165–166, 124–143 (this issue).
116. Westaway, R., Maddy, D., Bridgland, D., 2002. Flow in the lower continental crust as a mechanism for the Quaternary uplift of south-east England: constraints from the Thames terrace record. Quaternary Science Reviews 21, 559–603.
117. Westaway, R., Bridgland, D.R., Mishra, S., 2003. Rheological differences between Archaean and younger crust can determine rates of Quaternary vertical motions revealed by fluvial geomorphology. Terra Nova 15, 287–298.
118. Westaway, R., Demir, T., Seyrek, A., Beck, A., 2006. Kinematics of active left-lateral faulting in southeast Turkey from offset Pleistocene river gorges: improved constraint on the rate and history of relative motion between the Turkish and Arabian plates. Journal of the Geological Society of London 163, 149–164.
119. Westaway, R., Demir, T., Seyrek, A., 2008. Geometry of the Turkey-Arabia and AfricaArabia plate boundaries in the latest Miocene to Mid-Pliocene: the role of the Malatya-Ovacık Fault Zone in eastern Turkey. eEarth 3, 27–35.
120. Westaway, R., Bridgland, D.R., Sinha, R., Demir, T., 2009a. Fluvial sequences as evidence for landscape and climatic evolution in the Late Cenozoic: a synthesis of data from IGCP 518. Global and Planetary Change 68, 237–253.
121. Westaway, R., Guillou, H., Seyrek, A., Demir, T., Bridgland, D., Scaillet, S., Beck, A., 2009b. Late Cenozoic surface uplift, basaltic volcanism, and incision by the River Tigris around Diyarbakır, SE Turkey. International Journal of Earth Sciences 98, 601–625.
122. Wilkinson, T.J., 1997. The history of the Lake of Antioch: a preliminary note. In: Young, G.D., Chavalas, M.W., Averbeck, R.E. (Eds.), Crossing Boundaries and Linking Horizons: Studies in Honor of Michael C. Astour on His 80th Birthday. CDL Press, Betheseda, Maryland, pp. 557–576.
123. Wright, V.P., Tucker, M.E., 1991. Calcretes: an introduction. In: Wright, V.P., Tucker, M.E. (Eds.), Calcrete. International Association of Sedimentologists Special Publication, No. 2. Blackwell Scientific Publications, Oxford, pp. 1–22.
124. Yener, K.A., Edens, C., Harrison, T.P., Verstraete, J., Wilkinson, T.J., 2000. The Amuq Valley Regional Project. 1995–1998. American Journal of Archaeology 104, 163–220.
125. Yurtmen, S., Guillou, H., Westaway, R., Rowbotham, G., Tatar, O., 2002. Rate of strikeslip motion on the Amanos Fault (Karasu Valley, southern Turkey) constrained by K–Ar dating and geochemical analysis of Quaternary basalts. Tectonophysics 344, 207–246.

CHAPTER 7

THE NEGATIVE RELIEF OF LARGE RIVER FLOODPLAINS

John Lewin[a], Philip J. Ashworth[b]

[a] Institute of Geography and Earth Sciences, Aberystwyth University, Llandinam Building, Penglais, Aberystwyth, SY23 3DB, UK

[b] Division of Geography and Geology, School of Environment and Technology, University of Brighton, Lewes Road, Brighton, BN2 4GJ, UK

ABSTRACT

Large floodplains have multiple and complex negative relief assemblages in which depressions fall below local or general floodplain surfaces at a variety of scales. The generation and dynamics of negative relief along major alluvial corridors are described and compared. Such depressions are significant for the storage and passage of surface waters, the creation of a range of riparian, wetland, lacustrine and flowing-water habitats, and the long-term accumulation of organic materials.

Working on trunk channel remnants, drowned valleys and subsidence basins, fluvial processes modify floodplain negative relief through differential erosion and sedimentation. Effectively this takes place in three genetic zones: rheic, transitional and perirheic. We show that transitional zones marginal to active channels significantly diversify form complexes, and we demonstrate the diachronous nature of zonal processes and the complex nature and pace of depression modification and infilling. Four less well-understood sets of coupled phenomena are assessed: (i) floodplains associated with discontinuous river banks, (ii) the scales and types of scroll bar generation, (iii) factors underlying the contrasts between meander and braidplain surface relief, and (iv) the generation and function of large floodplain wetlands and lakes.

The survival likelihood of surface negative relief relates to geomorphological connectivity; this is described for each of the rheic, transitional and perirheic zones. The implications for contemporary aquatic system management are discussed. A key to understanding and managing negative relief on large river floodplains, and their associated ecologies and sedimentation, is to quantify *both* sedimentological and hydrological river-floodplain connectivity.

INTRODUCTION

Large floodplains possess significant relief that may be picked out by standing or flowing water in wet environments (respectively 'lentic' and 'lotic' environments in ecology) (Fig. 1). Depressions may also be distinguished by wetland and other ecological communities. The negative components of this relief are multi-scale; they may form enclosed depressions from m to km across, or they may be morphologically-connected linear systems of a variety of shapes and sizes. In depth, they can be up to or exceeding the bank full depth of their major trunk rivers, with water levels that fluctuate with hydrological regimes.

Citation: John Lewina, , Philip J. Ashworthb, The negative relief of large river floodplains, doi:10.1016/j.earscirev.2013.10.014

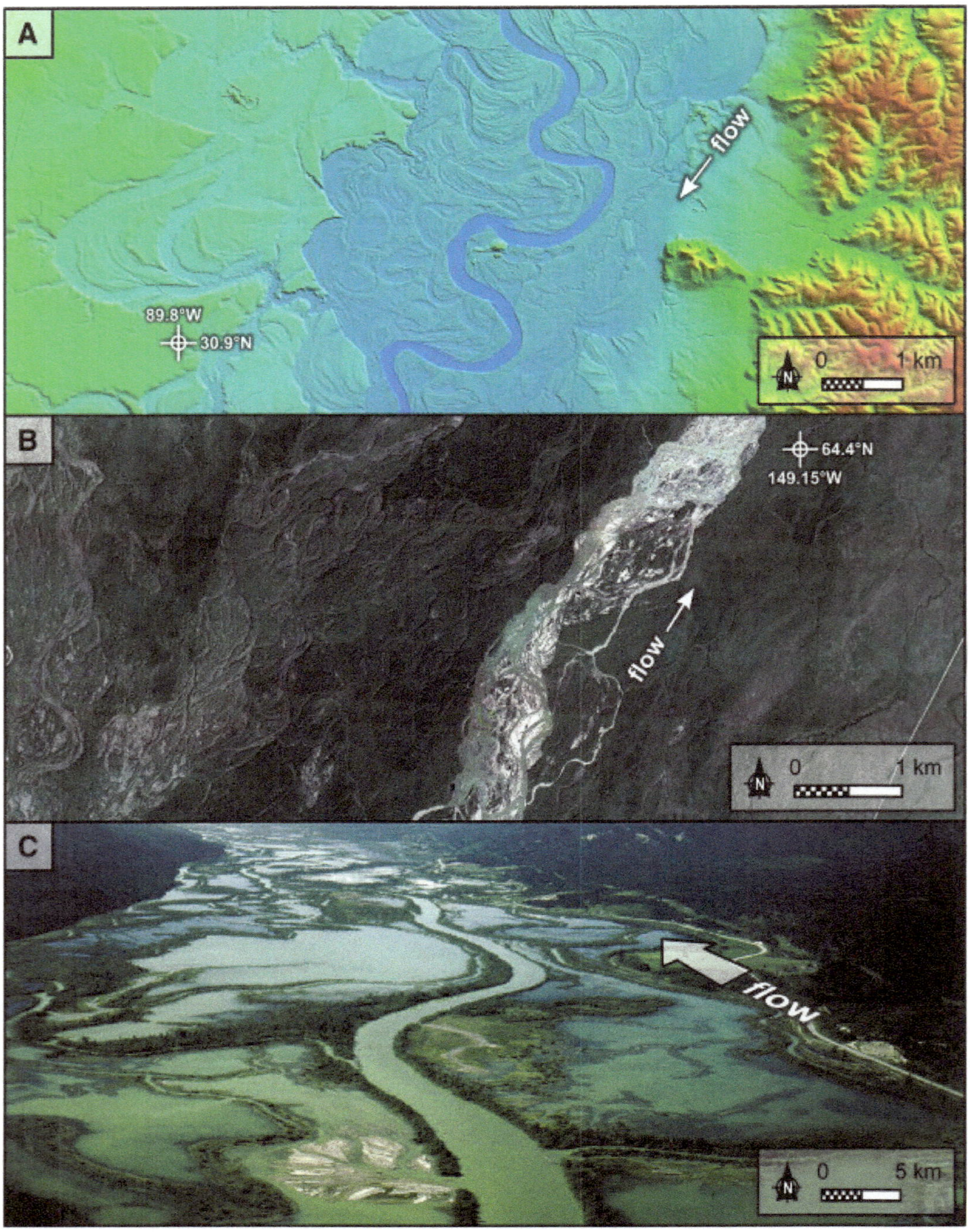

Figure 1: Negative relief elements in floodplain environments: (A) meandering Pearl River, Louisiana/Mississippi, USA. Digital elevation model (USGS DEM) from NE quadrant of Bogalusa East quadrangle, Louisiana, UTM 15 NAD83, Louisiana FEMA project — phase 1: Florida parishes, under USACE (2001) contract [3008910ne.dem]; (B) braided Nenana river, Alaska, USA. Image taken in 2006. Image from USDA-FSA aerial photography field office, Salt Lake City, Utah. Orthophoto mosaic of Fairbanks North Star Borough, Alaska; (C) anastomosing Columbia River, near Harrogate, British Columbia, Canada. Image courtesy of Derald Smith.

In recent decades, studies of the relief and form dynamics of larger floodplains have been facilitated by the wide availability of remote sensing imagery, by the use of digital elevation models (DEMs) and by the improved dating of sediments. Collectively these show that larger alluvial floodplains have great complexity of form, dynamics and age (Dunne and Aalto, 2013). For example, Trigg et al. (2012) used Landsat Pan-Sharpened, 15 m resolution imagery to 'reveal' extensive networks of floodplain channels on the Amazon (Fig. 2A–B). The floodplain 'channels' are typically up to 40 m wide and 10 s of m deep, and function to distribute contemporary floodwaters and sediment (Fig. 2B), though there is often no simple linkage with the geomorphology of the contemporary major trunk stream (Lewin and Ashworth, 2013). In origin, the depressions they occupy may be swales on point bars, partially infilled palaeochannels, tributary channels, or internally-generated floodplain drainage nets (Fig. 2B).

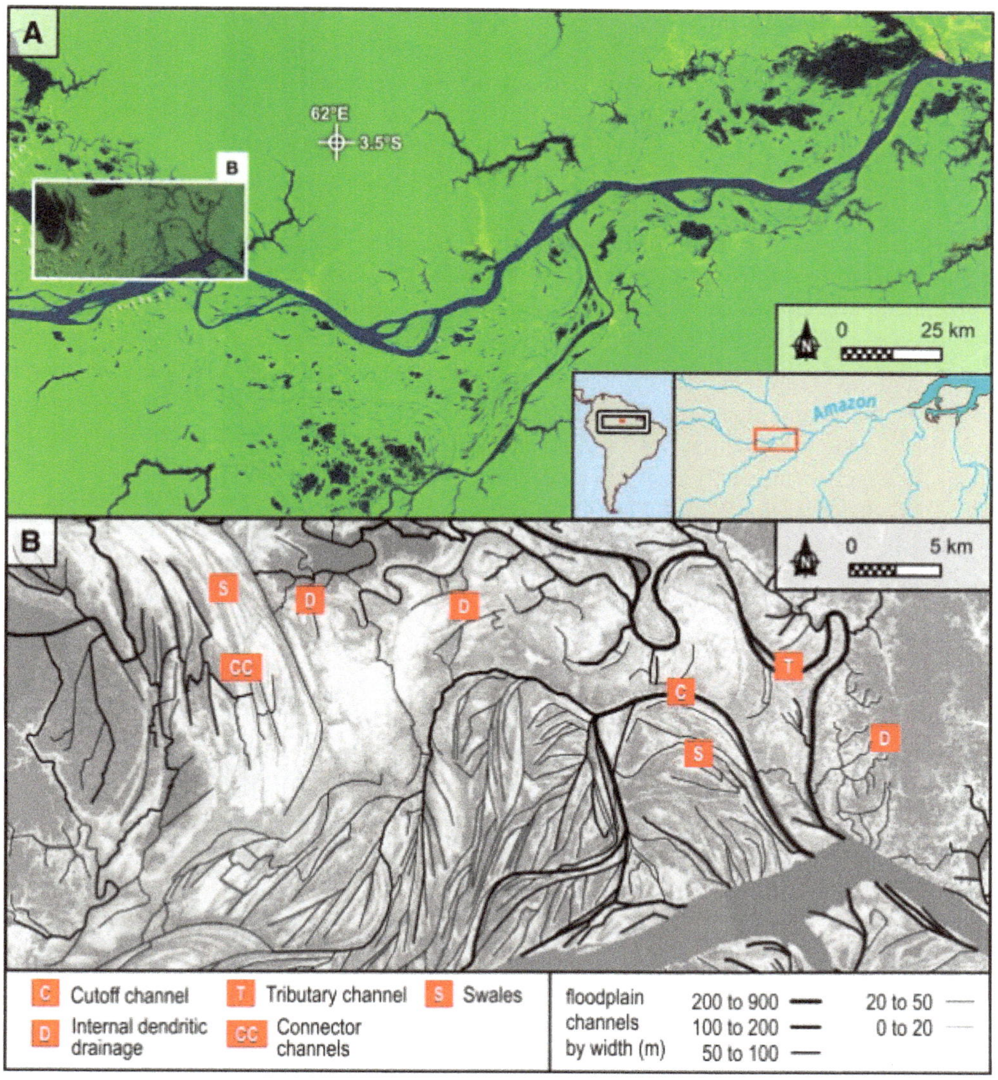

Figure 2: (A) 300 km reach of the Middle Amazon, centred on the municipality of Anori, Brazil, that was mapped byTrigg et al. (2012); (B) floodplain 'channels' mapped by Trigg et al. (2012) for the inset box in (A) using Pan-Sharpened Landsat imagery at spatial resolution of 15 m and local fieldwork. Channels are depicted proportional to their water widths; some swales and depressions are not mapped as water-filled. Labelling of the different types of floodplain drainage added by present authors. Note that in this transitional zone many of the floodplain channels are hydrologically-connected to the trunk stream, but that they may have a variety of origins in terms of geomorphological development. Landsat imagery in (A) courtesy of the U.S. Geological Survey. Image taken on 11 August 2001.

The diversity and heterogeneity in relief and form of large floodplains mean that the pioneering floodplain classification based on mainstream channel association by Nanson and Croke (1992) is less effective for large floodplains, because they are (i) plural (reflecting the activities of several channels and sub-systems, with a partial disconnect with trunk stream activity), (ii) complex (with zonal differences in processing and rates of activity), and (iii) diachronous (with forms that have developed over a range of timescales that include the effects of Quaternary and even earlier allogenic systems).

This paper focuses on the fluvial geomorphology that generates larger floodplain negative relief elements, (Table 1) concentrating on valley corridors beyond the reach of coastal and tidal processes where non-fluvial processes also come into play (e.g.,Fontana et al., 2008, Hijma and Cohen, 2011 and Blum et al., 2013). We (i) describe and classify negative relief forms, big and small; (ii) explain how negative-relief forms are differentiated zonally by mode of generation, dimensions and survival history; (iii) contrast the nature of floodplain relief in low-bank, meandering and braided environments; (iv) consider geomorphological processes for channelled and non-channelled flow in larger waterbodies; (v) document negative relief survival times; and (vi) show why it is important to understand the evolution and filling of negative relief for restoration and habitat management. Large rivers are defined loosely as the ones with channels greater than 100 m wide (cf. Miall, 2006, Latrubesse, 2008 and Ashworth and Lewin, 2012) and the forms examined here are meso- to macro-scale, from metres to kilometres across, within floodplains that are from c.10 to c.100 km in width. Their diachronous sub-aerial sedimentation surfaces incorporate boundaries ranging from Miall›s 3rd to 7th order (Miall, 2006, his Table 4.2). Extensive lake areas (Section 2), fluvial/lacustrine interactions (Section 7) and channelized flood flows (Trigg et al., 2012) are the highly significant responses to this relief along many large river systems.

Fluvial morphogenesis for large floodplains comes in several linked modes. First, there are features produced by active channel erosion (Section 3). Channel scour at first removes sediment, then as rivers relocate, accommodation space for sedimentation is opened up. Relict channel segments may nevertheless be left partially open for extended periods

(Section 8). On large river floodplains, these may formerly have been mainstream single or anabranching channels, or accessory and tributary ones (Ashworth and Lewin, 2012). As Gardner and Ashmore (2011) have argued, the scour by laterally migrating channels also creates a lower 'minimum surface' determining later deposit thickness for active systems. Over geological time periods, it is this 'combing' or 'working' depth (Paola and Borgman, 1991) that helps determine the preservation of sediments in the rock record both in-channel and as palaeoforms on and below the adjacent floodplain (Best and Ashworth, 1997, Brown et al., 2013 and Snedden, 2013).

Second, positive alluvial relief by sedimentation is created over a range of space and time scales (Table 2). Because the production and preservation of individual bedforms or sediment packages are spatially uneven (Bridge, 2003 and Horn et al., 2012), depressions are left at their margins and these may become long-lasting components of floodplain surfaces (3, 4, 5 and 6). Arising from such positive 'sediment occlusion', negative forms may be on a small scale, as in the swales between scroll ridges on point bars, or large-scale, as in the ponding of backswamps or 'impeded floodplains' (Latrubesse and Franzinelli, 2002) behind encircling levees or elevated channel belts, by blockage through fan extension, or between channel-belt alluvial ridges and bedrock valley sides. Such sedimentation may be observed in the short term using direct measurement or repeat imagery, but a timescale of centuries and millennia is required for understanding major floodplains. Essentially 'autogenic' processes (ones that are ongoing and associated with incident river regimes) overlap with 'allogenic' ones (resulting from environmental change). Palaeoform fragments in the form of fans, channel belts, islands and low terraces derived from earlier morphogenetic systems may be preserved on valley floors so as still to outline, or be incorporated into, regularly inundated areas.

Table 2: Meso- and macro-scale floodplain positive relief forms.

Genetic zone	Positive relief forms	Examples
(i)	1. Mid-channel bars (longitudinal, transverse, composite) 2. Side and point bars (including channel benches)	Moody et al., 1999; Ashworth et al., 2000; Santos and Stevaux, 2000; Best et al., 2003; Lunt and Bridge, 2004; Rice et al., 2009; Hooke and Yorke, 2011; Vietz et al., 2011
(ii)	3. Levees 4. Lateral accretion ridges and attached bars 5. Bed-material splays and channel plugs 6. Islands (commonly composites of 1, 4 and 5. May remain detached or incorporated into the general floodplain surface) 7. Overbank sedimentation, commonly infilling, in part, (ii) and (iii) in Table 1 8. Source-bordering aeolian dunes	Nanson, 1980, 1981; Brierley, 1991; Asselman and Middelkoop, 1995; Ferguson and Brierley, 1999; Aalto et al., 2003, 2008; Adams et al., 2004; Slingerland and Smith, 2004; Kemp and Spooner, 2007; Gurnell et al., 2012; Rozo et al., 2012
(iii)	9. Older abandoned valley floor islands 10. Substrate-erosional remnant islands 11. Abandoned valley-floor sides and alluvial terraces 12. Substrate-erosional valley sides	Latrubesse and Franzinelli, 2005; Bridgland and Westaway, 2008; Erkens et al., 2009; Blum et al., 2013; Lewin and Ashworth, 2013

Genetic zones

(i) Actively forming and emergent in-channel.

(ii) Transitional zone sedimentation being incorporated into floodplain surfaces and vegetated islands.

(iii) Palaeoforms in perirheic zone and at valley sides.

Thus a third major influence on inland valley floor forms, additional to the operation of current alluvial processes, concerns longer-term tectonics and prior geomorphological settings. Many large rivers have not currently infilled large-scale available sedimentation space in areas liable to contemporary flooding (e.g., Latrubesse and Franzinelli, 2005,Roy et al., 2011 and Rozo et al., 2012). Depressions on large river floodplains may be associated with foreland basins (peripheral, retroarc) including foredeep and backbulge, graben, half-graben and other fault complexes. Other neotectonic activities involving faulting, downwarping, and local river incision are also significant within large valleys (Mertes and Dunne, 2007 and Syvitski et al., 2012). Quaternary valley incision linked to sea level fluctuations has affected large rivers like the Amazon, in that case for some 2000 km inland. These have all provided opportunities for surface water ponding with only partial infilling by fluvial sedimentation.

Finally, there are ongoing post-formational activities. Overbank floodwater passage and groundwater drainage may dissect alluvial surfaces and connect prior depressions, so creating and maintaining internal dendritic drainage networks. Shoreline progradation, lateral encroachment of vegetation, and basal sedimentation operate to infill lakes. Covering all of these, fine sediment infilling and capping may eventually create a more planar floodplain. However, negative relief survives because infilling is neither immediate nor complete. This may be because erosion maintains internal drainage nets, or because sediment dispersal, by water or wind, has been inadequate to eradicate negative relief forms.

Floodplains have commonly been treated as alluvial sedimentation phenomena. Considerable progress has been made in quantifying the processes of floodplain accretion and sediment exchange (e.g. Pizzuto, 1987, Walling et al., 1996, Dietrich et al., 1999, Aalto et al., 2008, Day et al., 2008, Erkens et al., 2011, Van Dijk et al., 2012, Dieras et al., 2013, Khan et al., 2013 and Nicholas, 2013). Much less attention has been directed to understanding the origins, types and roles of floodplain negative relief and the impact this then has on sediment and organic sequestration and recycling (Aufdenkamp et al., 2011 and Bastviken et al., 2011).

Accordingly, this paper provides a different perspective: where has mineral sedimentation *not* been fully effective in levelling-up floodplain surfaces? The extended survival of negative-relief elements is important to different disciplines. At the event scale, it is of hydraulic significance for the guided passage of floodwaters, and for standing waterbody extent and dynamics (Trigg et al., 2013). These relate to impoundment; to replenishment by direct precipitation, overbank flows and groundwater rise; and to drainage and infiltration. Second, over seasons and years, biological habitats are created and maintained here in complex and diverse mosaics of freshwater, wetland and riparian ecosystems (Tockner and Stanford, 2002, Hohensinner et al., 2011 and van de Wolfshaar et al., 2011). This includes the generation of waterbodies that are involved in the transmission of diseases such as malaria (Smith et al., 2013). The different scales and relative heights of negative and positive relief, particularly on the very largest river floodplains, provide a range of sub-environments for different vegetation types to flourish and therefore to provide both floodplain stability and ecological diversity (Marchetti et al., 2013, Stevaux et al., 2013 and Valente et al., 2013). Third, negative relief elements that last for decades, and indeed very much longer, may accumulate organic deposits. These provide archival niches preserving a palaeoenvironmental record (Gibbard and Lewin, 2002), they operate as terrestrial carbon sinks and, over the long-term, give rise to hydrocarbon resources (Thomas, 2013).

Prevalence of negative relief in large rivers

Some of the world›s largest floodplains are characterised by extensive and water-filled negative-relief elements (e.g., Iriondo, 2004, Latrubesse et al., 2005, Ashworth and Lewin, 2012 and Latrubesse, 2010). For example, 55% of the floodplains on the 33 large rivers studied by Syvitski et al. (2012; their Table 1) contained deep floodplain depression zones. These include major negative relief associated with the Mompox depression in the Magdalena River system (9° 10′N, 74° 33′W), the Poyang depression and Lake Poyang (29° 07′N, 116° 17′E) and the Dong Ting at the intersection of the Yangtze and Xiangjiang rivers (28° 54′N, 112° 49′E). In South America, Sippel et al. (1992) reported that 11% of the mainstem Amazon floodplain in Brazil is covered by lakes (n = 6510) although only 10% of these lakes are larger than 2 km². In the Pantanal Basin in Brazil, accommodation space created by subsidence and increased humidity since the last glacial period has allowed the creation of an enormous seasonally flooded wetland dominated by alluvial sedimentation and is referred to as the world's largest continuous wetland (Alho et al., 1988, Assine and Silva, 2009 and Latrubesse, 2010). On the Paraná, Paira and Drago (2007) report a figure of nearly 50% for permanent or temporary lentic (lacustrine) waterbodies. Remote sensing imagery shows that many are humic or 'blackwater' lakes with minimal sediment input even during flood stage. On the Amazon near Manaus, open-water floodplain tracts include meander swales, ponded tributaries, and extensive backswamps behind channel-belt sediments (Latrubesse and Franzinelli, 2002 and Rozo et al., 2012). Hydraulic modelling of the Amazon flood wave along the main channel suggests that up to 30% of the main stem flow exchanges water with the floodplain (Richey et al., 1989). There are also small accessory sediment-rich channels (*paranás*) passing through the floodplain. These function especially at flood stage, when the system appears to operate as a giant elutriator allowing side channels to 'filter' water and finer sediment out onto the floodplain. These streams develop their own levees and may shift laterally (Mertes et al., 1996 and Rozo et al., 2012). Alluvial forms may be dominated by shallow lateral-accretion *várzea* derived from such accessory channel migration rather than from mainstream activity (Paira and Drago, 2007).

Table 1: Meso- and macro-scale floodplain negative relief forms.

Genetic zone	Negative relief forms	Examples
(i) Rheic	1. Main channels (single or anabranching) 2. Accessory through-channels 3. Tributary channels (including dammed tributaries with outlet channels)	Mertes et al., 1996; Sambrook Smith et al., 2006; Gupta, 2007; Iriondo et al., 2007; Latrubesse, 2008; Makaske et al., 2009; Ashworth and Lewin, 2012; Rozo et al., 2012; Lewin and Ashworth, 2013
(ii) Transitional	4. Channel margin slackwater zones 5. Bar-shelter backwaters* 6. Contiguous channel remnants (semi-detached cutoff arms*) 7. Tie channels 8. Internal drainage channel networks	Lewin, 1983; Rowland et al., 2005; Day et al., 2008; Constantine et al., 2010; Grenfell et al., 2012; Toonen et al., 2012; Trigg et al., 2012; Dieras et al., 2013
(iii) Perirheic	9. Cutoff palaeochannel segments* (cf. 1, 2 & 3 above) 10. Accretionary swales and irregular unsedimented voids* (cf. 4–8 above) 11. Large-scale flood basins occluded by channel-belt aggradation*	Panin et al., 1999; Paira and Drago, 2006; Werritty et al., 2006; Constantine and Dunne, 2009; Citterio and Piégay, 2009; Hoffmann et al., 2009; Sidorchuck et al., 2009; Gautier et al.,2010; Toonen et al., 2012

*occlusion-related.

Genetic zones.

(i) As components of geomorphologically active channel systems.

(ii) With hydrological and morphological connection to channels (one- or two-directional flows).

(iii) With morphological detachment from main channels (though with possible hydrological connection at high flows due to overbank inundation, tributary inflow, or groundwater rise) and limited geomorphological activity.

Cold environments can also have extensive floodplain relief but this is not simply created by flood flows during channel migration, but instead is associated with inherited glacial landscapes or present-day paraglacial and periglacial processes (Fig. 3A–B). Extensive thaw ponds and thermokarst features dominate such tundra environments but often these are disconnected from the main river channel except during snowmelt floods.

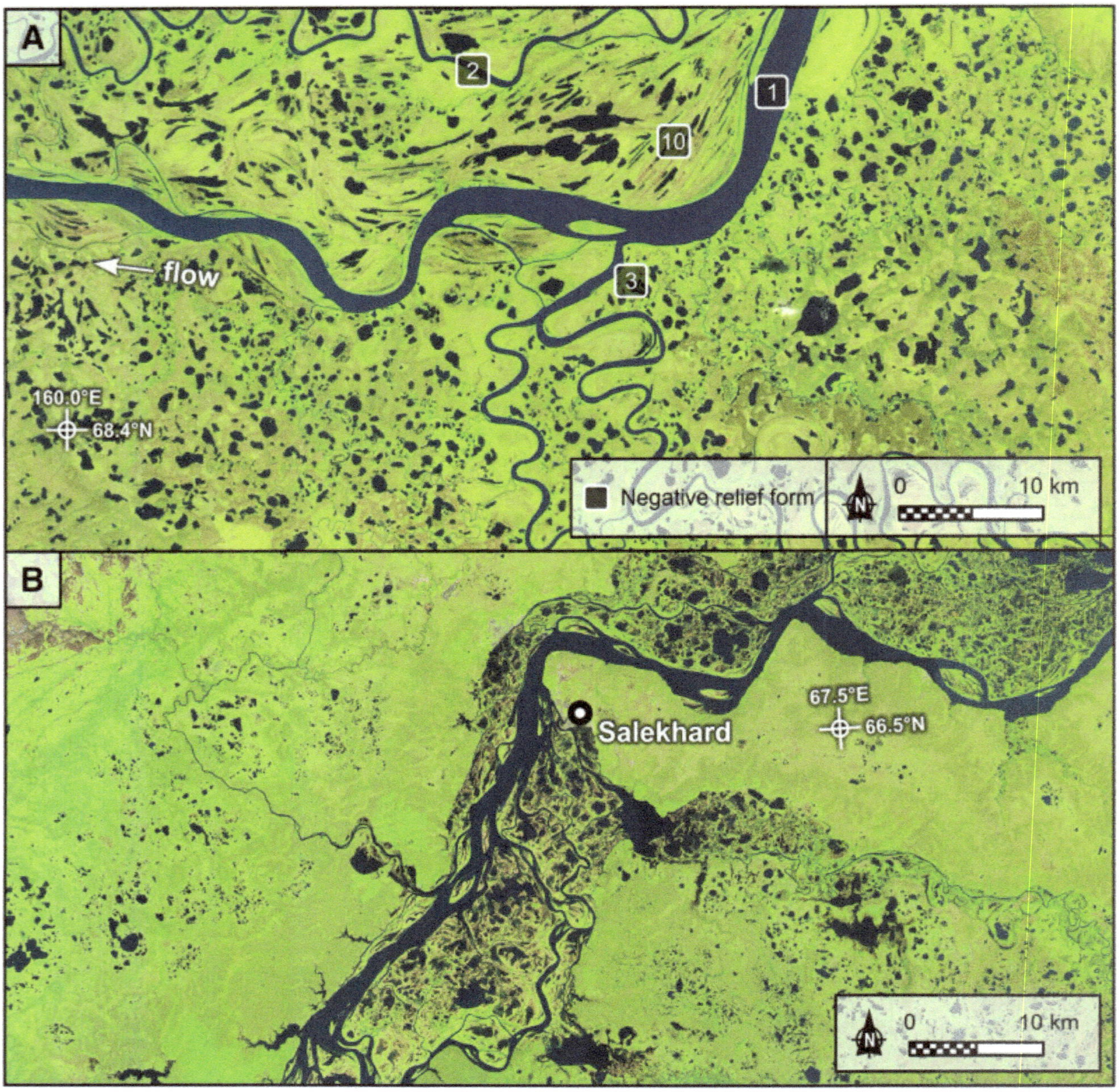

Figure 3: (A) River Kolymar, Russia (68° 24′ N, 160° 00′ E). Extensive thaw ponds and meandering systems (both main, accessory and tributary channels) with some lateral accretion ridges. Image taken on 27 July 2002; (B) River Ob (66° 32′ N, 67° 30′ E) near its mouth at Salekhard in NE Russia. Large thaw ponds up to 4 km wide. Image taken on 23 June 2011. Landsat imagery courtesy of the U.S. Geological Survey. Label numbers in this and later figures are as shown in Table 1 and Table 2.

Smaller floodplains with meandering rivers also have water-filled/wetland palaeochannels and backswamps, though in long-settled areas these may have been greatly modified because of sediment infilling following accelerated soil erosion (Lang et al., 2003 and James, 2013), or direct floodplain engineering and construction (Lewin, 2010, Lewin, 2013 and Hohensinner et al., 2011).

The prevalence of water-filled negative relief elements is not just governed by the hydrological and geomorphological connectivity between the river main-stem and floodplain drainage (Lewin and Ashworth, 2013). As Mertes (1997) noted, in some of the world's largest rivers the inundation of floodplains can be spatially and temporally variable and not just caused by overbank flooding events. As Fig. 4A–C illustrate, if the floodplain is essentially dry before overbank flooding from the river channel occurs, as the water rises the zone of flooding expands parallel to the river as levees are overtopped along the main channel and floodplain channels (Mertes, 1997). However, if water is already present on the floodplain – due to groundwater (or 'hyporheic' water), flooding of local tributaries, runoff from surrounding slopes, direct precipitation or antecedent water from prior floods – then local water ponding can resist the incursion of water across the surface of the floodplain (Fig. 4F). Saturation of parts of the floodplain may occur prior to the crest of the flood wave and the onset of overbank flooding (Fig. 4D–E). Thus, some of the world›s largest river floodplains are only partially inundated with water from main-stem rivers during floods (Mertes, 1997), and the local accumulation of floodplain water is not only governed by the morphology of floodplain relief but also the connectivity and interaction between surface and subsurface water. Subsurface hyporheic water connectivity, important both chemically and biologically (Castro and Hornberger, 1991 and Mertes, 1997), is only considered tangentially here.

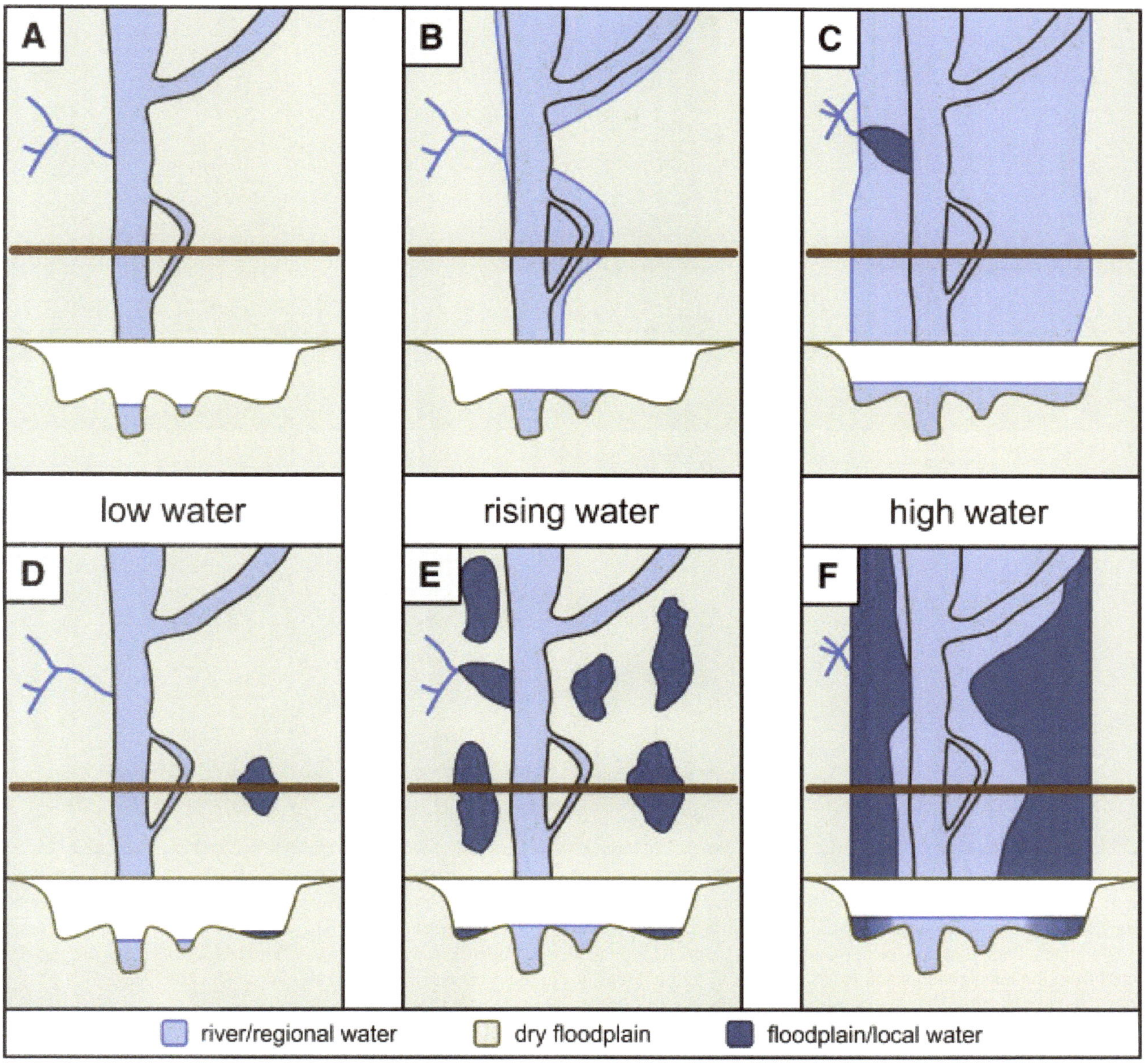

Figure 4: Inundation patterns for a dry (A–C) and saturated (D–F) floodplain at three different water levels. The upper graphs in each box are planform views of a typical river with cross-sections shown immediately below with a vertical exaggeration on the order of 100. Mixing of water types are shown for surface waters only. Redrawn from Mertes (1997).

NEGATIVE RELIEF ELEMENT GENERATION

The range of *fluvial* processes contributing to negative relief creation (Table 1) may usefully be resolved into three genetic zones that differ in terms of process activities and form replacement rates:

(i) Channel domains that are geomorphologically active. This rheic zone (that is, dominated by flowing surface water) includes main, secondary and accessory channels that have varying degrees of geomorphological activity and styles of channel pattern evolution (Fig. 5A).

Figure 5: Negative relief elements in the anabranching Rio Paraná, Argentina. (A) channel domains that are geomorphologically-connected and create elements such as topographic hollows, within-channel chutes and abandoned channels; (B) zones that are hydrologically-connected to the main channel and that may or may not receive influx of fines from suspended and washload; (C–D) perirheic zones that are mostly disconnected from the main channel and only fill during extreme floods or via sediment extrinsic to the fluvial system (see Table 3).

(ii) Geomorphologically transitional but hydrologically connected slackwater zones, backwaters and contiguous palaeochannel segments. These may or may not receive significant influxes of (generally finer) sediment to modify their forms (Fig. 5B).

(iii) Perirheic zones (Mertes, 1997) that are disconnected, both geomorphologically and in terms of hydrology, at all but extreme overbank flows or groundwater rise (Fig. 5C–D).

Most negative forms of fluvial origin are initiated in rheic domains (both main and other channels), they are transitionally but often considerably modified at rheic margins, and they may survive under slow infilling in perirheic environments. Boundaries may approximate those of abandoned channels, or may more closely be related to the margins of infilling sediment or occlusion bodies. The former reflect prior-channel dimensions such as width or radius of curvature. The latter are less likely to be simple elongate forms, and cover a considerable range of sizes, some marked by shelf-like inset fills, and lobate or digitate outlines. But the groups overlap: cutoff channels are plugged by occluding bed sediment (Constantine et al., 2010), and palaeochannels may be linked-up hydrologically with less regular between-bar depressions to give a beaded internal drainage network (van de Wolfshaar et al., 2011). Point bar scroll relief can also play a positive role in guiding chute development (Zinger et al., 2011) and the formation of 'bifurcate meander bends' (Grenfell et al., 2012), with swale irregularity itself having a feedback link to prior slow or rapid bar extension episodes. In addition to marginal sedimentation adjacent to channel flow, enclosure of negative relief areas takes place on a much larger scale in relation to valley-margin or tributary water-ponding beyond the reach.

NEGATIVE RELIEF IN RIVERS WITH LOW OR DISCONTINUOUS BANKS

Where rivers are laterally mobile, receding cut-banks on one side may not be rapidly or evenly balanced by sedimentation on the other because (i) point bars tend to be built at a lower elevation than cutbanks, and (ii) point bars tend to be shorter than the eroding portion of cutbanks because of channel curvature (Lauer and Parker, 2008). If a floodplain is in equilibrium this local imbalance should be evened-out primarily by overbank deposition (Dunne et al., 1998 and Aalto et al., 2008) and infill of negative relief elements such as abandoned stream courses or oxbow lakes. However, field studies show that diffuse sedimentation tends to be concentrated near the channel even on the largest river floodplains (Dietrich et al., 1999 and Aalto et al., 2008).

On the non-eroding river flank, the bank can be ill-defined or locally absent. Fine sediment and buoyant organic material deposition takes place at channel margins in slacker waters and flow-detached zones where erosion produces a widening channelway that is not fully occupied by laterally-relocating river stemflow. In some freely evolving meander trains (Fig. 6), this may be: (i) against concave banks from which channelways are receding, (ii) in bar-tail 'shelter' zones, and (iii) at the downstream end of convex bends that are translating down-valley (Lewin, 1983). Where unfilled, voids may remain as open water, and eventually these get incorporated into floodplains. Alternatively, fine sediment influx may infill these voids to produce concave valley benches (Nanson and Page, 1983) or counter-point bar deposits (Smith et al., 2009). A detailed study of the hydrodynamics of deposition suggests that this in-channel deposition may be particularly effective at well-below bankfull stage (Vietz et al., 2011). Such thick, but spatially limited bodies of organic-rich sediment, are quite unlike fine sediments deposited overbank, and may also have different sediment and biological signatures to palaeochannel fills that are topped-up only during out-of-channel flows (Moody et al., 1999). Stratified fills (or a lack of them) depend on sediment supply, with flood units that may be coarser than in palaeochannel fills deposited further away from mainstream activity (Toonen et al., 2012). Under anthropogenically modified conditions, where sediments and pollutants are fed into streams at lower discharges, these within-channel marginal sediment stores may become much more important (Asselman and Middelkoop, 1995 and Fryirs and Brierley, 2001).

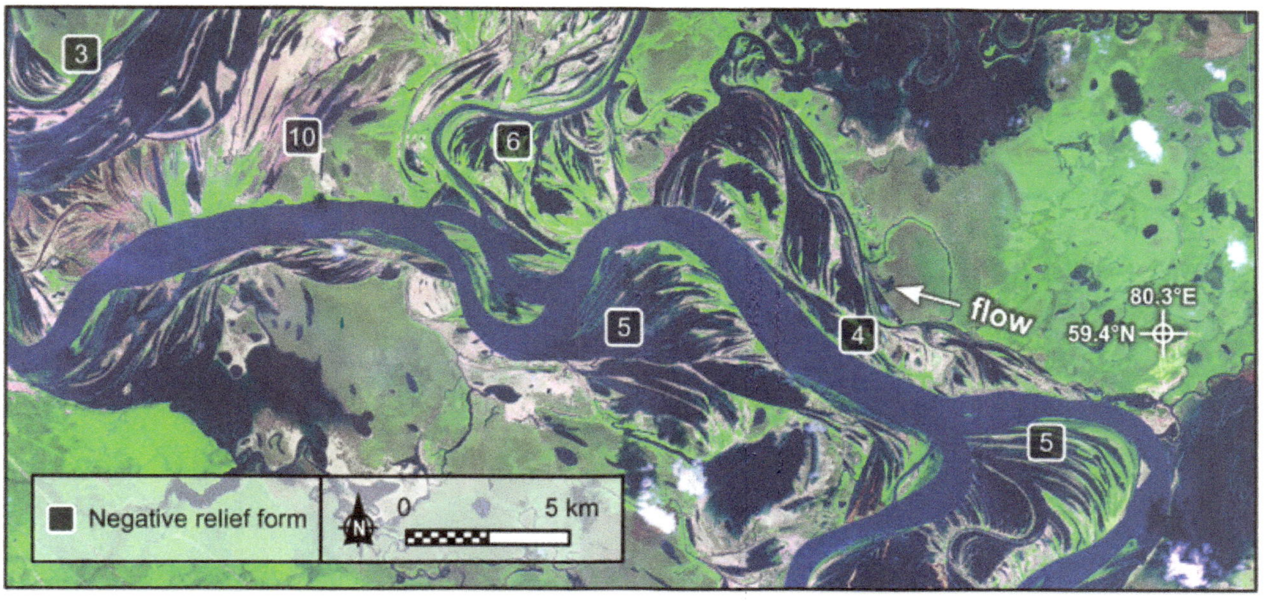

Figure 6: Point bar morphology on the River Ob (59° 23′N, 80° 02′ E). Note cutoff channels, arcuate swales and more irregular water-filled depressions. Landsat imagery courtesy of the U.S. Geological Survey. Image taken on 30 May 2011.

Lower 'bank' elevations may also form entry points for inundating water to back-up and pass out into extra-channel areas. Conversely, and especially during stage lowering and floodwater return, these locations may have inset scour channels at exit points for floodplain drainage. These channels can be extended by headward growth to form internal floodplain drainage networks. Floodwater headcuts may in some circumstances be an alternative 'bottom-up' mechanism for avulsive channel relocation or chute-cutoff generation (Gay et al., 1998, Tooth et al., 2007 and Zinger et al., 2011). To an extent, dryland rivers may also share some distinctive low-bank features, with aggradation and floodouts from intermittent flows, water transmission losses through percolation into alluvial deposits, and channel breakouts and splays (Tooth and Nanson, 2011 and Tooth, 2013).

NEGATIVE RELIEF CREATED BY SCROLL BAR DEVELOPMENT

Fig. 6 shows a reach of the River Ob in western Siberia. Depositional scrolls on the inside of currently-occupied and abandoned bends create an undulating topography that provides abundant sites for establishment of km-long narrow water bodies. Deposition follows lateral relocation of the river by erosion of the bank opposite, and scroll irregularity is a lagged reflection of the spatial and temporal pattern of bank erosion upstream and on the other side of the river. This gives the accreting channel margin a ridged and tooth comb appearance and there is a co-existing triple set of high-velocity main-channel stemflow, channel-side slackwater, and backwater arms. These have different degrees of intercommunication according to river stage. With low suspended sediment discharges but laterally mobile channels, there are greater opportunities for autochthonous organic deposition in irregular channel-side voids than in high suspended-sediment yield environments where the negative relief may be obliterated with fine sediment.

Point bar scrolls, despite their ubiquity and earlier pioneering research (Nanson, 1980 and Nanson, 1981), have attracted remarkably little attention as geomorphological elements contributing to positive and negative relief. Most detailed work has concentrated on understanding scroll bar evolution using either numerical models (e.g.,Howard, 1984, Blondeaux and Seminara, 1985, Parker and Johannesson, 1989, Sun et al., 1996, Parker et al., 2011, Motta et al., 2012 and Nicholas, 2013) or experimental work (e.g., Peakall et al., 2007, Dulal and Shimizu, 2010 and Van Dijk et al., 2012) (See alsoFig. 7A and E). Recent advances have also been made using high-resolution, three-dimensional seismic reflection data and drill cores that now provide unprecedented insights to ancient point bar and scroll morphology and sedimentology within hydrocarbon reservoirs that host some of the world›s largest petroleum reserves (e.g.,Smith et al., 2009, Hubbard et al., 2011, Fustic et al., 2012, Musial et al., 2012 and Moreton et al., 2013; see Fig. 7B–D). Pioneering numerical modelling work on the morphodynamics of the world›s largest anabranching rivers (Nicholas et al., 2012,Nicholas, 2013 and Nicholas et al., 2013) suggests that scroll-bar morphology can dominate the early stages of pattern development (Nicholas, 2013; his Fig. 3B, D, and F, p. 476) and that large negative relief elements may remain after channel abandonment (labels R in Fig. 7E). In the field, negative relief elements may be linear, but larger semi-rounded polyhedral forms are also possible where a combination of 'jerky' autogenic channel change and event-driven channel shifts are followed by only partial infilling.

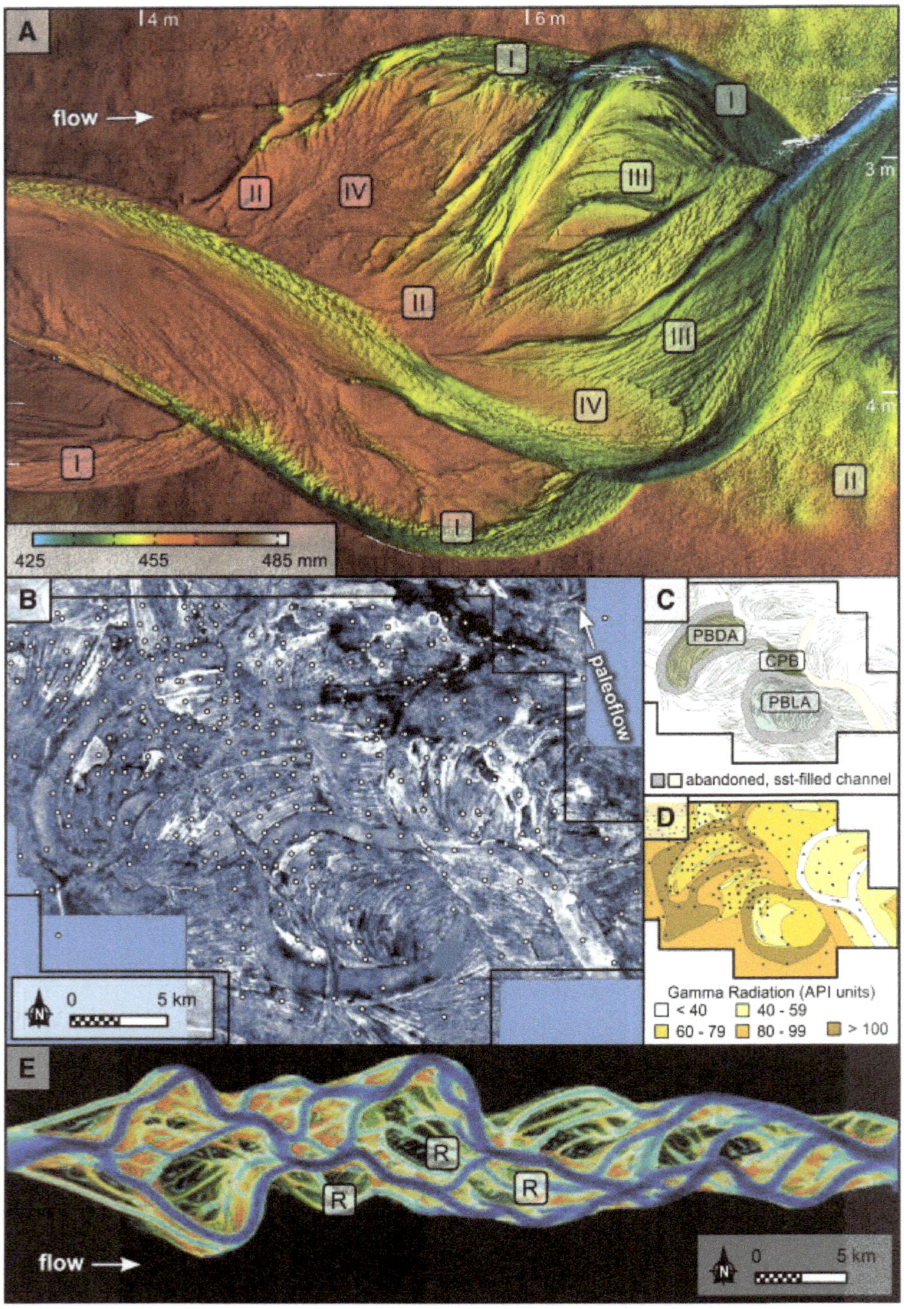

Figure 7: (A) Digital elevation model of the central region of a meandering experimental model after 191 h of run time. The channel cuts through its deposits and disconnects former active channels (I) by a plug bar (II). Locations of scroll ridges and swales (III), as well as bar accretion on the point bar complex (IV) are shown. From Van Dijk et al. (2012, their Fig. 6); (B–D) (B) Seismic time slice taken through the strata of interest, 8 ms (~ 8 m) from a marine flooding surface present near the top of the McMurray Formation. The white dots represent well locations. (C) Line-drawing trace of the main features in B. Some of the main depositional elements are highlighted, including abandoned channel, point bar associated with lateral accretion (PBLA), point bar associated with downstream accretion (PBDA), counter point bar (CPB), and sandstone-filled channel. (D) Gamma-radiation map, constructed from values measured from wireline logs (wells used indicated with black dots) at the same stratigraphic interval of seismic time slice. The map provides a proxy for averaged lithology across the depositional elements observed in B and C. For example, the abandoned channels filled with siltstone are associated with the highest average gamma-radiation values observed. (B–D) from Hubbard et al. (2011, their Fig. 2, p.1126) with the depositional environment interpreted as dominantly fluvial with some tidal influence. (E) Example of a simulated channel morphology for a large anabranching river (from Nicholas, 2013; his Fig. 3G, p. 476). Image shows the morphology at low flow (10000 $m^3 s^{-1}$). Label R shows abandoned scroll morphology as discussed in the text.

Individual ridges may prograde down-channel (Fig. 6) as a meander bend translates (Hooke and Yorke, 2011) or grow in arcuate form as the bend extends (Fig. 8A), with suspension material contributing to vertical growth (Rozo et al., 2012). Often, however, point bar scrolls accrete in a direction that is influenced by the local channel planform position at the time of maximum sediment transport, creating a mosaic of depositional packages (Fig. 8A), several scrolls wide and separated by an erosional-chute (labelled E in Fig. 8C) in between each patch of deposition. Individual scrolls on the Mississippi are up to 2 m high with erosional chutes 2–3 m deep (Fig. 8C). Elevated scrolls usually consist of finer traction or suspended load, and commonly have irregularly spaced negative-relief swales between them, though Fig. 8B–C show this is not always the case. Scroll relief is progressively smoothed away from the main channel as individual swales are filled with fine material (Fig. 8B). In planform, individual point bar scroll geometry is usually convex (e.g. Bridge et al., 1995 and Leclerc and Hickin, 1997), but counterpoint bars have scrolls that are concave and always locate down-river from or distal to the adjacent point bar (Smith et al., 2009).

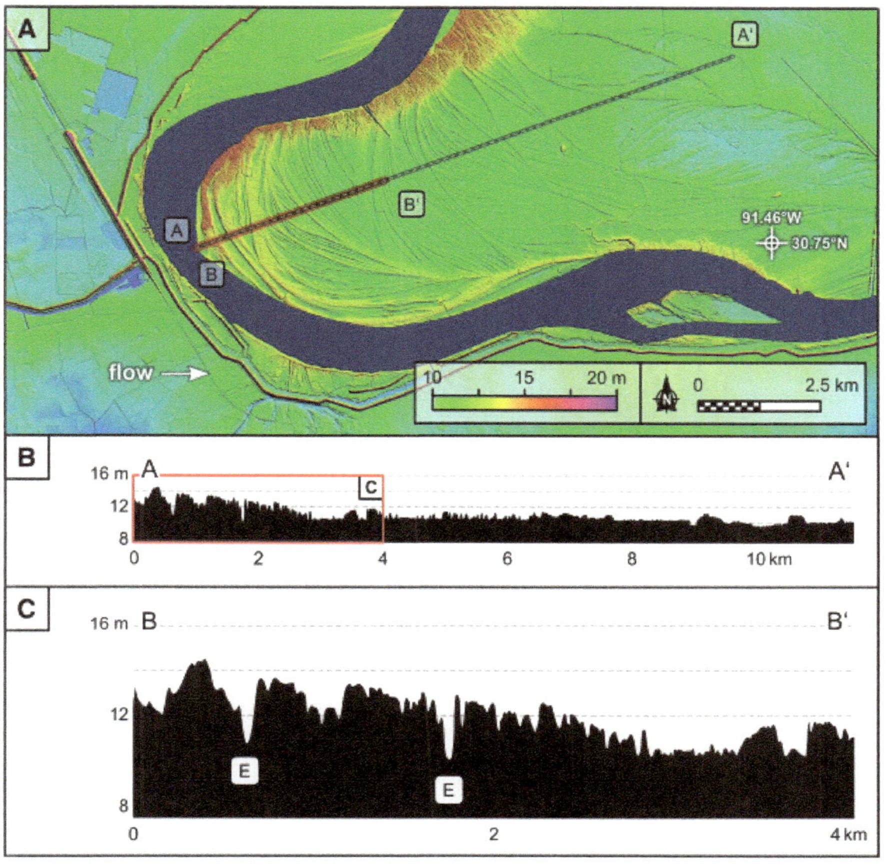

Figure 8: (A) LiDAR image of scroll bar morphology on the Mississippi at Morganza, Louisiana, USA. Digital elevation model (USGS DEM) from Task Area 11 – North, Louisiana, Louisiana/Federal Emergency Management Agency (FEMA) project – phase 3 of Louisiana. Image taken in 2004. Note the patchwork of different accretionary packages related to different channel planform configurations and separated by up to 3 m deep chutes (labelled E); (B) cross-section through the scroll bar showing the gradual filling of negative relief (swales) further away from the main channel, and higher sedimentation levels near the contemporary channel; (C) detailed cross-section showing ridges up to 1 m high and mostly symmetrical, that are separated by narrower swales.

Also on large and relatively straight rivers, compound sets of straight levee complexes may form as rivers oscillate from side to side and around islands (e.g., Rozo et al., 2012). These are rather different from lateral-accretion point bar scroll sets that are found on actively migrating secondary channels (Mertes et al., 1996 and Latrubesse and Franzinelli, 2002). Short-term levee growth has been documented for larger rivers byAalto et al., 2003 and Aalto et al., 2008. On large rivers transporting fine sediment, what may look like sets of accretionary bedforms may in fact be serial sets of extra-channel levees created as channels oscillate slightly in position over a timescale of many centuries.

In rivers with high suspended sediment loads, overbank and inner accretionary bank sedimentation, as well as bed-material point bars, closely track evolving planforms (Brooks, 2003, Page et al., 2003 and Aalto and Nittrouer, 2012). Floodplain surfaces may be more planar, some with ridges developed in finer sediment, accentuating relief as floodwaters are channelled along swales depositing material preferentially on growing ridges between them. Coarser bed-sediment point bars may be weakly developed or absent. By contrast, other meander plains may not only have both cutoff and quasi-regular point bar swales tracking former channel migrations, but also more complex sets of irregular depressions left when combing depth pockets remain unfilled.

For many meandering rivers, cutoff channels may dominate floodplain relief (Figs. 1A, and 6) with their lengths related to parent-channel sinuosity (Constantine and Dunne, 2009). Water-filled or wetland arcuate forms may be coupled with relatively dry and smaller-scale point bar scrolls and swales

(Fig. 9A–B). This applies to large tributaries of the Amazon, like the Purus and the Juruá, although, like the Amazon itself and its scroll-dominated plain, slight incision may mean that some alluvial forms occupy a low terrace (Latrubesse and Franzinelli, 2002 and Latrubesse and Kalicki, 2002). The Purus high-sinuosity courses developed neck cutoffs with little counterpoint voiding, though they may also have secondary channels and long multi-arc abandoned reaches (e.g., at 4° 42′ S, 66° 45′ W). Meander bend cutoffs through erosion of a new channel across the neck of a bend ('chute cutoff') are less common in large rivers (cf. Lewis and Lewin, 1983), but when this does occur it can trigger the rapid delivery of sediment into the river at rates that are one to five orders of magnitude larger than those produced by lateral migration of individual bends (Zinger et al., 2011). Shifting confluence zones (Rice et al., 2008) may also locally create deep scours and what Brown et al. (2013) call 'soft avulsions' over extended time periods.

Figure 9: (A) Meander morphology on the Rio Juruá (6° 31′ S, 68° 34′ W); cutoff domination, with ridges and swales. Image taken on 18 August 2011; (B) meander morphology on the Mississippi, USA (34° 39′ N, 90° 30′ W). Image taken on 2 November 2011. Landsat imagery courtesy of the U.S. Geological Survey.

MEANDER AND BRAIDPLAIN COMPARISONS

In meandering rivers, palaeochannel dimensions may be maintained after cutoffs or avulsions occur. Curving forms can retain their recognisable bank outline on floodplains until filled by subsequent sedimentation, often over long time periods (Panin et al., 1999,Werritty et al., 2006, Citterio and Piégay, 2009, Kleinhans et al., 2011, Grenfell et al., 2012 and Toonen et al., 2012). On braided rivers, the extra-channel equivalents of meandering system floodplains survive in the form of vegetated islands and surfaces (Reinfelds and Nanson, 1993, Thorne et al., 1993, Gurnell et al., 2009, Ashworth and Lewin, 2012 and Ham and Church, 2012). This may arise because of channel relocation, avulsive abandonment, incision, or in the longer-term in relation to glacial/interglacial cycles. Replacement by less extensive and confined channel styles has occurred in mid-latitudes with Quaternary transitions from cold to warm stage alluvial systems that were less laterally extensive (Gibbard and Lewin, 2002 and Lewin and Gibbard, 2010).

Braidplains exhibit more complex spatial and temporal relationships than in the distinctive three-stage sequence found in many meandering rivers (channel migration, cutoff and plugging, and overbank sediment infilling). Within active braided rivers, some channels are essentially geomorphologically inactive, although conveying water at high river stage (Ashmore, 1991 and Ashmore et al., 2011). Accommodation space is set by deep channel scour (Best and Ashworth, 1997), with convergent and divergent flows creating evolving bedforms with island development, and new channels created by avulsive switching of main-channel flow (Kleinhans et al., 2013). As switching occurs and channel scour wanes, bed sediment may still be fed through diminishing branches and deposited there in sheets and lobes, obliterating the outlines of the once major channels (Lunt and Bridge, 2004, Rice and Church, 2010 and Ashworth et al., 2011). For surviving alluvial surfaces, a critical difference between meander and braid plains is that channel abandonment has this longitudinally extensive waning, infilling and bed-material 'choking' process (Cheetham, 1979 and Leddy et al., 1993). This is coupled with ongoing reworking of high-flow forms, and multi-channel abandonment results in a latticed surface of partially infilled palaeochannels and islands, rather than that produced by the three-stage cutoff one of meandering rivers (Toonen et al., 2012). Relict depressions may be narrow, linear and beaded along the line of a former channel, reflecting a history of occluding bed-sediment infilling. Relict channels may also preserve former confluence/diffluence zones giving Y-shaped branching forms with deep intersect scour pools (Kleinhans et al., 2013). Additionally, positive relief forms may also incorporate surface depressions, as in the case of infilled bar-top hollows or the stacking of infilled scours produced in the lee of large dunes (Best et al. 2006).

Fig. 10A–B show reaches of the braided Brahmaputra and Indus rivers. These have multiple high-flow channels and bedforms exposed at low flows, together with more permanent islands. The positive relief forms (Table 2) identifiable include mid-channel and side bars within active channels; lateral accretion ridges; and channel-choking infill. As channels shift, positive and negative forms evolve interactively within the zone of active bedload transport (Fig. 11). The braidplains of large rivers may be completely or substantially reworked over decadal time periods (see location F in Fig. 11D). Although only ~ 600 m wide, Sambrook Smith et al. (2010) showed that the braidplain of the sandy South Saskatchewan River can be completely reworked in a 1:40 year flood, and on the same river Lunt et al. (2013) used geophysical surveys to show that during 'normal' flood years the lengths and widths of individual bars may be truncated by up to 90% before preservation in the subsurface. Fig. 12 illustrates the rapidity of channel change that is possible in some sandy braided rivers, and hence the rate of turnover from positive to negative relief and vice-versa. Vegetated bars can remain relatively immobile and are trimmed at the margins (label X in Fig. 12A) but unit bars (label U in Fig. 12A–B) migrating through anabranches at rates of up to 1.5 m day^{-1} are essentially transient features and usually only preserved as fragments in the subsurface (Parker et al., 2013). Up to 3 m deep scours can be created in the thalwegs and adjacent to resistant bank edges (label S in Fig. 12A–B) but these are filled rapidly by migrating dune bedforms (label D inFig. 12C). Likewise, new scours can be created, abandoned and start to fill in less than a year (label A in Fig. 12C).

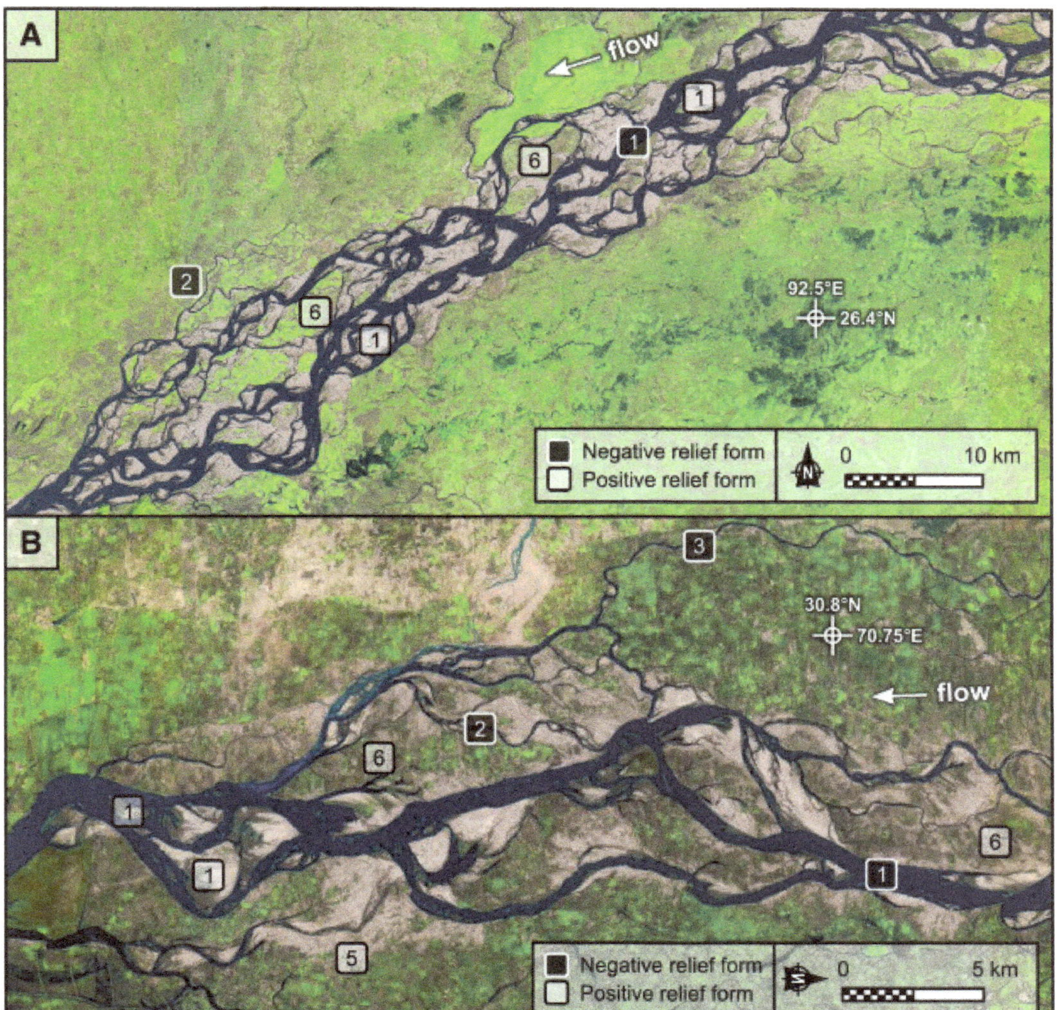

Figure 10: (A) Braiding morphology of the Brahmaputra River, India (26° 21′ N, 92° 07′ E) showing vegetated islands and multiple dry channels (at this low-flow stage). Dimensions of former main channels are not preserved. Image taken on 10 November 2011; (B) braiding on the Indus River, Pakistan (30° 36′ N, 70° 50′ E). Large bar splays in active channel, and secondary 'choking' channels. Image taken on 18 Decem-

ber 2010. Landsat imagery courtesy of the U.S. Geological Survey.

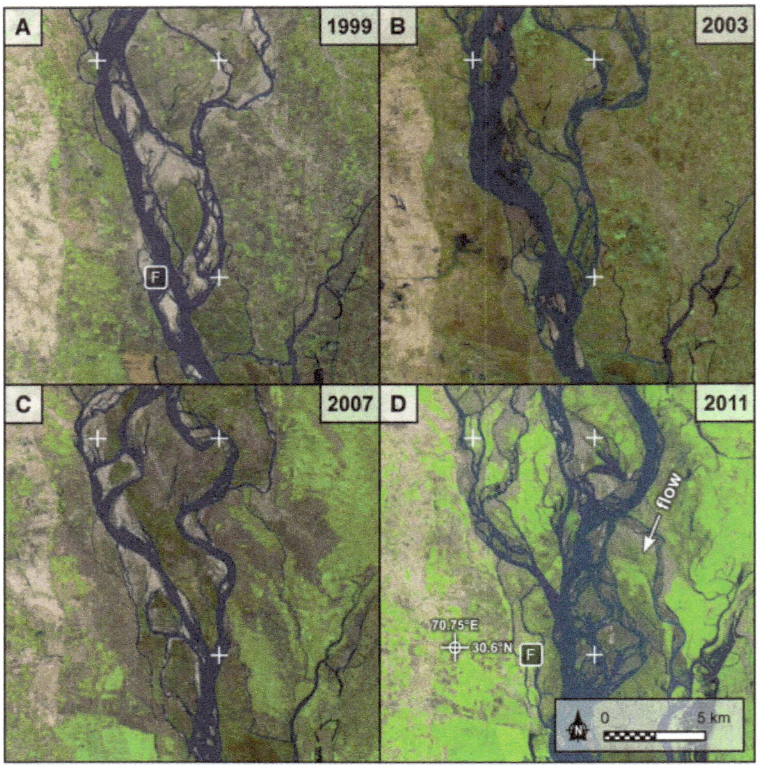

Figure 11: (A)–(D) Indus River at four time periods showing rapid sedimentation/switching on a large river. Note that Fig. 11C has been edited to remove the bands of black line that occur on Landsat 7 images due to a failing sensor. Crosses on image allow location of channels through different time periods. Label F shows a zone of the braidplain that was active channel in 1999 but then abandoned, and vegetated floodplain in 2011. Images taken on (A) 26 November 1999, (B) 27 April 2003, (C) 31 October 2007, (D) 2 October 2011. Landsat imagery courtesy of the U.S. Geological Survey.

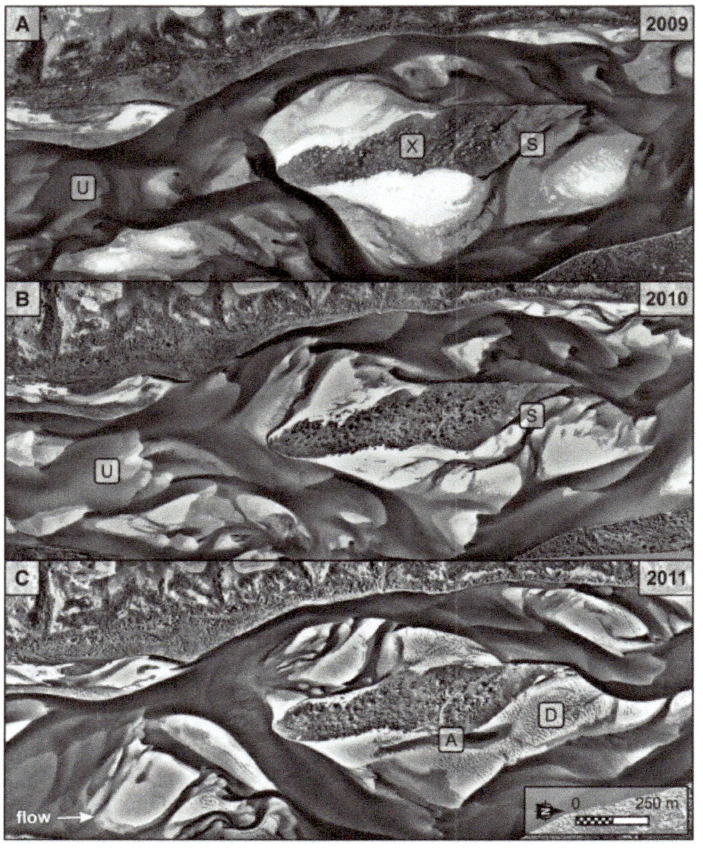

Figure 12: Aerial photographs of the South Saskatchewan River, Canada taken on (A) 23 October 2009, (B) 18 September 2010, and (C) 11 October 2011. Note the rapidity of channel change and therefore the rate of turnover of negative and positive relief. Labels are described in the text.

Thus the floodplain architecture resulting from braiding may actually have few width-preserved palaeochannel segments but rather depression 'strings' along their former alignments together with bedload fill outlines and drainage nets. The individual negative relief elements of braidplains are in general less pronounced, more complex in shape, and more frequent in number than in most meandering ones, though some of the latter (like the Ob, Fig. 6) have some of these characteristics as well.

Braided river process studies have commonly focused on 'early stage' processes: dune, bar and island development; bed scour and lateral accretion in active channels; and chutes, bifurcations and avulsions. Later-stage in-channel reworking and waning channel-infilling also contribute considerably to the relict topography of braidplains.*Relict* negative relief forms include isolated elongate scour holes (produced at former bends, bifurcating channels and channel convergence points) and extensive dendritic 'channel' talwegs, downstream of and between trimmed bar forms, and converging on the next set of scours (Doeschl et al., 2006, Lane, 2006 and Kleinhans et al., 2013). The trimming and anabranching patterning of low flow channels may only in part bring out the complexity of relict lobate/dendrite forms created at high flows (Fig. 13). In section, braided river deposits may be dominated by lower-order inclined and horizontal strata sets and, unlike meandering systems, have few distinguishable bounding surfaces that conform to the dimensions of abandoned main channels (Bridge and Lunt, 2006,Ashworth et al., 2011, Lunt et al., 2013 and Reesink et al., 2013). Quaternary braided-channel fills and surfaces may be equally lacking in major palaeochannel definition (Ashworth et al., 1999 and Lewin and Gibbard, 2010).

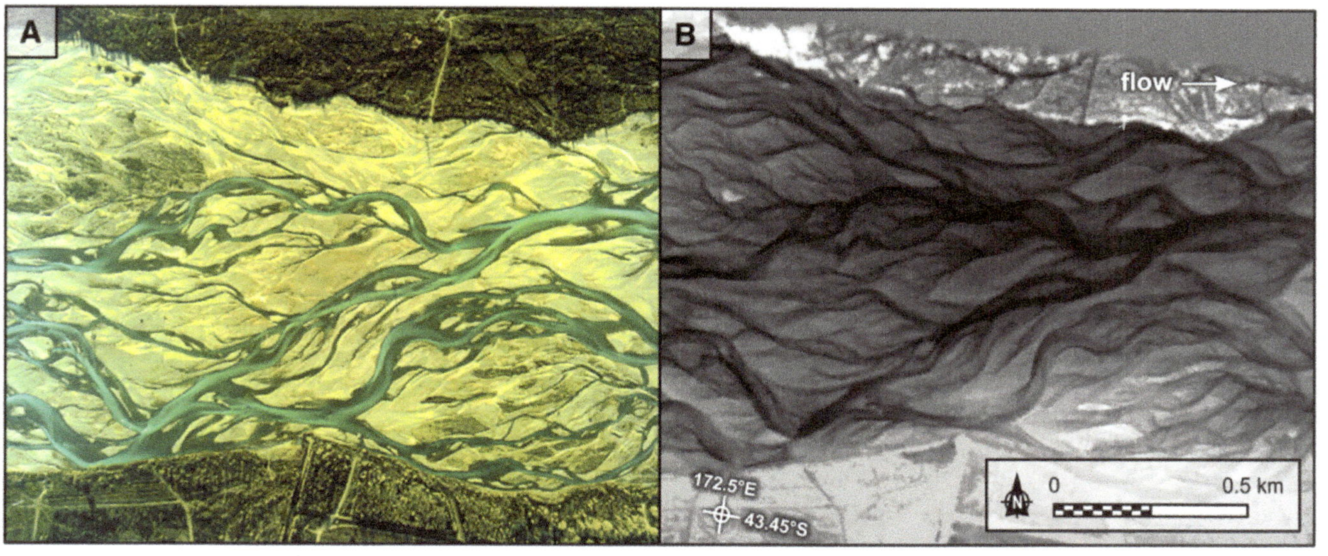

Figure 13: (A) Air photo of the Waimakariri; (B) DEM of the Waimakariri River, New Zealand (43° 26′ S, 172° 29′ E). Note the relatively simple braided pattern of water-filled channels in (A), but the complex pattern of relief with widespread negative elements in (B). This includes attached dendritic elements as well as longitudinally connected channels. Images courtesy of Stuart Lane. Further details on the image processing technique are in Lane et al., (2003) and Westaway et al. (2003).

CHANNELLED AND NON-CHANNELLED WATER IN MAJOR FLUVIAL WETLANDS

According to Fraser and Keddy (2005), five of the ten largest wetlands in the world are riverine. These border the Amazon, Congo, Paraná (including the Pantanal), Mississippi and Nile. At the present time, other wetlands in cold-climate taiga and tundra zones, as in the Mackenzie River catchment or West Siberian Lowlands, are generally string bogs along minor rivers, or raised interfluve bogs. Some floodplain-located bogs effectively form positive relief forms there that have developed entirely through vegetative growth. Major river floodplains also have large bodies of seasonally open water in occlusion-impeded floodplain drainage zones, or dammed tributary valleys (Ashworth and Lewin, 2012 and Latrubesse, 2010).

Many large tropical rivers replenish extensive associated waterbodies (Fig. 14A–C), together with seasonal flooding of forest and savannah bottomlands (Iriondo, 2004, Paira and Drago, 2006 and Paira and Drago, 2007). This arises in particular because sedimentation has been linear and localized in channel belts (Fig. 14A–C) and inadequate to infill intra-cratonic basins and the wide linear depressions that major rivers occupy (Mertes et al., 1996, Archer, 2005, Latrubesse and Franzinelli, 2005, Iriondo et al., 2007, Valente and Latrubesse, 2011 and Latrubesse et al., 2013). These include subsiding graben structures, valleys excavated in relation to lower relative sea levels, and tributaries back-ponded by major river aggradation.

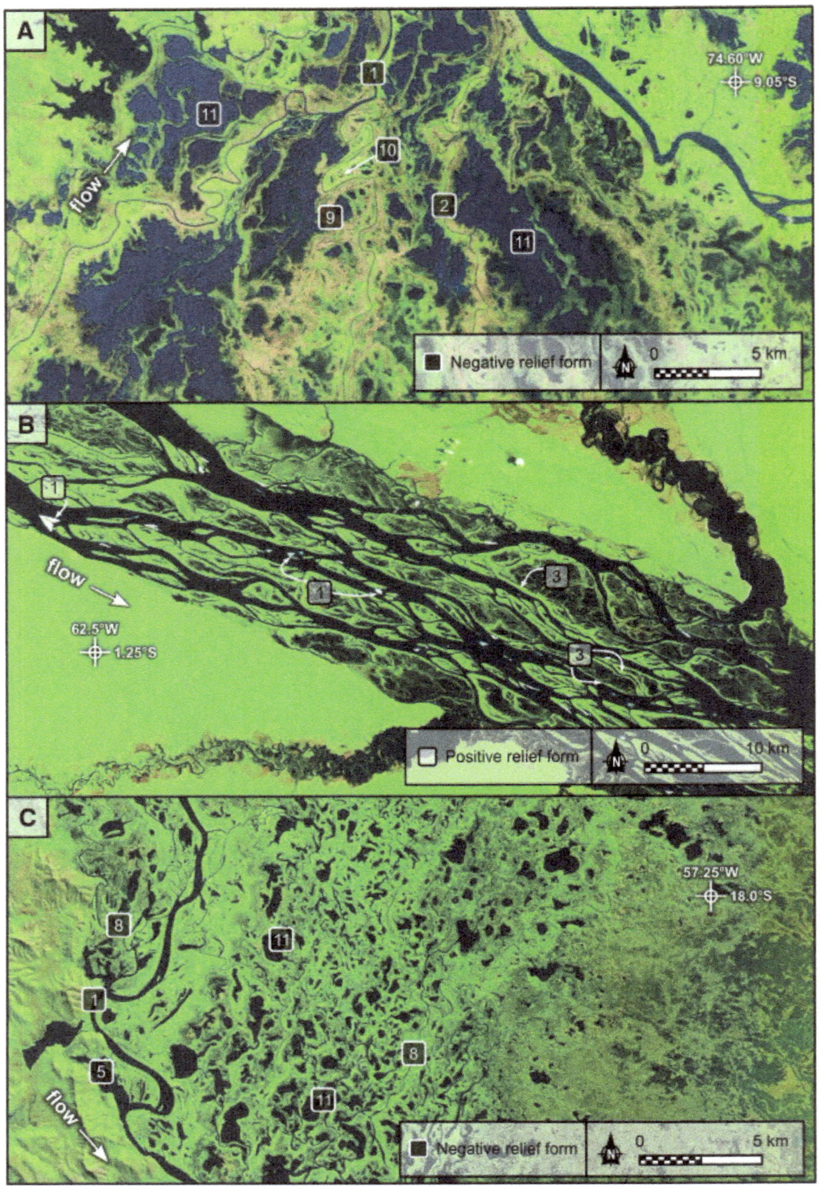

Figure 14: (A) Anastomosing River Magdalena, Columbia (9° 00′ S, 74° 46′ W). See also Smith (1986). Multiple anastomosing channels with levees and splays into flood basins — some reaches laterally mobile. Image taken on 12 January 2001; (B) anastomosing on the Rio Negro, Brazil (1° 11′ S, 62° 17′ W) (cf. Latrubesse and Franzinelli, 2005). Note numerous islands, extensive levee systems with occluded lakes and wetlands behind. Some lateral accretion ridges. Image taken on 9 October 2011; (C) Paraguay River in the Pantanal wetlands, south-west Brazil (18° 05′ S, 57° 26′ W). (cf. Assine and Soares, 2004 and Assine and Silva, 2009) showing an extensive and unconfined floodplain with swamps and lakes at the foot of the Taquari fan (cf. Makaske et al., 2012). Image taken on 24 May 2000. Landsat imagery courtesy of the U.S. Geological Survey.

Inter-channel waterbodies can also remain relatively sediment-free, despite the sediment-laden rivers flowing adjacent to them — even when there are open-water linkages between channels and ponded waters. With present-day large equatorial rivers, inundation waters may derive from runoff in lower parts of the catchments, and their levels may equilibrate with those in main channels that relate to floodwater inputs from distant headwaters. In practice there are complex adjustments of water level, but not a simple mainstream flood overflow involving dispersal of, and deposition from, sediment-rich waters (Fig. 4B). In such circumstances, there is little advection of sediment from the main river; there is narrow diffusive contact between the juxtaposed waterbodies concerned, and a restricted tendency for crevassing. For avulsion it is the existence of a hydraulic gradient or improved flow efficiency following diversion that is required; typically this occurs under aggrading channel or channel-belt conditions (Törnqvist and Bridge, 2002 and Slingerland and Smith, 2004). But if water levels in the extra-channel zone are similar to those in channels, then overflow, sediment dispersal and avulsion are less likely. A hydraulic barrier, as much as a bank of sediment, inhibits the lateral filling of broad valley troughs with sediment. Thus, subsiding environments like the wetlands of the lower Magdalena have quite narrow strings of mineral sediments within lacustrine fills (Smith, 1986). Such permanent waterbodies differ from those with markedly seasonal inundation deriving from mainstream overflow. Here advective transport may produce broad levees rather than narrow diffusive ones (Adams et al., 2004), whilst overbank deposition and avulsive relocation may distribute channel-belt sedimentation across wide valley floors (Valente and Latrubesse, 2011), decreasing the accommodation space for organic accumulation.

Marginal lakes and wetlands may have both channelled and non-channelled waters. On a small scale, an interesting but unappreciated process may be illustrated by the postglacial development of the Hay River that 'bridges' Lake Zama in Alberta, Canada (Fig. 15). In shallow water, the levee-lined trans-lacustrine channel has prograded to form a sinuous trail across the lake from upstream delta to exit point, together with some discontinued branches. These 'meanders' show very limited signs of lateral mobility, an activity commonly expected in the production of meandering planforms. But sinuosity here developed as part of initial downstream extension in a process quite unlike that of most meander models. Wetland river 'bridging' on a much larger scale can be seen on the Hwang He in China (29°54′N, 112°49′E), and in neotectonic basins on the Amazon (Latrubesse, 2010). Prograding levee arms are likely to be a major process for channel definition in shallow water across subsiding alluvial basins (where the rate of subsidence has been sufficient to 'drown' potential channel systems), or in the establishment of defined channels following avulsion into shallow inter-channel wetlands in anastomosing systems. Reviewing the establishment of new channels on floodplains following avulsion, Kleinhans et al. (2013) note the roles of incision or palaeochannel re-occupation, but neither may be the case for near-zero gradient lake/wetland channels.

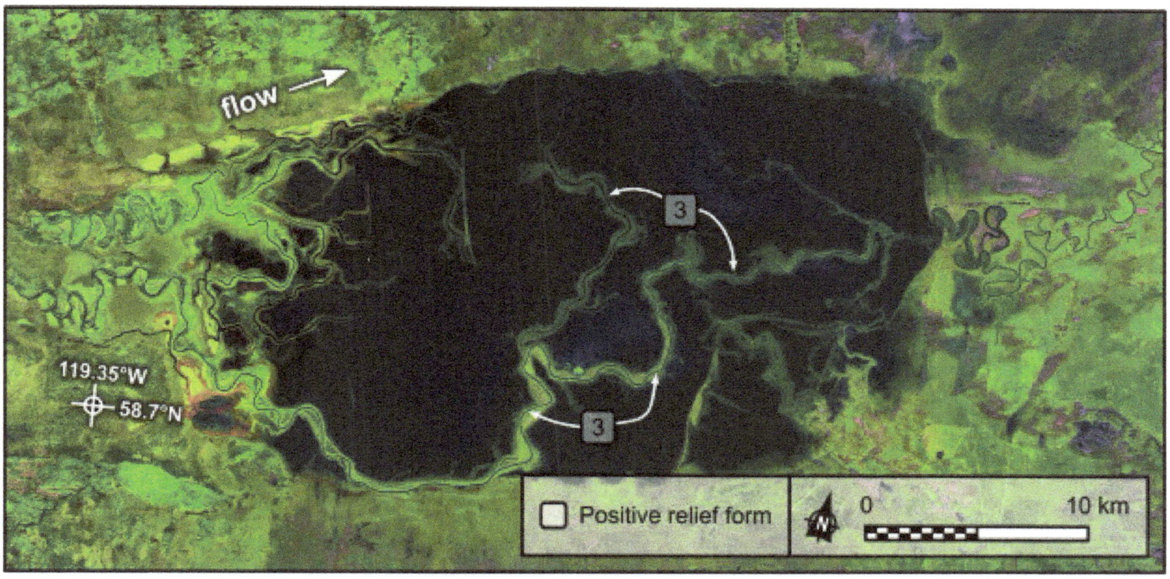

Figure 15: The Hay River and Lake Zama, Alberta (58° 48′ N, 119° 06′ W) showing 'snail-trail' of prograding channel across lake (including 'failed' or discontinued arms), with some lateral accretion in delta area. Image taken on 21 June 2008. Image courtesy of Government of Canada, Natural Resources Canada, Earth Sciences Sector, Mapping Information Branch, Centre for Topographic Information — Sherbrooke SPOT4.

'Rivers without banks' provide an important context both for contemporary aquatic biology and for the interpretation of ancient organic deposits. Differences between surface hydrological connectivity (with open water linkage), and geomorphological connectivity (involving sediment transfer), may allow negative water-filled relief to be long-lasting. With equilibrated water levels at high river stage, mobile biota can pass directly, and either way, between sediment-rich higher velocity channels and less-turbid backwater or lake environments through a kind of non-chemical osmotic transfer. This need not involve strong one-way overflow if extra-channel water levels are already high, though there may be subsequent falling-stage return flow when main channel levels drop. Such floodplain lacustrine environments can be remarkably sediment-free, long-lasting and without a chemically degrading influx of aerated water even where there are open-water connections. In the longer term, these conditions favour the production and preservation of freshwater organic deposits with little non-combustible content, as have many freshwater coal deposits (McCabe, 1984, McCabe and Parrish, 1992 and Thomas, 2013). There may, however, be lesser nutrient input from main rivers than might otherwise be expected. Iriondo (2004) has pointed to distinctions between sand wetlands (with relict features from wet and dry climate phases and the reworking of dunes) and mud wetlands (with fluvial sediment input and greater nutrient availability), whilstLatrubesse (2010) has also underlined the importance of geomorphological factors in differentiating the nutrient and primary productivity status of lacustrine environments.

In non-coastal areas (where other factors including beach barriers are also important), there are some geological circumstances that appear favourable to large-scale terrestrial organic sedimentation of eventual economic significance: large rivers with sandy sediment loading draining through cratonic troughs (in wet climates) that they incompletely fill by mineral sediment deposition. The limited amount of non-combustible mineral content in many associated organic sediments suggests low lateral advective sediment transfer, even though there may be hydrological connection. Ponded waters already present hold back lateral dispersal from mainstream floods. Lower sediment loadings might also suggest minimal tectonic uplift and erosion in continental headwaters, but for the largest rivers their derived sediment may already have been deposited in foreland basins leaving cratonic lower courses relatively poor in terms of supplied sediment from cratonic tributaries (Ashworth and Lewin, 2012).

SURVIVAL HISTORY OF NEGATIVE RELIEF

Negative relief may be infilled via a range of fluvial and non-fluvial sources such as channel accretion by bedforms, sediment fall-out from overbank flow, authigenic mineral precipitation associated with flood and groundwater evaporation, organic growth (Tooth and Nanson, 2011), aeolian dust and sand input (Bullard and McTainsh, 2003), and, more rarely, volcanic and marine sedimentary input (Marriott and Alexander, 1999). The rate of sedimentation as typically quantified from modern and ancient environments of different types varies by at least eleven orders of magnitude (Sadler, 1981) and decreases as a power-law function of the interval of time over which it is measured (Strauss and Sadler, 1989 and Tipper, 1983). The filling of negative relief by both autogenic and allocyclic-driven processes is episodic with non-depositional gaps of inactivity (hiatuses) and/or bursts of sedimentation (Ager, 1973 and Macklin et al., 2010). Within individual elements of filled negative relief, the preserved stratigraphic record may be contaminated or overprinted by internally-generated and variable autogenic processes (Jerolmack and Paola, 2007, Straub et al., 2009, Van De Wiel and Coulthard, 2010 and Wang et al., 2011a).

Direct fluvial input in rheic zones is through both bedload and finer sediment; in perirheic zones, infilling is largely by finer materials. The physical characteristics of sediment loads vary considerably. Up to 60% of the suspended sediment load of the middle Hwang He in China may be composed of sand-sized particles (Walling and Moorhead, 1989), thus supporting near-channel deposition and the creation of levees and alluvial ridges. High-latitude rivers may have low suspended loads resulting from lower runoff, lesser quantities of (fine) eroded soils from wetland-blanketed domains, and a proportionately greater channel reworking of (coarser) glacigenic and frost weathered material. Active lateral channel mobility of both meandering and braiding streams during peak melt discharges may create voids which are then only very slowly filled. In forest biomes floodplain depressions may also have a very long life where deforestation has not produced large quantities of fine sediment (Caselius, 1971). By contrast, in settled mid-latitudes soil erosion has contributed to blanket-like and accelerated fine overbank deposition, as for example in peak periods of agricultural activity in mediaeval Europe (Hoffmann et al., 2007 and Macklin et al., 2010). Where such sediment has moved to historically hasdepended on delivery pathways and storages (Lang et al., 2003 and James, 2013), with a proportion having yet to reach floodplains. Post-Pleistocene coarse sediment transport and channel mobility in temperate lowland rivers have been relatively restricted by interglacially-reduced stream power. In recent decades and centuries, complex response histories of river regulation have also involved channel engineering, floodplain inundation control, the trapping of fine sediment in reservoirs, and restricted channel dynamics because of a reduction in flows competent to transport bed material (Hudson et al., 2008 and Lewin, 2013).

This is true on a global scale and for large floodplains where it affects the sedimentation potential of negative relief elements. Over millennial timescales accelerated soil erosion has increased inputs of fine sediment (Syvitski et al., 2005 and Kundzewicz et al., 2009). Finer sediment from topsoil removal, towards the colloidal end of the river suspended sediment load, is more readily dispersed over floodplains far distant from channel margins unless moving in aggregate form (Walling and Moorhead, 1989). Secondly, and over the last 50 years in particular, large dam construction has reduced mainstream peak flows and transport-effective discharges, and trapped inputs of fine sediment from up-catchment (Nilsson et al., 2005 and Gupta et al., 2012). Although there are wide differences between quantitative estimates of the amounts of accelerated erosion and dam storage, and therefore effects on downstream floodplains, centuries of accelerated sediment supply in some systems appear likely to be succeeded by ongoing decades of decrease (Syvitski et al., 2005, Walling, 2008 and Wang et al., 2011b).

From a geomorphological perspective, and for quasi-natural floodplains, it is possible to begin constructing a qualitative framework for negative-relief 'survival-likelihood' (Table 3). In each of the domains initially recognised above (Table 1(i–iii)) negative relief is liable to differential elimination. To these should be added: (iv) under high river mobility, there may be reworking and elimination of whole floodplain assemblages, positive and negative elements alike; and (v) there may be relief preservation because of whole-assemblage isolation from mainstream reworking and infilling. That arises through river incision and terracing, or channel metamorphosis to a less active and laterally extensive state (Lewin and Macklin, 2003). Evidence for 'natural' survival rates is briefly discussed for each of these five groups in turn, although as yet global and corridor-variable survival rates have attracted limited study despite their biological significance.

Table 3: Natural negative-relief survival likelihood on large river floodplains (see text for explanation).

	Negative relief form	Survival likelihood
(i)	Active channel segments	Dependent on channel branch dynamics
(ii)	Transitional zones with surface-water channel connection	Low
(iii)	Perirheic depressions	High
(iv)	Autogenic assemblage recycling	Restricted on large rivers
(v)	Metamorphosis/incision preservation	Possible in the long-term

Segments of the Channel Network

Research has demonstrated that in both braided and anastomosing channel systems not all water-transmitting channels are geomorphologically active to the same degree (Makaske et al., 2009 and Ashmore et al., 2011). For example, typically about 40% of experimental braid channels have active sediment transport at any one time (Bertoldi et al., 2009 and Egozi and Ashmore, 2009). The same principle applies to plural large-river systems with mixed channel-patterning styles (Lewin and Ashworth, 2013). Sediment-transporting and eroding branches may be turbid and bed-abrasive, whereas geomorphologically moribund channels provide more sheltered habitats that are more suited as feeding and migration passages. How long moribund branches remain open has attracted limited documentation; most analyses concentrate on the causes and stability of bifurcation/avulsion (Huang and Nanson, 2007 and Kleinhans et al., 2011). The lifespan of bifurcations varies greatly — from channel abandonment caused by switching or plugging in a single flood event, to the 100s to 1000s of years it can take for a river to change its course on a floodplain or delta (Kleinhans et al., 2013).

Transitional Zones still with Surface-Water Channel Connection

Many floodplain channels in the transitional zone of large rivers are geomorphologically moribund but hydrologically conjoined to the main trunk stream. Channels can remain open so long as they can be scoured by clear water returning to the trunk stream during lower flows. If the supply of fine sediment is adequate, advective sediment transport is driven into connected backwater zones and high rates of deposition are likely there (Day et al., 2008 and Citterio and Piégay, 2009). Because so many large river systems anabranch dynamically in plural styles (Latrubesse, 2008 and Lewin and Ashworth, 2013), it follows that a range of partially connected or disconnected Y-junction channel forms should be expected, extending further away from trunk streams (Trigg et al., 2012 and Kleinhans et al., 2013). Infilling involves a whole range of negative relief elements of different origin (Dunne and Aalto, 2013 and Lewin and Ashworth, 2013) including mainstem bifurcations, formerly active channel segments still mainstem-connected at one end, branching-off accessory channels, downstream return channel junctions, and tributary input channels. These may undergo active erosion/sedimentation, or they may transmit passively, back-up, or pond water prior to slow sediment infilling. As yet there is little quantitative data for sedimentation rates for the filling of partially disconnected channel segments along large rivers. Latrubesse and Franzinelli (2002) note an abandoned channel belt on the Amazon that is still incompletely filled after c.1000 years. Trigg et al. (2012) conducted bed sonar surveys on 33 floodplain channels connected to the Amazon trunk stream that ranged from 40 to 900 m width. Their work showed a consistent relationship between depth of the floodplain channels and the water source. Channels with no local runoff input had a mean depth of 10.4 m, those with runoff input from the local catchment had a mean depth of 15.9 m and those channels strongly connected to the main river had a mean depth of 17.7 m. Clearly, despite the high sediment loads associated with many large rivers, infill rates are slow, at least in partially or seasonally, hydrologically-connected floodplain channels.

Perirheic Forms

By contrast, negative relief elements (and many lentic environments) that are distant from, and poorly linked to, main channels may have extended lives with very low sedimentation rates (Citterio and Piégay, 2009). This greatly depends on inundation patterns and depths, and upon floodwater sediment loadings. On some large rivers, such as the Strickland (Aalto et al., 2008), levee growth of up to 2 m can confine high flows to the main channel so that accretion on the floodplain is shut down for decades until a sufficiently large flood overtops the levee (Dunne and Aalto, 2013).

Field surveys have focused on 'oxbow' cutoff infilling (Werritty et al., 2006, Citterio and Piégay, 2009 and Toonen et al., 2012) and, more generally, on dispersed rates of overbank sedimentation (Asselman and Middelkoop, 1995, Walling et al., 1996, Törnqvist and Bridge, 2002, Benedetti, 2003 and Aalto and Nittrouer, 2012). Several overbank sedimentation rate models have been proposed (reviewed in Marriott, 1996) with deposition rates and the size of deposited material decreasing with distance from channels. In practice measured deposition rates are locally complex, with high rates near channel margins and in relation to floodplain microtopography. Domination of near-channel fine sediment accretion is consistent with those surveys that show that marginal and 'inner accretionary bank' accumulation on point bars rapidly follows channel migration (Steiger et al., 2001 and Gautier et al., 2010). Averaged over time, deposition rates (of sand in particular) may decrease abruptly away from channels, although this can lead to more rapid channelway aggradation and an enhanced likelihood for avulsion (Törnqvist and Bridge, 2002). Avulsive splays, dryland floodouts, depositional lobes or inland deltas on a range of scales may partially infill occluded basins (Latrubesse and Franzinelli, 2002, Ramonell et al., 2002, Assine and Silva, 2009 and Latrubesse et al., 2010, their Fig. 8; Tooth, 2013). Along the Amazon and the middle Paraná in particular, lakes behind channel belt barriers are only partially filled by sediment splays even when close to main channels.

Autogenic Assemblage Recycling

Only a few large trunk rivers like the Jamuna/Brahmaputra channel belt (Ashworth and Lewin, 2012) or the Indus (Fig. 11) recycle their entire floodplains over decades to centuries. Some, like reaches of the Paraná and Amur Rivers, have braided main channels set within wetlands and actively meandering accessory channel systems. The Lower Mississippi has actively reworked its floodplain so that few meander trains persist for longer than 120 years (Harmar and Clifford, 2006) although the river has been subject to multiple Holocene channel-belt avulsions. The Ganga exhibits decade-scale meander cycles (Gupta et al., 2013) within a longer-term history of valley-floor response to monsoonal oscillation (Roy et al., 2011). Other rivers like the Amazon (Mertes et al., 1996 and Rozo et al., 2012) have quasi-straight reaches that, without bend development, oscillate laterally over time to produce bordering multi-ridge accretionary zones of limited width. More laterally-remote wetlands and floodplains on large rivers frequently do not appear to be effectively reworked on a centuries to millennial timescale.

Metamorphosis or incisional preservation

Many rivers have terrace flights that preserve palaeoforms on their surfaces. On larger cratonic rivers, Quaternary depths of valley incision may be slight (Bridgland and Westaway, 2008), to the extent that palaeoforms relating to prior climatic regimes remain at near-channel level. With restricted lateral geomorphological reworking, elements that were geomorphologically active several thousand years previously may still form part of the inundated floodplain (Bravard, 1989, Sidorchuk, 2003, Busschers et al., 2007,Erkens et al., 2009, Latrubesse, 2010, Lewin and Ashworth, 2013 and Valente et al., 2013). Likewise, palaeovalley systems may preserve a significant morphological expression that is buried beneath Holocene delta plains (Busschers et al., 2007, Fontana et al., 2008 and Blum et al., 2013).

Management Considerations

In the last several centuries, geomorphological processing in major floodplains has been transformed through human activity, for example through flow regulation, channelization, land drainage and for flood protection (Hudson et al., 2008) (Fig. 16). The scale and nature of sediment recycling has been considerably modified to homogenize ecosystems (van de Wolfshaar et al., 2011) and to fossilize alluvial dynamics (Hohensinner et al., 2011). Under natural conditions, relief forms are diachronous and have different rates of element (re) formation. If particular negative forms are now to be preserved or restored/repaired in order to maintain hydromorphological quality, it must be appreciated that their 'normal' development is no longer in place. Like biological entities they are dynamic, albeit over varying and generally much longer timescales. Whilst allowing for short-term 'adjustments' (as evidenced by eroding banks or silting beds), many hydromorphic approaches to river systems appear to regard the morphological 'platform' as immutable until modified by human agency. This is not the reality that an analysis of negative relief forms reveals.

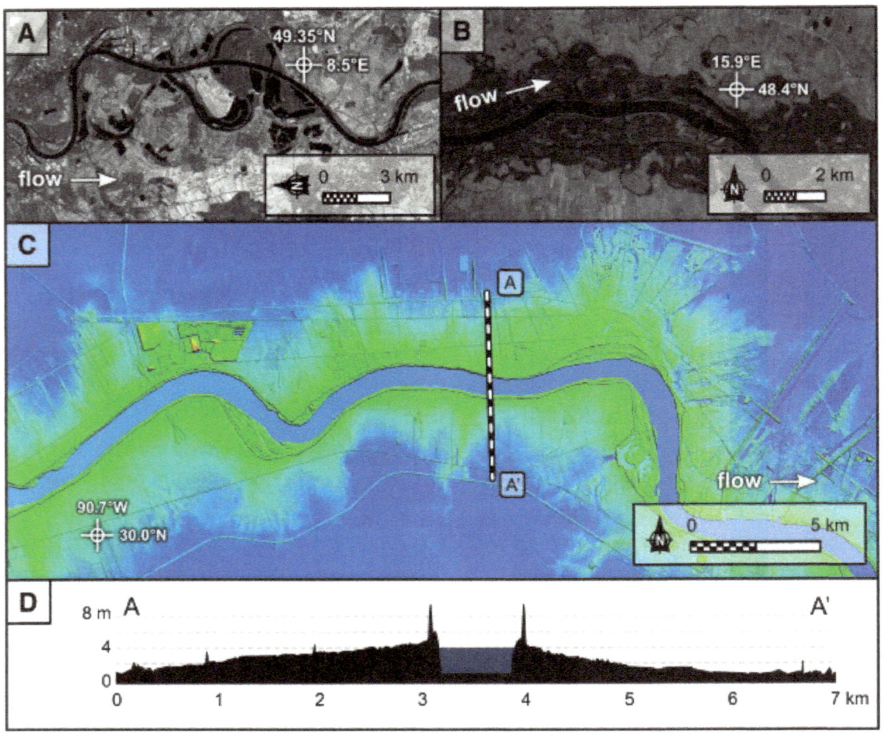

Figüre 16: (A) The channelized and formerly meandering Rhine, Germany (49° 25′ N, 8° 30′ E). Image taken on 15 April 2003; (B) the regulated, channelized and formerly anabranching Danube, Austria (48° 23′ N, 15° 51′ E). Image taken on 27 March 2003; (C) LiDAR image of the Mississippi between New Orleans and Baton Rouge with both artificial and natural flood embankments (D) showing detachment of river from its floodplain. Image from 2001 and courtesy of Digital Elevation Model (USGS DEM), SW quadrant of Reserve quadrangle, Louisiana, UTM 15 NAD83, Louisiana FEMA Project — Phase 1: Lake Maurepas/Lake. All Landsat imagery courtesy of the U.S. Geological Survey.

Three examples may be used to illustrate this. First, slackwater arms in rheic transition zones bordering contemporary channels are part of an evolving channel network. Their biological significance – for example as refuges, or for feeding or spawning – may lead to their artificial preservation or restoration. Ecologically useful though this may be, the long-term existence of such features is as components of dynamic systems that produce sequences of active and relict channels, which are then filled with sediment. Such arms may not individually have a long-term natural existence. Second, cutoff 'oxbow' lakes are produced through channel migration and then slowly infilled subsequently. These may have an extended 'life', but static preservation only of the floodplain set that currently exists is a distortion of their natural replacement sequence. Third, major wetlands may have existed for at least 10^4 years repeatedly during the Quaternary as occluding sediment bodies have built up or subsidence has occurred; once drained, artificial restoration would require both major engineering and attention to the intricacies of their former water drainage and sedimentation nets that evolved over millennia.

It may reasonably be argued that recognition of the ecological significance of the totality of negative relief forms, and preservation/restoration of some of them, are very worthwhile objectives (Hohensinner et al., 2011). But it has equally to be appreciated that in practice static conservation of form elements can only achieve 'half-natural' states. Opening-up water-filled voids and 'allowing the river to do the work' may underestimate both the timescale which such work will require and the varied cycles of formation and infilling that produced observed sets of features. The fidelity with which forms, conceived of as being static, can be artificially maintained is limited.

CONCLUSIONS

Negative relief on larger floodplains – important to hydrology, biology and geology, and thus to humans – is highly varied in terms of dimensions and dynamics. This paper shows that modification of formerly active channels and the preservation of negative relief elements on large floodplains is more complex than can be captured by a twofold channel/overbank descriptive framework. In *perirheic zones*, there are palaeoforms inherited from prior morphogenetic systems and ones related to tectonic depressions. Though geomorphologically relatively permanent and inactive, these may still be connected to active hydrobiological systems. By contrast, hydrologically-linked*transitional zones* at rheic margins are the scene for much varied geomorphologically activity. This zone performs especially significant transformational roles in modifying*rheic zone* channel forms, but ones that are more complex than the now relatively understood channel bar, cutoff and avulsion processes.

In specific terms, this paper has suggested that patterns of channel-margin sedimentation in partially-abandoned or flow-detached channel segments, and with low or discontinuous mainstream channel banks, are important. Geomorphologically-active*rheic zones* may occupy only part of the water system that provides hydrobiological diversity; branches and backwaters may or may not be historical hangovers now being modified by ongoing geomorphological processes. Prior surface water ponding across floodplains may aid negative relief preservation by inhibiting the lateral dispersal and sequestration of sediment, whilst vegetation may focus sediment sequestration. Braid and meander floodplains differ because of the ways their channels are abandoned and incorporated into floodplains as well as because of contrasts between the initial rheic patterns themselves. Sinuous channels prograding into extensive wetlands and shallow waters may form rather differently from those created by lateral meander mobility.

Still- and slack-water depressions and channel arms are both biologically significant yet also particularly sensitive to anthropogenic activities. These have included millennial-scale land cover changes, century-scale channel impoundment and engineering, and contemporary induced climatic change. At and beyond channel margins and away from mainstem flows, it is relatively easy to adopt sustainable engineering solutions for both hazard protection and economic development. But this may produce a reduction in geodiversity, which inevitably also leads on to a reduction of biodiversity. Although slow to operate (and over mixed timescales), the geomorphological processes and replacement cycles that generate floodplain negative relief variety continue to require further research. Future research needs to enhance: (i) the quantitative archive of large floodplain element hierarchies for a greater variety of large floodplains, especially using data derived from satellite imagery; (ii) using both field survey, historical archives, (maps and images) and geochronology, more comprehensive knowledge of the dynamics and survival spans under both quasi-natural and anthropogenically-modified conditions.

The management of negative relief on large river floodplains needs to quantify *both* the geomorphological and hydrological connectivity between river and floodplain. Understanding both ecosystems and the distribution of organic resources requires knowledge of the alluvial process context for negative relief formation and preservation.

ACKNOWLEDGEMENTS

PJA would like to thank the UK Natural Environment Research Council (NERC) for support under grant NE/E016022/1 (awarded to Ashworth, Hardy, Lane, Nicholas, Parsons and Sambrook Smith with Best) that funded fieldwork on the Rio Paraná, Argentina and was the catalyst for stimulating discussions between JL and PJA on the controls on large river dynamics. We are extremely grateful to Chris Simpson of Fulcrum Graphic Communications Inc., Vancouver, B.C., for his expertise in sourcing and producing the figures. Derald Smith kindly allowed us to use his image in Fig. 1C. Fig. 12was produced from aerial sorties commissioned by PJA, Jim Best and Greg Sambrook Smith. We are especially grateful to Kim Cohen, Edgardo Latrubesse, Andrew Nicholas and an anonymous referee for their considerable help in improving the text.

REFERENCES

1. Aalto, R., Nittrouer, C.A., 2012. 210Pb geochronology of flood events in large tropical river systems. Phil. Trans. R. Soc. A 370, 2040–2074. http://dx.doi.org/10.1098/rsta.2011.0607.
2. Aalto, R., Maurice-Bourgoin, L., Dunne, T., Montgomery, D.R., Nittrouer, C.A., Guyot, J.L., 2003. Episodic sediment accumulation on Amazonian floodplains influenced by El Niño/Southern Oscillation. Nature 425, 493–497. http://dx.doi.org/10.1038/ nature02002.
3. Aalto, R., Lauer, J.W., Dietrich, W.E., 2008. Spatial and temporal dynamics of sediment accumulation and exchange along Strickland River floodplains (Papua New Guinea) over decadal-to-centennial timescales. J. Geophys. Res. 113, F01S04. http://dx.doi.org/10.1029/2006JF000627.
4. Adams, P.N., Slingerland, R.L., Smith, N.D., 2004. Variations in natural levee morphology in anastomosed channel flood plain complexes. Geomorphology 61, 127–142.
5. Ager, D.V., 1973. The Nature of the Stratigraphic Record. Wiley, New York (114 pp.). Alho, C.J.R., Lacher, T.E., Gonçalves, H.C., 1988. Environmental degradation in the Pantanal ecosystem. Bioscience 38, 164–171.
6. Archer, D.R., 2005. Review of Amazonian depositional systems. In: Blum, M.D., Marriott, S.B., Leclair, S.F. (Eds.), Fluvial Sedimentology VII. Special Publication 35, International Association of Sedimentologists. Blackwell, Oxford, pp. 17–39.
7. Ashmore, P.E., 1991. How do gravel bed streams braid? Can. J. Earth Sci. 28, 326–341.
8. Ashmore, P., Bertoldi, W., Gardner, J.T., 2011. Active width of gravel-bed braided rivers. Earth Surf. Process. Landf. 36, 1510–1521. http://dx.doi.org/10.1002/esp.2182.
9. Ashworth, P.J., Lewin, J., 2012. How do big rivers come to be different? Earth Sci. Rev. 114, 84–107. http://dx.doi.org/10.1016/j.earscirev.2012.05.003.
10. Ashworth, P.J., Best, J.L., Peakall, J., Lorsong, J.A., 1999. The influence of aggradation rate on braided alluvial architecture: field study and physical scale modelling of the Ashburton River gravels, Canterbury Plains, New Zealand. In: Smith, N.D., Rogers, J. (Eds.), Fluvial Sedimentology VI, Special Publication. International Association of Sedimentologists, 28. Blackwell, Oxford, pp. 333–346.

11. Ashworth, P.J., Best, J.L., Roden, J.E., Bristow, C.S., Klaassen, G.J., 2000. Morphological evolution and dynamics of a large, sand braid-bar, Jamuna River, Bangladesh. Sedimentology 47, 533–555.

12. Ashworth, P.J., Sambrook Smith, G.H., Best, J.L., Bridge, J.S., Lane, S.N., Lunt, I.A., Reesink, A.J.H., Simpson, C.J., Thomas, R.E., 2011. Evolution and sedimentology of a channel fill in the sandy braided South Saskatchewan River and its comparison to the deposits of an adjacent compound bar. Sedimentology 58, 1860–1883. http:// dx.doi.org/ 10.1111/j.1365-3091.2011.01242.x.

13. Asselman, N.E.M., Middelkoop, H., 1995. Floodplain sedimentation: quantities, patterns and processes. Earth Surf. Process. Landf. 20, 481–499.

14. Assine, M.L., Silva, A., 2009. Contrasting fluvial styles of the Paraguay River in the northwestern border of the Pantanel wetland, Brazil. Geomorphology 113, 189–199. http://dx.doi.org/10.1016/j.geomorph.2009.03.012.

15. Assine, M.L., Soares, P., 2004. Quaternary of the Pantanal, west-central Brazil. Quat. Int. 114 (1), 23–34.

16. Aufdenkamp, A.L., Mayorga, E., Raymond, P.A., Melack, J.M., Doney, S.C., Alin, S.R., Aalto, R.E., Yoo, K., 2011. Riverine coupling of biogeochemical cycles between land, oceans and atmosphere. Ecol. Environ. 9, 53–60. http://dx.doi.org/10.1890/100014.

17. Bastviken, D., Tranvik, L.J., Downing, J.A., Crill, P.M., Enrich-Prast, A., 2011. Freshwater methane emissions offset the continental carbon sink. Science 331, 50. http:// dx.doi.org/10.1126/science.1196808. Benedetti, M.M., 2003. Controls on overbank deposition in the upper Mississippi River. Geomorphology 56, 271–290.

18. Bertoldi, W., Ashmore, P., Tubino, M., 2009. A method for estimating the mean bed load flux in braided rivers. Geomorphology 103, 330–340. http://dx.doi.org/10.1016/ j.geomorph.2008.06.014.

19. Best, J.L., Ashworth, P.J., 1997. Scour in large braided rivers and the recognition of sequence stratigraphic boundaries. Nature 387, 275–277.

20. Best, J.L., Ashworth, P.J., Bristow, C.S., Roden, J.E., 2003. Three-dimensional sedimentary architecture of a large, mid-channel sand braid bar, Jamuna River, Bangladesh. Journal of Sedimentary Research 73, 516–530.

21. Best, J., Woodward, J., Ashworth, P., Sambrook Smith, G., Simpson, C., 2006. Bar-top hollows: A new element in the architecture of sandy braided rivers. Sedimentary Geology 190, 241–255.

22. Blondeaux, P., Seminara, G., 1985. A unified bar-bend theory of river meanders. J. Fluid Mech. 112, 363–377.

23. Blum, M., Martin, J., Milliken, K., Garvin, M., 2013. Paleovalley systems: Insights from Quaternary analogs and experiments. Earth Sci. Rev. 116, 128–169. http:// dx.doi.org/ 10.1016/j.earscirev.2012.09.003.

24. Bravard, J.-P., 1989. La métamorphose des rivières des Alpes français à la fin du Moyen-Age et à l'époque moderne. Bull. Soc. Géog. Liège 25, 145–157. Bridge, J.S., 2003. Rivers and Floodplains: Forms, Processes and Sedimentary Record. Blackwell, Malden, USA (491 pp.).

25. Bridge, J.S., Lunt, I.A., 2006. Depositional models of braided rivers. In: Sambrook Smith, G.H., Best, J.L., Bristow, C.S., Petts, G.E. (Eds.), Braided Rivers; Process, Deposits, Ecology and Management, 36. International Association of Sedimentologists Special Publication, Blackwell, Oxford, pp. 11–50.

26. Bridge, J.S., Alexander, J., Collier, R.E.LL, Gawthorpe, R.L., Jarvis, J., 1995. Ground-penetrating radar and coring used to study the large-scale structure of point-bar deposits in 3-D. Sedimentology 42 (6), 839–852. Bridgland, D.R., Westaway, R., 2008. Climatically controlled river terrace staircases: a worldwide Quaternary phenomenon. Geomorphology 98, 288–315.

27. Brierley, G., 1991. Bar sedimentology of the Squamish River, British Colombia: definition and application of morphostratigraphic units. J. Sediment. Res. 61, 211–225. http:// dx.doi.org/10.1306/D42676D6-2B26-11D7-8648000102C1865D.

28. Brooks, G.R., 2003. Alluvial deposits of a mud-dominated stream: the Red River, Manitoba, Canada. Sedimentology 50, 441–458.

29. Brown, A.G., Toms, P.S., Carey, C.J., Howard, A.J., Challis, K., 2013. Late Pleistocene–Holocene river dynamics at the Trent-Soar conflu

30. Bullard, J.E., McTainsh, G.H., 2003. Aeolian-fluvial interactions in dryland environments: examples, concepts and Australia case study. Prog. Phys. Geogr. 27, 471–501.

31. Busschers, F.S., Kasse, C., van Balen, R.T., Vandenberghe, J., Cohen, K.M., Weerts, H.J.T., Wallinga, J., Johns, C., Cleveringa, P., Bunnik, F.P.M., 2007. Late Pleistocene evolution of the Rhine–Meuse system in the southern North Sea basin: imprints of climate change, sea level oscillation and glacio-isostacy. Quat. Sci. Rev. 26, 3216–3248.

32. Caselius, C., 1971. On the meandering rivers in the Lokka reservoir district, Finland. Fennia 90, 1–71. Castro, N.M., Hornberger, G.M., 1991. Surface-subsurface water interactions in an alluviated mountain stream channel. Water Resour. Res. 27, 1613–1621.

33. Cheetham, G.H., 1979. Flow competence in relation to stream channel form and braiding. Geol. Soc. Am. Bull. 90, 877–886.

34. Citterio, A., Piégay, H., 2009. Overbank sedimentation rates in former channel lakes: characterization and control factors. Sedimentology 56, 461–482. http:// dx.doi.org/ 10.1111/j.1365-3091.2008.00979.x.

35. Constantine, J.A., Dunne, T., 2009. Meander cutoff and the controls on the production of oxbow lakes. Geology 36, 23–26. http://dx.doi.org/10.1130/G24130A.1.

36. Constantine, J.A., Dunne, T., Piégay, H., Kondolf, G.M., 2010. Controls on the alluviation of oxbow lakes by bed-material load along the Sacramento River, California. Sedimentology 57, 389–407. http://dx.doi.org/10.1111/ j.1365-3091.2009.01084.x.

37. Day, G., Dietrich, W.E., Rowland, J.C., Marshall, A., 2008. The depositional web on the floodplain of the Fly River, Papua New Guinea. J. Geophys. Res. 113, F01S02. http:// dx.doi.org/10.1029/2006JF000622.

38. Dieras, P.L., Constantine, J.A., Hales, T.C., Piégay, H., Riquier, J., 2013. The role of oxbow lakes in the off-channel storage of bed material along the Ain River, France. Geomorphology. http://dx.doi.org/10.1016/j.geomorph.2012.12.024.
39. Dietrich, W.E., Day, G., Parker, G., 1999. The Fly River, Papua New Guinea: inferences about river dynamics, floodplain sedimentation and fate of sediment. In: Miller, A., Gupta, A. (Eds.), Varieties of Fluvial Form. Wiley, Chichester, pp. 345–376.
40. Doeschl, A.B., Ashmore, P., Davison, M., 2006. Methods for assessing exploratory computational models of braided rivers. In: Sambrook Smith, G.H., Best, J.L., Bristow, C.S., Petts, G.E. (Eds.), Braided Rivers: Process, Deposits, Ecology and Management. International Association of Sedimentologists Special Publication, 36. Blackwell, Oxford, pp. 177–197.
41. Dulal, K.P., Shimizu, Y., 2010. Experimental simulation of meandering in clay mixed sediments. J. Hydro Environ. Res. 4, 329–343. http://dx.doi.org/10.1016/j.jher.2010.05.001.
42. Dunne, T., Aalto, R.E., 2013. Large river floodplains. In: Shroder, J.F. (Ed.), Treatise on Geomorphology, vol. 9. Academic Press, San Diego, CA, pp. 645–678. http://dx.doi.org/ 10.1016/B978-0-12-374739-6.00258-X.
43. Dunne, T., Mertes, L.A.K., Meade, R.H., Richey, J.E., Forsberg, B.R., 1998. Exchanges of sediment between the flood plain and the channel of the Amazon River in Brazil. Geol. Soc. Am. Bull. 110, 450–467.
44. Egozi, R., Ashmore, P., 2009. Experimental analysis of braided channel pattern response to increased discharge. J. Geophys. Res. Earth Surf. 114, F02012. http://dx.doi.org/ 10.1029/2008JF001099.
45. Erkens, G., Dambeck, R., Vollenberg, K.P., Bouman, M.T.I.J., Boss, J.A.A., Cohen, K.M., Wallinga, J., Hoek, W.Z., 2009. Fluvial terrace formation in the northern Upper Rhine Graben during the last 20 000 years as a result of allogenic controls and autogenic evolution. Geomorphology 103, 476–495.
46. Erkens, G., Hoffmann, T., Gerlach, R., Klostermann, J., 2011. Complex fluvial response to Lateglacial and Holocene allogenic forcing in the Lower Rhine Valley (Germany). Quat. Sci. Rev. 30, 611–627. http://dx.doi.org/10.1016/j.quascirev.2010.11.019.
47. Ferguson, R.J., Brierley, G.J., 1999. Levee morphology and sedimentology along the lower Turos River, south-eastern Australia. Sedimentology 46, 627–648.
48. Fontana, A., Mozzi, P., Bondesan, A., 2008. Alluvial megafans in the Venetian–Friulian Plain (north-east Italy): evidence of sedimentary and erosive phases during Late Pleistocene and Holocene. Quat. Int. 189, 71–90.
49. Fraser, L.H., Keddy, P.A., 2005. The World›s Largest Wetlands: Ecology and Conservation. Cambridge University Press, Cambridge.
50. Fryirs, K., Brierley, G.J., 2001. Variability in sediment delivery and storage along river courses in Bega catchment, NSW, Australia: implications for re-mobilization of stored in channel sediment. Geomorphology 38, 237–265.
51. Fustic, M., Hubbard, S.M., Spencer, R., Smith, D.G., Leckie, D.A., Bennett, B., Larter, S., 2012. Recognition of down-valley translation in tidally influenced meandering fluvial deposits, Athabasca oil sands (Cretaceous), Alberta, Canada. Mar. Pet. Geol. 29, 219–232. http://dx.doi.org/10.1016/j.marpetgeo.2011.08.004.
52. Gardner, J.T., Ashmore, P.E., 2011. Geometry and grain-size characteristics of the basal surface of a braided river deposit. Geology 39 (3), 247–250. http://dx.doi.org/10.1130/G31639.1.
53. Gautier, E., Brunstein, D., Vauchel, P., Jouanneau, J.-M., Roulet, M., Garcia, C., Guyot, J.-L., Castro, M., 2010. Channel and floodplain sediment dynamics in a reach of the tropical meandering Rio Beni (Bolivian Amazonia). Earth Surf. Process. Landf. 35, 1838–1853. http://dx.doi.org/10.1002/esp.2065.
54. Gay, G.R., Gay, H.H., Gay, W.H., Martinson, H.A., Meade, R.H., Moody, J.A., 1998. Evolution of cutoffs across meander necks in Powder River, Montana, USA. Earth Surf. Process. Landf. 19, 651–662.
55. Gibbard, P.L., Lewin, J., 2002. Climate and related controls on interglacial fluvial sedimentation in lowland Britain. Sediment. Geol. 151, 187–210.
56. Grenfell, M., Aalto, R., Nicholas, A., 2012. Chute channel dynamics in large, sand-bed meandering rivers. Earth Surf. Process. Landf. 37, 315–331. http://dx.doi.org/ 10.1002/esp.2257.
57. Gupta, A. (Ed.), 2007. Large Rivers: Geomorphology Management. Wiley, Chichester. Gupta, H., Kao, S.-J., Dai, M., 2012. The role of mega dams in reducing sediment fluxes: a case study of large Asian rivers. J. Hydrol. 464–5, 447–458. http://dx.doi.org/10.1016/j.hydrol.2012.07.038.
58. Gupta, N., Atkinson, P.M., Carling, P.A., 2013. Decadal length changes in the fluvial planform of the River ganga: bringing a mega-river to life with Landsat archives. Remote Sens. Lett. 4, 1–9. http://dx.doi.org/10.1080/2150704X.2012.682658.
59. Gurnell, A.M., Surian, N., Zanoni, L., 2009. Multi-thread river channels: a perspective on changing European alpine river systems. Aquat. Sci. 71, 253–265.
60. Gurnell, A.M., Bertoldi, W., Corenblit, D., 2012. Changing river channels: the roles of hydrological processes, plants and pioneer fluvial landforms. Earth Sci. Rev. 111 (1–2), 129–141. http://dx.doi.org/10.1016/j.earscirev.2011.11.005.
61. Ham, D., Church, M., 2012. Morphodynamics of an extended bar complex, Fraser River, British Columbia. Earth Surf. Process. Landf. 37, 1074–1089. http://dx.doi.org/ 10.1002/esp.3231.
62. Harmar, O.P., Clifford, N.J., 2006. Planform dynamics of the Lower Mississippi River. Earth Surf. Process. Landf. 31, 825–843.
63. Hijma, M.P., Cohen, K.M., 2011. Holocene transgression of the Rhine river mouth area, The Netherlands/Southern North Sea: palaeogeography and sequence

stratigraphy. Sedimentology 58, 1453–1485. http://dx.doi.org/10.1111/j.1365-3091.2010.01222.x.

64. Hoffmann, T., Erkins, G., Cohen, K.M., Houben, P., Seidel, J., Dikau, R., 2007. Holocene floodplain sediment storage and hillslope erosion within the Rhine catchment. The Holocene 17, 105–118.

65. Hoffmann, T., Erkins, G., Gerlach, R., Klostermann, J., Lang, A., 2009. Trends and controls of Holocene floodplain sedimentation in the Rhine catchment. Catena 77, 96–106.

66. Hohensinner, S., Jungwirth, M., Muhar, S., Schmutz, S., 2011. Spatio-temporal habitat dynamics in a changing Danube River landscape 1812–2006. River Res. Appl. 27, 939–955. http://dx.doi.org/10.1002/rra.1407.

67. Hooke, J.M., Yorke, L., 2011. Channel bar dynamics on multi-decadal timescales in an actively meandering river. Earth Surf. Process. Landf. 36, 1910–1928. http://dx.doi.org/ 10.1002/esp.2214.

68. Horn, J.D., Fielding, C.R., Joeckel, R.M., 2012. Revision of Platte River alluvial facies model through observations of extant channels and barforms, and subsurface alluvial valley fills. J. Sediment. Res. 82, 72–91.

69. Howard, A.D., 1984. Simulation model of meandering. In: Elliott, C.M. (Ed.), River Meandering: Proceedings of the Conference Rivers› 83. American Society of Civil Engineers, New York, pp. 952–963.

70. Huang, H.Q., Nanson, G.C., 2007. Why some alluvial rivers develop an anabranching pattern. Water Resour. Res. 43, W07441. http://dx.doi.org/10.1029/2006WR005223.

71. Hubbard, S.M., Smith, D.G., Nielsen, H., Leckie, D.A., Fustic, M., Spencer, R.J., Bloom, L., 2011. Seismic geomorphology and sedimentology of a tidally influenced river deposit, Lower Cretaceous Athabasca oil sands, Alberta, Canada. Am. Assoc. Pet. Geol. 95, 1123–1145. http://dx.doi.org/10.1306/12131010111.

72. Hudson, P.F., Middlekoop, H., Stouthamer, E., 2008. Flood management along the Lower Mississippi and Rhine Rivers (The Netherlands) and the continuum of geomorphic adjustment. Geomorphology 101, 209–236. http://dx.doi.org/10.1016/ j.geomorph.2008.07.001.

73. Iriondo, M., 2004. Large wetlands of South America: a model for Quaternary humid environments. Quat. Int. 114, 3–9. http://dx.doi.org/10.1016/S1040-6182(03)00037-5.

74. Iriondo, M.H., Paggi, J.C., Parma, M.J. (Eds.), 2007. The Middle Paraná River. Springer, Verlag: Berlin. James, L.A., 2013. Legacy sediment: definitions and processes of episodically produced anthropogenic sediment. Anthropocene. http://dx.doi.org/10.1016/j.ancene.2013.04.001. Jerolmack, D., Paola, C., 2007. Complexity in a cellular model of river avulsion. Geomorphology 91, 259–270. http://dx.doi.org/10.1016/j.geomorph.2007.04.022.

75. Kemp, J., Spooner, N.A., 2007. Evidence for regionally wet conditions before the LGM in southeast Australia: OLS ages from a large palaeochannel in the Lachlan Valley. J. Quat. Sci. 22, 423–427. http://dx.doi.org/10.1002/jqs.1125.

76. Khan, N.S., Horton, B.P., McKee, K.L., Jerolmack, D., Falcini, F., Enache, M.D., Vane, C.H., 2013. Tracking sedimentation from the historic A.D. 2011 Mississippi River flood in the deltaic wetlands of Louisiana, USA. Geology. http://dx.doi.org/10.1130/G33805.1.

77. \Kleinhans, M.G., Cohen, K.M., Hoekstra, J., Ljmker, J.M., 2011. Evolution of a bifurcation in a meandering river with adjustable channel widths, Rhine delta apex, The Netherlands. Earth Surf. Process. Landf. 36, 2011–2027. http://dx.doi.org/10.1002/esp.2222.

78. Kleinhans, M.G., Ferguson, R.I., Lane, S.N., Hardy, R., 2013. Splitting rivers at their seams: bifurcations and avulsions. Earth Surf. Proc. Land. 38, 47–61. http://dx.doi.org/ 10.1002/esp.3268.

79. Kundzewicz, Z.W., Nohara, D., Tong, J., Oki, T., Buda, S., Takeuchi, K., 2009. Discharge of large Asian rivers — observations and projections. Quat. Int. 208, 4–10.

80. Lane, S.N., 2006. Approaching the system-scale understanding of braided river behaviour. In: Sambrook Smith, G.H., Best, J.L., Bristow, C.S., Petts, G.E. (Eds.), Braided Rivers: Process, Deposits, Ecology and Management. International Association of Sedimentologists Special Publication, 36. Blackwell, Oxford, pp. 107–135.

81. Lane, S.N., Westaway, R.M., Hicks, D.M., 2003. Estimation of erosion and deposition volumes in a large gravel-bed, braided river using synoptic remote sensing. Earth Surf. Process. Landf. 28, 249–271.

82. Lang, A., Bork, H.-R., Mäckel, R., Presto, N., Wunderlich, J., Dikau, R., 2003. Changes in sediment flux and storage within a fluvial system: some examples from the Rhine catchment. Hydrol. Process. 17, 3321–3334. http://dx.doi.org/10.1002/hyp.1389.

83. Latrubesse, E.M., 2008. Patterns of anabranching channels: the ultimate end-member adjustment of mega rivers. Geomorphology 101, 130–145. http://dx.doi.org/10.1016/ j.geomorph.2008.05.035.

84. Latrubesse, E.M., 2010. Amazon lakes. In: Bengtsson, L., Herschy, R.W., Fairbridge, R.W. (Eds.), Encyclopedia of Lakes and Reservoirs. Springer, Netherlands, pp. 13–26.

85. Latrubesse, E.M., Franzinelli, E., 2002. Holocene alluvial plain of the middle Amazon River Brazil. Geomorphology 44, 241–257. http://dx.doi.org/10.1016/S0169-555X(01) 00177-5.

86. Latrubesse, E.M., Franzinelli, E., 2005. The late Quaternary evolution of the Negro River, Amazon, Brazil: implications for island and floodplain formation in large anabranching tropical systems. Geomorphology 70, 372–397.

87. Latrubesse, E.M., Kalicki, T., 2002. Late Quaternary palaeohydrological changes in the upper Purus Basin, southwestern Amazonia, Brazil. Z. Geomorphol. Suppl. 129, 41–59.

88. Latrubesse, E.M., Stevaux, J.C., Sinha, R., 2005. Tropical rivers. Geomorphology 70, 187–206. http://dx.doi.org/10.1016/j.geomorph.2005.02.005.

89. Latrubesse, E.M., Cozzuel, M., da Silva-Caminha, S.A.F., Rigsby, C.A., Absy, M.L., Jaramillo, C., 2010.

The Late Miocene palaeogeography of the Amazon Basin and the evolution of the Amazon River system. Earth Sci. Rev. 99, 99–124. http://dx.doi.org/10.1016/j.earscirev.2010.02.005.

90. Latrubesse, E.M., Stevaux, J.C., Cremon, E.H., May, J.-H., Tatumi, S.H., Hurtado, M.A., Bezada, M., Argollo, J.B., 2013. Late Quaternary megafans, fans and fluvio–aeolian interactions in the Bolivian Chaco, Tropical South America. Palaeogeogr. Palaeoclimatol. Palaeoecol. 356–357, 75–88. http://dx.doi.org/10.1016/j.palaeo.2012.04.003.
91. Lauer, J.W., Parker, G., 2008. Net local removal of floodplain sediment by river meander migration. Geomorphology 96, 123–149.
92. Leclerc, R.F., Hickin, E.J., 1997. The internal structure of scrolled floodplain deposits based on ground-penetrating radar, North Thompson River, British Columbia. Geomorphology 21, 17–38.
93. Leddy, J.O., Ashworth, P.J., Best, J.L., 1993. Mechanisms of anabranch avulsion within gravel-bed braided rivers: observations from a scaled physical model. In: Best, J.L., Bristow, C.S. (Eds.), Braided Rivers. Geological Society of London Special Publication, 75, pp. 119–127.
94. Lewin, J., 1983. Changes of channel pattern and floodplains. In: Gregory, K.J. (Ed.), Background to Palaeohydrology. Wiley, Chichester, pp. 303–319.
95. Lewin, J., 2010. Medieval environmental impacts and feedbacks: the lowland floodplains of England and Wales. Geoarchaeology 25, 267–311.
96. Lewin, J., 2013. Enlightenment and the GM floodplain. Earth Surf. Process. Landf. 38, 17–29. http://dx.doi.org/10.1002/esp.3230.
97. Lewin, J., Ashworth, P.J., 2013. Defining large river channel patterns: alluvial exchange and plurality. Geomorphology. http://dx.doi.org/10.1016/j.geomorph.2013.02.024.
98. Lewin, J., Gibbard, P.L., 2010. Quaternary river terraces in England: forms, sediments and processes. Geomorphology 120, 293–311. http://dx.doi.org/10.1016/ j.geomorph.2010.04.002.
99. Lewin, J., Macklin, M.G., 2003. Preservation potential for Late Quaternary river alluvium. J. Quat. Sci. 18, 107–120. http://dx.doi.org/10.1002/jqs.738.
100. Lunt, I.A., Bridge, J.S., 2004. Evolution and deposits of a gravelly braid bar, Sagavanirktok River Alaska. Sedimentology 51, 415–432. http://dx.doi.org/10.1111/j.1365- 3091.2004.00628.x.
101. Lunt, I.A., Sambrook-Smith, G.H., Best, J.L., Ashworth, P., Lane, S.N., Simpson, C.J., 2013. Deposits of the sandy braided South Saskatchewan River: implications for the use of modern analogues in reconstructing channel dimensions in reservoir characterization. Bull. Am. Assoc. Pet. Geol. 97, 553–576. http://dx.doi.org/10.1306/09251211152.
102. Macklin, M.G., Jones, A.F., Lewin, J., 2010. River response to rapid Holocene environmental change: evidence and explanation in British catchments. Quat. Sci. Rev. 29, 1555–1576. http://dx.doi.org/10.1016/j.quascirev.2009.06010.
103. Makaske, B., Smith, D.G., Berendson, H.J.A., de Boer, A.G., Nielen-Kiezebrink, M.F., Locking, T., 2009. Hydraulic and sedimentary processes causing anastomosing morphology of the upper Columbia River, British Colombia, Canada. Geomorphology 111, 194–205.
104. Makaske, B., Maathuis, B.H.P., Padovani, C.R., Stolker, C., Mosselman, E., Jongman, R.H.G., 2012. Upstream and downstream controls of recent avulsions on the Taquari megafan, Pantanal, south-western Brazil. Earth Surf. Process. Landf. 37, 1313–1326. http://dx.doi.org/10.1002/esp.3278.
105. Marchetti, Z.Y., Latrubesse, E.M., Pereira, M.S., Ramonell, C.G., 2013. Vegetation and its relationship with geomorphologic units in the Parana River floodplain, Argentina. J. S. Am. Earth Sci. http://dx.doi.org/10.1016/j.jsames.2013.03.010.
106. Marriott, S.B., 1996. Analysis and modelling of overbank deposits. In: Anderson, M.G., Walling, M.G., Bates, P.D. (Eds.), Floodplain Processes. Wley, Chichester, pp. 63–93.
107. Marriott, S.B., Alexander, J., 1999. Introduction. In: Marriott, S.B., Alexander, J. (Eds.), Floodplains: interdisciplinary approaches. Geol. Soc. Lond. Spec. Publ., 163, pp. 1–13.
108. McCabe, P.J., 1984. Depositional environments of coal and coal-bearing strata. In: Rahmani, R.A., Flores, R.M. (Eds.), Depositional Environments of Coal and Coal Bearing Strata. International Association of Sedimentologists, Special Publication, 7, pp. 13–42.
109. McCabe, P.J., Parrish, J.T., 1992. Tectonic and climatic controls on the distribution and quality of Cretaceous coals. In: McCabe, P.J., Parrish, J.T. (Eds.), Controls on the Distribution and Quality of Cretaceous Coal. Geological Society of America, Special Paper, 267, pp. 1–15.
110. Mertes, L.A.K., 1997. Documentation and significance of the perirheic zone on inundated floodplains. Water Resour. Res. 33, 1749–1762.
111. Mertes, L.A.K., Dunne, T., 2007. The effects of tectonics, climatic history, and sea-level history on the form and behaviour of the modern Amazon River. In: Gupta, A. (Ed.), Large Rivers. Wiley, Chichester, pp. 115–144.
112. Mertes, L.A.K., Dunne, T., Martinelli, L.A., 1996. Channel-floodplain geomorphology along the Solimões–Amazon River, Brazil. Geol. Soc. Am. Bull. 108, 1089–1107.
113. Miall, A.D., 2006. Reconstructing the architecture and sequence stratigraphy of the preserved fluvial record as a tool for reservoir development: a reality check. Am. Assoc. Geol. Bull. 90, 989–1002.
114. Moody, J.A., Pizzuto, J.E., Meade, R.H., 1999. Ontogeny of a flood plain. Geol. Soc. Am. Bull. 111, 291–303.
115. Moreton, D., Carter, J., Peakcock, M., Rogola, B., 2013. Athabasca Oil Sands: Application of Integrated Technology in the Identification of Commercial Thermal and Mining Opportunities. International Petroleum Technology Conference, Paper 16755, Beijing, China. 13.

116. Motta, D., Abad, J.D., Langendoen, E.J., Garcia, M.H., 2012. A simplified 2D model for meander migration with physically-based bank evolution. Geomorphology 163–164, 10–25. http://dx.doi.org/10.1016/j.geomorph.2011.06.036.

117. Musial, G., Reynaud, J.-Y., Gingras, M.K., Féniès, H., Labourdette, R., Parize, O., 2012. Subsurface and outcrop characterization of large tidally influenced point bars of the Cretaceous McMurray Formation (Alberta, Canada). Sediment. Geol. 279, 156–172. http://dx.doi.org/10.1016/j.sedgeo.2011.04.020.

118. Nanson, G.C., 1980. Point bar and floodplain formation on the meandering Beatton River, northeastern British Columbia, Canada. Sedimentology 27, 3–29.

119. Nanson, G.C., 1981. New evidence of scroll-bar formation on the Beatton River. Sedimentology 28, 889–891. Nanson, G.C., Croke, J.C., 1992. A genetic classification of floodplains. Geomorphology 4, 459–486.

120. Nanson, G.C., Page, K., 1983. Lateral accretion of fine-grained concave benches on meandering rivers. In: Collinson, J.D., Lewin, J. (Eds.), Modern and Ancient Fluvial Systems. International Association of Sedimentologists Special Publication, 6, pp. 133–143.

121. Nicholas, A.P., 2013. Morphological diversity of the world›s largest rivers. Geology.GSA. http://dx.doi.org/10.1130/G34016.1.

122. Nicholas, A.P., Sandbach, S.D., Ashworth, P.J., Amsler, M.L., Best, J.L., Hardy, R.J., Lane, S.N., Orfeo, O., Parsons, D.R., Reesink, A.J.H., Sambrook Smith, G.H., Szupiany, R.N., 2012. Modelling hydrodynamics in the Rio Paraná, Argentina: an evaluation and intercomparison of reduced-complexity and physics based models applied to a large sand-bed river. Geomorphology 169–170, 192–211. http://dx.doi.org/10.1016/ j.geomorph.2012.05.014.

123. Nicholas, A.P., Ashworth, P.J., Sambrook Smith, G.H., Sandbach, S.D., 2013. Numerical simulation of bar and island morphodynamics in anabranching megarivers. Journal of Geophysical Research Earth Surface 118. http://dx.doi.org/10.1002/jgrf.20132.

124. Nilsson, C., Reidy, C.A., Dynesius, M., Ravenga, C., 2005. Fragmentation and flow regulation of the world›s large river systems. Science 308, 405–408. http://dx.doi.org/10.1126/ science.1107887.

125. Page, K.J., Nanson, G.C., Frazier, P.S., 2003. Floodplain formation and sediment stratigraphy resulting from oblique accretion on the Murrumbidgee River, Australia. J. Sed. Res. 73, 5–14.

126. Paira, A.R., Drago, E.C., 2006. Genetical, morphological and evolutional relationships of the floodplain lakes in the Middle Paraná River hydrosystem. Z. Geomorphol. 145, 207–228.

127. Paira, A.R., Drago, E.C., 2007. Origin, evolution, and types of floodplain water bodies. In: Iriondo, M.H., Paggi, J.C., Parma, M.J. (Eds.), The Middle Paraná River. Springer, Verlag: Berlin, pp. 53–81.

128. Panin, A.V., Sidorchuk, A.Yu., Chernov, A.V., 1999. Historical background to floodplain morphology: examples from the East European Plain. In: Marriott, S.B., Alexander, J. (Eds.), Floodplains: Interdisciplinary Approaches. Geological Society of London Special Publication, 163, pp. 217–229.

129. Paola, C., Borgman, L., 1991. Reconstructing random topography from preserved stratifi- cation. Sedimentology 38, 553–565.

130. Parker, G., Johannesson, H., 1989. Observations on several recent theories of resonance and overdeepening in meandering channels. In: Ikeda, S., Parker, G. (Eds.), River Meandering. Water Resource Monograph Series, 12. AGU, Washington, D.C., pp. 379–415.

131. Parker, G., Shimizu, Y., Wilkerson, G.V., Eke, E.C., Abad, J.D., Lauer, J.W., Paola, C., Dietrich, W.E., Voller, V.R., 2011. A new framework for modeling the migration of meandering rivers. Earth Surf. Process. Landf. 36, 70–86. http://dx.doi.org/10.1002/esp.2113.

132. Parker, N.O., Sambrook Smith, G.H., Ashworth, P.J., Best, J.L., Lane, S.N., Lunt, I.A., Simpson, C.J., Thomas, R.E., 2013. Quantification of the relation between surface morphodynamics and subsurface sedimentological product in sandy braided rivers. Sedimentology 60, 820–839. http://dx.doi.org/10.1111/j.1365-3091.2012.01364.x.

133. Peakall, J., Ashworth, P.J., Best, J.L., 2007. Meander-bend evolution, alluvial architecture, and the role of cohesion in sinuous river channels: a flume study. J. Sediment. Res. 77, 197–212. http://dx.doi.org/10.2110/jsr.2007.017.

134. Pizzuto, J.E., 1987. Sediment diffusion during overbank flows. Sedimentology 34, 301–317.

135. Ramonell, C.G., Amsler, M.L., Toniolo, H.A., 2002. Shifting modes of the Paraná River thalweg in its middle-lower reaches. Z. Geomorphol. Suppl. 129, 129–142.

136. Reesink, A.J.H., Ashworth, P.J., Sambrook Smith, G.H., Best, J.L., Parsons, D.R., Amsler, M.L., Hardy, R.J., Lane, S.N., Nicholas, A.P., Orfeo, O., Sandbach, S.D., Simpson, C.J., Szupiany, R.N., 2013. Scales and causes of heterogeneity in bars in a large multi-channel river: Rio Paraná. Argentina. Sedimentology. http://dx.doi.org/10.1111/sed.12092.

137. Reinfelds, I., Nanson, G., 1993. Formation of braided river floodplains, Waimakariri River, New Zealand. Sedimentology 40, 1113–1127.

138. Rice, S.P., Church, M., 2010. Grain-size sorting within river bars in relation to downstream fining along a wandering channel. Sedimentology 57, 232–251. http://dx.doi.org/ 10.1111/j.1365-3091.2009.01108.x.

139. Rice, S.P., Roy, A.G., Rhoads, B. (Eds.), 2008. River Confluences, Tributaries and the Fluvial Network. Wiley, Chichester. Rice, S.P., Church, M., Wooldridge, C.L., Hicken, E.J., 2009. Morphology and evolution of bars in a wandering gravel-bed river: lower Fraser river, British Columbia, Canada. Sedimentology 56, 709–736. http://dx.doi.org/10.1111/j.1365-3091.2008.00994.x.

140. Richey, J.E., Nobre, C., Deser, C., 1989. Amazon river discharge and climate variability: 1903–1985. Science 246, 101–103.

141. Rowland, J.C., Lepper, K., Dietrich, W.E., Wilson, C.J., Sheldon, R., 2005. Tie channel sedimentation rates,

owbow formation age and channel migration rate from optically stimulated luminescence (OSL) analysis of floodplain deposits. Earth Surf. Process. Landf. 30, 1161–1179. http://dx.doi.org/10.1002/esp.1268.

142. Roy, N.G., Sinha, R., Gibling, M.R., 2011. Aggradation, incision and interfluve flooding in the Ganga valley over the past 100,000 years: testing the influence of monsoonal precipitation. Palaeogeogr. Palaeoclimatol. Palaeoecol. http://dx.doi.org/10.1016/ j.palaeo.2011.08.012.
143. Rozo, M.G., Nogueira, A.C.R., Truckenbrodt, W., 2012. The anastomosing pattern and the extensively distributed scroll bars in the middle Amazon River. Earth Surf. Process. Landf. 14, 1471–1488. http://dx.doi.org/10.1002/esp.3249.
144. Sadler, P.M., 1981. Sediment accumulation rates and the completeness of stratigraphic sections. J. Geol. 89, 569–584.
145. Sambrook Smith, G.H., Best, J.L., Bristow, C.S., Petts, G.E. (Eds.), 2006. Braided rivers: process, deposits, ecology and management. International Association of Sedimentologists Special Publication, 36. Blackwell, Oxford.
146. Sambrook Smith, G.H., Best, J.L., Ashworth, P.J., Lane, S.N., Parker, N.O., Lunt, I.A., Thomas, R.E., Simpson, C.J., 2010. Can we distinguish flood frequency and magnitude in the sedimentological record of rivers? Geology 38, 579–582. http://dx.doi.org/10.1130/ G30861.1.
147. Santos, M.L., Stevaux, J.C., 2000. Facies and architectural analysis of channel sandy macroforms in the upper Paraná River. Quaternary International 72, 87–94.
148. Sidorchuck, A.Y., Panin, A.V., Borisova, O.K., 2009. Morphology of river channels and surface runoff in the Volga River basin (East European Plain) during the Late Glacial period. Geomorphology 113, 137–157.
149. Sidorchuk, A., 2003. Floodplain sedimentation: inherited memories. Glob. Planet. Chang. 39, 13–29. Sippel, S.J., Hamilton, S.K., Melack, J.M., 1992. Inundation area and morphometry of lakes on the Amazon River floodplain. Braz. Arch. Hydrobiol. 123, 385–400.
150. Slingerland, R., Smith, N.D., 2004. River avulsions and their deposits. Ann. Rev. Earth Planet. Sci. 32, 257–285.
151. Smith, D.G., 1986. Anastomosing river deposits, sedimentation rates and basin subsidence, Magdalena River, Northwestern Colombia, South America. Sediment. Geol. 46, 177–196.
152. Smith, D.G., Hubbard, S.M., Leckie, D.A., Fustic, M., 2009. Counter point-bar deposits: lithofacies and reservoir significance in the meandering modern Peace River and ancient McMurrary Formation, Alberta, Canada. Sedimentology 56, 655–669. http:// dx.doi.org/10.1111/j.1365-3091.2009.01050.x.
153. Smith, M.W., Macklin, M.G., Thomas, C.J., 2013. Hydrological and geomorphological controls of malaria transmission. Earth Sci. Rev. 116, 109–127. http:// dx.doi.org/ 10.1016/j.earscirev.2012.11.004.
154. Snedden, J.W., 2013. Channel-body basal scours: observations from 3D seismic and importance for subsurface reservoir connectivity. Mar. Pet. Sci. 39, 150–163. http:// dx.doi.org/10.1016/j.marpetgeo.2012.08.013.
155. Steiger, J., Gurnell, A.M., Petts, G.E., 2001. Sediment deposition along the channel margins of a reach of the Middle River Severn, UK. Regul. Rivers Res. Manag. 17, 443–460.
156. Stevaux, J.C., Corradini, F.A., Aquino, S., 2013. Connectivity processes and riparian vegetation of the upper Paraná River, Brazil. J. S. Am. Earth Sci. http:// dx.doi.org/10.1016/ j.jsames.2011.12.007.
157. Straub, K.M., Paola, C., Mohrig, D., Wolinsky, M.A., George, T., 2009. Compensational stacking of channelized sedimentary deposits. J. Sediment. Res. 79, 673–688. http:// dx.doi.org/10.2110/jsr.2009.070.
158. Strauss, D., Sadler, P.M., 1989. Stochastic-models for the completeness of stratigraphic sections. Math. Geol. 21 (1), 37–59. http://dx.doi.org/10.1007/BF00897239.
159. Sun, T., Meakin, P., Jossang, T., 1996. A simulation model for meandering rivers. Water Resour. Res. 32, 2937–2954.
160. Syvitski, J.P.M., Vorosmarty, C.J., Kettner, A.J., Green, P., 2005. Impact of humans on the flux of terrestrial sediment to the global coastal ocean. Science 308, 376–380.
161. Syvitski, J.P.M., Overeem, I., Brackenridge, G.R., Hannon, M., 2012. Floods, floodplains, delta plains — a satellite imaging approach. Sediment. Geol. 267–268, 1–14. http:// dx.doi.org/10.1016/j.sedgeo.2012.05.014.
162. Thomas, L., 2013. Coal Geology (2nd edition). Wiley-Blackwell, Chichester.
163. Thorne, C.R., Russell, A.P.G., Alam, M.K., 1993. Planform pattern and channel evolution of the Brahmaputra River, Bangladesh. In: Best, J.L., Bristow, C.S. (Eds.), Braided rivers: Special Publication of the Geological Society of London, 75, pp. 257–276.
164. Tipper, J.C., 1983. Rates of sedimentation, and stratigraphical completeness. Nature 302 (5910), 696–698. http://dx.doi.org/10.1038/302696a0.
165. Tockner, K., Stanford, J.A., 2002. Riverine flood plains: present state and future trends. Environ. Conserv. 29, 308–330. http://dx.doi.org/10.1017/S037689290200022X.
166. Toonen, W.H.J., Kleinhans, M.G., Cohen, K.M., 2012. Sedimentary architecture of abandoned channel fills. Earth Surf. Process. Landf. 37, 459–472. http://dx.doi.org/ 10.1002/esp.3189.
167. Tooth, S., 2013. Dryland fluvial environments: assessing distinctiveness and diversity from a global perspective. In: Shroder, J.F. (Ed.), Treatise on Geomorphology, vol. 9. Academic Press, San Diego, CA, pp. 612–644.
168. Tooth, S., Nanson, G.C., 2011. Distinctiveness and diversity of arid zone river systems, In: Thomas, D.S.G. (Ed.), Arid Zone Geomorphology: Process form and Change in Drylands, 3rd edition. Wiley, Chichester, pp. 269–300.
169. Tooth, S., Rodnight, H., Duller, G.A.T., McCarthy, T.S., Marren, P.M., Brandt, D., 2007. Chronology and controls

of avulsion along a mixed bedrock-alluvial river. Geol. Soc. Am. Bull. 119, 452–461.

170. Törnqvist, T.E., Bridge, J.S., 2002. Spatial variation of overbank aggradation rate and its in- fluence on avulsion frequency. Sedimentology 49, 891–905.

171. Trigg, M.A., Bates, P.D., Wilson, M.D., Schumann, G., Baugh, C., 2012. Floodplain channel morphology and networks of the middle Amazon River. Water Resour. Res. 48, W10504. http://dx.doi.org/10.1029/2012WR01188.

172. Trigg, M.A., Michaelides, K., Neal, J.C., Bates, P.D., 2013. Surface water connectivity dynamics of a large scale extreme flood. J. Hydrol. http://dx.doi.org/10.1016/j.jhydrol.2013.09.035.

173. Valente, C.R., Latrubesse, E.M., 2011. Fluvial archive of peculiar avulsive fluvial patterns in the largest Quaternary intracratonic basin of tropical South America: the Bananal Basin Central-Brazil. Palaeogeogr. Palaeoclimatol. Palaeoecol. http://dx.doi.org/ 10.1016/j.palaeo.2011.1.10.002.

174. Valente, C.R., Latrubesse, E.M., Ferreira, L.G., 2013. Relationships among vegetation, geomorphology and hydrology in the Bananal Island tropical wetlands, Araguaia River basin, Central Brazil. J. S. Am. Earth Sci. http://dx.doi.org/10.1016/j.jsames.2012.12.003.

175. Van De Wiel, M.J., Coulthard, T.J., 2010. Self-organized criticality in river basins: challenging sedimentary records of environmental change. Geology 38, 87–90. http:// dx.doi.org/10.1130/G30490.1.

176. van de Wolfshaar, K.E., Middlekoop, H., Addink, E., Winter, H.V., Nagelkerke, L.A.J., 2011. Linking flow regime, floodplain lake connectivity and fish catch in a large river- floodplain system, the Volga-Akhtuba floodplain (Russian Federation). Ecosystems 14, 920–934.

177. Van Dijk, W.M., van der Lageweg, W.I., Kleinhans, M.G., 2012. Experimental meandering river with chute cutoffs. J. Geophys. Res. 117, F03023. http://dx.doi.org/10.1029/ 2011FJ002314.

178. Vietz, G.J., Rutherfurd, I.D., Stewardson, M.J., Finlayson, B.L., 2011. Hydrodynamics and sedimentology of concave benches in a lowland river. Geomorphology 147 (8), 86–101. http://dx.doi.org/10.1016/j.geomorph.2011.07.033.

179. Walling, D.E., 2008. The changing sediment loads of the world›s rivers. Sediment Dynamics in Changing Environments, 325. IAHS Publ., pp. 323–328.

180. Walling, D.E., Moorhead, P.W., 1989. The particle size characteristics of fluvial suspended sediment: an overview. Hydrobiologia 176 (7), 125–149.

181. Walling, D.E., He, Q., Nicholas, A.P., 1996. Floodplains as suspended sediment sinks. In: Anderson, M.G., Walling, M.G., Bates, P.D. (Eds.), Floodplain Processes. Wiley, Chichester, pp. 399–440.

182. Wang, Y., Straub, K.M., Hajek, E.A., 2011a. Scale-dependent compensational stacking: an estimate of autogenic time scales in channelized sedimentary deposits. Geology 39 (9), 811–814. http://dx.doi.org/10.1130/G32068.1.

183. Wang, H., Saito, Y., Zhang, Y., Naishuang, B., Sun, X., Yang, Z., 2011b. Recent changes of sediment flux to the western Pacific Ocean from major rivers in East and Southeast Asia. Earth Sci, Rev http://dx.doi.org/10.1016/j.earscirev.2011.06.003.

184. Werritty, A., Paine, J.L., Macdonald, N., Rowan, J.S., McEwen, L.J., 2006. Use of multi-proxy records to improve estimates of flood risk: lower River Tay, Scotland. Catena 66, 107–119.

185. Westaway, R.M., Lane, S.N., Hicks, D.M., 2003. Remote survey of large-scale braided rivers using digital photogrammetry and image analysis. Int. J. Remote Sens. 24, 795–816.

186. Zinger, J.A., Rhoads, B.L., Best, J.L., 2011. Extreme sediment pulses generated by bend cutoffs along a large meandering river. Nat. Geosci. 4, 675–678. http://dx.doi.org/ 10.1038/NGEO1260.

CHAPTER 8

APPLICATION OF THE COASTAL HAZARD WHEEL METHODOLOGY FOR COASTAL MULTI-HAZARD ASSESSMENT AND MANAGEMENT IN THE STATE OF DJIBOUTI

Lars Rosendahl Appelquist[a], Thomas Balstrøm[b]

[a] UNEP Risoe Centre, Technical University of Denmark, UN City Copenhagen, Marmorvej 51, 2100 Copenhagen Ø, Denmark

[b] Department of Development and Planning, Aalborg University, A. C. Meyers Vænge 15, 2., 2450 Copenhagen SV, Denmark

ABSTRACT

This paper presents the application of a new methodology for coastal multi-hazard assessment and management in a changing global climate on the state of Djibouti. The methodology termed the Coastal Hazard Wheel (CHW) is developed for worldwide application and is based on a specially designed coastal classification system that incorporates the main static and dynamic parameters determining the characteristics of a coastal environment. The methodology provides information on the hazards of ecosystem disruption, gradual inundation, salt water intrusion, erosion and flooding and can be used to support management decisions at local, regional and national level, in areas with limited access to geophysical data. The assessment for Djibouti applies a geographic information system (GIS) to develop a range of national hazard maps along with relevant hazard statistics and is showcasing the procedure for applying the CHW methodology for national hazard assessments. The assessment shows that the coastline of Djibouti is characterized by extensive stretches with high or very high hazards of ecosystem disruption, mainly related to coral reefs and mangrove forests, while large sections along the coastlines of especially northern and southern Djibouti have high hazard levels for gradual inundation. The hazard of salt water intrusion is moderate along most of Djibouti's coastline, although groundwater availability is considered to be very sensitive to human ground water extraction. High or very high erosion hazards are associated with Djibouti's sedimentary plains, estuaries and river mouths, while very high flooding hazards are associated with the dry river mouths.

INTRODUCTION

Projected climate change will alter the environmental conditions along most of the world's coastlines and thereby the livelihoods of the local coastal populations (IPCC, 2014). According to the IPCC, the utilization of the coast has increased dramatically during the 20th century and this trend will continue during the 21st century, leading to a growth in the global coastal population from the current 1.2 billion to 1.8–5.2 billion by the 2080s depending on migration assumptions (IPCC, 2007a). Identifying climate-related hazards to coastal regions is therefore essential for managing potential hazards in due course, and this is especially a challenge in developing countries where data, expertise and economic resources are limited and coastal populations are growing rapidly.

To date, several different methodologies and approaches have been developed to assess and manage coastal climate

Citation: Lars Rosendahl Appelquista, Thomas Balstrømb, Application of the Coastal Hazard Wheel methodology for coastal multi-hazard assessment and management in the state of Djibouti, doi:10.1016/j.crm.2014.06.002

change vulnerability and coastal climate-related hazards. Generally, one can distinguish between index and indicator based methods, GIS-based decision support systems and dynamic computer models that are developed for different purposes and with different requirements for data and expertise (Ramieri et al., 2011). The index and indicator based methods such as the Coastal Vulnerability Index (CVI) and the Smartline system are to date the most realistic options for use in data-poor regions such as developing countries, but they cannot be directly used to identify a range of sector-specific hazards and management options and require relatively detailed input data (Ramieri et al., 2011, Thieler et al., 2000 and Sharples, 2014). The GIS-based decision support systems and dynamic computer models such as the DIVA model are very complex systems that would have to be combined with larger data-collection programs or be applied at very coarse resolution and also require highly specialized expertise (Global Climate Forum, 2013, Ramieri et al., 2011 and Mcleod et al., 2010).

The Coastal Hazard Wheel (CHW) framework is developed to address the gap in the current coastal vulnerability/hazard assessment methodologies and to offer a tool that can be used for combined assessment and management of the key climate-related coastal hazards in areas with limited data availability, resources and institutional capacity. The framework is especially designed to facilitate hazard screening and identification of hazard-hotpots at regional and national level but can also be applied as a first-line tool for hazard identification at local level prior to more detailed feasibility studies. The CHW framework makes use of publicly available geo-data and remote sensing information, and, depending on accuracy requirements, different levels of field verification. The system covers the hazards of ecosystem disruption, gradual inundation, salt water intrusion, erosion and flooding and can be used for development of hazard maps and hazard statics and subsequent identification of management options. The system can be applied for multi-hazard-assessments at three different steps depending on the data availability and accuracy requirements, namely:

- Step 1 that is designed for hazard assessments where data availability and accuracy requirements are relatively low. This step can generally be implemented based on remote sensing and publicly available data and is useful for hazard screening and for getting an initial picture of hazard presence in a cost-efficient manner.
- Step 2 that is designed for hazard assessments with moderate accuracy and this step generally requires additional field verification of the data obtained through remote sensing and public data sources.
- Step 3 that is designed for hazard assessments with moderate and locally focused accuracy and this step requires systematic and detailed field assessments at the local level, if possible supplemented with results from other locally focused assessment tools.

Generally, Step 1 and 2 are recommended for larger sub-regional, regional and national assessments, as it would require significant time and resources to implement Step 3 at this scale. Step 1–2 can therefore be used for broader hazard assessments, while Step 3 can be used for coastal stretches of specific interest or for detailed assessment of hazard-hotspots indentified at Step 1–2. Spot-assessments of a single coastal site can be carried out at any step depending on accuracy requirements, but it is important to be aware of the associated uncertainties at local level, unless the assessment is supplemented with more detailed results from other more locally focused assessment tools.

In this paper, the CHW framework is used to carry out a multi-hazard assessment at Step 1 (which can be thought of as a 'First Pass' assessment) for the full coastline of Djibouti and to develop a range of hazard maps along with national hazard statics. For the state of Djibouti, very little information is currently available on future climate-related hazards in coastal areas and it is therefore difficult for national planners and decision-makers to address and mitigate potential hazards. As systematic geophysical data collection until now has been limited and major challenges persist in downscaling regional climate models, this knowledge gap could potentially become a barrier for sound planning decisions. With a growing population and a possible future migration to coastal areas, the need for a robust decision-base for coastal planning becomes even more critical.

Together with developing relevant information for management of coastal climate change hazards in Djibouti, the assessment provides a good example of a CHW application in a data poor location for which the system is designed. The goal of this paper is therefore twofold, namely to showcase the practical procedure for applying the CHW methodology for national hazard assessments in a data poor location, and to provide relevant information on coastal hazards and management options for the coastline of Djibouti.

Since the assessment is carried out as a national hazard assessment at Step 1, it is designed to provide a good overview of where specific hazards are present and at what level the hazards are manifested for the full length of Djibouti's coastline. If a higher level of accuracy is needed, it might be necessary to supplement the assessment with addition field verification. At this stage, however, the assessment provides a good general picture of the coastal hazards for Djibouti. The result of the hazard assessment is presented as a series of five thematic overview maps for Djibouti, and along with this, the assessment has tested the possibility of developing detailed hazard layers for use in Google Earth. Whereas the overview maps are useful for getting a good general picture of the coastal hazards, the hazard layers can be used to support more detailed planning decisions at sub-regional level.

THE COASTAL HAZARD WHEEL (CHW) FRAMEWORK

The CHW framework is a hazard assessment and management methodology that can be used in areas with limited geophysical data availability and institutional capacity. The framework is based on a specially designed coastal classification system that incorporates the main geo-biophysical parameters determining the characteristics of a coastal environment and how this environment responds to the predicted changes in the global climate (IPCC, 2007b). The framework uses the coastal geological layout as basis, on which it adds the main dynamic parameters acting in the coastal environment, and it is designed to cover all generic coastal environments worldwide.

The framework provides information on the degree to which key climate change hazards are inherent in a particular coastal environment, defined as the hazards being an integrated part of the geo-biophysical properties of a coastal system when it is exposed to future climate change. The framework covers the inherent hazards of ecosystem disruption, gradual inundation, salt water intrusion, erosion and flooding, and each hazard evaluation is graduated into four different hazard levels based on a scientific literature review. The current version of the CHW framework includes 113 generic coastal environments and a total of 565 individual hazard evaluations.

When using the CHW, see Fig. 1, the user starts in the centre of the wheel and then moves outwards ending up with the inherent hazard evaluations in the outermost circles. Starting from the centre, the coastal classification parameters come in the following order where each category is represented by a new circle: Geological layout, wave exposure, tidal range, flora/fauna, sediment balance and storm climate. The special TSR category constitutes a group of highly dynamic geological environments. The inherent hazards then come in the following order: Ecosystem disruption, gradual inundation, salt water intrusion, erosion and flooding.

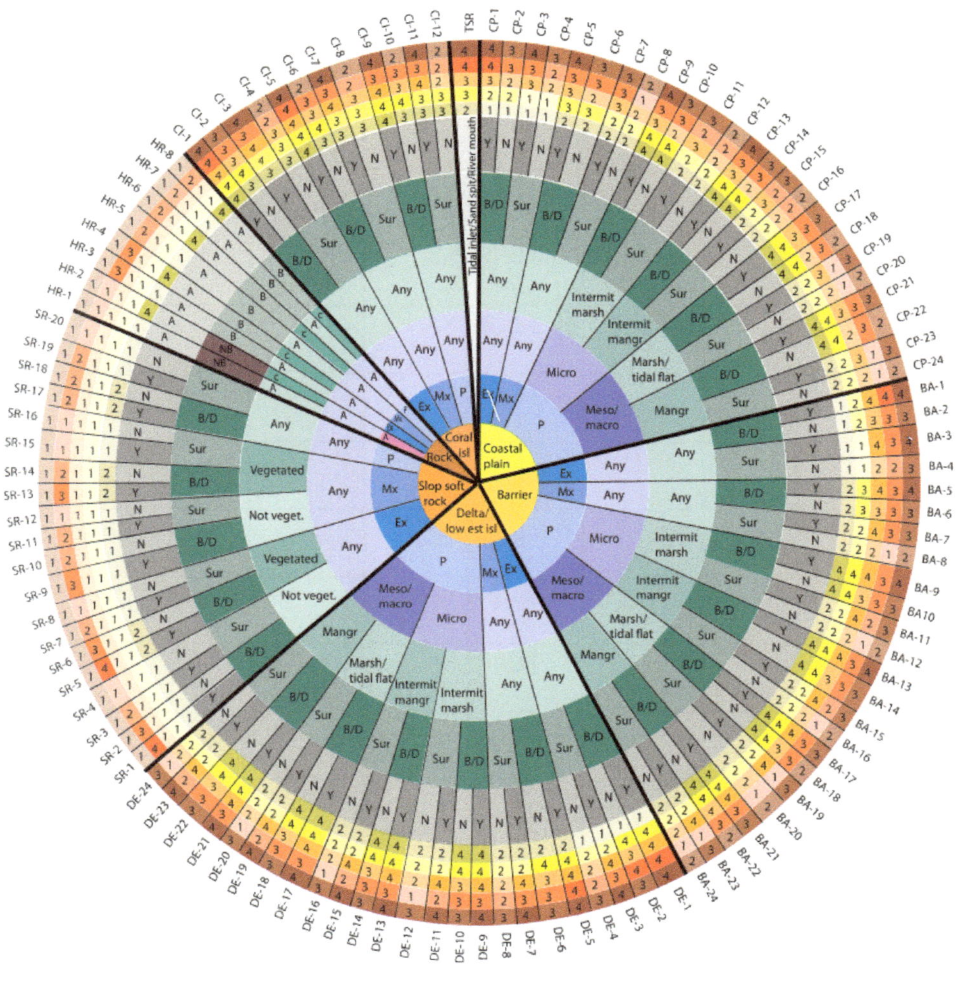

Figure 1: The Coastal Hazard Wheel consisting of six geo-biophysical classification circles, five hazard circles and the coastal classification codes. In the classification code, CP stands for coastal plain, BA for barrier, DE for delta, SR for sloping soft rock, HR for sloping hard rock, CI for coral island and TSR for tidal inlet/sand spit/river mouth (Rosendahl Appelquist, 2012).

In the practical application of the assessment framework, the user should make a new assessment every time any of the classification parameters change significantly. This can be done based on a combination of basic geo-biophysical data and remote sensing information, if possible combined with field verification. It is important that the user always is aware of human alterations of the natural environment that may have an impact on the coastal classification and thereby the hazard levels. A detailed description of the CHW methodology, assumptions and limitations can be found in Rosendahl Appelquist (2012).

The current version of the CHW operates with eight different generic coastal environments for rocky coastlines, and hence does not capture small variations and special rarities for these environments. Since the coastline of Djibouti is predominantly of rocky origin and includes some special features such as raised coral reef plains, it was decided to increase the detail of the wheel's rocky category so it spans over a total of 26 different coastal environments. In this way, the assessment for Djibouti will be able to provide relatively detailed hazard information, even when large parts of the coastline fall into the rocky coast category. The extended list of rocky coastal environments with their assigned hazard values is shown in Table 1. The list distinguishes between flat and sloping rocky coastlines, and the hazard values are derived from the rocky coast and coastal plain categories in the original CHW framework. Thus, the eight rocky types included in the original framework are not applied for this assessment.

Table 1: The extended list of rocky coastal categories

Coastal type	Geological layout	Wave exposure	Tidal range	Flora/ fauna	Sediment balance	Storm climate	Ecosystem disruption	Gradual inundation	Salt water intrusion	Erosion	Flooding
F-HR1	Flat hard rock	Exposed	Any	Any	No beach	Yes	1	3	2	1	3
F-HR2		Exposed	Any	Any	No beach	No	1	3	2	1	2
F-HR3		Exposed	Any	Any	Beach	Yes	1	3	2	2	3
F-HR4		Exposed	Any	Any	Beach	No	1	3	2	2	2
F-HR5		Exposed	Any	Coral reef	No beach	Yes	4	3	2	1	3
F-HR6		Exposed	Any	Coral reef	No beach	No	4	3	2	1	2
F-HR7		Exposed	Any	Coral reef	Beach	Yes	4	3	2	2	3
F-HR8		Exposed	Any	Coral reef	Beach	No	4	3	2	2	2
F-HR9		Moderately exposed	Any	Any	No beach	Yes	1	3	2	1	3
F-HR10		Moderately exposed	Any	Any	No beach	No	1	3	2	1	2
F-HR11		Moderately exposed	Any	Any	Beach	Yes	1	3	2	2	3
F-HR12		Moderately exposed	Any	Any	Beach	No	1	3	2	2	2
F-HR13		Moderately exposed	Any	Coral reef	No beach	Yes	4	3	2	1	3
F-HR14		Moderately exposed	Any	Coral reef	No beach	No	4	3	2	1	2
F-HR15		Moderately exposed	Any	Coral reef	Beach	Yes	4	3	2	2	3
F-HR16		Moderately exposed	Any	Coral reef	Beach	No	4	3	2	2	2
F-HR17		Protected	Any	None	Any	Yes	1	3	2	1	3
F-HR18		Protected	Any	None	Any	No	1	3	2	1	2
F-HR19		Protected	Any	Fringing marsh/ mangr	Any	Yes	3	3	2	1	3
F-HR20		Protected	Any	Fringing marsh/ mangr	Any	No	3	3	2	1	2
F-HR21		Protected	Any	Coral reef	Any	Yes	4	3	2	1	3
F-HR22		Protected	Any	Coral reef	Any	No	4	3	2	1	2
S-HR1	Sloping hard rock	Any	Any	Any	No beach	Any	1	1	1	1	1
S-HR2		Any	Any	Any	Beach	Any	1	2	1	2	1
S-HR3		Any	Any	Coral reef	No beach	Any	4	1	1	1	1
S-HR4		Any	Any	Coral reef	Beach	Any	4	2	1	2	1

CHARACTERISTICS OF DJIBOUTI'S COASTLINE

The coastline of Djibouti extends from the southern Red Sea and strait of Bab el Mandeb to the Somali border around the Gulf of Tadjoura (Bird, 2010). The general outline of Djibouti's coastline, along with political borders and larger cities is shown in Fig. 2. From the figure it can be seen that Djibouti shares its longest border with Ethiopia, while the coastal borders are shared with Eritrea and Somalia.

Figure 2: Overview map of Djibouti (CIA, 1996).

The geology of Djibouti is shaped by the tectonic trends of East Africa's Great Rift Valley which forms a complex landscape composed of high blocks and subsistence zones, mostly of volcanic and sedimentary rocks (Schlüter, 2006). The northern parts of Djibouti's coastline from the strait of Bab el Mandeb to the mouth of the Gulf of Tadjoura is characterized by raised coral reef plains interrupted by a rocky formation of basaltic lava, subordinate ignimbrites and rhyolites. Further south, a basaltic formation is located next to the city of Tadjoura, and after this, the coastline turns into alluvial deposits until ca. 10 km north of Ghoubet Bay. The coastline of Ghoubet Bay and the southern parts of the Gulf of Tadjoura is generally characterized by rocky formations of basalts, subordinate ignimbrites, rhyolites, silicic massifs and lava flows until just west of the city of Djibouti. From here to the Somali border, the coastline is composed of a mixture of eluvial and colluvial deposits, taluses, sheetfloods, dunes and beach deposits (Schlüter, 2006). Geomorphologic form elements such as barriers, spits and river mouths are found along the northern coastline facing the Gulf of Aden, the northern part of the Gulf of Tadjoura, as well as west and south of Djibouti City. In addition, Djibouti's coastline is characterized by having extensive stretches of fringing coral reefs. Generally, the Gulf of Aden basin is a young ocean formed by the rifting of Asia from Africa and is characterized by a well-defined continental margin, a small ocean basin and an active mid-ocean ridge. It is deepest in the middle with depths of 2000–2500 m and shallowest in the west with depths of less than 1000 m (Al Saafani, 2008).

The northern mountainous areas of Djibouti have an arid climate while the central and southern regions have semi-arid conditions. The country has two distinct seasons, namely the cold season spanning from October to April and the hot season spanning from June to September. Whereas the cold season is characterized by temperatures of 22–30 °C and increased humidity, the hot season has temperatures of 30–40 °C, violent dry winds, occasional sandstorms and is generally dry. The transition periods May–June and September–October have a variable wind climate and are generally dry. The precipitation pattern is very irregular with annual precipitation levels ranging from 50 to 215 mm, although 150 mm is rarely exceeded (Ministere de l'Habitat, l'Urbanisme, l'Environnement et de l'Amenagement du Territoire, 2012). Long dry periods can be followed by very intense rain, leading to catastrophic flooding events with damage to people and property. An example of this took place in 1994 where Djibouti received 360 mm rain over just two days (Ministere de l'Habitat, l'Urbanisme, l'Environnement et de l'Amenagement du Territoire, 2012). The wind regime in the Gulf of Aden is of monsoonal character with north-easterly winds during the winter months and south-westerly during the summer season (Ron Englebretson, 2002).

The tidal range along Djibouti's coastline is just at the transition between micro- and meso-tidal. The Red Sea has a semi-diurnal tidal regime with a tidal range of less than 1 m, while the Gulf of Aden generally has a mixed tidal regime with a tidal range exceeding 2 m (Jarosz and Murray, 2002). In the Ghoubet Bay, the tidal range is about 2 m and the tidal wave is generally one hour delayed compared to in the Gulf of Tadjoura (Salt Investment, 2008).

More than two-thirds of Djibouti's population of 865,000 live in the capital area which is located in a coastal setting facing the Gulf of Tadjoura and the Gulf of Aden (World Bank, 2011). The economy of Djibouti has been affected by political and economic instability as well as natural shocks such as droughts and floods, all damaging the country's competitiveness. Yet, recent developments in the marine and harbor industry have led to an increase in foreign direct investment. The large majority of Djibouti's rural population lives in infertile desert areas and is very susceptible to natural variations, especially water supply. Djibouti imports almost all its consumed cereal, and food aid makes up almost 10 percent of total imports (World Bank, 2011). The agricultural and industrial sectors represent the key livelihoods, although these are underdeveloped. The country is home to a large pastoralist population which lives on poor quality pasture lands and is vulnerable to climate change. Many pastoralist groups that rely on winter grazing grounds are already extremely vulnerable and are migrating to Sudan due to pasture degradation and increasing population pressure. The fishing sector constitutes a smaller source of livelihood with about one thousand people directly employed in this area (World Bank, 2011). Generally, 96.5 percent of the rural population lives below the poverty line.

Excessive pumping of groundwater and over-exploitation of surface waters are already placing significant stress on the limited water resources, and challenges in this area are only expected to increase with climate change. Furthermore, human pressures on coral reefs, estuaries and mangrove forests already affect the ecosystem services these systems provide (World Bank, 2011). With the current population growth of about 2 percent, the combined pressures from increasing population density and climate change will pose significant challenges to Djibouti's coastal areas in the years to come.

Climate change is to some degree already detectable in Djibouti. Over the past decades, the average temperature has been higher than normal and the period between 1991 and 2000 was one of the hottest decades on record. The absolute maximum monthly temperature has increased between 0.5 and 1.5 °C in the past three decades, while the minimum temperature has increased by 1.5 °C. Furthermore, there has been observed a significant decrease in rainfall in the months April–July, along with a significant increase for the months of January and October. It is projected that temperatures across Djibouti will increase by 0.6–2.4 °C by 2050, while the sea level is expected to rise between 8 and 39 cm compared to 1990 levels. Future precipitation patterns remain unclear although it is expected that critical rainfall periods are likely to be disrupted (World Bank, 2011).

DATA FOR THE HAZARD ASSESSMENT

the hazard assessment is carried out based on geophysical data available from relevant institutions and from the scientific literature. The complete list of data used for the assessment include the geological information and geological maps available in Schlüter (2006), the wave, tide and storm maps included in the original CHW framework paper and published by Masselink and Hughes (2003), supplementary data on wind conditions (Ron Englebretson, 2002), supplementary data on tidal range (Salt Investment, 2008,Jarosz and Murray, 2002 and Jarosz, 1997), the Reefbase database of global distribution of coral reefs (Reefbase, 2013), Google Earth satellite images with timeline and ground elevation functions (Google, 2013) and Bing Maps available in Esri's ArcGIS (Esri, 2012;Microsoft, 2013). The data used for the different classification components are briefly described below.

The geological layout

The geological layout is determined based on the geological map of Djibouti and Google Earth's satellite images and ground elevation function. Using these three data sources it is possible to classify the geological layout according to the CHW categories and a new classification is made every time the geological base material, geomorphology or slope changes significantly. The definitions for the different geological layouts are further described in Rosendahl Appelquist (2012) and it should be noted that several coastal types can occur at the same coastal stretch if they are located landwards of each other. This will e.g. be the case where a barrier is located landwards of a rocky or sedimentary mainland coast.

Large parts of Djibouti's coastline fall into one of the extended hard rock categories due to its volcanic, sedimentary or carbonate characteristics. Since it is not possible to determine the permeability and fractures of the carbonate rocks in the north and of the eluvial deposits in the south of Djibouti without additional field verification, it is assumed, that they have maintained most of their rocky characteristics. This assumption is also supported by the extensive presence of coral reefs shoreward of these deposits, as corals require a rocky structure to adhere to for proper habitat formation.

The slope of the coastline is determined using Google Earth's ruler and ground elevation functions. Generally, long stretches of Djibouti's coastline have a steep sloping profile, but especially the northern and southern coastal areas are characterized by extensive flat rocky coastlines. The geomorphology is determined through a visual inspection of Google Earth's satellite images, and form elements such as barriers, spits and dry river mouths can be relatively easily identified using a zoom level of about 10 km.

THE WAVE EXPOSURE

Since Djibouti is located in an area outside the swell/monsoon regions according to the maps in the original CHW framework paper (Rosendahl Appelquist, 2012 and Masselink and Hughes, 2003), the wave exposure is determined by the wind speeds and free fetch. As it has not been possible to acquire detailed wind data for the area, it has been decided to rely on the free fetch to determine the wave exposure. Since the wind is blowing from north-east and south-west in a monsoonal pattern, it is expected that the free fetch can be used as an acceptable proxy for the wave exposure for most of Djibouti's coastline. Google Earth is used for determining if the free fetch for a given coastal stretch is less than 10 km, 10–100 km or above 100 km, which are the defined boundaries for protected, moderately exposed and exposed coastlines in the CHW framework (Rosendahl Appelquist, 2012). As the wind is blowing from varying directions, the general maximum size of the water body in question is used as the free fetch length. Generally, the coastline facing the Gulf of Aden is classified as exposed, while the inner parts of the Gulf of Tadjoura are classified as moderately exposed, and the Ghoubet Bay is classified as protected.

The tidal range

Since Djibouti is situated just at the border between micro- and mesotidal regimes, supplementary data has been collected to determine the tidal conditions. The assessment therefore makes use of tidal data for the Red Sea, the Bab el Mandab Strait, the Gulf of Aden and Ghoubet Bay to determine the tidal conditions in more detail (Salt Investment, 2008, Jarosz and Murray, 2002 and Jarosz, 1997). Based on this data, the northernmost coast of Djibouti is assumed to have micro-tidal conditions while the remaining coastal stretches are assumed to be meso-tidal.

The flora/fauna

The flora/fauna of Djibouti's coastline is mainly related to the stretches with mangrove vegetation and the extensive coral reef systems. The mangrove areas are found in relation to the estuaries and bays in northern Djibouti, on the Moucha Islands and south of Djibouti City and can be easily identified using the satellite images available in Google Earth. The coral reef systems are identified using the Reefbase database of global distribution of coral reefs that provides a relatively detailed picture of the distribution of the coastal coral reefs (Reefbase, 2013).

THE SEDIMENT BALANCE

The sediment balance is an indication of whether a particular coastline has a balance, deficit or surplus of sediment over a time period of e.g. 10 years, and the sediment balance evaluation is carried out for all sedimentary coastal stretches based on Google Earth's historical satellite images. The sedimentary stretches includes the barriers, spits, coastal plains and sloping sedimentary coasts, and generally the quality of the available satellite images allows for a relatively detailed assessment. Due to the desert climate of Djibouti, however, most of the coastal stretches do not have a vegetation line that can be used for assisting the evaluation. The evaluation therefore relies mainly on the land-sea interface, which is affected by the high/low water levels at the time the satellite images were taken, and this adds some uncertainty to the assessment. Where there is any doubt about the sediment balance, it is assumed that the coastline has a balance/deficit as this gives the highest general hazard levels. For rocky coastlines, the CHW simply distinguishes between whether the coastline consists of bare rock or if some kind of beach environment is present. This is reflected in the special beach/no beach categories for these coastlines, which are assigned through a visual inspection of the coastline in Google Earth.

The storm climate

Since Djibouti is located outside areas influenced by tropical cyclones as indicated on the map in the original CHW framework paper (Rosendahl Appelquist, 2012 and Masselink and Hughes, 2003), the complete coastline is classified to be located in a non-tropical cyclone area. The extreme storm events present in Djibouti mainly relates to sand and rainstorms, sometimes associated with intense precipitation (Ministere de l'Habitat, l'Urbanisme, l'Environnement et de l'Amenagement du Territoire, 2006).

THE GIS PROCEDURE

The hazard assessment is carried out in ArcGIS, expanding on the assessment procedure suggested in the original CHW methodology paper (Rosendahl Appelquist, 2012). The assessment primarily makes use of the general functions available in ArcGIS, but uses the recent and historical satellite images and digital elevation model available in Google Earth as secondary data sources. Although other sources of historical satellite images and elevation models are available, Google Earth is selected as an easily accessible tool that is increasingly used for environmental management globally in areas ranging from marine, coastal and terrestrial assessments to wider science communication (Sawakuchi et al., 2012, Scheffers et al., 2012 and Whitmeyer et al., 2012).

First step in the assessment procedure is to create a geodatabase in ArcGIS that will contain all data on the coastal classification and subsequent hazard levels. Following this, and in order to have a relatively detailed and up-to-date digitized coastline of Djibouti that can be used for the assessment, a new line feature is created in the geodatabase referencing UTM Zone 38. The full coastline of Djibouti including the coastlines of backbarriers and estuaries is then digitized at the approximate Mean Sea Level (MSL) based on ArcGIS' hybrid Bing map, leaving gaps for river mouths and tidal inlets. The accuracy of the digitization is approximately 5–10 m, and the digitized coastline will be used as foundation for all the subsequent hazard maps.

To facilitate the assessment of the coastal slope and sediment balance, two supplementary line features are created in Google Earth. The first line feature that aims to support the slope assessment consists of a range of shore-parallel line sections, drawn landwards of the coastline in areas with a slope greater than 3–4 percent. In this way, the user can easily determine whether a specific coastal area is sloping or not when carrying out the assessment. The second line feature that aims to support the sediment balance evaluation is used to compare the coastline development between different historical satellite images. The line feature consists of a continuous line drawn at the approximate coastal vegetation line for all sedimentary stretches of Djibouti's coastline. Due to Djibouti's arid climate, however, many coastal stretches are not vegetated and hence it is generally necessary to draw the line at the approximate MSL. When the coastal classification is carried out, the sediment balance can then be evaluated by comparing the different historical satellite images available in Google Earth's historical image feature, looking at how the coast has been developing compared to the digitized, most recent coastline. Because the satellite images are captured at different tide levels and times of the year, there are some significant uncertainties related to this assessment, especially as it is not possible to use the vegetation line as proxy for coastal stability/retreat/advance. In practice, the methodology therefore simply assumes a balance/deficit whenever there is any uncertainty, and a sediment surplus is only assumed in areas with a coastal advance of at least several meters per year, which generally does not occur along Djibouti's coastline. As no detailed coastal sediment surveys or a clearly visible vegetation line are available, this approach is considered an acceptable proxy for evaluating the coastal sediment balance.

The coastal classification is carried out in ArcGIS on top of the digitized coastline of Djibouti. This is done using a polygon feature created in the geodatabase with the same coordinate system as the digitized coastline. The polygons are then used to split the coastline into smaller sections, each being classified based on the CHW classification system. The sections are stored in a so called linear referencing system that keeps track of the sections based on a simple measuring system defined along the coastline (Balstrøm, 2008). The practical classification is carried out by drawing a separate polygon for each coastal classification category along the coastline, based on a subjective evaluation of the classification parameters listed in the data section earlier. It should be noted that it is important to establish a snapping environment to make sure that the polygons are properly aligned with each other. The name of the coastal type in question is then written in the attribute table for each polygon in the ID field. Because the ID field in the attribute table only accepts numbers, the different coastal types have been assigned a number, which in the case of this assessment spans from 1 to 131 due to the expansion of the category for rocky coastlines. Since the coastal classification is carried out based on the CHW and a range of different data parameters, the user has to decide on an appropriate coastal type before each polygon is completed. Sometimes a coastline can maintain the same properties for many kilometers while at others it changes for every 50 m. This means that the length of each polygon varies significantly for the different parts of Djibouti's coastline.

The polygons are subsequently used to divide the initial digitized coastline into sections, each representing a specific coastal type. The hazard levels from the CHW system are then typed into a separate attribute table that is joined to the attribute table of the coastal classification file. Based on this, five different hazard maps are created for the respective hazards types and the different hazard levels are assigned a color code. The hazard maps are created based on a hybrid Bing map to optimize the visual readability, and the smaller villages along Djibouti's coastline are added to the maps manually. In addition to this, five separate hazard layers are developed for use in Google Earth to enable users to get a more detailed picture of the hazard presence.

RESULTS

The assessment results presented in this paper are a series of five national overview maps and some key hazard statistics. The overview maps are designed to provide a good general picture of the hazard hotspots and hazard distribution along Djibouti's coastline and are relevant for supporting coastal management decisions and climate change adaptation initiatives.

The national hazard map for ecosystem disruption is shown in Fig. 3, and for Djibouti, this hazard is mainly associated with the extensive coral reef systems and patchy mangrove forests. Most areas indicated to have a *very high* hazard of ecosystem disruption are related to the coral reef ecosystems, and these ecosystems extend for a full 50 km stretch of the northernmost part of Djibouti and then appear in more fragmented form on the remaining parts of the coastline in locations where physical conditions allow for coral growth. As coral reefs require a hard base to adhere to and a low level of dispersed sediment, they are normally present in areas with a rocky geological layout and a low level of fluvial sediment supply. The presence of river mouths south of Djibouti city therefore creates some gaps in an otherwise continuous coral reef system in this area. The mangrove habitats in Djibouti are generally considered to have a *high* hazard of ecosystem disruption, especially due to the limited sediment availability. These habitats are located in the bays and protected estuaries north of the Gulf of Tadjoura, but are also present on the Moucha Islands and in locations south of Djibouti City. The remaining part of the coastline is considered to have a *low* or *moderate* hazard level for ecosystem disruption and the hazards in these areas are considered minor compared to the hazards of the areas with coral reef and mangrove environments.

Figure 3: National hazard map for ecosystem disruption.

The national hazard map for gradual inundation is shown in Fig. 4 and this hazard is mainly related to the rocky and sedimentary plains, river mouths and barriers. The northern coastline of Djibouti from Eritrea until just north of the Gulf of Tadjoura is characterized by a low relief and several barrier systems that have a *high* and in some locations *very high* hazard of gradual inundation. The northern part of the Gulf of Tadjoura from Obock to just south of Tadjoura also has some low-lying areas with *high*gradual inundation hazards, while the remaining part of Gulf of Tadjoura and Ghoubet Bay generally have *low* and *moderate* hazard levels. The coastline around and south of Djibouti city, however, also has a *high* hazard of gradual inundation. It should be noted, however, that gradual inundation is a slow, long-term process and is therefore mainly relevant for longer-term coastal planning.

Figure 4: National hazard map for gradual inundation.

The national hazard map for salt water intrusion is shown in Fig. 5. The hazard of salt water intrusion is generally *high* or *very high* at the low-lying barriers and river mouths along Djibouti's coastline, while it is *low* or *moderate* along most of the remaining coastline. However, due to the very dry climatic conditions, all areas indicated to have*moderate* hazard levels can very easily move to *high* or *very high* hazard levels due to human extraction of water from the very limited freshwater reservoirs. The *moderate*hazard levels should therefore be seen as relative hazard levels compared to e.g. low-lying delta areas of Bangladesh and only applies when very little human ground water extraction takes place. One should therefore assume that coastal areas indicated to have a *moderate* hazard level will move to the *high* category as soon as any significant human ground water extraction takes place.

Figure 5: National hazard map for salt water intrusion.

The national hazard map for erosion is shown in Fig. 6 and this hazard is mainly related to the barriers, tidal inlets and river mouths as well as the flat and sloping sedimentary stretches along Djibouti's coastline. The coastline of the estuaries located between the Eritrean border and Gulf of Tadjoura has some sections with *high* and *very high* erosion hazards. In the northern part of the Gulf of Tadjoura, *very high* erosion hazards are related to the dry river mouths, while *high* erosion hazards are assigned to the sedimentary coastal plain. The remaining parts of the Gulf of Tadjoura and Ghoubet Bay generally have *low* and *moderate* erosion hazards, but some of the dry river mouths east and south of Djibouti city have *high* and *very high* erosion hazard levels.

Figure 6: National hazard map for erosion.

The national hazard map for flooding is shown in Fig. 7 and this hazard is especially related to the coastal stretches made of low-lying dry river mouths that are likely to be flooded during intense precipitation events. As most of these areas are completely dry during normal weather conditions, one might not be aware of the *very high* inherent flooding hazards of these locations. Some areas where these conditions are combined with human settlements are the eastern parts of Obock and west and south of Djibouti city. Other flooding hotspots include the barriers and spit systems in northern Djibouti facing the Gulf of Aden. The extensive coastal plains in northern Djibouti and the plains south of Djibouti City have a *moderate* flooding hazard which should also be kept in mind when planning human settlement.

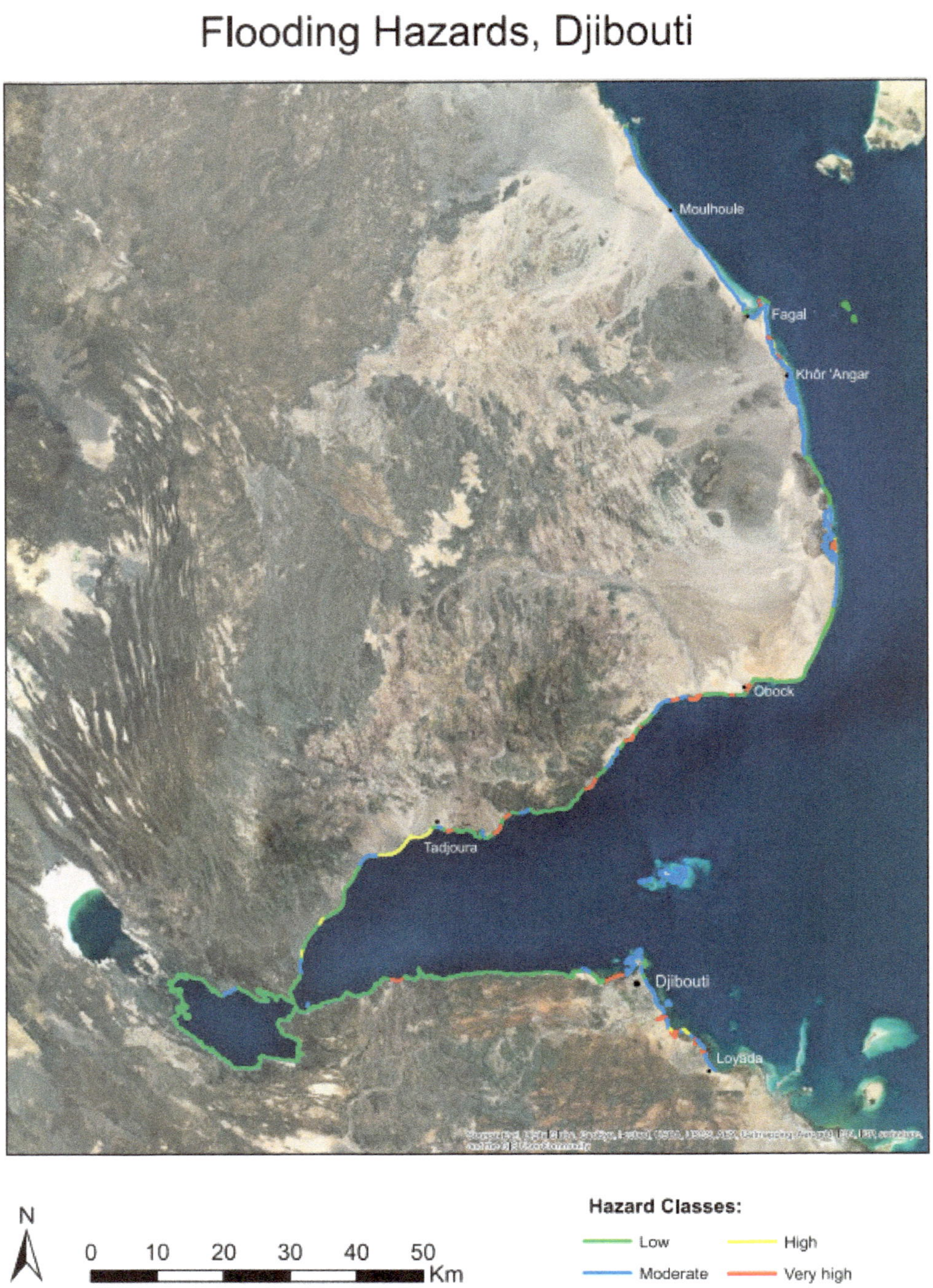

Figure 7: National hazard map for flooding.

Table 2 shows an overview of the hazard distribution in percent for the coastline of Djibouti. It can be seen that close to 50 percent of Djibouti's coastline has a *moderate,high* or *very high* hazard level for erosion and flooding. The hazard of gradual inundation is relatively widespread with about 65 percent of the coastline having *moderate*, *high* or*very high* hazard levels. About 50 percent of the coastline has a *moderate*, *high* or *very high* hazard level for salt water intrusion, while the hazard of ecosystem disruption is very prevalent with 60 percent of the coastline having a *moderate*, *high* or *very high* hazard level, and as much as 41 percent having a *very high* hazard level.

Table 2: The distribution of hazard levels in percentage for Djibouti's coastline

Hazard level	Ecosystem disruption	Gradual inundation	Salt water intrusion	Erosion	Flooding
Low	41	35	50	44	50
Moderate	11	16	40	42	38
High	7	48	9	5	4
Very high	41	1	1	8	8

Uncertainties and Limitations

The assessment is considered to provide a reasonably good picture of the climate change hazards for the coastline of Djibouti. Yet, there are a number of uncertainties and limitations that should be considered when using the assessment for management and planning purposes. Generally, one should be cautious about using the assessment to support local planning decisions, as the assessment is carried out as a Step 1 analysis based solely on published geophysical data and remote sensing information. However, the available data and remote sensing information is considered to be of relatively good quality, and the assessment should be sufficiently detailed and accurate for identifying hazard hotspots and for supporting national, regional and sub-regional planning decisions.

The hazard map for ecosystem disruption is considered to give a reasonable good indication of the future hazards to Djibouti's ecosystems under a changing climate. The main uncertainties are related to the current state of the extensive coral reef systems along Djibouti's coastline, and how they will respond to the changing ocean temperature and ocean acidity. But like most tropical coral reefs, they are likely to be at significant risk. The assessment framework is unable to cover coral reefs offshore of river mouths, spits and barriers as these would normally not occur because of the increased sediment load and lack of hard bottom substrate in these locations. Therefore, the assessment may underestimate the ecosystem hazards associated with smaller reef sections associated with these features e.g. north of Khor Angar. Yet, these are only small areas and probably insignificant compared to the uncertainty of the coral reef data currently available from global coral datasets.

The hazard map for gradual inundation is generally considered as relative robust for the full coastline of Djibouti. Some uncertainties may be related to the sediment supply and availability along the rocky stretches with beach environments, but those are not expected to influence the hazard levels significantly. Uncertainties related to this hazard are therefore mainly related to the rate of sea level rise, which is an uncertainty parameter that applies to all coastal areas globally.

The hazard map for salt water intrusion is generally considered to be surrounded by significant uncertainties as this hazard is influenced by a range of parameters other than the coastal dynamics and geomorphology. Due to the very dry climatic conditions in Djibouti, the freshwater aquifers are generally considered to be limited, and just a low level of human groundwater extraction can cause severe salt water intrusion if salty sea water is replacing the extracted freshwater. If this is combined with unusual low precipitation levels, the hazard of salt water intrusion can be *high* to *very high,* even when it is indicated as *moderate* on the hazard maps. The information given in the assessment should therefore be considered as a relative hazard level compared to other coastal environments. With human water extraction, the natural aquifer replenishment can quickly become overwhelmed leading to *high* and *very high* hazards of salt water intrusion in many locations. Thoughtful water management is therefore crucial along the coastline of Djibouti.

The hazard map for erosion is mainly surrounded by uncertainty in areas where it is difficult to determine the geological base material from the available geological maps and satellite images. This is mainly the case for the outer coastline of Djibouti slightly south of Khor Angar, the outer coastline of the estuary 25 km south of Khor Angar and the coastline south of Djibouti City. Yet, the hazard levels for erosion in these areas are considered to be relatively accurate, but additional field verification would be needed if the assessment should support local management decisions.

When using the hazard map for flooding, it is very important to be aware of the associated uncertainties due to the sudden and often dramatic nature of flooding events. Misjudgments and bad management decisions in relation to this hazard can lead to extensive property damage and in worst case loss of lives, and additional field verification may therefore be necessary in some locations to establish a more solid decision base. Yet, the assessment is considered to provide a reasonable reliable picture of the flooding hazards on a national, regional and sub-regional level. Since coastal flooding may arise from both ocean high water and intense precipitation and run-off, different uncertainties are associated with these conditions. Generally, the CHW framework covers the ocean-caused flooding hazards relatively well, but in the case of precipitation induced flooding, some uncertainties are present. As Djibouti generally has no real rivers but only dry river valleys that are occasionally flooded during extreme precipitation events, it is necessary to rely on visual observations of the geomorphology, using satellite images to identify the dry river beds. Most dry river beds have been identified and classified as river mouths in assessment, giving them a *very high* flooding hazard, but some of the smaller dry streams are difficult to detect using this methodology. However, as a Step 1 assessment, the maps provide a generally good picture of the flooding hazards and should be sufficient to support broader planning decisions.

Since limited information on current coastal hazards is available for Djibouti, it is not possible to make broader comparisons between existing conditions and the hazards indentified in the assessment. However, in some locations it is possible to compare current hazard conditions and assessment results. For ecosystem disruption, the hazards indentified in the hazard assessment can generally not be detected at this point in time. The coral reefs along Djibouti's coastline are currently in average to good condition and have not shown major signs of degradation due to climate-related impacts. The mangrove areas are under some pressure from human exploitation combined with natural fluctuations, but no consistent climate-related impacts are detectable (PERSGA/ALECSO, 2003). As the hazard of ecosystem disruption mainly relates to changes taking place over the next decades, this picture is expected. Little information is currently available on gradual inundation, while salt water intrusion already is a significant problem for Djibouti city, with salinity levels exceeding recommended WHO levels (Ministere de l'Habitat, l'Urbanisme, l'Environnement et de l'Amenagement du Territoire, 2006). The hazard assessment indicates moderate to high hazards for salt water intrusion for most of Djibouti city, which corresponds well with these conditions. There are no broader studies available for erosion, although it is a general concern for the coastline of Djibouti (Ministere de l'Habitat, l'Urbanisme, l'Environnement et de l'Amenagement du Territoire, 2006). With regards to flooding hazards, the alluvial areas of the city of Tadjoura experience regular flooding, while Djibouti city has experienced several catastrophic flooding events in relation to the Ambouli wadi

(Ministere de l'Habitat, l'Urbanisme, l'Environnement et de l'Amenagement du Territoire, 2012). This corresponds well with the assessment results, as these areas are indicated to have very high flooding hazards in the hazard assessment.

Planning and Management Perspectives

The assessment procedure outlined in the previous sections provides an example of how the CHW framework can be applied for national multi-hazard assessments. The maps developed in the assessment can be used for providing an overview of the hazard profile of the national coastline and for identifying hazard hotspots and where human activities may be affected by coastal hazards. The additional hazard layers for Google Earth can be used to support more detailed management decisions and to provide a first impression of the hazard presence at local level. Since few broader hazard assessments have been carried out in developing countries such as Djibouti, the CHW framework offers national planners a possibility for obtaining a picture of the hazard profile for the coastline, even when little geophysical data is available. The assessment for Djibouti can therefore function as a guide for carrying out similar assessments in other developing countries.

Since the hazards along Djibouti's coastline are of very different character and extension, a range of different measures and approaches are required to manage the hazards appropriately. Some coastal stretches have high hazard levels for several hazard types that to some degree are interrelated. It is therefore relevant to consider which measures can be used for addressing several hazards at the same time. Also, it is important to consider human use of the coastal area and the different ecosystem services the coastal systems provide, when deciding on hazard mitigation strategies.

Since ecosystem hazards are extensive along Djibouti's coastline, nation-wide measures are likely to be relevant for addressing this hazard type. Aside from the ethical aspects of preserving biological diversity, ecosystems are important for maintaining key ecosystem services for human society (Millennium Ecosystem Assessment, 2005). The significant threat to the country's coral reefs from climate change is therefore not only problematic seen from a biological diversity perspective but also constitutes a direct risk to the sustainability of the broader marine ecosystems and Djibouti's fishing sector. Likewise, the threat to the country's mangrove ecosystems is also related to coastal fisheries. As very little data is available on the health of these ecosystems and their role in the broader marine ecosystems, it is difficult to decide on appropriate management strategies. Initially, it might be appropriate to assess the direct human threats to these systems from unsustainable fishing methods, wastewater pollution, clearing of mangroves etc. but on the longer term, it is necessary to obtain more data on the state and dynamics of these systems. One way of doing this could be through a citizen science approach where local fishermen, tourists, etc. are involved in data collection as part of their normal activities through a simple, standardized data collection system. If such system is designed properly, it can be used for collecting significant amounts of temporal data which subsequently can be analyzed by scientists and coastal managers, providing a continuous indication of the state of the ecosystems. Along with providing a basis for implementing dynamic adaptation measures, such data collection systems may also increase the general knowledge about the ecosystems amongst coastal residents and increase their responsibility and ownership for the sustainability of these systems. Yet, these approaches have until now mainly been used in developed countries and have to be specifically designed so they fit into the conditions on the ground in Djibouti.

Although the hazard of gradual inundation is relatively widespread in Djibouti, it does not constitute an imminent threat to the coastal activities due to its slow, gradual nature. Yet, it is very important to consider this hazard for long-term planning decisions related to infrastructure development, human settlement etc. to avoid the need for costly relocation and adaptation measures at a later stage. The hazard maps developed in this assessment can be used to support such planning decisions and may be supplemented by more detailed data on isostatic uplift/subsidence and rate of sea level rise at a later stage.

The hazard of salt water intrusion is mainly related to the human extraction of groundwater because of the low precipitation levels in Djibouti. Careful water management and water conservation is therefore a key issue for all coastal areas and more detailed assessments may provide estimates of the amount of water that can be sustainably withdrawn from the different areas. Since the changes in precipitation levels with climate change are very uncertain, it may be relevant to establish a monitoring system to assess the temporal developments in water levels in wells and the possible salt water intrusion. For this purpose, a citizen science approach may also be relevant for broader data collection that can subsequently be analyzed by ground water specialists. From such monitoring it will over time be possible to see the impact of changing precipitation patterns and adapt dynamically to these changes.

As the hazard of erosion mainly is related to the low-lying dry river mouths, barriers and tidal inlets, it does not constitute a major nation-wide hazard in Djibouti. The most cost-efficient way of addressing this hazard is likely to minimize infrastructure development and human settlements in these erosion-prone locations and then implement some technical erosion control measures at specific locations if deemed necessary. Generally, it may also be relevant to make people aware of these erosion hazards if they are settling more permanently in these hotspot locations. Technical measures for tackling erosion include hard engineering approaches such as breakwaters, groins and sea walls, and soft measures as beach nourishment, but all these options come with a cost. It is therefore wise to consider erosion hazards early in the planning process to minimize the need for technical protection measures.

The flooding hazards related to the dry river beds pose a significant threat to human settlement in these locations. The best way to mitigate this hazard is to avoid any permanent settlement in these locations and over time assist inhabitants with a permanent relocation. As larger settlements are present in these hotspot areas in parts of Djibouti City, it may be necessary to consider different technical protection measures to manage the threat to the settlements. This could include the development of a levee system that directs the water away from the most densely populated areas or dams further upstream to absorb peak flows. With regards to flooding hazards for the extensive coastal plains of Djibouti, the most cost-efficient management approach is probably to create a small buffer zone along the coastline without human settlement, so the impact of a flooding event is limited.

Because the hazard management measures are highly dependent on the interrelationship between the natural coastal systems and human actives, it is important to consider the key goals of any management activity. For some measures, such as preserving coral reef and mangrove environments, maintaining the natural state is of direct benefit to human activities due to their important ecosystem services e.g. for coastal fisheries. In other cases, technical measures that modify the natural dynamics such as erosion and flood protection may be appropriate. Generally, it is recommended to consider climate change hazards in the early stages of all coastal planning processes to avoid damage to people and property, costly protection measures and unnecessary degradation of natural systems and associated services. The concept of working with nature, which aims at combining societal interests with natural dynamics is gaining increasing attention and can in many cases reduce planning costs, while at the same time maintaining the services provided by natural coastal systems.

CONCLUSION

The application of the CHW framework on the coastline of Djibouti has proved appropriate for a relatively detailed multi-hazard assessment for the full coastline. The extension of the CHW framework with additional categories for rocky coastlines has been appropriate for Djibouti's predominantly rocky coastline, and it may be considered to incorporate the extra categories in a future update of the CHW system. The national overview maps provide sufficiently detail and accuracy for supporting broader management decisions, while the hazard layers developed for Google Earth seems to be a useful supplement for supporting sub-regional and local planning. Some uncertainties are related to the geological layout and sediment balance evaluations, but the results are generally considered acceptable as a Step 1 assessment. For a more detailed assessment at Step 2 or 3, additional field verification is recommended to clarify some of these uncertainties. It may be considered to supplement the assessment with some dynamic data collection systems through a citizen science approach involving coastal residents. This would especially be relevant for the hazards of ecosystem disruption and salt water intrusion as uncertainties related to these hazards are difficult to address during a short field campaign. Generally, the assessment for Djibouti can be used as an example of a CHW application on a predominantly rocky coastline and the procedure should be replicable on other coastlines globally, yielding results of similar quality.

ACKNOWLEDGEMENTS

We would like to thank Kirsten Halsnæs and Rami Abu Salman for valuable discussions and advice and assistance with background material, and IFAD for a fruitful collaboration.

REFERENCES

1. Al Saafani, M.A.Y., 2008. Physical Oceanography of the Gulf of Aden. Goa University.
2. Balstrøm, T., 2008. Adaptions of ArcGIS' Linear Referencing System to the Coastal Environment. Esri Users Conference Proceedings, San Diego, CA.
3. Bird, E.C.F., 2010. Encyclopaedia of the World's Coastal Landforms. Springer.
4. CIA, 1996, CIA World Factbook 1996. Available online the 23-03-14 at <http://www.lib.utexas.edu/maps/cia96.html>.
5. Esri, 2012. ArcGIS ver. 10.1. Esri, Redlands, CA.
6. Global Climate Forum, 2013. DIVA Model. Available online the 19.11.13 at <http://www.diva-model.net/>.
7. Google, 2013. Google Earth. Available for download the 26-05-13 at <http://www.google.com/earth/index.html>.
8. IPCC, 2014. Climate Change 2014: Impacts Adaptation and Vulnerability, unedited accepted final draft report, IPCC.
9. IPCC, 2007a. In: Parry, M.L., Canziani, O.F., Palutikof, J.P., van der Linden, P.J., Hanson, C.E. (Eds.), Climate Change 2007: Contribution of the Working Group II to the Fourth Assessment Report of the Intergovernmental Panel on Climate Change. Cambridge University Press, Cambridge, UK.
10. IPCC, 2007b. In: Solomon, S., Qin, D., Manning, M., Chen, Z., Marquis, M., Averyt, K.B., Tignor, M., Miller, H.L. (Eds.), Climate Change 2007: The Physical Science Basis. Contribution of the Working Group I to the Fourth Assessment Report of the Intergovernmental Panel on Climate Change. Cambridge University Press, Cambridge, UK.
11. Jarosz, E., Murray, S.P., 2002. Tides in the Bab el Mandab Strait, The 2nd Meeting on the physical oceanography of sea straits, Villefranche 15–19th April,2002.
12. Jarosz, E., 1997. Tidal dynamics in the Bab el Mandab Stait. Louisiana State University.
13. Masselink, G., Hughes, M.G., 2003. Introduction to Coastal Processes and Geomorphology. Oxford University Press.
14. Mcleod, E., Poulter, B., Hinkel, J., Reyes, E., Salm, R., 2010. Sea-level rise impact models and environmental conservation: a review of models and theirapplications. Ocean Coastal Manage. 53, 507–517.
15. Microsoft, 2013. Bing Maps Hybrid. Available in Esri ArcGIS.
16. Millennium Ecosystem Assessment, 2005. Ecosystems and Human Well-being: Synthesis. Island press, Washington DC.
17. Ministere de l'Habitat, l'Urbanisme, l'Environnement et de l'Amenagement du Territoire, 2012. Seconde communication nationale de la Republique de Djibouti a la convention cadre des nations unis sur les changements climatiques, Djibouti.
18. Ministere de l'Habitat, l'Urbanisme, l'Environnement et de l'Amenagement du Territoire, 2006. Programme d'action national d'adaptation aux changements climatique, Djibouti.
19. PERSGA/ALECSO, 2003. Survey of Habitats in Djibouti and Plans for their Protection, PERSGA Technical Series No. 5. PERSGA, Jeddah.

20. Ramieri, E., Hartley, A., Barbanti, A., Santos, F.D., Gomes, A., Hilden, M., Laihonen, P., Marinova, N., Santini, M., 2011. Methods for Assessing Coastal

21. Vulnerability to Climate Change. European Environment Agency, European Topic Centre on Climate Change Impacts, Vulnerability and Adaptation.

22. Reefbase, 2013. Reefbase – A Global Information System for Coral Reefs. UNEP-WCMC.

23. Ron Englebretson, 2002. Arabian Sea/Gulf of Aden winds. NRL Monterey, Marine Meteorology Division.

24. Rosendahl Appelquist, L., 2012. Generic framework 59–74, http://link.springer.com/article/10.1007%2Fs11852-012-0218-z.

25. Salt Investment, 2008. Lake Assal Salt Project, Djibouti: Environmental Impact Assessment. Salt Investment.

26. Sawakuchi, A.O., Fornari, M., Tanaka, A.P.B., Giannini, P.C.F., Mendes, V.R., Guedes, C.C.F., Zular, A., Nascimento, D.R., 2012. Geomorphological analysis of coastal depositional systems in SE Brazil aided by Google Earth coupled with the integration of chronological and sedimentological data by means of a Google Fusion Table, Special Paper of the Geological Society of America, vol. 492, pp. 113–125.

27. Scheffers, A.M., Scheffers, S.R., Kelletat, D.H., 2012. The Coastlines of the World with Google Earth: Understanding Our Environment. Springer.

28. Schlüter, T., 2006. Geological Atlas of Africa. Springer.

29. Sharples, C., 2014. Smartline. Available online the 22-03-14 at <http://www.ozcoasts.gov.au/coastal/introduction.jsp>.

30. Thieler, E.R., Willams, J., Hammer-Klose, E., 2000. National Assessment of Coastal Vulnerability to Future Sea-level Rise. <http://woodshole.er.usgs.gov/ project-pages/cvi/> (accessed online at 19-11-13).

31. Whitmeyer, S.J., Bailay, J.E., Depaor, D.G., Ornduff, T., 2012. Google Earth & Virtual Visualizations in Geoscience. Geological Society of America.

32. World Bank, 2011. Vulnerability, Risk Reduction, and Adaptation to Climate Change, Djibouti. The World Bank Group.

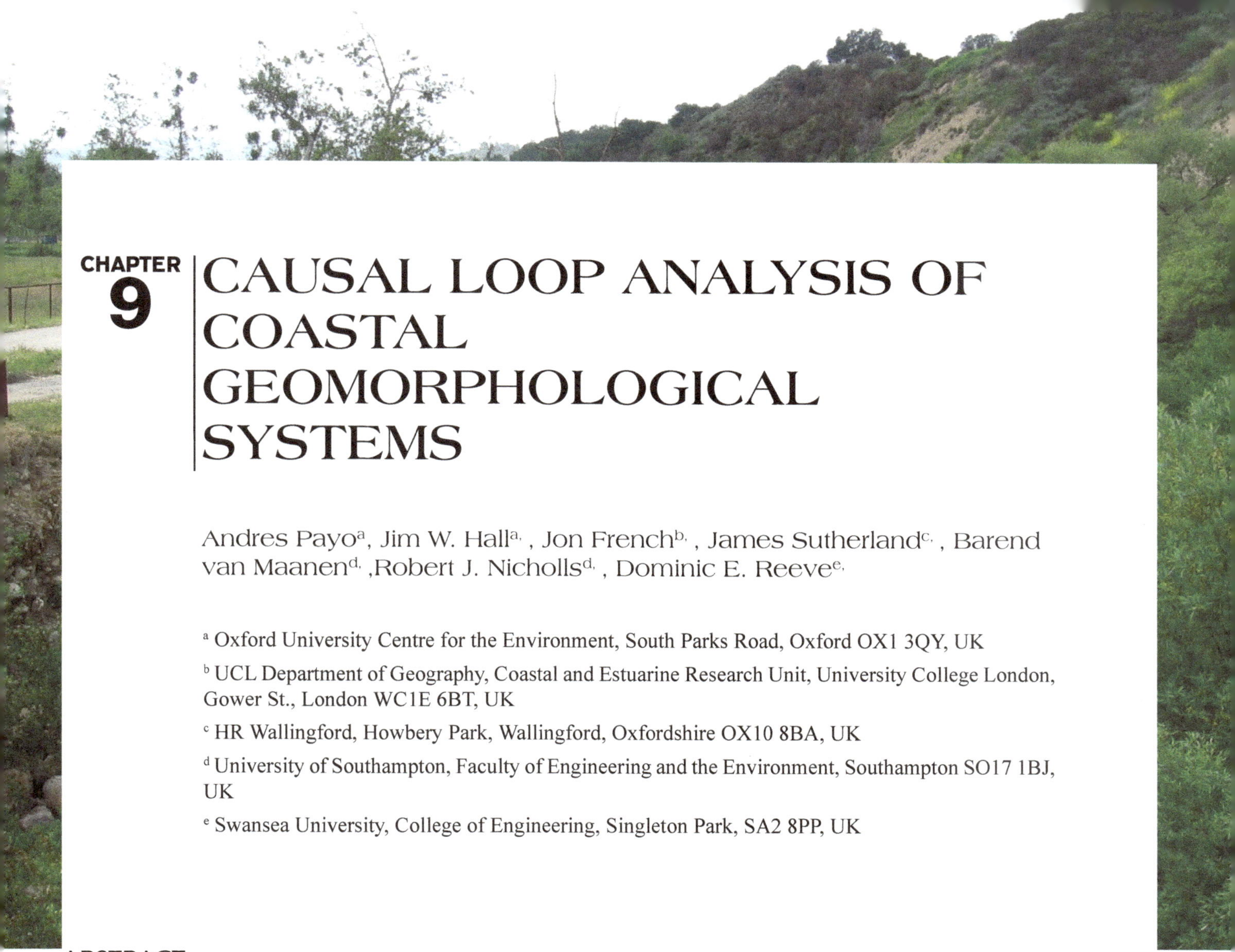

CHAPTER 9

CAUSAL LOOP ANALYSIS OF COASTAL GEOMORPHOLOGICAL SYSTEMS

Andres Payo[a], Jim W. Hall[a], , Jon French[b], , James Sutherland[c], , Barend van Maanen[d], ,Robert J. Nicholls[d], , Dominic E. Reeve[e],

[a] Oxford University Centre for the Environment, South Parks Road, Oxford OX1 3QY, UK

[b] UCL Department of Geography, Coastal and Estuarine Research Unit, University College London, Gower St., London WC1E 6BT, UK

[c] HR Wallingford, Howbery Park, Wallingford, Oxfordshire OX10 8BA, UK

[d] University of Southampton, Faculty of Engineering and the Environment, Southampton SO17 1BJ, UK

[e] Swansea University, College of Engineering, Singleton Park, SA2 8PP, UK

ABSTRACT

As geomorphologists embrace ever more sophisticated theoretical frameworks that shift from simple notions of evolution towards single steady equilibria to recognise the possibility of multiple response pathways and outcomes, morphodynamic modellers are facing the problem of how to keep track of an ever-greater number of system feedbacks. Within coastal geomorphology, capturing these feedbacks is critically important, especially as the focus of activity shifts from reductionist models founded on sediment transport fundamentals to more synthesist ones intended to resolve emergent behaviours at decadal to centennial scales. This paper addresses the challenge of mapping the feedback structure of processes controlling geomorphic system behaviour with reference to illustrative applications of Causal Loop Analysis at two study cases: (1) the erosion–accretion behaviour of graded (mixed) sediment beds, and (2) the local alongshore sediment fluxes of sand-rich shorelines. These case study examples are chosen on account of their central role in the quantitative modelling of geomorphological futures and as they illustrate different types of causation. Causal loop diagrams, a form of directed graph, are used to distil the feedback structure to reveal, in advance of more quantitative modelling, multi-response pathways and multiple outcomes. In the case of graded sediment bed, up to three different outcomes (no response, and two disequilibrium states) can be derived from a simple qualitative stability analysis. For the sand-rich local shoreline behaviour case, two fundamentally different responses of the shoreline (diffusive and anti-diffusive), triggered by small changes of the shoreline cross-shore position, can be inferred purely through analysis of the causal pathways. Explicit depiction of feedback-structure diagrams is beneficial when developing numerical models to explore coastal morphological futures. By explicitly mapping the feedbacks included and neglected within a model, the modeller can readily assess if critical feedback loops are included.

INTRODUCTION

Feedbacks and emergent behaviours in geomorphology have been long recognised (Schumm and Lichty, 1965 and Schumm, 1991) but it has been in recent decades when views of change, disturbance, response and recovery have expanded considerably (Phillips, 2009): conceptual frameworks emphasizing single-path, single-outcome trajectories of change have been supplemented – not replaced – by multi-path, multi-outcome perspectives. In this context, Phillips (2009) argues that any attempt to explore change and response studies should seek to identify potential feedbacks, determine their signs, and assess their relative importance. This is especially important when

Citation: Andres Payo, Jim W. Hall, , Jon French, James Sutherland , Barend van Maanen, Robert J. Nicholls, Dominic E. Reevee, Causal Loop Analysis of coastal geomorphological systems , doi:10.1016/j.geomorph.2015.07.048

studying coastal systems in which a broad range of feedback mechanisms drives the system's evolution. A feedback is a change to a component of the system that causes a knock-on effect that further alters the original change. A positive feedback amplifies the initial change. For example as waves erode the cliff, granular material will be released, which may abraid the shore platform, resulting in even more cliff erosion. Negative feedbacks have the opposite effect of the initial change. For example, as the shore platform is eroding it becomes wider and gentler diminishing the rate of mass wasting for the same given offshore wave energy flux. Identifying these feedbacks is the first step towards establishing their relevance at the spatial scales of geomorphological models (Lane, 2013). As an ever-greater number of feedbacks are identified and appreciated, the need to map them into a coherent framework is needed. Techniques for the formal assessment of the main feedbacks between coastal geomorphology and the drivers of change (i.e. climatic variability and human interventions) assist geomorphological modellers to explore how that variability might change during the 21st century and what this might mean for geomorphic processes, landforms and entire landscapes.

The concept of feedbacks has proved helpful in the idealized model domain, but extrapolation to the real world is complicated (i.e. Klocke et al., 2013). In geomorphology examples of qualitative stability assessment of the system based on the feedback loop structure go back to at least the early 1980s (Slingerland, 1981 and Phillips, 2006). The stability of the system (or conditions under which it is stable) can be determined if historical reconstructions or field observations identify the key system components and the positive, negligible, or negative links between them. This often takes the form of a directed graph, network model, or box-and-arrow diagram (Capobianco et al., 1999 and Townend, 2003) (Fig. 1). These can be translated into an interaction matrix, and the stability may be determined based only on a qualitative (+, −, 0) assessment.Payo et al. (2014) have taken this qualitative analysis forward by showing how the strength (i.e. not only the sign) of a single feedback loop (e.g. the cliff toe energy depletion feedback loop) can be assessed by reasoning on the current understanding of its causal pathway. Payo et al. (2014) used directed graphs (i.e. Lane, 2000) but limited their analysis to the main processes that control the morphodynamics of cliff and shore platforms.

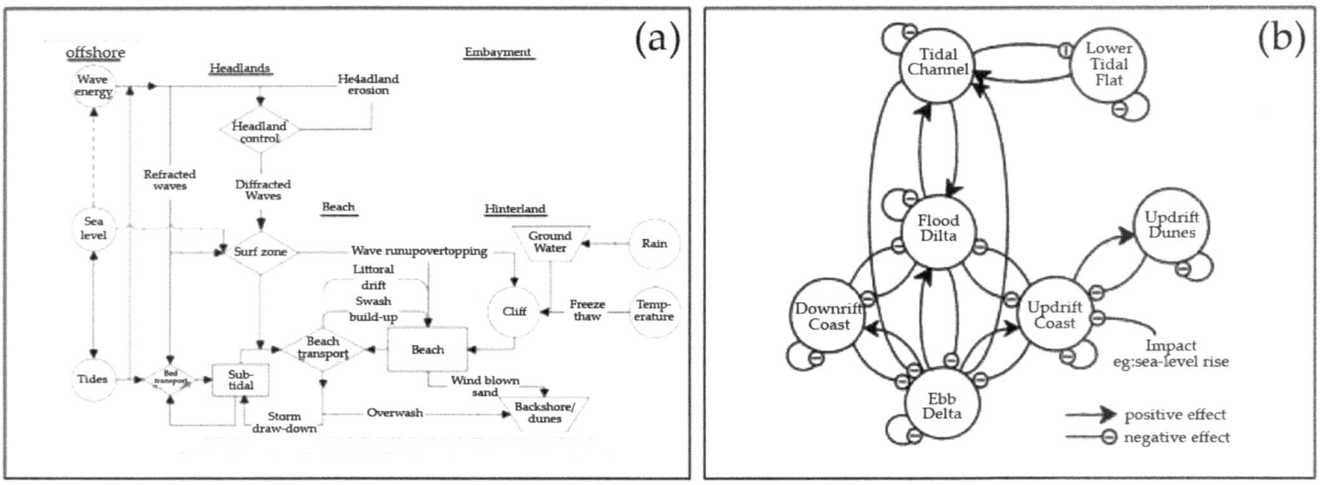

Figure 1: Examples of how stock and flow and causal loop diagrams have been used to represent coastal system functioning and behaviour for different purposes. (a) Stock and flow diagram of an embayment (Townend, 2003) and (b) causal loop diagram of the impact of sea-level rise on an inlet or lagoon entrance (Capobianco et al., 1999).

In this work, we extend the use of directed graphs to two different case studies: (1) processes that control the erosion–deposition behaviour of graded beds, and (2) processes that control the local alongshore sediment transport fluxes of sand-rich shorelines. These case studies are selected because they are illustrative of how directed graphs can be used to capture different types of causation, and because of their central role in the quantitative modelling of geomorphological futures. When modelling geomorphological futures at decadal and longer timescales, the modeller is likely to make informed decisions about how to numerically model the fundamentally different behaviour of mixed sediments versus uniformly graded ones. Le Hir et al. (2011) noted the need for any morphodynamic model of mixed sediment to add an active layer concept to deal with the erosion of the different fractions. Explanations of shoreline behaviours not captured by a purely diffusive 1-line approach (Ashton et al., 2001 and Van den Berg et al., 2011), and attempts to unify the quantitative modelling of a graded beach has been published elsewhere (van Rijn et al., 2007). However, to the authors› knowledge, no attempt to synthesize the feedback structure of these case studies in a unified way has yet been presented.

This manuscript is organised in four main sections. In Section 2, we present the rationale for this study and define the symbolic convention adopted in this work. The feedback structures for each of the study cases are presented in 3 and 4. To conclude, we highlight the benefits and limitations of our advocated qualitative modelling approach.

METHODOLOGY

Methodology of the varied representational approaches within the field of systems dynamics two predominate: Causal Loop Diagrams (CLDs) and Stock and Flow Diagrams (Lane, 2000). CLDs are a broad representations of the variables and feedback structure while, in contrast, Stock and Flow Diagrams are more detailed, discriminating both state and flow variables. These two forms are fairly standardised and while their benefits and limitations are generally understood (Morecroft, 1982), the preference for one over the other is still contested (Lane, 2000); both conventions have been used in the past for conceptualizing decadal to centennial coastal system dynamics (Fig. 1). For example,Townend (2003) used stock/flow system diagrams to represent coast and estuarine system behaviour and Capobianco et al. (1999) proposed the use of CLDs to assess the impact of sea-level rise on estuaries and adjacent coasts. From a model development perspective, it might be argued that stock and flow diagrams provide more information to guide the organisation and structure of model code developed from them. Here, however, we are interested in the identification of the feedback loop structure, and the use of directed graphs is favoured. While the most relevant literature at which the conceptualization presented here is built upon is cited, the authors acknowledge that is not possible to cite the entire significant body of literature.

Fig. 2 shows the symbolic convention used in this work. The hierarchy of levels is captured by a cluster of variables at each level (c.f. Phillips, 2012). The terms 'local' and 'global' are used in the broadest sense to refer to scale finer, shorter and more detailed, or coarser, longer and broader, respectively, than the scale of observation. The term 'scale' is also used in a broad sense, encompassing both spatial and temporal resolution and position within a hierarchy. For the sake of clarity, a minimal set of symbols is used to capture causality and feedback loop structure. This includes:

- Two types of variables: (1) state variables (stocks, levels, attributes) (e.g., beach width, dune volume, sea level, sediment size, threshold wind velocity for initiating sediment transport), and (2) rate variables (underlined) (flows) (e.g. rate of shoreline change, sediment transport rate).
- Positive (+), negative (−) or influence (+/−) links. Links connect two variables (e.g. X → Y) and represent the answer to the question if X increases, would Y increase or decrease compared to what it would otherwise have been? Links are positive if $dy/dx > 0$ or negative if $dy/dx < 0$. When the answer is not known or ambiguous it is represented as an influence link.
- Causal pathways. We can reason about the influence of one variable on another variable indirectly connected to it by examining the signs along their causal pathway (e.g., two negatives, whether adjacent or not, will act to reverse each other). Loops in a causal loop diagram indicate feedback in the system being represented. In this case, changes cascade through other factors so as to either amplify (reinforcing feedback; products of signs positive) or dampen the original change (balancing feedback; products of signs negative).

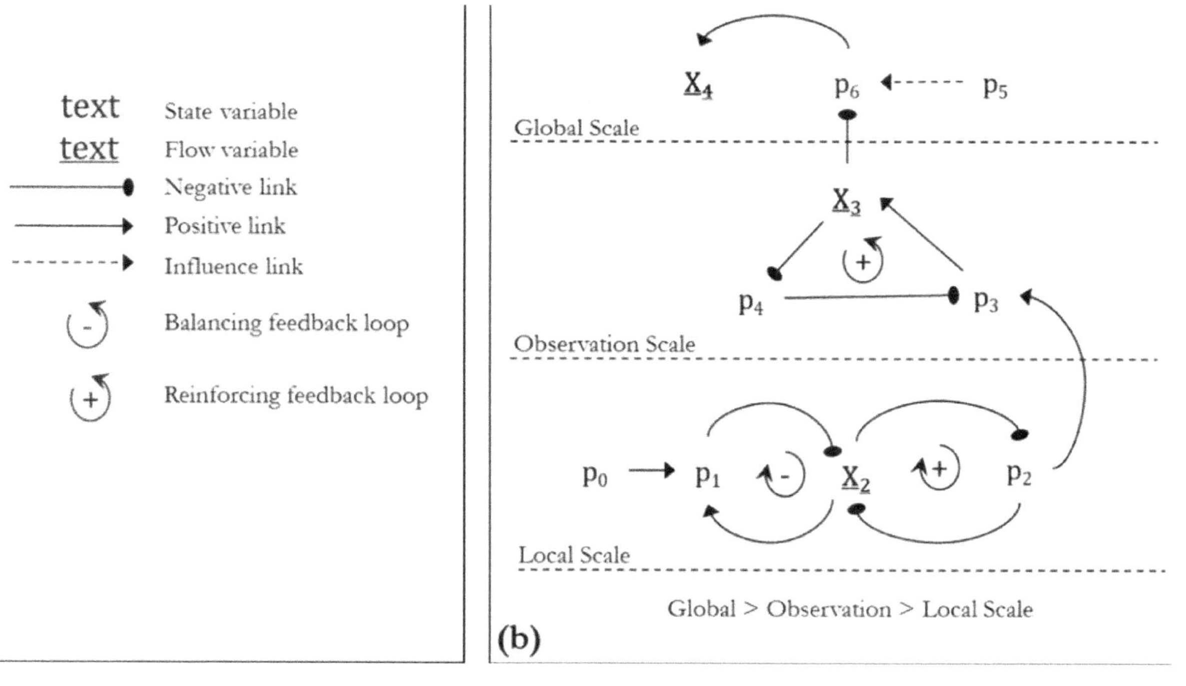

Figure 2: Convention used to represent the causal loop diagrams in this paper. (a) Symbolic convention and (b) example application of scale hierarchy at three scales (Global > Observation > Local). The directionality of the link is represented by the arrow and oval head for the positive and negative link respectively.

Mapping Graded Sediment Erosion–Accretion Behaviour

Building on field and laboratory observations of the active layer (Fig. 3), a conceptualization of the bottom elevation in terms of disequilibrium between erosion and accretion processes (Fig. 4), and the behaviour of a graded (i.e. mixed non-cohesive and cohesive) sediment bed (Fig. 5) can lead to a CLD that synthesizes how a generic active layer behaves (Fig. 6). Bottom elevation is defined as the elevation difference between the seabed/beach surface and a datum defined well below the active layer.

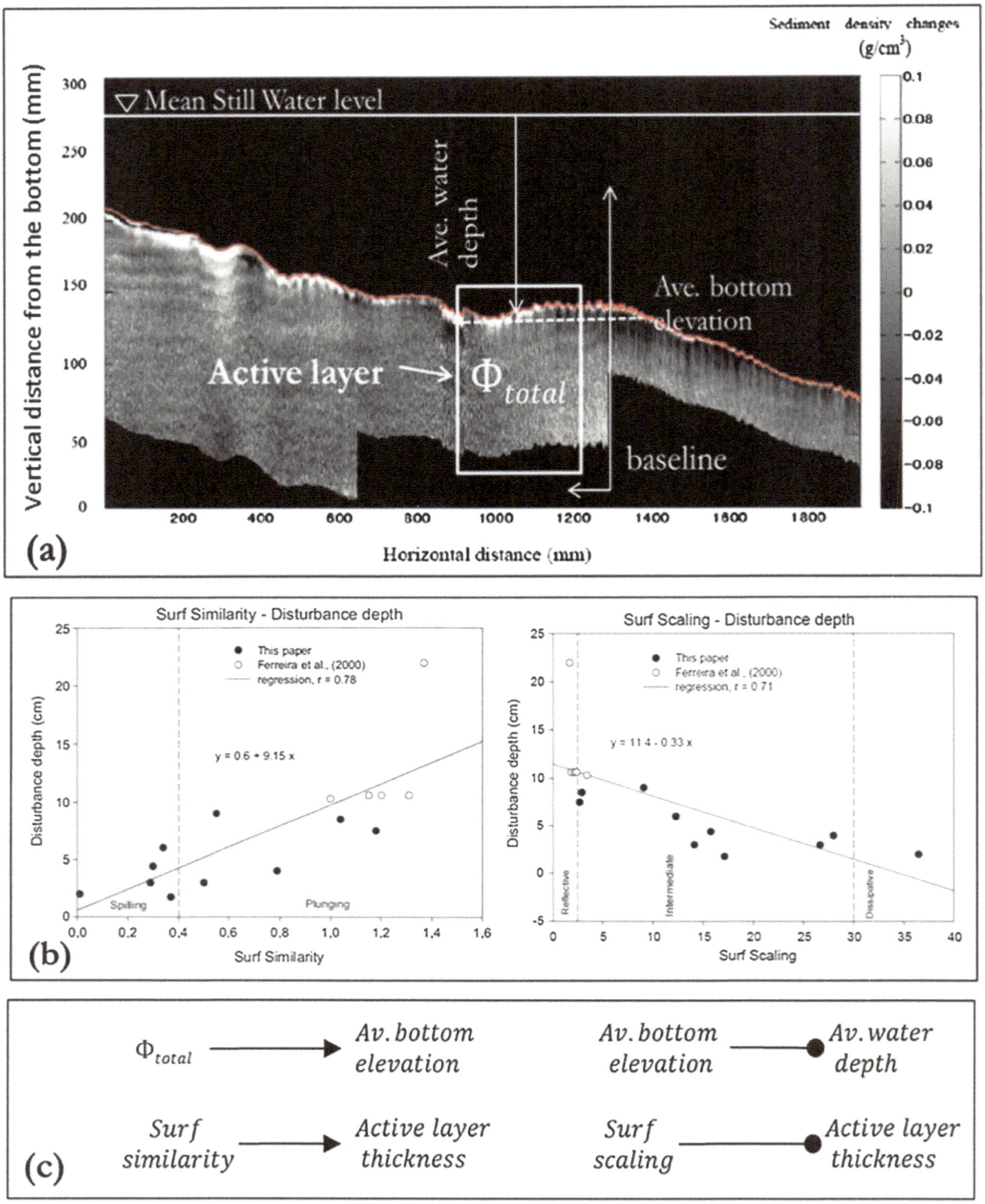

Figure 3: (a) Active layer represented as a white rectangle over an observed across-shore variation of a sand bed in a flume subject to regular wave forcing. Red line indicates the sediment surface after 5.5 h of wave forcing. Sediment density is represented relative to the initial density on a grey scale. Lighter (darker) colours indicate a decrease (increase) in density (modified from Yamada et al. (2013) published with permission of Journal of Coastal Research). (b) Field observation of the averaged active layer thickness over the entire surf-zone shows positive (negative) correlation with the surf similarity (surf scaling) parameter (fromAnfuso, 2005). (c) Above relationships represented using CLD symbolic convention.

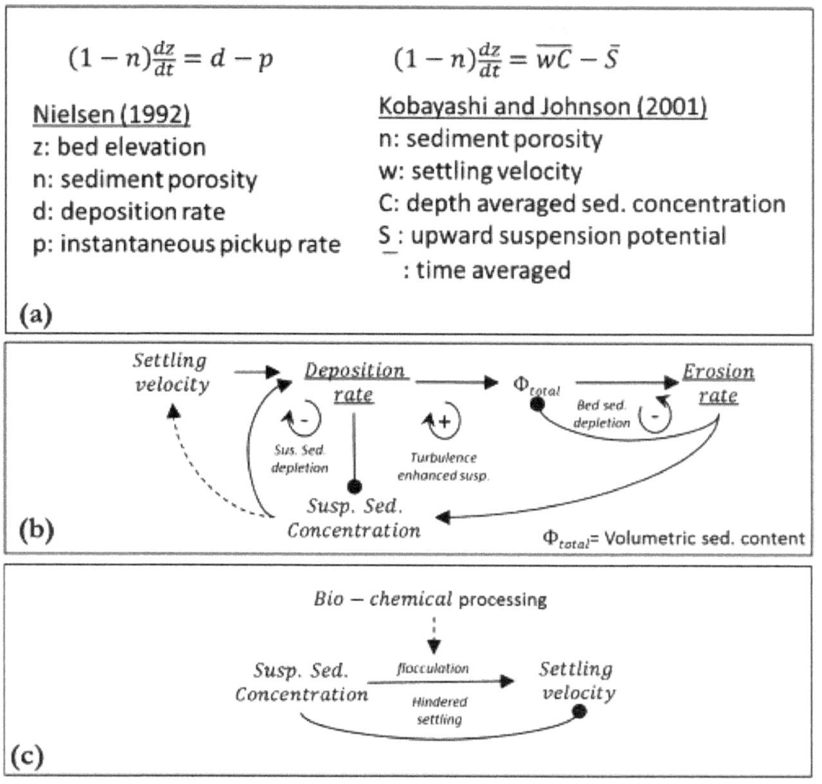

Figure 4: At the active layer scale, bed elevation can be understood as a balance between deposition and entrainment. (a) Examples of instantaneous and time-averaged parameterizations of deposition and entrainment rate. (b) Bed elevation change represented in CLD form including the supply feedbacks of suspended sediment and bed sediment. Eroded sediment from bed, at this scale, is transferred to the suspended sediment concentration. (c) For the mud fraction; flocculation and hindered settling also control the settling velocity of the mixture. No agreement exists on the sign of the bio-chemical processing influence on suspended sediment concentration, partly due to the limited field observations but also due to the wide range of potential interactions and circumstances involved.

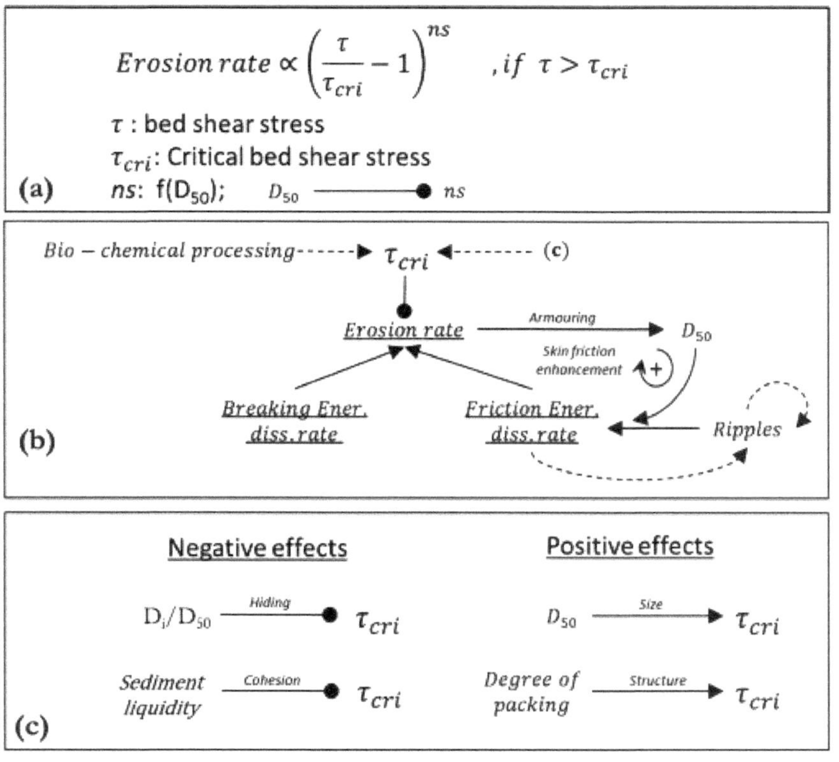

Figure 5: The volumetric sediment erosion rate is proportional to the difference between a critical shear stress and the actual bottom shear stress. (a) This expression can be generalized as a power function where the exponent, *ns*, is higher for smaller particles. (b) Erosion rate expression represented as a CLD. Bio-chemical processes influence the threshold shear stress but no agreement on the sign has emerged. The actual bottom shear stress is function of the energy dissipation rate due to wave breaking and bottom friction (i.e. wave and current). The armouring effect of an eroding-graded-bed reinforces erosion of bed material by enhancing the skin friction. (c) A graded bed also influences erosion either by increasing or decreasing the critical shear stress for erosion.

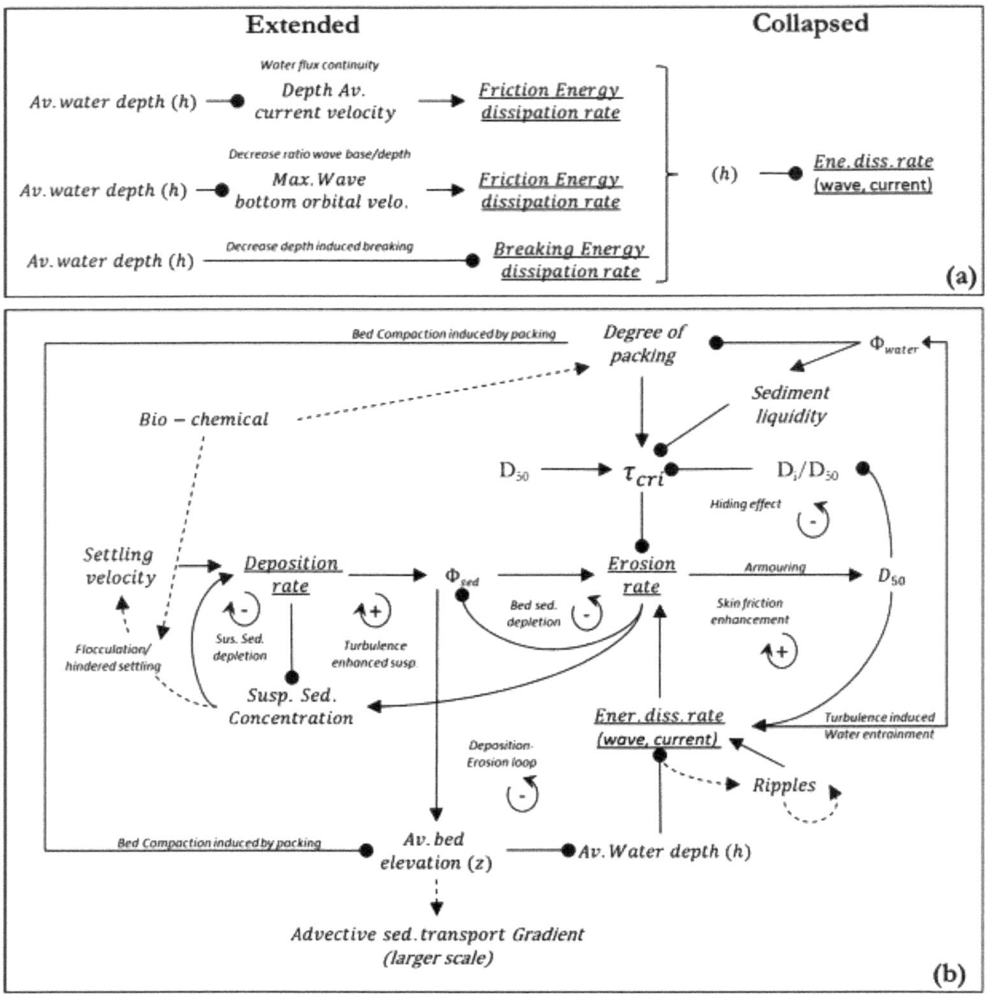

Figure 6: Synthesis of the erosional and depositional behaviour of bottom sediment at the active layer scale. (a) Causal pathway between bed elevation change and energy dissipation due to waves and currents is collapsed into a single negative effect. (b) Overall loop structure, showing the dynamical behaviour of local eroded/deposited sediments only (i.e. advected sediment not represented here but at higher scales as sediment transport gradients).

Feedback Structure

The active layer (Fig. 3a) represents the thickness (O(10 cm)) of the surficial sediment that is affected by hydrodynamic processes (essentially waves and currents) over timescales of a few minutes (mixing depth) or hours to the tidal cycle or several days (disturbance depth) (Harris and Wiberg, 1997). From field observations, Anfuso (2005)found that the thickness of the active layer averaged over the surf-zone correlates linearly with the surf scaling parameter (Guza and Inman, 1975), being greater for reflective sand beaches (20%–40% of the significant wave height at breaking H_s) and lower for more dissipative sand beaches (i.e. 1%–4% of H_s) (Fig. 3b). Yamada et al. (2013), based on a series of laboratory experiments, have observed how the active layer is not spatially homogeneous but varies across-shore for regular spilling breaker and intermediate beach states. The sediment content of the active layer is represented by the volumetric content of different fraction sizes [Φ_{total} = f(***Φcoarse***, ***Φsand***, ***Φmud***, ***Φwater***)]. By definition of the state variable total sediment volumetric content, a decrease (or increase) in bed volume sediment content decreases (or increases) the averaged bottom elevation relative to a datum defined well below the active layer (*d(Av. bottom elevation)/d*$_{total}$ *> 0*). A change in average bottom elevation has a negative effect on average water depth over the active layer (*d(Ave. water depth)/d(Av. bottom elevation) < 0*) (Fig. 3c).

The disequilibrium condition between sediment concentrations and shear stress (especially for finer sediment fractions) has been analysed by treating sediment entrainment and deposition independently (Houser and Barrett, 2010). Bed elevation change (dz/dt) can be estimated through the conservation of sediment as the difference between the instantaneous sediment entrainment and deposition (Nielsen, 1992) or as a balance of time-averaged upward suspension potential and the amount of sediment settling from the water column (Kobayashi and Johnson, 2001) (see Fig. 4a). Independently of the parameterization used, bed elevation change at the active layer scale can be represented as a balance of volumetric erosion and deposition rates (Fig. 4b). Both deposition and erosion rates are balanced by the depletion of depth-averaged suspended sediment and depletion of erodible sediment respectively; both can thus be depicted as balanced feedback loops. The settling velocity of the mixed fraction is influenced by the finer sediment fraction through flocculation and hindered settling (van Rijn, 2007). Flocculation increases the effective diameter and settling velocity, while hindered settling reduces the settling velocity due to flow and wake formation around the particles and an increase in the density and viscosity of the suspension. Flocculation is a dynamically active process that readily reacts to changes in hydrodynamically-generated turbulent shear stresses (τ) and Suspended Particulate Matter (SPM) concentration (Eisma, 1991 and Benson and French, 2007), together with salinity, mineralogy and biological stickiness (Manning et al., 2013). While the importance of bio-chemical processing processes on the suspended sediment behaviour has been long recognised, there is presently no agreement on the sign of their net effect on settling velocity; this is therefore represented as an influence link in Fig. 4c.

The erosional behaviour of graded sediments has been synthesized by van Ledden et al. (2004), and the extensive laboratory and field data have been reviewed separately byvan Rijn et al. (2007), and modelling approaches reviewed by Le Hir et al. (2011). Laboratory and field observations suggest that the erosion rate is proportional to the difference between the critical bed shear stress for erosion (τ_{crit}) and the actual bed shear stress (τ) to a power, *ns*, with *ns* being higher for smaller particles, inducing the bed armouring or increase in D_{50} (i.e. loss of finer fractions of active bed layer; Le Hir et al. (2011); Fig. 5a). The energy dissipation due to wave breaking and bottom friction (i.e. wave and current) has been found to be a robust proxy for near-bed shear stress due to wave breaking and bottom friction (Kobayashi et al., 2008 and Houser and Barrett, 2010). Ripple height and steepness are well known to increase the energy dissipation due to bottom friction but ripples are also influenced by the energy dissipation itself (Nielsen, 1992). Given that this synthesis focuses on the active layer scale, at which ripples are averaged into a time and space averaged bottom elevation, their self-organising behaviour and dependence on energy dissipation (represented as influences in Fig. 5b) are not explored further (see also the work of Coco et al. (2007a) for examples of the complexities of ripple behaviour). Bio-geochemical processes have been found to either increase or decrease the critical shear stress (Friedrichs and Perry, 2001a and Friedrichs, 2011) and are also represented as an influence link. For a uniform bed, the critical shear stress increases with the sediment size (e.g. D_{50}). For a non-uniform bed, the armouring of an eroding bed further influences bed erosion by: (1) increasing the critical shear stress of the finer particles and decreasing the critical shear stress of the coarser particles due to a hiding effect (ξ) and (2) increasing effective bed shear stress of the coarser fraction and decreasing that of the finer (skin friction modulation, λ) (Fig. 5b, c) (van Rijn, 2007). In particular, van Rijn (2007) found that $\xi \approx (D_i/D_{50})^{-1}$ and $\lambda \approx (D_i/D_{50})^{-0.25}$ for mixed sands. The critical shear stress to initiate sediment transport also depends on other factors that determine the vertical strength profile of the bed. van Ledden et al. (2004) proposed the use of the liquidity index and the degree of packing to explain the erosional behaviour of unconsolidated sediment mixtures (Fig. 5c). They reported a more gradual transition from non-cohesive to cohesive behaviour when the network structure becomes silt-dominated, and a sharp transition when the structure is sand-dominated. Critical shear stress increases with sediment size of mixed sediments when the mud fraction ranges from 20%–40%, and critical shear stress is dominated by cohesive forces when the mud content exceeds about 70% (Le Hir et al., 2011).

Erosion rate and deposition rate are balanced due to water flux continuity and depletion of wave and current transport capacity associated with water depth (and therefore bed level) changes (Fig. 6a). Due to water continuity, an increase in water depth (i.e. erosion of the bed) reduces the depth-averaged current velocity. Since energy dissipation due to bottom friction is proportional to the velocity amplitude, this leads to a decrease in energy dissipation rate. An increase in water depth, for a given wave height and period, decreases the maximum wave orbital velocity at the bottom (i.e. proxy for near-bed turbulence) until the wave base is reached. Additionally, an increase in water depth reduces depth-induced breaking. Collapsing all the above-mentioned pathways translates into a negative effect between the water depth and total (current plus wave) energy dissipation rate. Fig. 6b combines the active layer links discussed in this section into a single diagram. The effect of bed compaction induced by packing has been also included as a negative link between degree of packing and bed elevation.

UPWARD CAUSATION

We now illustrate how the proposed feedback structure of the erosional and depositional behaviour of a graded sediment bed can be used to capture an upward causation effect by which small-scale processes influence morphological changes at much larger time and space scales. To this end, we explore the elevation changes in tidal flats and their ability to keep pace with sea-level rise. In particular, we link the dominance of different loops with three distinct anticipated behaviours.

The ability of tidal flats to keep pace with sea-level rise is not only influenced by the sediment mass balance (i.e. accretion deficit or surplus) but also the elevation deficit that incorporates the additional effect of shallow subsidence by autocompaction and shrinkage of the sediment layer (Cahoon et al., 1995, Allen, 1999 and French, 2006). Compaction is largely driven by time evolution of the effective stress within the sediment column (Allen, 1999 and Brain et al., 2011), but also by shrinkage due to desiccation (Van Wijnen and Bakker, 2001). These processes can be represented (Fig. 7) by two causal pathways that influence the degree of packing (compaction) and both packing and sediment strength directly (desiccation and wetting). Compaction reduces the erosion rate by reducing sediment liquidity and increasing packing. The increase in critical shear stress shifts the balance between the local (i.e. non-advected) erosion/deposition rate and bed elevation (see grey thicker causal pathway in Fig. 7b). Compaction not only increases the critical shear stress but also reduces the bed elevation relative to water level. An increase in relative water level reduces the dissipation of wave and current energy, thereby further decreasing the erosion potential. While elevation deficit might be dominant on minerogenic tidal flats, the authors acknowledge that biological activity can further contribute to both the sediment budget and the erosion potential (e.g. diatom mucilage can significantly increase the critical shear stress (Paterson, 1997)).

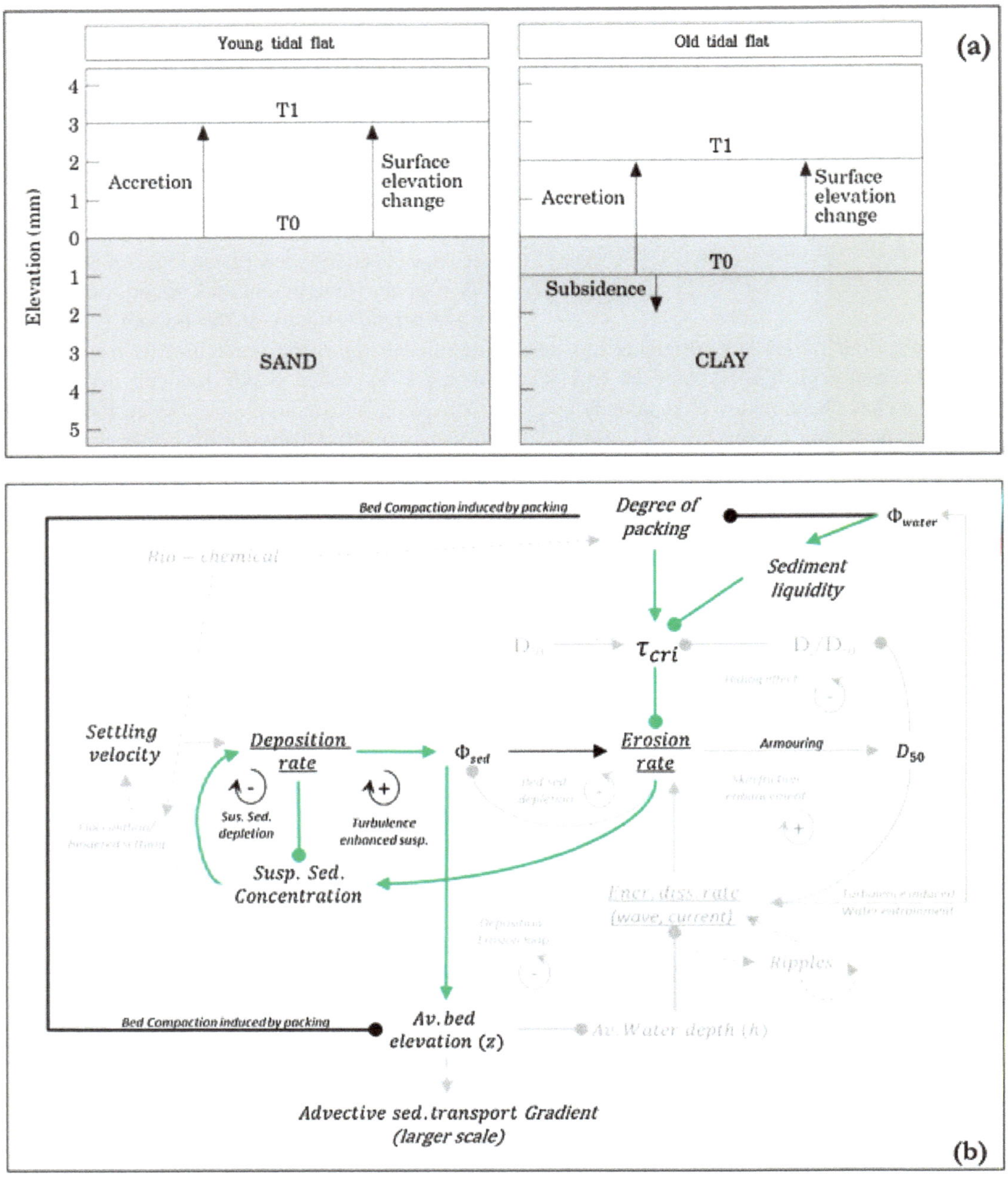

Figure 7: An elevation deficit within a tidal flat sedimentation system can be conceptualized as a direct and non-direct large-scale consequence of a set of processes at the active layer scale; a) shows the differences in elevation at initial stage (T0) and final stage (T1) of a young (sand-reach) and old tidal flat (clay-rich) (Van Wijnen and Bakker, 2001); b) direct effect of changes on seabed sediment volumetric water content (black) and non-direct effect (green) on averaged sea bed elevation. Direct effects are likely dominant in young tidal marshes and non-direct on old tidal flats.

More generally, the existence of multiple feedback loops in Fig. 7 means that the active layer can exhibit three qualitatively different behaviours. Fig. 8 sketches these as a function of the total bed shear stress and the time-averaged water depth, as a proxy for bed erosion potential. Since the bed stress reaches a minimum when water depth tends either to zero or close to tidal high water, the bed-shear stress variation with averaged water depth (black solid line in Fig. 8) must have a maximum between these two limits. For a bed layer of mono-sized sediment (i.e. no hiding effect, armouring or fine wash-over effect), the critical shear stress for entrainment is independent of the "bed memory" (i.e. history of maximum bed shear stress) and can be represented as a constant value. Where this straight line coincides with the bed shear stress it represents an equilibrium depth where bed-shear stress is just insufficient to erode the bed. A stable solution (dark circles) also occurs when a positive (negative) perturbation of the equilibrium depth is counteracted by increased (decreased) erosion. Conversely, instability (empty circles inFig. 8) occurs when a perturbation of the equilibrium depth causes evolution away from the stationary solution, either by further reducing the water depth (i.e. siltation) or towards a stable deeper water condition. For mixed sediments, the hiding effect, armouring and winnowing become active and the critical shear stress becomes dependent on the memory of the bed. At lower bed shear stress, the finer fractions are quickly washed out, such that a lag deposit of coarser grains armours the bed, hiding the remaining finer material from further erosion. The system might eventually achieve a stable state when armouring is the dominant effect, in which case the stable critical shear stress will depend on both the bed-shear history and the fraction of coarser material in the bed (i.e. higher shear stresses for mixed sediment with richer coarser fractions). For the example illustrated in Fig. 8, the behaviour of both mixed and single sized bed sediment, subject to the same initial critical shear stress is compared. Depending on the bed-shear stress history (shown here as a simple sigmoidal variation of bed shear–stress), the bed elevation (or equivalently averaged water depth) exhibits two equilibria — one stable and one unstable. These are closer together for the case of the non-graded bed (i.e. less resilience) than for the graded cases, and may even be disconnected from the shear stress due to elevated critical shear stresses (e.g. loss of sediment water content, or tidal inlets where the channel has reached a hard bottom).

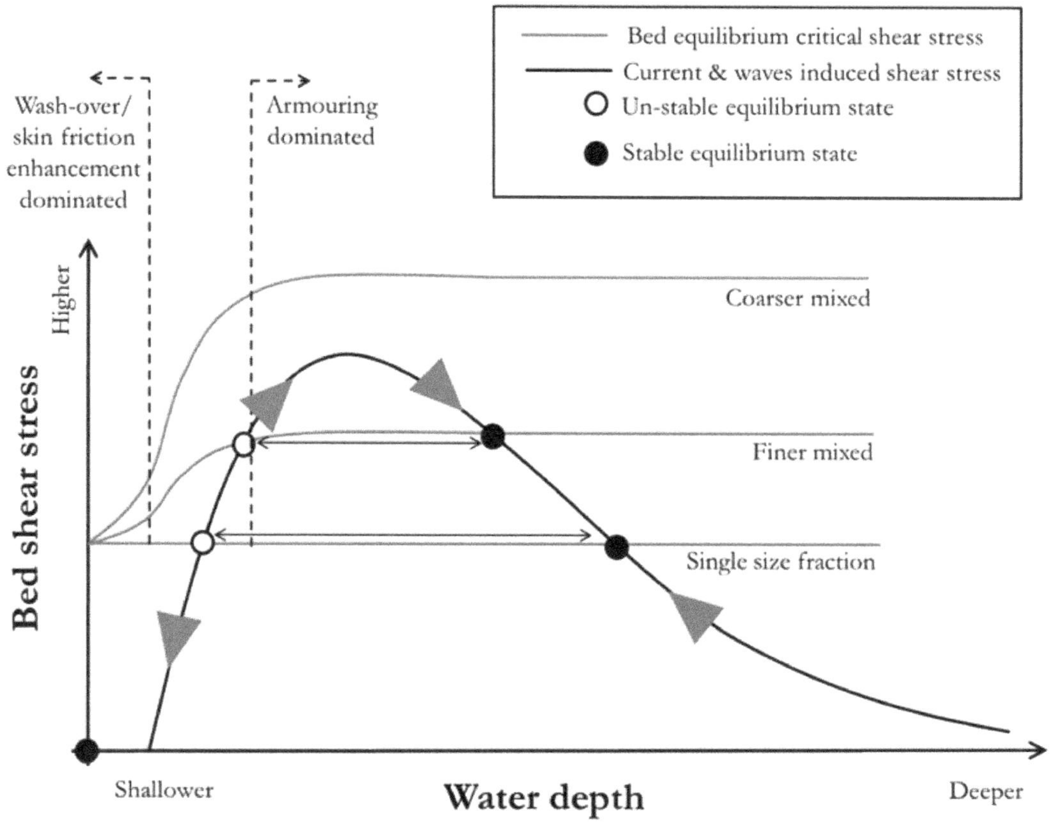

Figure 8: Qualitative stability diagram of active layer behaviour based on the feedback structure of erosion–deposition behaviour of graded beds. Solid black curve shows the total bed shear stress due to wave breaking-induced turbulence, wave stirring and currents, as a function of water depth. Equilibrium critical stress curves (solid grey lines) are shown for bed layers of a single-size-fraction and two mixed grainsized fractions. Assuming no changes on deposition rate, the bed is eroded if total bed shear stress is larger than critical shear stress and bed is accreted if total bed shear stress is lower. This bed's responses are represented by grey arrows that show system trajectories towards stable (filled circles) or away from unstable points (empty circles).

MAPPING SAND-RICH SHORELINE LOCAL PLANFORM BEHAVIOUR

In this section, we shift our attention to the landform scale and demonstrate how our current understanding of the processes that control the beach plan-form can be synthesized into a CLD.

Feedback Structure

On an open, long, sand-rich coast, wave-driven alongshore sediment transport tends to smooth the coastline if the angle between deep water wave crests and the shoreline is relatively small (Komar, 1998). However, for waves approaching at a large angle with respect to the shoreline, the gradients in alongshore drift may reinforce nearshore-bathymetry irregularities, rendering the rectilinear coast unstable (Ashton et al., 2001). This instability, termed High Angle Wave Instability (HAWI), has been used to explain persistent shoreline features over a large range of spatial scales, such as cuspate shorelines, alongshore sand waves, and flying spits (Ashton and Murray, 2006). Whereas earlier studies have pointed only to the alongshore drift as the cause of the instability, Ashton and Murray (2006) have shown that the instability mechanism involves both the surf and the shoaling zones, so that the link provided by the cross-shore sediment transport becomes crucial. Falqués et al. (2011) have shown that obliqueness of wave incidence has two effects on the alongshore drift: (i) a direct effect on the relative angle between the wave fronts and the shoreline, and (ii) an indirect effect on breaker height via the wave energy spreading during refraction. The direct effect is always stabilizing, and instability occurs only from the effects of wave energy spreading, which dominates at large incidence angles. In the following, we show how this system behaviour can be captured using CLDs.

Let us define the across-shore location of the shoreline and depth of closure at the observation point, x_{obs}, as, y_s, and y_c respectively, as shown in Fig. 9. The adjacent locations to the left and to the right of the observation point are x_l and x_r respectively, being the alongshore distance between each point to the observation point (equal to dx). The alongshore sediment transport, Q, is positive in the x direction or global alongshore coordinate, and the alongshore sediment transport gradient at the observation location, at a given time, is defined as:

$$\frac{dQ}{dx} = \frac{Q(x_r) - Q(x_l)}{dx} \quad (1)$$

where:

$$Q(x_r) \approx H_{b,r}^{5/2} \sin\left(2\left(\theta_{b,r} - \Phi_{s,r}\right)\right) = H_{b,r}^{5/2} \sin\left(2\alpha_{b,r}\right) \quad (2)$$

and:

$$Q(x_l) \approx H_{b,l}^{5/2} \sin\left(2\left(\theta_{b,l} - \Phi_{s,l}\right)\right) = H_{b,r}^{5/2} \sin\left(2\alpha_{b,l}\right) \quad (3)$$

are the Q values at the right and left side of the observation point, $(Hb\,l, Hb\,r)$ being the wave height at breaking, $(\,b\,l,\ b\,r)$ wave angle at breaking and $(\Phi s\,l, \Phi s\,r)$ the shoreline orientation at the left- and right-hand side of the observation point. Based on this plan-form schematization, an increase in $\frac{dQ}{dx}$ has a negative effect on the sediment volume at the observation point, and therefore on shoreline position. An advance of the shoreline position at x_{obs} increases the shoreline angle relative to the x axis at the left-hand side, $\Phi_{s,l} = \frac{y_s(x_{obs}) - y_s(x_l)}{dx}$, and decreases the shoreline orientation at the right-hand side, $\Phi_{s,r} = \frac{y_s(x_r) - y_s(x_{obs})}{dx}$. Similar reasoning applies to the orientation of the depth of closure to the left, $\Phi_{c,l} = \frac{y_c(x_{obs}) - y_c(x_l)}{dx}$ and right, $\Phi_{c,r} = \frac{y_c(x_r) - y_c(x_{obs})}{dx}$ of the observation point. Therefore, the effect of increasing y_s or y_c is always positive on the orientation of any bathymetric contour at the left-hand side and negative at the right-hand side. Next, we need to answer the question of what are the effects of a shoreline advance on the alongshore sediment transport gradient at each side of the perturbation?

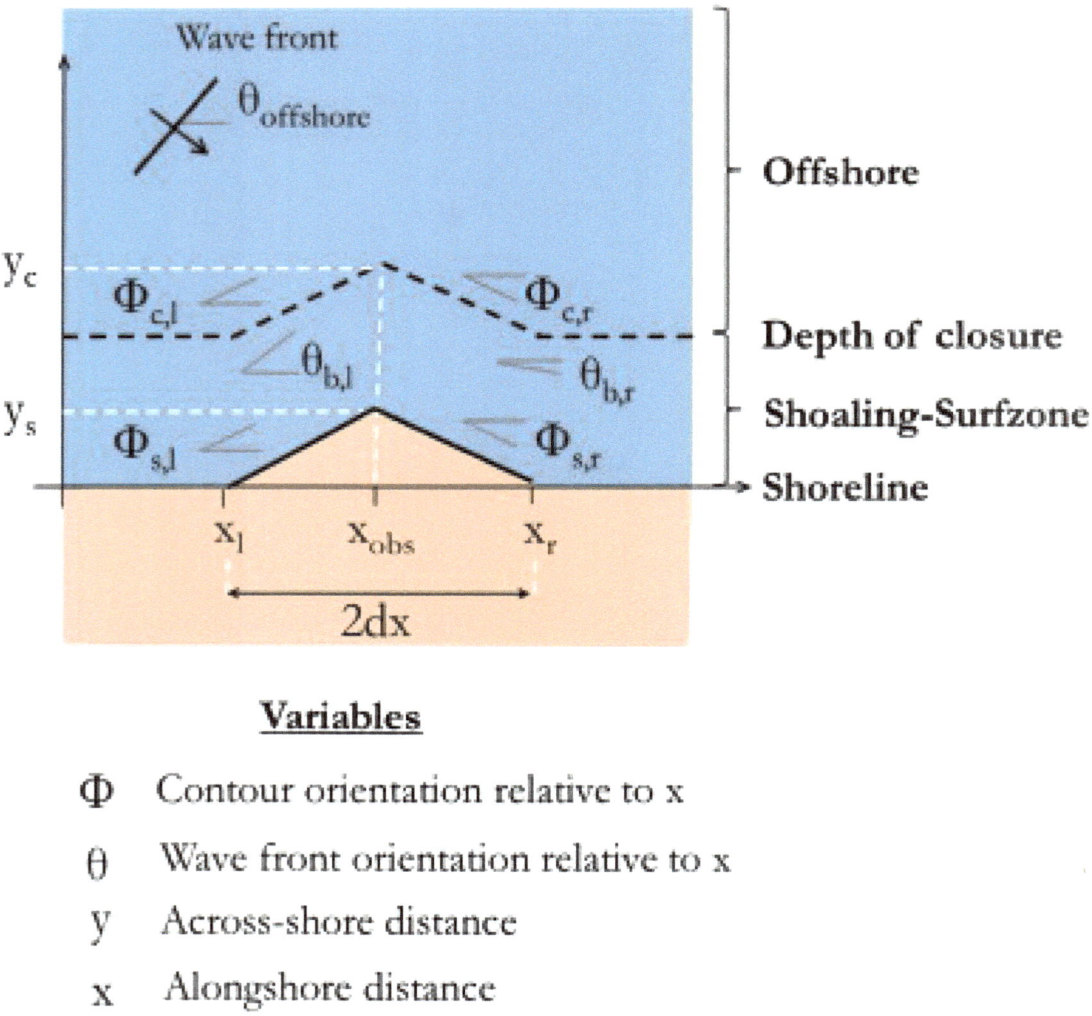

Figure 9: Open coast plan-form schematization of a perturbation on the shoreline position, y_s, and depth of closure, y_c, at the observation long-shore location, x_{obs}. A central difference scheme is used to obtain the alongshore sediment transport gradient at the observation point for any given offshore wave angle. The sub-indexes b, c, s, correspond with variables at breaking, closure and shoreline respectively and the sub-indexes obs, l and r correspond with variables at the observation point and left and right of observation point.

For waves at breaking (typically $\propto b \ll 45°$ because of refraction and therefore $sin2 \propto b \approx 2 \propto b = 2(\,b - \Phi s)$) an increase of shoreline orientation, Φs, has always a negative effect on the alongshore sediment transport. The latter applies to changes in shoreline orientation but not to other bathymetric contours or depth of closure as explained below. A causal pathway analysis (Fig. 10a) shows that an advance in shoreline position, always has a positive effect on the alongshore sediment transport gradient. In line with conventional Causal Loop Analysis (Morecroft, 1982), this follows from the sequence of signs: $+ - + - = +$ from left and $- - + + = +$ from the right. This is then collapsed into a positive link between shoreline positions and alongshore sediment transport gradient. However, the alongshore sediment transport gradient has also an indirect effect on shoreline position by changing the shoaling contours. While the alongshore sediment transport is driven by the wave angle at breaking, sediment is transported further offshore from the breaking point down to the depth of closure. This transport is explained by the spiral velocity profile resulting from vector addition of the alongshore and across-shore currents (Svendsen and Putrevu, 1994). Therefore, a perturbation of the shoreline propagates seaward to the shoaling zone, affecting the wave angle and height at breaking, and therefore the sediment transport gradient. As shown by Falqués et al. (2011), an increase in the offshore wave angle relative to the shoaling bathymetry contour always has a negative effect on wave height at breaking due to energy spreading. The actual magnitude of the effect will depend on the shoaling geometry (they assumed this to be parallel and rectilinear). Causal pathway analysis shows that an advance in depth of closure contour always has a negative effect on the alongshore sediment transport gradient, either from the left-hand side (+ – – + – = –) or right-hand side (– – – + + = –) due to a reduction in wave energy because of energy spreading (Fig. 10b). Energy spreading also affects the wave angle at breaking (i.e. very oblique waves arrive at the shore with smaller height and break in shallower water causing a decrease in $\propto b$) adding a negative effect on the alongshore sediment transport gradient (Fig. 10c). The resulting CLD for the open coast, including the existence of a local sink or source of sediment ($q(x_{obs},t)$) and the balanced feedback loop between erodible sediment volume, V_{sed}, and alongshore sediment transport gradient is shown in Fig. 10d. A reduction in V_{sed} reduces the outgoing (Q_r) sediment transport rate, relative to the potential sediment transport (Hanson and Militello, 2005), and therefore a positive effect on the alongshore sediment transport gradient. The resulting balanced feedback loop (termed "sediment depletion" in Fig. 10d) indicates that the alongshore sediment transport gradient can also be controlled by a lack of sediment and not only a reduction of transport capacity due to beach plan-form rotation (here termed "transport depletion").

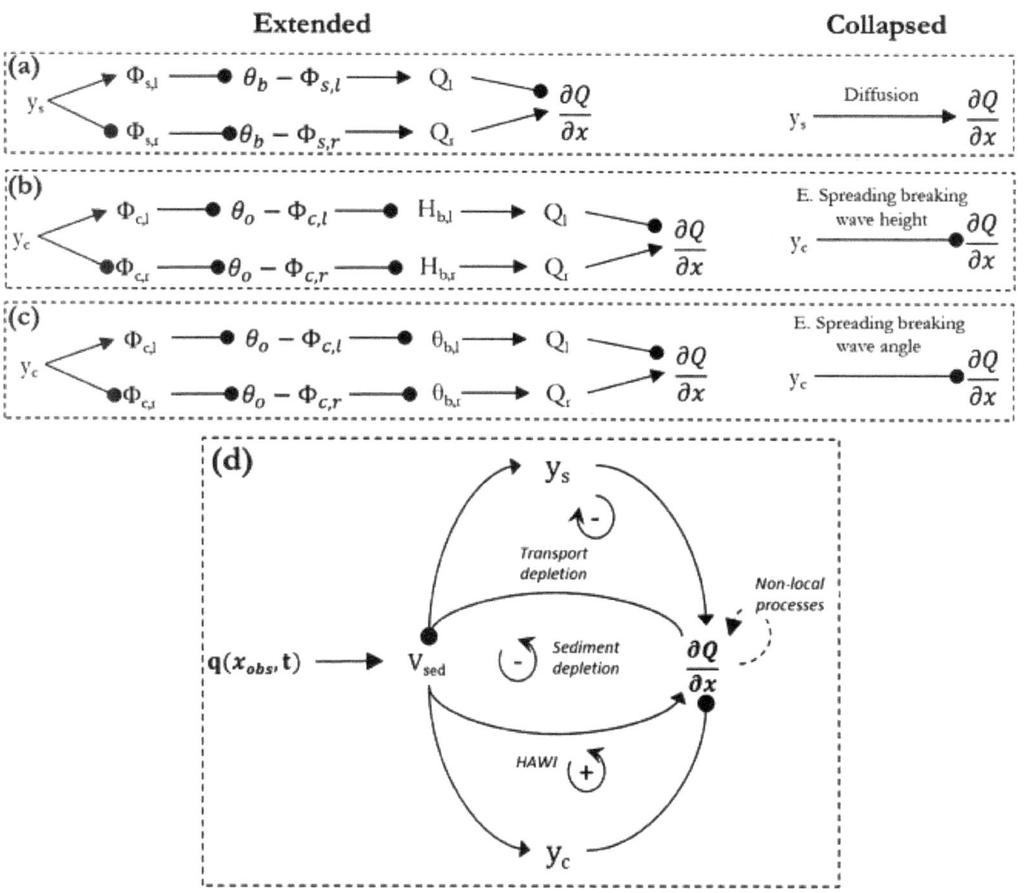

Figure 10: Translation of the mathematical description given by Falqués et al. (2011) of the quasi 2D processes involved in shoreline High Angle Wave Instability into CLDs. A perturbation of the shoreline and the depth of closure line leads unambiguously to an increase (a) and decrease (b, c) of the alongshore sediment transport gradient respectively. (d) Shows the overall beach volume (V_{sed}) behaviour, where $q(x_c,t)$ represent a source/sink of sediment at the observation location.

Downward Causation and Emergent Behaviour

The High Angle Wave Instability described above is also an example of how a local feedback can lead to both emergent landforms (capes) and emergent processes (wave shadowing). When a HAWI feedback loop dominates the local alongshore sediment transport overall feedback, it causes finite amplitude cuspate landforms to emerge from initial small perturbations. However, these features cause wave shadowing and influence the growth of adjacent features. This emergent process (wave shadowing) is unrelated to the initial instability and ends up controlling the alongshore sediment transport gradient at the smaller scale O(100 m). As noted by Coco and Murray (2007), this emergent non-local process implies that detailed analyses of the local surfzone processes that result in bulk alongshore transport will not lead to direct insights about how the large-scale shoreline features develop, nor will analyses confined to the analysis of the shoreline instability that causes features to grow initially. When the shoreline smoothing induced by small angle wave, HAWI and shadowing are included in a numerical modelling of initially straight shorelines under strongly asymmetric wave climate, cuspate spits develop and exhibit a striking array of non-local interactions (Ashton and Murray, 2006). Notably, these emergent spit features can dominate the alongshore sediment transport gradient behaviour for a considerable distance through shadowing effects.

The effect of wave shadowing (a non-local process) on the local feedback structure shown in Fig. 10d is acknowledged as an influence link between the alongshore sediment transport and itself. An unsigned link is used since wave shadowing can induce both an increase and a decrease of the local alongshore sediment transport gradient at different locations along the shoreline. This is best illustrated by reasoning how changes on the shoreline position at either the exposed shoreline, shoreline in the lee of the spit or shadowed shoreline might influence the alongshore sediment transport (Fig. 11). At both sides, the exposed side and in the lee of the spit, the schematization of shoreline shown on Fig. 9 is still valid, but since the local orientation might differ significantly from the global orientation, the definition of Q as positive along the global x-axis is no longer valid. In practice this is resolved by interpreting on which type of shoreline (exposed, in the lee of the spit or shadowed shoreline) the local shoreline is located (x_c) (e.g. Ashton and Murray, 2006). Since the directed graph shown in Fig. 10d is not spatially explicit, the type of local shoreline cannot be interpreted but is at least acknowledged as an influence link.

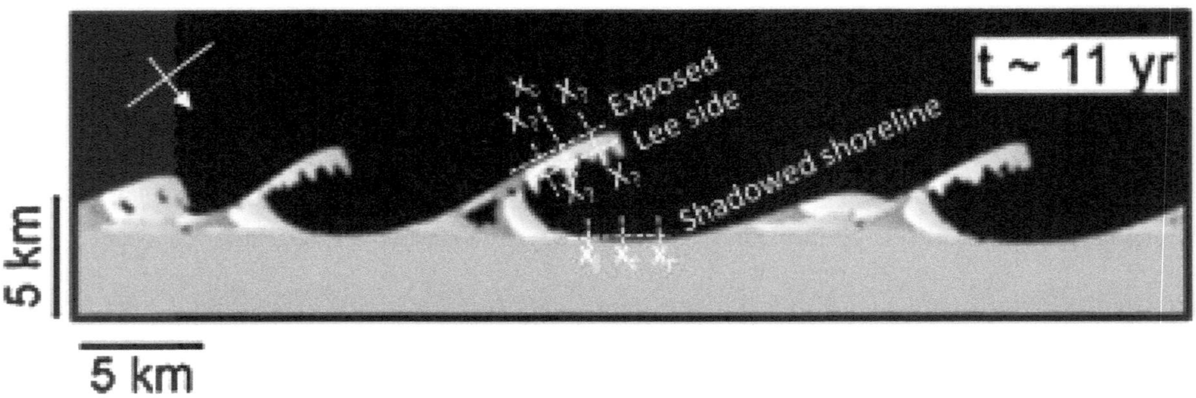

Figure 11: Numerical result showing a snapshot plan view of an initially almost rectilinear beach after 11 years of strongly asymmetric wave forcing (more waves approaching from the left than the right). Arrow shows approximate direction of incoming waves approaching from the left.

Modified from Coco and Murray (2007).

Linking model Simplifications to Model Behavioural Validity

An important application of CLDs is in revealing the limitations introduced by assumptions made in the formulation of a numerical process-based model. For example, let us assume that a modeller aims to reproduce the observed alongshore migration of a foreland such as the one shown in Fig. 12. Foreland dynamics are a product of a bidirectional wave climate and a simple segmented linear coastline (Burningham and French, 2014) so, in principle, both diffusive and anti-diffusive feedback loops might be active. A one-line model based on an assumed equilibrium profile is selected. To speed up the computation, the wave propagation algorithm is simplified by assuming the bathymetry is rectilinear and parallel to the shoreline. This ignores the wave shoaling that occurs between the wave breaking point and the depth of closure, which is needed to close the HAWI reinforcing feedback loop in the CLD. The resulting model will thus behave in purely diffusive way. The GENCADE model Hanson and Kraus (2011) will behave in this way. Only if the bathymetry is not simplified and use made of a numerical wave propagation model (Van den Berg et al., 2011), is the alongshore wave energy spreading that closes the feedback loop included (with the effect of significantly increasing the computation time). In contrast, the raster-based Coastal Evolution Model of Ashton and Murray (2006) explicitly encodes this behaviour in its otherwise more highly parameterized numerical scheme. CLDs are non-software specific and therefore cannot be used to assess if models with similar assumptions and simplifications but different parameterization will reproduce the same behaviour. However, analysis of the feedback factors (see above) provides a robust basis for comparative evaluation of alternative model formulations.

Causal Loop Analysis is non-software specific and, in advance of the selection or formulation of a quantitative model, provides a means of identifying the important system behaviours that should emerge at the scales being simulated. While many features and some of the feedbacks depicted in a CLD appear self-evident, taken together the qualitative insights into aggregate system behaviour can prove illuminating and, as in the simple example presented above, these can be fundamental to the explanation of the patterns of mesoscale coastal morphological change. Indeed, it can be argued that a lack of attention to anticipated system behaviours and a desire to obtain quantitative predictions irrespective of the robustness of the underlying model has led to weak and/or incomplete parameterizations that persist beyond their intended scope. The 'Bruun rule' is an obvious example that is noteworthy, not only for its continued application despite widespread criticism of its underlying assumptions (Cooper and Pilkey, 2004) but also for is embedded use in models that purport to offer more sophisticated insights into mesoscale coastal behaviour (French et al., 2016-in this issue).

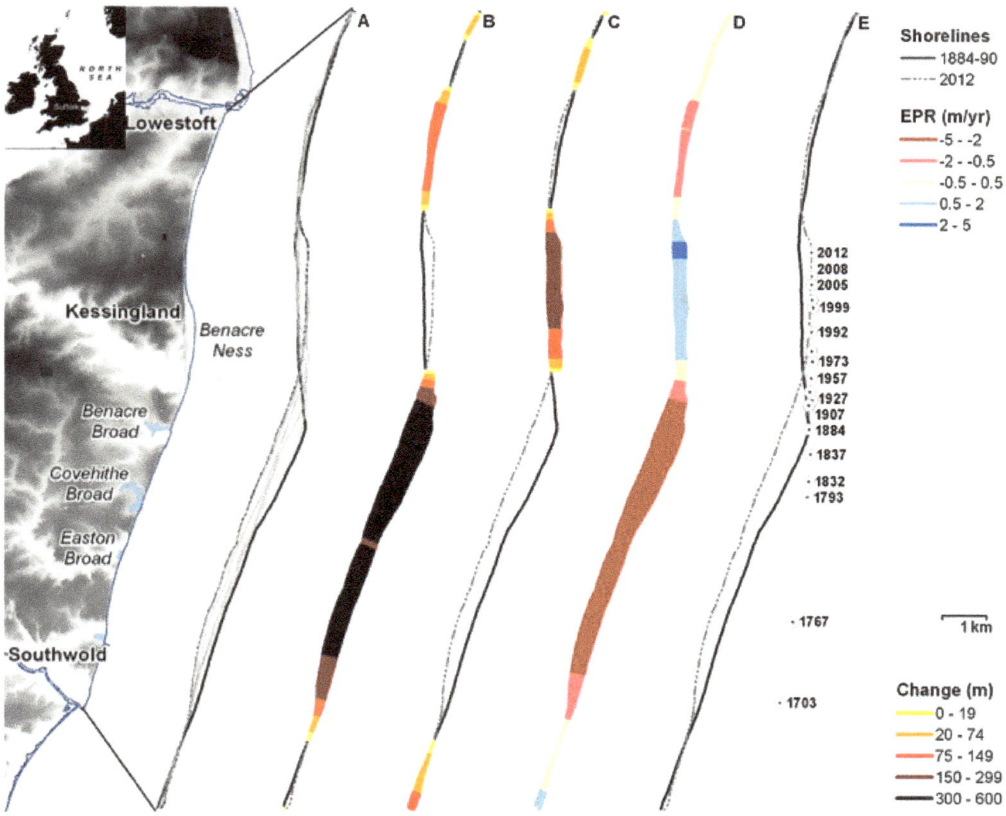

Figure 12: Shoreline change analysis of the north Suffolk coast, showing historic migration of Benacre Ness foreland. A) Shoreline change envelope. B, C) net shoreline recession and advance. D) Historical end-point rate change. E) Historical change in the position of Benacre foreland.

Source: Burningham and French (2014).

DISCUSSION AND CONCLUSIONS

The preceding analysis has highlighted the challenge, as well as the importance, of mapping some of the many feedbacks that govern coastal morphodynamics at decadal and longer timescales. Following Morecroft (1982), CLDs that represent systems as directed graphs are advocated as a means of distilling the cumulative effect of individual feedbacks that have been derived elsewhere by conceptualizing the parts of the system and simulating or observing their interaction. Essentially, this constitutes a knowledge formalisation process, in which a system-level understanding is synthesized from relevant scientific literature and observational data.

The case studies analysed herein are certainly not exhaustive but are intended to illustrate the advantages of applying a formal approach at two contrasting scales: (1) the erosion–accretion behaviour of graded (mixed) sediment beds, and (2) the local alongshore sediment fluxes of sand-rich shorelines. These case study examples are chosen on account of their central role in the quantitative modelling of geomorphological futures and as they illustrate different types of causation (upward causation, downward causation and emergent processes). To our knowledge, no similar qualitative behavioural modelling has been attempted for these geomorphological phenomena.

In this first example, we conceptualize the sediment bed elevation as a disequilibrium between erosion and accretion processes, and the local alongshore gradient as a disequilibrium between diffusive and anti-diffusive processes. This conceptualization aligns naturally with the pioneering work by Phillips (2009), whose framework recognises the possibility of multiple response pathways or trajectories, as well as multiple outcomes. In our case study of graded sediment beds, we show how up to three different outcomes (a null response, and two disequilibrium states) can be derived from a simple qualitative stability analysis. In the case of sand-rich shoreline planform behaviour, we show how two fundamentally different shoreline responses (diffusive and anti-diffusive) can be triggered by small changes of the shoreline cross-shore position. These responses have been described elsewhere using different numerical approaches (e.g.Ashton et al., 2001 and Van den Berg et al., 2011).

The graded sediment study case also illustrates how a direct and non-direct upward causation can be mapped into a simple directed graph (Fig. 7). Upward causation, by which small-scale processes directly cause larger-scale, longer-term landscape evolution, has long been used by geomorphologists to understand large-scale behaviour (Murray et al., 2014). As an example, the sedimentation–erosion balance of intertidal sediment has implications for the evolution of minerogenic-tidal flat morphology and the occurrence of a transition pathway to tidal saltmarsh (Friedrichs and Perry, 2001b and Fagherazzi et al., 2007). We have shown how a non-direct upward causation, the elevation deficit, can also be mapped as a directed graph. While not all feedbacks can be represented (either due to lack of understanding or for clarity seek), it can be acknowledged as an un-signed link as we have done for the feedback between ripple behaviour and bottom energy dissipation (e.g. Coco et al., 2007b).

The sand-rich local planform behaviour has been used to test how downward causation and emergent processes and landforms can be mapped as directed graphs. The growth of an initial small perturbation can be explained by the dominance of the HAWI feedback loop over the balancing feedbacks of the local alongshore sediment transport gradient. We have shown how the effect of wave shadowing, induced when the initial perturbation becomes large enough, is location-specific (exposed shoreline, lee of the spit, shadowed shoreline). Despite the directed graph not being spatially explicit, this downward causation can be represented as an un-signed link on the alongshore sediment transport itself. Spatially explicit numerical models overcome this issue by "interpreting the coastline" at each location (e.g. Ashton et al., 2001).

We believe that explicit depiction of feedback-structure diagrams is beneficial when exploring geomorphic futures with numerical models. System analysis and depiction of feedback structures is the first step of numerical modelling (e.g. Capobianco et al., 1999), but it is more often than not implicit and hence inaccessible. By explicitly mapping the feedbacks included and neglected within the model, the user is better positioned to assess whether or not relevant feedback loops are included. As an example, we have shown how model simplifications that appear intuitively independent (e.g. assuming bathymetry is rectilinear and parallel to the shoreline for speedy wave propagation of 1-line models) can limit modelled outcomes (e.g. a failure to resolve a critical HAWI loop in the case of shore planform evolution). The problem of how to assess the relative importance of a given feedback loop is the natural extension of the work presented here (e.g. Klocke et al., 2013 and Payo et al., 2014).

ACKNOWLEDGEMENTS

We would like to acknowledge the useful discussions with Professor Brad Murray and Dr Laurent Amoudry during the preparation of this manuscript. This work was funded by the Natural Environment Research Council (NERC) as part of the Integrating COAstal Sediment SysTems (iCOASST) project (NE/J005541/1), with the Environment Agency as an embedded project stakeholder. The authors gratefully acknowledge helpful discussions with other iCOASST project team members and with the invited attendees of the iCOASST International Conference on Simulating Decadal Coastal Morphodynamics, held from 15 to 17 October 2013 in Southampton, UK.

REFERENCES

1. Allen, J., 1999. Geological impacts on coastal wetland landscapes: some general effects of sediment autocompaction in the Holocene of northwest Europe. The Holocene 9 (1), 1–12.
2. Anfuso, G., 2005. Sediment- activation depth values for gentle and steep beaches. Mar. Geol. 220 (1), 101–112.
3. Ashton, A.D., Murray, A.B., 2006. High-angle wave instability and emergent shoreline shapes: 1. Modeling of sand waves, flying spits, and capes. J. Geophys. Res. Earth Surf. 111 (F4), F04011.
4. Ashton, A., Murray, A.B., Arnoult, O., 2001. Formation of coastline features by large-scale instabilities induced by high-angle waves. Nature 414 (6861), 296–300.
5. Benson, T., French, J.R., 2007. InSiPID: a new low-cost instrument for in situ particle size measurements in estuarine and coastal waters. J. Sea Res. 58 (3), 167–188.
6. Brain, M.J., Long, A.J., Petley, D.N., Horton, B.P., Allison, R.J., 2011. Compression behaviour of minerogenic low energy intertidal sediments. Sediment. Geol. 233 (1–4), 28–41.
7. Burningham, H., French, J.R., 2014. Travelling forelands: complexities in drift and migration patterns. J. Coast. Res. 102–108.
8. Cahoon, D.R., Reed, D.J., Day Jr., J.W., 1995. Estimating shallow subsidence in microtidal salt marshes of the southeastern United States: Kaye and Barghoorn revisited. Mar. Geol. 128 (1), 1–9.
9. Capobianco, M., DeVriend, H.J., Nicholls, R.J., Stive, M.J.E., 1999. Coastal area impact and vulnerability assessment: the point of view of a morphodynamic modeller. J. Coast. Res. 15 (3), 701–716.
10. Coco, G., Murray, A.B., 2007. Patterns in the sand: from forcing templates to selforganization. Geomorphology 91 (3–4), 271–290.
11. Coco, G., Murray, A.B., Green, M.O., 2007a. Sorted bed forms as self-organized patterns: 1. Model development. J. Geophys. Res. Earth Surf. 112 (F3).
12. Coco, G., Murray, A.B., Green, M.O., Thieler, E.R., Hume, T., 2007b. Sorted bed forms as self-organized patterns: 2. Complex forcing scenarios. J. Geophys. Res. Earth Surf. 112 (F3).
13. Cooper, J.A.G., Pilkey, O.H., 2004. Sea-level rise and shoreline retreat: time to abandon the Bruun Rule. Glob. Planet. Chang. 43 (3), 157–171.
14. Eisma, D., 1991. Particle size of suspended matter in estuaries. Geo-Mar. Lett. 11 (3), 147–153.
15. Fagherazzi, S., Palermo, C., Rulli, M., Carniello, L., Defina, A., 2007. Wind waves in shallow microtidal basins and the dynamic equilibrium of tidal flats. J. Geophys. Res. Earth Surf. 112 (F2).
16. Falqués, A., Calvete, D., Ribas, F., 2011. Shoreline instability due to very oblique wave incidence: some remarks on the physics. J. Coast. Res. 27 (2), 291–295.
17. French, J., 2006. Tidal marsh sedimentation and resilience to environmental change: exploratorymodelling of tidal, sea-level and sediment supply forcing in predominantly llochthonous systems. Mar. Geol. 235 (1), 119–136.
18. French, J., Payo, A., Murray, A.B., Orford, J., Elliot, M., Cowell, P.J., 2016. Appropriate complexity for the prediction of coastal and estuarine geomorphic behaviour at decadal to centennial scales. Gemorphology(iCOASST Special Issue) 256, 3–16 (in this issue).
19. Friedrichs, C., 2011. 3.06 Tidal Flat Morphodynamics: a Synthesis. Friedrichs, C.T., Perry, J.E., 2001a. Tidal salt marsh morphodynamics: a synthesis. J. Coast.Res. 7–37.
20. Friedrichs, C.T., Perry, J.E., 2001b. A review of tidal salt marsh morphodynamics. AGU Spring, Meeting Abstracts, 1.
21. Guza, R.T., Inman, D.L., 1975. Edge waves and beach cusps. J. Geophys. Res. 80 (21), 2997–3012.

22. Hanson, H., Kraus, N.C., 2011. Long-term evolution of a long-term evolution model. J. Coast. Res. 118–129.

23. Hanson, H., Militello, A., 2005. Representation of nonerodible (hard) bottom in twodimensional morphology change models. ERDC, CHL CHETN-IV-63. US Army Corps of Engineers.

24. Harris, C.K., Wiberg, P.L., 1997. Approaches to quantifying long-term continental shelf sediment transport with an example from the Northern California STRESS midshelf site. Cont. Shelf Res. 17 (11), 1389–1418.

25. Houser, C., Barrett, G., 2010. Divergent behavior of the swash zone in response to different foreshore slopes and nearshore states. Mar. Geol. 271 (1), 106–118.

26. Klocke, D., Quaas, J., Stevens, B., 2013. Assessment of different metrics for physical climate feedbacks. Observational, Theoretical and Computational Research on the Climate System 41(5) pp. 1173–1185.

27. Kobayashi, N., Johnson, B.D., 2001. Sand suspension, storage, advection, and settling in surf and swash zones. J. Geophys. Res. Oceans 106 (C5), 9363–9376.

28. Kobayashi, N., Payo, A., Schmied, L., 2008. Cross-shore suspended sand and bed load transport on beaches. J. Geophys. Res. Oceans 113 (C7), C07001.

29. Komar, P.D., 1998. Beach Processes and Sedimentation, 1998. Prentice Hall, Upper Saddle River, NJ. Lane, D.C., 2000. Diagramming conventions in system dynamics. J. Oper. Res. Soc. 51 (2), 241–245.

30. Lane, S.N., 2013. 21st century climate change: where has all the geomorphology gone? Earth Surf. Process. Landf. 38 (1), 106–110.

31. Le Hir, P., Cayocca, F., Waeles, B., 2011. Dynamics of sand and mud mixtures: a multiprocess-based modelling strategy. Cont. Shelf Res. 31 (10), S135–S149.

32. Manning, A.J., Spearman, J.R., Whitehouse, R.J.S., Pidduck, E.L., Baugh, J.V., Spencer, K.L., 2013. Flocculation dynamics of mud: sand mixed suspensions. Sediment Transport Processes and Their Modelling Applications.

33. Morecroft, J.D., 1982. A critical review of diagramming tools for conceptualizing feedback system models. Dynamica 8 (1), 20–29.

34. Murray, A.B., Coco, G., Goldstein, E.B., 2014. Cause and effect in geomorphic systems: complex systems perspectives. Geomorphology 214, 1–9.

35. Nielsen, P., 1992. Coastal Bottom Boundary Layers and Sediment Transport. 4. World scientific. Paterson, D., 1997. Biological mediation of sediment erodibility: ecology and physical dynamics. Cohesive sedimentspp. 215–229.

36. Payo, A., Hall, J., Dickson, M., Walkden, M.A., 2014. Feedback structure of cliff and shore platform morphodynamics. J. Coast. Conserv. 1–13.

37. Phillips, J.D., 2006. Deterministic chaos and historical geomorphology: a review and look forward. Geomorphology 76 (1), 109–121.

38. Phillips, J.D., 2009. Changes, perturbations, and responses in geomorphic systems. Prog. Phys. Geogr. 33 (1), 17–30.

39. Phillips, J.D., 2012. Synchronization and scale in geomorphic systems. Geomorphology 137 (1), 150–158.

40. Schumm, S.A., 1991. To Interpret The Earth : Ten Ways to be Wrong. Cambridge University Press, Cambridge.

41. Schumm, S.A., Lichty, R.W., 1965. Time, space, and causality in geomorphology. Am. J. Sci. 263 (2), 110–119.

42. Slingerland, R., 1981. Qualitative stability analysis of geologic systems, with an example from river hydraulic geometry. Geology 9 (10), 491–493.

43. Svendsen, I.A., Putrevu, U., 1994. Nearshore mixing and dispersion. Proc. R. Soc. Lond. Ser. A Math. Phys. Sci. 445 (1925), 561–576.

44. Townend, I., 2003. Coast and estuary behaviour systems. Coastal Sedimentpp. 1–14. Van den Berg, N., Falqués, A., Ribas, F., 2011. Long-term evolution of nourished beaches under high angle wave conditions. J. Mar. Syst. 88 (1), 102–112.

45. van Ledden, M., Van Kesteren, W., Winterwerp, J., 2004. A conceptual framework for the erosion behaviour of sand–mud mixtures. Cont. Shelf Res. 24 (1), 1–11.

46. van Rijn, L.C., 2007. Unified view of sediment transport by currents and waves. III: Graded beds. J. Hydraul. Eng. 133 (7), 761–775.

47. van Rijn, L.C., Walstra, D.-J.R., van Ormondt, M., 2007. Unified view of sediment transport by currents and waves. IV: Application of morphodynamic model. J. Hydraul. Eng. 133 (7), 776–793.

48. Van Wijnen, H., Bakker, J., 2001. Long-term surface elevation change in salt marshes: a prediction of marsh response to future sea-level rise. Estuar. Coast. Shelf Sci. 52 (3), 381–390.

49. Yamada, F., Tateyama, R., Tsujimoto, G., Suenaga, S., Long, B., Pilote, C., 2013. Dynamic monitoring of physical models beach morphodynamics and sediment transport using X-ray CT scanning technique. J. Coast. Res., SI 65, 1617–1622.

CHAPTER 10

CATASTROPHIC IMPACT OF EXTREME FLOOD EVENTS ON THE MORPHOLOGY AND EVOLUTION OF THE LOWER JÖKULSÁ Á FJÖLLUM (NORTHEAST ICELAND) DURING THE HOLOCENE

Edwin R.C. Baynes, Mikaël Attal, Andrew J. Dugmore, Linda A. Kirstein, Kathryn A. Whaler

School of GeoSciences, University of Edinburgh, Edinburgh EH8 9XP, UK

ABSTRACT

The impact of extreme flood events is rarely considered in studies of long-term landscape evolution, despite the potential for catastrophic landscape change in a short period of time. Here, we use an integrated approach of geomorphological mapping, topographic analysis and geophysical surveys to identify and quantify the impact of extreme flood events (jökulhlaups) along the Jökulsá á Fjöllum, Iceland, where evidence for the action of such floods is widespread on microspatial to macrospatial scales. The apex of the 28-km-long Jökulsárgljúfur canyon is characterised by a complex network of palaeo-flood channels and large vertical knickpoints such as Dettifoss (54 m high) and Hafragilsfoss (20 m high). Downstream, the Forvoð valley contains large terraces of boulder-rich deposits (50 m thick, > 3 km long). Near the outlet of the canyon is Ásbyrgi, a dry canyon (3 km long, 1 km wide, up to 90 m deep) with eroded cataracts and scabland morphology immediately upstream and ~ 90 m above the current river channel. Topographic analysis and electrical resistivity tomography surveys show that 0.144 km^{3}of rock was eroded from Ásbyrgi during its formation ~ 10,000 years ago, and just 4% of this eroded volume is currently filled with sediment deposits, up to 5 m thick. Deposited boulders across the canyon floor of Ásbyrgi demonstrate that the discharge of the jökulhlaup that formed the canyon was at least 39,000 m^3 s^{-1}. We present a model for the evolution of the lower Jökulsá á Fjöllum and the Jökulsárgljúfur canyon during various stages of an extreme flood event. Reconstruction of the early Holocene flood event includes the initiation and development of different canyons before the capture of all floodwater within one canyon at the end. We tie the evolution of the lower Jökulsárgljúfur canyon to established chronology of flood events during the Holocene farther upstream and highlight the dominant impact of extreme flood events over background processes in this landscape.

INTRODUCTION

Extreme flood events are characterised by the release of a large volume of water over the landscape in a short period of time. Such events occur in a range of environments and can be triggered by glacial lake outbursts (e.g., Baker et al., 1993), landslide or moraine dam failures (e.g., Dunning et al., 2006), or by subglacial volcanic eruptions (e.g.,Bjornsson, 2009 and Dunning et al., 2013). Extreme flood events are common over geological timescales, and the potential for geomorphic change during such events is great owing to high peak discharges, potentially over 10^6 m^3 s^{-1} (Baker, 2002). Previous work has identified the impact of extreme flood events in the evolution of a range of terrestrial environments such as

Citation: Edwin R.C. Baynes, Mikaël Attal, Andrew J. Dugmore, Linda A. Kirstein, Kathryn A. Whaler, Catastrophic impact of extreme flood events on the morphology and evolution of the lower Jökulsá á Fjöllum (northeast Iceland) during the Holocene, doi:10.1016/j.geomorph.2015.05.009

the Channeled Scabland of northwestern USA following the draining of Glacial Lake Missoula (Bretz, 1923), the Tsangpo gorge of southeastern Tibet (Montgomery et al., 2004), and the Transbaikalia and Altai Mountains of Siberia (Carling et al., 2009a and Margold et al., 2011); it has also been suggested that such floods could have played a key role in the evolution of the Straits of Gibraltar (Garcia-Castellanos et al., 2009), the English Channel (Gupta et al., 2007), and the surface of Mars (Warner et al., 2010 and Warner et al., 2013). Despite this, current landscape evolution models do not consider the impact of extreme flood events in controlling bedrock landscape morphology (Carling et al., 2009b). Detailed quantitative studies of the impact of extreme flood events on the landscape are therefore required.

Glacial outburst floods, termed 'jökulhlaups', occur regularly in Iceland owing to the location of large ice caps atop active volcanoes (e.g., Bjornsson, 2002), which makes Iceland a globally important place to study the impact of extreme flood events. Previous work on Icelandic jökulhlaups includes the interpretation of deposited sediments (e.g.,Maizels, 1997, Duller et al., 2008 and Marren et al., 2009), the reconstruction of the hydraulic conditions (e.g., Baker et al., 1993, Alho et al., 2005, Alho et al., 2010,Carrivick, 2006 and Carrivick, 2007), and the geomorphic impact of jökulhlaups in proglacial areas close to the floodwater source (e.g., Magilligan et al., 2002, Carrivick et al., 2004 and Dunning et al., 2013). Our current understanding of canyon formation and bedrock erosion processes during extreme flood events is limited, especially in distal areas, and is based on studies such as that of the Channeled Scabland in Washington, USA (e.g., Baker and Kale, 1998) and a small number of studies in Idaho, USA (Lamb et al., 2008, Lamb et al., 2014 and Lamb and Dietrich, 2009) where the main motivation was to use the terrestrial landscape to infer the formation mechanisms of morphologically similar canyons on Mars. Building on a recent work in the upper reaches of the Jökulsárgljúfur canyon, Iceland (Baynes et al., 2015), the aim of this study is to reconstruct the bedrock landscape evolution of the lower Jökulsá á Fjöllum, in particular the impact of extreme flood events that are known to have occurred since deglaciation (Thórarinsson, 1950, Sæmundsson, 1973, Tómasson, 1973, Eliasson, 1974, Eliasson, 1977, Sigbjarnarson, 1996, Waitt, 2002, Kirkbride et al., 2006 and Baynes et al., 2015). This objective is achieved through documenting an inventory of landscape features within the Jökulsá á Fjöllum that are characteristic of the work of extreme floods, establishment of the chronology of floods, and assessment of the geomorphic impact of these extreme flood events during the Holocene using topographic analysis and Electrical Resistivity Tomography (ERT) surveys.

STUDY AREA

The Jökulsá á Fjöllum is one of Iceland's largest rivers, draining much of the 8100 km^{2}Vatnajökull ice cap in the south of the island and flowing 206 km north across central Iceland to the Arctic Ocean (Fig. 1A). The Jökulsá á Fjöllum has experienced multiple jökulhlaups of varying magnitude since the Last Glacial Maximum, with peak discharge for the largest flood estimated at 0.9×10^6 m^3 s^{-1} (Alho et al., 2005 and Carrivick et al., 2013). Jökulhlaups occur along the Jökulsá á Fjöllum as a result of either subglacial volcanism beneath Vatnajökull from one or more of the Kverkfjöll, Grimsvötn, or Bárðarbunga volcanic centres (Bjornsson, 2009) or the release of floodwater from an ice-dammed lake to the south of Kverkfjöll (Bjornsson, 2002) (Fig. 1B). Attempts have been made to identify and interpret the impact of jökulhlaups in the Jökulsá á Fjöllum in recent history (Isaksson, 1985 and Russell and Knudsen, 2002) and during the Holocene (Thórarinsson, 1950, Sæmundsson, 1973, Tómasson, 1973, Eliasson, 1974, Eliasson, 1977, Sigbjarnarson, 1996, Waitt, 2002, Carrivick et al., 2004, Kirkbride et al., 2006 and Baynes et al., 2015). Much of the recent work on this river has focussed on the geomorphic impact and sedimentary evidence of jökulhlaups close to the floodwater source (e.g., Carrivick et al., 2004, Carrivick, 2007 and Marren et al., 2009) and on modelling the hydraulic conditions of the floods in mid-stream to upstream reaches (e.g.,Alho et al., 2005, Carrivick, 2006, Carrivick, 2007 and Carrivick et al., 2013). Baynes et al. (2015) identified evidence for large-scale, rapid canyon cutting within the Jökulsárgljúfur canyon during three erosive periods in the Holocene, using cosmogenic nuclides concentrations to date the exposure of fluvially sculpted bedrock surfaces.

Three distinct reaches are identified within the Jökulsárgljúfur canyon, each exhibiting evidence for extreme flood events (Fig. 1C). The first reach is the main study area ofBaynes et al. (2015) at the head of the canyon. There, the Jökulsá á Fjöllum becomes deeply incised into the surrounding terrain, with three large waterfalls over a 5-km-long reach: these are Selfoss (13 m high), Dettifoss (54 m high), and Hafragilsfoss (20 m high); the canyon was carved through the retreat of these waterfalls during extreme floods. Downstream of this reach is the Forvoð valley, where widespread evidence for deposition of large volumes of sediment during extreme floods is present. At the lower end of the Jökulsárgljúfur canyon, additional evidence is present for the erosive impact of extreme flood events with the Klappir scabland area and Ásbyrgi canyon, a large dry cataract now disconnected from the current course of the Jökulsá á Fjöllum. This area contains outstanding preservation of large-scale fluvial landforms that have not undergone any alteration since their formation and therefore offer an excellent opportunity to quantify the impact of extreme floods. Downstream of this reach, the Jökulsá á Fjöllum flows for 18 km over a large depositional sandur plain to the coast. The geology of the area is characterised by young (< 0.8 Ma) basalt lava flows stacked on top of each other, ranging in structure from regular, near-vertical columns with metre-scale joint spacing to blocky, rubbly lavas with a centimetre to decimetre scale jointing. The ages of abandoned bedrock surfaces show significant canyon formation occurred at Ásbyrgi ~ 10,000 years ago and at the head of the Jökulsárgljúfur canyon ~ 5000 and ~ 1500 years ago through the upstream migration of large knickpoints such as Dettifoss and Selfoss (Baynes et al., 2015).

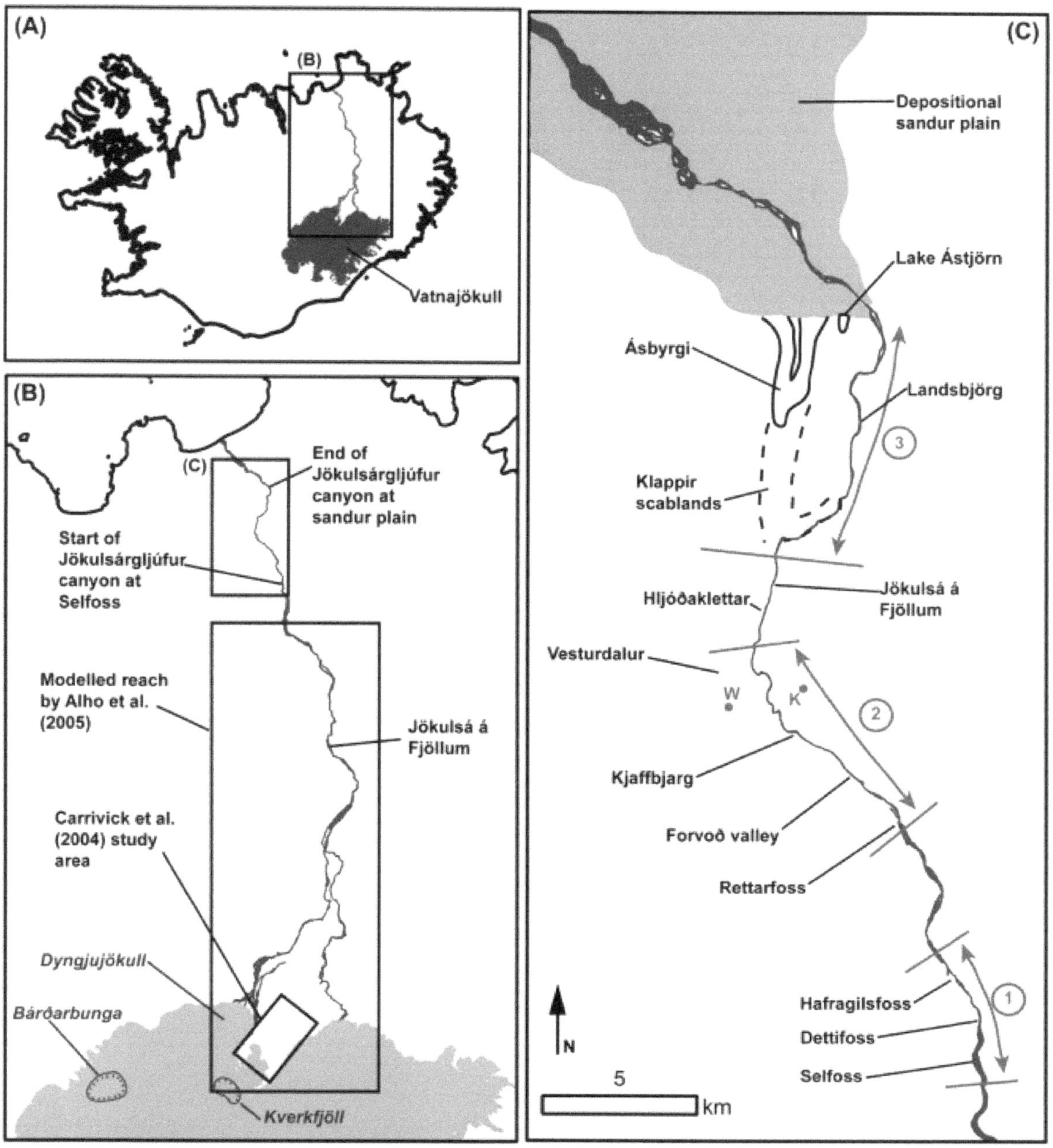

Figure 1: (A) Outline of Iceland with Vatnajökull ice cap (grey-shaded area), the source of the floodwaters, and the course of the present day Jökulsá á Fjöllum draining to the north coast. (B) Zoomed in map showing the areas studied by previous authors (Carrivick et al., 2004 and Alho et al., 2005) in the upper and middle reaches of the Jökulsá á Fjöllum as well as the Jökulsárgljúfur canyon, the focus of this study. Kverkfjöll and Bárðarbunga volcanic centres are highlighted. Grimsvötn volcano is located ~ 25 km south of Bárðarbunga, just beyond the extent shown in the map. (C) The Jökulsárgljúfur canyon is divided into three sections for the study: (1) the apex of the Jökulsárgljúfur canyon between Selfoss and Hafragilsfoss; (2) the Forvoð valley, containing depositional landforms; and (3) the lowermost section of the Jökulsárgljúfur canyon, with the Ásbyrgi horseshoe and the Klappir scabland system. The grey points labelled 'W' and 'K' indicate the location of the sedimentary sequences discussed in Waitt (2002) and Kirkbride et al. (2006), respectively.

MORPHOLOGICAL AND SEDIMENTOLOGICAL EVIDENCE FOR EXTREME FLOODS ALONG THE JÖKULSÁ A FJÖLLUM DOWNSTREAM OF SELFOSS

Carrivick et al. (2004) created a list of key criteria to identify the occurrence of extreme floods in bedrock channels, from macroscale landforms such as cataracts and anastomosing channels to microforms such as potholes and flutes (Table 1). Notably, many of the landforms listed in Table 1 are not exclusive to the action of extreme flood events, and the presence of these landforms within a landscape should not necessarily lead to the conclusion that an extreme flood event has taken place (Carrivick et al., 2013). However, considering the landscape as a whole and how multiple landforms are 'associated' to each other across a range of spatial scales can give an insight into the magnitude of the events that formed them (Carling et al., 2009c). We use the criteria inTable 1 to document erosional and depositional landforms in the study landscape using field observations and aerial photographs. The following sections describe this evidence in each of the three study reaches (Fig. 1C): (i) Selfoss to Hafragilsfoss, (ii) the Forvoð valley, and (iii) Ásbyrgi and the Klappir scablands.

Table 1: Table of geomorphological evidence used for identifying the impact of extreme flood events in bedrock channels (adapted from Carrivick et al., 2004); modifications include the addition of boulder terraces in 'macroscale landforms' and recent studies, e.g., Lamb et al., 2008 and Lamb et al., 2014

Scale	Geomorphological feature for distinguishing bedrock-channelled jökulhlaups	Selected references
Macroforms	– Anastomosing channel pattern of valley-wide palaeo-channels cut into bedrock	Kehew and Lord (1986); Rudoy (2002); Waitt (2002); Gupta et al. (2007); Baynes et al. (2015)
	– Deep trench-shaped valleys	Kehew and Lord (1986); Lamb et al., 2008 and Lamb et al., 2014
	– Cataracts	Baker (1973); Kehew and Lord (1986); Rudoy (2002); Lamb et al., 2008 and Lamb et al., 2014;Baynes et al. (2015)
	– Flow overspilling previous drainage divides	Shakesby (1985); Kehew and Lord (1986)
	– Scoured surface	Kehew and Lord (1986); Lamb et al., 2008 and Lamb et al., 2014
	– Boulder terraces	Baker (1973); O'Conner (1993)
Mesoforms	– Streamlined residuals	Baker (1988); Komar (1984); Wiedmer et al. (2010)
	– Obstacle and iceblock marks	Baker (1973); Fay (2002)
	– Wash limits	Maizels (1995)
	– Boulder surfaces and boulder bars	Baker (1973); O'Conner (1993)
	– Dunes	Baker (1973); Maizels (1995); Carling (1996);Wiedmer et al. (2010)
	– Bars	Carling et al. (2002)
	– Kettled surfaces	Fay (2002)
	– Slackwater deposits	Baker (1973); Baker and Bunker (1985)
Microforms	– Potholes, flutes, furrows, obstacle marks, and grooves	Hancock et al. (1998); Whipple et al. (2000);Richardson and Carling (2005); Wilson and Lavé (2014)

Selfoss to Hafragilsfoss

From the apex of the Jökulsárgljúfur canyon at Selfoss to ~ 5 km farther downstream, the Jökulsá á Fjöllum becomes deeply incised into the surrounding terrain (Fig. 2). Exposed in the canyon wall ~ 4 km downstream of its head is a volcanic conduit that brought lava to the surface in a fissure eruption about 8.5 ka BP (Eliasson, 1974). This event provides an independent constraint on the maximum age for the formation of the canyon upstream of the fissure and indicates that at least 4 km of the canyon was cut in the last 8.5 ka. In this section, a clear pattern of multiple palaeo-channels has been cut into bedrock, including the Sanddalur overspill channel (200 m wide) that contains a 20-m-high cataract, a dry vertical waterfall characteristic of erosion during jökulhlaups (Carrivick et al., 2004, Lamb et al., 2008 and Lamb et al., 2014) (Fig. 3A), and a 50-m-high cataract where the channel rejoins the western wall of the main canyon. The vertical headwalls of the three waterfalls in the active channel are also characteristic of the migration of knickpoints in columnar basalt environments during large floods (Lamb and Dietrich, 2009 and Baynes et al., 2015) (Fig. 3B). Further macroscale evidence for the action of extreme flooding is the relative size of the contemporary river compared to the size of the canyon. With the exception of the 500-m reach immediately downstream of Dettifoss, the Jökulsá á Fjöllum does not fill the canyon floor, even during regular annual spate stages (peak annual discharge from 1973–1979 was ~ 500 $m^3 s^{-1}$ at Grimsstadir, 25 km upstream of Selfoss; Schunke, 1985) (Fig. 3C). This underfit suggests that the canyon was formed when the flow in the river was significantly greater. Three distinct strath terrace levels are present within the canyon, indicating the palaeo-location of the river bed (Baynes et al., 2015). All of these terraces, and the contemporary river bed, correspond to the top of lava flows. Despite small-scale fluting (on the scale of tens of centimetres) and submeter scale scouring on the strath terraces, evidence is limited for widespread vertical incision of the channel into the lava flows through abrasion. This fact demonstrates that the dominant mechanism of canyon erosion is the upstream propagation of knickpoints through the toppling and subsequent transportation of bedrock columns, once the flow depth has surpassed a threshold value (Baynes et al., 2015).

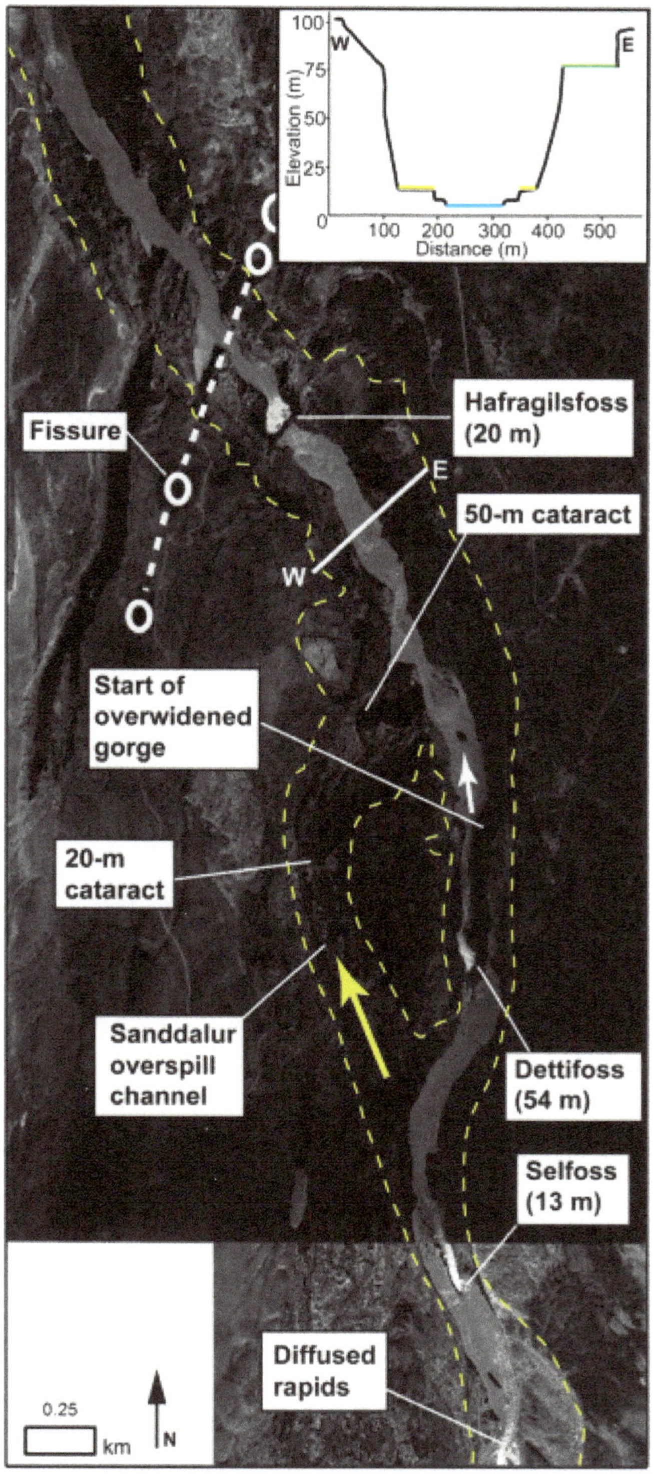

Figure 2: Aerial photograph showing the upper 5 km of the Jökulsárgljúfur canyon where the Jökulsá á Fjöllum is deeply incised with three vertical waterfalls; Selfoss (13 m in height), Dettifoss (54 m in height), and Hafragilsfoss (20 m in height) (adapted from Baynes et al., 2015). The dashed yellow lines indicate the areas where evidence for erosion during extreme flood events is clear. The 200-m-wide Sanddalur overspill channel contains a 20-m vertical cataract and a 50-m vertical cataract where it rejoins the main canyon. The fissure that erupted 8.5 ka ago (Eliasson, 1974) is highlighted with white circles, and a cross section of the gorge across the line from W to E is inset. With the exception of the 500 m of canyon immediately downstream of Dettifoss, the Jökulsá á Fjöllum does not fill the canyon, suggesting that the flow was much greater when the canyon was formed. Aerial photograph source: Landmælingar Íslands.

Figure 3: Evidence for the impact of extreme flood events in the upper reach of the Jökulsárgljúfur canyon. (A) Looking upstream (south): the Sanddalur overspill channel contains clear fluvially sculpted surfaces and a 20-m dry cataract. (B) Looking upstream toward Dettifoss (54 m in height): the Jökulsárgljúfur canyon contains three large vertical waterfalls formed by the upstream retreat of knickpoints through the toppling of basalt columns. (C) Looking upstream toward Hafragilsfoss (20 m in height) from the location of the fissure that erupted 8.5 ka ago (Eliasson, 1974): the canyon cuts through the fissure, providing an independent constraint on the age of the canyon as all of the erosion has occurred since the fissure erupted. Strath terraces, indicating the palaeo-location of the river bed, can be seen on the edges of the canyon. Long stretches of these terraces have been exposed at the same time, including the Sanddalur overspill channel (Baynes et al., 2015).

Forvoð valley to Vesturdalur

Nine kilometres downstream of the apex of the Jökulsárgljúfur canyon is the Forvoð valley, which contains landforms that testify to the action of extreme flood events in erosional and depositional contexts (Fig. 4A). Downstream of the Rettarfoss waterfall, the river is incised in a relatively narrow valley (20 m wide); 48 m above the current river channel on the eastern side of the valley is an extensive, heavily scoured bedrock surface with relief of up to a few metres (Fig. 4B); this surface was likely formed and then abandoned during an extreme flood (Waitt, 2002), and the high amplitude relief may be the result of efficient plucking promoted by the small size of the basaltic columns and intense fracturing, making blocks with size rarely exceeding 30 cm available for transport. Downstream of the slot canyon, the valley widens and landforms associated with deposition rather than erosion are evident. Boulder surfaces and boulder bars are defined by Carrivick et al. (2004) as 'mesoform' evidence for the action of extreme floods (Table 1), although they should also be considered on the macroscale here as they extend on both sides of the Forvoð valley for > 3 km and some of the deposits are up to 50 m thick (Fig. 5). On the western side of the river, boulder-rich deposits up to 47 m thick (some blocks > 1 m in diameter) are found above the current river level; and two clear boulder-rich terraces are located on the eastern side of the valley, 5 m and 21 m above the river bed, respectively (Fig. 5). Such extensive, thick, and coarse deposits are likely associated with extreme palaeo-flow conditions (Wohl, 1992). We suggest that the boulder-rich sediment deposited in the Forvoð valley is a result of floodwaters losing energy as they pond behind the forced narrowing caused by the bedrock constriction at Kjaffbjarg (Fig. 4A). Subsequent stages of the flood, or subsequent floods, have reworked the boulder-rich deposits in the valley, incising through them but preserving the terrace surfaces high on the valley sides.

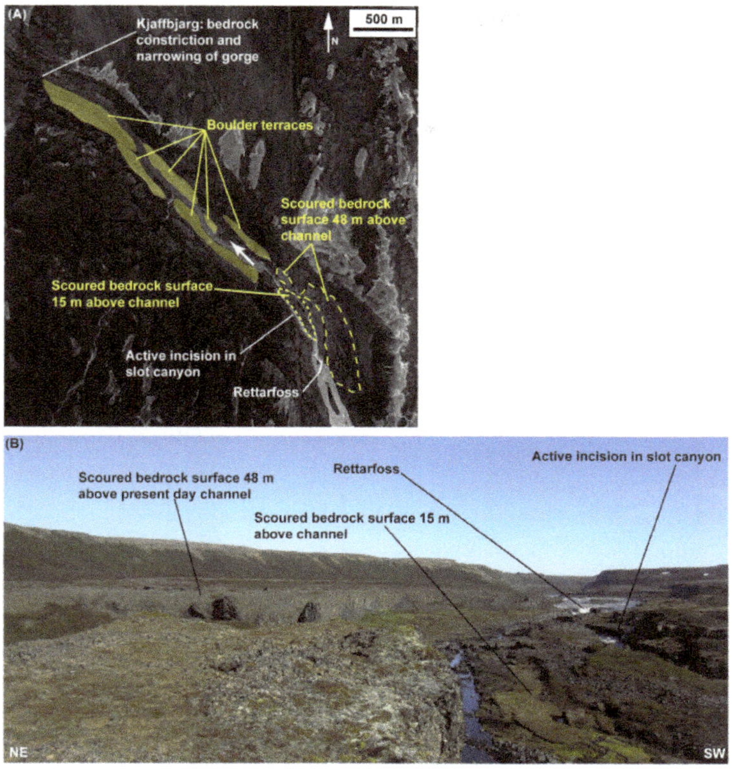

Figure 4: (A) Aerial photograph of the Forvoð valley. Rettarfoss, the waterfall in the upstream part of a narrow, actively incising slot canyon, is highlighted. The dashed yellow lines indicate the heavily scoured bedrock surface 48 m above the current river channel and a lower bedrock terrace abandoned in the 1950s (Vatnajökulsþjóðgarður National Park Tourist Information). The yellow-shaded areas identify the large boulder terraces, thought to have been deposited during an extreme flood event, possibly owing to a backwater effect as a result of water ponding up behind the narrow bedrock constriction at Kjaffbjarg (also highlighted). Aerial photograph source: Landmælingar Íslands. (B) Looking upstream toward Rettarfoss waterfall, showing the upper part of the Forvoð valley and the strath terraces identified in (A). The surface is heavily scoured owing to the rubbly nature of the bedrock. The columns are thin, not well developed, and fractured such that joint spacing between blocks rarely exceed 30 cm, making blocks easily plucked and transported.

Figure 5: Within the Forvoð valley, large boulder-rich terraces are exposed on both sides of the valley. On the eastern side of the valley, two terraces made up of sediment containing abundant large boulders are located 5 and 21 m above the current river

level as seen in (A). The white arrow indicates the direction of flow of the river. On the western side of the valley, one large fill terrace is present that is 47 m thick and contains boulders with diameter > 1 m, shown in picture (B) (taken looking upstream); the scoured bedrock terrace from Fig. 4is shown in the background of the photograph. For scale, the fence posts on the top of the deposits are ~ 0.5 m high.

Downstream of the Kjaffbjarg bedrock constriction, the Jökulsá á Fjöllum flows within a deeply incised scabland area at Vesturdalur (Fig. 1C) before flowing along the eastern edge of another post-glacial volcanic fissure at Hljóðaklettar (Eliasson, 1974 and Waitt, 2002). Vesturdalur is a key location of previous studies that have identified extreme flood events along the Jökulsá a Fjöllum. Waitt (2002) and Kirkbride et al. (2006) identified sedimentary sequences containing sandy flood deposits from this location (Fig. 1C).Waitt (2002) identified up to 16 sandy flood layers high above the west side of the Jökulsá á Fjöllum thought to have been laid down between 8000 and 4000 years ago, constrained by the presence of H4 and H3 tephra layers in the sequence that were deposited following eruptions of Hekla volcano ~ 3800 YBP and ~ 2900 YBP, respectively (Kirkbride et al., 2006). Two flood layers in a sequence on the eastern side of the valley, corresponding to the layers at the top of the sequence identified by Waitt (2002), were dated by Kirkbride et al. (2006) to 5020 and 4610 cal. YBP. This sedimentary evidence suggests that multiple large flood events affected this part of the canyon during the mid-Holocene.

Ásbyrgi and Klappir Scablands

Perhaps the most striking evidence for erosion during extreme floods along the Jökulsá á Fjöllum can be found at Ásbyrgi canyon and the Klappir scablands. Ásbyrgi is a horseshoe-shaped canyon (3 km long, 1 km wide, up to 90 m deep), which is disconnected from the current river that now flows in a deeply incised canyon at Landsbjörg to the east (Figs. 1C, 6). Between Ásbyrgi and the main Jökulsárgljúfur canyon is Lake Ástjörn, a small cataract now filled with water, that exhibits the same amphitheatre shape as Ásbyrgi albeit on a smaller scale (250 m wide). Upstream of Ástjörn is a narrow scabland tract leading from the main Jökulsárgljúfur canyon but hanging ~ 60 m above the modern river. Ásbyrgi has been cut into a succession of lava flows with a northward dipping surface (slope: 0.025). The Klappir scablands are a flood-scoured area of bedrock ridges and pools that clearly mark out the route of the floodwaters into the head of Ásbyrgi (Figs. 6/7B). At the 'upstream' (southern) end of the area are four smaller (100 m wide, 10 m high) amphitheatre-shaped cataracts that also open toward the north (indicating flow direction from the south), which are located high (~ 90 m) above the current course of the river (Fig. 7). These cataracts include plunge pools at the base of the headwall featuring sediment ridges that could be interpreted as push-bars (Carling et al., 2009c); these bars show no obvious reworking since their formation (Fig. 7A). At the rim of Ásbyrgi, large-scale potholes (up to 10 m in depth) and flutes are clearly visible (Fig. 7C), and several notches (3–5 m in height) have been cut into the rim of the vertical headwall of the canyon (Fig. 8A). The exposure age of bedrock in one of these notches has been put at between 7.2 and 12.5 ka, indicating that Ásbyrgi and the Klappir scablands were formed during an extreme flood event in the early Holocene, shortly after deglaciation (Baynes et al., 2015).

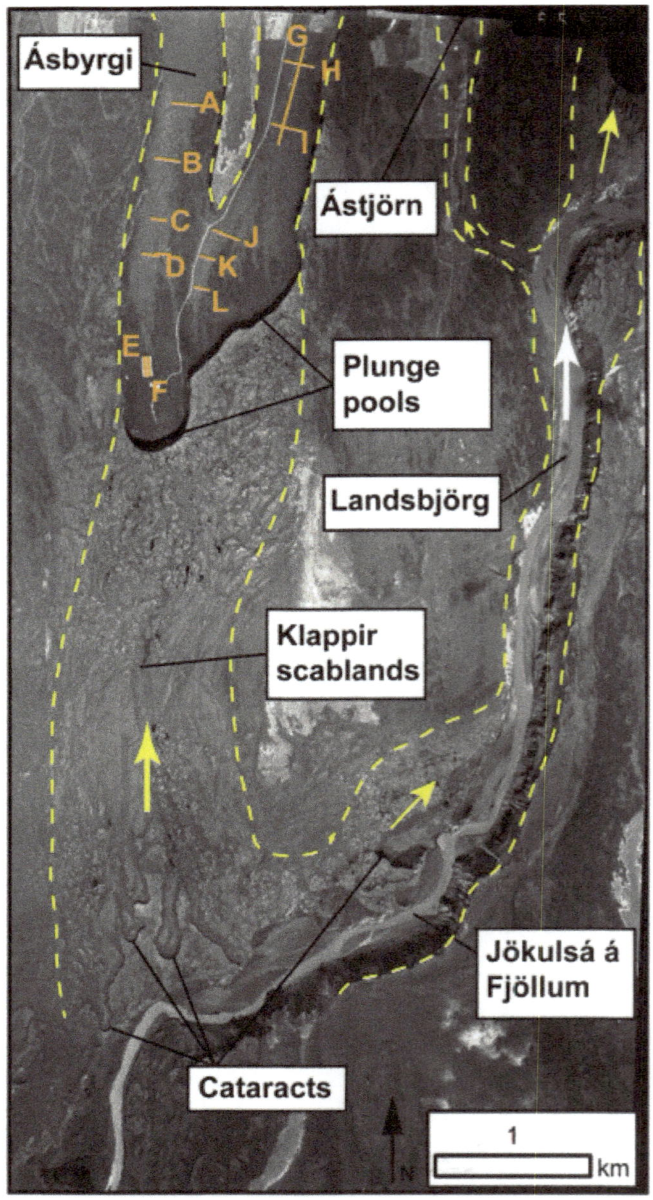

Figure 6: Aerial photograph of the Klappir scablands and Ásbyrgi canyon. The present day course of the Jökulsá á Fjöllum is within the deeply incised main Jökulsárgljúfur canyon that flows to the east at Landsbjörg. Ásbyrgi is a large (3-km-long, 1-km-wide) horseshoe-shaped canyon cut into a northward-dipping lava succession with an island preserved between the two main channels. The yellow dashed lines indicate the areas that have been inundated by floodwaters, with the Klappir area of scabland topography with bedrock ridges and pools clearly visible. At the upstream limit of the scablands, four dry cataracts provide additional evidence for erosion during an extreme flood event in this area. The locations of the plunge pools at the apex of the western and eastern Ásbyrgi canyons are shown. The orange lines within Ásbyrgi indicate the location of the electrical resistivity tomography (ERT) surveys in the canyon floor, with letters corresponding to the profiles in Fig. 12.

Aerial photograph source: Landmælingar Íslands (adapted from Baynes et al., 2015).

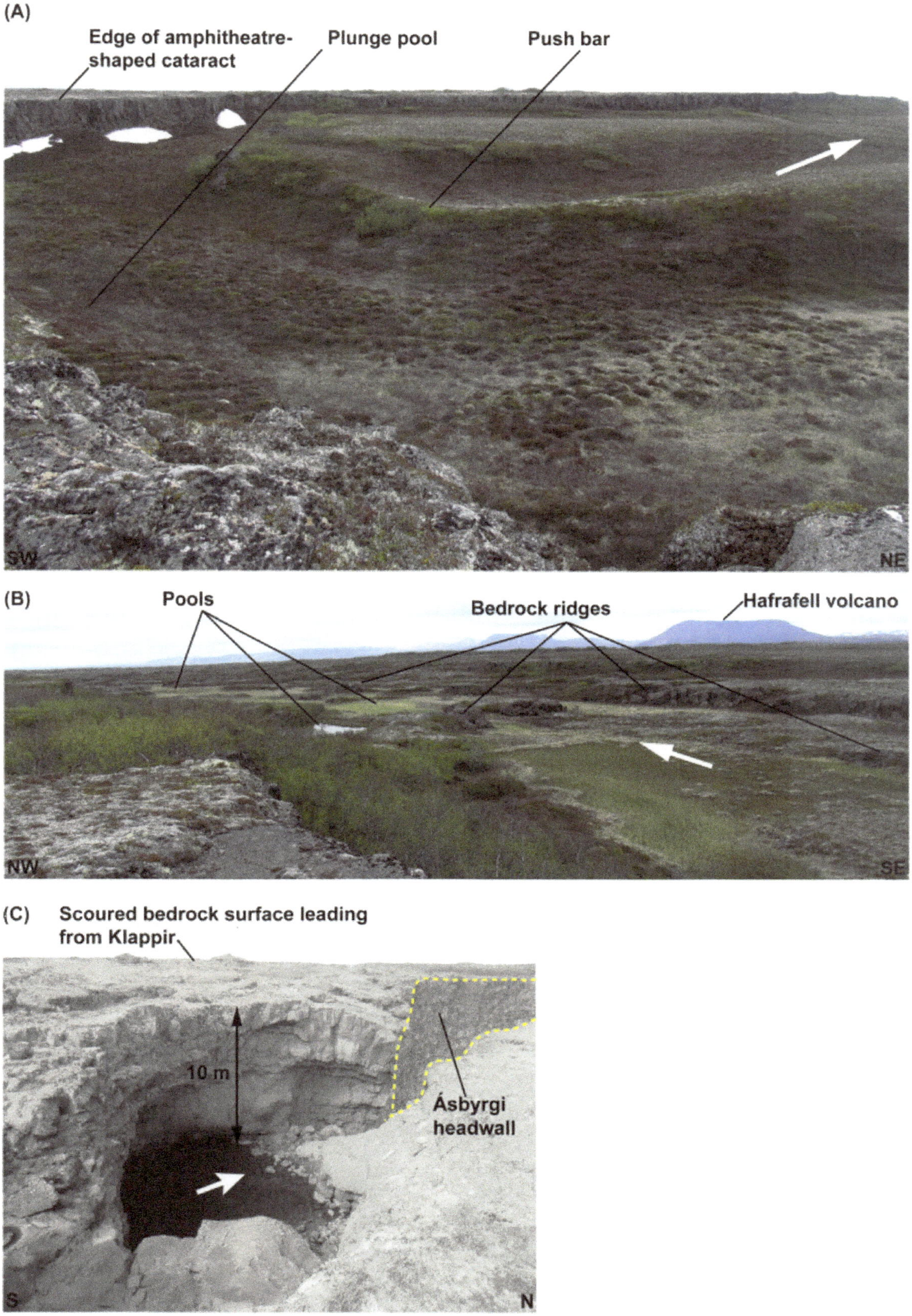

Figure 7: Landforms in the Klappir scabland area suggest the action of powerful flows. In the south are three dry cataracts ~ 100 m wide with a 5–10 m vertical cliff at their head (A). Within the cataracts are arcuate sediment ridges that could be interpreted as push-bars preserved in pristine condition, showing no evidence for subsequent fluvial reworking. Downstream of the cataracts is an area with distinctive scabland morphology (B). Characterised by a series of bedrock ridges and pools (local relief ~ 5 m), this area clearly marks the route of the flood waters, as can also be seen in the aerial photograph (Fig. 6). White arrows indicate the direction of the floodwaters. Hafrafell, a nearby table-top volcano that erupted subglacially, is shown. The eruption age of Hafrafell has been dated to 11,100 ± 2200 years, thought to coincide with deglaciation in this area (Licciardi et al., 2007). (C) Megascale fluvial bedrock erosion landforms at the scoured rim of Ásbyrgi. Looking west into a 10-m-deep pothole within one of the eroded notches identified in Fig. 8, with flow direction to the north (white arrow). In the background of the photograph is the 90-m-high vertical headwall of Ásbyrgi (edge of rim highlighted in yellow).

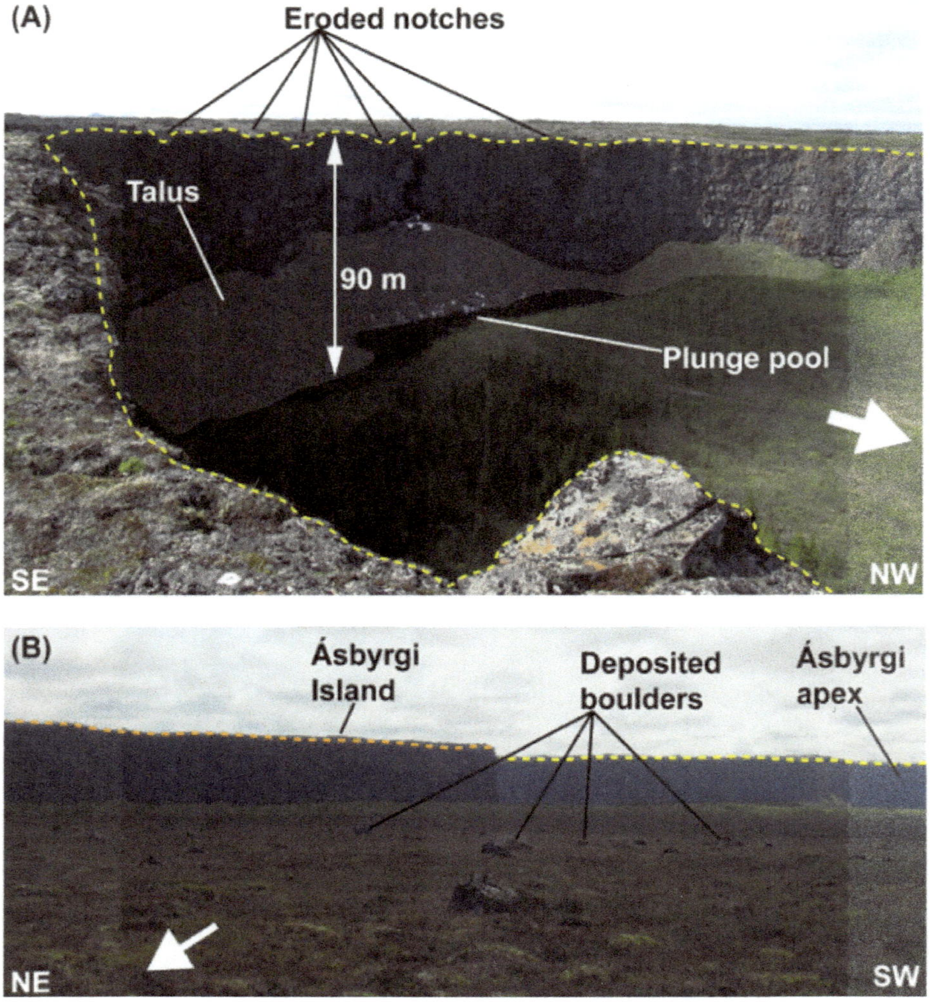

Figure 8: Ásbyrgi canyon is a horseshoe-shaped cataract cut into a northward-dipping lava succession. The vertical headwall at the apex of Ásbyrgi is 90 m high with a large plunge pool at the base and a pile of sediment deposited immediately downstream (A). Several eroded notches are present along the rim of the canyon (canyon edge highlighted with yellow dashed line). At the base of the headwall, talus deposits resulting from rockfalls show no evidence for reworking by overland flow. (B) Looking into the western canyon of Ásbyrgi from near the canyon outlet. The rim of the main canyon is highlighted in yellow and the rim of the Ásbyrgi Island is shown in orange. The floor of Ásbyrgi is covered in sediment including large boulders (some > 3 m in diameter). No evidence for recent flow within the canyon exists so it is hypothesised that these boulders were deposited following transportation during an extreme flow. The source of the boulders is impossible to determine, but we suggest that they were initially part of the lava succession into which Ásbyrgi has been cut rather than transported from farther upstream. From this we propose that the floodwaters that formed Ásbyrgi were powerful enough to transport boulders of 3 m in diameter. White blocky arrows indicate palaeo-flow direction (to the north).

The horseshoe of Ásbyrgi is made up of two parallel channels that have eroded back and coalesced (Fig. 6). Between the two parallel channels is 'Eyjan', or 'Island', a bedrock outcrop rising to the same elevation as the lava surface around the main rim of Ásbyrgi. The western canyon retreated farther south, and its headwall marks the location of the highest cliffs (90 m) in Ásbyrgi. At the base of the headwalls of the western and eastern canyons are large relict plunge pools (Fig. 8B). The floor of Ásbyrgi is covered in sediment, with many large boulders (some > 3 m in diameter) found on the surface of the deposits (Fig. 8B). The maximum measured boulder size can be used to calculate the minimum discharge of the palaeo-flood that transported them (e.g., Costa, 1983, Clarke, 1996 and Stokes et al., 2012). Caution should be employed when using such a method as different equations can give different estimates of flood discharge, and there are issues with the collection of the boulder size data and the interpretation of the resulting estimates (see discussion in Stokes et al., 2012). Within these caveats, we used the method described by Stokes et al. (2012) to calculate a rough estimate of the minimum flood discharge that would be required to transport the largest boulders in Ásbyrgi. The largest measured boulder in the eastern Ásbyrgi channel (diameter = 1.49 m) gives a minimum palaeo-discharge estimate of 12,000 $m^3 s^{-1}$. In the western channel, where the diameter of the largest measured boulder is 3.75 m, the minimum discharge estimate is 39,000 $m^3 s^{-1}$ (see Supplementary information for sensitivity analysis and full list of parametres used).

Small-scale fluvially sculpted bedforms on the top surface of the Island between the two eroded channels that make up the Ásbyrgi 'horseshoe' provide evidence that, pre-flood, the river flowed over the lava surface into which Ásbyrgi has been eroded (Fig. 9). Surveys from across the Island indicate a palaeo-flow direction that is consistently from the south (Fig. 9). These surfaces were formed before Ásbyrgi was eroded as the canyon walls cut straight across some of the landforms (Fig. 9B), and we propose that they were not formed during the flood as they are substantially smaller in scale (relief in the order of a few tens of centimetres) than the flutes, furrows, and potholes found at the rim of Ásbyrgi (relief in the order of a few metres, up to 10 m; Fig. 7C; Richardson and Carling, 2005). Similar-scale fluvial surfaces to those found on the Island are found on the eastern rim of Ásbyrgi and on the western rim of the modern canyon to the east (Fig. 6). During the last glacial period, the Icelandic ice sheet extended beyond the north coast of Iceland, covering the area containing Ásbyrgi and the Jökulsárgljúfur canyon (Norðdahl, 1990,Hubbard et al., 2006 and Licciardi et al., 2007). During the retreat of the ice sheet across the central highlands, the discharge of the proto-Jökulsá á Fjöllum was likely greater owing to enhanced glacial ablation during deglaciation. Upstream of the Jökulsárgljúfur canyon, the Jökulsá á Fjöllum is at present a large braided river system (sometimes > 1 km in width) flowing on a bedrock substratum; it is possible that the river developed such morphology all the way to the coast before the canyons were eroded. The fluvial surfaces on Ásbyrgi Island, on the eastern rim of Ásbyrgi and the western rim of the main canyon, indicate the palaeo-course of this system. Fluvial sediment is lacking on these surfaces, possibly because the sediment would have been entrained and transported during the early stages of the jökulhlaup before the surfaces were abandoned by the upstream propagation of the canyon headwalls.

Figure 9: Evidence for palaeo-flow on Ásbyrgi Island. (A) The morphology of the flutes and furrows allows the identification of the flow direction of the palaeo-river that formed them. Flutes and furrows (solid black lines) are parallel to the flow direction; dashed lines represent

the crests of upstream-facing convex surfaces that are perpendicular to the palaeo-flow (Wilson and Lavé, 2014). Flow-direction is to the north. Local relief of bedrock surface is ~ 30 cm. (B) The landforms atop the Island between the two Ásbyrgi canyons were formed before the canyon was carved because the flutes and furrows lead right to the edge of the vertical walls (white dashed line). Photograph looking southeast on eastern edge of the Island. For scale, dimensions of GPS unit are 17 × 9 × 4 cm. C. Aerial photograph (source: Landmælingar Íslands) of the apex of Ásbyrgi Island (outline shown by dashed white line) showing the palaeo-flow direction of the fluvially sculpted bedrock features from 27 survey locations (total measurements = 182). All of the sites indicate a palaeo-flow direction broadly to the north.

VOLUME OF ROCK ERODED FROM ÁSBYRGI AND SEDIMENT DEPTH

As demonstrated in Section 3.3, the evidence for extreme flood events at Ásbyrgi and the Klappir scabland area immediately upstream is clear. The landscape is disconnected from the course of the present day Jökulsá á Fjöllum, which now flows in a deeply incised canyon to the east at Landsbjörg (Fig. 6). The Klappir scablands and Ásbyrgi contain landforms preserved in pristine condition, unburied and with no evidence for fluvial modification through erosion since their formation, suggesting abandonment following the event that carved Ásbyrgi. Klappir and Ásbyrgi therefore provide an unusually good opportunity to examine the impact of a single extreme flood event in eroding bedrock and then in depositing sediment. Combined topographic analysis and near-surface geophysics surveys were used to evaluate the volume of bedrock eroded from Ásbyrgi and the thickness of sediment deposited during the waning stages of the flood.

The volume of rock eroded from Ásbyrgi was quantified using high-resolution topographic data based on a total station survey and a 1.8-m resolution digital elevation model (DEM; source: TanDEM-X collected on 2 September 2012). A 'pre-flood' surface was constructed by interpolating the elevation values from around the outer rim and the Island surface across the top of the canyon. An initial estimate of the rock eroded from Ásbyrgi during formation was made through the subtraction of the 'pre-flood' surface from the 'present day' DEM (Fig. 10A), giving a total of 0.139 km^3. However, this is an underestimate of the true amount of rock eroded from Ásbyrgi as the floor of the canyon is completely covered with sediment. An assessment of the sedimentary thickness was carried out using electrical resistivity tomography (ERT) surveys across the canyon floor. The ERT surveys are nondestructive and provide greater spatial coverage than point measurements when multiple profiles are collected. The ERT is an established method for imaging the near subsurface and has been used for a wide range of applications, including detecting the bedrock-sediment interface (Hsu et al., 2010 and Chambers et al., 2012), aquifer characterisation (Doetsch et al., 2012), detection of subsurface cavities (Martinez-Lopez et al., 2013), rockwall retreat rates (Siewert et al., 2012), and permafrost depth and structure (You et al., 2013). The ERT surveys were carried out across transects A–L shown in Fig. 6, with 25 electrodes at 5-m spacing, allowing electrode spacings ranging from 5 to 40 m. By increasing the spacing between the electrodes, the current penetrates deeper, building up a data section that can be interpreted in terms of lateral and depth variations in electrical resistivity. Some of the transects were built up from multiple surveys in order to cover a longer distance than the 120 m possible in a single 25-electrode survey, such as the long transect along the middle of the eastern canyon (transect G, 680 m long; Fig. 6).

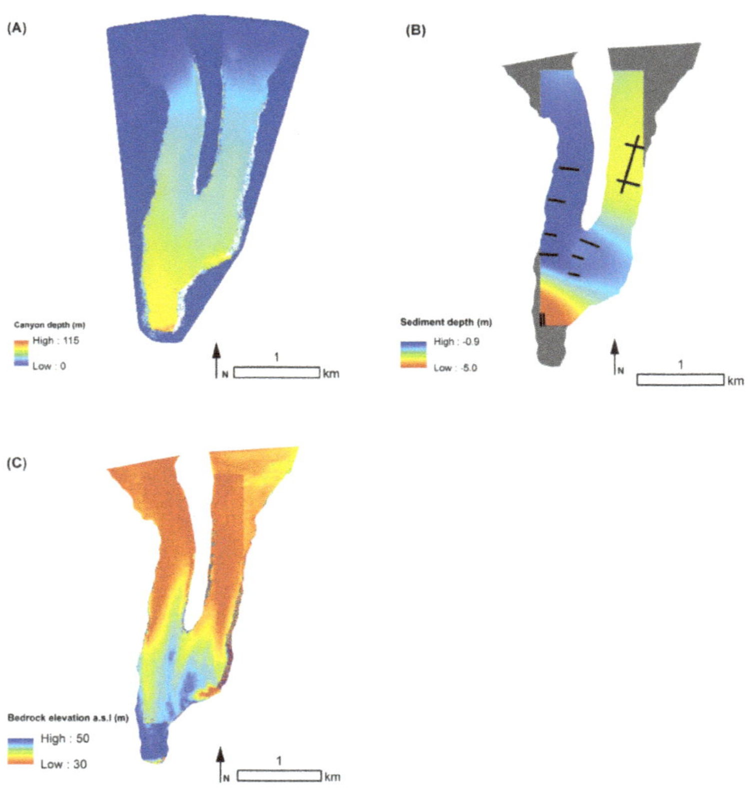

Figure 10: (A) Depth of Ásbyrgi canyon calculated by subtracting the DEM of the 'pre-flood' top surface interpolated from elevation values around the outer rim and the Island and the DEM of the present day canyon. The total volume of rock eroded between the two DEMs, with-

out accounting for the deposited sediment in the floor of Ásbyrgi, is 0.139 km^3. (B) Interpolated sediment depth to the floor of Ásbyrgi from the ERT data using the 'spline with barriers' function in ArcGIS. This surface is used to calculate a minimum estimate of the sediment deposited within Ásbyrgi through a comparison with the DEM of the canyon floor (0.005 km^3). The areas at the exit of the canyon are not considered owing to the spatial coverage of the ERT surveys (black lines) and the processing extent of the interpolation algorithm. C. Elevation of bedrock above sea level. Beyond the processing extent of the interpolation algorithm, the bedrock elevation is represented by the DEM of the canyon floor.

Different inversion methods are available in the 'res2Dinv' software (res2Dinv version 3.4; Geotomo, 2001). The conventional least squares method minimises the square of the difference between the measured and the calculated apparent resistivity values and produces a model with smooth resistivity variations (Loke et al., 2003). However, the technique is not perfectly appropriate when the subsurface contains sharp boundaries between resistivity interfaces as the smoothing of the boundaries between layers makes their localisation difficult. We therefore employed a 'robust iterative inversion' to model our survey data, whereby the absolute changes in the resistivity values are minimised (Claerbout and Muir, 1973). This approach produces models of the subsurface with sharp interfaces between different subsurface structures that have different resistivity values (Loke et al., 2003) and was deemed most appropriate because we expect to see a sharp boundary between the sediment deposits and the basalt bedrock beneath; all images presented here have been produced using this method (Fig. 11). The model iterations were stopped when the percentage misfit between the measured and the calculated apparent resistivity was < 5% or no further improvement to the fit was possible with further iterations. In the case of transect E, no further improvement to the fit occurred after five iterations, when RMS error was 5.9%.

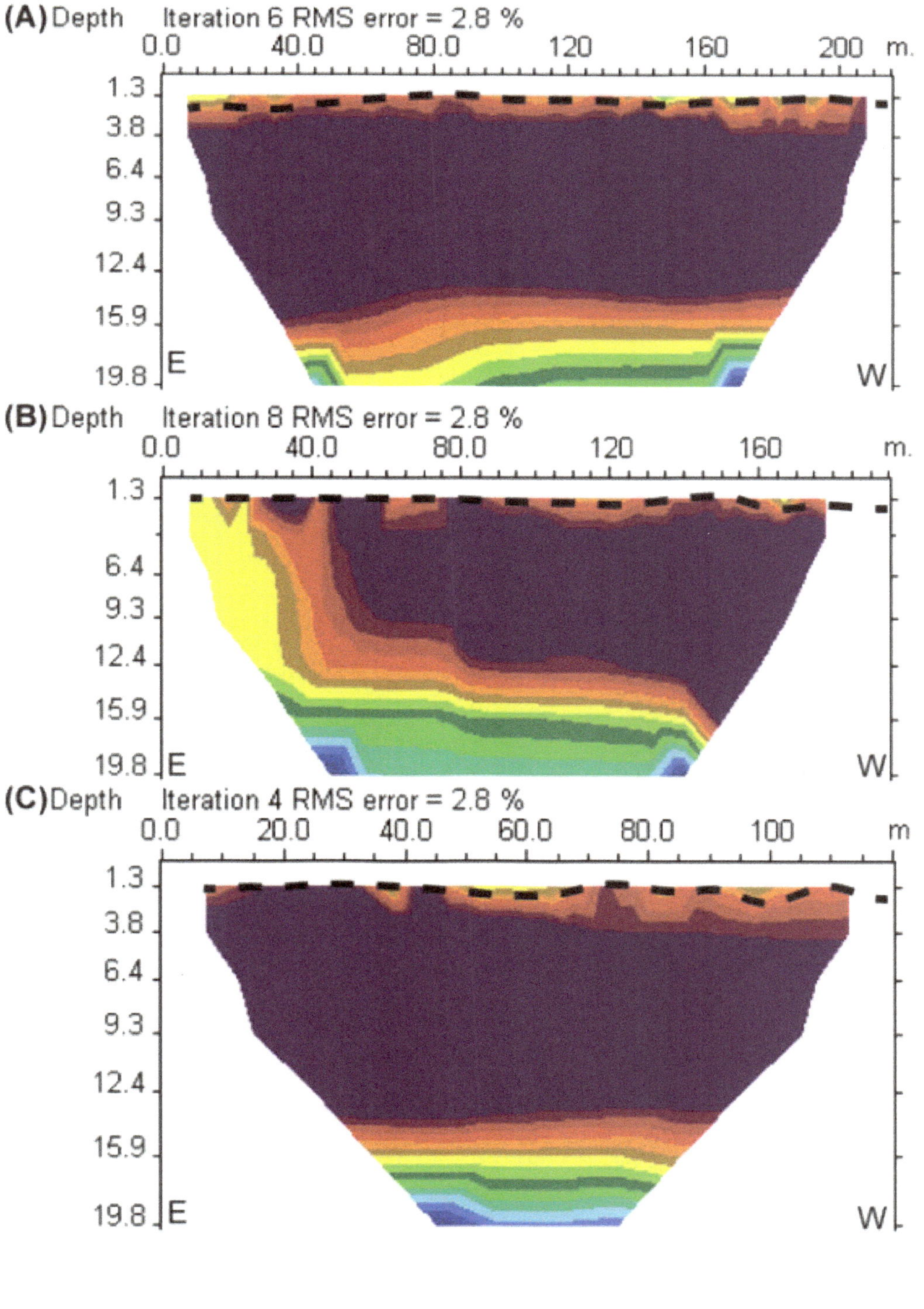

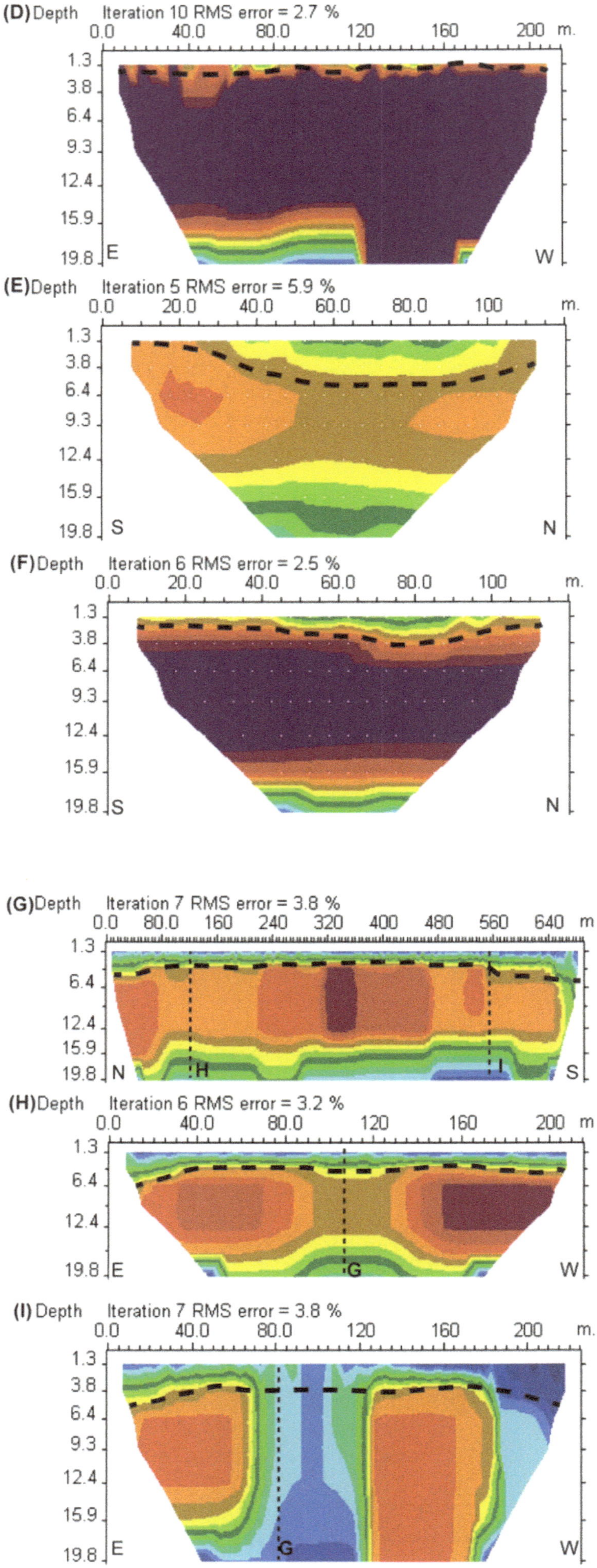

(D) Depth
Iteration 10 RMS error = 2.7 %
0.0 40.0 80.0 120 160 200 m.
1.3
3.8
6.4
9.3
12.4
15.9
19.8
E
W
(E) Depth
Iteration 5 RMS error = 5.9 %
0.0 20.0 40.0 60.0 80.0 100 m.
1.3
3.8
6.4
9.3
12.4
15.9
19.8
S
N
(F) Depth
Iteration 6 RMS error = 2.5 %
0.0 20.0 40.0 60.0 80.0 100 m.
1.3
3.8
6.4
9.3
12.4
15.9
19.8
S
N
(G) Depth
Iteration 7 RMS error = 3.8 %
0.0 80.0 160 240 320 400 480 560 640 m
1.3
6.4
12.4
15.9
19.8
N
H
I
S
(H) Depth
Iteration 6 RMS error = 3.2 %
0.0 40.0 80.0 120 160 200 m
1.3
6.4
12.4
19.8
E
G
W
(I) Depth
Iteration 7 RMS error = 3.8 %
0.0 40.0 80.0 120 160 200 m.
1.3
3.8
6.4
9.3
12.4
15.9
19.8
E
G
W

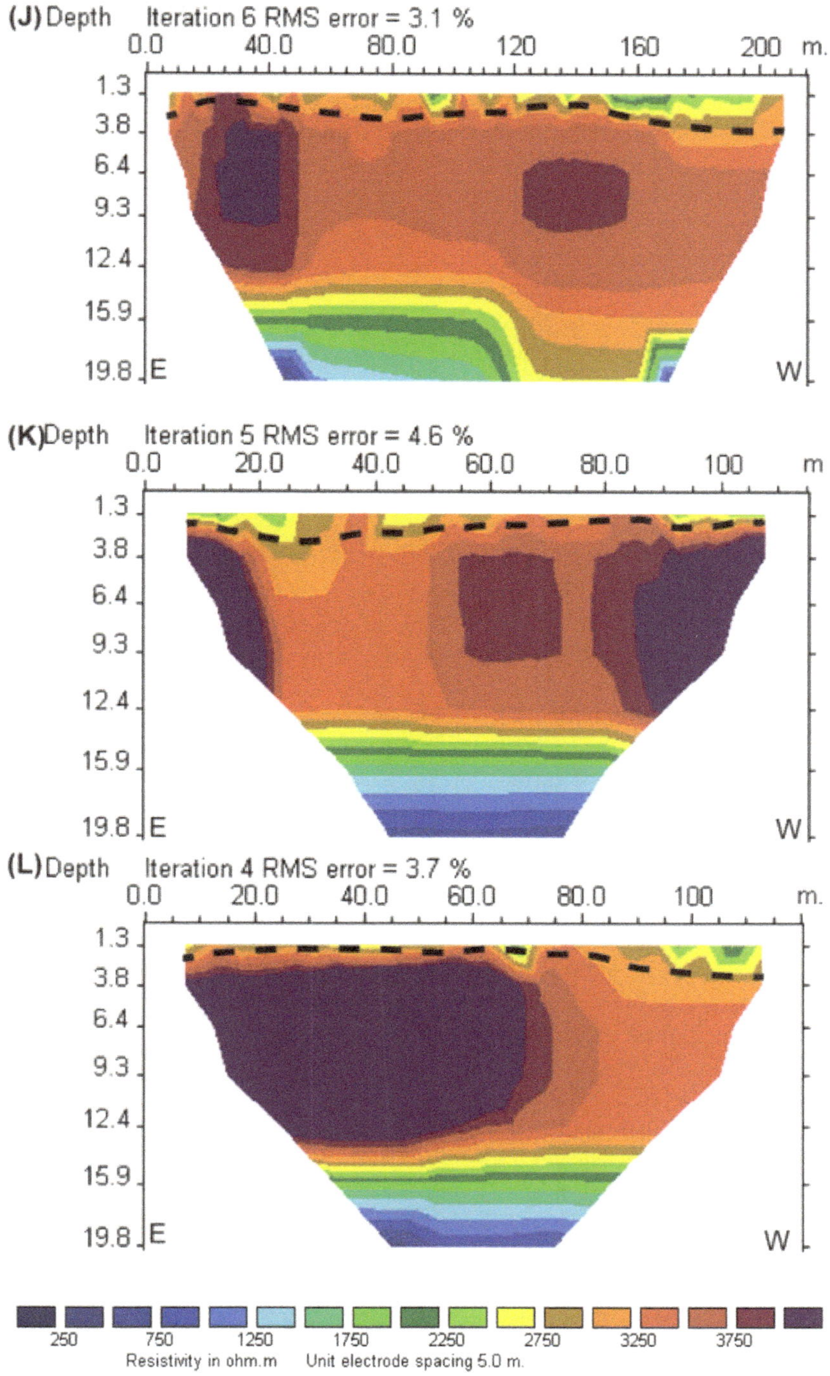

Figure 11: Electrical resistivity tomography imaging of the subsurface. Labels (A–L) refer to the location of each transect shown in Fig. 6. Surveys A–D are from the western canyon; they show a depth to the bedrock-sediment interface of ~ 1 m. Surveys E and F are parallel to each other from the field close to the apex of Ásbyrgi; they show a sediment thickness of ~ 5 m. Survey G is a longitudinal survey along the middle of the eastern canyon with surveys H and I also from the eastern canyon, each showing a uniform sediment thickness of ~ 3 m. Surveys J–L are from the region between the two main canyons and have a sediment depth of ~ 1.5 m. The letters at the edges of each profile (bottom) indicate the orientation of the transects. Vertical dashed lines and corresponding labels on (G), (H), and (I) indicate the location where the transects cross each other.

Broadly, sedimentary deposits have the lowest resistivity and igneous rocks the highest (Telford et al., 1990). We therefore interpret the bedrock-sediment interface in each of our profiles as the sharp horizontal downward transition from regions of low to high resistivities (Fig. 11). The typical range of resistivity for basalt is large: 10^1–1.3×10^7 Ωm (Telford et al., 1990) because of a number of factors, including the water content in fractures and pore space. The resistivity of dry (0% water content) basalt is 1.3×10^7 Ωm, whereas basalt with 0.95% water content typically has a much lower resistivity of 4×10^4 Ωm (Telford et al., 1990). The peak resistivity in each of our surveys is up to 3.7×10^3 Ωm, which implies that the basalt in our study area has a water content > 1%. This is to be expected as the rocks are located in a coastal region with a wet climate. We are confident that the transition to high resistivity found a few metres below the surface is the top of bedrock (Fig. 11). The layer of lower resistivity at the base of each of the surveys is interpreted to represent the water table owing to its broadly consistent depth at ~ 15 m across all surveys.

The ERT surveys show that the sediment is ~ 1 m thick across the floor of the western gorge (Figs. 11A–D) and 3 m thick in the eastern gorge (Figs. 11G–I). Owing to forest cover, only two surveys were carried out in a field near the apex of the western gorge, but these show that the sediment in this region of the canyon is ~ 5 m thick (Figs. 11E–F). We hypothesise that this is because of the survey location on top of the pile of sediment immediately downstream of the plunge pool. These surveys were parallel to each other and have produced a similar subsurface morphology despite a slight difference in the peak resistivity values, indicating reproducibility of the results. The three surveys undertaken in the region between the two main channels indicate a sediment depth of ~ 1.5 m in this region (Fig. 11J–L).

Sediment depths were interpolated across the canyon floor using the 'Spline with Barriers' function in ArcGIS (Fig. 10B). Owing to the limits on the spatial coverage of the ERT surveys, the interpolated surface does not cover the canyon floor in the apex of the western channel of Ásbyrgi or in some of the areas at the exit of the western and eastern canyons. The volume of sediment within Ásbyrgi was estimated by subtracting the interpolated surface from the DEM of the canyon floor topography, giving a volume of 0.005 km^3, making up < 4% of the total volume of rock eroded from Ásbyrgi at 0.144 km^3. This is a minimum estimate as the interpolated surface does not cover the entire floor of Ásbyrgi, although the additional sediment located beyond the interpolated surface is unlikely to cause a significant increase in the estimate of total rock eroded. Fig. 10C shows the bedrock surface elevation above sea level, created by subtracting the interpolated sediment depth (Fig. 10B) from the DEM of the canyon floor. The area close to the apex of Ásbyrgi is affected by the presence of trees that are picked up by the DEM (the highest elevation areas in blue) but farther north, near to the outlet of the two canyons, the elevation of the bedrock surface above sea level is very similar. This observation suggests that when the two canyons were retreating, before they coalesced, the vertical knickpoints at the headwall of the canyon were the same height.

DISCUSSION

Some of the features described in Section 3, such as boulder erratics, are not exclusive to the action of extreme flood events and individually should not be used as evidence for the action of extreme flood events (Carling et al., 2009c and Carrivick et al., 2013). However, as multiple different landforms across all scales of the Carrivick et al. (2004)criteria are found in three distinct and very different reaches of the Jökulsárgljúfur canyon, we suggest that the evidence for extreme flood events is unequivocal in this landscape. Combining this with the identification of three significant periods of canyon cutting by Baynes et al. (2015) at ~ 10,000, ~ 5000 and ~ 2000 years ago, the following sections reconstruct the landscape evolution of the lower Jökulsá á Fjöllum during the Holocene.

Model of Formation of Ásbyrgi and Klappir during a flood ~ 10,000 years Ago

The presence of fluvially sculpted surfaces on the top of Ásbyrgi Island as well as strath terraces above the Jökulsárgljúfur canyon to the east suggests that during the retreat of the last Icelandic ice sheet, a major river flowed from the south over the northward-dipping lava surface into which the canyons have been eroded. This proto-Jökulsá á Fjöllum may have been substantially wider than the modern river channel as the discharge may have been higher because of increased meltwater generated during a major period of deglaciation. It possibly generated a large braided river system with multiple active channels on the lava substrate similar to the present day Jökulsá á Fjöllum upstream of Selfoss. This palaeo-river system could have simultaneously occupied, and fluvially sculpted, the surface at the top of Ásbyrgi Island and the surface close to the present day main canyon (Fig. 12A). Alternatively, the palaeo-river system could have been similar in size to the present day Jökulsá á Fjöllum and could have migrated the 2.5 km across the lava surface, sculpting the two bedrock surfaces at different times (Fig. 12A).

During the initial phases of the early Holocene jökulhlaup, the floodwaters spread across the Klappir area and the area to the east, over what would become the course of the modern day river. The eastern floodwaters split, and two canyons (the origins of the main Jökulsárgljúfur canyon at Landsbjörg and lake Ástjörn) began to be incised through the plucking and toppling of large basalt blocks and columns at the lava flow front (Fig. 12B). The floodwaters in the Klappir area also began incising at the lava flow front with two canyons forming close to each other (the beginnings of the modern Ásbyrgi canyon) (Fig. 12B). Upstream of these four main canyons, the Klappir area began to be sculpted into the ridge and pool scabland morphology seen today, with the smaller cataracts starting to be formed under a similar process to the main canyons to the north.

Fig. 12C shows the proposed locations of the canyons midway through the jökulhlaup, with the floodwaters flowing into the Ástjörn canyon captured owing to the upstream migration of the head of the main Jökulsárgljúfur canyon farther east. We propose that the jökulhlaup had no further impact on the scabland tract leading to Lake Ástjörn, which is now exposed ~ 60 m above the modern channel (Fig. 12C). The two canyons of Ásbyrgi were also still retreating at the mid flood stage (Fig. 12C); and at some point during the latter stages of the flood, the two canyons coalesced to form the horseshoe-shaped canyon seen today (Fig. 12D). Based on the maximum size of boulders deposited across the canyon floor, calculations suggest that the discharge of the jökulhlaup that eroded Ásbyrgi was at least 39,000 $m^3 s^{-1}$, although it may have been greater than this magnitude. The perfect preservation of landforms in the Klappir scablands and the maintained vertical headwall of Ásbyrgi suggest that the floodwaters were diverted from this area at the end of the flood, and we propose that this occurred through the capture of the waters because of the retreat of the headwall of the main Jökulsárgljúfur canyon to the east (Fig. 12D). The ERT profiles reveal that the sediment in the canyon floor is just a thin veneer only a few metres thick over the bedrock surface, filling < 4% of the total volume. We propose that this sediment was deposited across the canyon floor of Ásbyrgi during the latter stages of the flood when the waning floodwaters were no longer powerful enough to transport the sediment load (Fig. 12D).

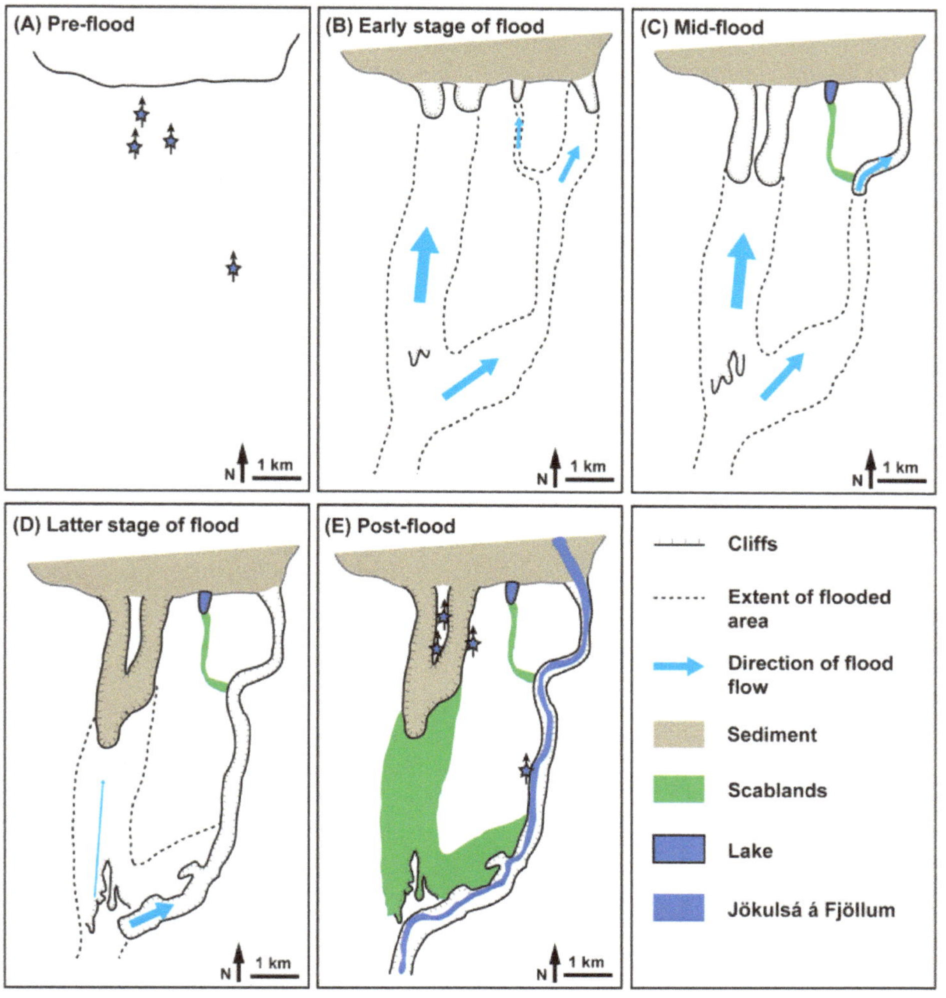

Figure 12: Proposed macroform evolution of the lower Jökulsárgljúfur canyon during an extreme flood event in the early Holocene. (A) Before the flood, the precise course of the Jökulsá á Fjöllum is impossible to determine; but fluvially sculpted surfaces on the top of Ásbyrgi Island, the eastern rim of Ásbyrgi, and the western rim of the main canyon at Landsbjörg indicate that a river once flowed here before the canyons were formed. The locations of the fluvially sculpted surfaces are shown by the blue stars (also shown in E for comparison), with direction of palaeo-flow shown by the black lines. (B) In the early stages of the flood, the floodwaters follow the course of the pre-flood river and also spread to the east. At the northern limit of the lava surface, four canyons begin to be incised. Through time, the floodwaters flowing into the canyon that currently contains Lake Ástjörn are captured by the faster retreat of the canyon to the east (C) while the two Ásbyrgi canyons continue to retreat until they coalesce. The western canyon of Ásbyrgi continues to retreat and, eventually, the large canyon to the east retreats far enough to also capture the floodwaters flowing across the Klappir scablands into Ásbyrgi. During the waning flow, a thin layer of sediment is deposited in the bottom of Ásbyrgi (D). The headwall in the main canyon continues to retreat, disconnecting Ásbyrgi and Klappir from the course of the Jökulsá á Fjöllum leading to the outstanding preservation of the landforms (E). Subsequent floods along the Jökulsá á Fjöllum are channelled in the main canyon, although some potential minor reoccupation of Klappir may have occurred, which stripped some of the soils (Waitt, 2002). The main canyon at Landsbjörg is drawn here assuming that all the erosion in this main canyon occurred during the early Holocene flood, although additional reworking of the canyon morphology during later floods cannot be ruled out.

Waitt (2002) proposed that the eroded scabland area immediately upstream of Ásbyrgi was reoccupied during the late Holocene flood ~ 1500 years ago as the soil in this area lacks the H3 (~ 2900 YBP), H4 (~ 3800 YBP), and H5 (~ 6000 YBP) tephra layers, while the soil beyond the scabland limits do contain them. This observation suggests that the soils in the scabland area were washed away after the deposition of the H3 layer, most likely during the late Holocene flood. However, the exposure age from the eroded notch at the rim of Ásbyrgi (9850 ± 2650 years from a sample in a notch a couple of metres under the original surface of the lava flow; Baynes et al., 2015) suggests that any flow through here during the mid- and late Holocene was not powerful enough to cause any significant bedrock erosion (i.e., not enough to 'reset' the concentration of cosmogenic nuclides). Thus, we can be confident that the carving of Ásbyrgi represents the impact of an early Holocene flood event. The effect of any mid- or late Holocene floodwaters that overtopped the scablands and flowed into Ásbyrgi on the sediments deposited across the canyon floor is unknown, but the presence of the eroded boulders (from the early Holocene flood) and the thin layer of canyon floor deposits suggest that at least some of the material was preserved. The loss of additional material through aeolian processes is unlikely because of the morphology of the canyon and the vegetation cover.

Over time, overland flow into a canyon with a vertical headwall should act to diffuse the knickpoint through abrasion and plucking of small blocks (Lamb et al., 2014). As the Ásbyrgi headwall is vertical and contains no evidence for diffusion since its formation, we propose that the Klappir scablands and Ásbyrgi were formed during a single extreme flood event. The floodwaters were diverted at the end of the flood preventing further fluvial activity that could have diffused the canyon headwall or reworked the landforms present on the Klappir scablands. Fig. 12E shows the state of the landscape at the present day, which is likely to be very similar to that of the immediate aftermath of the early Holocene flood, although the morphology of the main canyon at Landsbjörg may have been altered after the early Holocene flood owing to subsequent modification during moderate and large floods in the mid- and late Holocene.

Evolution of lower Jökulsá á Fjöllum during mid-late Holocene floods

While we hypothesise that the knickpoint at the head of the main Jökulsárgljúfur canyon retreated at least as far as to capture the floodwaters flowing into Ásbyrgi, we have no evidence to suggest the exact position of the knickpoint at the end of the early Holocene flood. Waitt (2002) stated that the canyon already existed before the eruption of a fissure at Hljodaklettar ~ 9000 years ago, as some of the cinder cones lie within the canyon (Fig. 1C). This chronology supports the theory that an early-Holocene flood, pre-fissure eruption, initiated formation of the Jökulsárgljúfur canyon and that erosion through headwall retreat proceeded at least as far as Hljodaklettar. We suggest that the 16 floods identified by Waitt (2002) and Kirkbride et al. (2006) at Vesturdalur have contributed to the upstream propagation of the knickpoint(s) from Hljodaklettar to the current apex of the Jökulsárgljúfur canyon. The two youngest floods dated by Kirkbride et al. (2006), as well as a late Holocene extreme flood that several authors agree has taken place (Sæmundsson, 1973, Tómasson, 1973, Helgason, 1987 and Waitt, 2002), led to significant erosion within the upper 5 km of the canyon (Baynes et al., 2015). Additional erosion of the downstream reach of the main canyon during the mid- and late Holocene floods cannot be ruled out, but we believe that this is minimal because of the absence of active, or relict, knickpoints within this part of the canyon. An abandoned terrace on the east side of the canyon at Landsbjörg represents a historical position of the river bed (Fig. 6), but the age of formation and abandonment of this terrace is not currently known.

CONCLUSIONS

Our work documents widespread evidence for bedrock erosion during extreme flood events in the lower Jökulsá á Fjöllum in northern Iceland. Multiple discrete phases of extreme flooding have occurred during the Holocene, leaving a lasting legacy on the landscape morphology in three distinct reaches. Evidence for erosion during extreme floods is clear at Dettifoss and Ásbyrgi, while evidence for deposition is found in the Forvoð valley. Ásbyrgi, unaltered since formation, contains a thin veneer of sediment in the floor of the canyon documented using an ERT survey; sediment fills < 4% of the total 0.14 km^3 volume of material that was eroded during an early Holocene extreme flood event, with reconstructed discharge of at least 39,000 $m^3\ s^{-1}$. During this flood, coincident erosion was occurring in what is now the main Jökulsárgljúfur canyon through upstream migration of the canyon headwall. The canyon head retreated far enough to capture the floodwaters flowing across the Klappir scablands into Ásbyrgi; all future flow of the Jökulsá á Fjöllum and all subsequent floods were channelled within the main canyon at Landsbjörg to the east and caused significant erosion farther upstream, although a small-scale overtopping over Klappir during later floods cannot be ruled out. The overall contribution of extreme flooding along the Jökulsá á Fjöllum during the Holocene has been the formation of a 28-km-long, up to 100-m-deep canyon in < 10 ka. This highlights the importance of extreme flood events in the erosion of bedrock landscapes, with discrete high-magnitude events having the potential to cause catastrophic landscape change that can be preserved over millennial timescales.

ACKNOWLEDGEMENTS

We thank the Vatnajökulsþjóðgarður National Park for allowing access, survey permission, and providing logistical support. We thank Katie Whitbread for her assistance in the field and Noel Gourmelen for obtaining the high-resolution DEM from*TanDEM-X* and the DLR for providing this data (grant numbers: gourmele_IDEM_GEOL0141 and XTI_GLAC0296). We are grateful to Paul Carling, an anonymous reviewer, and the Editor for their time and constructive suggestions that have improved this manuscript. This work was funded by NERC PhDStudentshipNE/H525270/1 to E.R.C.B., a grant from the Carnegie Trust for the Universities of Scotland to M.A., and National Science Foundation grant 1249313 to A.J.D.

REFERENCES

1. Alho, P., Russell, A.J., Carrivick, J.L., Kayhkö, J., 2005. Reconstruction of the largest Holocene jökulhlaup within Jökulsá á Fjöllum, NE Iceland. Quat. Sci. Rev. 24, 2319–2334.
2. Alho, P., Baker, V.R., Smith, L.N., 2010. Paleohydraulic reconstruction of the largest Glacial Lake Missoula draining(s). Quat. Sci. Rev. 29, 3067–3078.
3. Baker, V.R., 1973. Palaeohydrology and sedimentology of Lake Missoula flooding in eastern Washington. Geol. Soc. Am. Spec. Pap. 144, 1–79.
4. Baker, V.R., 1988. Flood erosion. In: Baker, V.R., Kochel, R.C., Patton, P.C. (Eds.), Flood Geomorphology. Wiley, New York, pp. 81–95.
5. Baker, V.R., 2002. The study of superfloods. Science 295, 2379–2380.
6. Baker, V.R., Bunker, R.C., 1985. Cataclysmic late Pleistocene flooding from glacial lake Missoula: a review. Quat. Sci. Rev. 4, 1–41.
7. Baker, V.R., Kale, V.S., 1998. The role of extreme floods in shaping bedrock channels. In:Tinkler, K.J., Wohl, E.E. (Eds.), Rivers Over Rock: Fluvial Processes in Bedrock Channels. American Geophysical Monograph 107, pp. 153–165.
8. Baker, V.R., Benito, G., Rudoy, A.N., 1993. Paleohydrology of Later Pleistocene superflooding. Altay Mountains, Siberia. Science 259, 348–350.
9. Baynes, E.R.C., Attal, M., Niedermann, S., Kirstein, L.A., Dugmore, A.J., Naylor, M., 2015.
10. Erosion during extreme flood events dominates Holocene canyon evolution in North-East Iceland. Proc. Natl.

Acad. Sci. 112 (8), 2355–2360.

11. Björnsson, H., 2002. Subglacial lakes and jökulhlaups in Iceland. Glob. Planet. Chang. 35, 255–271.
12. Björnsson, H., 2009. Jökulhlaups in Iceland: sources, release and drainage. In: Burr, D.M., Carling, P.A., Baker, V.R. (Eds.), Megaflooding on Earth and Mars. Cambridge University Press, Cambridge, pp. 50–64.
13. Bretz, J.H., 1923. The channeled scabland of the Columbia Plateau. J. Geol. 31, 617–649.
14. Carling, P.A., 1996. Morphology, sedimentology and palaeohydraulic significance of large gravel dunes, Altai Mountains, Siberia. Sedimentology 43, 647–664.
15. Carling, P.A., Kirkbride, A.D., Parnachov, S., Borodavko, P.S., Berger, G.W., 2002. Late Quaternary catastrophic flooding in the Altai Mountains of south-central Siberia: a synoptic overview and an introductionto flood deposit sedimentology. In: Martini, P.I., Baker, V.R., Garzon, G. (Eds.), Flood and Megaflood Processes and Deposits:
16. Recent and Ancient Examples. Special Publications of the International Associationof Sedimentologists. Blackwell Science, Oxford, pp. 17–35.
17. Carling, P.A., Martini, I.P., Herget, J., Borodavko, P., Parnachov, S., 2009a. Megaflood sedimentary valley fill: Altai Mountains, Siberia. In: Burr, D.M., Carling, P.A., Baker, V.R.(Eds.), Megaflooding on Earth and Mars. Cambridge University Press, Cambridge, pp. 243–264.
18. Carling, P.A., Herget, J., Lanz, J.K., Richardson, K., Pacifici, A., 2009b. Channel-scale erosional bedforms in bedrock and in loose granular material: character, processes and implications. In: Burr, D.M., Carling, P.A., Baker, V.R. (Eds.), Megaflooding on Earth and Mars. Cambridge University Press, Cambridge, pp. 13–32.
19. Carling, P.A., Burr, D.M., Johnsen, T.F., Brennand, T.A., 2009c. A review of open-channel megaflood depositional landforms on Earth and Mars. In: Burr, D.M., Carling, P.A., Baker, V.R. (Eds.), Megaflooding on Earth and Mars. Cambridge University Press,Cambridge, pp. 33–49.
20. Carrivick, J.L., 2006. Application of 2D hydrodynamic modelling to high-magnitude outburst floods: an example from Kverkfjöll, Iceland. J. Hydrol. 321, 187–199.
21. Carrivick, J.L., 2007. Hydrodynamics and geomorphic work of jökulhlaups (glacial outburst floods) from Kverkfjöll volcano, Iceland. Hydrol. Process. 21, 725–740
22. Carrivick, J.L., Russell, A.J., Tweed, F.S., 2004. Geomorphological evidence for jökulhlaups from Kverkfjöll volcano, Iceland. Geomorphology 63, 81–102.
23. Carrivick, J.L., Tweed, F.S., Carling, P.A., Alho, P., Marren, P.M., Staines, K., Russell, A.J., Rushmer, E.L., Duller, R., 2013. Discussion of 'Field evidence and hydraulic modelling of a large Holocene jökulhlaup at Jökulsá á Fjöllum channel, Iceland' by Douglas Howard, Sheryl Luzzadder-Beach and Timothy Beach, 2012. Geomorphology 201,512–519.
24. Chambers, J.E., Wilkinson, P.B., Wardrop, D., Hameed, A., Hill, I., Jeffrey, C., Loke, M.H.,Meldrum, P.I., Kuras, O., Cave, M., Gunn, D.A., 2012. Bedrock detection beneath riverterrace deposits using three-dimensional electrical resistivity tomography. Geomorphology 177–178, 17–25.
25. Claerbout, J.F., Muir, F., 1973. Robust modelling with erratic data. Geophysics 38 (5),826–844.
26. Clarke,A.O., 1996. Estimating probable maximum floods in the Upper Santa Ana Basin,Southern California, from stream boulder size. Environ. Eng. Geosci. 2, 165–182.
27. Costa, J.E., 1983. Paleohydraulic reconstruction of flash-flood peaks from boulder deposits in the Colorado Front Range. Geol. Soc. Am. Bull. 94 (8), 986–1004.
28. Doetsch, J., Linde, N., Pessognelli, M., Green, A.G., Günther, T., 2012. Constraining 3-D electrical resistance tomography with GPR reflection data for improved aquifer characterization. J. Appl. Geophys. 78, 68–76.
29. Duller, R.A., Mountney, N.P., Russell, A.J., Cassidy, N.C., 2008. Architectural analysis of avolcaniclastic jökulhlaup deposit, southern Iceland: sedimentary evidence forsupercritical flow. Sedimentology 55, 939–964.
30. Dunning, S.A., Rosser, N.J., Petley, D.H., Massey, C.R., 2006. Formation and failure of theTsatichhu landslide dam, Bhutan. Landslides 3, 107–113.
31. Dunning, S.A., Large, A.R.D., Russell, A.J., Roberts, M.J., Duller, R., Woodward, J., Meriaux,A.-S., Tweed, F.S., Lim, M., 2013. The role of multiple glacier outburst floods inproglacial landscape evolution: the 2010 Eyjafjallajökull eruption, Iceland. Geology41 (10), 1123–1126.
32. Eliasson, S., 1974. Eldsumbrot i Jökulsárgljúfrum. Nature 44, 52–70.
33. Eliasson, S., 1977. Molar um Jökulsárhlaup og Ásbyrgi. Nature 47, 160–179.
34. Fay, H., 2002. Formation of kettle holes following a glacial-outburst flood (jökulhlaup),Skeiðarársandur, southern Iceland. In: Snorasson, A., Finnsdottir, H.P., Moss, M. (Eds.), The Extremes of the Extremes: Extraordinary Floods. IAHS Publication. vol. 271. IAHS Press, Oxford, pp. 205–210.
35. Garcia-Castellanos, D., Estrada, F., Jimenez-Munt, I., Gorini, C., Fernandez, M., Verges, J., De Vicente, R., 2009. Catastrophic flood of the Mediterranean after the Messinian salinity crisis. Nature 462, 778–781.
36. Geotomo, 2001. Res2Dinv Electrical Resistivity Tomography processing software. Version3, 4.
37. Gupta, S., Collier, J.S., Palmer-Felgate, A., Potter, G., 2007. Catastrophic flooding origin of shelf valley systems in the English Channel. Nature 448, 342–345.
38. Hancock, G.S., Anderson, R.S., Whipple, K.X., 1998. Beyond power: bedrock river incisionprocesses and form. In: Tinkler, K.J., Wohl, E.E. (Eds.), Rivers Over Rock: Fluvial Processes in Bedrock Channels. American Geophysical Union Monograph 107,pp. 35–60.
39. Helgason, J., 1987. Jarðfræðirannsóknir á Vatnasviði Jökulsár á Fjöllum við Möðrudal. Report OS-87005/VOD-01: Orkustofnun Reykjavík (Available here:

http://www.os.is/gogn/Skyrslur/OS-1987/OS-87005.pdf Accessed October 6, 2014).

40. Hsu, H.-L., Yanites, B.J., Chen, C.-C., Chen, Y.-G., 2010. Bedrock detection using 2D electricalresistivity imaging along the Peikang River, central Taiwan. Geomorphology 114,406–414.

41. Hubbard, A., Sugden, D., Dugmore, A.J., Norddahl, H., Petursson, H.G., 2006. A modellinginsight into the Iceland Last Glacial Maximum ice sheet. Quat. Sci. Rev. 25, 2283–2296.

42. Isaksson, S.P., 1985. Stórhlaup í Jökulá á Fjöllum á fyrri hluta 18. Aldar. Náttúrufræðingurinn 54 (4–5), 165–191.

43. Kehew, A.E., Lord, M.L., 1986. Origin and large-scale erosional features of glacial-lake spillways in the northern Great Plains. Geol. Soc. Am. Bull. 97, 162–177.

44. Kirkbride, M.P., Dugmore, A.J., Brazier, V., 2006. Radiocarbon dating of mid-Holocenemegaflood deposits in the Jökulsá á Fjöllum, north Iceland. The Holocene 16 (4),605–609.

45. Komar, P.D., 1984. The lemniscate loop-comparisons with the shapes of streamlined landforms. J. Geol. 92, 133–146.

46. Lamb, M.P., Dietrich, W.E., 2009. The persistence of waterfalls in fractured bedrock. Geol. Soc. Am. Bull. 121 (7–8), 1123–1134.

47. Lamb, M.P., Dietrich, W.E., Aciego, S.M., DePaolo, D.J., Manga, M., 2008. Formation of Box Canyon, Idaho, by Megaflood: implications for seepage erosion on Earth and Mars. Science 320 (5879), 1067–1070.

48. Lamb, M.P., Mackey, B.H., Farley, K.A., 2014. Amphitheater-headed canyons formed by megaflooding at Malad Gorge, Idaho. Proc. Natl. Acad. Sci. 111 (1), 57–62.

49. Licciardi, J.M., Kurz, M.D., Curtice, J.M., 2007. Glacial and volcanic history of Icelandic table mountains from cosmogenic 3He exposure ages. Quat. Sci. Rev. 26, 1529–1546.

50. Loke, M.H., Acworth, I., Dahlin, T., 2003. A comparison of smooth and blocky inversion methods in 2D electrical imaging surveys. Explor. Geophys. 34, 182–187.

51. Magilligan, F.J., Gomez, B., Mertes, L.A.K., Smith, L.C., Smith, N.D., Finnegan, D., Garvin, J.B.,2002. Geomorphic effectiveness, sandur development, and the pattern of landscape response during jökulhlaups: Skeiðarársandur, southeastern Iceland. Geomorphology 44, 95–113.

52. Maizels, J.K., 1995. Sediments and landforms of modern proglacial terrestrial environments.In: Menzies, J.E. (Ed.), Modern Glacial Environments. Butterworth-Heinemann, Oxford,pp. 365–416.

53. Maizels, J.K., 1997. Jökulhlaup deposits in proglacial areas. Quat. Sci. Rev. 16, 793–819.

54. Margold, M., Jansson, K.N., Stroeven, A.P., Jansen, J.D., 2011. Glacial Lake Vitim, a 3000- km3 outburst flood from Siberia to the Arctic Ocean. Quat. Res. 76, 393–396.

55. Marren, P.M., Russell, A.J., Rushmer, E.L., 2009. Sedimentology of a sandur formed by multiple jökulhlaups, Kverkfjöll, Iceland. Sediment. Geol. 213, 77–88.

56. Martinez-Lopez, J., Rey, J., Duenas, J., Hidalgo, C., Benavente, J., 2013. Electrical tomographyapplied to the detection of subsurface cavities. Journal of Cave and Karst Studies 75 (1), 28–37.

57. Montgomery, D.R., Hallet, B., Yuping, L., Finnegan, N., Anders, A., Gillespie, A., Greenberg, H.M., 2004. Evidence for Holocene megafloods down the Tsangpo River gorge, southeastern Tibet. Quat. Res. 62, 201–207.

58. Norðdahl, H., 1990. Late Weichselian and Early Holocene deglaciation history of Iceland.Jökull 40, 27–50.

59. O›Conner, J., 1993. Hydrology, hydraulics and geomorphology of the Bonneville Flood. Geol. Soc. Am. Spec. Pap. 274, 1–83.

60. Richardson, K., Carling, P.A., 2005. A typology of sculpted forms in open bedrock channels.Geol. Soc. Am. Spec. Pap. 392, 1–108.

61. Rudoy, A.N., 2002. Glacier-dammed lakes and geological work of glacial superfloods in the late Pleistocene, Southern Siberia, Altai Mountains. Quat. Int. 87, 119–140.

62. Russell, A.J., Knudsen, Ó., 2002. Jökulhlaup deposits at the Ásbyrgi Canyon, northernIceland: sedimentology and implications for flow type. In: Snorasson, A., Finnsdóttir, H.P., Moss, M. (Eds.), The Extreme of the Extremes: Extraordinary Floods.Proceedings Symposium at Reykjavik, Iceland, July 2000, 271 271. IAHS Publication,pp. 107–112.

63. Sæmundsson, K., 1973. Straumrákaðarklappir I kringum Ásbyrgi. Nature 43, 52–60.

64. Schunke, E., 1985. Sedimenttransport und fluviale Abtragung der Jökulsá á Fjöllum imperiglazialen Zentral-Island [Sediment transport and fluvial erosion of Jökulsá Fjöllum in periglacial Central Iceland]. Erdkunde 39 (3), 197–205.

65. Shakesby, R.A., 1985. Geomorphological effects of jökulhlaups and ice-dammed lakes, Jotunheimu, Norway. Nor. Geol. Tidsskr. 39, 1–16.

66. Siewert, M.B., Krautblatter, M., Christiansen, H.H., Eckerstorfer, M., 2012. Arctic rockwall retreat rates estimated using laboratory-calibrated ERT measurements of talus cones in Longyeardalen, Svalbard. Earth Surf. Process. Landf. 37, 1542–1555.

67. Sigbjarnarson, G., 1996. Norðan Vatnajökuls III. Eldstöðvar og hraun frá nútíma. Nature 65, 199–212.

68. Stokes, M., Griffiths, J.S., Mather, A., 2012. Palaeoflood estimates of Pleistocene coarse grained river terrace landforms (Rio Almanzora, SE Spain). Geomorphology 149–150, 11–26.

69. Telford, W.M., Geldart, L.P., Sheriff, E.R., 1990. Applied Geophysics. Second Edition. Cambridge University Press, Cambridge.

70. Thórarinsson, S., 1950. Jokulhlaup og eldgos á jökulvatnasvæði Jökulsár á Fjöllum. Nature 20, 113–133.

71. Tómasson, H., 1973. Hamfarahlaup i Jökulsá a Fjöllum. Nature 43, 12–34.

72. Waitt, R.B., 2002. Great Holocene floods along the Jökulsá á Fjöllum, north Iceland. In:Martini, P.I., Baker, V.R., Garzon, G. (Eds.), Flood and Megaflood Processes and Deposits: Recent and Ancient Examples. Special Publications of International Association of Sedimentologists, pp. 37–51.

73. Warner, N.H., Gupta, S., Kim, J.-R., Lin, S.-Y., Muller, J.-P., 2010. Retreat of a giant cataract in a long-lived (3.7–2.6 Ga) Martian outflow channel. Geology 38 (9), 791–794.

74. Warner, N.H., Sowe, M., Gupta, S., Dumke, A., Goddard, K., 2013. Fill and spill of giant lakes in the eastern Valles Marineris region of Mars. Geology 41 (6), 675–678.

75. Whipple, K.X., Hancock, G.S., Anderson, R.S., 2000. River incision into bedrock: mechanicsand relative efficacy of plucking, abrasion and cavitation. Geol. Soc. Am. Bull. 112 (3), 490–503.

76. Wiedmer, M., Montgomery, D.R., Gillespie, A.R., Greenberg, H., 2010. Late Quaternary megafloods from Glacial Lake Atna, Southcentral Alaska, U.S.A. Quat. Res. 73,412–424.

77. Wilson, A., Lavé, J., 2014. Convergent evolution of abrading flow obstacles: Insights fromanalogue modelling of fluvial bedrock abrasion by coarse bedload. Geomorphology 208, 207–224.

78. Wohl, E., 1992. Gradient irregularity in the Herbert Gorge of Northeastern Australia. EarthSurf. Process. Landf. 17, 69–84.

79. You, Y., Yu, Q., Pan, X., Wang, X., Guo, L., 2013. Application of electrical resistivity tomography in investigating depth of permafrost base and permafrost structure in Tibetan Plateau. Cold Reg. Sci. Technol. 87, 19–26.

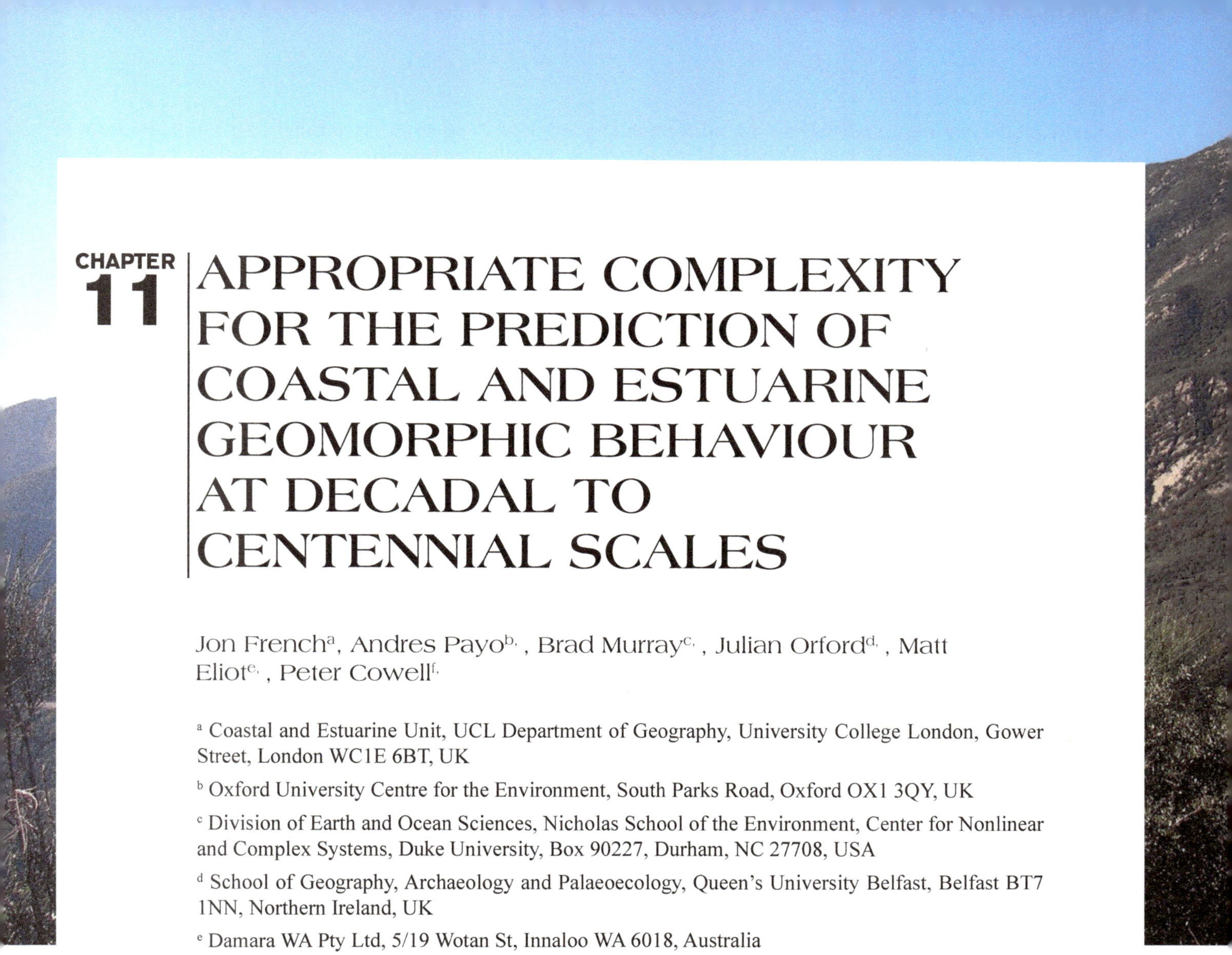

CHAPTER 11 APPROPRIATE COMPLEXITY FOR THE PREDICTION OF COASTAL AND ESTUARINE GEOMORPHIC BEHAVIOUR AT DECADAL TO CENTENNIAL SCALES

Jon French[a], Andres Payo[b], Brad Murray[c], Julian Orford[d], Matt Eliot[e], Peter Cowell[f]

[a] Coastal and Estuarine Unit, UCL Department of Geography, University College London, Gower Street, London WC1E 6BT, UK

[b] Oxford University Centre for the Environment, South Parks Road, Oxford OX1 3QY, UK

[c] Division of Earth and Ocean Sciences, Nicholas School of the Environment, Center for Nonlinear and Complex Systems, Duke University, Box 90227, Durham, NC 27708, USA

[d] School of Geography, Archaeology and Palaeoecology, Queen's University Belfast, Belfast BT7 1NN, Northern Ireland, UK

[e] Damara WA Pty Ltd, 5/19 Wotan St, Innaloo WA 6018, Australia

[f] School of Geosciences, University of Sydney, NSW 2006, Australia

ABSTRACT

Coastal and estuarine landforms provide a physical template that not only accommodates diverse ecosystem functions and human activities, but also mediates flood and erosion risks that are expected to increase with climate change. In this paper, we explore some of the issues associated with the conceptualisation and modelling of coastal morphological change at time and space scales relevant to managers and policy makers. Firstly, we revisit the question of how to define the most appropriate scales at which to seek quantitative predictions of landform change within an age defined by human interference with natural sediment systems and by the prospect of significant changes in climate and ocean forcing. Secondly, we consider the theoretical bases and conceptual frameworks for determining which processes are most important at a given scale of interest and the related problem of how to translate this understanding into models that are computationally feasible, retain a sound physical basis and demonstrate useful predictive skill. In particular, we explore the limitations of a primary scale approach and the extent to which these can be resolved with reference to the concept of the coastal tract and application of systems theory. Thirdly, we consider the importance of different styles of landform change and the need to resolve not only incremental evolution of morphology but also changes in the qualitative dynamics of a system and/or its gross morphological configuration. The extreme complexity and spatially distributed nature of landform systems means that quantitative prediction of future changes must necessarily be approached through mechanistic modelling of some form or another. Geomorphology has increasingly embraced so-called 'reduced complexity' models as a means of moving from an essentially reductionist focus on the mechanics of sediment transport towards a more synthesist view of landform evolution. However, there is little consensus on exactly what constitutes a reduced

---

Citation: Jon French, Andres Payo, Brad Murray, Julian Orford, Matt Eliot , Peter Cowell, Appropriate complexity for the prediction of coastal and estuarine geomorphic behaviour at decadal to centennial scales, doi:10.1016/j.geomorph.2015.10.005

complexity model and the term itself is both misleading and, arguably, unhelpful. Accordingly, we synthesise a set of requirements for what might be termed 'appropriate complexity modelling' of quantitative coastal morphological change at scales commensurate with contemporary management and policy-making requirements: 1) The system being studied must be bounded with reference to the time and space scales at which behaviours of interest emerge and/or scientific or management problems arise; 2) model complexity and comprehensiveness must be appropriate to the problem at hand; 3) modellers should seek a priori insights into what kind of behaviours are likely to be evident at the scale of interest and the extent to which the behavioural validity of a model may be constrained by its underlying assumptions and its comprehensiveness; 4) informed by qualitative insights into likely dynamic behaviour, models should then be formulated with a view to resolving critical state changes; and 5) meso-scale modelling of coastal morphological change should reflect critically on the role of modelling and its relation to the observable world.

INTRODUCTION

Landform behaviour is intrinsically complex due to the nature of the feedbacks between morphology and sediment transport and the range of scales over which these operate (Schumm and Lichty, 1965). Geomorphological systems are also complicated on account of the multiplicity of connected morphological components within the landform complexes that constitute the broader landscape (Werner, 1999 and French et al., 2016). Morphodynamic complexity arises in several ways, including the residual influence of previous states (*state dependence*, or *inheritance*; Wright and Short, 1984 and Favis-Mortlock, 2013); the interplay between self-regulation (or *equilibrium tendency*; Howard, 1965, Thorn and Welford, 1994 and Orford et al., 2002) and self-forcing (which leads to*thresholds* and *complex response*; Schumm, 1973 and Brunsden and Thornes, 1979), and the non-linear nature of many of the functional linkages between system components (see, for example, Wright and Thom, 1977, Cowell and Thom, 1994 and Murray et al., 2008). Predicting such complex non-linear behaviour beyond the short-timescales at which we can tightly specify governing physics and boundary conditions continues to present major difficulties. From the perspective of understanding the impacts of contemporary climate change, relevant time scales span decades and, potentially, centuries. Corresponding spatial scales are less clear-cut. In a coastal context, management planning is increasingly engaged with regional shoreline behaviour at scales of the order of 10^2 km (e.g. Stive et al., 1991, Mulder et al., 2011 and Nicholls et al., 2013). However, there is still a demand for improved prediction of changes likely to occur locally, especially in the context of proposed engineering or management schemes. At extended spatial scales, the complicated nature of landscapes becomes problematic, since much of our modelling capability is restricted to the consideration of individual landforms. This leads naturally to the question of whether landscape evolution is best understood through the coupling of specialised landform-scale models or through the development of more tightly integrated models that are able to simulate morphological evolution at whole landscape scales; this is explored further byvan Maanen et al. (2016).

As Thieler et al. (2000) have argued in the context of beach behaviour modelling, the transition from models intended to advance and articulate scientific understanding to those capable of application to societal problems has not been a smooth one. Indeed, widespread engineering application of shoreline change models based on the equilibrium shoreface profile (Dean, 1991) has provoked intense criticism from geoscientists concerned at the weak theoretical and empirical support for this concept as well its neglect of the broader-scale geological context (e.g. Pilkey et al., 1993, Young et al., 1995 and Cooper and Pilkey, 2004a). Moreover there is considerable scepticism over whether quantitative prediction of shoreline change is actually possible at multi-decadal scales (Cooper and Pilkey, 2004b and Pilkey et al., 2013) and whether expert judgement or more qualitative modelling approaches (e.g. Cooper and Jay, 2002) might be the best way to bring scientific understanding of coastal behaviour to bear on management problems. Predictions of morphological change at the coast are increasingly important, however, since coastal landforms provide a physical template that not only accommodates diverse ecosystem functions and human activities (Murray et al., 2008), but also mediates flood and erosion risk (Sayers et al., 2002 and Narayan et al., 2012).

This position paper arises from a need to formulate an overarching theoretical framework for a programme of mesoscale coastal behaviour model development being undertaken in the Integrating Coastal Sediment Systems (iCOASST) project (Nicholls et al., 2012). In it, we unpack the problem of how to deliver such predictions into a series of issues pertaining to our conceptualisation of geomorphological systems at the time and space scales of interest and the translation of geomorphological process understanding into models that deliver the insights demanded by managers and policy makers. Firstly, we revisit the well-worked question of how to define the relevant scales at which to seek quantitative predictions of landform change within an age defined by historical interference with natural sediment systems and also by the increasing prospect of significant changes in climate and ocean forcing. Secondly, we consider the theoretical bases for determining which processes are most important at a given scale of interest and the related problem of how to represent the processes of interest into models that are computationally feasible, retain a sound physical basis and demonstrate useful predictive skill (French and Burningham, 2013). Specifically, we explore the limitations of a primary-scale approach (de Vriend, 1991) and the extent to which these can be resolved with reference to ideas drawn from complex-systems theory (Werner, 1999 and Werner, 2003). Thirdly, we consider the nature of the change to be modelled and the particular need to resolve not only incremental evolution of morphology but also changes in either the gross configuration (e.g. barrier breakdown; Orford, 2011) or the dynamic nature of system operation (e.g. a shift between estuary flood and ebb dominance; Dronkers, 1986). We note that whilst geomorphology has increasingly embraced so-called 'reduced complexity' models as a means of moving away from an essentially reductionist focus on the mechanics of sediment transport towards a more synthesist view of landform evolution at broader scales (Murray and Paola, 1994,Coulthard et al., 2002, Paola, 2002, Brasington and Richards, 2007 and Murray, 2007), there appears to be little formal consensus how to define a reduced complexity model or what constitutes an appropriate level of complexity. Accordingly, we identify a set of requirements for what might be termed 'appropriate complexity modelling' of quantitative coastal morphological change at a mesoscale that is commensurate with contemporary management and policy-making requirements.

RELATING SCALE TO THE DEMANDS OF COASTAL MANAGEMENT

As Schumm and Lichty (1965) convincingly demonstrated, the scale at which we approach geomorphological phenomena introduces – indeed imposes – choices to do with the relationship between cause and effect, the levels of abstraction that are relevant and the modes of explanation and prediction that are possible. Within coastal geomorphology, as in other areas of the discipline, nested temporal hierarchies have been proposed to accommodate disparate styles of research that range from reconstructions of past coastal and estuarine evolution over extended geological timescales to interactions between fluid mechanics, sediment movement and bedforms at timescales measured in seconds. Terminology varies, with significant differences between the geoscience and engineering communities (e.g. Kraus et al., 1991, Stive et al., 1991, Fenster et al., 1993, Cowell and Thom, 1994 and Komar, 1999). Almost all schemes emphasise the correlation between temporal and spatial scale, and invariably include one or more areas of study that lie comfortably within the realm of geophysical fluid dynamics and process geomorphology, and which encompass both the fundamentals of sediment transport under the influence of waves and tides and the effect of intermittent events on landform morphology. At the other end of the spectrum, geological studies are primarily descriptive and rely on palaeoenvironmental evidence to infer past coastal dynamics. A particularly active area of study concerns recent historical timescales at which various forms of observational evidence, including instrument records and systematic monitoring, can inform explanations for documented coastal morphological change. This is also the scale at which humans have sought to manage and constrain natural shoreline dynamics, such that the term 'engineering scale' is also commonly applied (e.g. Cowell and Thom, 1994).

Whilst these kinds of classification are typically applied to the past, they can also inform our approach to the future (Gelfenbaum and Kaminsky, 2010). Coastal stakeholders worldwide increasingly demand more reliable and more quantitative assessment of likely changes in coastal morphological response to human interventions and climate change, not least to quantify the damage and adaptation costs (e.g., Hinkel et al., 2014 and Kousky, 2014). Despite inconsistencies in terminology, there is a broad consensus that the relevant time scales here extend from a few decades to a century or more. Such a time frame is clearly determined in part by human lifespans, political horizons and the extent to which these condition societal actions more generally and strategic coastal management and planning in particular. As Nicholls et al. (2013) note, a more strategic approach emerged after the 1970s under separate paradigms of coastal zone management and shoreline management. The latter has dominated the management of coastal geohazards in many countries, including the USA, UK and Netherlands (Mulder et al., 2011 and Mitsova and Esnard, 2012) and a recent tendency has been for the timescales considered to extend beyond those associated with the lifespan of engineering structures (typically a few decades) to the 2100 horizon adopted initially by climate change science. The current IPCC synthesis (Stocker et al., 2013) incorporates results from multi-century projections of climate change. These have important implications for the vulnerability of major cities (Nicholls, 2011), especially since even under stabilised conditions of radiative forcing, it would take 200 to 400 years for global mean sea-level rise to return to its mid-20th century rate (Jevrejeva et al., 2012). Timescales in excess of a century are also relevant to certain critical infrastructure — notably nuclear power stations at coastal sites, given that generation and decommissioning schedules extend well beyond 100 years.

Timescales of relevance to shoreline management do, of course, also reflect intrinsic aspects of coastal system behaviour. Variability in marine forcing and coastal response is high at shorter timescales associated with periodic tidal and intermittent wave and surge-driven processes. Given that the management of progressive change is typically the primary concern (Cowell et al., 2003a and Van Rijn, 2011), it clearly makes sense to filter in some way phenomena predominantly associated with sub-annual scales. There is also much interest in disaggregating secular trends in coastal behaviour from significant interannual variability arising from dominant modes of atmospheric behaviour such as ENSO and NAO (e.g. Storlazzi and Griggs, 2000, Esteves et al., 2011 and Burningham and French, 2013). This further justifies a focus on multi-decadal to century timescales, though even at these scales 18.6 year nodal tidal variation can be significant in macro-tidal estuaries (Townend et al., 2006) and in some open coastal settings (e.g. Gratiot et al., 2008).

The spatial dimensions associated with the timescales considered above are largely implicit, with an underlying assumption that extended timescales require consideration of broader spatial scales. The oft-assumed correlation between time and space scales is weakened by the fact that the range of circumstances encountered in coastal management planning is such that spatial scales appropriate for understanding, intervention and planning become diffuse and arbitrary (Capobianco et al., 1998). Moreover, it is clear that not all geophysical phenomena conform to such a simple scaling relationship, surge-generated extreme water levels, tsunamis and isostatic adjustment being example exceptions.

Translation of basic time and space geomorphic principles into applied coastal understanding has not been an easy discourse. This is exemplified by the situation in Britain where, prior to the 1990s, coastal protection was seen as an expensive and essentially reactive approach to coastal change with a virtual disregard for geomorphological concepts. Initial progress came with the implementation of a first generation of 42 Shoreline Management Plans (SMPs) for England and Wales (around 7000 km of open coast). These were nested in a two-tier hierarchy of major coastal cells (median length approximately 550 km) and sub-cells (mean length 100 km; Motyka and Brampton, 1993; see also Fig. 1). At the sub-cell scale, each SMP divided the coast into finer-scale management units, for which alternative strategic defence options were evaluated for a somewhat arbitrary 50-year epoch. A second generation of SMPs, initiated in 2006, incorporated more substantive geomorphological thinking, largely emanating from the DEFRA-funded FutureCoast project (Burgess et al., 2002 and Cooper and Jay, 2002). This departed from the previous place-specific analysis of coastal problems in favour of a broader context that embraced the spatial and temporal scales by which the coast and its human problems had evolved and (considering anticipated change in climate, or more specifically, mean sea level) would likely evolve. FutureCoast identified a three-tier spatial hierarchy of 424 local shoreline response systems, 108 shoreline behaviour systems and 23 coastal behaviour systems. The latter translated into 22 Phase 2 SMPs that are rather larger than their predecessors, ranging from approximately 30 to 950 km (median 170 km), subdivided into around 1500 policy units. The principal tract remained the wave-sediment cell though in reality it was the sediment pathway that dominated the cell definition. The overall conditional status of each cell was evaluated against two coastal behavioural types, swash-aligned and drift-aligned shorelines. These two dynamic types were a conceptual recognition of sediment supply and, tacitly in terms of wave exposure, longshore sediment transport. They were in effect time-averaged statements of temporal and spatial variation in the sediment pathway, and given the alongshore interconnection of such structures, a means of establishing a qualitative forecasting position for multiple future epochs (0 to 20; 20 to 50; and 50 to 100 years). A small number of similarly-founded SMPs have been implemented in Scotland, although the more indented nature of the Scottish coast and weaker littoral-drift systems mean that sediment cell-based management plans might not be the best way forward (Hansom et al., 2004).

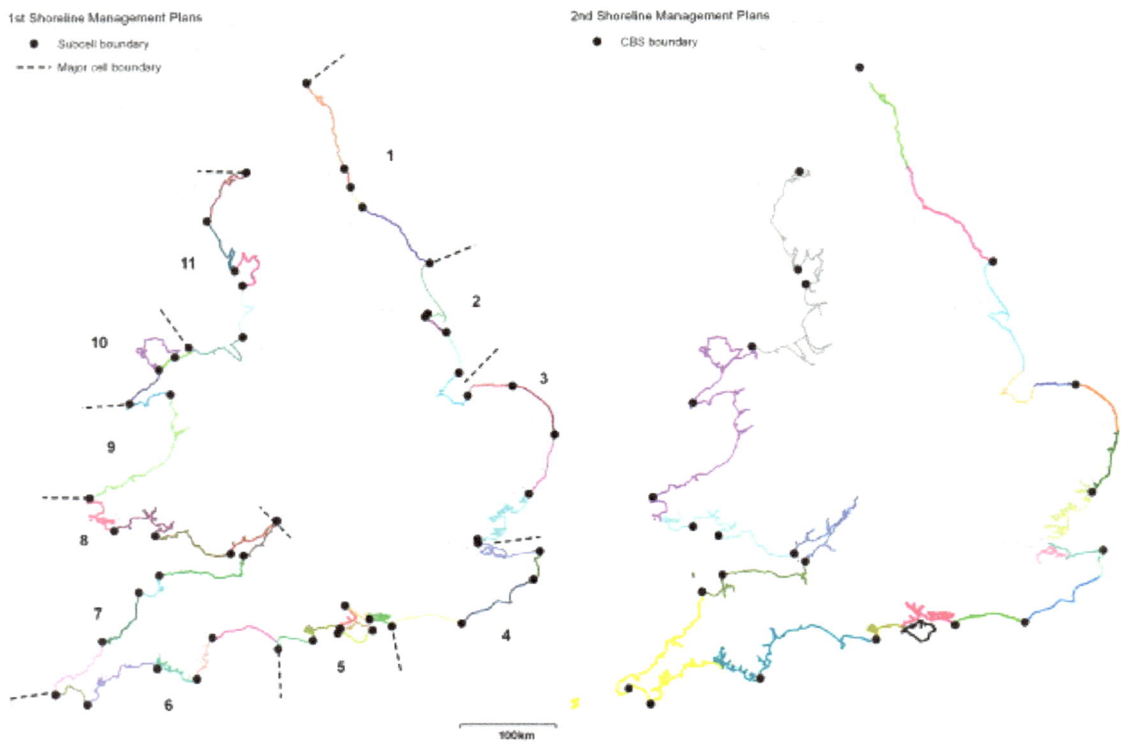

	Cells	Subcells (~SMP1)	CBS	SBS	SMP2
N	11	47	23	108	22
Min length (km)	296	33	60	6.9	29
Max length	663	388	439	217	946
Median length	550	100	207	38	171

CBS Coastal Behaviour Systems
SBS Shoreline Behaviour Statements
SMP1, SMP2 Shoreline Management Plans (1st and 2nd generation)

Figure 1: Comparison of Motyka and Brampton (1993) mapping of major cells and sub-cells for England and Wales (which broadly correspond to phase 1 SMPs) and the coastal behaviour systems derived from the FUTURECOAST project (Burgess et al., 2002), used for the phase 2 SMPs.

Several issues arose from this type of analysis: i) the understanding of what drift- and swash-alignment meant in terms of coastal behaviour was not widely understood beyond academic geomorphology; ii) the interaction of estuaries with the open coast was not given real prominence, given the difficulty of developing pathways for finer non-beach grade sediment (Cooper and Pontee, 2006); and iii) sediment pathways, as articulated by supply changes, were not recognised as the major time and spatial scale determinant of coastal evolution. These limitations meant that the structure of these pathways and their dynamics (including decaying impacts of past climate shifts, as well as potential future inputs due to climate change) were not easily translated into individual SMPs. There was a major requirement in for geomorphic 'expertise' to translate scientific understanding of coastal behaviour into an underlying formulation of larger-scale process and expected outcomes that, if unrecognised by planners, meant the loss of both SMP function and the ability to devise realistic scenarios of coastal evolution.

The British experience of incorporating geomorphological principles into shoreline management has thus been mixed. On the positive side, an explicit spatial up-scaling has moved planners to consider a more integrated coastal perspective, whereby local problems have to be considered in the context of broader-scale sediment pathways. However, management plans remain somewhat truncated in their time-scale perspectives. Much of this limitation must be tied back to lack of understanding of larger-scale coastal behaviour (e.g. shoreline status and sediment cells (see Cooper and Pontee, 2006)). Such limitations are especially apparent at the critical multi-decadal to centennial scales over which systems-level behaviour is seen to emerge as a complex function of a set of uncertain forcings, whose uncertainty appears to increase into the 21st century. Quantitative modelling of such outcomes has thus remained largely restricted to local spatial scales, and at a reductionist level that is unsuited to problems arising at a mesoscale. The FutureCoast project (Burgess et al., 2002), and the revised SMPs that it underpinned, formalise a wealth of knowledge that could inform predictive modelling, but which has still to be taken up in any substantive manner.

The UK is something of an outlier in terms of the extent to which scientific knowledge pertaining to a large proportion of its open coast has been formalised and systematically incorporated into successive generations of strategic management plans. It is instructive to compare this situation with that in the USA, where a more complex interplay of federal and state government and stakeholders (e.g. Kousky, 2014) has precluded analysis and management at a national scale. However, this is partly a question of scale, and similarly cohesive assessments of coastal vulnerability are emerging at a state level, one of the most obvious initiatives being the Louisiana Coastal Masterplan (Coastal Protection and Restoration Authority, 2012 and Peyronnin et al., 2013), which covers 100,000 km^2. This project is of particular interest in that it marks a departure from analysis of historic change and qualitative synthesis of coastal behavioural tendencies in favour of quantitative predictive modelling. To this end, an integrated suite of mesoscale sediment balance and ecohydrological models is used to evaluate nearly 400 coastal protection and restoration projects and to capture the essential trends in coastal behaviour necessary to inform strategic planning over a 50 year timescale (Meselhe et al., 2013).

MATCHING UNDERSTANDING TO MESOSCALE PROBLEMS

The challenge is to meet the demand for long-term morphological simulations with theoretical and mechanistic knowledge based on sound physical principles and empirical knowledge. Progress on this front is still hindered by the lack of a generally accepted up-scaling theory. In the following, the rationale for a more robust approach to up-scaling and limitations of three proposed conceptualizations; the primary scale relationship, the coastal tract and the hierarchical approach, are presented.

Up-Scaling on the Basis of the Primary Scale Relationship

It is clear from the preceding analysis that the scales of shoreline management, whilst partly defined by dominant system forcings, are also substantially imposed by timescales of policy-making that is increasingly conducted within an overarching paradigm of climate change science. Both imply an entry into coastal system behaviour at a scale that is rather far removed from that at which much of our primary knowledge resides. This gives rise to an up-scaling problem, since much of our understanding of the mechanisms of landform change is obtained at the relatively small temporal spatial scales of experimental studies. This is not to say that we do not have access to historical and geological insights into the evolutionary behaviour of coastal environments at scales from centuries to millennia (Woodroffe and Murray-Wallace, 2012). However, much of the mechanistic detail that is pertinent to such scales continues to elude us. Such a predicament is by no means unique to geomorphology. In ecology, for example, it is widely acknowledged that we know a lot about organisms, less about populations and even less about entire ecological landscapes (e.g. Urban, 2005). As noted above, the scales of interest for shoreline management necessitate the disaggregation of progressive trends from extreme variability, some of which extends into the mesoscale range. From a sediment budget perspective, this boils down to the resolution of tiny residual fluxes from much larger gross transports. Some of these fluxes, for example cross-shore exchange between lower and upper shoreface or, estuarine sediment sinks, cannot be resolved directly through measurement. Long-term geological context can guide our understanding of both the mode/direction of coastal change under specified forcing and also help to constrain the fundamental driver of sediment supply (e.g.McNinch, 2004). However, there is an ultimate information limit associated with historic data, since these can only represent a small portion of the environmental state space. Furthermore, there has been massive variation in terms of sediment availability and processes, especially over the later Holocene, which makes it difficult to extrapolate with confidence. Together with sheer extent of human interference in natural sediment systems (Syvitski and Kettner, 2011 and Knight and Harrison, 2014) this means that the past is not always a good guide to the future and this is especially so of systems that exhibit the possibility of divergent future states and emergent behaviours (Dearing et al., 2010; see also Section 4 below). Rather, future coastal behaviour is best understood and predicted through modelling that is able to capture the non-linearities evident in many aspects of geomorphological system behaviour.

In the absence of a formal aggregation theory, and based on the seminal work ofSchumm and Lichty (1965) on time and space causality in geomorphological systems, the largely implicit relationship between temporal and spatial scale has afforded a convenient basis for specifying quantitative modelling approaches (e.g. de Vriend, 1991 and de Vriend et al., 1993). As de Vriend (1991) observes, such a 'primary-scale relationship' implies that "the explanation of a phenomenon in coastal behaviour is primarily sought in physical processes in a similar scale range". As our process understanding improves, however, there are a number of issues that mitigate against uncritical use of the primary-scale relationship as an aggregation theory. For example, bar formation at sub-annual timescales may influence longer-term shoreline stability through the cumulative effect of repeated cycles of formation and dissipation. For example, the seasonal arrival and migration of mud banks significantly increases the storm and tsunami resistance of villages on the southwest coast of Brazil and in India (e.g. Elgar and Raubenheimer, 2008 and Holland et al., 2009). A particularly impressive example occurred in 1998 at Cassino Beach (Fig. 2), southern Brazil, where deposition of fluid mud on the beach during a single storm event caused a total absence of breaking waves along 13 km of shoreline for nearly 14 months (Holland et al., 2009). Anthony and Dolique (2004) document a slightly different interplay between the interannual morphodynamics of sandy beaches in Cayenne, French Guiana, and migrating mud banks fed by sediment supply from the Amazon. This mud bank migration drives beach rotational responses that are completely unrelated to any variability in deepwater wave climate, with extreme instances resulting in muting of normal ocean beach dynamics for several years until mud dispersal.

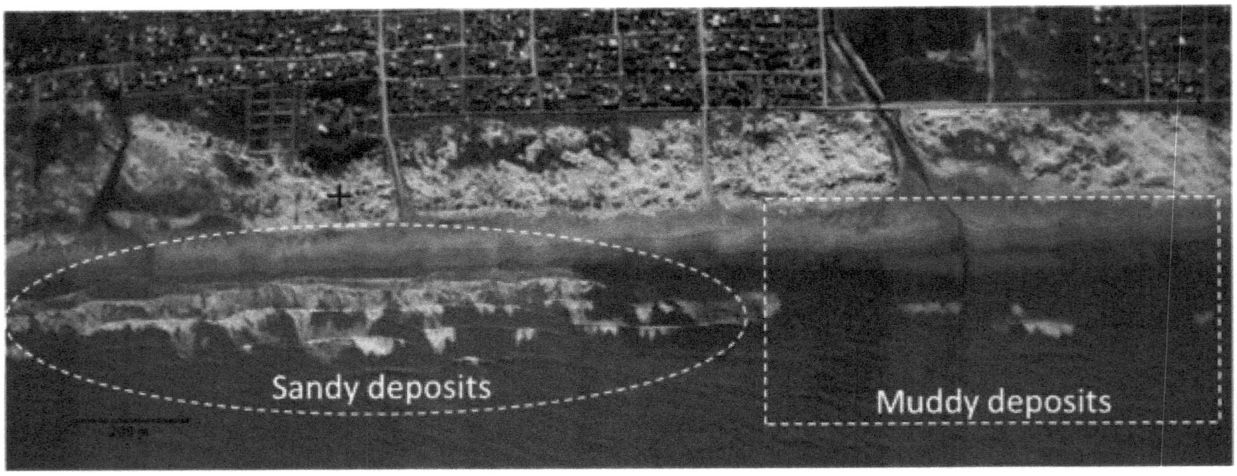

Figure 2: Example of how seasonal timing and non-linear energy transfer might influence long-term shoreline morphological evolution. Airborne image showing differential wave breaking due to ephemeral shallow water mud deposits off Cassino beach in southern Brazil.

From Holland et al. (2009).

Various attempts have been made to formalise alternative bases for applying our knowledge of coastal processes to modelling of morphological change at broader scales. One of the earliest comprehensive treatments is that of de Vriend et al. (1993), who identify three distinct approaches to longer-term modelling of coastal morphology. Firstly, input reduction methods are based around the premise that one can capture the essence of complicated forcings by simpler representative inputs. A classic example is the use of a 'morphological tide' (i.e. a characteristic tidal cycle than can be continuously repeated to achieve an effect similar to a real sequence of individual tides) to model long-term bed evolution using hydrodynamic and sediment transport models (e.g. Latteux, 1995). In effect, this up-scales observational data in a way that permits application of reductionist models at extended space and, especially, time scales. The second approach, model reduction (or process filtering; see Capobianco et al., 1998) endeavours to reformulate the model around processes relevant at the scale of interest whilst averaging or neglecting finer scale phenomena. A distinction is made here between the explicit representation of finer-scale processes and the inclusion of their effects (Murray, 2007). This is akin to the turbulence closure schemes used in fluid dynamics; an example from coastal morphodynamics is the use of tide-averaging to resolve the long-term sediment fluxes and bathymetric evolution of tidal basins (e.g. Di Silvio et al., 2010). A third approach side-steps the scaling problem by using a simplified model that exhibits behaviour that is consistent with observation or theory. Examples of such a behaviour-oriented approach include highly aggregated (0-dimensional) models for the evolution of coupled tidal delta, inlet channel and tidal flat systems (Stive et al., 1998 and Kragtwijk et al., 2004), and the shoreface translation model of Cowell et al. (1999). With regard to the two upscaling approaches, a variety of formal methods exist although these are often heavily influenced by place-specific considerations such that generalisation is difficult (de Vriend et al., 1993). In contrast, no general procedure for aggregation exists and substantial reliance is placed on the imagination and ingenuity of the modeller in elucidating the essential system behaviours and devising appropriate modelling analogues. Such models need not be directly founded on physical principles and many are essentially phenomenological (cf Urban, 2005). As, such, their nature and explanatory and predictive power depend greatly on the conceptual framework within which they are formulated.

The Coastal Tract

Perhaps the most fully developed conceptual framework is that of the coastal tract (Cowell et al., 2003a). This introduces a composite morphology of interacting landscape components within which a hierarchical cascade of processes drive coastal morphological change. At any level within the hierarchy, morphological change is driven not only by processes that are naturally observed at the associated range of time and space scales but also by the accumulated residual effects of higher order processes acting at smaller scales (Fig. 3). The tract concept is important in that it explicitly sets out to define a spatially-extended meta-morphology that encompasses not only the coast, or upper shoreface, but also connected back-barrier (including estuarine) environments as well as the lower shoreface. The tract constitutes the lowest order within a temporal scale hierarchy and constrains, via the operation of internal boundary conditions, the operation of higher order processes. As a framework for aggregation of processes into mesoscale models, the tract concept fulfils a number of roles. First, it seeks to resolve progressive (low order) changes, which tend to be associated with the more pressing shoreline management problems, from higher order variability. Second, it nests the processes and sub-systems associated with such changes within a geological framework that provides a degree of closure of the sediment budget that, together with external climate and ocean forcing, is a primary driver of landform change. Such a geological context is necessarily region-specific. Third, it presents a method for defining a hierarchy of morphodynamic behaviours that take place within the tract, and a protocol for mapping site-specific real-world problems onto this, to facilitate the formulation of appropriately dimensioned and scaled models (Cowell et al., 2003b).

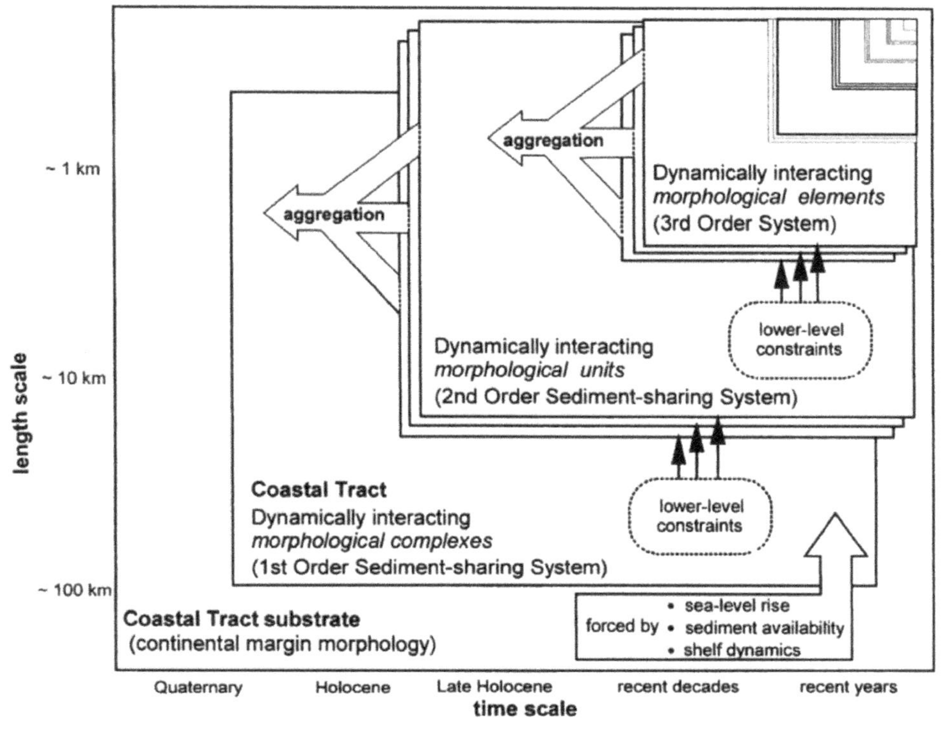

Figure 3: The nesting of the coastal tract cascade highlights the needs to aggregate smaller scale processes up to larger scales of interest as well as the role of lower-order constraints that act at each level within the cascade.

Reproduced from Cowell et al. (2003a).

However, there remains the issue of how best to aggregate the effects of finer scale processes as we move up the scale hierarchy (Fig. 3). In part, this can be accomplished through careful structuring of the scale hierarchy. Accordingly, in their 'tract cascade',Cowell et al. (2003a) invoke three orders of landform complex, morphological unit and morphological element to accommodate the spatial and functional complexity that is evident in nature. This discourages overly simplistic aggregation and also better accommodates place-specific circumstances. That said, regional application will invariably involve subjective modification of the detail, and there is scope for different assignments of specific landforms (tidal flats, saltmarshes, dunes etc.) within the absolute scale hierarchy that they present (see, for example, French et al., 2016).

The tract cascade is defined with explicit reference to hierarchy theory (Simon, 1962,O›Neill et al., 1986, Haigh, 1987 and Werner, 1999). Formal interpretations of hierarchy theory have been most numerous in ecology and ecosystem science (Müller, 1997), where it has long been appreciated that spatially-distributed environmental systems are naturally organised such that behaviour emerges on distinct temporal and spatial scales under the influence and constraint of relatively lower and higher levels (e.g. Valentine and May, 1996 and Hay et al., 2001). A key aspect of hierarchy theory that is incorporated into the tract cascade is the asymmetric nature of the vertical coupling between levels. Vertical coupling is conditioned primarily by process rates and higher levels (lower orders in the scheme of Cowell et al., 2003a) exert constraints whereas lower levels (high orders of behaviour) influence the focal level through their accumulated residual effects.

Landforms as Self-organised, Hierarchical Systems

A generalised analysis has been presented by Werner, 1999 and Werner, 2003 with reference to the self-organising properties of geomorphological systems (Hallet, 1990 and Coco and Murray, 2007), as well as ideas drawn from complex systems science. Werner›s hierarchical treatment is consistent with the primary-scale relationship and the coastal tract approaches, but generalizes similar ideas as an appropriate way to analyse any system in which the effects of processes occurring on specific time and space scales propagate over a wide range of scales. The hierarchical modelling approach arises from complex systems perspectives, including especially the idea of 'emergent phenomena' originally espoused in the physics community but subsequently finding widespread applications in the earth and environmental sciences. From this viewpoint, no single scale range exists in which the native processes are the 'fundamental' causes of phenomena at much larger scales (as is often implicitly or explicitly assumed when we attempt to construct numerical models starting with grain-scale interactions to address behaviours on much larger scales). Rather, as we move upward through the scales, the non-additive interactions between myriad degrees of freedom in one scale range collectively give rise to effectively new variables and interactions at much larger scales — for example, the way molecular motions give rise to macroscopic variable such as density and pressure (Murray et al., 2008). It is increasingly accepted that it is the interactions between these emergent variables that most directly cause the phenomena that occur on commensurate scale. For example, interactions between macroscopic fluid variables are generally viewed as the cause of phenomena such as water waves.

The hierarchical perspective also highlights the fact that cause and effect propagate downward through the scales as well as upward (Werner, 1999, Werner, 2003 and Murray et al., 2014). Structures and behaviours that arise from emergent interactions on relatively large scales dictate the context within which the much smaller scale processes occur (for example, the way water waves influence molecular motions). In addition, the emergent variables and interactions in one scale range may be effectively independent of details of the much smaller scale processes. Although the larger scale processes would not occur if the smaller scale ones did not exist, the effects from the small scale processes that are important on the larger scale might not depend on all the properties of the small scale system. Formal downscaling analyses have been limited in geomorphology, but Eliot et al. (2013) present a hierarchical landform analysis framework that incorporates an evidence-based downscaling process. This approach can potentially address a number of the issues that confound up-scaling of geomorphic modelling, including sediment volume closure, external influences, geomorphic state-dependency (see Section 4 below) and the elucidation of dominant processes and feedbacks (see also Payo et al., 2015). It may also provide a basis for identifying or discriminating between alternative up-scaling pathways from more reductionist sediment transport modelling.

Taken together, these realizations – that analysing emergent interactions can provide the most direct explanations of phenomena in a given scale range (and possibly the most reliable predictions (e.g. Murray, 2007 and Murray, 2013), and that cause and effect propagate at least as directly downward through the scales as upward – imply that smaller scale processes are not necessarily more 'fundamental' than larger-scale ones. Thus, when devising numerical models, it would seemingly be most effective to search for interactions at the 'primary scale' of the phenomena we are interested in.

As noted above, however, there are limitations to the straightforward implementation of a hierarchical approach when developing a model of mesoscale landscape phenomena. Many of the accepted, well-tested parameterizations available currently are based on the observation of processes on laboratory scales and relatively few parameterizations addressing mesoscale interactions have been developed thus far. Future work may help improve this situation, but at this stage, lacking a formal method that prescribes how to determine which effects of much smaller scale processes are most relevant for the scale of interest and how to most effectively represent those effects, we must rely heavily on scientific creativity to mint new parameterizations. Parameterizations, if they are to ultimately become widely accepted, should have a rational basis, although initially minted parameterizations (sometimes called 'rules') are not likely to have a high degree of quantitative accuracy. Ideally, time and rigorous field testing will determine which are the most effective mesoscale parameterizations, and strategic observation campaigns will improve the empirical grounding of those parameterizations. On the other hand, there is always the danger that convenience and a desire to predict regardless of the robustness of the underlying model may lead weak parameterizations to endure beyond their 'expiry date' in the science community. The so-called 'Bruun rule', in its original sense that involved only the shoreface (Bruun, 1962 and Schwartz, 1967), is perhaps the most obvious case in point here. Although the conservation of mass assumptions employed in the in the original Bruun have been extended to encompass the broader coastal profile active over longer timescales (Wolinsky and Murray, 2009), making an 'extended' Bruun rule appropriate for evaluating a sea-level rise-driven component of coastline change, the original, restricted Bruun rule continues to be applied in inappropriate contexts (see also, Bruun (1988)). This particular example is noteworthy not only for the continued application of a highly idealised model of beach response to sea-level rise despite widespread criticism of its underlying assumptions, but also the fact that this model continues to be embedded in models that purport to offer far more sophisticated insights into mesoscale coastal behaviour (Cooper and Pilkey, 2004a).

One mesoscale parameterization for modelling wave-dominated coastlines that benefits from some history of use, is the treatment of alongshore sediment flux as a relatively simple function of shoreline orientation relative to wave-approach angles, or a distribution of them (Komar, 1998; Ashton et al., 2001 and Hanson and Kraus, 2011). This has empirical support (Ashton and Murray, 2006a and Moore et al., 2013), and further refinements can be expected (not least to answer some of the criticisms levelled at the rather empirical approaches to the quantification of alongshore transport rate;Cooper and Pilkey, 2004c). As Lazarus et al. (2011) have demonstrated, gradients in net alongshore sediment flux appear to be responsible for cumulative change on wave-dominated coastlines on scales greater than a few kilometres and over years to decades. Their analysis of high resolution shoreline change data for 130 km of the predominantly sandy North Carolina coast shows that, whilst processes and patterns of shoreline change occurring on scales of a kilometre or less over storm time scales are fascinating in their own right and can adversely impact human development, they do not necessarily directly contribute to long-term coastline change and the spatial distribution of chronic erosion on this coast.

Analysing mesoscale coastline change in terms of gradients in alongshore sediment flux (treated relatively simply, without explicitly simulating processes on much smaller scales) provides an example of the hierarchical systems perspective applied to the meso-scale. Alongshore sediment flux can be related to flux of alongshore momentum entering the surf zone (Longuet-Higgins, 1970, Fredsoe and Deigaard, 1992 and Ashton and Murray, 2006b). The commonly used semi-empirical CERC equation takes almost the same form as this alongshore (Komar and Inman, 1970 and Komar, 1998). How these fluxes change as a function of shoreline orientation relative to offshore wave approach angles can drive the evolution of large-scale coastline shape (e.g. Ashton et al., 2001) and how that shape responds to changing forcing (Moore et al., 2013). The basic physics relating fluxes of energy and momentum to shoreline orientation do not depend on the details of how sediment is transported along the shore—it could hypothetically take place all in the surf zone, or all in the swash zone, and produce the same large-scale result.

Large-scale morphodynamic interactions thus arise from the basic relationship between momentum fluxes and shoreline orientation (Murray and Ashton, 2013); large-scale coastline shapes dictate large-scale patterns of alongshore sediment flux, and the patterns of sediment determine changes in coastline shape. These morphodynamic interactions can involve positive feedbacks that cause coastline undulations to increase in amplitude, as well as finite-amplitude interactions between growing coastline features that lead to self-organised coastline shapes (including cuspate capes and flying spits;Ashton and Murray, 2006a). In the case of self-organised rhythmic coastline patterns, it is clear that the emergent structures arise most directly from the large-scale interactions, rather than complicated details of processes on surf zone scales or smaller; interactions at the scale of coastline patterns determine local shoreline orientations and therefore processes on smaller scales, which can be interpreted as cause and effect propagating directly downward through a range of scales (Murray et al., 2014). However, these findings are not limited to coastlines with large-scale rhythmic patterns; large-scale morphodynamic interactions will drive or influence coastline change on any wave-dominated coastline (even where boundary conditions such as headlands that compartmentalize the coastline influence shoreline orientations).

THE IMPORTANCE OF CHANGES IN SYSTEM STATE

Much of geomorphology is concerned with the elucidation or prediction of incremental changes in either process rates or morphology. These aspects of geomorphological system functioning can be increasingly resolved by reductionist modelling that is grounded in hydrodynamic and sediment transport principles (Roelvink and Reniers, 2012 and Villaret et al., 2013). As the scale of investigation expands to encounter the self-organisational tendencies discussed in the preceding section, qualitative changes in state are encountered. These include changes in a critical aspect of system dynamics (e.g. a shift from flood-dominance to ebb-dominance in an estuary) as well as changes in gross configuration (as in the breaching and detachment and degradation of a spit). At the mesoscales of primary interest here, both kinds of state change constitute important aspects of landform behaviour that must be captured in any quantitative model.

Phillips (2014) presents a slightly different schematisation of state changes, classified on the basis of their network properties with reference to graph theory. It is beyond the scope of the present paper to fully enumerate all these types; a subset, however, is prevalent enough in coastal and estuarine settings to merit immediate attention from landform behaviour modellers. The most straightforward case involves a sequential transition between discrete states, as in the classic tidal flat, lower saltmarsh, upper saltmarsh sequence. A second case involves a sequence that repeats in a cyclical manner; examples are some circumstances of tidal flat–saltmarsh alternation (e.g. Pedersen and Bartholdy, 2007 and Singh Chauhan, 2009) or bypassing cycles that involve growth, detachment, migration and reattachment of inlet sediment shoals (Burningham and French, 2006). Other important modes of state change involve either divergent or convergent evolution. Divergence is of particular interest in that it implies the existence of multiple evolutionary pathways that may culminate in alternative stable states. An important example in the present context is the potential for evolution towards either wave-regimes or tide-dominated intertidal sedimentation (Fagherazzi and Wiberg, 2009 and Kirwan et al., 2010). Here, state changes may simultaneously encompass both changes in configuration (e.g. replacement of tidal flat by saltmarsh or vice-versa) and shifts in process dynamics (e.g. a shift from estuary sediment import to export; French et al., 2008).

State change has been widely recognised in the context of beach morphodynamic regime. This is thought to exhibit state transitions between distinct beach states varying from nil at the regime continuum ends (reflective and dissipative) to annual to decadal variation in state in the middle or intermediate states (Wright and Short, 1984). However, for multi-decadal modelling of state change one needs to upscale to 'state' definitions that can accommodate aspects of wider (time and space) coastal change and associated sediment paths. This may entail some subjective judgments about what precisely constitutes a discrete state. This problem is neatly illustrated with reference to recent work on gravel barrier dynamics. A tradition of coastal barrier recognition in the context of drift-alignment, and swash alignment (Orford et al., 2002) can provide two potential end member states that have been interpreted as mutually exclusive attractors (Orford et al., 1996). However, in the context of long-term sediment supply reduction or loss, then two more states are worth defining: 'segmented barriers' where underlying terrestrial basements have taken control of residual sediment volumes into a headland and bay configuration; and 'barrier breakdown' which the thrust of residual sediment is in a dominant onshore direction given then loss of alongshore sediment pathway. An extension to $n > 2$ states enables the structuring of a transition matrix by which characteristics and statistical significance of transition probabilities are potentially delimited using Markovian-based models.

Once putative states are defined, the question of state-change timing has to be considered. There is evidence of such changes in the form of a switch from drift to swash alignment at 10^2 to 10^3 yr scales (Orford and Jennings, 2007). Such shifts can be regarded as one way unless major forcing conditions occurred (notably changing sea level so as to revitalize sediment supply). The transition from drift to swash alignment was not necessarily due to catastrophic forcing changes. Isla and Bujalesky (2000) have offered a mechanism through what is termed as barrier cannibalization that defines a slow state change potentially measured at centennial plus scale. The inevitably of a one-way state change in barrier structure has been challenged by Orford and Jennings (2007) with reference to mid- to late-Holocene coastal barrier changes on the south coast of England (Sussex) as reconstructed from related estuary palaeoecological changes. These show how open and closed phases of estuary conditions (freshwater–brackish variability) can be related to state change between swash and drift aligned coastal tracts at multi-centennial scales. Such changes appear to occur under the influence of a relatively slow rate of sea-level change (around 5 m in 5 ka to 1 mm yr^{-1}) and restricted spatial freedom of barrier morphology movement due to lithified basement and cliff morphology. On open coasts, temporal and spatial changes between swash and drift alignment are therefore sensitive to changes in sediment supply rates and by changes in near-shore water depth as lower beachfaces react to reducing supply rates.

Efforts to model state change should also assess the conditional probabilities by which system components would act against state change. The principal components are those that control sediment supply per se, as well as the conditional control imposed by barrier resilience. Orford (2011) explored the time scales by which resilience could be defined, thus de facto imposing conditional probabilities on an absence of state change. Our present understanding indicates that this control operates at sub-annual to multi-annual scales and as such would be likely filtered from any treatment of mesoscale behaviour at multi-decadal to centennial scales. What might be more effective as a conditional control of state change are catastrophic storm events. However, in temperate climates (even with the possibility of extra tropical hurricane incursion) significant changes in the incidence of extreme events (susceptibility) have not been evident in the 20th century (Orford, 2011). This would suggest that in some forcing regimes, and in the case of coastal barriers at least, present state is well-tuned to quite a broad a range of high magnitude events, none of which stand-out as being so extreme as to cause major state change in terms of landform breakdown (Orford and Anthony, 2011).

From the above, it is clear that state change is an important aspect of coastal behaviour that becomes more, not less significant as we approach the mesoscale. Gross changes in configuration are clearly part of the normal sequence of landform evolution in some contexts, notably within the estuarine intertidal zone. On the open coast, they may be associated with abrupt changes in system functioning, possibly with major implications for flood and erosion risk (for example, a persistent barrier breach and the creation of new inlet and estuary). However, the inherent resilience of many coastal sediment systems means that such changes may not be particularly common at decadal scales. That said, there are clearly implications for mesoscale modelling in that models have to potentially capture not only subtle shifts in behaviour that can ultimately be reduced to the direction of a mediating sediment flux (especially in the case of estuaries), but also the appearance or disappearance of discrete landforms and or changes in the structure of the interactions in a broader-scale complex of landforms. Certainly, as idealised modelling by Slott et al. (2006) shows, responses to climate change will by no means by restricted to uniform retreat as envisaged by some applications of the Bruun model, and reconfiguration of more complex planform geometries is possible under plausible changes in wave climate.

Whilst capturing the existence or potential for such behaviour is clearly crucial, it is not always clear how the range of discrete states can be anticipated. Whilst some 'pressure points' may be obvious from local knowledge and previous research (e.g. the potential for breaching of a barrier beach to create a new inlet; Hartley and Pontee, 2008) other dynamic state changes may be harder to discern a priori. Alternatively, some insights may be obtained qualitative modelling of systems depicted as network graphs (Capobianco et al., 1999 and Phillips, 2012). Causal loop analysis (CLA) is a particularly promising technique for elucidating qualitative aspects of system dynamic behaviour that has hitherto seen little application in geomorphology. Payo et al. (2015) extend CLA to the analysis of coastal geomorphological systems and demonstrate its potential to reveal the existence of alternative equilibria in advance of quantitative modelling. Divergent configurational states can also be captured in the Coastal and Estuarine System Mapping methodology proposed by French et al. (2016).

TOWARDS AN APPROPRIATE COMPLEXITY FOR MESOSCALE MODELLING

The preceding sections have set out a case for focusing quantitative models of coastal change on a mesoscale that is bounded in time by the roughly decadal time frame required to exclude seasonal and interannual variability and, at the upper end, by a timescale of one or more centuries that frames contemporary climate-change impact debates. At these time scales, coastal morphology exerts a significant mediating effect on the vulnerability of human communities and assets to erosion (especially on open coasts) and flooding (especially in estuaries) (see also Lazarus et al., 2015). Spatial scales emerge less obviously from application of hierarchy theory to natural geomorphological systems and only in an arbitrary manner from the needs of shoreline management planning. Moreover, as shown schematically in Fig. 4, the association between time and space scales is not quite as tight as textbook conceptualisations imply.

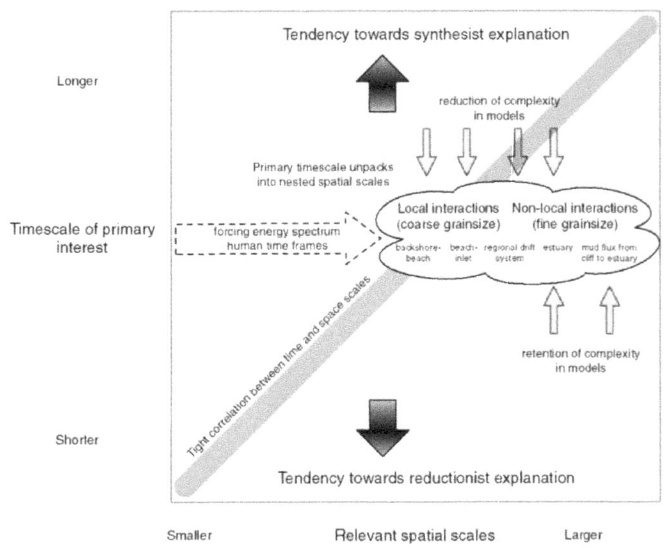

Figure 4: Schematic representation of linkages between a timescale of primary interest (determined with reference to intrinsic forcing pe-

riodicities as well as engineering, social and political timeframes) and the associated multiple spatial scales of interaction in coupled coast–estuary–shelf systems (the latter being strongly associated with distinct sediment size fractions). From a modelling perspective, broadly synthesist and reductionist approaches can be brought to bear on various aspects of this coupled system.

Much of the discussion thus far has been concerned with how to deal with complexity that arises due to non-linear behaviour and the way that this complicates scaling of our predictive ability along the time axis. In fluvial geomorphology, there is considerable scope to reduce the complexity of the problem in that many of the most problematic aspects of fluid-sediment interaction can be dispensed with in favour of simplified routing algorithms that effectively replace hydraulics with hydrology (e.g. Coulthard et al., 2002 and Nicholas, 2009). Similarly the dynamics of sediment transport can be approximated by slope-dependent functions that dispense with much of the physics that gives rise to non-linear effects that accumulate erroneously and which are sensitive to initial conditions that are hard to specify. In this sense the 'reduced complexity' label may well be appropriate in that difficult problems are made tractable largely by replacing complex phenomena by simplified parameterizations. Of course, as Nicholas and Quine (2007) observe, this form of abstraction is virtually ubiquitous in modelling.

For coastal problems a different set of considerations emerges. Most fundamentally, the dependence of sediment transport on geophysical flows cannot be quite so readily parameterised as a simple function of topography. Tidal flows, for example, arise from pressure rather than topographic gradients and drive sediment movements that emerge as tiny residuals of opposing gross fluxes. The residual water and sediment movements are much less amenable to robust parameterisation, which favours the retention of more of the hydrodynamic complexity in models that aim to resolve morphological change. This particularly applies to estuaries and to systems dominated by cohesive sediments. The sediment transport pathways that drive morphological change are highly grain-size dependent (e.g. Bass et al., 2007) and the interaction between cohesive and non-cohesive sediment pathways can be quite complex, even involving opposing residual fluxes in some tidal inlets (e.g. Van der Kreeke and Hibma, 2005).

There is also the question of how comprehensively to model coupled systems that are complex in their own right and exhibit different scales of behaviour and strengths of horizontal coupling. Change on open coasts is typically mediated by beach grade material (sands, gravels) and littoral drift systems that tend to have a strong serial dependency. Even in the case of a dependence on more distant sources, the filtering effect (e.g. via slower shelf transport rates) acts to decouple sources from their ultimate sinks, such that the problem reduces to one of shorter-range transfers with serial dependency (i.e. dominant direction of littoral drift system). Some estuarine systems also participate in short-range coupling with adjacent coast via tidal deltas and inlet bypassing processes (e.g. Burningham and French, 2006) and the evolution of estuary morphology can exert a considerable influence on the open coast (e.g. Stive et al., 1991 and Fitzgerald et al., 2006). Other estuaries and inlets are only weakly coupled via inlet dynamics to up- and down-drift coasts, such that much longer-range transfers of cohesive material determine their sediment budget and morphological evolution. A comprehensive model of coupled estuary and coastal behaviour at decadal timescales thus has to incorporate spatial scales of interaction that range from a relatively small multiple of the spatial resolution up to potentially hundreds of kilometres to handle different pathways within the shelf-scale sediment system (Fig. 4). In the case of, estuaries, fusion of reductionist and synthesist approaches may offer the best basis for prediction, especially where the timescale of interest lies in the decadal to centennial range (Thornhill et al., 2015).

How best to approach the problem of deciding upon an appropriate model complexity therefore? Building on the preceding synthesis of ideas, we see a number of requirements for successful predictive modelling at a mesoscale.

1. The system being studied must be bounded with reference to the time and space scales at which behaviours of interest emerge and/or scientific or management problems arise. Natural timescales may be revealed more-or-less objectively through analysis of the forcing energy spectrum, supplemented by increasingly sophisticated data-driven analysis of coastal change (Tebbens et al., 2002, Kroon et al., 2008 and Lazarus et al., 2011), although it is clear that timescales will invariably be societally imposed in many instances. The concept of a low-order coastal tract that contains a hierarchy of higher-order sediment-sharing morphological units (Cowell et al., 2003a) provides a vital conceptual framework through which we can bound the system in a way that specifies key sediment exchanges and constraints on supply. These include contemporary terrestrial sources, exchanges between coast and inner shelf, and the magnitude of active stores that integrate a long history of time-varying inputs and outputs and estuarine sinks. However, we here advocate a slightly different approach to the treatment of scales within the tract hierarchy in favour of relative rather than absolute time scales allied to a broader spectrum of spatial scales. Identification of nested littoral sediment cells is not sufficient to resolve the web of interactions that drive coupled coastal and estuarine behaviour and a more sophisticated mapping of morphological components and influence (including sediment) pathways is an essential first step in establishing the configuration of the system, and especially the multiplicity of spatial interactions that need to be modelled within a specified time frame. This mapping should also identify the various geological (e.g. Burningham, 2008 and Cooper et al., 2012) and human constraints (structures, interventions in the sediment regime) that constrain evolution towards a natural equilibrium. The Coastal and Estuarine System Mapping methodology advocated by French et al., 2016) provides a basis for deriving these constraints and interactions in a consistent and transparent way based on formalisation of current knowledge.
2. Levels of model complexity and comprehensiveness must be appropriate to the problem at hand. Whilst is clear that progress is more likely to be made through models that are predominantly synthesist, (in the sense of Paola, 2002), the nature of marine forcing, especially in estuaries, means that it is frequently appropriate to retain a degree of hydrodynamic complexity that

can only be obtained from a primarily reductionist approach. Accordingly, we see great potential in fusing these approaches rather than deploying them in isolation as end members of a modelling spectrum. This can be achieved in two ways. Firstly, the power of reductionist coastal area models can be harnessed to extend the spatial scale of shelf sediment transport (e.g. Barnard et al., 2013 and Bian et al., 2013). Again, this possibility reflects the observation that spatial and temporal scales are not quite as tightly intertwined as conventional scale-based classifications of geomorphic phenomena typically assume (see also Fenster et al., 1993). Crucially, this allows us to use the proven explanatory power of computational hydrodynamics both to generate large-scale pathways and also to resolve potentially important local cross-shore exchanges that are below the limits of direct observation. Secondly, hybrid model architectures can be devised to retain physically complete representations where these are necessary to generate behaviours we expect to be important at the scale of interest but which otherwise reduce complexity in favour of a more synthesist approach. Exploratory work of this kind has been undertaken in estuaries (Hinwood and McLean, 2002 and Dearing et al., 2006) and Fig. 5 illustrates how this vision is being implemented in the Estuary SpaTial LandscapE Evolution Model (ESTEEM;Thornhill et al., 2015) that is being developed as part of the iCOASST Project in the UK (Nicholls et al., 2012).

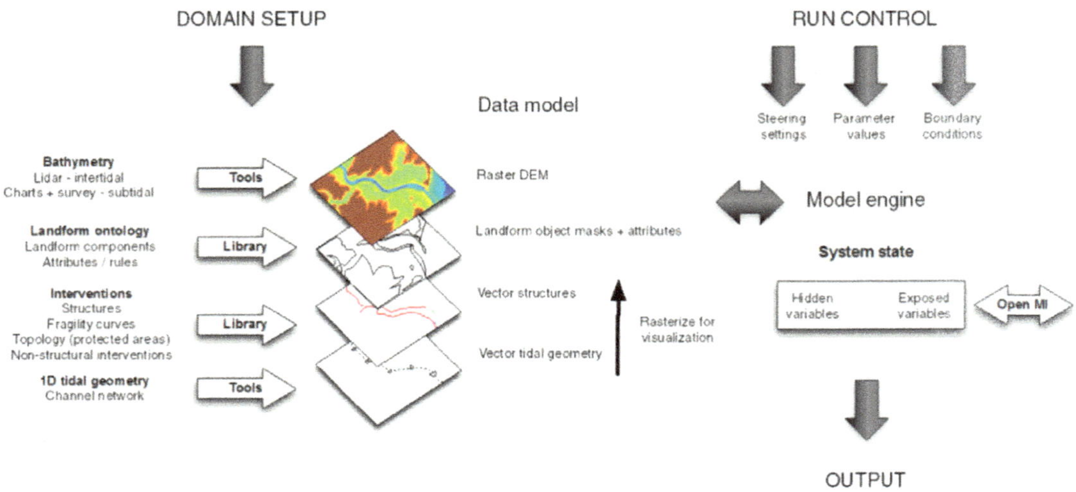

Figure 5: Schematization of Estuary SpaTial LandscapE Evolution Model (ESTEEM) model code architecture (Thornhill et al., 2015), showing fusion of low-dimensional but essentially reductionist representation of tidal hydrodynamics and channel sediment flux, parameterized ('reduced complexity') intertidal flat sedimentation, and largely rule-based saltmarsh sedimentation.

3. Modellers should seek a priori insights into what kind of behaviours are likely to be evident at the scale of interest and the extent to which the behavioural validity of a model may be constrained by its underlying assumptions and its comprehensiveness (i.e. the range of processes included). Since such behaviours emerge at the scale of interest from non-additive interactions between more fundamental components, they are often not easily predictable. However, as Payo et al. (2015) show, qualitative mathematical modelling of alternative functional representation of system structure can provide invaluable insights into sets of indicative behaviours and the processes that are most important in driving them. Anticipation of modelled behaviours is not purely of intellectual interest — it can guide evaluation of model performance by suggesting appropriately objective criteria against which to evaluate model performance. This can be problematic with models of landform evolution, since conventional point-based accuracy metrics struggle to accommodate subjective judgements of morphological similarity (Bosboom and Reniers, 2014). Eliot et al. (2013) advocate enhanced validation with reference to a multi-axis observational space that takes account of a broader suite of distinct processes than conventional assessments of hydrodynamic and sediment transport models. Finally, if we know the behaviours involved, model output indicators can be chosen that more closely map on to user needs (Van Koningsveld et al., 2005).
4. Informed by qualitative insights into likely dynamic behaviour, models (and compositions of models) should then be formulated with a view to resolving critical state changes. Here too, appropriate complexity is required. Dynamic changes, such as changes in the sign of a cohesive sediment flux, may well depend on more complex representation of shallow water tidal asymmetry. Divergent end states, such as transitions between wave and tide-dominated intertidal landform states may be accommodated more parametrically but a mechanistic treatment of some form is essential (e.g. Mariotti and Fagherazzi, 2013) — such behaviours are not easily resolved at all in simpler rule-based schemes such as SLAMM (Clough et al., 2010). But as we engage more fully with the meso-scale, we also need models that can handle configurational change. Model coupling (Warner et al., 2008 and Van Maanen et al., 2016) is a way forward here, especially when carried out within a conceptual framework that identifies potential changes in state (French et al., 2016). Indeed, the ability to dynamically assemble sets of component coastal and estuary models on the fly to

handle the appearance or disappearance of individual landforms or landform complex may well be a factor favouring externally coupled compositions of disparate models rather than more tightly integrated models (see also Voinov and Shugart, 2013).

5. Meso-scale modelling of coastal morphological change should reflect critically on the role of modelling and its relation to the observable world. Conventionally, models have provided a means of articulating and testing theories in a well-worked (though occasionally critiqued; Oreskes et al., 1994 and Thieler et al., 2000) validation framework. As Manson (2007) observes, this is accommodated within both the classical axiomatic and normal conceptions of science (Fig. 6). As the ambition of our modelling grows, however, we encounter problems involving too many components and interactions (size, comprehensiveness) to be amenable to such a treatment. At this point we enter territory that has become the realm of complex systems science in which the relation between model and data becomes somewhat different. This relates, in part to the 'complexity paradox' of Oreskes (2004) under which the more we strive to incorporate real world complexity in increasingly sophisticated and comprehensive models, the more we struggle to make sense of the model. In this situation, the model becomes almost as complex as reality and, arguably, reveals as much about the consequences of adopting a particular view of how the world works as it does about the world itself.

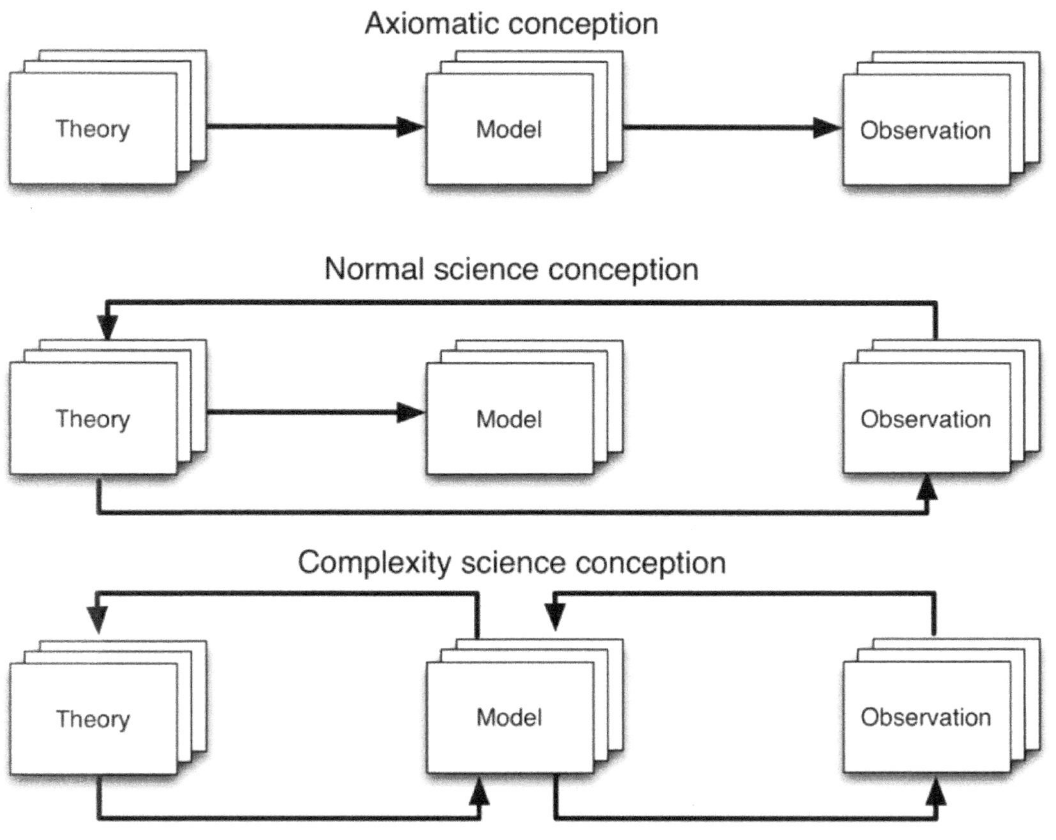

Figure 6: Axiomatic, normal-science and complexity science conceptions of the relationships between theory, models and observables (based on Manson, 2007). The last of these more closely corresponds to many of the modelling endeavours considered in the present paper.

Whilst the preceding observations are undoubtedly of academic interest, they do have implications, especially for comprehensive simulations using compositions of coupled models to advance coastal science as well as to inform management decision-making. At the very least, it necessitates a subtler approach to the question of validation. Rather than conventional point-based metrics in a calibration–validation sequence, we should seek metrics that diagnose qualitatively correct behaviours and distributions of outcomes within probabilistic uncertainty frameworks (e.g. Murray and Paola, 1996 and Murray, 2003). Where validation in the traditional sense is not feasible, inter-model comparison to test specific aspects of highly parameterised system behaviour against that modelled using more physically-complete (e.g. computational fluid dynamics) models is likely to play a larger role in evaluating more synthesist model codes (Nicholas and Quine, 2007). A potentially fruitful approach to the alignment of model complexity and observable reality is to derive meta-models that parameterise and reproduce behaviours of finer-scale models. Urban et al. (1999) demonstrate this in the context of landscape ecology and the problem of scaling knowledge of trees, through stands, to whole forests. Fine-scale models generate behaviours that can be statistically abstracted into simpler meta-models that generate essentially consistent results since they share a conceptual foundation and empirical base. Moreover, the meta-models can be tuned to meet the needs of specific applications, thereby mitigating some of the pitfalls of comprehensive broader-scaling modelling outlined above. This track does not appear to have been widely followed in geomorphology, but would seem to have potential in the coastal contexts considered here.

As models and their outputs become more complex, clear communication of modelling goals, approaches and capabilities to stakeholder audiences also becomes critically important. As Hall et al. (2014), observe, some of the best known, but worst understood models have emanated from climate science. In a coastal context, relatively simple models (notably those found on the original 'Bruun rule' of beach response to sea-level rise) have generated a broader scepticism (e.g. Pilkey et al., 2013) concerning the practical feasibility of quantitatively modelling coastal process and morphological change. As we acknowledge above, this is partly a consequence of the continued use of weak mesoscale parameterizations in the form of the shoreface profile of equilibrium. But it also arises through uncritical translation of scientific models, intended as exploratory tools, to applied ones. If we wish to avoid a similar fate for the emerging generation of meso-scale coastal change models, we clearly need to supplement our technical modelling effort with effective strategies for achieving a mutual understanding of key concepts, how complicated systems operate, and how scientific model predictions can be reconciled with non-modellers› knowledge of environmental systems and behaviours (often accumulated informally over many years, even generations, of first-hand experience). As Voinov and Gaddis (2008) have argued (albeit in the context of participatory catchment modelling), it is critical to engage with stakeholders early on and gain acceptance of the modelling methodology well ahead of the presentation of model results.

CONCLUDING REMARKS

It should be fundamental to any modelling endeavour to match the level of mechanistic understanding with scale of the problem. In the context of understanding and managing coastal change, that scale of interest is often as much determined by applications and stakeholder needs than by any intrinsic organisational property that the geomorphological systems involved are dependent upon. One of the major challenges of applied coastal modelling is that highly abstract models, useful for exploring specific aspects of system behaviour and honing our scientific understanding, are typically rather far removed from stakeholder perceptions of how the world works and, moreover, do not generate the kind of quantitative predictions that management problems increasingly demand. Whilst there is certainly scope for changing the kind of questions that we demand of models there remains a real demand for quantitative prediction of coastal morphodynamic behaviour, even if this is ultimately translated into semi-qualitative outcomes — such as the probabilistic assessment of potential changes in state.

As we engage with the evolution of coastal morphological changes at the decadal to centennial mesoscales of greatest human interest, it is clear that model complexity must be adjusted away from the tendency towards reductionism that continues to characterise much existing work in coastal morphodynamics in favour of more synthesist principles. A crucial step here is the recognition that analysing emergent mesoscale interactions can provide the most direct explanations of phenomena and, quite likely, the most reliable predictions. Whilst this is often carried out as a prelude to what has come to be known, in geomorphology at least, as 'reduced complexity' modelling, the arguments that we have advanced in this paper reinforce the view that smaller scale processes are not necessarily more 'fundamental' than larger-scale ones.

Reduced complexity modelling has proved especially popular in the context of landscape evolution under the influence of hydrological processes at the catchment scale. However, coastal and estuarine hydrodynamic forcings are harder to parameterise than their hydrological equivalents. Moreover, as we have argued here, the nesting of time and space scales is not quite as tight as is commonly envisaged. For these reasons, we advocate not a reduced complexity approach, per se, but an 'appropriate complexity' one that balances reductionism and synthesism to capture the behaviours that provide the explanatory and predictive capabilities we need. Aside from a neater philosophical rationale, such an approach would also dispense with the negative connotations that 'reducing complexity' conveys to lay audiences who may pose simple questions but who undoubtedly assume that solutions must address real world complexity and be technologically sophisticated.

ACKNOWLEDGEMENTS

This work was funded by NERC (http://www.nerc.ac.uk) as part of the Integrating COAstal Sediment SysTems (iCOASST) project (NE/J005541/1), with additional support from the Environment Agency (http://environment-agency.gov.uk) as an embedded stakeholder. The authors gratefully acknowledge helpful discussions with other iCOASST project team members and with the invited attendees of the iCOASST International Conference on Simulating Decadal Coastal Morphodynamics, held from 15 to 17 October 2013 in Southampton, UK. Dr. Helene Burningham (UCL) is also thanked for her invaluable assistance with the production of Fig. 1. The paper has also benefited from comments by J.A.G. Cooper and one other anonymous reviewer.

REFERENCES

1. Anthony, E.J., Dolique, F., 2004. The influence of Amazon-derived mud banks on the morphology of sandy, headland-bound beaches in Cayenne, French Guiana: a short- to long-term perspective. Mar. Geol. 208, 249–264.
2. Ashton, A., Murray, A.B., 2006a. High-angle wave instability and emergent shoreline shapes: 2. Wave climate analysis and comparisons to nature. J. Geophys. Res. 11, F04012.
3. Ashton, A., Murray, A.B., 2006b. High-angle wave instability and emergent shoreline shapes: 1. Modeling of sand waves, flying spits, and capes. J. Geophys. Res. 11, F04012.
4. Ashton, A., Murray, A.B., Arnoult, O., 2001. Formation of coastline features by large-scale instabilities induced by high-angle waves. Nature 414, 296–300.
5. Barnard, P.L., Erikson, L.H., Elias, E.P.L., Dartnell, P., 2013. Sediment transport patterns in San Francisco Bay Coastal System from cross validation of bedform asymmetry and modeled residual sediment flux. Mar. Geol. 345, 72–95.
6. Bass, S.J., McCave, I.N., Rees, J.M., Vincent, C.E., 2007. Sand and mud flux estimates using acoustic and

optical backscatter sensors: measurements seaward of the Wash, southern North Sea. In: Balson, P.S., Collins, M.B. (Eds.), Coastal and Shelf Sediment Transport. Geological Society of London, Special Publications 274, pp. 25–35.

7. Bian, C., Jiang, W., Greatbatch, R.J., 2013. An exploratory model study of sediment transport sources and deposits in the Bohai Sea, Yellow Sea, and East China Sea. J. Geophys. Res. Oceans 118, 5908–5923.
8. Bosboom, J., Reniers, A.J.H.M., 2014. Displacement-based error metrics for morphodynamic models. Adv. Geosci. 39, 37–43.
9. Brasington, J., Richards, K.S., 2007. Reduced-complexity, physically-based geomorphological modelling for catchment and river management. Geomorphology 90, 171–177.
10. Brunsden, D., Thornes, J.B., 1979. Landscape sensitivity and change. Trans. Inst. Brit. Geogr. NS 4, 463–484.
11. Bruun, P., 1962. Sea-level rise as a cause of shore erosion. Proc. Am. Soc. Civ. Eng. J. Waterw. Harb. Div. 88, 117–130.
12. Bruun, P., 1988. The Bruun rule of erosion by sea-level rise: a discussion on large-scale two- and three-dimensional usages. J. Coast. Res. 4, 627–648.
13. Burgess, K., Orford, J., Dyer, K., Townend, I., Balson, P., 2002. FUTURECOAST — the integration of knowledge to assess future coastal evolution at a national scale. Proceedings 28th International Conference on Coastal Engineering. World Scientific vol. 3, pp. 3221–3233.
14. Burningham, H., 2008. Contrasting geomorphic response to structural control: the Loughros estuaries, northwest Ireland. Geomorphology 97, 300–320.
15. Burningham, J., French, J.R., 2006. Morphodynamic behaviour of a mixed sand-gravel ebbtidal delta: Deben estuary, Suffolk, UK. Mar. Geol. 225, 23–44.
16. Burningham, H., French, J.R., 2013. Is the NAO winter index a reliable proxy for wind climate and storminess in northwest Europe? Int. J. Climatol. 33, 2036–2049.
17. Capobianco, M., Stive, M.J.F., Jimenez, J.A., Sanchez-Arcilla, A., 1998. Towards the definition of budget models for the evolution of deltas. J. Coast. Conserv. 4, 7–16.
18. Capobianco, M., de Vriend, H.J., Nicholls, R.J., Stive, M.J.F., 1999. Coastal area impact and vulnerability assessment: the point of view of the morphodynamic modeller. J. Coast. Res. 15, 701–716.
19. Clough, J.S., Park, R.A., Fuller, R., 2010. SLAMM 6 Beta Technical Documentation. Warren Pinnacle Consulting Inc. (49 pp.).
20. Coastal Protection and Restoration Authority, 2012. Louisiana's Comprehensive Master Plan for a Sustainable Coast. CPRA, Baton Rouge (190 pp. http://www.coastalmasterplan.la.gov).
21. Coco, G., Murray, A.B., 2007. Patterns in the sand: from forcing templates to self organization. Geomorphology 91, 271–290.
22. Cooper, J.A.G., Pilkey, O.H., 2004a. Alternatives to the mathematical modeling of beaches. J. Coast. Res. 20, 641–644.
23. Cooper, J.A.G., Pilkey, O.H., 2004b. Sea level rise and shoreline retreat: time to abandon the Bruun Rule. Glob. Planet. Chang. 43, 157–171.
24. Cooper, J.A.G., Pilkey, O.H., 2004c. Longshore drift: trapped in an Expected Universe. J. Sediment. Res. 74, 599–606.
25. Cooper, J.A.G., Jackson, D.W.T., Dawson, A.G., Dawson, S., Bates, C.R., Ritchie, W., 2012. Barrier islands on bedrock: a new landform type demonstrating the role of antecedent topography on coastal evolution. Geology 40, 923–926.
26. Cooper, N.J., Jay, H., 2002. Predictions of large-scale coastal tendency: development and application of a qualitative behavior-based methodology. J. Coast. Res. Spec. Issue 36, 173–181.
27. Cooper, N.J., Pontee, N.I., 2006. Appraisal and evolution of the littoral 'sediment cell' concept in applied coastal management: experiences from England and Wales. Ocean Coast. Manag. 49, 498–510.
28. Coulthard, T.J., Macklin, M.G., Kirkby, M.J., 2002. A cellular model of Holocene river basin and alluvial fan formation. Earth Surf. Process. Landf. 27, 269–288.
29. Cowell, P.J., Thom, B.G., 1994. Morphodynamics of coastal evolution. In: Carter, R.W.G., Woodroffe, C.D. (Eds.), Coastal Evolution: Late Quaternary Shoreline Morphodynamics. Cambridge University Press, Cambridge, pp. 33–86.
30. Cowell, P.J., Roy, P.S., Cleveringa, J., de Boer, P.L., 1999. Simulating coastal tracts with the shoreface translation model. SEPM Spec. Publ. 62, 165–175.
31. Cowell, P.J., Stive, M.J.F., Niedoroda, A.W., de Vriend, H.J., Swift, D.J.P., Kaminsky, G.M., Capobianco, M., 2003a. The coastal-tract (part 1): a conceptual approach to aggregated modeling to lower-order coastal change. J. Coast. Res. 19, 812–827.
32. Cowell, P.J., Stive, M.J.F., Niedoroda, A.W., Swift, D.J.P., de Vriend, H.J., Buijsman, M.C., Nicholls, R.J., Roy, P.S., Kaminsky, G.M., Cleveringa, J., Reed, C.W., de Boer, P.L., 2003b. The coastal-tract (part 2): applications of aggregated modeling to lowerorder coastal change. J. Coast. Res. 19, 828–848.
33. de Vriend, H.J., 1991. Mathematical modelling and large-scale coastal behaviour. J. Hydraul. Eng. 26, 727–753.
34. de Vriend, H.J., Capobianco, M., Chesher, T., de Swart, H.E., Latteux, B., Stive, M.J.F., 1993.
35. Approaches to long-term modelling of coastal morphology: a review. Coast. Eng. 21, 225–269.
36. Dean, R.G., 1991. Equilibrium beach profiles: characteristics and applications. J. Coast. Res. 7, 53–84.
37. Dearing, J., Richmond, N., Plater, A., Wolf, J., Prandle, D., Coulthard, T., 2006. Modelling approaches for coastal simulation based on cellular automata: the need and potential. Philos. Trans. R. Soc. A Math. Phys. Eng. Sci. 364, 1051–1071.

38. Dearing, J.A., Ademola, K.B., Anette, R., Sander Van Der, L., Billie, L.T., 2010. Complex land systems: the need for long time perspectives to assess their future. Ecol. Soc. 15, 21.

39. Di Silvio, G., Dall'Angelo, C., Bonaldo, D., Fasolato, G., 2010. Long-term model of planimetric and bathymetric evolution of a tidal lagoon. Cont. Shelf Res. 30, 894–903.

40. Dronkers, J., 1986. Tidal asymmetry and estuarine morphology. Neth. J. Sea Res. 20, 117–131.

41. Elgar, S., Raubenheimer, B., 2008. Wave dissipation by muddy seafloors. Geophys. Res. Lett. 35, L07611.

42. Eliot, M., Stul, T., Eliot, I., 2013. Revisiting landforms in coastal engineering. Coasts and Ports 2013: 21st Australasian Coastal and Ocean Engineering Conference and the 14th Australasian Port and Harbour Conference. Engineers Australia, Barton, A.C.T., pp. 733–738.

43. Esteves, L.S., Williams, J.J., Brown, J.M., 2011. Looking for evidence of climate change impacts in the eastern Irish Sea. Nat. Hazards Earth Syst. Sci. 11, 1641–1656.

44. Fagherazzi, S., Wiberg, P.L., 2009. Importance of wind conditions, fetch, and water levels on wave-generated shear stresses in shallow intertidal basins. J. Geophys. Res. 114, F03022.

45. Favis-Mortlock, D., 2013. Non-linear dynamics, self-organization, and cellular automata models. In: Wainwright, J., Mulligan, M. (Eds.), Environmental Modelling: Finding Simplicity in Complexity. Wiley, Chichester, pp. 349–367.

46. Fenster, M.S., Dolan, R., Elder, J., 1993. A new method of predicting shoreline positions from historical data. J. Coast. Res. 9, 147–171.

47. Fitzgerald, D.M., Buynevich, I., Argow, B., 2006. Model of tidal inlet and barrier island dynamics in a regime of accelerated sea level rise. J. Coast. Res. SI 39, 789–795.

48. Fredsoe, J., Deigaard, R., 1992. Mechanics of Coastal Sediment Transport: Advanced Series on Ocean Engineering. World Science, Singapore (392 pp.).

49. French, J.R., Burningham, H., 2013. Coasts and climate: insights from geomorphology. Prog. Phys. Geogr. 37, 550–561.

50. French, J.R., Benson, T., Burningham, H., 2008. Tidal and meteorological forcing of suspended sediment flux in a muddy mesotidal estuary. Estuar. Coasts 31, 843–859.

51. French, J.R., Burningham, H., Thornhill, G., Whitehouse, R., Nicholls, R.S., 2016. Conceptualizing and mapping coupled estuary, coast and inner shelf sediment systems. Geomorphology. 256, 17–35.

52. Gelfenbaum, G., Kaminsky, G.M., 2010. Large-scale coastal change in the Columbia River littoral cell: an overview. Mar. Geol. 273, 1–10.

53. Gratiot, N., Anthony, E.J., Gardel, A., Gaucherel, C., Proisy, C., Wells, J.T., 2008. Significant contribution of the 18.6 year tidal cycle to regional coastal changes. Nat. Geosci. 1, 169–172.

54. Haigh, M.J., 1987. The holon: hierarchy theory and landscape research. In: Ahnert, F. (Ed.), Geomorphological Models: Theoretical and Empirical ModelsCatena Supplement 10. Catena Verlag, pp. 181–192.

55. Hall, D.M., Lazarus, E.D., Swannack, T.M., 2014. Strategies for communicating systems models. Environ. Model. Softw. 55, 70–76.

56. Hallet, B., 1990. Spatial self-organization in geomorphology: from periodic bedforms and patterned ground to scale-invariant topography. Earth-Sci. Rev. 29, 57–75.

57. Hansom, J.D., Lees, G., McGlashan, D.J., John, S., 2004. Shoreline management plans and coastal cells in Scotland. Coast. Manag. 32, 227–242.

58. Hanson, H., Kraus, N.C., 2011. Long-term evolution of a long-term evolution model. J. Coast. Res. Spec. Issue 59, 118–129.

59. Hartley, L., Pontee, N., 2008. Assessing breaching risk in coastal gravel barriers. Proc. Inst. Civ. Eng. Marit. Eng. 161, 143–152.

60. Hay, G., Marceau, D.J., Dubé, P., Bouchard, A., 2001. A multiscale framework for landscape analysis: object-specific analysis and upscaling. Landsc. Ecol. 16, 471–490.

61. Hinkel, J., Lincke, D., Vafeidis, A.T., Parrette, M., Nicholls, R.J., Tol, R.S.J., Marzeion, B., Fettwies, X., Leverman, A., 2014. Coastal flood damage and adaption coasts under 21st century sea-level rise. PNAS 111, 3292–3297.

62. Hinwood, J.B., McLean, E.J., 2002. Modelling the evolution of an estuary. Proceedings Littoral 2002 — The Changing Coast, Porto, Portugal, pp. 245–252.

63. Holland, K.T., Vinzon, S.B., Calliari, L.J., 2009. A field study of coastal dynamics on a muddycoast offshore of Cassino beach, Brazil. Cont. Shelf Res. 29, 503–514.

64. Howard, A.D., 1965. Geomorphological systems — equilibrium and dynamics. Am. J. Sci. 263, 302–312.

65. Isla, F.I., Bujalesky, G.G., 2000. Cannibalisation of Holocene gravel beach-ridge plains, northern Tierra del Fuego, Argentina. Mar. Geol. 170, 105–122.

66. Jevrejeva, S., Moore, J.C., Grinsted, A., 2012. Sea level projections to AD2500 with a new generation of climate change scenarios. Glob. Planet. Chang. 80-81, 14–20.

67. Kirwan, M.L., Guntenspergen, G.R., D'Apaos, A., Morris, J.T., Mudd, S.M., Temmerman, S., 2010. Limits on the adaptability of coastal marshes to rising sea level. Geophys. Res. Lett. 37, L23401.

68. Knight, J., Harrison, S., 2014. Limitations of uniformitarianism in the Anthropocene. Anthropocene 5, 71–75.

69. Komar, P.D., 1998. Beach Processes and Sedimentation. Prentice Hall, New Jersey (544 pp.).

70. Komar, P.D., 1999. Coastal change — scales of processes and dimensions of problems. In:

71. Kraus, N.C., McDougal, W.M. (Eds.), Coastal Sediments'99 vol. 1. American Society of Civil Engineers, New York, pp. 1–17.

72. Komar, P.D., Inman, D.L., 1970. Longshore sand transport on beaches. J. Geophys. Res. 75, 5914–5927.

73. Kousky, C., 2014. Managing shoreline retreat: a US perspective. Clim. Chang. 124, 9–20.

74. Kraus, N.C., Larson, M., Kriebel, D.L., 1991. Coastal Sediments '91. American Society of Civil Engineers, Seattle, pp. 527–587.

75. Kragtwijk, N.G., Zitman, T.J., Stive, M.J.F., Wang, Z.B., 2004. Morphological response of tidal basins to human intervention. Coast. Eng. 51, 207–221.

76. Kroon, A., Larson, M., Moller, I., Yokiki, H., Rozynski, G., Cox, J., Laroude, P., 2008. Statisticalanalysis of coastal morphological data sets over seasonal to decadal time scales. Coast. Eng. 55, 581–600.

77. Latteux, B., 1995. Techniques for long term morphological simulation under tidal action. Mar. Geol. 126, 129–141.

78. Lazarus, E.D., Ashton, A., Murray, A.B., Tebbens, S., Burroughs, S., 2011. Cumulative versus transient shoreline change: dependencies on temporal and spatial scale. J. Geophys. Res. 116, F02014.

79. Lazarus, E.D., Ellis, M.A., Murray, A.B., Hall, D.M., 2015. An evolving research agenda for human–coastal systems. Geomorphology. http://dx.doi.org/10.1016/j.geomorph. 2015.07.043.

80. Longuet-Higgins, M.S., 1970. Longshore currents generated by obliquely incident waves: 1. J. Geophys. Res. 75, 6778–6789.

81. Manson, S.M., 2007. Challenges in evaluating models of geographic complexity. Environ. Plan. B Plan. Des. 34, 245–260.

82. Mariotti, G., Fagherazzi, S., 2013. A two-point dynamic model of the coupled evolution of channels and tidal flats. J. Geophys. Res. Earth Surf. 118, 1387–1399.

83. McNinch, J.E., 2004. Geologic control in the nearshore: shore oblique sandbars and shoreline erosional hotspots, Mid-Atlantic Bight, USA. Mar. Geol. 211, 121–141.

84. Meselhe, E., McCoquodale, J.A., Sheldon, J., Dortch, M., Brown, T.S., Elkan, P., Rodrique, M.D., Schindler, J.K., Wang, Z., 2013. Ecohydrology component of Louisiana's 2012 Coastal Master Plan: mass-balance compartment model. J. Coast. Res. SI 67, 16–28.

85. Mitsova, D., Esnard, A.-M., 2012. Holding back the sea: an overview of shore zone planning and management. J. Plan. Lit. 27, 446–459.

86. Moore, L.J., McNamara, D., Murray, A.B., Brenner, O., 2013. Observed changes in hurricanedriven waves explain the dynamics of modern cuspate shorelines. Geophys. Res. Lett. 40, 5867–5871.

87. Motyka, J.M., Brampton, A.H., 1993. Coastal Management: Mapping of Littoral Cells. Wallingford Report SR 328. Hydraulics Research Ltd, Wallingford, UK (102 pp.).

88. Mulder, J.P.M., Hommes, S., Horstman, E.M., 2011. Implementation of coastal erosion management in The Netherlands. Ocean Coast. Manag. 54, 888–897.

89. Müller, F., 1997. State-of-the-art in ecosystem theory. Ecol. Model. 100, 135–161.

90. Murray, A.B., 2003. Contrasting the goal, strategies, and predictions associated with simplified numerical models and detailed simulations. In: Iverson, D., Wilcock, P. (Eds.), Prediction in GeomorphologyGeophysical Monograph 135. American Geophysical Union, pp. 151–165.

91. Murray, A.B., 2007. Reducing model complexity for explanation and prediction. Geomorphology 90, 178–191.

92. Murray, A.B., 2013. Which models are good (enough), and when? In: Shroder, J. (Editor in Chief), Baas, A.C. (Ed.), Treatise on Geomorphology, vol. 2. Academic Press, San Diego, pp. 50–58.

93. Murray, A.B., Paola, C., 1994. A cellular model of braided rivers. Nature 371, 54–57.

94. Murray, A.B., Paola, C., 1996. A new quantitative test of geomorphic models, applied to a model of braided streams. Water Resour. Res. 32, 2579–2587.

95. Murray, A.B., Ashton, A., 2013. Instability and finite-amplitude self-organization of largescale coastline shapes. Philos. Trans. R. Soc. A Math. Phys. Eng. Sci. 371, 20120363.

96. Murray, A.B., Lazarus, E., Ashton, A., Baas, A., Coco, G., Coulthard, T., Fonstad, M., Haff, P.K., McNamara, D., Paola, C., Reinhardt, L., 2008. Geomorphology, complexity, and the emerging science of the Earth's surface. Geomorphology 103, 496–505.

97. Murray, A.B., Coco, G., Goldstein, B., 2014. Cause and effect in geomorphic systems: complex systems perspectives. Geomorphology 214, 1–9.

98. Narayan, S., Hanson, S., Nicholls, R.J., Clarke, D., Willems, P., Ntegeka, V., Mobaliu, J., 2012. A holistic model for coastal flooding using system diagrams and the source-pathwayreceptor (SPR) concept. Nat. Hazards Earth Syst. Sci. 12, 1431–1439.

99. Nicholas, A.P., 2009. Reduced-complexity flow routing models for sinuous single-thread channels: intercomparison with a physically-based shallow-water equation model. Earth Surf. Process. Landf. 34, 641–653.

100. Nicholas, A.P., Quine, T.A., 2007. Crossing the divide: representation of channels and processes in reduced-complexity river models at reach and landscape scales. Geomorphology 90, 318–339.

101. Nicholls, R.J., 2011. Planning for the impacts of sea level rise. Oceanography 24, 144–157.

102. Nicholls, R.J., Bradbury, A., Burningham, H., Dix, J., Ellis, M., French, J.R., Hall, J.W., Karunarathna, H.U., Lawn, J., Pan, S., 2012. iCOASST — integrating coastal sediment systems. Coastal Eng. Proc. 33, 100 (sediment).

103. Nicholls, R.J., Townend, I.H., Bradbury, A.P., Ramsbottom, D., Day, S.A., 2013. Planning for long-term coastal change: experience from England and Wales. Ocean Eng. 71, 3–16.

104. O'Neill, E.G., DeAngelis, D.L., Waide, J.B., Allen, T.F.H., 1986. A Hierarchical Concept of Ecosystems. Princeton University Press, Princeton (253 pp.).

105. Oreskes, N., 2004. Beyond the ivory tower: the scientific consensus on climate change. Science 306, 1686–1690.

106. Oreskes, N., Scrader-Frechette, K., Belitz, K., 1994. Verification, validation, and confirmationof numerical models in the earth sciences. Science 263, 641–646.

107. Orford, J.D., 2011. Gravel-dominated coastal-barrier reorganisation variability as a function of coastal susceptibility and barrier resilience. Coastal Sediments '11. In: Wang,

108. P., Rosati, J.D., Roberts, T.M. (Eds.), Proceedings of the International Conference onCoastal Sediments 2011. CD-ROM published by World Scientific Publishing Corp. and East Meets West Productions, Corpus Christi, Texas, pp. 1257–1270.

109. Orford, J.D., Anthony, E.J., 2011. Extreme events and the morphodynamics of coastal gravel barriers: strengthening uncertain ground. Mar. Geol. 290, 41–45.

110. Orford, J.D., Carter, R.W.G., Jennings, S.C., 1996. Control domains and morphological phases in gravel-dominated coastal barriers. J. Coast. Res. 12, 589–605.

111. Orford, J.D., Forbes, D.J., Jennings, S., 2002. Organisational controls, typologies and timescales of paraglacial gravel-dominated coastal systems. Geomorphology 48, 51–85.

112. Orford, J.D., Jennings, S.C., 2007. Variation in the organisation of gravel-dominated coastal systems: evidence from Nova Scotia and Southern England. Coastal Sediments '07. New Orleans. American Society of Civil Engineers vol. 1, pp. 434–448.

113. Paola, C., 2002. Quantitative models of sedimentary basin infilling. Sedimentology 41 (Suppl. 1), 121–178.

114. Payo, A., Hall, J.W., French, J.R., Sutherland, J., van Maanen, B., Nicholls, R.J., Reeve, D.E.,

115. 2015. Causal Loop Analysis of coastal geomorphological systems. Geomorphology. http://dx.doi.org/10.1016/j.geomorph.2015.07.048.

116. Pedersen, J.B.T., Bartholdy, J., 2007. Exposed salt marsh morphodynamics: an example from the Danish Wadden Sea. Geomorphology 90, 115–125.

117. Peyronnin, N., Green, M., Richards, C.P., Owens, A., Reed, D., Chamberlain, J., Groves, D.G., Rhinehart, W.K., Belhadjali, K., 2013. Louisiana's 2012 Coastal Master Plan: overviewof a science-based and publicly informed decision-making process. J. Coast. Res. Spec. Issue 67, 1–15.

118. Phillips, J.D., 2012. Synchronization and scale in geomorphic systems. Geomorphology 137, 50–58.

119. Phillips, J.D., 2014. State transitions in geomorphic responses to environmental change. Geomorphology 204, 208–217.

120. Pilkey, O.H., Young, R.S., Riggs, S.R., Smith, A.W., Wu, H., Pilkey, W.D., 1993. The concept of shoreface profile of equilibrium: a critical review. J. Coast. Res. 9, 255–278.

121. Pilkey, O.H., Young, R., Cooper, J.A.G., 2013. Quantitative modeling of coastal processes: a boom or bust for society? In: Baker, V.R. (Ed.), Geological Society of America Special Paper 502. Rethinking the Fabric of Geology, pp. 135–144.

122. Roelvink, D., Reniers, A., 2012. A Guide to Modeling Coastal Morphology. World Scientific, New Jersey (292 pp.).

123. Sayers, P.B., Hall, J.W., Meadowcroft, I.C., 2002. Towards risk-based flood hazard management in the UK. Proc. Inst. Civ. Eng. 150, 36–42.

124. Schumm, S.A., 1973. Geomorphic thresholds and complex response of drainage systems. In: Morisawa, M. (Ed.), Fluvial Geomorphology. State University of New York Publications, Geomorphology, Binghampton, pp. 299–310 (Binghamton, New York). Schumm, S.A., Lichty, R.W., 1965. Time, space, and causality in geomorphology. Am. J. Sci.263, 110–119.

125. Schwartz, M., 1967. The Bruun theory of sea-level rise as a cause of shore erosion. J. Geol. 75, 76–92.

126. Simon, H.A., 1962. The architecture of complexity. Proc. Am. Philos. Soc. 106, 468–482.

127. Singh Chauhan, P.P., 2009. Autocyclic erosion in tidal marshes. Geomorphology 110, 45–57.

128. Slott, J.M., Murray, A.B., Ashton, A.D., Crowley, T.J., 2006. Coastline response to changingstorm patterns. Geophys. Res. Lett. 33, L18404.

129. Stive, M.J.F., Roelvink, D.J.A., de Vriend, H.J., 1991. Large-scale coastal evolution concept. Proceedings of the 22nd International Conference on Coastal Engineering. American Society of Civil Engineers, New York, pp. 1962–1974.

130. Stive, M.J.F., Capaobianco, M., Wang, B., Ruol, P., Buijsman, M.C., 1998. Morphodynamics of a tidal lagoon and adjacent coasts. 8th International Conference on Physics of Estuariesand Coastal Seas. The Hague. Rotterdam, Balkema, pp. 397–407.

131. Stocker, T.F., Qin, D., Plattner, G.-K., Tignor, M., Allen, S.K., Boschung, J., Nauels, A., Xia, Y., Bex, V., Midgley, P.M. (Eds.), 2013. IPCC, 2013: climate change 2013: the physical science basisContribution of Working Group I to the Fifth Assessment Report of the Intergovernmental Panel on Climate Change. Cambridge University Press, Cambridge (1535 pp.).

132. Storlazzi, C.D., Griggs, G.B., 2000. Influence of El Nino-Southern Oscillation (ENSO) events on the evolution of California's shoreline. Geol. Soc. Am. Bull. 112, 236–249.

133. Syvitski, J.P.M., Kettner, A., 2011. Sediment flux and the Anthropocene. Philos. Trans. R. Soc. Lond. A Math. Phys. Eng. Sci. 369, 957–975. Tebbens, S.F., Burroughs, S.M., Nelson, E.E., 2002. Wavelet analysis of shoreline change n

134. the Outer Banks of North Carolina: an example of complexity in the marine sciences. Proc. Natl. Acad. Sci. U. S. A. 93 (Suppl. 1), 2554–2560.

135. Thieler, E.R., Pilkey, O.H., Young, R.S., Bush, D.M., Chei, F., 2000. The use of mathematical models to predict beach behavior for U.S. coastal engineering. J. Coast. Res. 16, 48–70.

136. Thorn, C.E., Welford, M.R., 1994. The equilibrium concept in geomorphology. Ann. Assoc. Am. Geogr. 84, 666–696.

137. Thornhill, G., French, J.R., Burningham, H., 2015. ESTEEM — a new 'hybrid complexity' model for simulating estuary morphological change at decadal to centennial scales. In: Wang, P., Rosati, J.D., Cheng, J. (Eds.), Proceedings Coastal Sediments 2015. World Scientific (13 pp.).

138. Townend, I.H., Wang, Z.B., Rees, J.G., 2006. Millennial to annual volume changes in the Humber estuary. Proc. R. Soc. A 463, 837–854.

139. Urban, D.L., 2005. Modeling ecological processes across scales. Ecology 86, 1996–2006.Urban, D.L., Acevedo, M.F., Graman, S.L., 1999. Scaling fine-scale processes to large-scale processes using models derived from models: meta-models. In: Mladenoff, D., Baker, W. (Eds.), Spatial Modelling of Forest Landscape Change: Approaches and Applications. Cambridge University Press, Cambridge, pp. 70–98.

140. Valentine, J.W., May, C.L., 1996. Hierarchies in biology and paleontology. Paleobiology 22, 23–33.

141. Van Koningsveld, M., Davidson, M.A., Huntley, D.A., 2005. Matching science with coastal management needs: the search for appropriate Coastal State Indicators. J. Coast. Res. 213, 399–411.

142. Van der Kreeke, J., Hibma, A., 2005. Observations on silt and sand transport in the throat section of the Frisian Inlet. Coast. Eng. 52, 159–175.

143. Van Maanen, B., Nicholls, R.J., French, J.R., Barkwith, A., Bonaldo, D., Burningham, H., Murray, A.B., Payo, A., Sutherland, J., Thornhill, G., Townend, I.H., van der Wegen, M., Walkden, M.J.A., 2016. Simulating mesoscale coastal evolution for decadal coastal management: A new framework integrating multiple, complementary modelling approaches. Geomorphology. 256, 68–80.

144. Van Rijn, L.C., 2011. Coastal erosion and control. Ocean Coast. Manag. 54, 867–887.

145. Villaret, C., Hervouet, J.-M., Kopmann, R., Merkel, U., 2013. Morphodynamic modelling using the Telemac finite-element system. Comput. Geosci. 53, 105–113.

146. Voinov, A., Gaddis, E.J.B., 2008. Lessons for successful participatory watershed modeling: aperspective from modeling practitioners. Ecol. Model. 216, 197–200.

147. Voinov, A., Shugart, H.H., 2013. 'Integronsters', integral and integrated modelling. Environ. Model. Softw. 39, 149–158.

148. Warner, J.C., Perlin, N., Skyllingstad, E.D., 2008. Using the Model Coupling Toolkit to couple earth system models. Environ. Model. Softw. 23, 1240–1249.

149. Werner, B.T., 1999. Complexity in natural landform patterns. Science 284, 102–104.

150. Werner, B.T., 2003. Modeling landforms as self-organized, hierarchical dynamical systems. In: Wilcock, P.R., Iverson, R.M. (Eds.), Prediction in Geomorphology: Geophysical Monograph 135. American Geophysical Union Press, Washington, D.C., pp. 133–150.

151. Wolinsky, M.A., Murray, A.B., 2009. A unifying framework for shoreline migration: 2. Application to wave-dominated coasts. J. Geophys. Res. 114, F01009. http://dx.doi. org/10.1029/2007JF000856.

152. Woodroffe, C.D., Murray-Wallace, C.V., 2012. Sea-level rise and coastal change: the past as a guide to the future. Quat. Sci. Rev. 54, 4–11.

153. Wright, L.D., Short, A.D., 1984. Morphodynamic variability of surf zones and beaches: a synthesis. Mar. Geol. 56, 93–118.

154. Wright, L.D., Thom, B.G., 1977. Coastal depositional landforms: a morphodynamic approach. Prog. Phys. Geogr. 1, 412–459.

155. Young, R.S., Pilkey, O.H., Bush, D.M., Thieler, E.R., 1995. A discussion of the Generalized Model for Simulating Shoreline Change (GENESIS). J. Coast. Res. 11, 875–886.

CHAPTER 12

DATA-DRIVEN AND HYBRID COASTAL MORPHOLOGICAL PREDICTION METHODS FOR MESOSCALE FORECASTING

Dominic E. Reeve[a], Harshinie Karunarathna[a], Shunqi Pan[b], Jose M. Horrillo-Caraballo[a], Grzegorz Różyński[c], Roshanka Ranasinghe[d]

[a] College of Engineering, Swansea University, Wales, UK

[b] School of Engineering, Cardiff University, Wales, UK

[c] Institute of Hydro-Engineering, Polish Academy of Sciences, Gdańsk, Poland

[d] UNESCO-IHE/Delft University of Technology, The Netherlands

ABSTRACT

It is now common for coastal planning to anticipate changes anywhere from 70 to 100 years into the future. The process models developed and used for scheme design or for large-scale oceanography are currently inadequate for this task. This has prompted the development of a plethora of alternative methods. Some, such as reduced complexity or hybrid models simplify the governing equations retaining processes that are considered to govern observed morphological behaviour. The computational cost of these models is low and they have proven effective in exploring morphodynamic trends and improving our understanding of mesoscale behaviour. One drawback is that there is no generally agreed set of principles on which to make the simplifying assumptions and predictions can vary considerably between models. An alternative approach is data-driven techniques that are based entirely on analysis and extrapolation of observations. Here, we discuss the application of some of the better known and emerging methods in this category to argue that with the increasing availability of observations from coastal monitoring programmes and the development of more sophisticated statistical analysis techniques data-driven models provide a valuable addition to the armoury of methods available for mesoscale prediction. The continuation of established monitoring programmes is paramount, and those that provide contemporaneous records of the driving forces and the shoreline response are the most valuable in this regard. In the second part of the paper we discuss some recent research that combining some of the hybrid techniques with data analysis methods in order to synthesise a more consistent means of predicting mesoscale coastal morphological evolution. While encouraging in certain applications a universally applicable approach has yet to be found. The route to linking different model types is highlighted as a major challenge and requires further research to establish its viability. We argue that key elements of a successful solution will need to account for dependencies between driving parameters, (such as wave height and tide level), and be able to predict step changes in the configuration of coastal systems.

Citation: Dominic E. Reeve, Harshinie Karunarathna, Shunqi Pan, Jose M. Horrillo-Caraballo, Grzegorz Różyński, Roshanka Ranasinghe, Data-driven and hybrid coastal morphological prediction methods for mesoscale forecasting, doi:10.1016/j.geomorph.2015.10.016

INTRODUCTION

Planning and development on our shorelines is increasingly undertaken within the framework of structured shoreline management plans that require the consideration of morphological change over a window of up to 100 years into the future. Methods to perform this have been scarce and predictions have been made on an ad hoc, case by case, basis. The absence of a consistent predictive framework has provided the motivation to develop morphological models that can provide useful mesoscale (of the order of 10^1 to 10^2 km length scales and 10^1 to 10^2 year timescales) estimates of coastal morphological change. The hurdles to developing such models are significant. The deterministic process models that have proved useful for predicting short-term storm response encounter difficulties when applied to mesoscale problems. Not only does the increased number of time steps required lead to an unacceptable accumulation of numerical errors, as well as huge increases in computational time, but the approach has difficulties in reliably reproducing some of the broad morphological tendencies observed in practice.

In this paper we discuss some of the methods that have been attempted to make mesoscale predictions of coastal change and then propose an alternative approach, based on using information contained in the growing amount of observational evidence gathered in coastal monitoring programmes. This approach has gained the epithet 'data-driven' modelling and is demonstrated through a number of applications to study sites from around the world. We also provide a demonstration of how data-driven methods can be combined with reduced complexity models as an example of how a good observational database can be combined with elements of physical understanding to form the basis of prediction. Accordingly, we argue for the importance of maintaining and extending long records of good quality observations, such as those gathered at Duck in the USA, Lubiatowo in Poland and the Channel Coastal Observatory in the UK, and the intention to expand the observational network to cover a wider range of shoreline types and exposures. Further, methods that combine the observational evidence with elements of our understanding of the key physical processes seem to show some promise, and go some way towards addressing the criticism that purely data-driven methods are based entirely on historical measurements.

One question that arises naturally from scale categorisation is whether forecasting methods can (or should) be developed at each scale, or whether the detailed process knowledge at small scales should simply be extended into the description of larger scale processes, at consequent computational cost. Section 2 provides some background to the different types of methods that have been developed for predicting mesoscale coastal morphology. Section 3 presents a range of data-driven techniques together with a selection of applications to specific sites. The merging of data-driven and mechanistic approaches in hybrid models is discussed in Section 4. The paper concludes withSection 5.

BACKGROUND

Our coastlines change as a consequence of the aggregation of forces due to winds, waves and tides and the consequent movement of sediments. The net result of continual sediment transport is an alteration of the shape or morphology of the shoreline. De Vriend et al. (1993) concluded over 20 years ago that understanding this process was one of the most challenging issues confronting coastal engineers and managers. Despite improvements in process understanding, computational power and monitoring techniques, our understanding of coastal morphodynamics remains limited. Unfortunately there are practical limitations to simply expanding the process models that have worked quite well for short-term prediction to the mesoscale problem. These arise partly through the potential for errors associated with the numerical approximation of derivatives to accumulate and eventually dominate the procedure, rendering the solution useless (e.g. Lilly, 1965). There are theoretical arguments that suggest there may be an inherent uncertainty or limit of predictability in the equations used for coastal process modelling due to their strong nonlinearity. In practical terms this means that the computations for very similar starting conditions will at some point diverge to very different, but equally valid, solutions. This type of behaviour has been termed "chaos" after the pioneering work of Lorenz (1963) into dynamical systems. The consequence of apparently deterministic equations of motion supporting chaos is that even if our set of morphological prediction equations are solved perfectly, we cannot be sure that our predictions will be perfect because of the limited accuracy of the initial conditions. BothBaas (2002) and Southgate et al. (2003) discuss a number of methods developed in the discipline of nonlinear processes and explain how these can be applied in the context of coastal engineering. To date, it has not been established whether the equations for morphodynamic evolution support chaos or not. From a pragmatic point of view, it seems only prudent to assume that uncertainties in initial conditions, measurement errors and numerical errors are likely to limit the period over which useful predictions can be obtained through deterministic process modelling. Some progress has been made using the 'brute force' approach of running process models over areas of several square kilometres and for periods up to 5 years or so as reported by Lesser (2009). The primary computational constraint is that the hydrodynamic and morphodynamic time scales are quite different. The currently preferred technique for addressing this is the morphodynamic acceleration method, in which the bed level changes that occur during a single hydrodynamic time step are multiplied by a factor so that the morphodynamic updating step does not have to be computed for each short hydrodynamic step. Detailed discussion of the technique in a range of applications can be found in Lesser et al. (2004), Jones et al. (2007), van der Wegen and Roelvink (2008), Roelvink et al. (2009),Ranasinghe et al. (2011) and Dissanayake et al. (2012) amongst others. Thence, there is good motivation to find alternative means to make mesoscale predictions that are required to inform coastal planning.

Coastal management, design of coastal structures, and flood risk all depend upon our understanding of how the shoreline changes, especially at decadal and longer timescales. Engineered structures such as groynes, artificial headlands and detached breakwaters are used as means to control the movement of sediment on the beach or near the coast. Despite such measures, changes in the prevailing conditions can lead to dramatic variations in coastal morphology and hence flood and erosion risks. Gaining insight into the physical processes that govern mesoscale morphological evolution (French et al., in this issue; van Maanen et al., in this issue), is crucial for the successful design of coastal defence systems and formulation of shoreline management strategy. From a forecasting perspective, various classes of approach are possible:

a) Process-based modelling: These models include explicit representations of physical processes to describe the complex interaction between waves, tides, sediment transport, coastal defence structures and the resulting morphological and shoreline changes. This approach can be successful for short-term forecasting, such as single or multiple storm events, often at a limited spatial scale associated with specific engineering schemes. It becomes less feasible for longer term simulations and larger domains for the reasons already mentioned above. Nevertheless, process-based modelling is capable of providing valuable insights into complex processes, thus improving the level of understanding of those processes, as demonstrated in the reviews by de Vriend et al. (1993), Nicholson et al. (1997), Roelvink (2006), Pan et al. (2010) and others.

b) Data-driven modelling: This relatively new class of approach is discussed in detail inSection 3, so only a brief outline is given here. In essence, data-driven models use measurements of past conditions at a site, together with sophisticated statistical techniques, to identify patterns of behaviour that are then extrapolated into the future to form a forecast.

c) Hybrid Modelling: This covers models in which simplifications to the governing equations or the forcing, or both, are made in order to make mesoscale forecasting tractable. Such approaches have also been termed 'reduced complexity methods' or 'behaviour-oriented' models. This class also includes approaches that combine two or more modelling concepts to form hybrids, such as combining empirical equilibrium models with wave sequences to introduce an evolutionary element (e.g. Yates et al., 2009 and Splinter et al., 2014). In some situations information from complex process models is used to provide parameterised data to simpler models, such as one-line or N-line models, in order to retain the primary coastal processes. An example of this type of approach includes the use of Unibest-CL + with parameterizations derived from a number of simulations using the model suite Delft3D reported by van Koningsveld et al. (2005). Huthnance et al. (2008) provide a brief survey of hybrid models developed for estuary and coastal inlet morphology prediction with a range of complexity, including the Analytical Emulator described by Manning (2007), the Hybrid Regime model of HR Wallingford (2006), the SandTrack approach of Soulsby et al. (2007), ASMITA described by Wang (2005), and the inverse techniques ofKarunarathna et al. (2008), that have demonstrated applications with encouraging results.

d) Probabilistic modelling: This class of modelling is used to quantify uncertainties and is included as a separate approach because the input data and output quantities are distinct from those of deterministic models. Specifically, descriptions of the probability distribution functions and correlation properties of the drivers are required as input, and the output is (sample) statistics of the dependent variables. Many probabilistic models are essentially deterministic models run many times over to create a Monte Carlo simulation, in order to assess the levels of uncertainty in the predictions. For example, Monte Carlo simulations have been presented by Vrijling and Meijer (1992) for a port development, by Dong and Chen (1999) to investigate the effect of storms on an open coast, by Lee et al. (2002) for forecasting cliff erosion,Wang and Reeve (2010) to assess the performance of a detached breakwater scheme and Callaghan et al. (2013) to estimate storm erosion. The output from these models can provide an indication of the uncertainty in the predictions of deterministic models but are difficult to validate in the sense that there are rarely sufficient measurements to verify that the statistics of the model output correspond to the statistics of the measurements. To address this deficiency other probabilistic approaches have been developed to provide an independent check of Monte Carlo results. These are quite complex and have been developed for 1-line model conditions only to date. In one of the first examples, Reeve and Spivack (2004)provided analytical expressions for the first two moments of beach position in the case of a beach nourishment. Subsequently Dong and Wu (2013) formulated a Fokker–Planck equation for the probability distribution of shoreline position, andReeve et al. (2014) presented a closed-form analytical solution for the mean beach position near a groyne, subject to random wave attack.

DATA-DRIVEN METHODS

Most models are driven by data through initial or boundary conditions so the term 'data-driven' might at first seem rather all-encompassing. However, the term 'data-driven method' refers to techniques that rely solely on the analysis of measurements, without invoking knowledge of physical processes. Attribution of particular behaviours found through the analysis to particular physical processes is through inference. At its most basic, a data-driven method involves analysing a sequence of measurements of forcing variables or coastal state indicators in order to find evidence of trends, cycles or other smoothly varying modes of change. The term 'data driven model' is best reserved for the process of making a forecast of future coastal state formulated around an extrapolation of the patterns found from analysis. Changes in morphological regime are unlikely to be captured unless the past records also contain such episodes. Much of the literature covers the application of methods to finding patterns in measurements and interpreting these in terms of physical processes. There is less work reported on the use of data-driven methods for forecasting.

The origins of data-driven modelling are difficult to pinpoint but evolved from the application of prognostic analysis techniques and a recognition that many coastal datasets seemed to exhibit coherent patterns of temporal behaviour that could be extrapolated to form a prediction. The extent of signal extraction has improved as the sophistication of statistical methods has increased. One of the most basic problems encountered in coastal data analysis is that measurements usually provide poor resolution in time and are often intermittent, thereby making interpolation very error-prone. Spatial sampling is often much better with surveying along fixed profiles at regular intervals becoming a norm in monitoring schemes. Analysis methods based on Fourier analysis (in time) are therefore difficult to use and alternatives have been sought, with Empirical Orthogonal Function analysis being one of the most widely used. Early analyses with Empirical Orthogonal Functions, (EOF), such as those of Winant et al. (1975) and Aranuvachapun and Johnson (1978), relied mostly on the records of beach profile measurements. In contrast, the later studies of Wijnberg and Terwindt (1995) andReeve et al. (2001) extended analyses to nearshore and offshore morphology respectively. Notable exceptions are the analyses of Aubrey and Emery (1983), Solow (1987) and Ding et al. (2001) of data from long-established national tide-gauge networks, used for tidal harmonic decomposition and surge analysis.

The link between data-driven methods and process understanding is not a one-way relationship. For example, we might well expect from dynamical considerations that a beach subjected to seasonally changing wave conditions would show a seasonal signature in its response. This deduction can be tested through data-driven analysis such as that reported by Haxel and Holman (2004) who used EOF analysis to isolate seasonal patterns in beach alignment. Both Haxel and Holman (2004) and Thomas et al. (2010)were also able to identify longer interannual patterns in beach plan shape which were attributed to the El Niño–La Niña atmospheric oscillation, something that could be tested in process modelling.

National beach monitoring programmes, such as those run by the US Army Corps at Duck, Carolina, the Polish Academy of Sciences Lubiatowo Coastal Research Station, and the New Forest District Council›s Channel Coastal Observatory in the UK, (New Forest District Council, 2014), have spurred a rapid expansion in the type and sophistication of statistical methods that have been used for analysis. For example, Singular Spectrum Analysis (SSA) was employed to trace forced and self-organised components of shoreline change (Różyński et al., 2001), Principal Oscillation Patterns were used to derive a data-driven model of changes in nearshore bathymetry byRóżyński and Jansen (2002) and Canonical Correlation Analysis (CCA) was employed by Różyński (2003) to study interactions between bars on the multi-barred beach at Lubiatowo while Larson et al. (2000) used it to analyse the links between beach profile changes and wave conditions at Duck, North Carolina.

Hsu et al. (1994) described one of the first studies to use data-driven methods for forecasting beach levels. They combined beach level measurements with an indicator of wave conditions, (Irribarren number), in an EOF analysis so as to forecast beach levels on the basis of the prevailing wave conditions. The technique was moderately successful but did not find widespread use. CCA has one advantage over the other methods in that it explicitly establishes linked patterns of behaviour between two variables, thereby opening the possibility of forecasting one variable on the basis of the other. This means it is ideally suited to morphological prediction where beach response is strongly dependent upon hydrodynamic conditions. The thinking behind this is thus: our predictive capability for waves and tides is generally greater than it is for coastal morphology; if we can establish a link between the hydrodynamics and the beach morphology then this, together with hydrodynamic forecasts, might be used as an effective predictor of coastal morphology. Horrillo-Caraballo and Reeve, 2008 and Horrillo-Caraballo and Reeve, 2010 demonstrated how CCA could be employed to analyse beach profile and wave measurements at a sandy beach at Duck, North Carolina and a mixed sand/gravel beach at Milford-on-Sea on the south coast of the UK. Forecasts of beach profiles at both sites, based on the correlations between waves and beach profiles, had a quality useful for planning purposes over a period of about a decade. In a parallel, but mathematically quite similar, development artificial neural networks have also found application in coastal sciences. Very often presented as a method in which an artificial 'brain' is first 'trained' on a range of input and output data and then used to make forecasts using new input data, this method is in essence a multiple regression technique. One of the earliest applications in coastal engineering was by van Gent and van den Boogaard (2001), who used measurements obtained in laboratory experiments to create a neural network to predict forces on vertical structures. A similar process was used to create a wave overtopping calculator for coastal structure design in the EurOtop Manual (2007), and many subsequent applications for engineering design purposes. The use of neural networks for coastal morphology was discussed by Southgate et al. (2003), and Pape et al., 2007 and Pape et al., 2010 have described the application of a variety of neural net algorithms to the problem of predicting the movement of a nearshore bar over the period of several years. The core assumption of data-driven models is that information on the evolution of a system can be extracted from signals that are manifestations of that evolution.

In the following subsection a description of the Lubiatowo coastal station and dataset is provided as this forms much of the basis of the discussion of the different data-driven methods that follows. In the subsequent subsections the different analysis techniques are described briefly, following which the process of applying this sequence of data-driven methods to the data is illustrated to demonstrate how the individual methods can be applied and what extra information becomes available at each stage.

Lubiatowo Coastal Station

The IBWPAN Lubiatowo Coastal Research Station is located on the shores of the southern Baltic Sea, near Gdansk, Poland (Fig. 1). The beach is sandy and approximately north facing. Tides are extremely small (a few centimetres) and most water level variations arise as a result of surges due to the passage of low pressure atmospheric depressions, and local wind and wave set up.

Figure 1: Location of IBW PAN Coastal Research Station at Lubiatowo.

The coastal station was constructed in the 1970s and consisted of 8 measuring towers from which measurements including waves, water levels and seabed elevations were made, (Fig. 2 left panel). Storm damage and economic constraints have meant that the most offshore towers have not been replaced so that now there are two towers remaining (Fig. 2 right panel).

Figure 2: (Left) 1974–8 measuring towers operational, erected for joint COMECON experiments in 1974 and 1976; (Right) 2008–2 towers remaining.

Measurements performed and archived at the station include:

- Routine monthly records of beach topography and shoreline configuration along 27 geodetically referenced cross-shore profiles, since 1983 (Fig. 3);

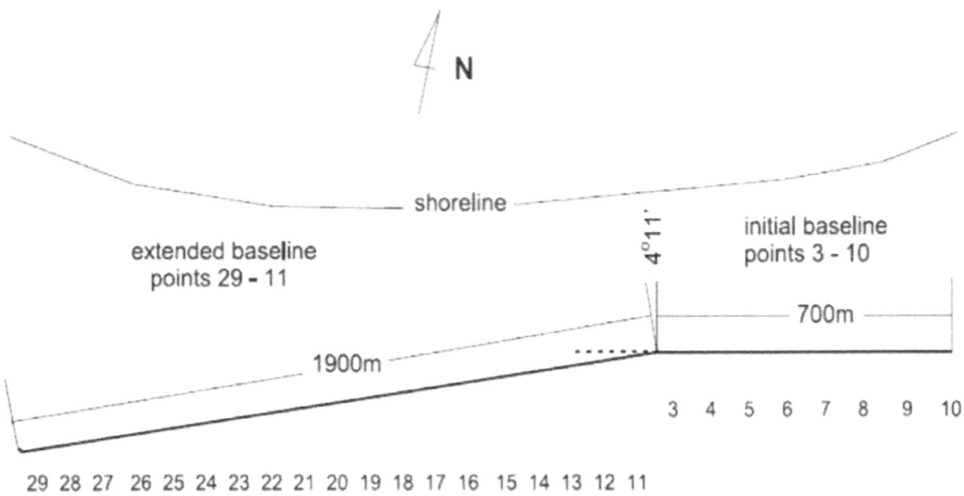

Figure 3: Location of survey profiles along the baseline at Lubiatowo.

- Routine annual/bi-annual records of nearshore topography along the same profiles; cross-shore range ca. 1000 m;
- Records of wave height and wave driven currents within the surf zone (wave gauges, current metres) during field experiments;
- Records of deepwater wave height, period and direction by a waverider — during field campaigns;
- Records of wind speed and direction as supplementary measurements.

The Lubiatowo site provides a very good example of a beach with multiple bars. Its minimal tides mean that this process can be discounted as a major factor when trying to understand the evolution of the shoreline. On beaches with multiple bars the outer bars control the evolution of inner bars by altering hydrodynamic regimes, through wave breaking and energy dissipation that affect inner bars. During moderate storms they are much less active (waves may steepen but do not break), so almost all energy dissipation occurs over inner bars. Thus, apparently simple non-tidal, longshore uniform beaches with multiple bars are highly nonlinear and therefore difficult for classical process based modelling. That is why statistical techniques are a useful option to further our understanding of such cases.

Fig. 4 shows the nearshore seabed evolution over a few years. The inner bars are fairly stable features; they are well-defined in the time mean, whereas the outer region of the seawardmost bar is much less homogeneous in its behaviour.

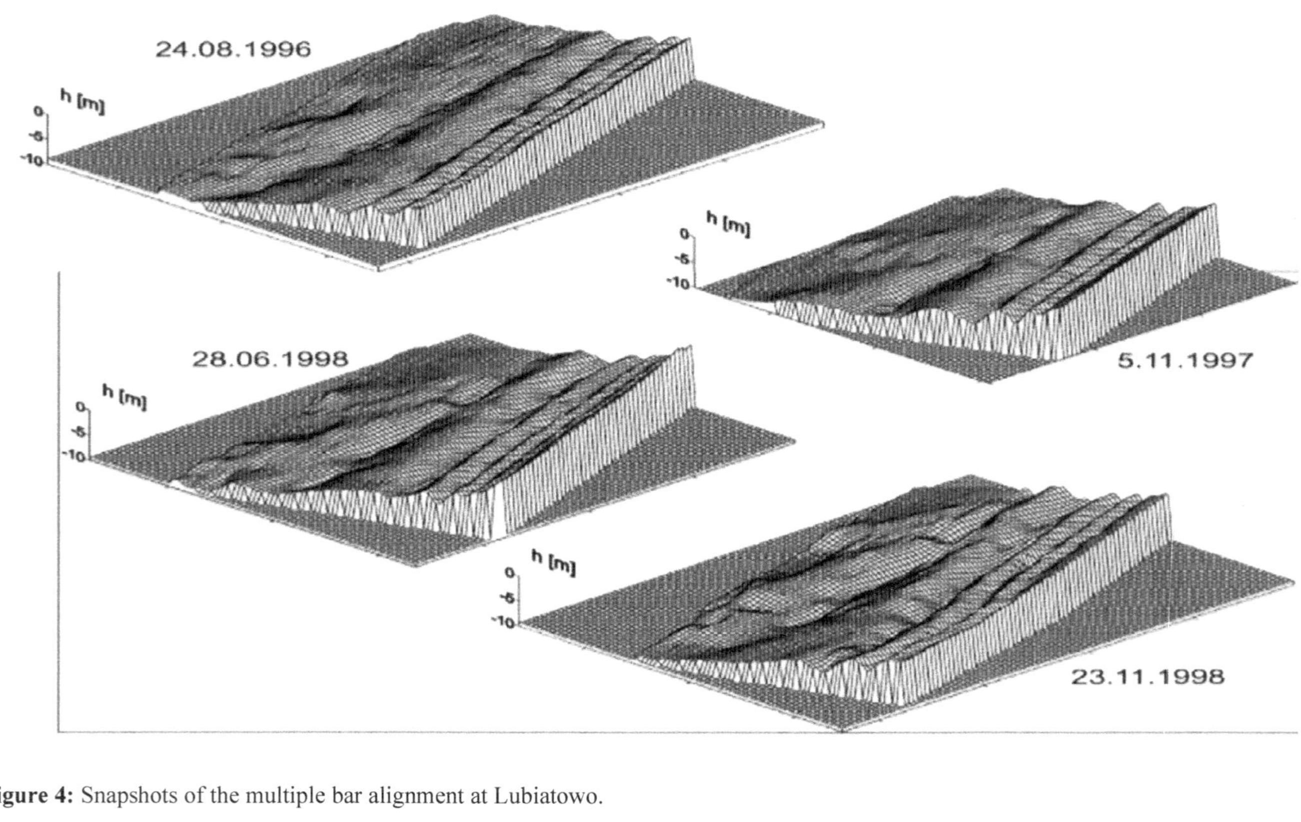

Figure 4: Snapshots of the multiple bar alignment at Lubiatowo.

Methods Based on Signal Covariance Structures

Data-driven techniques based on the covariance structure of analysed signals often assume that all information about statistical properties of those signals is contained by that structure. This is equivalent to the assumption that all random variables of the signal are normal (Gaussian). However approximate this assumption may be, it is usually exact enough to be accepted. The most important consequence of normality is that if random variables, centred about their mean values, are orthogonal, then they are uncorrelated and, by virtue of their normality, independent. Methods that fall into this category include Empirical Orthogonal Functions, Complex Principal Component Analysis, Canonical Correlation Analysis, Singular Spectrum Analysis and Multi-channel singular spectrum analysis. In the following subsections these analysis techniques are described briefly, following which the process of applying this sequence of data-driven methods to the data is illustrated to demonstrate how the individual methods can be applied and what extra information becomes available at each stage.

Signal Covariance Methods

Theoretical Background

The common idea of all techniques investigating signal covariance structure is based on breaking down the covariance matrix into eigenvectors, which are orthonormal. Particular variants of this approach are related to whether the analysis is done in the spatial or temporal domain or in both. When the covariance matrix is computed in the spatial domain, the associated method is termed the Empirical Orthogonal Function or Principal Component Analysis. Its formalism is simple and well known. Let us represent the measured seabed depths, centred about their mean depths as hxt, where x ranges between 1 and nx and t between 1 and nt. The terms of the symmetric covariance matrixA^{EOF} can then be expressed as:

$$a_{i,j} = \frac{1}{n_x n_t} \sum_{t=1}^{n_t} h_{i,t} h_{j,t}. \tag{1}$$

The matrix A^{EOF} possesses a set of positive eigenvalues λn and a set of corresponding eigenfunctions enx, defined by the matrix equation:

$$A^{EOF} e_n = \lambda_n e_n. \tag{2}$$

We can directly see from this equation that the sum of all eigenvalues is equal to the trace of the matrix, i.e. the global signal variance. Each of the n eigenvalues represents a portion of that variance and they are usually rearranged in decreasing order of magnitude. The magnitude of the ordered eigenvalues typically decreases quite rapidly and the number of eigenvalues that are significant can be determined using an empirical 'rule of thumb' proposed by North et al. (1982). Due to the orthonormality of eigenvectors, the principal components cnt are obtained from:

$$c_{nt} = \sum_{x=1}^{n_x} h_{xt} e_{nx} \tag{3}$$

The principle components encapsulate the time variation in the measurements. The key merit of EOF decomposition is the separation of the temporal and spatial patterns within the signal. The spatial patterns (eigenvectors) are usually plotted in the spatial domain. Since the eigenvectors are scaled to unit length regions where they have relatively large amplitude indicate areas of significance of a particular eigenvector. The principal components can reveal trends, oscillations and other types of behaviour. Studied jointly with the associated eigenvectors these can assist in interpreting the patterns within the signal.

Given that we might expect, from consideration of the physical processes, a linkage between two variables, it is reasonable to ask whether there are methods that could test the strength of such a linkage. One such technique is Canonical Correlation Analysis (CCA). It was introduced by Hotelling (1935) and used in climatology (Glahn, 1968).Larson et al. (2000) were the first to apply the method in a coastal engineering context. The method consists in analysing two data sets in form of vector time series Yt,y and Zt,zwhere t indicates observations in time of spatial variations y and z. The number of observations must be the same for each data set but the number of spatial points do not need to be equal. Subtracting mean values for each spatial location centres the data sets. Finally, we construct linear combinations of Y and Z respectively to define new variables that are maximally correlated. Additionally, a scaling is introduced such that these new variables have unit variances and are orthogonal. The CCA method then proceeds by finding, through matrix algebra, the (linear regression) relationship between the predictor field Y and predictand Z. Linear regression methods have been used in coastal engineering in order to investigate dependencies and establish empirical relationships between variables (e.g., Larson and Kraus, 1995 and Hedges, 2001).

In practice, the performance of regression analysis methods is subject to the characteristics of the observations, the level of noise in the data, and the choice of regression function and technique. In using regression for prediction it should always be borne in mind that the regression relationship is based on past measurements, so that if the system being measured undergoes a major change, or a new driver suddenly becomes important, predictions based on historic data may no longer be reliable, and may give misleading results (Freedman, 2005).

Application of EOF and Canonical Correlation Analysis

An example of an application of the EOF method is presented below, for four neighbouring cross-shore profiles, (profiles 4, 5, 6 and 7) at Lubiatowo. The seabed is usually characterised by 4 alongshore bars, but during the measurements the number of crests varied between 3 and 6. Fig. 5 presents the average profiles, which are very much uniform alongshore. The spatial structures of the three most important eigenvectors, associated with the greatest eigenvalues of the covariance matrix and their respective principal components are plotted in Fig. 6.

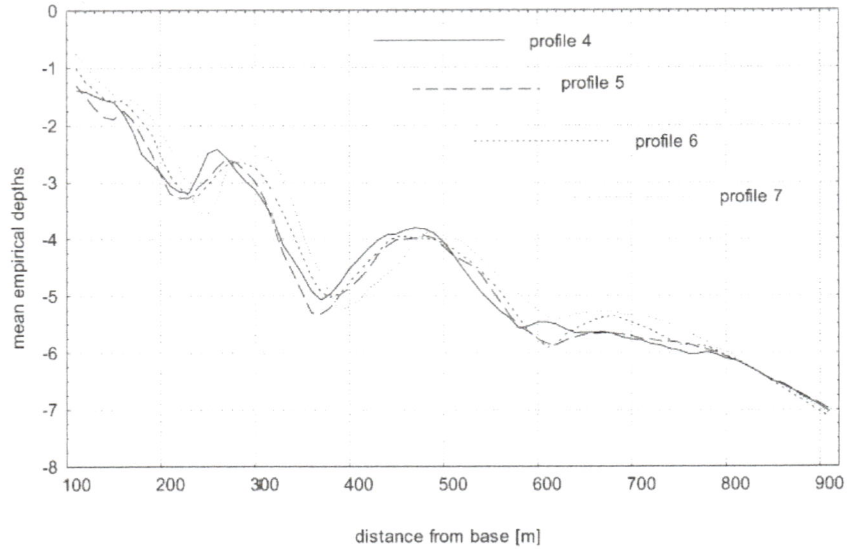

Figure 5: Mean profiles 4, 5, 6 and 7, Lubiatowo, Poland 1987–1998, (from Różyński and Jansen, 2002).

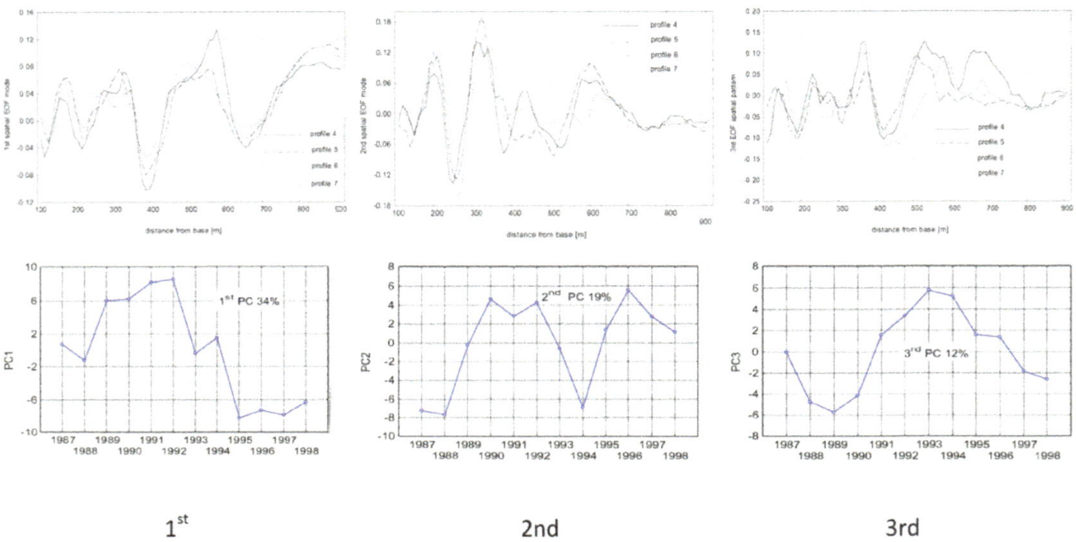

Figure 6: EOF eigenvectors (top) and principal components (bottom) at Lubiatowo, Poland, 1987–1998, (fromRóżyński and Jansen, 2002).

The first three EOFs explain 65% of the total variance (1st EOF: 34%, 2nd EOF: 19% and 3rd EOF: 12%). The 1st EOF, is very consistent for all four profiles; it displays similar amplitudes of its extremes all over the study area, highlighting high alongshore uniformity of the Lubiatowo beach. The extremes coincide with average positions of crests and troughs (see Fig. 5 and Fig. 6). This implies that the 1st EOF describes vertical oscillations of the whole system, which are expected to be a bit more significant in the area of outer bars. The 2nd EOF, is also fairly well alongshore uniform and is most pronounced for the area of inner bars 1 and 2. The most substantial changes, related to the 3rd EOF occur for the area stretching from the offshore slope of bar 2 up to the trough between bar 3 and 4. It may therefore describe interactions between inner bars 1 and 2 and outer ones 3 and 4 in the form of cycles with the period of approximately 12 years. The 3rd EOF shows dissimilarities from one profile to the next, indicating that interactions amongst inner and outer bars are not always uniform in the alongshore direction. The presence of independent oscillations in the inner bars suggests that the inner and outer bars may form separate sub-systems in their long-term evolution.

When using CCA the choice of predictor and predictand requires some care, as noted byRóżyński (2003). Some of the results of a CCA analysis are shown in Fig. 7. The standard deviations of measurements are drawn as a thick solid line at the top. We can see they have minima over bar crests and troughs (the latter except for the trough between bars 3 and 4), whereas the maxima are associated with onshore and offshore bar slopes. Such a pattern indicates that crests and troughs are fairly firmly fixed. Firm positions of crests are quite surprising at first glance, but together with less surprising firm positions of troughs they imply that bar oscillations occur in fairly narrow bounds, because it looks unlikely for crests to travel far enough to take positions of troughs and vice versa. However, bar oscillations are strong enough to cause very pronounced changes in depth over bar slopes. In other words, the slope nearer a crest at one time becomes the slope nearer a trough at some other time, which can only occur in connection with dramatic changes in depth. It should be remembered though, the bars do not oscillate as rigid structures and the records of bed topography only reflect joint effects of movements of each and every sediment grain.

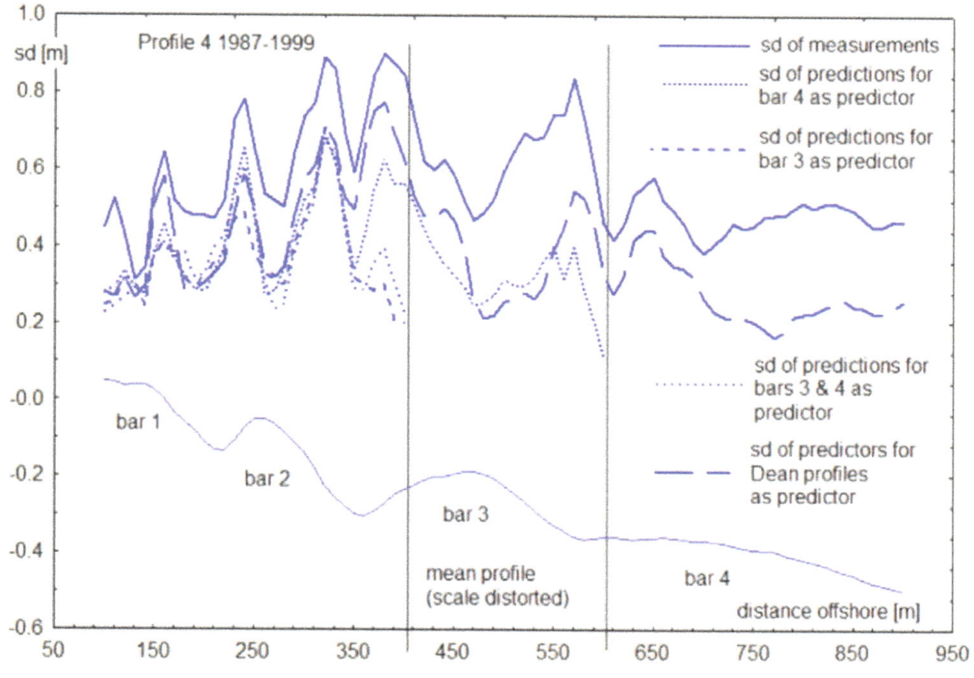

Figure 7: CCA analysis of outer and inner bar interactions at Lubiatowo, Poland (Różyński 2003).

Since the EOF analysis suggested the existence of the subsystem of inner and outer bars, an initial choice was to select part or the whole outer subsystem as a predictor and the rest as a predictand, so the first analysis was done for bar 4 as the predictor (610–900 m) and bars 1, 2 and 3 as the predictand (100 to 600 m), in order to assess the dependence of inner bars on the outermost bar. The standard deviations of predictions are plotted in Fig. 7 as a dense dotted line. This line shows that roughly speaking prediction errors are proportional to standard deviations of measurements, so positions of crests and troughs are predicted with better accuracy than onshore and offshore bar slopes.

The 2nd analysis took positions of bars 1 and 2 as predictand (100 – 400 m) and bar 3 as predictor (410–620 m). The dense dashed line of prediction standard deviations is very similar to the previous one. The results hardly change when the outer bars 3 and 4 are used as a predictor (410 to 900 m). Standard deviations of predictions are shown as a dotted line and are again very close to previous results. It can be therefore concluded that predictands and predictors must share a common feature, so that practically the same results are obtained for various combinations of predictor and predictand. They also point out the upper threshold (roughly 60%) of the variability of inner bars that corresponds to the variability of outer bars.

While helpful, such results do not relate directly to empirical understanding. One means of injecting such knowledge into the CCA approach is to try to link the observed seabed changes to the seabed equilibrium profile, expressed by Dean›s coefficients derived upon least square fit to the measurements. These define equilibrium profiles, which may serve as the predictor to measurements treated as the predictand, over the entire stretch 100–900 m. The CCA analysis performed for such a pair of predictor–predictand illustrates to what extent equilibrium profiles control different portions of beach profiles. The line of standard deviations of predictions using the equilibrium profiles is plotted as intermittent line in Fig. 7 and matches both the patterns of previous analyses as well as the line of standard deviations of measurements. Różyński (2003) found that the average prediction error over the entire profile is very similar to results found with the previous predictors. Thus from the analysis over the entire line 100–900 m with the Dean profiles as predictors the outcome of partial CCA computations over the profile subsets can be deduced very accurately.

The above results demonstrate inner bars 1 and 2 depend much less on equilibrium profiles than the outermost bar 4. Since changes in equilibrium profiles depict vertical profile variability, it may therefore be concluded that the outermost bar is dominated by vertical fluctuations, whereas for inner bars vertical movements play less crucial role. It confirms the EOF analysis, where the 1st pattern, expounding vertical oscillations of the whole seabed, was most pronounced in the vicinity of the outermost bar. This conclusion also illustrates the synergistic effect that using two methods for the same data set and obtaining consistent results may have.

It might be considered that while using the measurement of seabed elevation in one location to predict the location of the seabed in another is helpful to support physical explanations of the observed behaviour it doesn›t help much in terms of practical prediction, (if you are measuring the seabed level at one point it is probably quicker and cheaper to go and measure the elevation at another nearby point while you have kit and crew mobilised). The CCA method is ideally suited for use in a forecasting role to predict one variable on the basis of another one. This works best when there is a physically plausible link between the two series such as wave energy and shoreline position. While the CCA makes no use of any physical process understanding, a user can improve its performance by using informed judgement in the choice of predictor and predictand. One possible approach is as follows: given a sequence of beach profile measurements and wave conditions we split this into two sections; the first section will be used for the CCA analysis to compute the regression matrix, while the second section will be used to assess 'forecasts' made on the basis of the regression matrix; in order to make the number of beach profiles (approximately monthly) and wave measurements (approximately 3 hourly) the same some aggregation method is required for the waves.Larson et al. (2000) suggested that as any of the measured profile changes were likely to be the result of the wave conditions between this and the previous profile survey this aggregating could be achieved by creating the probability density function, (pdf), of wave heights of the waves in the interval; forecasts could be made on a monthly basis by using the pdf of the wave heights between the starting point and a month hence, and then estimating the corresponding beach profile using this pdf combined with the regression matrix. Larson et al. (2000) investigated the performance of this approach for different wave parameters as well as inshore and offshore wave conditions. Horrillo-Caraballo and Reeve (2008) extended Larson et al.›s (2000) study with additional measurements from Duck and investigated how the choices made in describing the wave climate could influence the quality of predictions made on the basis of the CCA results. They found the best performance was achieved using an empirical description of wave heights and suggested that resampling techniques could be useful in quantifying the forecast uncertainties. In a subsequent investigation Horrillo-Caraballo and Reeve (2010) applied the same technique to measurements taken at a shingle beach in Milford-on-Sea, UK. Results were as good, if not better, than for the Duck site, demonstrating that the CCA can find patterns of linked behaviour in two distinct datasets, irrespective of the details of the underlying physical processes, which are quite different on sandy and shingle beaches. Fig. 8 shows an example of the forecasts made at Milford over a 3 month period using the CCA regression together with offshore wave measurements. The root mean square error in the forecast profile is of the order of 0.3 m which, for a 3 month forecast compares favourably with the quality of forecast obtained from process-based models over the same time frame.

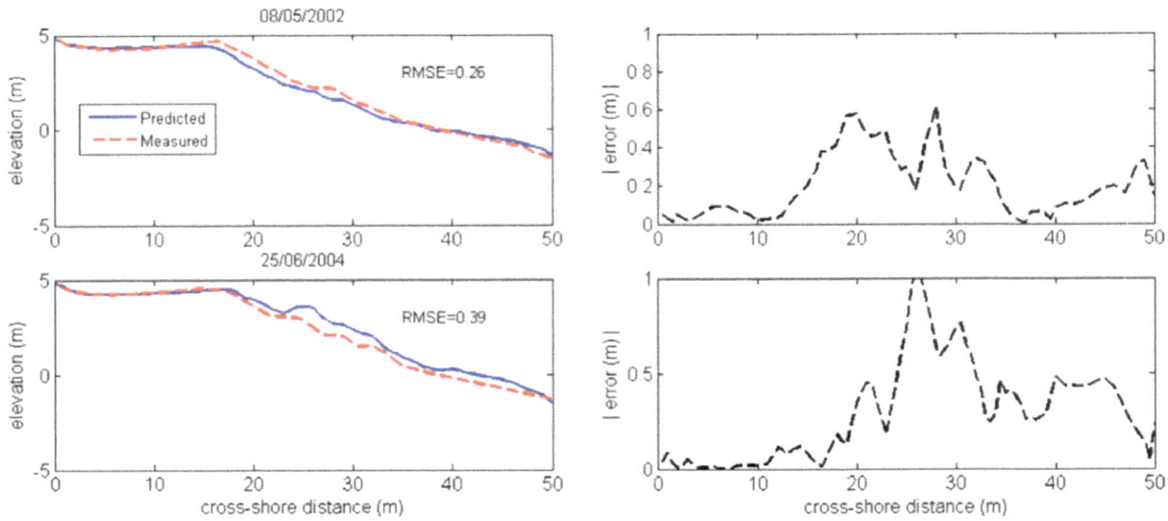

Figure 8: Comparison of hindcast profiles and actual for May 2002 and June 2004 (left panel). Cross-shore distribution of error (right hand panel). Hindcasts were made over a period of 6 months — the gap between successive profile measurements.

The right hand panels (Fig. 8) show the distribution of the error across the profile, which reaches a maximum across the intertidal area, a feature also found for predictions at Duck. This reflects the fact that both beaches are micro-tidal and therefore their response will reflect the influences of tides. This effect is absent from the CCA regression and is likely to be one of the main sources of the discrepancies. These results also indicate that any model of the mesoscale evolution of these beaches should account for both tides and waves for best results. The method is not limited to beach profiles and may also be applied to other coastal indicators such as beach plan shape. De Alegría-Arzaburu et al. (2010) describe an application of CCA to derive forcing–response relations between the wave climate and shoreline position on a macrotidal gravel barrier located in the southwest of the U.K. that is known to rotate in response to variations in the prevailing wave direction. The link between wave direction and beach plan shape orientation was isolated from the measurements purely from the statistical analysis without any direct physical process knowledge. More recently the CCA method was used to relate offshore wave conditions to beach plan shape at three morphologically distinct areas along the North Kent coast; one in an open exposure, one sheltered behind a breakwater and subject to diffracted waves, and one bounded by shore normal groynes. Using CCAReeve and Horrillo-Caraballo (2014) showed that the beach shape in each of the three locations could be forecast to a useful degree of accuracy from the same offshore wave conditions, but with different regression coefficients. This result demonstrates that if there is a link in the mutual patterns of behaviour between two datasets, the CCA will find them irrespective of the details of the physical processes and irrespective of their relative geographical locations. The significance of such results is that it provides a means of guiding decisions on the density and location of wave buoys for coastal management. On the basis of the locations studied so far this suggests that a network of offshore rather than inshore buoys might be a more economical means of routine monitoring of coastal wave conditions.

Returning again to the multiple bars at Lubiatowo, it seems we are able to understand the bathymetric variations to a moderate degree within linear methods, but this leaves a significant component unexplained. One possibility, mentioned in the Introduction, is that the complicated observed behaviour is a manifestation of a chaotic system. That is, a system that is governed by deterministic equations but, nevertheless, can exhibit seemingly unpredictable behaviour. There is a large literature on this subject but for our purposes we are interested in the associated time series analysis methods. The germane point from our perspective is that while it is now straightforward to analyse and identify chaotic behaviour from the solution of a set of nonlinear equations known to show chaotic behaviour, given only the output and an analysis that confirms chaotic behaviour, it is still virtually impossible to deduce what the governing equations were, or indeed how many governing equations there are. A technique called Singular Spectrum Analysis (SSA), can be used to assist in separating noise from underlying irregular but smooth changes in a signal. It can also give some insight into the dimension of the problem, that is, how many governing equations there might be producing the observed behaviour.Broomhead and King (1986) showed how the method, based on deep mathematical results, could be applied to the output of the Lorenz equations. Vautard et al. (1992)provide a useful summary of SSA and its applications to adaptive filtering and noise reduction. The method is based on analysing the lagged covariance matrix of the a time series xi, where $1 \leq i \leq nt$. To perform the analysis the user needs to define the number MLwhich is the window length or embedding dimension and its value needs to be chosen with care. It is, in effect, your estimate of the dimension of the system producing the output being analysed. The covariance matrix is symmetric, so its eigenvalues λk are all positive, and the corresponding eigenvectors are orthogonal. The eigenvectors form the time-invariant part of the SSA decomposition, whereas the variability of a given system is contained in principal components (PCs).

The PCs are time series of length nt − ML and are orthogonal to each other as MLconsecutive elements of the original time series are needed to compute one term of every PC. In one of the first applications of the SSA method to coastal morphodynamicsRóżyński et al. (2001) analysed 16 years of shoreline position measurements (1983–1999), sampled at equal 100 m intervals upon a monthly basis along a 2.8 km shoreline segment at Lubiatowo. No systematic behaviour could be deduced from the observations, but the SSA results suggested that the shoreline exhibits standing wave behaviour with periods lasting several decades found in the western part of the coastal segment, approximately 16 years in the middle part of the segment, and ~ 8 years in the central-eastern part of the segment. A more sophisticated version of the SSA, multi-channel SSA (MSSA), in which covariances in both time and space are computed, was subsequently applied to the same dataset by Różyński (2005). Some of the results of this analysis are shown in Fig. 9. The reconstructed components shown in Fig. 9 are analogous to the eigenfunctions of EOF analysis.

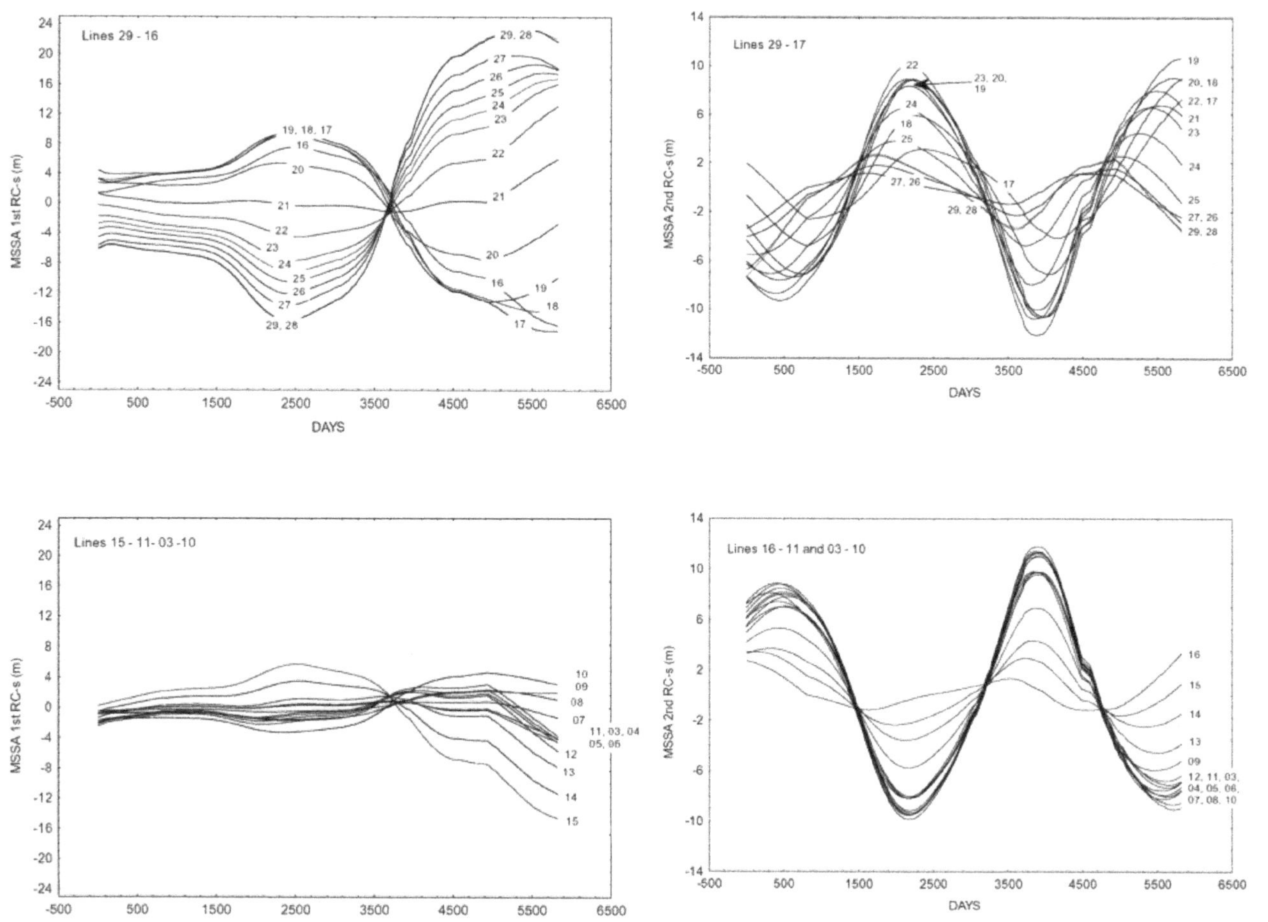

1st 2nd

Figure 9: Reconstructed components (1st — Left; 2nd — Right) using MSSA, Lubiatowo 1983–1999, (Różyński, 2005).

Fig. 9 (left panels) shows the 1st reconstructed component. Western profiles are shown on the left (lines 17 to 29), eastern on the right (lines 11–16 and 03–10). They feature a long shoreline standing wave with a period of several decades. The amplitude of that wave on the eastern sector is much less pronounced, so it had not been detected by the ordinary SSA method. Fig. 9 (right panels) demonstrates trajectories of the 2nd MSSA reconstructed components. They can also be interpreted as standing waves with a period of 7–8 years, with nodes at profiles 27–26 and 16–15. The antinodes can be observed at profiles 22–19 and 03–06, and a corresponding wavelength of 1000 to 1300 m can be deduced. Since this pattern was not as strong as the 1st reconstructed component in the western sector, it had not been detected by the ordinary SSA method either. Thus, the MSSA had confirmed the detection of the same shoreline standing waves as the SSA and found some extra patterns as well.

As another example of how findings from data-driven analyses can then initiate further research on the dynamics the detection of this 8 year oscillation prompted extensive studies into the coupling of beach response with a similar periodic component of the North Atlantic Oscillation (NAO) and its winter index (NAOWI) (Różyński, 2010). It was found that this coupling is transmitted by the wave climate during winters.

In concluding this section we summarise the methods based on the analysis of covariance structures within measurements as:

- .Always begin with fairly simple methods like EOF or SSA, where covariances in either space or time are considered;
- .When the amount of data is sufficient, and the results of Step 1 are inconclusive, a more advanced tool (eg. MSSA, CPCA) may be able to extract just a few dominant patterns;
- .When the coastal system is complicated by other additional processes, for example tides or stratification, there may be 2D spatial patterns or migrating seabed forms that may require advanced tools (CPCA, MSSA);
- .Beach nourishment schemes are usually frequently monitored, so they can be a perfect subject for data-driven analyses;
- .When hydrodynamic information is available, every effort should be made to use this in the analysis or interpretation.

Other Methods

We conclude our discussion of data-driven methods with a section on two of the more recent techniques which perhaps have more in common with Fourier decomposition than covariance methods. Namely, wavelets and empirical mode decomposition (EMD). Both methods have been developed to isolate underlying features of a time series that may be obscured by high frequency noise, but have yet to be used in a predictive scheme for coastal morphology.

Wavelets are conceptually similar to Fourier decomposition, the main difference being that the functions used in the expansion, the wavelets, are non-zero only in a finite interval, rather than extending to infinity. Just like Fourier analysis there are continuous and discrete versions; the continuous version being more appropriate for theoretical developments while the discrete version being more suitable for practical application to time series of measurements taken at discrete points in time or space. Further, Fourier transforms implicitly assume periodicity of the analysed signal, and hence stationarity of the studied signal. Although they can provide useful information about a signal, it is frequently not enough to characterise signals whose frequency content changes over time. Hence, Fourier analysis is not an ideal tool for studying signals that are non-stationary. By contrast, due to the localisation property wavelet transforms enable us to obtain orthonormal expansions of a signal that is intermittent and nonstationary.

The discrete wavelet transform, (DWT), has been used widely in studies where there is a need to examine intermittent or variable phenomena in long data series, such that byMallat (1989) in signal analysis, by Kumar and Foufoula-Georgiou (1997) in geophysics, by Iyama and Kuwamura (1999) on earthquake analysis and by Smallwood (1999) for shock analysis. One attractive property of wavelets is time (or space)–scale localisation so that the signal variance at a particular scale is related to a particular time (location). In contrast to Fourier analysis, which uses harmonic sine and cosine functions, wavelet analysis is based on a 'mother wavelet' function. Some examples of wavelet functions that have been employed are shown in Fig. 10. One useful analogy for thinking about the difference between Fourier and wavelet analysis is to imagine the result of analysing a piece of piano music with both. Fourier analysis tells you how much of each note was played, whereas wavelet analysis will tell you what notes were played and when. Some of the first applications of wavelet analysis in a coastal engineering context were the studies by Short and Trembanis (2004) who analysed beach levels with continuous wavelets, and Różyński and Reeve (2005) who analysed water level and current records over the period of a storm using DWT. Li et al. (2005) developed an adapted maximal overlap wavelet transform to investigate the multiscale variability of beach levels at Duck, while Esteves et al. (2006) applied DWT to investigate seasonal and interannual patterns of shoreline changes in Rio Grande do Sul, Southern Brazil. The application of wavelet analysis to identify elements of predictive morphological models that required improvement was suggested by French (2010). One variant of DWT, wavelet packet transforms which yields improved resolution of the scale intervals of the variability than DWT, were applied by Reeve et al. (2007) to the Duck beach profile dataset. In much the same way that Rozynski›s MSSA analysis of the Lubiatowo site confirmed and extended the earlier SSA analysis so Reeve et al.›s (2007) wavelet packet analysis of the Duck site confirmed their DWT results but also lead to the finding that over 25% of the temporal variance in beach levels was attributable to interannual periods (16 to 22 months).

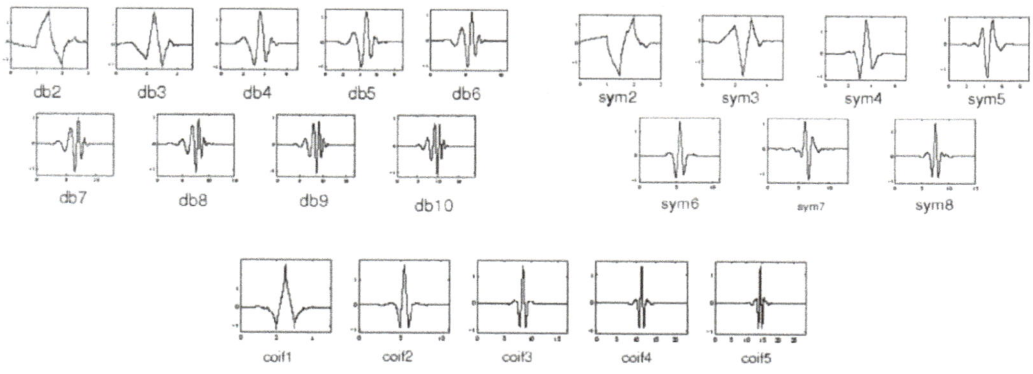

Figure 10: Orthogonal wavelets: near anti-symmetric dbN (top left), near symmetric symN (top right) and near symmetric coifN (bottom centre).

Pruszak et al. (2012) examined extended records of shoreline and dune foot positions at Lubiatowo, covering 25 years of observations between 1983 and 2008.

Using the coif5 wavelet function they found that the shoreline standing wave, detected as the 1st reconstructed component by the MSSA study, is also imprinted in the variations of dune foot, (see Fig. 11, top panel). Moreover, they were able to assess its lengthL $\approx$ 3700 m and period T $\approx$ 30 years as well as the amplitudes which, unsurprisingly, were more pronounced for the shoreline wave. A comparison of the outputs of the MSSA and DWT analyses (Fig. 9 and Fig. 11), demonstrates clearly that both MSSA and DWT methods captured the most important morphological patterns, but the greater number of records analysed in the DWT analysis allowed for a more precise description of its parameters.

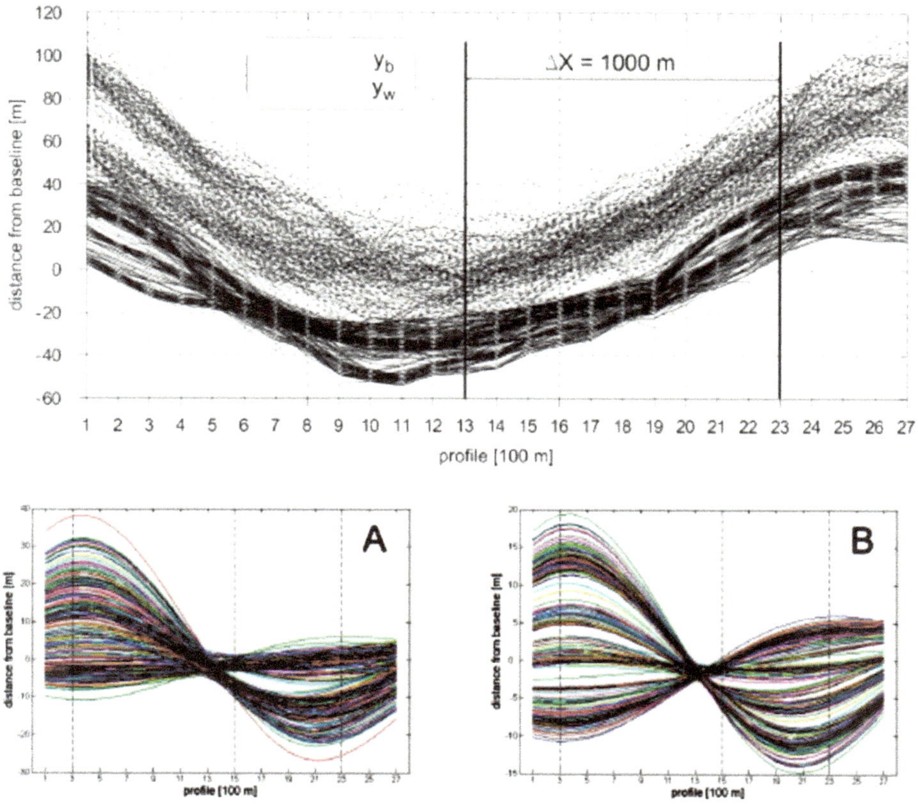

Figure 11: Extended (1983–2008) measurements of shoreline (yb) and dune foot (yw) positions, Lubiatowo, Poland (top panel). Shoreline (bottom left) and dune foot (bottom right) standing wave fixed by the DWT method using coif5 wavelet function (bottom panels) (Pruszak et al., 2012).

Empirical mode decomposition was introduced by Huang et al. (1998) as an adaptive method suitable for analysing all types of signal, containing nonlinear and non-stationary components. In common with other decomposition methods discussed above, EMD expresses a signal as the sum of functions termed Intrinsic Mode Functions, (IMFs). The IMFs are defined by the zero crossings of the signal and represent an oscillation embedded in the data. The signal being analysed must fulfil three rather general constraints: (1) the signal has at least two extrema: one maximum and one minimum; (2) the characteristic time scale is defined by the time lapse between the extrema; (3) if the signal has no extrema but contains inflexion points, it can be differentiated once or more times to reveal the extrema and the final results can be obtained by integration(s) of the components.

The decomposition of a signal into the IMFs is called the sifting process. In this, oscillations are removed successively from the original signal, creating a sequence of functions that iterate towards a signal that meets the conditions for an IMF. The first time through this procedure yields one IMF which contains high frequency content of the original signal. This IMF is then removed from the original signal and the process is repeated to find the next IMF. A sequence of IMFs can thus be computed from a single signal. The process is completed by imposing a halting criterion, such as when the magnitude of the standard deviation computed from two consecutive sifting results falls below a specified value.

The EMD method has been applied to beach profiles at Lubiatowo. Fig. 12 top left shows an example record of profile 4 from 1987 together with its IMFs. The purpose of this decomposition is illustrated in Fig. 13. The ultimate goal of the study was to assess the rates of wave energy dissipation in fully developed wave breaking regime to identify the areas of erosion and accumulation. To do so, the linear trend and low-frequency components were recombined to arrive at a smooth but monotonic trend. From this short term departures of wave energy dissipation rates from the constant rate valid for the Dean profile could be estimated. As a result, Różyński and Lin (2015) demonstrated that not only could the EMD method be used to extract smooth monotonic profiles reminiscent of equilibrium profiles but it was also possible to incorporate the hydrodynamic information to identify areas of the cross-shore profile prone to erosion and accumulation during storms.

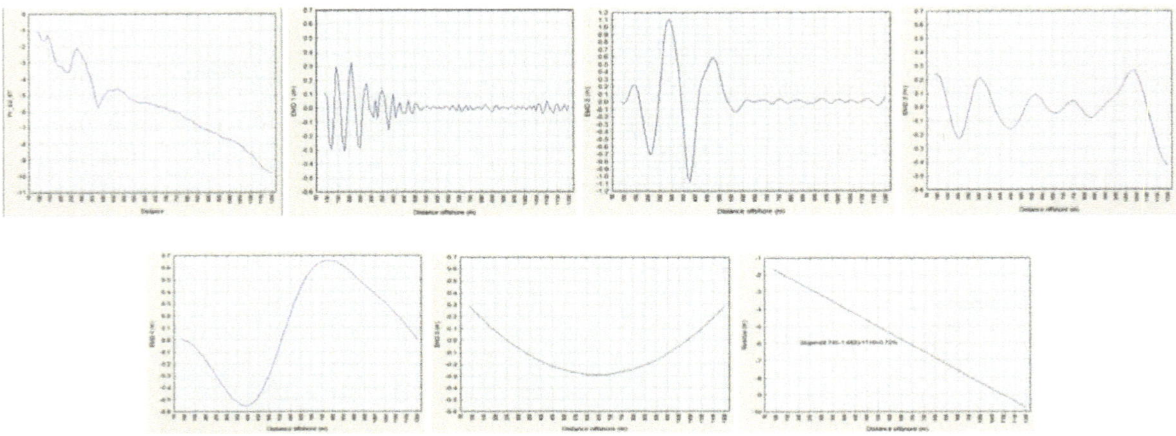

Figure 12: Decomposition of cross-shore profile with multiple bars with EMD method: surveyed profile (top left), IMFs (short oscillations top centre left, bars top centre right, low frequency modes top right and bottom left and centre, trend bottom right).

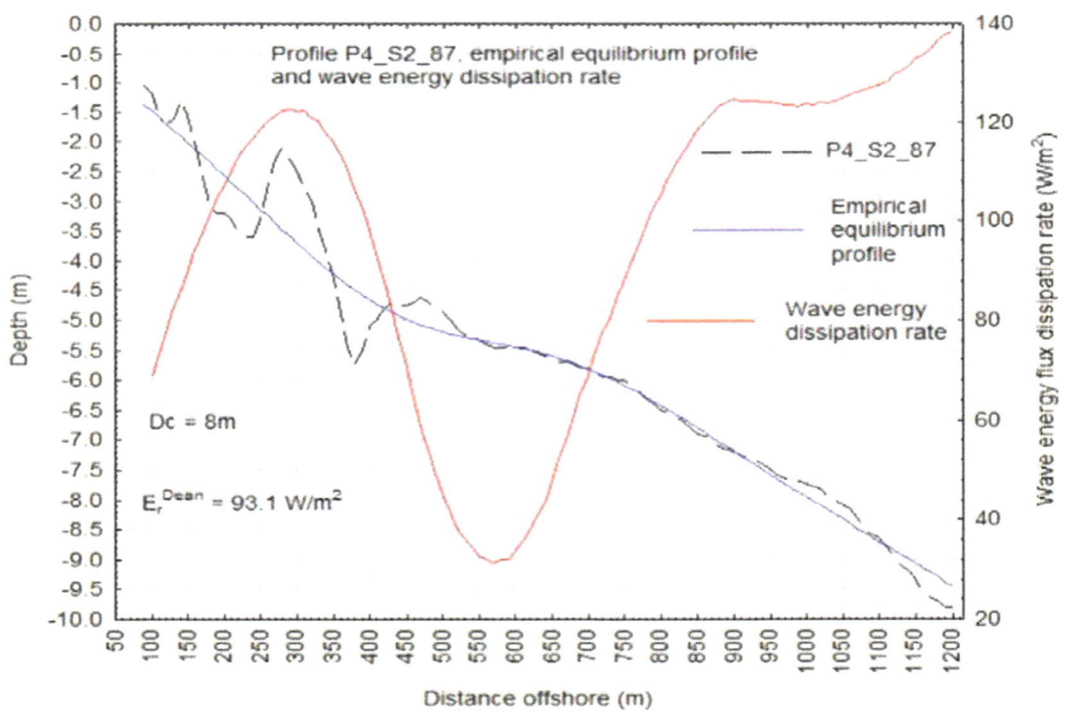

Figure 13: Raw profile (intermittent line), EMD monotonic trend (solid line) and wave energy dissipation rates (red line) evaluated using monotonic trend: rates greater/lower than 93 W/m^2 indicate erosion/deposition areas, seabed beyond D_c = 8 m should not develop wave breaking regimes, Różyński and Lin (2015).

HYBRID MODELS

As we have seen in the last section purely statistical methods are data hungry and do not make use of physical understanding (except indirectly through the choice of quantities being analysed). On the other hand, process models have difficulties with mesoscale forecasting for the reasons already mentioned in Section 1. There are also arguments that a scale-based approach to modelling would be a more successful way forward, and these chime with the wavelet type of analysis that is very much based on the concept of scale rather than correlation. Such arguments generally advance along the lines that morphological models developed upon our understanding of processes at small (human) scale do not contain all the necessary ingredients to encapsulate observed meso- and macro-scale variations. Murray (2007) provides an eloquent exposition of these qualitative arguments. They are supported by observations of shoreline evolution,Murray et al. (2015), and the difficulties experienced by process models in simulating observed mesoscale behaviours. This leaves us facing something of a dichotomy: Is it possible to scale process models up to mesoscale or is there a fundamental difficulty with this approach? The deterministic reductionist approach would argue that we should be able to predict at all scales if all the relevant processes are included in the model. By comparing model output with observations the performance of the model can be tested and the absence of relevant processes in the model gauged. If there are discrepancies then further experiments/observations are required to find and identify the missing process. The development of beach profile modelling provides an interesting lesson in this regard. In the 1980s and 1990s many beach profile models were developed for predicting beach response to storms, such as: HR Wallingford›s COSMOS model, Nairn and Southgate (1993); UNIBEST-TC of Delft Hydraulics, Reniers et al. (1995); CROSMOR2000 from the University of Utrech, Van Rijn and Wijnberg (1996); and many others. Roelvink and Broker (1993) and Van Rijn et al. (2003) noted that while very good performance was achieved by these models when compared against laboratory experiments of beach response to storm waves, these same models were not able to reproduce the observed post-storm beach recovery. The development of the XBeach code (Roelvink et al., 2009) sparked renewed interest in profile modelling. Subsequent development of the code, such as that reported by Jamal et al. (2014), have included the slightly longer time scale process of infiltration, (amongst other processes), to reproduce post-storm beach building. The point being that to identify a missing process from observations that has only a slightly longer time scale than a wave period, and to build this into a process model, has taken several decades.

For mesoscale modelling there is also a practical problem in that as time and space scales get larger so do the monitoring requirements. There is another, more undecidable, problem in that the system is nonlinear; nonlinear systems can exhibit chaos and emergent behaviour, which is predictable if you know the governing equations but if not, as discussed earlier not even methods such as SSA can determine the equations solely from observations. As a result a different class of model has emerged; termed variously 'reduced complexity', 'behaviour-oriented' or 'hybrid'. These terms encapsulate the idea behind the approach to create a model that distils the essence of mesoscale evolution without the difficulties associated with process models nor the data requirements of data-driven methods. It is fair to highlight a distinction between hybrid models and the other forms of model; hybrid models have been very much focussed on developing understanding as opposed to creating accurate forecasts.

In hybrid models, elements of two or more model types are combined. Some elements of the physics are eliminated in order to reduce computational costs and simplify the dynamics on the assumption that the broad scale morphological changes will be captured. They use simplified governing equations that exhibit the behaviour of the application. Good examples of such models in a coastal context are: the empirical equilibrium beach profile model of Dean (1977) and beach plan shape model of Hsu and Evans (1989); 1- or N-line shoreline evolution models pioneered by Pelnard-Considere (1956) and subsequent developments due to Hanson and Kraus (1989), Hanson et al. (2003), Dabees and Kamphuis (1998); the shoreline evolution model of Karunarathna and Reeve (2013); the cross-shore profile evolution models of Stive and de Vriend (1995)and Niedoroda et al. (1995); the estuary and tidal inlet models proposed by Stive et al. (1998), Karunarathna et al. (2008) and Spearman (2011); and the equilibrium shoreline models described by Yates et al. (2009) and Splinter et al. (2014). Hybrid models have been developed on the basis of physical intuition, hypothesis and mathematical considerations. By their very nature they tend to be process or scenario specific and, as yet, there is no well-defined or standardised methodology of deriving them from first principles. They will normally take as a starting point either a guiding physical principle (such as conservation of mass), or a specific physical process (such as transport or diffusion). An early example of such a model is due to Cowell et al. (1995) which relies only on sea level change and changes in geometry to predict shoreface translation, rather than hydraulic forcing.

Diffusion Models

One of the most widely used type of hybrid models is based on diffusion-type equations. Indeed the model proposed by Pelnard-Considere (1956) is such an equation for describing the evolution of the beach plan shape on the basis of laboratory experiments. In order to account for cross-shore sediment transport, 1-line models have been extended to track two or more contours along the beach. This group of models is known as multi-line models or 'N-line' models, and have been developed successively by Perlin and Dean (1983), Dabees and Kamphuis (1998), Hanson and Larson (2000) andShibutani et al. (2009). In contrast to 1-line models, N-line models simulate changes in cross-shore beach profile as well as beach plan form. However, they require detailed information on both cross-shore sediment transport rates and cross-shore distribution of alongshore sediment transport rates. This is often implemented as the rate of cross-shore transport being based on the principle that any profile deviation from the equilibrium profile will result in cross-shore exchange of sediment between contours as in the formulations proposed by Steetzel and de Vroeg (1999) and Hanson et al. (2003). While a diffusion type equation for longshore transport can be derived from physical principles, and the assumption of small wave angles, the same is not true for models of cross-shore transport. Nevertheless, one and two dimensional advection–diffusion type formulations have been used in dynamical beach evolution models, with appropriate initial and boundary conditions. In these models, sea bottom bathymetry change is assumed to take place as a result of the collective effects of diffusive and non-diffusive processes.

Diffusion has the effect of smoothing irregularities in the sea bed. However, smoothing is not the only morphological response of a seabed and other morphological changes have to be included in some way. A mathematically convenient way to achieve this is by including a source function in the equation which is an aggregation of changes driven by physical processes other than diffusion. The natural question that arises is of course how the source function should be specified. One way to investigate this is to take observations at two distinct times and determine the source function that is required by predicting the earlier observation to the second observation with the simplified model. This process amounts to what is termed an inverse problem; one is using observations to determine the parameter and function values in an equation. Many of the simplified morphological evolution equations take the form of diffusion-type equations and some progress has been made recently in developing methods to solve inverse problems of this nature. In fact, because diffusion-type models have been proposed for beach plan shape, beach profile and beach area problems the methods have a level of generality. If the source term does in fact represent the aggregation of coherent morphological processes it may be possible to hypothesise particular causal physical mechanisms on the basis of the shape of the source function in different settings. If this is indeed the case then it provides a natural way in which to extend the simplified model by including the newly identified process. In any case, if the inversion is carried out with sequential observations it is possible to build up a time sequence of source terms and analyse these for patterns of behaviour which could be extrapolated into the future to allow predictions of morphological evolution.

Here we discuss the method as applied to beach plan shape and profiles to illustrate how the inversion method can be used to understand mesoscale morphological changes. For beach plan shape the idea is to use the 1-line model as the governing equation, with observations of the beach position to determine the corresponding values of the diffusion coefficient and forcing function. The first such approach with this method was Reeve and Fleming (1997) who used a constant diffusion coefficient and a specific form of forcing function. These restrictions were subsequently relaxed in developments of the method by Spivack and Reeve, 1999 and Spivack and Reeve, 2000. Based on this initial work and extending the form of 1-line equation as proposed by Larson et al. (1997),Karunarathna and Reeve (2013) developed an advection–diffusion model to simulate beach plan shape evolution relative to a fixed line of reference:

$$\frac{\partial y}{\partial t} = \frac{\partial}{\partial x}\left(K(x,t)\frac{\partial y}{\partial x}\right) + S(x,t) \tag{4}$$

Eq. (4) describes the variation of shoreline position y(x,t) defined relative to a fixed reference line at longshore location x at time t. K(x,t) is interpreted as the space- and time-dependent diffusion coefficient which relates the response of the shoreline to the incoming wave field through longshore transport. S(x,t) is a space- and time-dependent source function which describes all processes that contribute to shoreline change other than longshore transport by incident waves, (including tides, wave induced currents and anthropogenic impacts).

Solving Eq. (4) to find K and S simultaneously, given a sequence of observations of beach plan shape is a challenging mathematical problem. Karunarathna and Reeve (2013) used an approximate, two step approach to determine K and S for Colwyn Bay Beach in the UK, from a series of historic shoreline surveys and wave measurements. (Shorelines were measured every 6 months from 2001 to 2006 and waves for the same duration, at five locations along the beach, were derived from hindcasting). This site was selected because Colwyn Bay beach has a clearly defined longshore extent and exhibits substantial morphodynamic variability; shoreline recession of 0.69 m/year has been observed at unprotected parts of the beach. There are also good records of waves and beach positions. In addition there are some rock groynes on the beach that provide the potential for interrupting longshore drift and causing some cross-shore transport; a signature that ought to be picked up in the source function.

The time mean diffusion coefficient was computed using the equation:

$$K(x,t) = \frac{2Q_0}{D_c}. \tag{5}$$

The depth of closure, Dc, was determined using the formula due to Hallermeier (1981)and Q_0 was computed from the time history of wave conditions and the CERC equation due to US Army Corps of Engineers (1984) to generate a time sequence of K(x,t) along the beach. The time mean diffusion coefficient $\overline{K}(x)$ was obtained by time averaging K(x,t) over intervals between successive beach surveys. The source function was determined as the inverse solution to Eq. (4).

The resulting longshore variation of the time mean diffusion coefficient for Colwyn Bay is shown in Fig. 14.

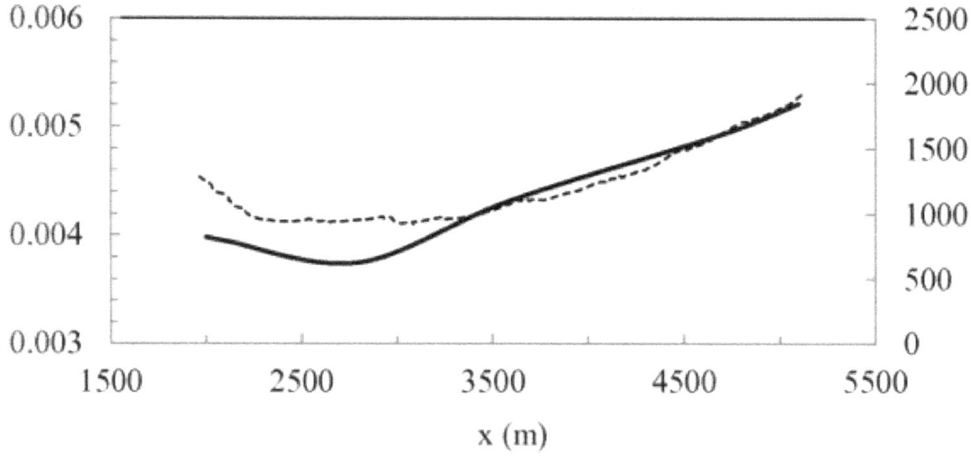

Figure 14: Longshore variation of mean diffusion coefficient along Colwyn Bay (dark line) and the mean shoreline (dotted line), (Karunarathna and Reeve, 2013). Left axis shows diffusion coefficient and the right axis the shoreline position relative to a reference line.

The distribution of K shows that the lefthand side of the beach is less mobile than the righthand side. It could reflect a grading in sediment size or wave exposure. The sediment shows little sign of alongshore grading, but an analysis of the nearshore wave climate confirms that the righthand end of the beach is exposed to larger and slightly more oblique waves.

Fig. 15 shows the envelope of all source functions recovered from historic shoreline measurements along the beach. The envelope of shoreline change, determined directly from measured shoreline surveys, is also shown in the figure. The two envelopes are distinctly different. The source function relates to the residual of cross-shore wave transport processes. Its largest values coincide with the region of the beach which has rock groynes. Therefore, the difference between the two envelopes provides an estimate of the alongshore transport contribution to shoreline change.

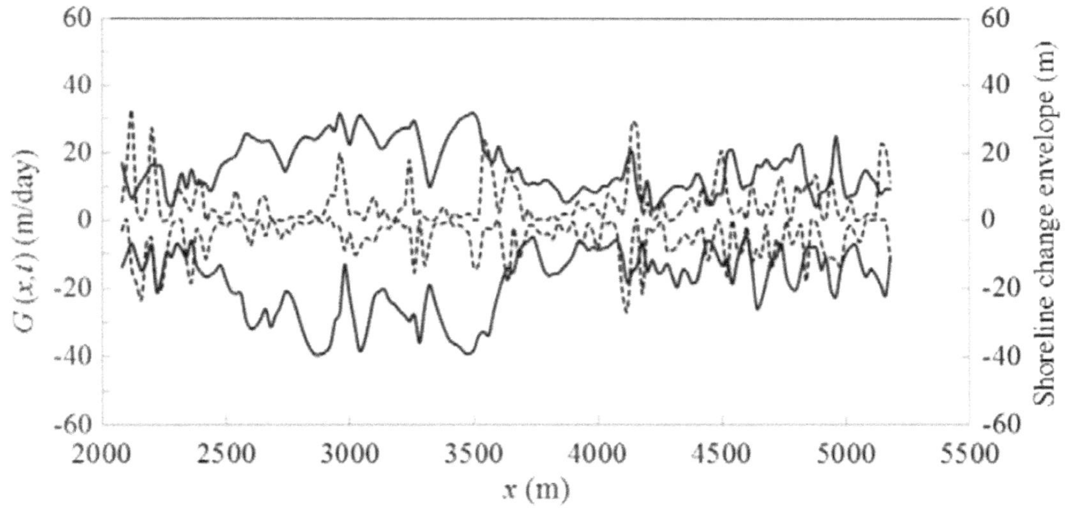

Figure 15: Spatial distribution of time varying source function envelope (dotted lines), measured shoreline change envelope (dark line) (Karunarathna and Reeve, 2013).

The combination of a hybrid morphodynamic shoreline model, data, and an inverse technique has provided an approximate means of distinguishing between the contributions of cross-shore and alongshore transport to shoreline change. Further, the mean diffusion coefficient can be related to the longshore variations in wave climate, thus giving some tangible physical interpretation to the somewhat abstract quantities K and S.

Stive et al. (1991) and Niedoroda et al. (1995) proposed an advection–diffusion model for predicting cross-shore beach profile change, of the form given in Eq. (6):

$$\frac{\partial h(x,t)}{\partial t} = \frac{\partial}{\partial x}\left(K(x)\frac{\partial h(x,t)}{\partial x}\right) + S(x,t). \tag{6}$$

The cross-shore position of the profile is described as a function of profile depth where beach profile evolution is assumed to take place as a result of the collective effects of diffusive and non-diffusive processes. In Eq. (6), K is a cross-shore varying diffusion coefficient and S is a time and space dependent source function. The specification of the diffusion coefficient and source term was discussed as being likely to be site dependent but little practical guidance on how these should be determined in practice was given.Karunarathna et al. (2009) demonstrated how the method developed by Spivack and Reeve (2000) could be used to retrieve the diffusion coefficient and source term from a sequence of historic beach profile measurements, taking K and S as a sum of time varying and time averaged components as in a Reynolds expansion. Assuming that there are no drastic changes to the morphodynamic forcing in future, the sequence of Kand S can be analysed for patterns which can be extrapolated to formulate suitable future values to be used in forecasts made with Eq. (6). Karunarathna et al. (2011) established a correlation between incident wave conditions and the source function, using CCA, which was then used to predict future source terms. The predicted source functions were then used, with the mean distribution of K calculated previously, to determine future changes in beach profiles at three monthly intervals. A comparison of measured and predicted beach profile change using the above method is shown in Fig. 16 (lefthand panels), and the absolute errors in Fig. 16 (righthand panels).

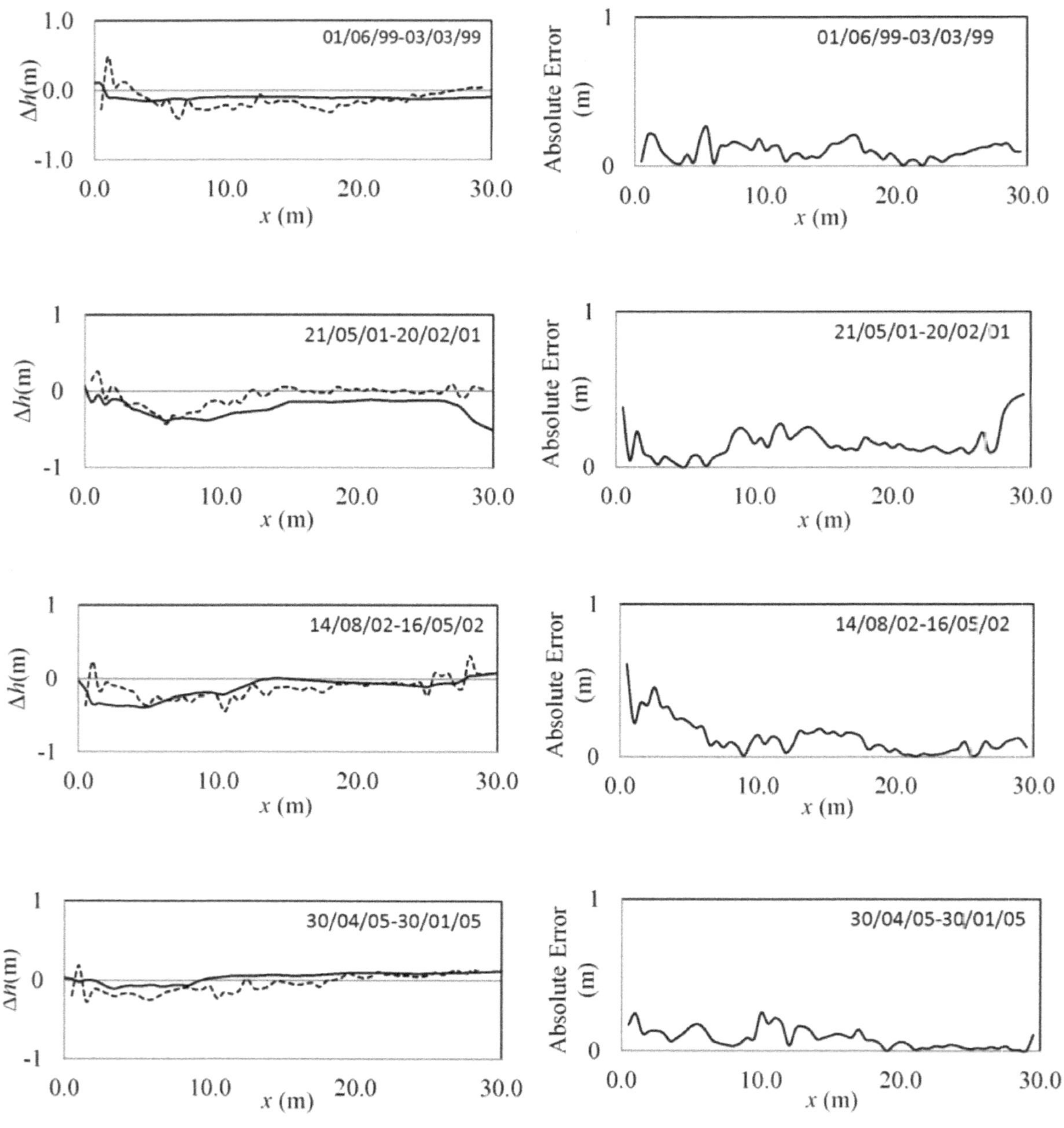

Figure 16: Comparison of measured (dark line) and computed (broken line) cross-shore profile change by the diffusion model [left panel] and absolute errors between the two [right panel] (Karunarathna et al. 2011).

A comparison by eye suggests that the error in forecasts using this method is less than the errors obtained by forecasting evolution directly from the profile data and a CCA regression. A quantitative assessment of the comparative performance is the subject of current research. Karunarathna et al. (2012) also noted that the long term mean profile was well-represented by Dean's equilibrium profile and, on the basis that the Milford-on-Sea mean profile represents an equilibrium, showed that the mean diffusion coefficient should follow an $x^{1/3}$ distribution. The computed coefficient did indeed follow this distribution to a good degree, demonstrating a clear link between the heuristically defined diffusion process and the sediment grain size.

Despite the dynamical simplicity of hybrid modelling approaches, an encouraging level of performance has been found by different researchers using a variety of techniques. The absence of detailed exactitude in such models may be troubling to reductionists, and the authors have some empathy with this. However, given that many researchers and practitioners are at ease using equilibrium-type models for development of conceptual arguments and design, there is perhaps a case for revisiting what metrics are most appropriate for gauging the performance of mesoscale models as opposed to those for small-scale models. This is reinforced by the fact that hybrid models are data-dependent and thus site- or scenario-specific and as a result, the accuracy of model predictions will be dependent on the accuracy and resolution of the data as well as model assumptions.

Other Models

In addition to the approaches described above, there are a number of other techniques that have been developed and applied to investigate coastal morphodynamics. Amongst these are: stochastic methods that treat the coast as a stochastic variable, driven by random forcing, for example the Monte Carlo methods proposed by Vrijling and Meijer (1992), Lee et al. (2002), Wang and Reeve (2010) and Callaghan et al. (2013); models that investigate the nature of shoreline instabilities due to the interaction of waves and shoreline as described by Ashton et al. (2001) and Idier et al. (2011); network or systems models that attempt to describe the behaviour of the coast as a system of linked elements like those of Baas (2002), Reeve and Karunarathna (2009) and French and Burningham (2011). The last of these types of models is discussed in detail by Payo et al. (in this issue).

One issue potentially of relevance to mesoscale coastal change that has gained prominence in recent years is that of wave chronology and its significance, or otherwise, to morphological evolution. The importance of the order in which storm episodes occur, or storm sequencing or 'chronology', was raised by Southgate (1995) and investigated on a more formal basis with Monte Carlo simulation by Dong and Chen (1999). They concluded that chronology effects could be significant in the short to medium term but became less significant over longer periods. Indeed, for periods of storm groups, in a 1-line model study of beach response near a groyne Reeve (2006) showed that while using time-averaged wave conditions yielded the same net transport as the corresponding time varying conditions, the final beach configuration was dependent on the sequence of storms. Walton and Dean (2011) showed analytically that the final planform shape of the shoreline may depend on the wave sequence despite wave conditions being spatially uniform. More recently, Reeve et al. (2014) demonstrated analytically that for the simple case of waves varying randomly in angle only, the timescale of changes in beach position are directly proportional to the timescale of changes in wave direction. Taking a storm time scale of one week this means that we would expect any chronology effects in the beach position to be negligible after a period of a few time scales, say a fortnight to a month. In an analysis of pocket beaches Turki et al. (2012) found that to explain the observed behaviour of several pocket beaches near Barcelona over the period of years it was necessary to explicitly include an element of beach memory of antecedent wave conditions in their hybrid model. Suffice to say this is an active area of research, and that current understanding indicates that the direct impact of chronology effects is likely to be at the smaller scale end of mesoscale, although this will depend on the structure of correlation between the different driving variables.

This brings us back to the statistical description of morphology and its forcing. This section concludes with a discussion of how some ideas from turbulence could be applied to hybrid modelling. For the purposes of this discussion we consider the 1-line equation for the evolution of a beach in response to waves under the small angle assumption. This is a simple morphological model but illustrates the stochastic approach which can also be applied to more complicated descriptions. If the shoreline position and wave conditions are considered to be random variables, with $y = \langle y \rangle + y'$ and $K = \langle K \rangle + K'$ where $\langle \rangle$ denotes an ensemble average and ' denotes the fluctuation about the ensemble average then we may write the ensemble average of the 1-line equation as:

$$\frac{\partial \langle y \rangle}{\partial t} = \langle K \rangle \frac{\partial^2 \langle y \rangle}{\partial x^2} + \left\langle K' \frac{\partial^2 y'}{\partial x^2} \right\rangle \tag{7}$$

where $\langle y' \rangle = \langle K' \rangle = 0$, $\langle\langle y \rangle\rangle = \langle y \rangle$ and $\langle\langle K \rangle\rangle = \langle K \rangle$. This assumption is the basis of 'Reynolds Averaging' in fluid mechanics, and is used to develop theories of turbulence. An important aspect of this assumption is that the fluctuations average to zero over the chosen averaging period. That is, the overall motion has a separation of scales of change that means such an averaging can be performed unambiguously. Assuming that this is the case it is clear from Eq. (7) that the ensemble average shoreline position depends on the ensemble averaged wave forcing, $\langle K \rangle$, as might be expected, but also on the correlations between the fluctuations in wave forcing and the second derivative of the beach plan shape; in essence a form of 'morphodynamic turbulence'.

To solve Eq. (7) it is necessary to specify the last term on the right hand side in some way. This is not straightforward and is analogous to the 'turbulence closure problem' in fluid mechanics. One approximation would be to assume the correlation term is negligible; another would be to parameterise it in terms of the ensemble quantities, for example by replacing it with a term proportional to $\langle y \rangle$ or its spatial gradient. Yet another approach, open to us in this age of computational modelling, is to use Monte Carlo simulation. Thus, if wave conditions are considered random variables driving a stochastic shoreline response, then by repeating simulations of shoreline evolution with multiple but independent realisations of wave conditions a corresponding set of realisations of likely shoreline responses can be generated. Ensemble statistics of shoreline position can be calculated from the realisations directly, without the need to specify the turbulence term in Eq. (7).

This might seem like a winning strategy which neatly bypasses the problems of 'turbulence closure'. It can certainly be a powerful method but it comes with drawbacks. For instance, irrespective of the number of realisations generated, these can only form a small subset of all possible outcomes. That is, the ensemble averages calculated from Monte Carlo simulation are sample statistics, calculated from the sample comprising the set of generated realisations, and can only ever be an approximation to the true solution. Further, Monte Carlo simulations can only ever be as good as the accuracy of the statistical characteristics of the input variables. If the statistics of the input variables do not accurately represent the statistics of the variables in nature, both in terms of distribution and correlation properties, then the resulting outputs will be unreliable. Unfortunately, good statistics requires very extensive observations against which to fit statistical distributions. One such example is the Dutch coast and Li et al. (2014) describe a recent modelling investigation that incorporates dependencies between variables based on observed data. Most valuable of all are datasets that contain contemporaneous records of the driving forces and the shoreline response, since these facilitate cross-correlation analysis as well as testing process model predictions.

DISCUSSION AND CONCLUDING REMARKS

Data-driven methods constitute a rigorous means of identifying trends, cycles and other patterns of behaviour. When used for prediction, by extrapolating such patterns, they are unlikely to be able to predict changes in coastal system state or configuration unless such changes are captured in the measurement record used for the pattern analysis. It may also have occurred to the reader that the hybrid morphological prediction methods described in the preceding sections really constitute a wide range of tools, rather than a seamless and integrated system of models ready to be used by the practitioner. In the view of the authors this is indeed the case; reflects the current state of understanding and is perhaps indicative of the fact that we are at a crossroads where, to make improvements in process modelling necessary for mesoscale forecasting, we need to develop a better understanding of the processes that dominate mesoscale evolution of coastal morphology. This understanding may come from direct modification of process models through, for example, experimentation with different parameterisation schemes or time compression algorithms, or through establishing empirical relationships with statistical techniques or through the development of hybrid models designed to capture mesoscale features of morphological evolution. Whichever one, or combination of these, transpires to be successful it will require more measurements; specifically, the continuation of current coastal monitoring programmes and perhaps the initiation of new programmes to extend monitoring to coastal types and exposures not represented by the current programmes. Long-term monitoring will underpin all forecasting methods, model development and validation; requiring significant and continued investment. Developments in monitoring technology such as video, LIDAR, and UAV systems, mean that much larger areas can now be monitored quickly.

The uncertainties in modelling the medium and long term that arise from the accumulation of rounding errors, the potential nonlinear sensitivity to initial condition, and limitations of our understanding of the key processes, mean that at least for the present, we must accept that predictions with such techniques must be uncertain. These uncertainties can be estimated through statistical methods such as Monte Carlo simulation. This would appear to be one possible area for further research that has the potential to provide bounds on the uncertainty of predictions and also to provide insight into which elements of the problem contribute most to the overall uncertainty. The importance of long term measurements and their role in guiding the development of our understanding cannot be understated. These provide an extremely valuable resource for investigating the statistical nature of the quantities we are trying to forecast. Furthermore, the longer the record the greater the chance of observing infrequent erosion events, breaching of barriers or dunes and inlet closures that can all lead to radical changes in the morphodynamics of the shoreline.

From a practical perspective medium to long term forecasts of coastal evolution are being demanded for coastal management. In such situations an answer of some form has to be provided and the selection of forecasting approach depends very often on the availability and quality of measurements at a site. This is likely to remain the case for the foreseeable future, while the existing monitoring programmes continue to build the database of measurements to an extent that allows some of the uncertainties to be reduced or eliminated and our understanding to improve. As such, having the flexibility of a variety of methods to make forecasts seems both apposite and persuasive.

Finally, the prospect of linking some of the many different approaches is appealing and would help to develop a system-wide capability for forecasting coastal change. Linking data-driven models with hydraulic quantities has always been difficult and challenging, due to the fact that contemporaneous measurements are very rare. However, some progress is being made in this regard, for instance, combining CCA with simple wave transformation calculations as proposed by Valsamidis et al. (2013), or combining EMD and wave transformation to investigate details of the impact of wave breaking over a multi-barred beach as proposed by Różyński and Lin (2015). The route to linking different model types, (data-driven with hybrid, hybrid with process-based and so on), is not clear at this moment and requires further research to establish its viability; which if demonstrated, will require the elucidation of practical methods of implementation.

To conclude, the quantity, quality and duration of observations are the vital factors when considering options for mesoscale morphological predictions and when assessing the accuracy of the model output. It is important to bear in mind the limitations of any model. Data-driven methods can play an important role by providing forecasts that are site specific but independent of any particular process-based modelling package. Process models provide a flexible methodology that is not site specific, but do require intensive computational resources and can be fraught with difficulties associated with numerical stability, accumulation of rounding errors and sensitivity to small changes in boundary conditions. Hybrid models often provide the optimum means of forecasting given constraints on observational records and physical process understanding at all but the best monitored sites. However, the outputs from such models can be frustratingly devoid of detail. The increasing availability of more high quality measurements provides the promise of improved forecasts, better modelling methodologies, reduced uncertainties and a more complete understanding of the mechanisms driving mesoscale coastal morphodynamics.

ACKNOWLEDGEMENTS

This work was funded by the Natural Environment Research Council (NERC) as part of the Integrating COAstal Sediment SysTems (iCOASST) project (NE/J005541/1), with the Environment Agency as an embedded project stakeholder. The authors gratefully acknowledge the helpful discussions with other iCOASST project team members and with the invited attendees of the iCOASST International Conference on Simulating Decadal Coastal Morphodynamics, held from 15 to 17 October 2013 in Southampton, UK.

REFERENCES

1. de Alegría-Arzaburu, A.R., Pedrozo-Acuña, A., Horrillo-Caraballo, J.M., Masselink, G.,Reeve, D.E., 2010. Determination of wave–shoreline dynamics on a macrotidal gravelbeach using Canonical Correlation Analysis. Coast. Eng. 57, 290–303.
2. Aranuvachapun, S., Johnson, J.A., 1978. Beach profiles at Gorleston and Great Yarmouth.Coast. Eng. 2, 201–213.
3. Ashton, A., Murray, A.B., Arnoult, O., 2001. Formation of coastline features by large-scale instabilities induced by high-angle waves. Nature 414, 296–300.
4. Aubrey, D.G., Emery, K.O., 1983. Eigen analysis of recent United States sea levels. Cont. Shelf Res. 2, 21–33.
5. Baas, A.C.W., 2002. Chaos, fractals and self-organization in coastal geomorphology: simulating dune landscapes in vegetated environments. Geomorphology 48, 309–328.
6. Broomhead, D.S., King, G.P., 1986. Extracting qualitative dynamics from experimental data. Physica D 20, 217–236.
7. Callaghan, D.P., Ranasinghe, R., Roelvink, D., 2013. Probabilistic estimation of storm erosion using analytical, semi-empirical, and process based storm erosion models. Coast. Eng. 82, 64–75.
8. Cowell, P.J., Roy, P.S., Jones, R.A., 1995. Shore-face translation model: computer simulation of coastal sand body response to sea level rise. Math. Comput. Simul. 33, 603–608.
9. Dabees, M.A., Kamphuis, J.W., 1998. ONELINE, a numerical model for shoreline change. Proceedings of the ASCE 27th International Conference on Coastal Engineering, ASCE, Vicksburg USA, pp. 2668–2681.
10. De Vriend, H.J., Capobianco, M., Chesher, T., de Swart, H.E., Latteux, B., Stive, M., 1993. Approachesto long-term modelling of coastal morphology: a review. Coast. Eng. 21, 225–269.
11. Dean, R.G., 1977. Equilibrium beach profiles, US, Atlantic and Gulf Coasts. Department of Civil Engineering Ocean Engineering Report No. 12. University of Delaware, Delaware (45 pp.).
12. Ding, X.L., Chao, J., Zheng, D.W., Chen, Y.Q., 2001. Long-term sea-level changes in Hong Kong from tide-gauge records. J. Coast. Res. 17, 749–754.
13. Dissanayake, D.M.P.K., Ranasinghe, R., Roelvink, D., 2012. The morphological response of large tidal inlet/basin systems to sea level rise. Clim. Chang. 113, 253–276.
14. Dong, P., Chen, H., 1999. A probability method for predicting the time-dependent long term beach erosion. Coast. Eng. 36, 243–261.
15. Dong, P., Wu, X.Z., 2013. Application of a stochastic differential equation to the prediction of shoreline evolution. Stoch. Env. Res. Risk A. 27, 1799–1814.
16. Esteves, L.S., Williams, J.J., Dillenburg, S.R., 2006. Seasonal and interannual influences on the patterns of shoreline changes in Rio Grande do Sul, southern Brazil. J. Coast.Res. 22, 1076–1093.
17. EurOtop Manual, 2007. In: Allsop, N.W.H., Pullen, T., Bruce, T., van der Meer, J.W.,Schüttrumpf, H., Kortenhaus, A. (Eds.), Overtopping Manual; Wave Overtopping of Sea Defences and Related Structures — Assessment Manual (Available at www.overtopping-manual.com (Accessed 3 September 2015)).
18. Freedman, D.A., 2005. Statistical Models: Theory and Practice. Cambridge University Press, Cambridge, UK (458 pp.).
19. French, J.R., 2010. Critical perspectives on the evaluation and optimization of complex numerical models of estuary hydrodynamics and sediment dynamics. Earth Surf. Process. Landf. 35, 174–189.
20. French, J., Burningham, H., 2011. Qualitative mathematical modelling of coupled coastand estuary morphodynamics: a modified Boolean network approach. American Geophysical Union, Fall Meeting 2011 (abstract #EP33A-0888).
21. van Gent, M., van den Boogaard, H., 2001. Neural network modeling of forces on vertical structures. Proceedings of the 27th International Conference of Coastal Engineering. ASCE, Vicksburg, USA, pp. 2096–2123.
22. Glahn, H.R., 1968. Canonical correlation analysis and its relationship to discriminant analysis and multiple regression. J. Atmos. Sci. 25, 23–31.
23. Hallermeier, R.J., 1981. A profile zonation for seasonal sand beaches from wave climate. Coast. Eng. 4, 253–277.
24. Hanson, H., Kraus, N.C., 1989. GENESIS: generalised model for simulating shoreline change. Technical Report CERC-89-19. Coastal Engineering Research Station, US Army Corps of Engineers, Vicksburg, USA (185 pp.).
25. Hanson, H., Larson, M., 2000. Simulating coastal evolution using a new type N-line model. Proceedings of the ASCE 27th International Conference on Coastal Engineering. ASCE,Vicksburg USA, pp. 2808–2821.
26. Hanson, H., Aarninkhof, S., Capobianco, M., Jimenez, J.A., Larson, M., Nicholls, R., Plant, N.,Southgate, H.N., Steetzel, H.J., Stive, M.J.F., De Vriend, H.J., 2003. Modelling coastal evolution on early to decadal time scales. J. Coast. Res. 19, 790–811.
27. Haxel, J.H., Holman, R.A., 2004. The sediment response of a dissipative beach to variationsin wave climate. Mar. Geol. 206, 73–99.
28. Hedges, T.S., 2001. Regression and dimensional analysis in coastal engineering: some pitfalls. Proceedings of the ICE - Water and Maritime Engineering 148,pp. 219–225.
29. Horrillo-Caraballo, J.M., Reeve, D.E., 2008. An investigation of the link between beach morphology and wave climate at Duck, North Carolina, USA. J. Flood Risk Manage. 1, 110–122.

30. Horrillo-Caraballo, J.M., Reeve, D.E., 2010. An investigation of the performance of a datadriven model on sand and shingle beaches. Mar. Geol. http://dx.doi.org/10.1016/ j. margeo.2010.03.010.
31. Hotelling, H., 1935. The most predictable criterion. J. Educ. Psychol. 26, 139–142. HR Wallingford, 2006. Thames Estuary 2100: a hybrid morphological prediction. Technical Note EP 6, p. 7 (24 pp.).
32. Hsu, J.R.C., Evans, C., 1989. Parabolic bay shapes and applications. Proc. of the Institution of Civil Engrs. 87 (2), 557–570.
33. Hsu, T.-W., Ou, S.-H., Wang, S.-K., 1994. On the prediction of beach changes by a new 2-D empirical eigenfunction model. Coast. Eng. 23, 255–270.
34. Huang, N.E., Shen, Z., Long, S.R., Wu, M.C., Shih, H.H., Zheng, Q., Yen, N.-C., Tung, C.C., Liu, H.H., 1998. The empirical mode decomposition and the Hilbert spectrum for nonlinear and non-stationary time series analysis. Proc. R. Soc. Lond. A 454, 903–995.
35. Huthnance, J.M., Karunarathna, H., Lane, A., Manning, A.J., Norton, P., Reeve, D.E., Spearman, J., Soulsby, R.L., Townend, I.H., Wolf, J., Wright, A., 2008. Development of estuary morphological models. R & D Technical Report FD2107/TR, Joint Defra/Environment Agency Flood and Coastal Erosion Risk Management R & D Programme
36. (71 pp.).
37. Idier, D., Falqués, A., Ruessink, B.G., Garnier, R., 2011. Shoreline instability under low-angle wave incidence. J. Geophys. Res. 116, F04031.
38. Iyama, J., Kuwamura, H., 1999. Application of wavelets to analysis and simulation of earthquake motions. Earthq. Eng. Struct. Dyn. 28, 255–272.
39. Jamal, M.H., Simmonds, D.J., Magar, V., 2014. Modelling gravel beach dynamics with XBeach. Coast. Eng. 89, 20–29.
40. Jones, O.P., Petersen, O.S., Kofoed-Hansen, H., 2007. Modelling of complex coastal environments: some considerations for best practice. Coast. Eng. 54, 717–733.
41. Karunarathna, H., Reeve, D.E. 2013. A hybrid approach to model shoreline change at multiple timescales. Cont. Shelf Res. 66, 29–35. doi.org/http://dx.doi.org/10.1016/j. csr. 2013.06.019
42. Karunarathna, H., Horrillo-Caraballo, J.M., Ranasinghe, R., Short, A., Reeve, D.E., 2012. An analysis of the cross-shore beach morphodynamics of a sandy and composite gravel beach. Mar. Geol. 299, 33–42.
43. Karunarathna, H., Horrillo-Caraballo, J.M., Spivack, M., Reeve, D.E., 2011. Analysis of key parameters in a diffusion type beach profile evolution model. Cont. Shelf Res. 31,98–107.
44. Karunarathna, H., Reeve, D.E., Spivack, M., 2008. Long term morphodynamic evolution of estuaries: an inverse problem. Estuar. Coast. Shelf Sci. 77, 386–395.
45. Karunarathna, H., Reeve, D.E., Spivack, M., 2009. Beach profile evolution as an inverse problem. Cont. Shelf Res. 29, 2234–2239.
46. Kumar, P., Foufoula-Georgiou, E., 1997. Wavelet analysis for geophysical applications. Rev. eophys. 35, 385–412.
47. Larson, M., Kraus, N.C., 1995. Prediction of cross-shore sediment transport at different spatial and temporal scales. Mar. Geol. 126, 111–127.
48. Larson, M., Capobianco, M., Hanson, H., 2000. Relationship between beach profiles and waves at Duck, North Carolina, determined by canonical correlation analysis. Mar. Geol. 163, 275–288.
49. Larson, M., Hanson, H., Kraus, N.C., 1997. Analytical solutions of one-line model for shoreline change near coastal structures. J. Waterw. Port Coast. Ocean Eng. 123, 180–191.
50. Lee, E., Meadowcroft, I.C., Hall, J., Walkden, M., 2002. Coastal landslide activity: a probabilistic simulation model. Bull. Eng. Geol. Environ. 61, 347–355.
51. Lesser, G. 2009. An Approach to Medium-Term Coastal Morphological Modeling, (PhDthesis). Delft University of Technology, ISBN 978-0-415-55668-2, (255 pp.).
52. Lesser, G., Roelvink, J.A., Van Kester, J.A.T.M., Stelling, G.S., 2004. Development and validation of a three-dimensional morphological model. Coast. Eng. 51, 883–915.
53. Li, L., Lark, M., Reeve, D.E., 2005. The multi-scale variability of beach profiles at Duck, N. C., USA, 1981–2003. Coast. Eng. 52, 1133–1153.
54. Li, F., van Gelder, P.H.A.J.M., Ranasinghe, R., Callaghan, D.P., Jongejan, R.B., 2014. Probabilistic modelling of extreme storms along the Dutch coast. Coast. Eng. 86, 1–13.
55. Lilly, D.K., 1965. On the computational stability of numerical solutions of time-dependent non-linear geophysical fluid dynamics problems. Mon. Weather Rev. 93, 11–25.
56. Lorenz, E.N., 1963. Deterministic non-periodic flow. J. Atmos. Sci. 20, 130–141.
57. Mallat, S., 1989. A theory for multi-resolution signal decomposition: the wavelet representation. IEEE Trans. Pattern Anal. Mach. Intell. 11, 674–693.
58. Manning, A.J., 2007. Morphological modelling scenario comparisons using the Analytical Emulator for WP2.7 FD2107. HR Wallingford Tech. Report TR166 (24 pp.).
59. Murray, A.B., 2007. Reducing model complexity for explanation and prediction. Geomorphology90, 178–191.
60. Murray, A.B., Lazarus, E.D., Moore, L.J., Lightfoot, J., Ashton, A.D., McNamara, D.E., Ells, K., 2015. Decadal scale shoreline change arises from large-scale interactions, while small-scale changes are forgotten: observational evidence. Proceedings of Coastal Dynamics 2015, (10 pp.), ISBN: 978-981-4689-96-0 (dvd, San Diego), World Scientific,New York.
61. Nairn, R.B., Southgate, H.N., 1993. Deterministic profile modelling of nearshoreprocess. Part 2. Sediment transport and beach profile development. Coast. Eng. 19, 58–96.

62. New Forest District Council, 2014. http://www.channelcoast.org (Accessed 1 July 2014). Nicholson, J., Broker, I., Roelvink, J.A., Price, D., Tanguy, J.M., Moreno, L., 1997. Intercomparison of coastal area morphodynamic models. Coast. Eng. 31, 97–123.
63. Niedoroda, A.W., Reed, C.W., Swift, C.W., Arato, H., Hoyanagi, K., 1995. Modelling shorenormal large-scale coastal evolution. Mar. Geol. 126, 181–199.
64. North, G.R., Bell, T.L., Cahalan, R.F., Moeng, F.J., 1982. Sampling errors in the estimation of empirical orthogonal functions. Mon. Weather Rev. 110, 699–706.
65. Pan, S., Reeve, D.E., Davidson, M., O'Connor, B., Vincent, C., Dolphin, T., Wolf, J., Thorne, P., Bell, P., Souza, A., Chesher, T., Johnson, H., Leadbetter, A., 2010. Larger-scale morphodynamic impacts of segmented shore-parallel breakwaters on coasts and beaches: an overview of the LEACOAST2 Project. 78, 35–43.
66. Pape, L., Kuriyama, Y., Ruessink, B.G., 2010. Models and scales for cross-shore sandbar migration. J. Geophys. Res. 115, F03043, doi :http://dx.doi.org/10.1029/2009JF001644
67. Pape, L., Ruessink, B.G., Wiering, M.A., Turner, I.L., 2007. Recurrent neural network modeling of nearshore sandbar behaviour. Neural Netw. 20, 509–518.
68. Pelnard-Considere, R., 1956. Essai de Theorie de l'Evolution des Formes de Ravage en Plages de Sables et de Galets. Societe Hydrotechnique de France, IV'eme Journee de L'Hydraulique Question III, Rapport, 1, 74-1-10 (in French).
69. Perlin, M., Dean, R.G., 1983. A numerical model to simulate sediment transport in the vicinity of coastal structures. MR83-10, Coastal. Engineering Research Centre, US Army Engineers Waterways Experiment Station, Vicksburg, MS (199 pp.).
70. Pruszak, Z., Schönhofer, J., Różyński, G., 2012. Spatiotemporal analysis of variations of certain morphological beach forms. Proceedings of the 33rd International Conference on Coastal Engineering, ASCE, Vicksburg USA (10 pp, available online at https://journals. tdl.org/icce/index.php/icce/article/view/6603).
71. Ranasinghe, R., Swinkels, C., Luijendijk, A., Roelvink, D., Bosboom, J., Stive, M., Walstra,D.J.R., 2011. Morphodynamic upscaling with the MORFAC approach: dependencies and sensitivities. Coast. Eng. 58, 806–811.
72. Reeve, D.E., 2006. Explicit expression for beach response to non-stationary forcing near a groyne. J. Waterw. Port Coast. Ocean Eng. 132, 125–132.
73. Reeve, D.E., Fleming, C.A., 1997. A statistical-dynamical method for predicting long term coastal evolution. Coast. Eng. 30, 259–280.
74. Reeve, D.E., Horrillo-Caraballo, J.-M., 2014. Linking regional wave conditions to local beach profile change. Marit. Eng. 167, 29–41.
75. Reeve, D.E., Karunarathna, H., 2009. On the prediction of long-term morphodynamic response of estuarine systems to sea level rise and human interference. Cont. Shelf Res. 29, 938–950.
76. Reeve, D.E., Spivack, M., 2004. Evolution of shoreline position moments. Coast. Eng. 51, 661–673.
77. Reeve, D.E., Li, Y., Lark, M., Simmonds, D.J., 2007. Analysis of long-term temporal variability of beach profiles at Duck using wavelet packets. Coast. Eng. 54, 401–415.
78. Reeve, D.E., Li, B., Thurston, N., 2001. Eigenfunction analysis of decadal fluctuations in sandbank morphology. J. Coast. Res. 17, 371–382.
79. Reeve, D.E., Pedrozo-Acuña, A., Spivack, M., 2014. Beach memory and ensemble average of the shoreline evolution near a groyne. Coast. Eng. 86, 77–87.
80. Reniers, A.J.H.M., Roelvink, J.A., Walstra, D.J.R., 1995. Validation study of UNIBEST-TC model. Report H2130. Delft Hydraulics, Delft, The Netherlands (72 pp.). Roelvink, J.A., 2006. Coastal morphodynamic evolution techniques. Coast. Eng. 53,277–287.
81. Roelvink, J.A., Broker, I., 1993. Cross-shore profile models. Coast. Eng. 21, 163–191.
82. Roelvink, J.A., Reniers, A., van Dongeren, A., van Thiel De Vries, J., McCall, R., Lescinski, J., 2009. Modelling storm impacts on beaches, dunes and barrier islands. Coast. Eng.56, 1133–1152.
83. Różyński, G., 2003. Data-driven modeling of multiple longshore bars and their interactions. Coast. Eng. 48, 151–170.
84. Różyński, G., 2005. Long term shoreline response of a non-tidal, barred coast. Coast. Eng.52, 79–91.
85. Różyński, G., 2010. Long-term evolution of Baltic Sea wave climate near a coastal segment in Poland; its drivers and impacts. Ocean Eng. 37, 186–199.
86. Różyński, G., Jansen, H., 2002. Modeling nearshore bed topography with principal oscillation patterns. J. Waterw. Port Coast. Ocean Eng. 128, 202–215.
87. Różyński, G., Lin, J.-G., 2015. Data-driven and theoretical beach equilibrium profiles: implications and consequence. J. Waterw. Port Coast. Ocean Eng. 141. http://dx.doi. org/10.1061/(ASCE)WW.1943-5460.0000304.
88. Różyński, G., Reeve, D.E., 2005. Multi-resolution analysis of nearshore hydrodynamicsusing discrete wavelet transforms. Coast. Eng. 52, 771–792.
89. Różyński, G., Larson, M., Pruszak, Z., 2001. Forced and self-organized shoreline response for a beach in the southern Baltic Sea determined through singular spectrum analysis. Coast. Eng. 43, 41–58.
90. Shibutani, Y., Kuroiwa, M., Matsubara, Y., Kim., M., Abualtaef, M., 2009. Development of Nline numerical model onsidering the effects of beach nourishment. J. Coast. Res. SI 56, 554–558.
91. Short, A.D., Trembanis, A.C., 2004. Decadal scale patterns in beach oscillation and rotation Narrabeen Beach, Australia—time series, PCA and wavelet analysis. J. Coast. Res. 20, 523–532.
92. Smallwood, D.O., 1999. Characterization and simulation of gunfire with wavelets. Shock. Vib. 6, 59–72.

93. Solow, A., 1987. The application of eigenanalysis to tide-gauge records of relative sea level. Cont. Shelf Sci. 7, 641–692692.

94. Soulsby, R.L., Mead, C.T., Wood, M.J., 2007. Development of a Lagrangian morphodynamic model for sandy estuaries and coasts. Report TR150. HRWallingford, Oxon., UK (56 pp.).

95. Southgate, H.N., 1995. The effects of wave chronology on medium and long term coastal morphology. Coast. Eng. 26, 251–270.

96. Southgate, H.N., Wijnberg, K.M., Larson, M., Capobianco, M., Jansen, H., 2003. Analysis of field data of coastal morphological evolution over yearly and decadal timescales: part 2 Non-linear techniques. J. Coast. Res. 19, 776–789.

97. Spearman, J., 2011. The development of a tool for examining the morphological evolution of managed realignment sites. Cont. Shelf Res. 31, S199–S210. http://dx.doi.org/10. 1016/j.csr.2010.12.003.

98. Spivack, M., Reeve, D.E., 1999. Recovery of a variable coefficient in a coastal evolution equation. J. Comp. Physiol. 151, 585–596.

99. Spivack, M., Reeve, D.E., 2000. Source reconstruction in a coastal evolution equation. J. Comp. Physiol. 161, 169–181.

100. Splinter, K.D., Turner, I.L., Davidson, M.A., Barnard, P., Castelle, B., Oltman-Shay, J., 2014. A generalized equilibrium model for predicting daily to interannual shoreline response.

101. J. Geophys. Res. Earth Surf. 119. http://dx.doi. org/10.1002/2014JF003106. Steetzel, H.J., de Vroeg, J.H., 1999. Application of a multilayer approach for morphological modelling. Coastal Sediments' 99ASCE, Vicksburg, USA, pp. 2206–2218.

102. Stive, M.J.F., De Vriend, H.J., 1995. Modelling shoreface profile evolution. Mar. Geol. 126, 235–248.

103. Stive, M.J.F., Capobianco, M., Wang, Z.B., Ruol, P., Buijsman, M.C., 1998. Morphodynamics of a tidal lagoon and the adjacent coast. In: Dronkers, J., Scheffers, M. (Eds.), Physics of Estuaries and Coastal Seas. Balkema, Rotterdam, pp. 397–407.

104. Stive, M.J.F., Nicholls, R.J., de Vriend, H.J., 1991. Sea level rise and shore nourishment. Coast. Eng. 16, 147–163.

105. Thomas, T., Phillips, M.R., Williams, A.T., 2010. Mesoscale evolution of a headland bay: beach rotation processes. Geomorphology 123, 129–141.

106. Turki, I., Medina, R., Gonzalez, M., 2012. Beach memory. Proceedings of the 33rd International Conference on Coastal EngineeringASCE, Vicksburg USA (10 pp., available online at https://journals.tdl.org/icce/index.php/icce/issue/view/361).

107. US Army Corps of Engineers, 1984. Shore protection Manual. Dept. of the Army, Waterways Experiment Station, Corps of Engineers, Coastal Engineering Research Center, Washington, DC. (656 pp).

108. Valsamidis, A., Różyński, G., Reeve, D.E., 2013. Data-driven analysis of wave–seabed interactions at a non-tidal beach with multiple bars. Mar. Geol. 344, 144–154.

109. Van der Wegen, M., Roelvink, J.A., 2008. Long-term morphodynamic evolution of a tidal embayment using a two-dimensional, process-based model. J. Geophys. Res. 113, C03016. http://dx.doi.org/10.1029/2006JC003983.

110. Van Koningsveld, M., Stive, M.J.F., Mulder, J.P.M., 2005. Balancing research efforts and management needs: a challenge for coastal engineering. Proceedings of the 29th International Conference of Coastal Engineering. World Scientific, New York,pp. 2985–2997.

111. Van Rijn, L.C., Wijnberg, K.M., 1996. One-dimensional modelling of individual waves and wave-induced longshore currents in the surf zone. Coast. Eng. 28, 121–145.

112. Van Rijn, L.C., Walstra, D.J.R., Grasmeijer, B., Sutherland, J., Pan, S., Sierra, J.P., 2003. The predictability of cross-shore bed evolution of sandy beaches at the time scale of storms and seasons using process-based profile models. Coast. Eng. 47, 295–327.

113. Vautard, R., Yiou, P., Ghil, M., 1992. Singular spectrum analysis: a toolkit for short, noisy, chaotic signals. Physica D 158, 95–126.

114. Vrijling, J.K., Meijer, G.J., 1992. Probabilistic coastline position computations. Coast. Eng. 17, 1–23.

115. Walton, T.L., Dean, R.G., 2011. Shoreline change at an infinite jetty for wave time series. Cont. Shelf Res. 31, 1474–1480.

116. Wang, Z.B., 2005. Application of semi-empirical long-term morphological models ESTMORPH and ASMITADelft Hydraulics, Delft, p. Z2839 (34 pp.).

117. Wang, B., Reeve, D.E., 2010. Probabilistic modelling of long-term beach evolution near segmented shore-parallel breakwaters. Coast. Eng. 57, 732–744.

118. Wijnberg, K.M., Terwindt, J.H.J., 1995. Extracting decadal morphological behaviour from high-resolution, long-term bathymetric surveys along the Holland coast usingeigenfunction analysis. Mar. Geol. 126, 301–330.

119. Winant, C.D., Inman, D.L., Nordstrom, C.E., 1975. Description of seasonal beach changes using empirical eigenfunctions. J. Geophys. Res. 80, 1979–1986. http://dx.doi.org/10.1029/JC080i015p01979.

120. Yates, M.L., Guza, R.T., O'Reilly, W.C., 2009. Equilibrium shoreline response: observations and modelling. J. Geophys. Res. 114(C09014), doi :http://dx.doi.org/10.1029/ 2009JC005359.

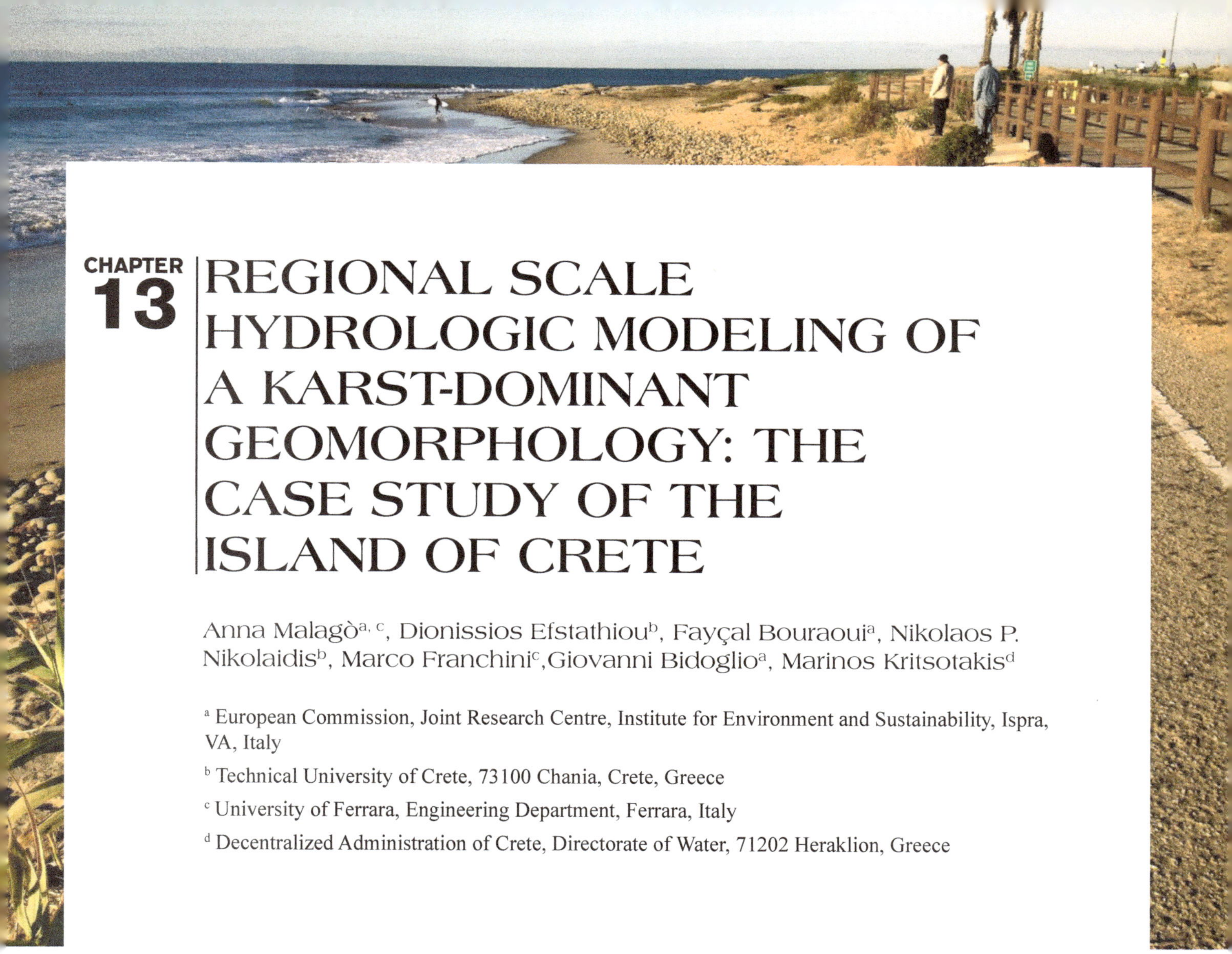

CHAPTER 13

REGIONAL SCALE HYDROLOGIC MODELING OF A KARST-DOMINANT GEOMORPHOLOGY: THE CASE STUDY OF THE ISLAND OF CRETE

Anna Malagò[a, c], Dionissios Efstathiou[b], Fayçal Bouraoui[a], Nikolaos P. Nikolaidis[b], Marco Franchini[c], Giovanni Bidoglio[a], Marinos Kritsotakis[d]

[a] European Commission, Joint Research Centre, Institute for Environment and Sustainability, Ispra, VA, Italy

[b] Technical University of Crete, 73100 Chania, Crete, Greece

[c] University of Ferrara, Engineering Department, Ferrara, Italy

[d] Decentralized Administration of Crete, Directorate of Water, 71202 Heraklion, Greece

SUMMARY

Crete Island (Greece) is a karst dominated region that faces limited water supply and increased seasonal demand, especially during summer for agricultural and touristic uses. In addition, due to the mountainous terrain, interbasin water transfer is very limited. The resulting water imbalance requires a correct quantification of available water resources in view of developing appropriate management plans to face the problem of water shortage.

The aim of this work is the development of a methodology using the SWAT model and a*karst-flow model* (KSWAT, Karst SWAT model) for the quantification of a spatially and temporally explicit hydrologic water balance of karst-dominated geomorphology in order to assess the sustainability of the actual water use. The application was conducted in the Island of Crete using both hard (long time series of streamflow and spring monitoring stations) and soft data (i.e. literature information of individual processes). The KSWAT model estimated the water balance under normal hydrological condition as follows: 6400 Mm^3/y of precipitation, of which 40% (2500 Mm^3/y) was lost through evapotranspiration, 5% was surface runoff and 55% percolated into the soil contributing to lateral flow (2%), and recharging the shallow (9%) and deep aquifer (44%). The water yield was estimated as 22% of precipitation, of which about half was the contribution from spring discharges (9% of precipitation). The application of the KSWAT model increased our knowledge about water resources availability and distribution in Crete under different hydrologic conditions. The model was able to capture the hydrology of the karst areas allowing a better management and planning of water resources under scarcity.

Citation: Anna Malagò, Dionissios Efstathiou, Fayçal Bouraoui, Nikolaos P. Nikolaidis, Marco Franchini, Giovanni Bidoglio, Marinos Kritsotakis, Regional scale hydrologic modeling of a karst-dominant geomorphology: The case study of the Island of Crete, doi:10.1016/j.jhydrol.2016.05.061

INTRODUCTION

"Karst" identifies a specific geological landscape and morphology formed by the dissolving action of water on soluble carbonate rocks such as primarily limestone, but also marble, dolomite, and gypsum. These rocks are mechanically strong but chemically soluble with high degree of secondary porosity. As a consequence, the hydrological cycle provides the primary source of energy for karst formation because water is the solvent that dissolves carbonate rocks and then carries the ions away in solution (Williams, 2004).

The process of dissolution ('karstification') leads to the development of caves, sinkholes, springs and sinking streams that are typical features of a karst system. With progressing karstification, groundwater flow in the karst aquifer develops from a flow in an interconnected fissure network to a flow concentrated in several large pipes, interconnected cavities and cave systems (EC, 2003 and EC, 2004). The downstream end of a karst system usually is a spring where the underground conduit reaches the surface as an output point from an extensive network of groundwater conduits (Smart and Worthington, 2004).

In Europe, soluble carbonate rocks are widespread in Western, Southern and Eastern part covering 35% of whole Europe (Daly et al., 2002), so that the karst processes are significant components of the physical geography of Mediterranean basins. In particular, limestones reach great thickness in Spain, southern France, Italy, the Balkan Peninsula, Turkey and in many islands in the Mediterranean (Crete, Majorca and Sicily). As a consequence, karst aquifers and springs are an important source of water supply for Mediterranean countries and special strategies are required to manage the quantity and quality of their waters.

Bakalowicz (2015) pointed out the importance to study the karst aquifer functioning and the local geological evolution in order to manage in realistic and sustainable way the water resources. The monitoring and management of these resources are recognized in Europe as an essential issue and the European Union prompted the creation of COST Actions 620 and 621 to develop a comprehensive methodology for risk assessment and for the sustainable management of karst systems (EC, 2003 and EC, 2004). Unfortunately, most countries are lacking behind in monitoring the discharge of springs or wells and the exploitation of karst aquifers in generally is inappropriate (Bakalowicz, 2015).

In this context, large scale hydrologic models are essential tools for watershed management at regional scale. Regional scale models with an appropriate discretization of watershed can adequately account for the spatial heterogeneity improving water predictions (Wooldridge and Kalma, 2001). A variety of karst models have been developed and applied to karst watersheds (Nikolaidis et al., 2013). Recently, Hartmann et al. (2015) presented for the first time the simulations of groundwater recharge in Europe with a grid-hydrological model (VarKarst-R) pointing out the importance of a characterization of subsurface heterogeneity.

Baffaut and Benson (2009) modified the SWAT (Soil and Water Assessment Tool) model to simulate faster aquifer recharge in a Missouri karst watershed (SWAT-B&B) modifying deep groundwater recharge equations, increasing the hydraulic conductivity of sinkholes simulated as ponds and losses from streams. After that, Yachtao (2009) further modified the SWAT model in order to improve the simulation of water quality and quantity in the Opequon Creek watershed (USA). The author introduced two new parameters for simulating the hydrology and nitrate transport in a sinkhole and losses from sinking streams. More recently, Wang and Brubaker (2014) proposed a non-linear modification of groundwater algorithm in SWAT (ISWAT) improving the recession and low-flow simulation. Nikolaidis et al. (2013) developed a reservoir model approach linked with SWAT to simulate karst's behavior and the recharge of springs of Koiliaris basin in Crete adding five new parameters.

All these studies indicate that a specific parameterization of aquifer discharge, return flow, stream losses and sinkholes is required in karst watersheds. In addition, Wang et al. (2014) showed that a specific database of karst SWAT soils should be used to assess the influence of soil hydrological process in a karst region.

In this paper the SWAT model was integrated with a *karst-flow model* (Nikolaidis et al., 2013 and Tzoraki and Nikolaidis, 2007). Karst-subbasins were defined in order to reproduce more accurately the water balance at regional scale.

This integrated model, KSWAT model, was applied to the Island of Crete, one of the most intensively managed Mediterranean islands, where the major water use is irrigation (84.5% of the total consumption) and the main water source is karst aquifers. The growing water demand of the region makes the rational management of water resources extremely important in view of a sustainable development. Consequently, the specific objective of this study is the development of a large scale methodology for the quantification of a spatially explicit water balance of karst-dominated geomorphologies using the SWAT model in order to assess the sustainability of water use in the Island of Crete and potentially to other areas.

MATERIALS AND METHODS

Study Area

The Island of Crete occupies the southern part of Greece and is the largest island of Greece and the fifth in the Mediterranean. The Island covers an area of 8336 km^2 and is divided into prefectures, including from east to west: Lasithi (1810 km^2), Heraklion (2626 km^2), Rethymno (1487 km^2) and Chania (2342 km^2) (Fig. 1a). The maximum length of the Island is 269 km and the maximum width 60 km. Four main mountains run west to east: the White Mountains in the west (2453 m), Idis mountain (2456 m) in the center, Asterousian (1280 m) in south Heraklion and Dikti (2148 m) in the east (Baltas and Tzoraki, 2013).

The Island of Crete is characterized by a dry semi-humid Mediterranean climate with dry and warm summers and humid and relatively cold winters where mean annual rainfall decreases from west to east and from north to south, but increases with altitude (MEDIWAT, 2013). Annual precipitations are highly variable ranging between 300 mm in coastal areas and 2000 mm in headwaters in White Mountains. The mean annual temperature ranges from 18.5° in the west to 20° in the south of island and decreases with altitude.

The mountainous areas, in particular in the western part, have mountainous climate. As a consequence, Crete contains sub-regions with very different hydrological characteristics.

Crete has about 2550 km^2 of agriculture land, about 30% of whole Crete, with more than 1100 km^2 in Heraklion, and 3800 km^2 of pasture (45% of total areas of Crete). The main crops are olives, grapes, and the main vegetables crops are tomatoes, cucumbers, onions, potatoes, watermelons and melons. The demand for irrigation water is high (about 360 Mm^3/y), while only 47% (1200 km^2) of agricultural land is irrigated. In Heraklion the irrigated area is around 600 km^2, followed by Lasithi and Chania with around 300 km^2, while a very small area in Rethymno is irrigated (Agriculture statistics of Greece, 2005).

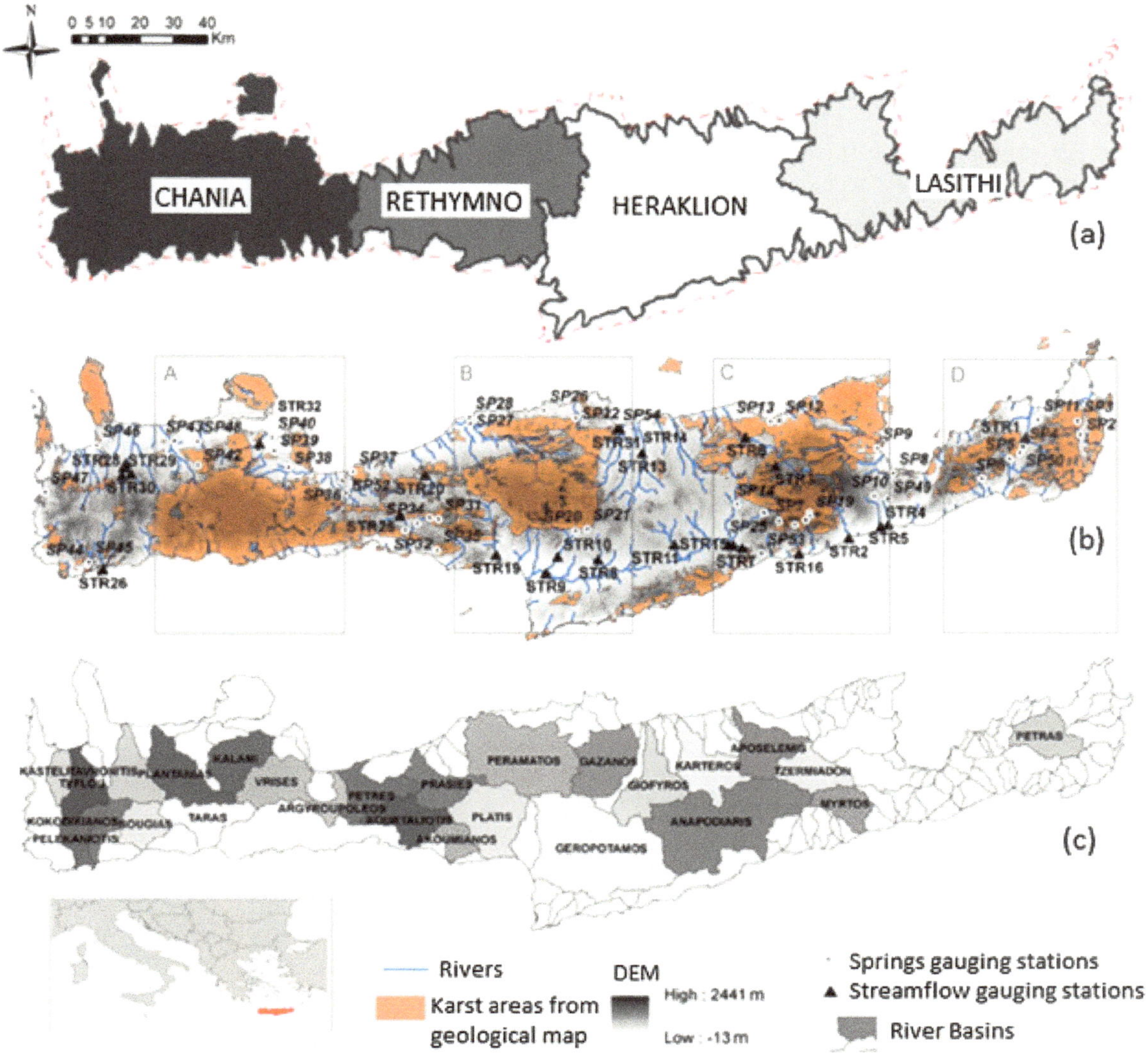

Figure 1: Map of prefectures (a); the main carbonate rocks of Crete (karst areas from geological map) with springs and streamflow gauging stations (b) A, White Mountainous; B, Idi; C, Dikti; D, Sitia; map of the main river basins (c).

The Karst Geology and Hydrology of Island of Crete

The geology of Crete is composed of carbonate rocks (limestone, marble and dolomite) which allow water to penetrate, creating major karst formation (Baltas and Tzoraki, 2013). More than 30% of carbonate rocks cover the total area of the Island and the major carbonate rocks of Crete are located in the White mountains (Lefka Ori), Idi, Dikti and Sitia (Fig. 1b). The total karst area covers about 2730 km^2 and the water contribution to the karst aquifer is estimated around 2000 Mm^3/y which discharges out in many springs (Chartzoulakis et al., 2001).

There are 47 gauged springs in Crete with relative large flow, which are subdivided in three main classes: freshwater springs, brackish water springs and undersea springs. Most of springs are karst springs and refer to the same karst hydrogeological system (Lefka Ori, Idi, Dikti and Sitia) discharging around 500 Mm^3/y freshwater into rivers. The most important springs are Stylos, Platanos and Kourtaliotis that discharge 85 Mm^3/y in the Kalami/Koiliaris basin, 67 Mm^3/y in Plantanias Basin and 38 Mm^3/y in Kourtaliotis basin, respectively (Fig. 1c).

Big brackish springs, located on coastal areas, include Almiros-Heraklion that discharges around 235 Mm^3/y, Almiros-Agios-Nikolaos (83 Mm^3/y) and brackish springs of lower discharge include Almiros-Mallia (3.2 Mm^3/y), Georgioupolis, Grammatikaki and Malavra. Brackish springs discharge directly about 285 Mm^3/y in the sea. Submarine discharges can be found in the southern part of Lefka Ori, Souda bay, Bali bay, Mallia bay, Elounta, Skinia, and the eastern part of the limestones of Zakros (MEDIWAT, 2013).

Streamflow and Springs Measurements

Stream and spring discharge measurements were provided by the bureau of Water Resources at the Decentralized Administration of the Region of Crete that deals with data elaboration and the assessment of their accuracy. The stream measurements at 25 permanent locations were conducted using a continuous graphic record of the water depth. For a comprehensive description of continuous measurements of water depth, the reader can refer to Rantz et al. (1982).

These measurements were then converted to streamflow rate with an established rating curve (streamflow, m^3/s vs water depth, m) calculated every year for each monitoring station following the procedure described in Buchanan and Somers (1976). The streamflow-water depth relationship was adopted to calculate the streamflow at daily time step, and then aggregated at monthly time step. The available dataset for this study included 22 gauging stations because the streamflow stations STR30 and STR29 were excluded due to their uncertain localization and drain areas. STR5 was also excluded because observed data were lacking during the study period 1980–2009.

The spring's discharges measurements were directly calculated by measuring velocity once a month on surveyed cross-sectional channel geometry at 47 permanent locations. The only exception was the Almiros spring for which only the calculated monthly discharge was available. Table 1 summarizes the streamflow and springs data utilized in this study.

Table 1: Description of available streamflow and spring discharges measurements in the Crete Island provided by the bureau of Water Resources at the Decentralized Administration of the Region of Crete. The symbol # represent the number

Type of measure	# gauging stations	Type of location	Type of measures	Frequency	Period extension	# Calibrated gauging stations	# Validated gauging stations
Streamflow	22	Permanent gauging station	Water depth	Continuous measurements on a graphic record (Rantz et al., 1982)	1980–2009 (30years; #5120 data entries)	15	7
Spring Discharge	47	Permanent gauging station	Water velocity	Once a month	1983–2005 (23years; #8015 data entries)	47	–

Approach for the Prediction of Karst Water Resources

The proposed karst modeling approach links the process-based model SWAT and a*karst-flow model* (Nikolaidis et al., 2013). Fig. 2 represents the sequential steps of our approach that involved the setup of SWAT model, the identification of karst subbasins in which a modified version of SWAT model was applied, the calibration of streamflow at selected gauging stations (step-wise calibration), the regionalization of calibrated parameters for ungauged subbasins, the spring's discharges calibration outside the SWAT model using the *karst-flow model* and finally the introduction of calibrated discharge of springs as a point sources in the SWAT model. The overall procedure included the use of hard and soft data. Hard data are defined as long-term, measured time series, typically at a point within a watershed, while soft data are defined as information on individual processes within a budget that may not be directly measured within the study area, generally from literature information (Arnold et al., 2015). In this study long time series of 22 streamflow and 47 of spring monitoring stations were used as hard data, while for instance literature information of the extension of karst recharge areas, as soft data. This approach enables the dialog between local experts, experimentalists and modelers, increases the reliability of the model results in ungauged areas and helps to better constrain model parameters (Seibert and McDonnell, 2002).

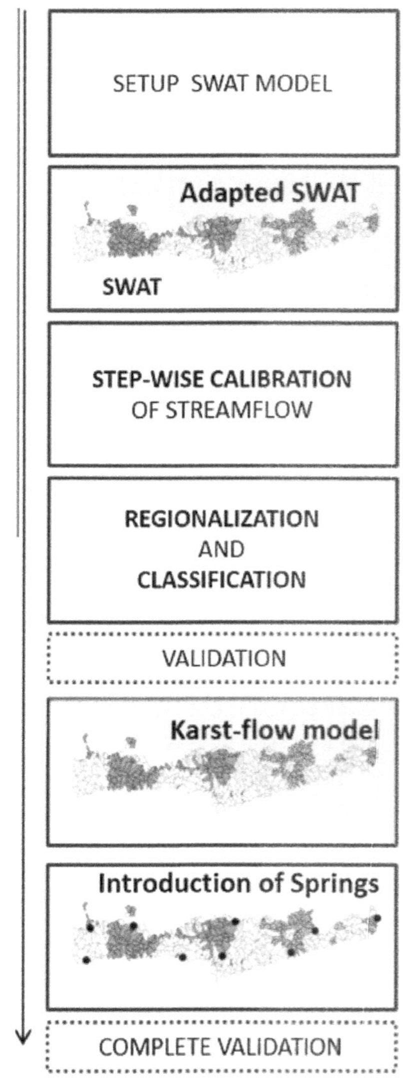

Figure 2: Karst SWAT modeling flowchart.

The SWAT Model

The SWAT model (Arnold et al., 1998) model is a continuous-time, semi-distributed, process based river basin model, originally developed to predict the long-term impact of climate and land use management practices on water, sediment, and agricultural chemical yields in large complex basins. SWAT uses the water balance approach to simulate watershed hydrologic partitioning. In particular, the water balance is calculated considering the snow, soil, shallow aquifer and deep aquifer components.

The watershed is first subdivided into subbasins and each subbasin into hydrologic response units (HRUs) that are a function of soil type, land use and land slope. The main components of the model are: weather, hydrologic mass balance, soil temperature and soil properties, plant growth, nutrients cycling, and sediments yield. SWAT runs on a daily time step. For specific aspects of SWAT, Neitsch et al. (2010) provide a complete description in the theoretical documentation of the model.

Adapting SWAT Model to Simulate Karst Processes

SWAT2012 (v.622) uses two linear reservoirs to partition groundwater into two aquifer systems (Fig. 3a): a shallow aquifer which contributes baseflow to streams (GWQ_1) and a deep aquifer which can also contribute baseflow to streams (GWQ_2). The remaining portion in the deep aquifer (DA_RCHRG – GWQ_2) can be considered lost from the system. The recharge for a specific day is calculated as a linear function of the daily seepage, the recharge of the previous day and the groundwater delay. Daily seepage includes seepage through the soil profile, through ponds or wetlands and loosing streams (tributaries and main channels). All seepage losses are added together and assumed to travel vertically to the aquifer with the same velocity. The baseflow from the shallow aquifer GWQ_1 is calculated through the aquifer recharge (GW_RCHRG, mm H_2O), the deep aquifer recharge (DA_RCHRG, mm H_2O) and the baseflow recession constant (ALPHA_BF, 1/day). Similarly, the baseflow from the deep aquifer (GWQ_2) depends on the deep aquifer recharge and on the deep-baseflow recession constant (ALPHA_BF_D, 1/day).

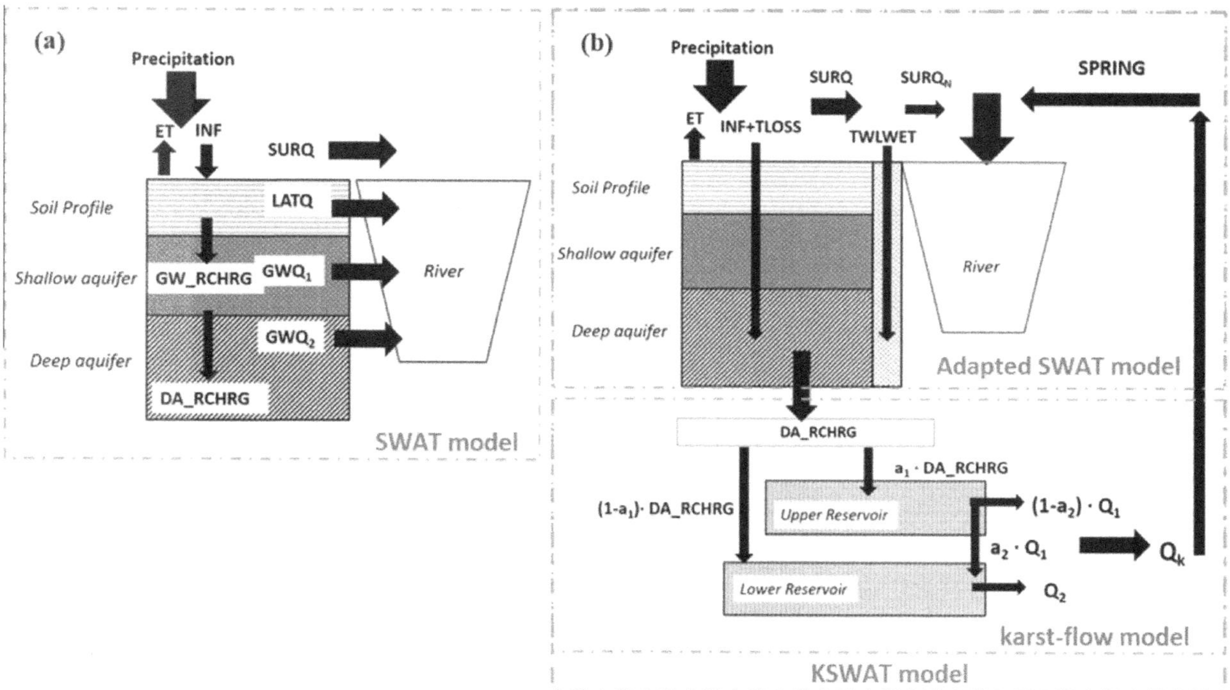

Figure 3: Comparison between the classic SWAT model configuration and the adapted SWAT model linked with the karst-flow model. In (a): DA_RCHRG, the deep aquifer recharge; ET, evapotranspiration; GW_RCHRG, the shallow aquifer recharge; GWQ_1, baseflow from shallow aquifer; GWQ_2, baseflow from deep aquifer; INF, infiltration in the soil; LATQ, Lateral flow; SURQ, surface runoff. In (b): a_1, fraction of DA_RCHRG to the upper reservoir; a_2, fraction of flow from upper to lower reservoir; DA_RCHRG, amount of direct recharge of deep aquifer from several subbasins (inlet of the karst-flow model); ET, evapotranspiration; INF, infiltration in the soil; Q_1, outlet of upper reservoir; Q_2, outlet of lower reservoir; Q_k, calibrated spring's discharge; SURQ, surface runoff; $SURQ_N$, surface runoff excluding TLOSS and TWLWET; TLOSS, tributary stream losses; TWLWET, losses from the bed of wetlands.

In order to simulate the specific characteristics of karst aquifers (i.e. fast infiltration, movement of water in cave systems) and to calculate the contribution of the karst areas to streamflow as spring flow, the SWAT model structure (Fig. 3a) was adapted as shown in Fig. 3b.

This new SWAT Model structure (Fig. 3b) was the result of the combination of two main studies: Baffaut and Benson, 2009 and Nikolaidis et al., 2013 and represents the concept of the karst model and its hydrological pathways. Hereafter, this model configuration is called KSWAT and combines an *adapted SWAT model* and a *karst-flow model*.

In particular, the *adapted SWAT model* considers fast infiltration through caves and sinkholes up to the deep-aquifer, while the *karst-flow model* represents the fast movement of water in subterranean conduits and the lower movement in narrow fractures through the interconnection of reservoirs structure.

The *adapted SWAT model* consists in representing sinkholes by wetlands with small drainage area and a large hydraulic conductivity at the bottom of the wetlands and loosing streams were represented by tributary channels with high hydraulic conductivity in the stream bed.

All the percolation in the soil profile, stream losses and seepage from the bottom of wetlands directly recharges the deep aquifer (DA_RCHRG, mm H_2O). This is achieved by setting the deep aquifer percolation fraction (RCHRG_DP) to 1, imposing minimum groundwater delay (GW_DELAY equal to 1) and setting the groundwater coefficient of capillarity rise ("revap") 0.1 in order to avoid that water moves from shallow into the overlying unsaturated zone while the baseflow from shallow and deep aquifers was imposed negligible (Table 2).

Table 2: Parameters and parameter ranges used in the calibration (in alphabetic order), in the *adapted SWAT model*and in the *karst-flow model*. In column "Process" the information about the order and group of calibration processes: 1, snow process; 2, runoff process; 3, lateral flow process; 4: groundwater process; 5: karst process

Model	Parameter	Description	Process	Range of calibration values	Range of values in karst-subbasins
Karst-flow model	a1	Fraction of inflow to upper reservoir	5	0–1	
Karst-flow model	a2	Fraction of upper reservoir to lower	5	0–1	
SWAT/adapted SWAT model	ALPHA_BF	Baseflow recession constant [1/d]	4/5	0–1	0.1
SWAT	ALPHA_BF_D	Baseflow recession constant for deep aquifer [1/d]	4	0	0
Adapted SWAT model	CH_K (1)	Effective hydraulic conductivity in tributary channel alluvium [mmh−1]	5	–	300
Adapted SWAT model	CH_K (2)	Effective hydraulic conductivity in the main channel [mmh−1]	5	–	5–500 a
SWAT	CN2	SCS runoff curve number for moisture condition II	2	−15% to +15%	–
SWAT	EPCO	Plant evaporation compensation factor	3	0.01–1	–
SWAT	ESCO	Soil evaporation compensation factor	3	0.01–1	–
SWAT/Adapted SWAT model	GW_DELAY	Groundwater delay [d]	4/5	0–500	1
SWAT/Adapted SWAT model	GW_REVAP	Groundwater 'revap' coefficient	4/5	0.02–2	0.1
SWAT	GWQMN	Threshold depth of water in the shallow aquifer required for return flow to occur [mm]	4	0–1000	–
Karst-flow model	kl	Upper reservoir recession constant [1/d]	5	0–5	–
Karst-flow model	ku	Lower reservoir recession constant [1/d]	5	0–5	–

SWAT	PLAPS	Precipitation laps rate [mm/km]	1	0–100	–
Karst-flow model	Qk0	Initial karstic flow [m3/day]	5	0–3×105	–
SWAT/adapted SWAT model	RCHRG_DP	Groundwater recharge to deep aquifer [fr]	4/5	0–1	1
SWAT	REVAPMN	Threshold depth of water in the shallow aquifer for revap to occur [mm]	4	0–500	–
SWAT	SFTMP	Snowfall temperature [°C]	1	−5 to +5	–
SWAT	SMFMN	Minimum melt rate for snow on Dec 21 [mm°C−1d−1]	1	0–10	–
SWAT	SMFMX	Minimum melt rate for snow on Jun 21 [mm°C−1d−1]	1	0–10	–
SWAT	SMTMP	Snow melt base temperature	1	−5 to +5	–
SWAT	SNOEB	Initial snow water content [mm]	1	0	–
SWAT	SOL_AWC	Available water capacity of the soil layer [fr]	3	−25% to +25%	–
SWAT	SOL_K	Saturated hydraulic conductivity [mmh−1]	3	−25% to +25%	–
SWAT	TIMP	Snow pack temperature lag factor	1	0.01–1	–
SWAT	TLAPS	Temperature laps rate [°C/km]	1	−10 to 0	–
Adapted SWAT model	WET_FR	Fraction of subbasin area that drain into wetlands	5	–	0.1–1b
Adapted SWAT model	WET_K	Hydraulic conductivity through bottom of wetland [mmh−1]	5	–	300b
Adapted SWAT model	WET_NSA, WET_MXSA	Surface area of wetlands at normal water level and at maximum water level [ha]	5	--	20b
Adapted SWAT model	WET_NVOL, WET_MXVOL, WET_VOL	Volume of water stored in wetlands when filled to normal, maximum water level and initial volume [104m3]	5	–	300b

[a]CH-K (2) was manually calibrated in the final calibration.

[b]Wetlands parameters were setting based on geological map extension.

The spring's discharge was simulated with a *karst-flow model* developed in "Excel" environment by Nikolaidis et al. (2013). The input was the daily deep aquifer recharge (DA_ RCHRG, mm) calculated by the SWAT model.

The *karst-flow model* is composed by an upper reservoir with faster response that represents wide conduits in a karst system and a lower reservoir with a slower response that simulates narrow fractures (Kourgialas et al., 2010).

The *karst-flow model* uses the deep aquifer recharge of the surrounding subbasins and involved at the same time the simulation and calibration of springs discharges. The procedure required the contribution/exclusion of deep aquifer recharge of nearest subbasins and the manually adjustment of five parameters of *karst-flow model*minimizing the Normalized Root Mean Square Errors.

These five parameters included: Q_{ko}, the initial karst flow (m^3/day), a_1 the fraction of deep groundwater discharge entering the upper reservoir, a_2 the fraction of flow from the upper reservoir discharge entering the lower reservoir and k_u and k_l were recession constants ($1/d$) for the upper and lower reservoirs (Table 2). For further details of *karst-flow model*equations, the reader can refer to Nikolaidis et al. (2013) who provided a complete description.

The outputs of the *karst-flow model* were then introduced in SWAT as point sources with aggregated values at monthly time step.

This modified SWAT was applied only in the karst-subbasins identified by three dominant soils that were strongly correlated to the karst areas of the Geological Map (Fig. 4b and c). According to the European Soil Database these are Leptosols (European Soil Portal, 2014), poorly developed and shallow soils over hard rock and comprise of very gravelly or highly calcareous material.

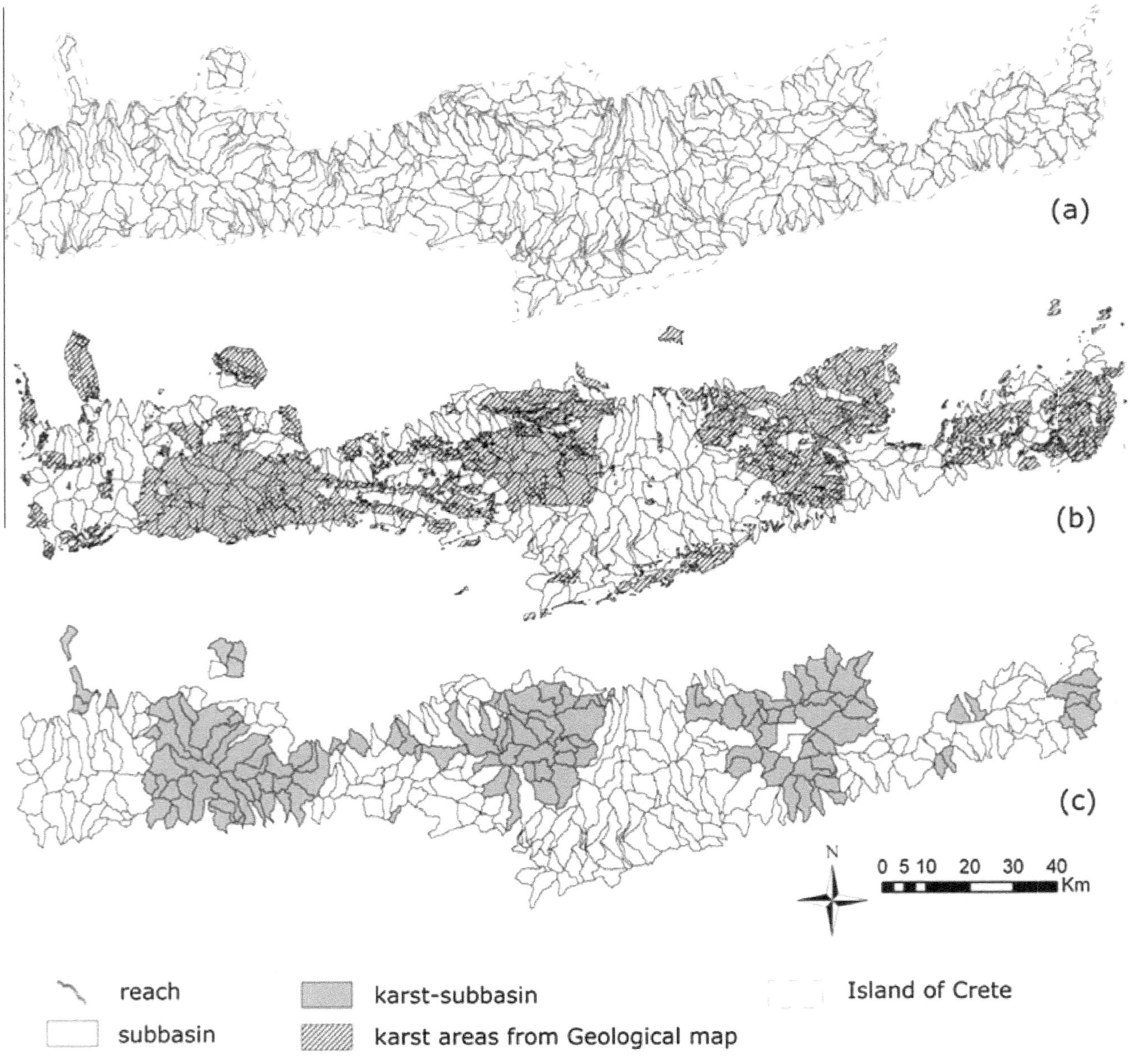

Figure 4: Rivers and subbasins (a); definition of karst geological features: karst areas from geological map (b) and the subbasins with karst soils (karst subbasins) (c).

However, the use of the "dominant karst soils" sometimes was limiting since in some cases the size of the subbasins (about 20 km^2) and the fragmentation of soils inside the subbasin did not allow identifying it as "karst subbasin". This occurred for instance in the south-central part of the Island in Kourtaliotis Basin (corresponding to Kourtaliotis river basin in Fig. 1c), where this limitation was overcome using literature information (i.e.Steiakakis et al., 2011) and the geological map. Thus, both karst-soils extension and the geological map were used to identify the karst-subbasins where the KSWAT model was applied. It is noteworthy that the extension of wetlands inside each karst-subbasin was defined using the geological map, generally covering the entire subbasins where the karst-soils are dominant, or only a percentage of subbasin area as in the case of the Kourtaliotis Basin. The resulted area covered by the karst-subbasins was estimated around 2600 km^2.

Model Setup

The SWAT model requires pedological, climatological, topographical and land use data. Subbasins were delineated using the ArcSWAT interface with a Digital Elevation Model of 25 m pixel size (EU-DEM, xxxx and Bashfield and Keim, 2011). The Digital Elevation Map was obtained from a Pan-European elevation data at 1 arc-second (EU-DEM). Subbasins and streams were defined using a drainage area threshold of 1000 ha resulting in 352 subbasins with an average area of 19 km^2 covering 6700 km^2 (Fig. 4a).

Land cover was derived from a 1 km raster map built from the combination of CAPRI (Britz, 2004), SAGE (Monfreda et al., 2008), HYDE 3 (Klein and van Drecht, 2006) and GLC (Bartholome and Belward, 2005) for the year 2005. Land use was obtained from theAgriculture statistics of Greece (2005). For each subbasin the "dominant land cover approach" was used in order to reduce the complexity and the computational level of the model. Six classes of dominant land cover were defined: arable land, pasture, forest, urban area, water and range grasses. Each subbasin with dominant arable land cover was further subdivided in three parts (15–35–50% of subbasin area) in order to distribute more accurately the crops reported by the Agricultural census. Soil type and characteristics were defined using a 1 km soil raster map, obtained from the Harmonized World Soil Database (HWSD) (FAO, 2008).

From the combination of land use, dominant soils, and single slope for each subbasin the final configuration resulted in 502 HRUs (Hydrological Response Units) with an average area of 13 km^2. The management practices for each crop included planting, fertilization, irrigation and harvesting. The simulated timing of plant sowing and harvesting were implemented in SWAT through daily heat unit concept. In this study the heat units for each crop were calculated by Bouraoui and Aloe (2007) using the PHU (Potential Heat Units) program (PHU, 2007), developed at Texas Agricultural Experiment Station. The PHU program calculates the total number of heat units required to bring a crop to maturity. For each crop the crop growing season (winter versus spring crop), the base growing temperature (°C), the optimum growing temperature (°C), the dry down fraction, the time to maturity (number of days between planting and harvesting) were considered as input in the PHU program. These attributes were related to the different European climatic zone.

The amount of manure and mineral fertilization applied were retrieved from the Common Agricultural Policy Regionalized Impact (CAPRI) agro-economic model (Britz and Witzke, 2008). It provided at NUTS 2 administrative level (Nomenclature of Territorial Units for Statistics) the mineral and organic nitrogen and phosphorus fertilizer application estimated from sectoral statistics and animal production levels.

A manual irrigation was applied selecting the irrigated HRUs for a total irrigated area around 25,000 ha in Chania, 9300 ha in Rethymno, 60,000 ha in Heraklion and 26,000 ha in Lasithi. The total abstractions for irrigation use in each Prefecture was 89 Mm^3/y in Chania, 35 Mm^3/y in Rethymno, 170 Mm^3/y in Heraklion and 66 Mm^3/y in Lasithi, respectively. Both irrigated areas and volumes were obtained from the Agriculture statistics of Greece (2005).

According to the National statistics, the sources of abstraction for irrigation uses were set from deep aquifer and shallow aquifer, and from springs. In the model the abstraction for irrigation from springs (38 Mm^3/y) was implemented as outside-watershed source in the schedule management plan and the corresponding abstractions were subtracted from the daily deep aquifer recharge (DA_RCHRG, m^3/s) (outside SWAT), so that the calibration of springs discharge in the *karst-flow model* was performed using the net of the abstractions.

The abstractions for drinking water, industry and olive mills (40 Mm^3/y) were introduced in SWAT by subtracting water from the deep aquifer and shallow aquifer at monthly time step, while the abstractions for drinking water from the springs (24 Mm^3/y) were subtracted from the daily deep aquifer recharge (DA_RCHRG) (outside SWAT).

The climate data used in this study include 69 stations with daily data for precipitation and 21 stations for temperature from 1961 to 2009. Monthly statistics of solar radiation, wind speed and relative humidity were calculated using the pan European high-resolution gridded daily data set EFAS-METEO, acronym of European Flood Awareness System – METEO (Ntegeka et al., 2012) for 29 stations uniformly distributed over the modeled territory.

The subbasins were split into elevation bands, and snow cover and snowmelt were simulated separately for each elevation band. In this study steep subbasins were subdivided into 4 classes of elevation.

Calibration of Streamflow

For the calibration and validation of the Crete SWAT model 22 stations of streamflow were used. Based on the simplest test technique (Klemeš, 1986), 15 of them were used for calibration and 7 for validation covering the period 1980–2009. The selected subbasins for calibration were headwaters and were uniformly distributed in the Island. The streamflow was calibrated in the 15 selected subbasins in different steps ("step-wise calibration") according to the involved hydrological process, providing a calibrated parameter set NOP (Near Optimal Parameters set). The obtained NOP set was then transposed in ungauged subbasins using the "hydrological similarity approach".

The step-wise calibration is a procedure that allowed calibrating the different components of the streamflow according to different hydrologic processes underpinning each calibration objective. This procedure proposed by Pagliero et al., 2015 and Malagò et al., 2015 was performed using the SWAT-CUP (SWAT Calibration Uncertainty Procedures) programs and the SUFI-2 algorithm (Abbaspour, 2008).

The calibration period covered 30 years from 1980 to 2009 with 10 years of warm up. Among several parameters in SWAT, 19 parameters (see parameters that refer to SWAT and SWAT/Adapted SWAT model in Table 2) were chosen for the streamflow calibration as the most sensitive and representative of each hydrological process involved in the calibration. The parameters and their initial ranges (Table 2) were selected based on preliminary model runs, literature reviews (van Griensven at al., 2006) and sensitivity test. It is noteworthy however that the parameters in karst-subbasins were not changed keeping the settings of the *adapted SWAT model* configuration as described in Table 2.

In this work, since the streamflow data was only available at monthly resolution, the streamflow was not divided into the main components (surface runoff, lateral flow, baseflow). However, the calibration was still performed step wise:

Step 1: Calibration of total monthly flow in order to control the timing of runoff signal by adjusting the snow parameters by minimizing the coefficient of determination r^2;

Step 2: Calibration of total monthly flow using only the parameters that regulate the surface runoff process by minimizing the coefficient of determination r^2;

Step 3: Calibration of total monthly flow using only the parameters that regulate the lateral flow process by minimizing the coefficient of determination r^2;

Step 4: Calibration of total monthly flow using only the parameters that regulate the baseflow process by minimizing the Nash Sutcliffe Efficiency, NSE (Nash and Sutcliffe, 1970);

Step 5: Final step, calibration of total monthly flow using all previous parameters together with a reduced uncertainty ranges by minimizing the NSE coefficient.

The coefficient of determination r^2 (Taylor, 1990) was adopted as objective function for the first three steps and the NSE coefficient (Nash and Sutcliffe, 1970) was selected for the fourth and final step as described below. The selection of the objective functions is the result of several tests since the differences between objective functions on parameter sets could be quite striking (Abbaspour et al., 2015). The calibration was performed using 1000 simulation runs at each step.

The Regionalization and Classification of Ungauged Subbasins

The Final Near Optimal Parameter set (NOP) of calibrated subbasins (called "donors") was transposed to ungauged subbasins using the hydrological similarity approach. The measure of hydrologic similarity between subbasins is translated into subbasin characteristics which are common for gauged and ungauged watersheds. These characteristics that linked the subbasin characteristics with the hydrological responses have been performed in the literature (Shamir et al., 2005 and Oudin et al., 2010). However, for a comprehensive enlightenment of hydrological similarity approach the reader can refer to Wagener et al., 2007 and Sawicz et al., 2011. The calibration for the parameters of the karst-subbasins were kept unchanged using the settings of the*adapted SWAT model.*

The hydrological similarity approach was performed using the PLSR (Partial Least Squares Regression) that allowed identifying similar subbasins based on the correlation between the watershed characteristics and the discharge characteristics. FollowingMalagò et al. (2015), this correlation was performed using 17 gauged independent basins using the R package "pls" (Mevic and Wehrens, 2007). The gauging stations selected to perform the PLSR analysis involved stations from the group selected for the calibration and some from the validation group in order to identify a group of stations with almost 8 years of continuous data from 1988 to 1995 that allowed calculating robust statistical indices.

For the selected subbasins 20 independent variables representing the subbasin characteristics ("c" matrix) and 2 dependent flow variables ("q" matrix) were used in the PLSR analysis as the responses and regressors ($q \sim c$) (Malagò et al., 2015). The independent variables included river length (km), minimum, maximum and average elevation (m), median slope (%), content of clay and sand in the soil, percentage of different classes of land cover (%), percentage of karst area, mean annual precipitation (mm), minimum, maximum and average temperature (°C), annual potential evapotranspiration (mm) and average number of days with precipitation. The dependent variables included the mean annual discharge (mm/y) and the corresponding coefficient of variation. The PLSR analysis defined latent variables that were used to perform the classification of subbasins.

Due to the limited number of potential donors (15 subbasins, 4% of total number of subbasins), it was necessary to define "hydrological regions" and then the "classification" of similar subbasins was performed inside each region. For a review of regionalization techniques, the reader can refer to Parajka et al. (2005).

Using the latent variables, the regionalization was performed with the Ward's hierarchical cluster analysis (Ward, 1963) based on the Euclidean distances and the classification of subbasins in donor's classes was pursued inside each hydrological region according to the K-Nearest-Neighbor using the R package "supclust" (Dettling and Maechler, 2012). The classification of each subbasin allowed the automatic assignation of the NOP of "donor subbasins" to the classified ungauged subbasins.

In order to verify the strength of the parameters regionalizationintheungaugedsubbasinsthemodelperformance was evaluated using the percent bias (PBIAS; Gupta et al., 1999) between monthly observed and simulated streamflow at seven gauging stations. This was an intermediate validation that anticipated a final complete validation performed in all gauging stations after the introduction of springs in the model as described in the next paragraphs.

The Calibration of Springs and Delineation of Karst Recharge Area of each Spring

The springs were calibrated using the daily recharge of deep aquifer (DA_RCHRG) from the *adapted SWAT model*, except for Spring SP54 (Almiros Heraklion) that was calibrated using DA_RCHRG at monthly time step. The springs were calibrated manually changing parameters of the *karst-flow model* in reasonable ranges: the initial karst flow (m³/day), the fractions a_1, a_2 and the recession constants k_u and k_l (1/*d*) for each reservoir as show in Table 2. This process ended when the Normalized Root Mean Square Error (NRMSE) calculated using the simulated and observed values for each spring was minimized.

42 springs out of 47 were calibrated having hypothesized that some springs drained the same karst area. This assumption was strongly correlated to the neighborhood location of these springs that didn't allow distinguishing the contribution of deep aquifer recharge from the same karst-subbasins; hence SP12 and SP13, SP44 and SP45, SP4 and SP50, SP7, SP18 and SP19 were grouped into SP12-13, SP44-45, SP18-19-7 and SP4-50.

The springs were entered as points sources in the model and the associated calibrated monthly Q_k (m³/s) were implemented as input data in SWAT. For each spring the karst recharge areas (inside and outside the hydrological boundary) was defined as sum of subbasins areas whose deep aquifer recharge contributes to a specific spring.

The subbasins that contributed to spring flow through their deep aquifer recharge in the*karst-flow model* were not necessarily hydrologically connected. The linkage between these subbasins and the springs represented the movement of water through the cave systems whose spatial extent and organization was different from the surface subbasin hydrologic boundaries.

The performances of the *karst-flow model* were evaluated by visual comparison of measured and simulated hydrographs of springs and using performance indicators including the Normalized Root Mean Square Error (NRMSE), the percent bias (PBIAS;Gupta et al., 1999), the coefficient of determination r^2 multiplied by the slope of the regression line between simulated and observed values (br²; Krause et al., 2005) to overcome the intrinsic weakness of each indicator (Bennett et al., 2013).

Final Calibration and Complete Validation

After the introduction of calibrated springs in the SWAT model as point sources, an additional manual calibration was necessary to adjust the water discharge including peaks and baseflow since the addition of the spring contribution modified the already calibrated discharge. In particular, the effective hydraulic conductivity in main channel (mm/hr) in Messera, Anapodiaris and in Petras Basins was adjusted (Table 2).

Finally, the complete SWAT model was validated using the whole dataset of streamflow gauging stations in order to evaluate the model performance after the introduction of the springs. Similarly to calibrated streamflow and spring's discharges, the complete SWAT model was validated comparing monthly observed and simulated streamflow in the period 1983–2009 calculating the NRMSE (%), PBIAS (%), the Nash Sutcliffe efficiency (NSE), br² and r^2.

RESULTS AND DISCUSSION

Hydrological Simulation

The performance indicators obtained comparing the monthly observed and simulated streamflow for the 22 stream gauging stations after calibration, regionalization and the introduction of the calibrated springs are summarized in Table 3.

Table 3: Comparison of observed and simulated streamflow in the period 1983–2009 after the introduction of the springs (complete validation): in the first two columns the code and name of the stations; in the third column the drain area (km²); the *% data* represents the % of observed monthly values in the period 1983–2009 (324 total months); the observed and simulated volumes are respectively *Mm³/year OBS* and *SIM;* the performance indicators between monthly observed and simulated steamflow are the normalized root mean square error (NRMSE%), the percent bias (PBIAS%), the Nash Sutcliffe Coefficient (NSE), the coefficient of determination (r^2) and the coefficient of determination multiplied by the slope of the regression line between simulated and observed values (br²)

Station	Name	Drain area (km²)	% data	Mm³/ year OBS	Mm³/ year SIM	NRMSE%	PBIAS%	NSE	r^2	br²
STR32	Koiliaris	132.6	21	180.97	182.74	16.5	3.3	0.68	0.69	0.62
STR19R	Platis	209.6	69	50.15	46.30	11.1	2.4	0.63	0.63	0.47
STR7R	AnapodiarisH	522.7	54	30.13	31.55	6.9	16.5	0.78	0.8	0.73
STR13	Giofiros	186.3	69	22.24	22.94	12.5	25.4	0.53	0.66	0.65
STR20	Prasanos	101.9	54	17.83	22.17	13.2	2.8	0.23	0.36	0.27
STR9R	Geropotamos	394.5	69	17.21	33.04	32.2	183.4	−3.94	0.46	0.24
STR3	Katavothres	21.35	54	16.63	11.56	11	−32.7	0.53	0.62	0.3
STR26	Kakodikianos	78.11	77	15.89	19.02	11.6	31.6	−0.87	0.25	0.23

STR27	Sebreniotis	28.48	80	15.83	12.45	8.5	−17.9	0.58	0.6	0.38
STR11	Anapodiaris Pla	89.51	47	11.89	12.91	13.4	−2	0.64	0.65	0.53
STR10	Koutsoulidis	132.2	54	11.53	13.07	5.7	17.2	0.83	0.84	0.82
STR2R	Mirtos	96.34	84	10.87	11.31	15.3	39.9	−0.35	0.29	0.26
STR6R	Aposelemis	204.8	69	10.69	26.46	14.4	158.7	−0.05	0.63	0.5
STR28	Rumatianos	22.06	76	6.75	7.18	16.3	36.9	−0.06	0.26	0.18
STR8	Litheos	41.98	77	6.39	6.13	20.6	8.2	−0.31	0.26	0.2
STR14R	Gazanos	186.8	54	6.32	11.86	36.9	111.8	−3.05	0.23	0.17
STR1	Patelis	84.57	56	5.91	6.12	12.1	13.7	0.48	0.57	0.51
STR4	Kalamafkianos	36.14	10	5.41	3.57	40.7	10.4	−1.33	0.32	0.3
STR15	Baritis	105.5	44	4.08	3.06	13.2	61.7	−0.04	0.19	0.12
STR25	Agios Vasiliou	35.64	74	2.37	4.62	28	103.4	−2.38	0.26	0.2
STR16	Arvis	26.53	80	1.66	1.14	11.1	11.2	0	0.32	0.23
STR12R	Iniotis	105.9	44	1.56	2.21	23.7	57.3	−0.81	0.16	0.12

[R]Gauging stations used in the validation of regionalization technique.

It is noteworthy that about 64% of the calibrated gauging streamflow stations reached satisfactory PBIAS% (values in the range ± 25%), while only 40% had NSE greater than 0.5 (see Moriasi et al., 2007 for performance rating as reference). Furthermore, 50% and 36% of calibrated streamflow gauging stations reached respectively r^2 and br^2 larger than 0.5.

Fig. 5 shows some examples of comparison of hydrographs. Even though some of the statistical indicators did not score very high, the hydrographs comparison illustrates that the model was able to reproduce accurately the monthly variations over a long period of simulation even in subbasins influenced by excessive water pumping, such as STR16.

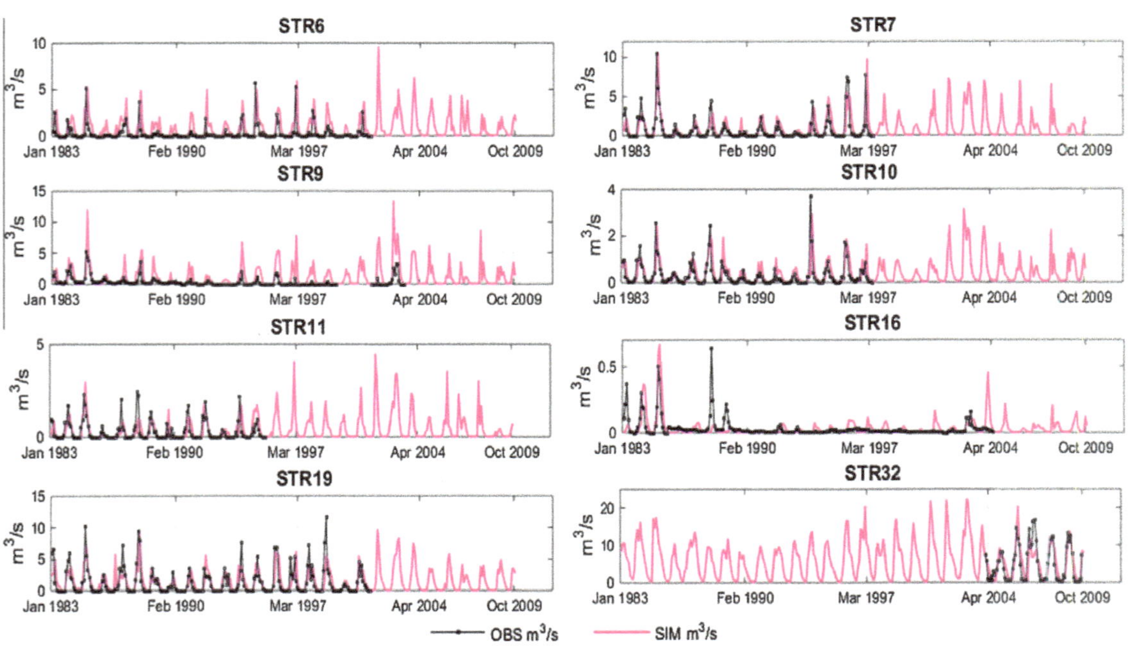

Figure 5: Some examples, comparison between monthly simulated and observed streamflow values from 1983 to 2009 after the allocation of springs in the SWAT model.

This gauging station was fed by precipitation and discharge of spring SP16. After the year 1990, the spring's discharge was diverted to supply domestic uses reducing drastically the streamflow in particular during the summer and autumn.

It was estimated that the percentage of streamflow gauging stations that reached satisfactory PBIAS (PBIAS < ±25%, Moriasi et al., 2007) increased from 59% to 64% adding spring's discharges in the SWAT model highlighting the significant influence of springs in the water balance of Crete. Fig. 6 shows the PBIAS cumulative curves of performance indicators calculated between monthly and observed streamflow before and after the introduction of springs as point sources in the SWAT model.

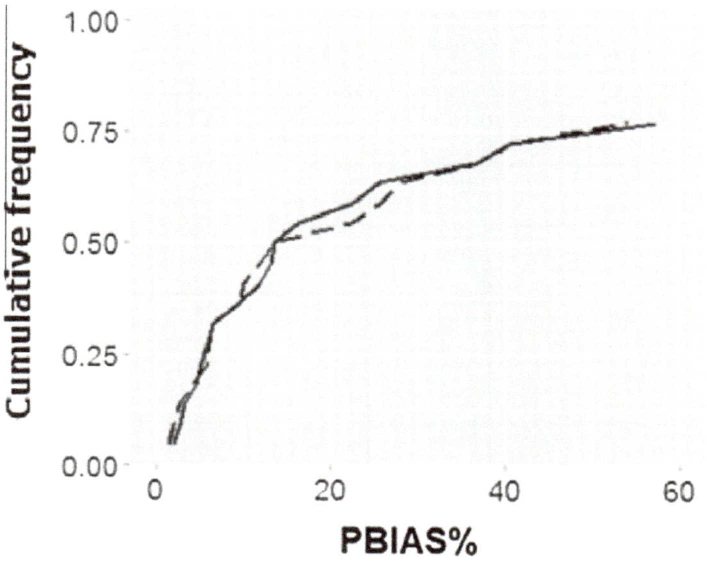

Figure 6: Cumulative frequency curves of percent bias (PBIAS, %) obtained between monthly and observed streamflow before (only adapting SWAT model) and after the introduction of springs (KSWAT).

In particular, streamflow discharge of gauging stations STR32 (Koiliaris), STR16 (Avli), and STR10 (Koutsoulidis) were strongly influenced by the discharge of springs SP41 (Stylos), SP14-SP25 (Migilisi and Emparos) and SP20 (Zaros), respectively. For STR32 the PBIAS decreased to 3% and the NSE increased to 0.68, for STR16 the PBIAS% reached 11%, and finally for STR10 the PBIAS% and NSE were also satisfactory reaching 17% and 0.83 respectively (Table 3).

These results demonstrated that the *karst-flow model* correctly simulated the discharge of springs increasing the SWAT model performance. Fig. 7 shows some examples of comparison between daily observed and simulated spring flows (only for SP54 monthly comparison), and Table 4 summarizes the comparison between monthly and observed long-term annual discharge on the period 1983–2009 and the performance indicators for all springs. It is noteworthy that for more than 70% of the calibrated springs the PBIAS% is within the range of ±25%, indicating that the performance of the *karst-flow model* was satisfactory (Moriasi et al., 2007). NSE was positive for 80% of calibrated springs, whiler^2 and br^2 had a wide range of values with around 50% of the values larger than 0.4.

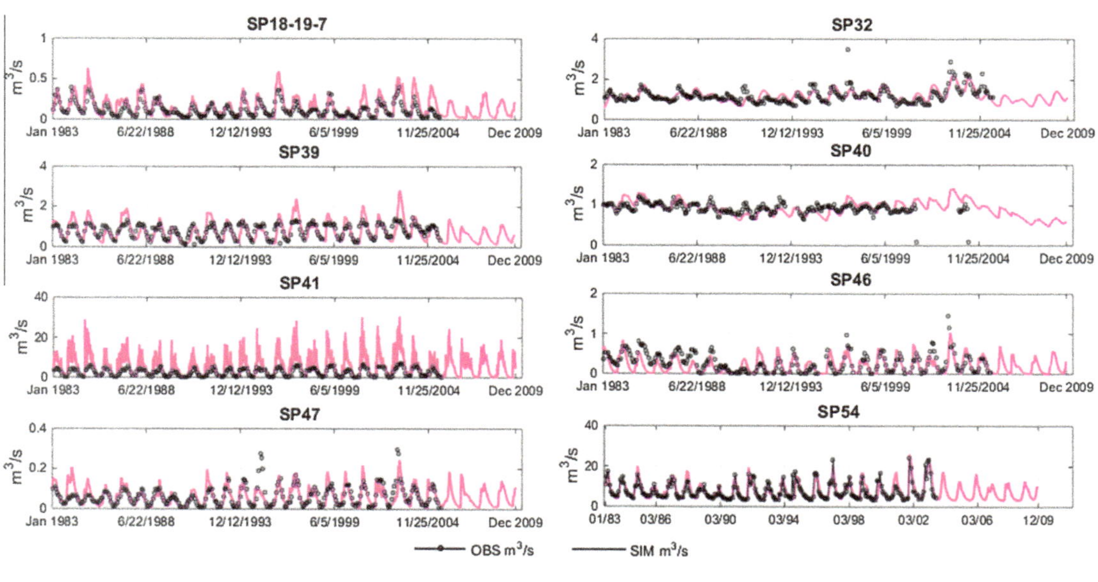

Figure 7: Some examples, comparison between daily simulated and observed spring's values from 1983 to 2009 (monthly comparison for SP54).

Table 4: Comparison of observed and simulated spring discharges in the period 1983–2009: in the first two columns the code and name of the stations; in the third the karst area that contributes of each spring (km^2) and involves the area inside the hydrological basin and outside; the % *data* represents the percentage of observed values (once a month) in the period 1983–2009 (324 total months); the observed and simulated volumes are respectively *Mm^3/year OBS* and *SIM;* the performance indicators calculated between daily observed and simulated spring discharges are the normalized root mean square error (NRMSE%), the percent bias (PBIAS%), the Nash Sutcliffe Coefficient (NSE), the coefficient of determination (r^2) and the coefficient of determination multiplied by the slope of the regression line between simulated and observed values (br^2)

Station	Name	Estimated karst recharge area (km^2)	% data	Mm^3/year OBS	Mm^3/year SIM	NRMSE%	PBIAS%	NSE	r^2	br^2
SP54a	Almiros-Heraklion	323.48	77	235.72	236.59	9.8	0.7	0.77	0.8	0.8
SP41	Stylos	185.01	79	85.81	168.85	66	102.7	−4.8	0.64	0.31
SP9	Almiros-Agios-Nikolaos	155.42	80	83.22	83.73	15.9	3.9	0.22	0.54	0.56
SP43	Platanos/kalamionas-Agia	110.18	10	67.11	67.43	23.8	6.1	0.18	0.55	0.59
SP32	Kourtaliotis	103.37	84	37.99	37.08	9.9	−0.8	0.44	0.43	0.45
SP42	Meskla	114.2	84	30.7	30.94	17.4	11.2	0.31	0.32	0.38
SP40	Kourbos	53.32	68	29.85	28.68	14.6	2.1	−0.61	0.12	0.12
SP38	Vrisses	20.21	77	26.95	26.53	13.4	−3	0.48	0.41	0.51
SP39	Armenoi	78.48	84	24.79	24.09	26.1	1.2	−0.08	0.5	0.5
SP37	Petres	26.2	70	13.77	13.76	15.2	2	0.34	0.26	0.39
SP44_45	Elliniki-Konto Kinigi	40.32	84	13.5	13.52	9.5	5.3	0.44	0.32	0.45
SP34	Spilianos	34.07	84	12.65	11.24	10.7	−6.3	0.36	0.21	0.36
SP36	Mousela Xalikouti	47.95	77	11.95	11.26	9.4	18.9	0.24	0.36	0.41
SP46	Drapania	40.9	75	9.87	6.35	14.9	−21.9	0.16	0.21	0.32
SP48	Therisos	12.6	42	9.68	9.35	16.6	−3	0.32	0.18	0.32
SP28	Geropotamos	25.58	64	7.75	7.81	14.2	−2.5	0.13	0.13	0.22
SP2	Zakros	28.45	80	5.4	5.77	9.5	−1.8	0.62	0.59	0.63
SP3	Chochlakies	34.75	49	5.33	5.47	17.3	2.3	0.34	0.38	0.46
SP27	Panormos Almiro Nero	16.74	69	4.68	4.65	13.1	4	0.07	0.49	0.5
SP18_19_7	Kria Vrisi-Simis-Kefalovrisi Viannou	35.14	84	4.65	4.74	20.6	26.7	0.17	0.55	0.67
SP22	Fodele	22.45	84	3.78	3.99	8.7	−3.2	0.21	0.13	0.22
SP52	Argiroupoli	24.23	44	3.67	3.54	18.7	18.7	0.09	0.51	0.6
SP49	Kefalovris/Kalamoukas-Ierapetra	17.91	84	3.54	3.59	10	−4.8	0.41	0.33	0.46
SP12_13	Almiros Mallia-Grammatikaki	22.89	80	3.51	3.43	12.7	10.4	0.49	0.48	0.56

SP6	Archon Stavro-choriou – Sitia	28.91	83	3.26	3.23	12.5	−1.7	0.38	0.49	0.51
SP20	Zaros – Votomos	26.38	10	3.15	3.03	32.4	−0.1	−0.06	0.04	0.04
SP30	Spili	14.21	21	3.07	3.1	30.4	−2	−0.85	0.1	0.11
SP8	Kalo Chorio	14.61	76	3.05	3.1	14.3	1.2	0.29	0.26	0.3
SP11	Zou	29.54	48	2.14	2.14	13.2	−14.7	0.51	0.35	0.52
SP47	Sfinari	32.57	84	1.96	1.99	12.8	−0.2	0.42	0.43	0.49
SP31	Agia Fotia-Spili	14.21	35	1.94	1.97	21.4	−9.7	0.29	0.26	0.36
SP25	Emparos	16.14	36	1.53	1.52	13	11.5	−0.11	0.01	0.02
SP26	Seises	2.5	80	1.53	1.23	16.7	−20.9	0.05	0.3	0.4
SP4_50	Agios Georgios Sitia-Sikia Sitia	18.27	42	1.52	1.49	23.8	103.4	−1.06	0.26	0.35
SP21	Gergeri	41.8	17	1.49	1.86	14.8	−8.3	0.57	0.57	0.63
SP35	Ligres	8.35	17	1.46	1.41	14.7	14	0.41	0.46	0.47
SP5	Lithines-Sitia	18.73	81	1.38	1.31	11.9	−3.6	0.54	0.42	0.56
SP14	Avli Migilisi	18.85	4	1.15	1.02	50.7	43.8	−1.04	0.41	0.62
SP51	Loutraki	33.22	58	1.09	1.19	19.1	27.2	−0.1	0.42	0.49
SP10	Petikou	3.22	80	1.01	1.02	8.1	11.5	0.62	0.56	0.63
SP16	Mega Vrisi	28.97	34	0.65	0.64	20.3	32.7	0.04	0.1	0.16
SP53	Agios Georgios-Viannou	3.69	33	0.11	0.2	19.1	7	0.22	0.37	0.39

[a]For SP54 the performance indicators were calculated using monthly values.

However, the *karst-flow model* markedly overestimated the discharge for springs SP4-50 (Agios Georgios Sitia-Sikia Sitia), SP14 (Avli Migilisi), SP16 (Mega Vrisi), SP41 (Stylos) and SP51 (Loutraki) maybe due to the quality of the observed data.

Concerning SP41, the most studied spring in Crete (Nikolaidis et al., 2013, Moraetis et al., 2010 and Kourgialas et al., 2010), was intentionally calibrated to overestimate the flow in order to take into account the contribution of ungauged springs. Nikolaidis et al. (2013) reported that Stylos spring has two permanent springs at elevation +17 m AMSL (with 85 Mm^3/year of volume), but there was also an ungauged intermittent spring (Anavreti at elevation +24 m AMSL). Both contributed to Koiliaris discharge. The authors argued that these springs have a total average discharge around 154 Mm^3/y (2007–2010), and this volume was in agreement with the total overestimated volume of Stylos in this work (150 Mm^3/y, for the same period 2007–2010, 170 Mm^3/y considering 1983–2009). Finally, it is noteworthy that the spring SP46 (Fig. 7) had two different hydrological behaviors: observed data were higher prior to year 1988, then the values decreased sharply highlighting changes in the regime of discharge. This may be explained by a change in the method of measurement or more probably by an excessive pumping of water as occurred for the STR16 spring.

The estimated recharge areas of karst springs

Fig. 8 shows some schematic examples of recharge areas of springs obtained considering the subbasins that contribute with their deep aquifer recharge to the calibration of springs. Table 4 summarized also the karst recharge areas (km^2) for each spring or group of springs. The karst recharge areas of springs cover karst-subbasins and karst areas for a total of 1928 km^2, about the 70% of total karst areas in Crete. The main karst recharge area drains more than 300 km^2 into Almiros Heraklion Spring (SP54), followed by the karst recharge area of Almiros-Agios-Nikolaos Spring (SP9) that cover about 155 km^2, and 130 km^2 of karst recharge area of springs in Koiliaris Basin.

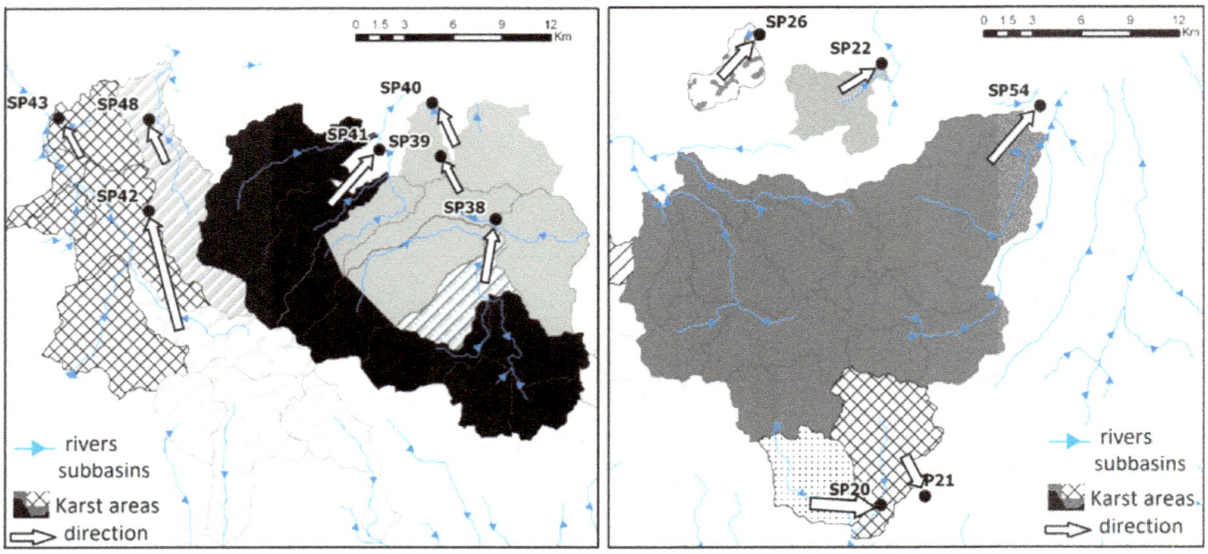

Figure 8: Examples of estimated karst recharge area of selected springs.

To the best of our knowledge this work was the first estimation of karst recharge areas of all gauged springs in the Island of Crete, providing valuable information for water resources management. The only exception was the karst area drained by Almiros Spring and the knowledge of extended areas (outside the hydrological boundary of basin) of springs in Koiliaris Basin.

Bonacci and Fistanic (2006) pointed out the difficulties to define the karst area drained by Almiros Spring due to complex geology and hydrogeology. There were different assumptions of its boundaries and size which varies from 300 km^2 (Arfib et al., 2000) to 500 km^2 (Lambrakis et al., 2000). However, Bonacci (1995) and Bonacci and Ljubenkov (2005) estimated the karst recharge area of Almiros of about 300 km^2, similar to our finding.

The area outside the Koiliaris Basin that contributed to the karst recharge area of Stylos Spring and other ungauged springs in Koiliaris Basin was estimated around 80 km^2 (7 subbasins that contribute with 70% of their total area). As a consequence, more than 60% of karst recharge area of Koiliaris springs was outside the boundary of the hydrological basin. This result was confirmed by Nikolaidis et al. (2013) that estimated an extended karst area of springs in Koiliaris Basin of 79 km^2 based on hydrologic modeling and geologic cross-sections.

Albeit this approach may be criticized by the fact that karst recharge areas should be only delineated based on hydrogeological considerations derived, i.e. by tracer tests, it has to be taken into account that there are limits in the use of tracer studies in karst areas such as in Koiliaris basin. In a geological karst area which discharges a large volume of water there is a risk of high degree of dilution of tracers (Knithakis, 1995) which in turn limits the usefulness of the methods in delineating the extend area. In addition, breakthrough curves are highly dependent on the flow conditions, often necessitating the repetition of the tracing experiment under low flow and high flow conditions as observed byGoldscheider, 2005 and Ravbar et al., 2011.

Spatial and Temporal Variation of Hydrological Components in Crete

The mean annual precipitation ranges from 550 mm (driest year) to 1400 mm (wettest year) during the period 1983–2009, with a decreasing gradient from west to east. The precipitation was higher at high elevation, reaching more than 2500 mm in the wet years (1996, 1997, 2001 and 2003) and around 1000 mm in the central north areas, while in the plain areas and along the south cost the average annual rainfall was below 1000 mm.

The mean calculated actual evapotranspiration of Crete varied from 340 mm/y (driest year) to 390 mm/y (wettest year) reaching the maximum values long the north coast and, in particular in Karteros, Aposelemis, Tzermiadon, Myrtos and Anapodiaris basins (east-central part of Crete).

Large values of surface runoff were simulated in the western part of Crete and in particular in Tavronitis Basin with maximum value in wet years, such as in 1997 and 2003 (more than 1100 mm). Minor contributors to the water balance include lateral flow and baseflow. The baseflow reached high values in Kourtaliotis, Akoumianos and Platis basins (south-central Crete). The annual and spatial variations of deep aquifer recharge were strongly correlated to the precipitation, as a consequence in zones with high precipitation the deep aquifer recharge reached the largest values since the main mountainous systems were associated with the most important karst systems, allowing water to penetrate directly to the deep aquifer as showed in Fig. 9. In particular, Fig. 9illustrates how precipitation, evapotranspiration and recharge of the deep aquifer change during the year. Precipitation and deep aquifer recharge changed during the year decreasing from January (~1120 and 463 Mm^3/month respectively) to August (~18 and 62 Mm^3/month respectively) and increasing from September (~163 and 63 Mm^3/month respectively) to December (~1250 and 416 Mm^3/month respectively). During the wettest months, the west part of Crete reached the highest values of precipitation and also the highest value of deep aquifer recharge. The evapotranspiration increased from January (~240 Mm^3/month) to April (~320 Mm^3/month), in particular in subbasins long the cost, and then decreased until August (~54 Mm^3/month) and increased from September (~71 Mm³/month) to December (~240 Mm³/month). These long-term spatially and monthly variations could be very useful for planning and implementing conservations measures and programs and evaluating their performance.

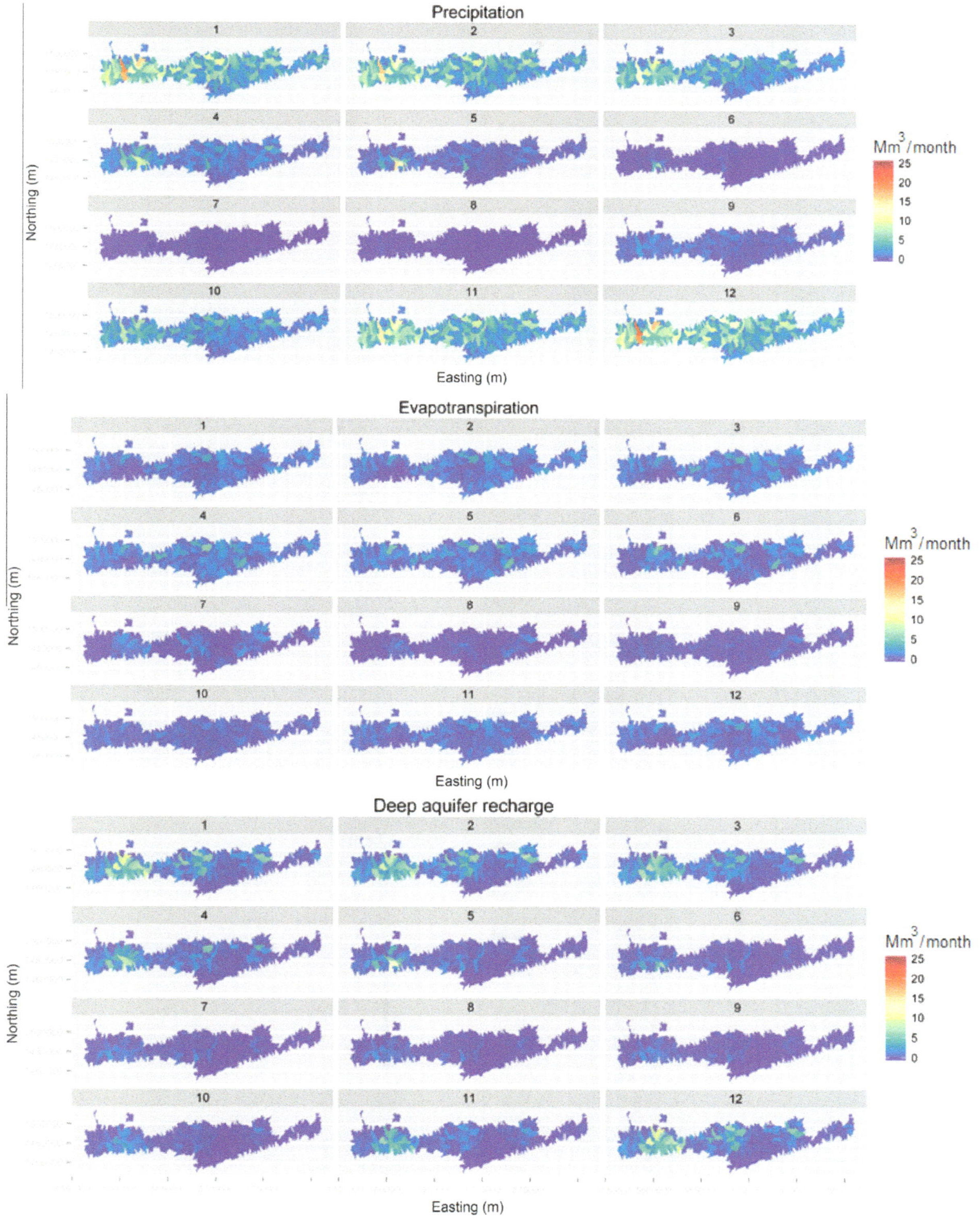

Figure 9: Maps of monthly variations of precipitation, evapotranspiration and the deep aquifer recharge (Mm³/month).

The Estimated Water Balance of Crete

Chartzoulakis et al. (2001) and many other studies (RGC, 2002, Vardavas et al., 2004,MEDIWAT, 2013, Koutroulis et al., 2013 and Baltas and Tzoraki, 2013) pointed out the importance of the evapotranspiration on the water balance of the whole Crete giving infiltration a secondary role. Conversely to these findings, the Crete SWAT model has allowed the estimation of the water balance of Crete resulting in significantly different estimates. Fig. 10 summarizes the percentage of flow contribution of the water balance main components with respect to the long-term average volumes and different hydrological conditions during the period 1983–2009. In addition, Fig. 11 represents the previous results following the KSWAT model approach providing a more readable and systematic information of the fluxes.

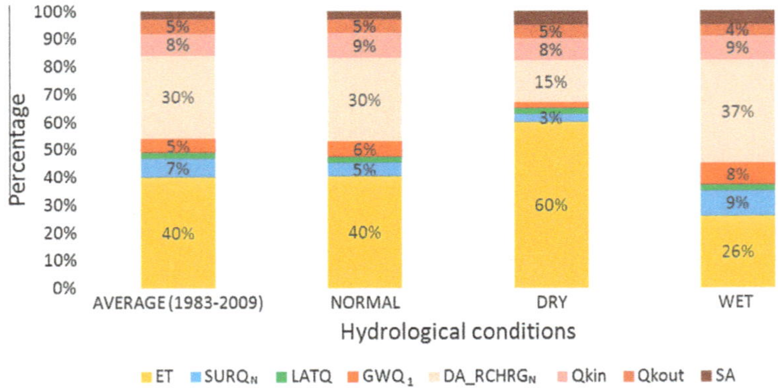

Figure 10: Annual water balance in different hydrological conditions (in percentage), average of 27 years (total volume of precipitation 6370 Mm^3/y), normal year (total volume of precipitation 6400 Mm^3/y), dry year (total volume of precipitation 3700 Mm^3/y) and wet year (total volume of precipitation 9600 Mm^3/y). ET, evapotranspiration; $SURQ_N$, surface runoff excluding losses from tributaries and wetlands; LATQ, lateral flow; GWQ_1, baseflow; DA_RCHRG_N, amount of water recharge in the deep aquifer without the contribution for springs that discharge inside (Qkin) and outside (Qkout) Crete Watershed; SA, shallow aquifer storage.

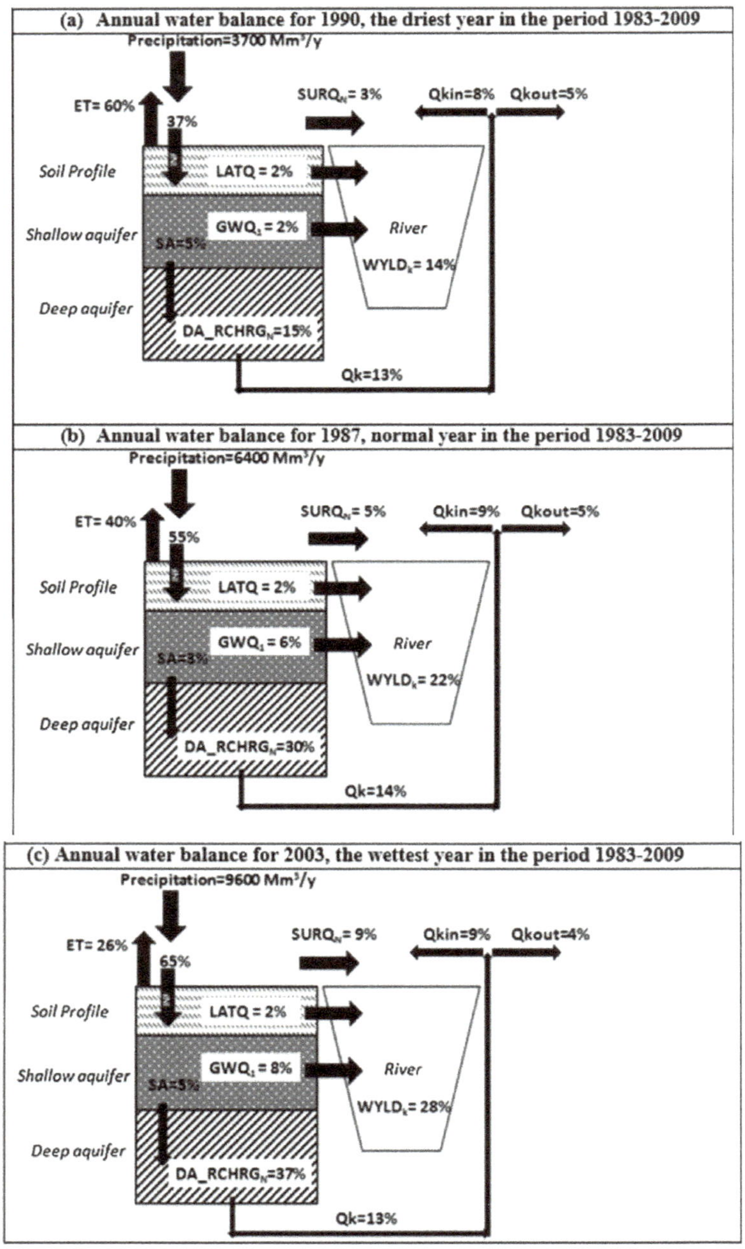

Figure 11: Annual water balance schematization in different hydrological conditions, dry (a), normal (b) and wet (c) year. ET, evapotranspiration; INF, infiltration in the soil; $SURQ_N$, surface runoff excluding the losses from tributaries and wetlands; LATQ, lateral flow; GWQ_1, baseflow; DA_RCHRG_N, amount of water recharge in the deep aquifer without the contribution for springs that discharge inside (Qkin) and outside (Qkout) Crete; $WYLD_k$, freshwater contribution; SA, shallow aquifer storage.

The evapotranspiration volume ranged from around 2250 Mm^3/y in the driest year to 2500 Mm^3/y in the wettest year exhibiting a small variation in terms of absolute values but with a significant variation with respect to the total volume of precipitation (Fig. 10). In the wettest year the main component of hydrological processes was the deep aquifer recharge, while in the driest year the evapotranspiration had the main role. From dry to wet year the surface runoff increased from 95 Mm^3/y to more than 800 Mm^3/y and the total net water available in the deep aquifer increased from 550 Mm^3/y to around 3500 Mm^3/y in dry and wet year, respectively.

As a consequence, during the wet conditions there was high infiltration, but also the surface runoff was larger than that during driest and normal hydrological condition. On the other hand, in dry years there was also significant percolation to the deep aquifer (1000 Mm^3/y). This result was also highlighted by Hartmann et al., 2014 and Hartmann et al., 2015 pointing out that karst regions might be more resilient to climate change in terms of both flooding and droughts. Furthermore, the authors demonstrated the existence a nonlinear relationship between precipitation and recharge rate indicating that the recharge of deep aquifer is more sensitive to a decrease than to an increase of precipitation (Hartmann et al., 2014). This was confirmed also in this study where the total deep aquifer recharge was estimated 28% of precipitation in driest year, while was 50% in wettest year departing of 16% and 6% from normal conditions, respectively.

Concerning the percentage of water balance components (Fig. 11), it was estimated that from the total precipitation of 6400 Mm^3/y about 40% was lost through evapotranspiration, 55% infiltrates and only 5% was surface runoff. The surface runoff including lateral flow (LATQ), baseflow (GWQ_1) and spring's discharge (Qkin) contribution to streamflow represents 22% of total precipitation. The total amount of deep aquifer recharge was around 44% of which 14% contributed as springs discharge to streamflow (Q_{in} = 9%) or sea (Q_{out} = 5%) (Fig. 11b). These percentages were valid for all years in normal hydrological conditions (for instance year 1987), but they changed in extreme hydrological conditions.

Considering the driest year (year 1990, Fig. 11a), it was estimated that from the total precipitations of 3700 Mm^3/y about 60% was lost through evapotranspiration, 37% infiltrated and only 3% was surface runoff.

The amount of deep aquifer recharge decreased with respect to the normal hydrological condition to 28% of which 13% contributed as springs discharge to streamflow (Q_{in} = 9%) or sea (Q_{out} = 4%). The total contribution to freshwater decreased to 14% with respect to the normal condition.

In the wettest year (year 2003, Fig. 11c), considering a total volume of precipitation around 9600 Mm^3/y, evapotranspiration was estimated around 26%, 65% was infiltration and 9% was surface runoff. The deep aquifer recharge reached the highest percentage of 50% and the water yield increased to 28%. Our work has shown that the evapotranspiration ranged between 2250 Mm^3/y in a driest year and of 2500 Mm^3/y in wettest year, corresponding to 60% and 26% of total precipitation, respectively. These results suggested the importance of karst water supply during dry periods.

Strengths and Challenges of the karst Modeling Approach

The modeling approach proposed in this study allowed achieving satisfactory simulations of streamflow and springs discharges at monthly and daily time step respectively across the whole Crete Island. These temporal resolutions are inherently linked to the available input data (Table 1), the nature of the simulated hydrologic processes and the modeling objectives (Baffaut et al., 2015). Conversely to previously studies that focus mainly on annual or long-term mean annual prediction (i.e. Malard et al., 2015), here the monthly simulations of streamflow have allowed capturing the seasonal variations and gathering precious information for a better water resources management. Similarly, the daily simulated springs discharges provided information of each karst recharge area and the spring hydrograph, if further investigated, can also express some geometrical and hydraulic characteristics of aquifers (Fiorillo, 2014).

Furthermore, the spatial and temporal (annual and monthly) distribution of simulated water resources (i.e. precipitation, deep aquifer recharge, evapotranspiration and runoff) and the mean annual water balances in different hydrological conditions are important benefits of the present work.

Despite these strengths, the modeling approach suffers from some limits. For instance, the monthly time step of streamflow has limited the application of step-wise calibration not allowing a reliable estimation of each component of the streamflow. Furthermore, in data scarce areas, where only annual time step is available and spatial data has coarse resolutions, the reproducibility of the KSWAT model needs some adaptations that should be evaluated case by case. Likewise, the knowledge of springs discharges and their position is vital and in extremis they could be approximated using literature information. In this context, soft data has an important role and should be seen as injections of some insightful common sense into the automatic or semiautomatic calibration procedure (Seibert and McDonnell, 2002). For instance, Malard et al. (2015) used 42 alpine and peri-alpine specific springs discharges from literature as references sites in Switzerland for karst recharge assessments, and Hartmann et al. (2015) used 40 independent observations of mean annual recharge both from field and modeling studies over Europe and Mediterranean.

Finally, albeit the identification of karst recharge areas of each spring is a valuable aspect of the KSWAT model, the associated uncertainty was not object of this study and this topic deserves further investigations that may involve climatic and topographic descriptors as pointed out in Hartmann et al. (2015).

CONCLUSION

In this study the SWAT model was adapted and integrated with a karst-flow model to simulate the karst water resources in the Island of Crete. The KSWAT model allowed calibrating the daily spring discharges in 47 gauging stations, estimating their karst recharge areas and their introduction in the SWAT model increased the performance of streamflow prediction in karst-subbasins. These springs contributed significantly to total discharge with 300 Mm³/y in dry hydrological condition, ensuring ecological services (Malard et al., 2015), and 850 Mm³/y in wet periods. In addition, the seasonal variation of volume of springs suggests that these valuable sources should be conserved and preserved in particular from April to September when available volumes are the lowest and agriculture and tourism demand increases.

The analysis of the water balance also showed that water resources are not homogeneously distributed in Crete and change significantly in different hydrological conditions. In particular, the western part of Crete has a surplus of water resources with respect to the eastern part where there is low water availability and high demand. Messara Valley (Geropotamos Basin) is an example of intensively managed basin in south-eastern Crete where an overexploitation of groundwater is occurring in a large number of wells since 1984 that continue to decrease groundwater levels at alarming proportions (Kritsotakis and Tsanis, 2009).

In this context, the present study provides a methodology and a tool for the integrated water management of Crete and other similar areas, by providing detailed spatially distributed hydrologic balances and accurate estimation of water availability using both hard and soft data.

ACKNOWLEDGEMENTS

The authors are grateful to the reviewers for their precious remarks, which made it possible to improve the quality of the paper. Funding for the work was provided by the EU FP7-ENV-2009 Project SoilTrEC "Soil Transformations in European Catchments" (Grant# 244118). The work of Dionissios Efstathiou has been co-financed by the European Union (European Social Fund – ESF) and Greek national funds through the Operational Program Education and Lifelong Learning of the National Strategic Reference Framework (NSRF) – Research Funding Program: Heracleitus II. Investing in knowledge society through the European Social Fund. The shapefiles of monitoring points of streamflow and springs, as well as the databases of the related time series are published on ResearchGate.

REFRENCES

1. Abbaspour, K.C., 2008. SWAT-CUP2: SWAT Calibration and Uncertainty Programs –A User Manual. Department of Systems Analysis, Integrated Assessment and Modelling (SIAM), Eawag, Swiss Federal Institute of Aquatic Science and Technology, Duebendorf, Switzerland.
2. Abbaspour, K.C., Rouholahnejad, E., Vaghefa, S., Srinivasan, R., Yang, H., Kløve, B.,2015. A continental-scale hydrology and water quality model for Europe: Calibration and uncertainty of a high-resolution large-scale SWAT model. J. Hydrol. 524 (1), 733–752.
3. Agriculture Statistics of Greece, 2005. PEIPAIATM-2009-PIREAS. Arfib, B., de Marsily, G., Ganoulis, J., 2000. Pollution by seawater intrusion into a karst system: new research in the case of the Almyros source (Heraklion, Crete,
4. Greece). Acta Carsologica 29 (1), 15–31.
5. Arnold, J.G., Youssef, M.A., Yen, H., White, M.J., Sheshukov, A.Y., Sadeghi, A.M., Moriasi, D.N., Steiner, J.L., Amatya, D.M., Skaggs, R.W., Haney, E.B., Jeong, J., Arabi, M., Gowda, P.H., 2015. Hydrological processes and model representation: impact of soft data on calibration. Trans. ASABE 58 (6), 1637–1660. http://dx.doi.org/10.13031/trans.58.10726.
6. Arnold, J.G., Srinivasan, R., Muttiah, R.S., Williams, J.R., 1998. Large area hydrologic modeling and assessment: Part I. Model development. J. Am. Water Resour. Assoc. 34 (1), 73–89.
7. Baffaut, C., Benson, V.W., 2009. Modeling flow and pollutant transport in a karst watershed with SWAT. Trans. ASABE 52, 469–479.
8. Baffaut, C., Dabney, S.M., Smolen, M.D., Youssef, M.A., Bonta, J.V., Chu, M.L., Guzman,J.A., Shedekar, V., Jha, M.K., Arnold, J.G., 2015. Hydrologic and water quality modeling: spatial and temporal considerations. Trans. ASABE 58 (6), 1661– 1680. http://dx.doi.org/10.13031/trans.58.10714.
9. Bakalowicz, 2015. Karst and karst groundwater resources in the Mediterranean.Environ. Earth Sci. 74, 5–14.
10. Baltas, E., Tzoraki, O., 2013. Water Resources Management on the Island of Crete: Lessons Learnt. In: United Nations International Year of Water Cooperation (Eds.), Free Flow – Reaching Water Security Through Cooperation, Tudor Rose. p. 285–9.
11. Bartholome, E., Belward, A.S., 2005. GLC2000: A new approach to global land cover mapping from Earth observation data. Int. J. Remote Sens. 26 (9), 1959–1977.
12. Bashfield, A., Keim, A., 2011. Continent-wide DEM creation for the European Union. In: 34th International Symposium on Remote Sensing of the Environment, Sydney, Australia, 10–15 April 2011 <http://www.isprs.org/proceedings/2011/ ISRSE-34/211104015Final00143.pdf> (accessed 21st January 2015); data available at http://www.eea.europa.eu/data-and-maps/data/eu-dem#tabmetadata (accessed 21st January 2015).
13. Bennett, N.D., Croke, B.F.W., Guariso, G., Guillaume, J.H.A., Hamilon, S.H., Jakeman, A.J., Marsili_libeli, S., Newham, L.T.H., Norton, J.P., Perrin, C., Pierce, S.A., Robson, B., Voinov, A.A., Fath, B.D., 2013. Characterising performance of environmental models. Environ. Model. Softw. 40, 1–20.
14. Bonacci, O., 1995. Brackish karst spring Pantan. Acta Carsologica 24, 97–107.
15. Bonacci, O., Fistanic I., 2006. Contribution to hydrological analysis of the coastalkarst spring Almyros (Creete, Greece). In: Conference on Water Observation and information system for decision support, Balwois (Ohrid, Republic of Macedonia). Radic (Eds.), UNDP-Agency of Skopje. p. 191–2.
16. Bonacci, O., Ljubenkov, I., 2005. Nova saznanja o hidrologiji rijeke Krke. Hrvatske Vode 13 (52), 265–281.
17. Bouraoui, F., Aloe, A., 2007. European Agrochemicals Geospatial Loss Estimator: Model development and Applications, EUR – Scientific and Technical Research

Series, ISSN 1018-5593. Office for Official Publications of the EuropeanCommunities, Luxembourg.

18. Britz, W., 2004. CAPRI Modeling System Documentation. Final Report of the FP5 Shared Cost Project CAP-STRAT "Common Agricultural Policy Strategy for Regions, Agriculture and Trade." QLTR-2000-00394. Univ. Bonn, Bonn,
19. Germany.
20. Britz, W., Witzke, P., 2008. CAPRI Model Documentation 2008: Version 2. In: Britz, W., Witzke, P. (Eds.). Bonn: University of Bonn, Institute for Food and Resource Economics.
21. Buchanan, T.J., Somers, W.P., 1976. Discharge measurements at gaging stations. Techniques of Water-Resources Investigations of the United States Geological Survey. Washington, D.C.: USGS. Book 3, Chapter A8.
22. Chartzoulakis, K.S., Paranychianakis, N.V., Angelakis, A.N., 2001. Water resources management in the island of Crete, Greece, with emphasis on the agricultural use. Water Pol. 3 (3), 193–205.
23. Daly, D., Dassargues, A., Drew, D., Dunne, S., Gldscheider, N., Neale, S., Popescu, C., Zwhalen, F., 2002. Main concepts of the "European Approach" for (karst) groundwater vulnerability assessment and mapping. Hydrogeol. J. 10 (2), 340–345.
24. Dettling, M., Maechler, M., 2012. Supclust: Supervised clustering of predictor variables such as genes. R package version 1.0-7.
25. EC, 2003. Vulnerability and risk mapping for the protection of carbonate (Karst)aquifers 2003. In: Zwahlen, F. (Ed.), COST-620 Action Final Report. EuropeanCommission, DG-Research, p. 320.
26. EC, 2004. The main coastal karstic aquifers of southern Europe 2004. In: Calaforra, J.M., (Ed.), COST-621 Action, Groundwater Management of Coastal KarsticAquifers, European Commission, DG-Research, 126.
27. EU-DEM, Metadata <http://www.eea.europa.eu/data-and-maps/data/eu-dem#tabmetadata>.European Soil Portal, 2014 <http://eusoils.jrc.ec.europa.eu/ESDB Archive/ESDB/>(accessed April 2014).
28. FAO, 2008. Harmonized World Soil Database (version 1.0). FAO, Rome. Italy andIIASA, Laxenburg, Austria.
29. Fiorillo, F., 2014. The recession of spring hydrographs, focused on karst aquifers.Water Resour. Manage. 28, 1781–1805.
30. Goldscheider, N., 2005. Fold structure and underground drainage pattern in thealpine karst system Hochifen-Gottesacker. Eclogae Geol. Helv. 98 (1), 1–17.http://dx.doi.org/10.1007/s00015-005-1143-z.
31. Gupta, H.V., Sorooshian, S., Yapo, P.O., 1999. Status of automatic calibration forhydrologic models: comparison with multilevel expert calibration. J. Hydrol. Eng. 4 (2), 135–143.
32. Hartmann, A., Gleeson, T., Rosolem, R., Pianosi, F., Wada, Y., Wagener, T., 2015. A large-scale simulation model to assess karstic groundwater recharge over Europe and the Mediterranean. Geosci. Model Dev. 8, 1729–1746. http://dx.doi. org/10.5194/gmd-8-1729-2015.
33. Hartmann, A., Mudarra, M., Andreo, B., Marín, A., Wagener, T., Lange, J., 2014. Modeling spatiotemporal impacts of hydroclimatic extremes on groundwater recharge at a Mediterranean karst aquifer. Water Resour. Res. 50 (8), 6507–6521. http://dx.doi.org/10.1002/2014WR015685.
34. Klein, G.K., van Drecht, G., 2006. HYDE 3: Current and historical population and land cover. In: Integrated Modeling of Global Environmental Change. An Overview of IMAGE 2.4.93-111. Netherlands Environmental Assessment Agency (MNP), Bilthoven, The Netherlands.
35. Klemeš, V., 1986. Dilettantism in hydrology: transition or destiny. Water Resour.Res. 22 (9), 177S–188S.
36. Knithakis, M., 1995. Hydrogeological Investigation of Rethymnon Region (Prefecture of Rethymnon). Institute of Geological and MineralogicalInvestigations (IGME), Rethymnon.
37. Kourgialas, N.N., Karatzas, G.P., Nikolaidis, N.P., 2010. An integrated framework for the hydrologic simulation of a complex geomorphological river basin. J. Hydrol. 381, 308–321.
38. Koutroulis, A.G., Tsanis, I.K., Jacob, D., 2013. Impact of climate change on water resources status: a case study for Crete Island, Greece. J. Hydrol. 479, 146–158. http://dx.doi.org/10.1016/j.jhydrol.2012.11.055.
39. Krause, P., Boyle, D.P., Base, F., 2005. Comparison of different efficiency criteria for hydrological model assessment. Adv. Geosci. 5, 89–97.
40. Kritsotakis, M., Tsanis, I.K., 2009. An integrated approach for sustainable water resources management of Messara Basin, Crete, Greece. EWRA 27 (28), 15–30.
41. Lambrakis, N., Andreou, A.S., Polydoropoulos, P., Georgopoulos, E., 2000. Nonlinear analysis and forecasting of a brackish karstic spring. Water Resour. Res. 36 (4),875–884.
42. Malagò, A., Pagliero, L., Bouraoui, F., Franchini, M., 2015. Comparing calibrated parameter sets for the Scandinavian Peninsula and for the Iberian Peninsula. Hydrol. Sci. J. 60 (5), 949–967. http://dx.doi.org/10.1080/02626667.2014.97833.
43. Malard, A., Sinreich, M., Jeannin, P., 2015. A novel approach for estimating karst groundwater recharge in mountainous regions and its application. Hydrol. Process. http://dx.doi.org/10.1002/hyp.10765.
44. MEDIWAT, 2013. State of the art of Water Resources in Mediterranean Islands. Available at: <http://www.mediwat.eu/sites/default/files/D.3.1.1.pdf>.
45. Mevic, B., Wehrens, R., 2007. The pls package: principal component analysis and partial least squares regression in R. J. Stat. Softw. 18 (2), 1–23.
46. Monfreda, C., Ramankutty, N., Foley, J., 2008. Farming the planet: 2. Geographic distribution of crop areas, yields, physiological types, and net primaryproduction in the year 2000. Global Biogeochem. Cycles 22, GB1022.

47. Moraetis, D., Efstathiou, D., Stamati, F., Tzoraki, O., Nikolaidis, N.P., Schnoor, J.L.,Vozinakis, K., 2010. High frequency monitoring for the identification ofhydrological and biochemical processes in a Mediterranean river basin. J.Hydrol. 389, 127–136.

48. Moriasi, D.N., Arnold, J.G., Van Liew, M.W., Bingner, R.L., Harmel, R.D., Veith, T.L., 2007. Model evaluation guidelines for systematic quantification of accuracy in watershed simulations. Trans. ASABE 50, 885–900.

49. Nash, J.E., Sutcliffe, J.V., 1970. River flow forecasting through conceptual models. Part I: a discussion of principles. J. Hydrol. 10, 282–290.

50. Neitsch, S.L., Arnold, J.G., Kiniry, J.R., Srinivasan, R., Williams, J.R., 2010. Soil and Water Assessment Tool Input/Output File Documentation Version 2009,Grassland. Soil and Water Research Laboratory, Agricultural Research Service and Blackland Research Center, Texas Agricultural Experiment Station, College Station, Texas.

51. Nikolaidis, N.P., Bouraoui, F., Bidoglio, G., 2013. Hydrologic and geochemical modeling of a karstic Mediterranean watershed. J. Hydrol. 477, 129–138.

52. Ntegeka, V., Decloedt, L.C., Willems, P., Monbaliu, J., 2012. Quantifying the impact of climate change from inland, coastal and surface conditions. In: Klijn, F., Schweckendiek, T. (Eds.), Comprehensive Flood Risk Management – Research for Policy and Practise, CRC Press, Taylor & Francis Group, Leiden, The

53. Netherlands; In: Proceedings of the 2nd European Conference on Flood Risk Management FLOODRisk 2012, Rotterdam, The Netherlands, 19–23 November 2012.

54. Oudin, L., Kay, A., Andréassian, V., Perrin, C., 2010. Are seemingly physically similar catchments truly hydrologically similar? Water Resour. Res. 46, W11558. http://dx.doi.org/10.1029/2009WR008887.

55. Pagliero, L., Bouraoui, F., Willems, P., Diels, J., 2015. Large-scale hydrological simulations using the soil water assessment tool, protocol development, and application in the Danube Basin. J. Environ. Qual. 43, 145–154.

56. Parajka, J., Merz, R., Bloschl, G., 2005. A comparison of regionalisation methods for catchment model parameters. Hydrol. Earth Syst. Sci. 9, 157–171.

57. PHU 2007. POTENTIAL HEAT UNIT (PHU) PROGRAM download <http://swat.tamu. edu/software/ potential-heat-unit-program/> (accessed 21st January 2007). Rantz, S.E., et al., 1982. Measurement and Computation of Streamflow: Volume 1. Measurement of Stage and Discharge. U.S. Geological Survey Water-Supply Paper 2175.

58. Ravbar, N., Engelhardt, I., Goldscheider, N., 2011. Anomalous behaviour of specific electrical conductivity at a karst spring induced by variable catchment boundaries: the case of the Podstenjšek spring, Slovenia. Hydrol. Process. 25(13), 2130–2140. http://dx.doi. org/10.1002/hyp.7966.

59. RGC, 2002. Integrated Water Resources Management on Crete. Study Made Under the Department of Water Resources Management (in Greek). Sawicz, K., Wagener, T., Sivapalan, M., Troch, P.A., Carillo, G., 2011. Catchment classification: empirical analysis of hydrologic similarity based on catchment function in the eastern USA. Hydrol. Earth Syst. Sci. 15, 2895–2911.

60. Seibert, J., McDonnell, J.J., 2002. On the dialog between experimentalist and modeler in catchment hydrology: use of soft data for multicriteria model calibration. Water Resour. Res. 38 (11), 23.1–23.14.

61. Shamir, E., Iman, B., Morin, E., Gupta, H.V., Sorooshian, S., 2005. The role of hydrograph indices in parameter estimation of rainfall–runoff models. Hydrol. Process. 19, 2187–2207.

62. Smart, C.C., Worthington, S.R.H., 2004. Springs. In: Gunn, J. (Ed.), Encyclopedia of Caves and Karst Science. Fitzroy Dearbon, London, pp. 699–703.

63. Steiakakis, E., Monopolis, D., Vavadakis, D., Lambrakis, N., 2011. Effective infiltration assessment in Kourtaliotis karstic basin (S. Crete). Adv. Res. Aquatic Environ., 485–492

64. Taylor, R., 1990. Interpretation of the correlation coefficient: a basic review. J. Diagn. Med. Sonog. (January/February), 35–39

65. Tzoraki, O., Nikolaidis, N.P., 2007. A generalized framework for modeling the hydrologic and biogeochemical response of a Mediterranean temporary river basin. J. Hydrol. 346, 112–121.

66. van Griensven, A., Meixner, T., Grunwald, S., Bishop, T., Diluzio, M., Srinivasan, R., 2006. A global sensitivity analysis tool for the parameters of multi-variable catchment model. J. Hydrol. 324 (1–4), 10–23. http:// dx.doi.org/10.1016/j. jhydrol.2005.09.008.

67. Vardavas, I., Chartzoulakis, K., Papamastorakis, D., Xepapadeas, A., Spanoudaki, K., Zacharioudakis, G., Kritsotakis, M., Bertaki, M., 2004. Report of Water Resources of Crete, 2004. Technical Report.

68. Wagener, T., Sivapalan, M., Troch, P.A., Woods, R.A., 2007. Catchment classification and hydrologic similarity. Geogr. Comp. 1 (4), 901–931.

69. Wang, X., Liang, Z., Wang, J., 2014. Simulation of runoff in Karst-influenced Lianjiang Watershed using the SWAT Model. Sci. J. Earth Sci. 4 (2), 85–92.

70. Wang, Y., Brubaker, K., 2014. Implementing a nonlinear groundwater module in the soil and water assessment tool (SWAT). Hydrol. Process. 28 (9), 3388–3403.

71. Ward, J.H., 1963. Hierarchical grouping to optimize an objective function. J. Am. Statist. Assoc. 58, 236–244.

72. Williams, P.W., 2004. Karst evaluation. In: Gunn, J. (Ed.), Encyclopedia of Caves and Karst Science, vol. 475. Fitzroy Dearbon, London, p. 8.

73. Wooldridge, S.A., Kalma, J.D., 2001. Regional-scale hydrological modelling using multiple-parameter landscape zones and a quasi-distributed water balance model. Hydrol. Earth Syst. Sci. Discuss. 5 (1), 59–74.

74. Yachtao, G.A., 2009. Modification of the SWAT Model to Simulate Hydrologic Processes in a Karst Influenced Watershed MS Thesis. Virginia-Tech, Department of Biosystems Engineering, Blacksburg.

CHAPTER 14

IMPACT OF CLIMATE CHANGE ON UK ESTUARIES: A REVIEW OF PAST TRENDS AND POTENTIAL PROJECTIONS

Peter E. Robins[a], Martin W. Skov[b], Matt J. Lewis[b], Luis Giménez[b], Alan G. Davies[a], Shelagh K. Malham[a], Simon P. Neill[b], James E. McDonald[c], Timothy A. Whitton[a], Suzanna E. Jackson[b], Colin F. Jago[a,b, c]

[a] Centre for Applied Marine Sciences, Bangor University, LL59 5AB, United Kingdom

[b] School of Ocean Sciences, Bangor University, LL59 5AB, United Kingdom

[c] School of Biological Sciences, Bangor University, LL57 2UW, United Kingdom

ABSTRACT

UK estuarine environments are regulated by inter-acting physical processes, including tidal, wave, surge, river discharge and sediment supply. They regulate the fluxes of nutrients, pollutants, pathogens and viruses that determine whether coastlines achieve the Good Environmental Status (GEnS) required by the EU's Marine Strategy Directive. We review 20th century trends and 21st century projections of changes to climatic drivers, and their potential for altering estuarine bio-physical processes. Sea-level rise will cause some marine habitats to expand, and others diminish in area extent. The overall consequences of estuarine morphodynamics to these habitat shifts, and vice-versa, are unknown. Increased temperatures could intensify microbial pathogen concentrations and increase public health risk. The patterns of change of other climatic drivers are difficult to predict (e.g., river flows and storm surges). Projected increased winter river flows throughout UK catchments will enhance the risks of coastal eutrophication, harmful algal blooms and hypoxia in some contexts, although there are spatial variabilities in river flow projections. The reproductive success of estuarine biota is sensitive to saline intrusion and corresponding turbidity maxima, which are projected to gradually shift landwards as a result of sea-level rise. Although more-frequent flushing events in winter and longer periods of drought in summer are predicted, whereby the subsequent estuarine mixing and recovery rates are poorly understood. With rising estuarine salinities, subtidal species can penetrate deeper into estuaries, although this will depend on the resilience/adaptation of the species. Many climate and impact predictions lack resolution and spatial cover. Long-term monitoring and increased research, which considers the catchment-river-estuary-coast system as a whole, is needed to support risk predicting and mitigatory strategies.

INTRODUCTION

Estuaries are process centres for coastal hydrological, biogeochemical and biological cycles and their biological productivity rivals those of tropical rainforests and coral reefs (Cai, 2011). The high productivity combines with the provisioning of important ecosystem services, such as flood protection and providing recreational space, to maintain a Good Environmental Status (GEnS) so that they are socio-economically valuable and sustainable (Barbier et al., 2011 and Elliott et al., 2015). Estuaries are also biologically important; many in the UK are biodiversity rich and most host a series of habitats that are important nursery areas for commercial fisheries species (Beck et al., 2001), or feeding

Citation: Peter E. Robins, Martin W. Skov, Matt J. Lewis, Luis Giménez, Alan G. Davies, Shelagh K. Malham, Simon P. Neill, James E. McDonald, Timothy A. Whitton, Suzann E. Jackson, Colin F. Jago, , Impact of climate change on UK estuaries: A review of past trends and potential projections, doi:10.1016/j.ecss.2015.12.016

and breeding areas for terrestrial populations, such as birds (Sharps et al., 2015). Many estuarine habitats have statutory protection and all have roles in prolific estuarine foodwebs and the cycling of organic material (Kaiser et al., 2005). The ecological functions of providing habitat and nursery areas are attributed to the discontinuity in productivity, sediment characteristics, and turbidity, when compared with the surrounding coasts (Power et al., 2000). Estuaries are known to trap, filter and recycle suspended particulate matter (Ittekkot, 1988), as well as components that are potentially harmful to human health, such as nutrients and pathogenic viruses. In the UK, estuarine processes are directly affected by intrinsic anthropogenic activities, such as aquaculture and shellfish fisheries, and indirectly by activities such as catchment agriculture, forestation/deforestation, urbanisation and coastal management decisions (Attrill, 1998).

Estuaries are commonly defined as "*a semi-enclosed coastal body of water which has free connection to the open sea, extending into the river as far as the limit of tidal influence and within which seawater is measurably diluted with freshwater derived from land drainage*" (Dyer, 1997). This definition has shortcomings: tidal influence often exceeds the salt water intrusion in the river, and the seaward region of fresh water influence (ROFI) may extend offshore, such as in the Severn Estuary. Thus, estuaries can be further classified based on salinity structure (well mixed through to stratified), tidal range (microtidal to hypertidal) and geomorphology (Appendix 2). A review of 155 UK 'estuaries' classified sites by their geomorphology and tidal type (Fig. 1; Appendix 1;Davidson et al. (1991)). Mesotidal estuaries (2 m < tidal range < 4 m) are typical in the eastern UK, whereas macrotidal (4 m < tidal range < 6 m) or hypertidal (tidal range > 6 m) estuaries occur in the west, and are generally connected to catchments that respond quickly to rainfall events (Brown et al., 1991).

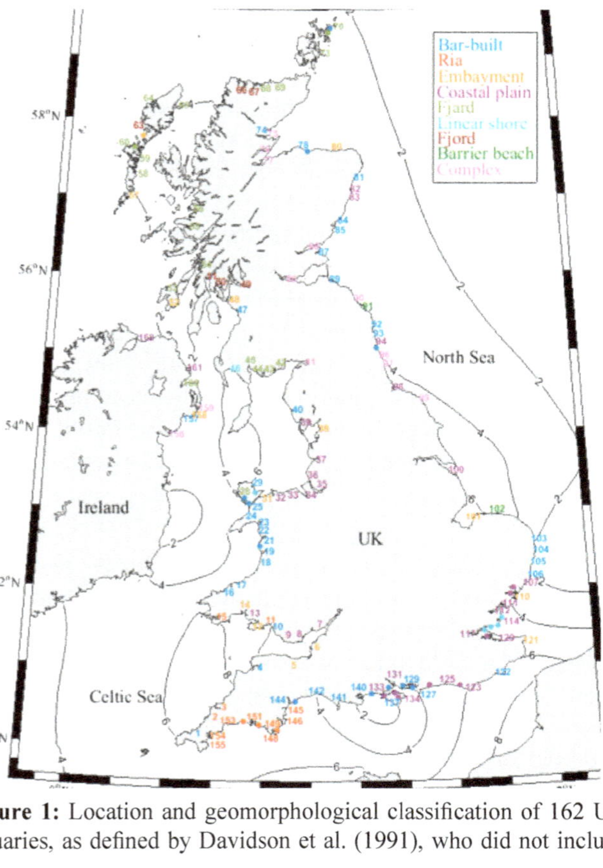

Figure 1: Location and geomorphological classification of 162 UK estuaries, as defined by Davidson et al. (1991), who did not include estuaries in Northern Ireland. Tidal type can be inferred from tidal range (m) contours (calculated from a 3D ocean model, simulated and validated by Lewis et al., 2014), classified by Davies (1964) as: microtidal (tidal range < 2 m), mesotidal (2 m–4 m), macrotidal (4 m–6 m), and hypertidal (>6 m).

The Intergovernmental Panel on Climate Change (IPCC) is 'virtually certain' that global temperatures will increase during the 21st century, with climate models projecting UK shelf seas to warm by between 1.5 °C and 4 °C by 2100 (Jenkins et al., 2009). In turn, global mean sea-levels are projected to rise by between 44 and 74 cm, depending on our future carbon dioxide emissions (IPCC, 2013), with UK mean sea-level closely matching the global trend (Woodworth et al., 2009). Climate models also predict changes to precipitation pattern in the UK, with more inter-regional and inter-seasonal variability, albeit with high model uncertainty and less confidence than sea-level projections (IPCC, 2013); but with high impact, such as the damaging storm clusters experienced during the winter of 2013/2014 along the western coasts of the UK (Wadey et al., 2014). It follows that the long-term climate-driven impacts that UK estuaries face include a deterioration in water quality and alterations to habitat composition, which could ultimately affect their ecological integrity (Kennish, 2002).

In this paper, we review current understanding of how anthropogenic climate change will impact UK estuaries and their complex and important environments. Trends suggest that, over the 21st century, estuarine ecosystems will be mostly impacted by habitat loss, as a consequence of climate change and urban expansion. Other projected high priority problems include excessive nutrient and sewage inputs to estuaries, leading to eutrophication, hypoxia and anoxia (Kennish, 2002). Further damaging problems will arise from over-fishing, chemical contaminant fluxes in urbanised regions, freshwater diversions, the introduction of invasive species, and coastal subsidence and erosion. Although it is difficult to quantify, these impacts on estuarine ecosystems are at least in part linked to climate change.

We amalgamate current knowledge from peer-reviewed literature on climate change trends and projections for the UK, and discuss how climate change has and will impact upon both physical and ecological processes in UK estuaries. This review paper is structured as follows: In Section 2, we summarise past trends and future projections of UK coastal climate change, relevant to physical estuarine processes; Sections 3 and 4review observed trends and projected impacts of climate change to the physical and ecological processes in UK estuaries, respectively; finally, discussions and conclusions are summarised in Sections 5 and 6, respectively.

Climate Changes to Physical Forcing

We focus on sea surface temperature (SST), sea-level rise and atmospheric shifts as the main climate drivers that influence UK estuarine environments and ecosystems, and their interactions. Table 1 summarises observed trends in the drivers over the past century and projected changes for the 21st century, and estimates the level of confidence associated with each projection. Fig. 2 illustrates the key expected spatial and seasonal variability across the UK, according to the UK Climate Change Projections Report (UKCP09; Lowe et al., 2009).

Table 1: Observed trends and future projections of climate change to UK estuaries. The level of confidence (low, medium, or high) has been ascribed, based on the amount of evidence and the level of agreement of the evidence

Process	Observed trend for the UK	*CONFIDENCE* and information sources	Projected 21st century change for the UK	*CONFIDENCE* and information sources
Sea surface temperature	+0.7 °C (1971–2010)	*HIGH* IPCC (2013) Jenkins et al. (2009)	+1.5 °C to +4 °C	*HIGH* Lowe et al. (2009)
Sea-level rise	+1 to +3 mm yr^{-1} during last century	*HIGH* Cazenave and Nerem (2004) Church et al. (2004) Woodworth and Blackman (2004) Church and White (2006) Menendez and Woodworth (2010) IPCC (2013)	+0.44 m to +0.74 m (Steric and eustatic processes) +1.9 m (Ice sheet melt)	*HIGH* Worth et al. (2006) Lowe et al. (2009) Woodworth et al. (2009) Lewis et al. (2011) IPCC (2013) *LOW* Jevrejeva et al., 2014 Knight et al., 2015
Storm surge	Pole-ward shift of storm tracks. Increased intensity/ decreased frequency extra-tropical cyclones.	*MEDIUM* McCabe et al. (2001) *HIGH* Paciorek et al. (2002) Geng and Sugi (2003) Woodworth and Blackman (2004) Woodworth et al. (2009) Menendez and Woodworth (2010)	Centennial changes in extreme water levels are marginally significant (i.e., are of the same order as the natural climatological variability).	*LOW* Bijl et al. (1999) Alexandersson et al. (2000) Wakelin et al. (2003) Worth et al. (2006) Debernard and Røed (2008) Lowe et al. (2009) Howard et al. (2010) Lewis et al. (2011)
	High variability in storm surges Clustering of extreme sea-levels	*MEDIUM* Allan et al. (2009) *LOW* Wadey et al. (2014).		
River flows	1969–2008: Increased flow during autumn and winter, decreased flow during spring, no trends during summer Rainfall (river flows) occurring in clustered events	*HIGH* Keef et al. (2009) Hannaford and Buys (2012) Hannaford (2015) Watts et al. (2015) *LOW* Jenkins et al. (2008) Burt and Ferranti (2012) Jones et al. (2013) Watts et al. (2015)	Increase in winter mean flow by up to 25% Decrease in summer mean flow by 40–80%	*MEDIUM* Fowler and Kilsby (2007) Jenkins et al. (2009) Prudhomme and Davies (2009a; 2009b) Fowler and Wilby (2010) Christierson et al. (2012) Prudhomme et al. (2012) Smith et al. (2013) Kay et al. (2013a,b)

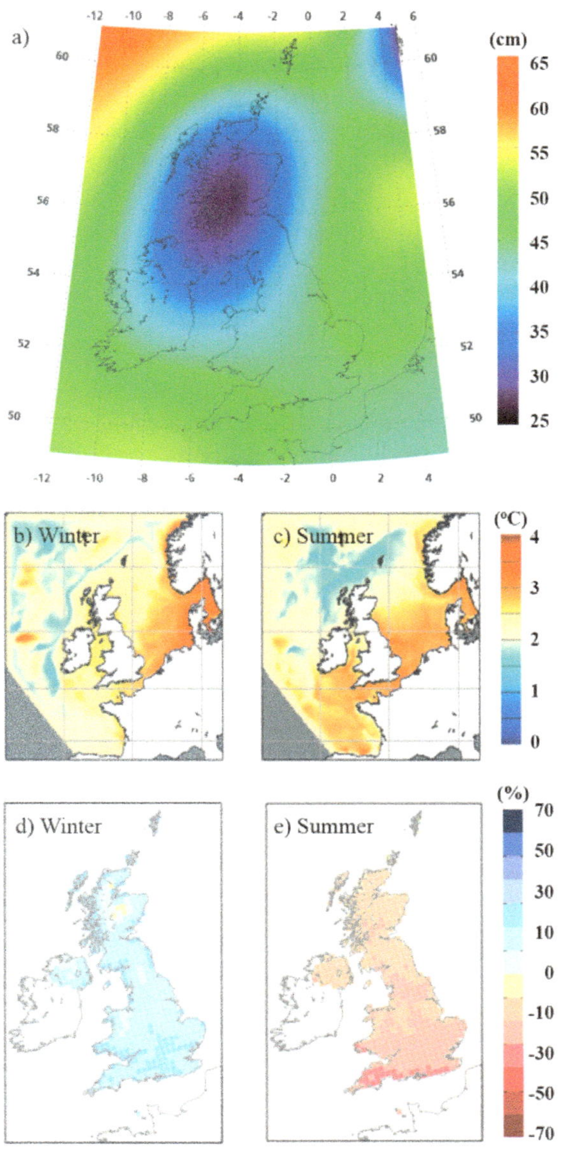

Figure 2: A selection of UKCP09 projected changes to UK climate drivers over the 21st century, showing spatial and seasonal variability (all figures published by Lowe et al. (2009)). (a) Shows relative sea-level change (cm) projections by 2100 (combined land movements and absolute sea-level change estimates from the central simulation (out of 11 ensemble members) for the IPCC medium emissions scenario). The middle panels show simulated seasonal mean trends in SST (°C), from (1961–1990) to (2070–2098), for (b) winter and (c) summer. Bottom panels show projected changes in mean precipitation (%) during (d) winter and (e) summer, at the 50% probability levels, for the 2080s under the IPCC medium emissions scenario.

TEMPERATURE

The warming of the global climate system is unequivocal (Jenkins et al., 2009). Data reveals that SST has warmed since the 1970s, at a rate of 0.11 °C per decade, and there is some evidence that SST has warmed since the 1870s (IPCC, 2013). Despite considerable year-to-year and even decade-to-decade variability, Jenkins et al. (2009)have shown that there is a clear UK coastal SST increase of about 0.7 °C over the past three decades, based on several long-term and short-term data sets.

Climate models project global mean SST to increase by 0.6–2 °C by 2100, depending on emission scenarios (low, medium, or high) and taking into account model uncertainty and seasonal variability (IPCC, 2013). Over the UK shelf edge regions and northern North Sea, the UKCP09 report projected temperature increases of 1.5–2.5 °C to occur this century, with larger increases of 2.5–4 °C in the Celtic, Irish and southern North Sea – a trend of 0.3 °C per decade (Lowe et al., 2009), e.g., Fig. 2b and c. However, a potential shutdown of the North Atlantic current (Gulf Stream) would result in much lower winter SST in the UK – but consensus among current climate model projections is that this is considered a low-probability, high-impact event (Wood et al., 2003). There has been a 1–3 °C rise in river water temperatures over UK and Europe during the past 100 years; since river water temperatures closely positively correlate with air temperatures (Whitehead et al., 2009).

Sea-level Rise

Global mean sea level has risen at a rate of 1–2 mm yr^{-1} since 1900 (Church et al., 2004 and IPCC, 2013), with an apparent change in rate to 3 mm yr^{-1} during the past 30 years, determined from satellite altimetry measurements (Cazenave and Nerem, 2004 and Church and White, 2006) and analysis of tide gauges worldwide (Woodworth and Blackman, 2004 and Menendez and Woodworth, 2010). Climate models, assuming different emission scenarios, projected with "medium confidence" that global mean sea levels will increase by 44–74 cm by 2100 (IPCC, 2013); an increase on the 18–59 cm previously projected by IPCC (2007). Alternative estimates based on vast ice sheet melt suggest, with low confidence, that sea level could rise by 1.9 m by 2100 (Jevrejeva et al., 2014), although recent studies suggest this trend is increasingly more likely (Knight et al., 2015).

Sea-level rise is not spatially uniform, with variability due to steric and eustatic processes, as well as vertical land movements such as isostatic adjustment (e.g.,Shennan and Horton, 2002). It is this relative mean sea-level which is most important for coastal communities (Lewis et al., 2011). In the UK, observed sea level trends are broadly consistent with the global average (Woodworth et al., 2009). Although, the effects of vertical land movements after the last ice age will augment this trend in the far southwest UK (by 1.2 mm yr^{-1}) and reduce it in parts of Scotland (by +2 mm yr^{-1}) (Shennan and Horton, 2002). Therefore, estuaries in southeast England are most at risk from the combined effects of land sinking and sea levels rising (Davidson et al., 1991), e.g., Fig. 2a. What is clear from recent analysis is that relative mean sea level around the UK is rising, albeit at a spatial varying rate, which could have significant consequences on estuarine systems (e.g., Robins et al., 2014), and shall be considered in Sections3 and 4.

Storm Surges

Analysis of atmospheric observations, and application of this data to models, show that the second half of the twentieth century experienced a pole-ward shift in the majority of storm tracks across the Northern Hemisphere (McCabe et al., 2001). Further, there was an increase in intensity, but a decrease in frequency, of extra-tropical cyclones (Paciorek et al., 2002 and Geng and Sugi, 2003). However, storm surge incidents in the UK were unrelated with changes to extra-tropical cyclones: extreme water levels were driven by a rise in mean sea-level, rather than an increase in storminess (Woodworth et al., 2009 and Menendez and Woodworth, 2010). Therefore, it is uncertain whether the UK storm surge frequency has changed beyond natural levels of variability (Allan et al., 2009) – yet, some limited evidence is now emerging of clustering of extreme sea-levels (Wadey et al., 2014). For instance, there was an exceptional number of extreme high waters during the 2013/14 winter in the UK (Wadey et al., 2014).

Mean sea-level rise will change the dynamics of storm surge generation and propagation (Mcinnes et al., 2003) and estuarine flood risk and inundation (Senior et al., 2002). Models indicate future centennial changes in extreme water levels, which pose the greatest flood risk to estuaries, will be moderate and of the same order as the natural climatological variability (Worth et al., 2006, Debernard and Røed, 2008 and Lewis et al., 2011). The uncertainty within atmospheric models is high due to the parameterization of sub grid-scale processes, such as cloud formations and inter-decadal variability (Howard et al., 2010 and Lowe et al., 2009), and quantification of decadal variability of storm surge climates (Alexandersson et al., 2000 and Wakelin et al., 2003); therefore, future changes to the storm surge climate in the UK appear to be unresolved at present.

Rainfall and River Flow

River flows represent the integrated response of all hydrometeorological processes acting upon a catchment, and are therefore a key indicator of potential impacts of climate change to UK estuaries (Hannaford, 2015). Atmospheric warming has been shown byWatts et al. (2015) to intensify the hydrological cycle; although there is a complex, non-linear process leading from increasing temperatures, through changes in precipitation to river flow response (Hannaford, 2015). Past trends show that, while annual average rainfall in the UK is unchanged since the 1960›s, winter rainfall has intensified and increasingly occurred in clustered events (Burt and Ferranti, 2012, Jones et al., 2013 and Watts et al., 2015). For the same period, river-flows have increased in winter also, especially in upland western areas; autumn flows rose and spring flows decreased slightly, while there was no consistent change in the summer (Hannaford and Buys, 2012, Hannaford, 2015 and Watts et al., 2015). For past extreme events, data from extensive UK river-gauge networks show that the relationship between precipitation and river flow is spatially variable and dependent on catchment characteristics (Keef et al., 2009).

Many studies have investigated potential changes to future UK river flow rates (e.g., Kay et al., 2009, Prudhomme and Davies, 2009a and Prudhomme and Davies, 2009b), including impacts to groundwater (e.g., Jackson et al., 2011), and more recent studies have analysed the latest UKCP09 future river flow data series (1 km-resolution and catchment-bias-corrected; Prudhomme et al., 2012). The scientific consensus on projected future UK precipitation patterns is the "wetter winters and drier summers" signal (Arnell, 2003 and Christierson et al., 2012), with some spatial variability in the UK to this pattern. Indeed, the UKCP09 Report explains that little change is expected in the median precipitation amount by 2100, but significant changes in the trends of winter and summer precipitation are expected (Jenkins et al., 2009) (Fig. 2d and e). The report generally found a slight increase in the winter mean precipitation and a decrease in summer mean precipitation for the western UK (Jenkins et al., 2009). River catchment models indicate river flows are projected to reduce in summer (by 40–80%) and increase in winter (by up to 25%), again particularly in mountainous regions of western UK (Fowler and Wilby, 2010, Christierson et al., 2012 and Prudhomme et al., 2012). These projections were biased through inability to capture monthly precipitation climates (Prudhomme and Davies, 2009a, Prudhomme and Davies, 2009b and Smith et al., 2013). Furthermore, when future precipitation patterns are projected through to future river flow rates there is considerable uncertainty in GCM projections (Prudhomme and Davies, 2009a), as well as downscaling and catchment model uncertainties (Kay et al., 2009 and Prudhomme and Davies, 2009b).

CLIMATE IMPACTS ON THE PHYSICAL ESTUARINE ENVIRONMENT

Flooding and Inundation

Increased estuarine flood risk and inundation occurrence are caused primarily by a combination of increased sea-level (sea-level rise, storm surge and changes to the wave climate) and changes to incidences of high river flow events (Lewis et al., 2013 and Quinn et al., 2014). Estuarine flooding can also be the volumetric combination of storm surge and extreme river flow that can drastically increase flood hazard (e.g.,Maskell et al., 2013). Further, there is evidence for a positive relation in some UK estuaries between storm surge and peak river flows (Samuels and Burt, 2002, Senior et al., 2002 and Svensson and Jones, 2002), suggesting that surges should to be taken into account for flood risk estimation (Keef et al., 2009 and Zheng et al., 2013). Both sea-level and river flow climates in the UK appear to be changing due to climate change, potentially resulting in increased flood risk to UK estuaries (King, 2004). Potential negative impacts of estuarine flooding to ecosystem goods and services, in the absence of mitigation, include: damage to infrastructure and population displacement (Hanson et al., 2011 and Nicholls et al., 2011), increased erosion and loss of land to the sea (Zhang et al., 2004), pollution and increased human health risk, and significant loss of wetland habitats and ecosystems (Day et al., 1995 and Nicholls et al., 1999).

Modelling approaches are important for predicting flood risk, assessing impact, and mitigation decision-making – where risk is generally defined as the probability of the occurrence multiplied by the consequences. It is important to simulate the nonlinear coupling of surge-wave-river events, since static (sea-level-only) events will underestimate the cost significantly (Prime et al., 2015). Using published climate change scenarios, a UK panel (King, 2004) found that a combination of sea-level rise and increased storminess will allow storm surges to propagate much further inland, so that UK estuaries will be subject to both higher water levels and more energetic wave attack. However, Worth et al. (2006) simulated no significant increased storm surge on the east coast of the UK over the next century. Although the likelihood of extreme sea-level rise (up to 1.9 m over the 21st century) due to rapid ice sheet melt is apparently low (Jevrejeva et al., 2014), the consequence would be an increase to the projected risk; for the case of the UK›s largest estuary – the Severn, the projected flood risk has been reported by Quinn et al. (2013) to increase by 30%.

A key consideration for estuarine flood risk and inundation occurrence, under accelerating rates of sea-level rise is the potential for geomorphological features and flood depths to adjust, through sediment transport (i.e., the sediment flux into or out of an estuary), estuarine morphology (erosion and accretion), and sediment compaction (Dickson et al., 2007 and Shennan et al., 2012). Sediment infilling will cause flood risk to increase incrementally over long (decadal) timescales, in which case mitigation measures can be foreseen and employed in advance to reduce risk. However, based on analysis of historical data, Pye and Blott (2006) showed that morphological changes on the Suffolk coast (around Blyth, Ore-Alde-Butley, and Deben Estuaries) over the past 50 years have actually been lower than historically. Instead, rapid morphological changes (and potential increased flood risk) will result from extreme surge/wave/river events (Horrillo-Caraballo et al., 2013). Interestingly, Dissanayake et al. (2015) explain that the initial storm in a cluster of storms has most morphological impact on the coastline. In addition to these natural trends, shoreline protection has greatly reduced the input of new sediment sources, resulting in much contemporary coastal erosion (Nicholls et al., 2013). The capacity for estuaries to respond to sea-level rise will depend on landward constraints to transgression, available accommodation space and available sediment supply (marine and fluvial) (Nicholls et al., 2013). Rossington et al. (2007) predicted the maximum rate of sea-level rise that UK estuaries can experience without losing all intertidal areas. Their results suggest that some estuaries are more vulnerable to sea-level rise (e.g., Southampton Water) and others are more resilient (e.g., the Ribble), depending on estuary area, sediment supply, sediment transport potential and human interference. Some studies of intertidal and saltmarsh areas suggest low resilience to sea-level rise (e.g., Orford and Pethick, 2006, French, 2006 and Marani et al., 2007).

Projected increases in UK average temperatures will increase evapotranspiration throughout many UK catchments, which may alter vegetation types and reduce soil moisture – all of which could reduce runoff into estuaries and potentially dampen flood risk slightly (Boorman, 2003). As an example, modelled catchment runoff from IPCC climate change scenarios in the Ouse-Humber Estuary showed increased evapotranspiration (5% in winter, 15% in summer), although total runoff generally increased by up to 30% in winter due to projected increases in rainfall (Boorman, 2003).

From our review, it is clear that uncertainty is high in predicting future changes to physical drivers in UK estuaries, due to the spatial variability associated with projected changes to rainfall patterns. Furthermore, considering GCM model uncertainty, downscaling uncertainties, hydrological model uncertainties and land use changes, as well as variability of the climate system (e.g., Kay et al., 2009, Prudhomme and Davies, 2009a,Prudhomme and Davies, 2009b and Christierson et al., 2012), prediction of future estuary changes requires further research to determine confidence in future river flow rates. Furthermore, surge–rainfall association requires further investigation (Zheng et al., 2014) and their implications to biology, water quality and sediment transport processes (Lowe et al., 2009 and Robins et al., 2014).

Hydrodynamics and Mixing

In the UK, the majority of the main estuaries are relatively shallow, and are classified as either mesotidal, macrotidal or hypertidal, in terms of tidal forcing (Davidson et al., 1991 and Prandle, 2009). The strong tidal forcing produces a vertically well-mixed water column, though weak vertical stratification is sometimes evident, e.g., approaching slack high water in the Conwy (Howlett et al., 2015), or more commonly during the ebb tide depending on the river flow (Pye and Blott, 2014). Typically, the tide becomes asymmetric upon entering the estuary due to shallow water frictional effects – resulting in a shorter, stronger flooding phase and a longer, weaker ebbing phase (Dronkers, 1986 and Moore et al., 2009). Influences of waves on hydrodynamics and mixing are usually small within the estuary due to wave dissipation over the UK shelf, particularly over offshore sand banks (Carini et al.,

2015), and prevented from penetrating into bar-built estuaries because of the constricted entrance. Of high importance for hydrodynamics and mixing are variabilities in river flows – particularly due to the nature and frequency of extreme flow events (Struyf et al., 2004). Rivers in the UK tend to rise very quickly after rainfall, especially in the north and west, but flowrates reduce more slowly; as such, UK estuaries are often in a state of non-equilibrium in terms of stratification and salt balance, and nutrient/material fluxes; processes which are critical for ecosystem goods and services (Prandle, 2009).

Consequently, projected alterations to sea-levels and river flows due to climate change could have significant implications for estuarine functioning (Whitehead et al., 2009). Modelling in the Conwy by Robins et al. (2014) shows interaction of sea-level rise with maximal projected changes in river flow over the 21st century will potentially increase the saline intrusion length. Sea-level rise combined with long periods of low flow conditions (drought) simulated the saline intrusion and turbidity maximum to be pushed further up-estuary from today's limit and retaining river-borne substances in the estuary for extended periods. On the other hand, increased river flows simulated the saline intrusion to be pushed down-estuary, and potentially flush nutrients and pollutants offshore (Robins et al., 2014). These findings relate to similar studies in the US by Yang et al. (2015). The extent of the saline pumping effect is dependent on the strength and frequency of river flow events, which are projected to be more variable in the UK with climate change. Correspondingly, Prandle and Lane (2015) investigated 96 UK estuaries and concluded that a sea-level rise of 1 m will increase the saline intrusion length by >7% in deep estuaries and by >25% in estuaries shallower than 10 m. Further, a change in river flow of 25% (either increase or decrease) will have significant effects on both vertical mixing and salinity intrusion. Accordingly, estuarine recovery after rainfall events restores a steady state salt distribution (MacCready, 1998 and Hetland and Geyer, 2004). However, few observations of estuarine tidal-pumping/recovery have been documented in the UK (e.g., Simpson et al., 2001) and there is generally a lack of knowledge about how estuarine hydrodynamics react to extreme climatic events for UK case studies.

Estuarine hydrodynamics in the majority of UK systems is further complicated by the presence of estuarine fronts – small-scale regions which exhibit significant changes in hydrographic variables (O›Donnell, 1993). Estuarine fronts have significant roles in estuarine circulation, and affect sediment grain size distribution (Neill, 2009) and larval dispersal (Robins et al., 2012). Since estuarine fronts are driven by horizontal salinity gradients (controlled by river flow and tidal range), it is anticipated that the role of fronts in estuarine processes will become increasingly important within the context of 21st century climate change, when extremes in river flow, coupled with increased estuarine tidal range due to sea-level rise, are likely.

Sediment Transport and Morphology

The development of UK estuaries has typically resulted from the latest post-glacial rise in sea level (of the order hundreds of metres), which inundated the UK coastline and drowned river mouths (Dyer, 1997). However, Tessier et al. (2012) have demonstrated that the majority of morphological development across northwest European estuaries has occurred following centennial-scale periods of major climate change (enhanced storminess), in addition to the (lesser) continual effect of relative sea-level rise, given an available sediment supply. In the present climate, the morphological development of such systems, other than impacts caused by human activities, has been determined by the changing relative importance of the river, wave and tidal influences (Prandle, 2009), with tidal flow asymmetry being considered as the main factor influencing net sediment transport and hence being responsible for the stability of the system (Hoitink et al., 2003). Both the magnitude and duration of the flood and ebb tides affect the morphological outcome and feedback effects can keep an estuary in a state of dynamic equilibrium, oscillating between morphological states (Pethick, 1994), subject to the influence of sea-level rise. According to Lanzoni and Seminara (1998) and more recently Pye and Blott (2014), deep or bar-built estuaries tend to generate ebb-current dominance, whereas in shallow or wide-mouthed estuaries, flood-current dominance occurs, which will affect sediment transport pathways. But classification of flood/ebb flow asymmetry (e.g.,Friedrichs and Aubrey, 1988) does not necessarily give an accurate indication of the net sediment transport.

State-of-the-art, process-based, sediment transport models are believed to have the correct general behaviour over the wide range of wave-current conditions encountered on site. However, the morphological models in which these processes are embedded are capable of predicting only the statistics defining channel geometry and meandering on time scales of decades or longer. The seabed is inherently unstable and morphological modelling outcomes are capable of reconstructing only the distribution of channels and sand flats that characterise many shallow tidally-dominated estuaries. Small bathymetric irregularities introduced at the start of a numerical simulation evolve into larger scale bed features. McCann et al. (2011) demonstrated how a tidal channel/sand flat network appears spontaneously and exhibits 'self-organisation' linked to the formation of ebb-dominant channels and flood-dominant shoals. In McCann's simulations sediment was exported seaward in the deeper channels, while net transport was shown to be landward (i.e., up-estuary) on the tidal flats.

Robins and Davies (2010) attempted to systematise this general behaviour, classifying estuary regions as 'flood-dominant' if the asymmetry in the tide causes a net, up-estuary, sediment accumulation (because the stronger flood flow may be above the threshold of sediment motion for longer than the ebb flow). They referred to the opposite behaviour as 'ebb-dominant'. They demonstrated that an estuary is likely to become more flood-dominant as the tidal wave travels landward, while providing clear evidence of the role of ebb channels in promoting ebb dominance in net sediment transport. Their modelling suggested that, for idealised estuary shapes characterising mesotidal systems found along the UK coast, net flood-dominance in sediment transport occurs in regions of shallower water (e.g., tidal flats and salt marshes in the upper estuary) where $a/h > 1.2$ (a = local tidal amplitude, h = local, cross-sectionally averaged, water depth), while net ebb-dominance occurs in regions of deeper water (Robins and Davies, 2010). However, the magnitude of net ebb-dominance was less pronounced than the net flood-dominance. Sea-level rise caused the threshold value of a/h to be reduced, with the estuary shifting from ebb-dominance towards flood-dominance, with less transport overall. Reducing the sediment grain size increased both the gross and net sediment transport but did not significantly affect the transport pattern. Importantly, the inclusion of an ebb channel promoted ebb-dominance in the channel. Applying these outcomes to a realistic case (the Dyfi Estuary), Robins and Davies (2010) concluded that, due to a projected sea-level rise of 1 m in the next 100 years, greater flood-dominance (i.e., net sediment accumulation into the estuary) but less sediment transport overall would occur. In a similar vein, Pye and Blott (2014) postulated generally that sediment accumulation would continue until the creation of a central channel and high flats, which would switch the system from flood-to ebb-dominance. Pye and Blott (2014) also stated that the effects of extreme river events and storm surges on the long-term sediment estuarine budget remain unclear and require further research.

The mobilisation of historic, consolidated, heavy metal contaminants in semi-industrialised UK estuaries (e.g., from mudflats and saltmarshes in the Humber and Mersey) poses a serious risk to water quality and human health (Lee and Cundy, 2001). On the other hand, a potential positive outcome is that this re-mobilisation, which is often a result of extreme stormy weather, accelerates sediment exchange with the open sea (e.g., the exchange of mercury-rich fine sediments from the Mersey Estuary into Liverpool Bay (Harland et al., 2000)).

Fluxes of Nutrients

Biogeochemical cycles vary over both time and space. There is a requirement in the UK to further understand such variability and nutrient coupling and processing in different catchment types to aid development of mitigation actions under climate change. Temporal changes include diurnal cycling and event based changes between high and low flow conditions, seasonal, decadal and longer. Spatially, estuarine biogeochemical cycles form a continuum from freshwater to marine, transporting and processing many forms of macronutrients with discontinuities in the retention and storage of macronutrients (Statham, 2012). Estuaries, therefore, play a crucial role in biogeochemical cycling of nutrients, which in turn influence the growth and resilience of plants and animals (Nedwell et al., 1999; Malham et al., 2014).

The impact of nutrient loading in UK estuaries depends on the physical and geological setting (Jickells et al., 2000) and the physical and chemical conditions prevalent at any particular time. Importantly, projections for the UK suggest that winter river flows will increase by between 4% and 9% by 2020, with summer decrease of roughly 11% (1–32% range), based on IPCC climate change projections (Whitehead et al., 2009). Nutrient projections suggest a possible rise of macronutrient concentrations (such as nitrogen and phosphorus) under climate change conditions, with consequent impacts of augmented eutrophication, hypoxia and harmful algal blooms (Tappin, 2002 and Statham, 2012). Projected 21st century sea-level rise, combined with reduced river flows (projected during summer in the UK), could heighten the likelihood of estuarine nutrient trapping, in between river flushing events (Robins et al., 2014). Conversely, nutrient retention will be less likely during winter if projected increases in river flow frequency occur, which could counteract the effects of sea-level rise to some extent (Struyf et al., 2004). Nutrification in summer is more prone to cause eutrophication than in winter: only summer production is nutrient limited; winter production is light limited. Thus, the dry summer effect has the greatest potential for augmenting eutrophication.

However, current understanding, let alone prediction, of biogeochemical processes in estuaries is highly complex and requires further study (Jickells et al., 2000, Andrews et al., 2006, Najjar et al., 2010 and Statham, 2012). Overall, more is known about fluxes of inorganic than organic nutrients (Jarvie et al., 2012). Further research is needed on bacterial cycling of organic nutrients (Asmala et al., 2014), nutrient reactivity (Valdemarsen et al., 2014), and on nutrient speciation and interaction with fresh and salt water under different pH and dissolved organic matter conditions (Morgan et al., 2012). Catchment-estuary model coupling methods are improving (Uncles, 2003 and Huang et al., 2013), and there is currently a UK drive in data generation (e.g., NERC Macronutrients Cycles Programme), which will prove valuable for model validation.

CLIMATE IMPACTS ON ESTUARINE ECOSYSTEMS

Human health: Pollutants and Pathogenic Microorganisms

By 2100, an increase in average temperatures, together with an increase in winter rainfall and longer drought periods will ultimately have impacts upon public health risk to UK estuaries, associated with water-borne and food-borne diseases caused by microbial pathogens (McMichael et al., 2006). There are over 100 recognised types of microbial pathogen found in contaminated water used for recreation and as potable water (Rose et al., 2001), with the potential for food-borne transmission via seafood (Marques et al., 2010). Projected increased precipitation and flooding may result in untreated sewage discharge via combined sewer overflows (CSOs) and increased agricultural runoff (Perciasepe, 1998). In addition, toxic pollutants, ammonia, organic solids and oxygen-demanding substances are also discharged into the water system (Friedman-Huffman and Rose, 1999 and Stachel et al., 2004) and may increase concurrently with high precipitation events. There are precedents for precipitation events preceding outbreaks of water-borne disease (Curriero et al., 2001 and Rose, 2005); for example, outbreaks of the protozoan pathogen *Cryptosporidium* have been observed following extreme precipitation events, which reduces the efficiency of water treatment processes resulting in increased human health risk (Rose et al., 2001). In addition, faecal indicator organism concentrations have been found to be higher during a winter of high precipitation, when compared with the rest of that year (Lipp et al., 2001), and high precipitation also results in an increase in the viral load of estuarine environments, particularly after CSOs, where a significant increase in Adenovirus and Norovirus concentrations have been observed in receiving waters (Rodríguez et al., 2012). It is well established that increased rainfall may influence the concentrations of toxic metals, organic chemicals, algal toxins, and human pathogen contaminants in seafood (Marques et al., 2010). Conversely, drought conditions will result in decreased river flow and lower concentrations of microbial contaminants (Rose et al., 2001). Sea-level rise may lead to the loss of coastal wetlands, which filter nutrients, microbial agents, and chemical agents (McMichael et al., 2006). However, the implications of sea-level rise on the fate of microbial, nutrient and chemical agents and their interactions with the wider ecological dynamics of the system are poorly understood, and could be addressed by integrated catchment-river-estuary models (Rose et al., 2001).

The cholera model perhaps represents the best-studied mechanism for understanding the role of climate on infectious disease transmission (Lipp et al., 2002). *Vibrio cholerae*(the bacterial species responsible for cholera) and other *Vibrio* spp. are endemic in marine and coastal environments and demonstrate strong associations with warmer sea surface temperatures that drive their seasonality and ecological distribution (Rose et al., 2001). Furthermore, favourable environmental conditions (dependent on nutrients and temperature) promote the formation of algal blooms that act as a reservoir for *Vibrio* spp. and can cause reversion from a quiescent to infectious state (Rose et al., 2001). Future temperature shifts, therefore, have the potential to expand the prevalence of *V. cholera* at both temporal and geographical scales (Lipp et al., 2002). However, cholera outbreaks have disappeared from much of the developed world (including the UK), and the endemicity of cholera in the tropics and subtropics exemplifies the favourable environmental conditions that drive their ecological distribution. In contrast to cholera, which is autochthonous in estuarine systems and favours dry weather and warmer temperatures, the prevalence of other bacterial and viral pathogens of allochthonous origin are governed by different conditions. For example, faecal indicator organisms demonstrate enhanced survival in estuarine environments at lower temperatures and their abundance has a strong correlation with high precipitation events (Lipp et al., 2001 and Malham et al., 2014). Further research is required to understand the projected effects of future climate scenarios on estuarine pathogen load and subsequent human health risk.

Estuarine Habitats

There is little evidence-based research on the effects of climate change drivers on estuarine habitats, except for the effects of temperature and sea-level rise (Mieszkowska et al., 2013 and Fagherazzi et al., 2012). Here, we highlight current knowledge for the most researched habitats, the majority of which are intertidal.

Many UK salt marshes were lost to agricultural expansion and coastal development in the 18th and 19th centuries (Doody, 2008). The total area of 40,522 ha in England and Wales was still declining, by <100 ha y^{-1} at the last survey (1989–2009; EA, 2011), mainly due to losses in the southeast associated with isostatic adjustment. Other geographical areas had marsh expansion over the same period (EA, 2011) and it is not known whether these patterns of change are associated with emergent climate change. In principle, climate change might be both 'good' and 'bad' for salt marshes. A moderate rise in the annual average temperature can boost vegetation productivity and favour marsh expansion, although a greater frequency of dry summer spells could depress plant growth through desiccation and evaporation-driven increases in sediment salinity (Gedan and Bertness, 2010). Increased frequency of high rainfall events is likely to boost riverine sediments supply to marshes, which is a key stimulant of marsh vertical growth (Fagherazzi et al., 2013). Conversely, increased storminess could accelerate erosion of marshes located in wide estuaries, where the fetch, and thus potential wave energy, is higher. Evidence suggest marshes are well-capable of keeping pace with sea-level rise if given the space to transgress inland, and if exceptional land subduction and depressed sediment supplies are absent (Fagherazzi et al., 2012). Coastal squeeze remains the greatest threat to salt marshes in terms of sea-level rise, the elimination of which is linked to policy decisions, such as shoreline management planning. Marsh vertical and horizontal growth is strongly dependent on sediment supply, which, if diminished, can switch marshes from accreting to eroding and preclude their keeping pace with sea-level rise (Bouma et al., 2014). Section 2 showed sediment supply is governed by hydrodynamics, determining for instance the up-estuary transport of marine sediments, as well as the down-estuary transport of catchment and riverine particles. Tendencies of increased fluvial transport of sediments bode well for the capacity of marshes to keep pace with sea-level rise. Emergent evidence for the past 70 years show that many marshes on the west coast of England and Wales have undergone dynamic changes in areal extent, but the collective area has expanded, particularly since the 1960s (Martin Skov, *pers. comm.*). These changes are not linked to isostatic adjustment, which is commonly invoked as a cause for national variation in marsh erosion rates (EA, 2011). Future projections of changes to estuarine function might do well to carefully consider the causes for such regional differences in ecosystem change, and the implications to estuarine management.

Rocky shores in the UK are predominantly located on open coasts, although they do occur in some estuaries. The following findings are mainly from coastal shores, but many are likely to be transferable to estuarine settings. The area cover of intertidal rocky shores is likely to decline with sea-level rise: as many shorelines increase in steepness with elevation, the distance between the low and high tidal mark, and thus the intertidal area, will decrease with sea-level rise (Jackson and McIlvenny, 2011). Temperature rise is already having some effects. Throughout the UK, a number of rocky shore species have been gradually moving north in response to rise in temperature, punctuated by cool years in which the advance has ceased, or reversed (Mieszkowska et al., 2013). The tendency is for species that are intolerant to warming to retreat northwards to remain in cooler conditions, while Mediterranean/Lusitanian species are coming up from the south, facilitated by an increasingly benign climate (Mieszkowska et al., 2013). Similar changes are occurring in other estuarine habitats, such as *Sabellaria* reefs (Frost et al., 2004), as well as in terrestrial and marine systems (Hiddink and Hofstede, 2008). Changes to community composition brings with it the potential for change in biodiversity and ecological functioning (organic production, degradation, habitat provisioning), as altered composition and biodiversity changes the interactions between species and their ecological roles (Balvanera et al., 2006). Rocky shore habitats are competed for by algae and grazing molluscs, and the relative dominance of these groups is determined by environmental conditions: increased wave exposure and higher temperatures favour animals (Hawkins et al., 2008). Thus, predictions are that algae will become less prevalent in a warming and more hydrologically energetic climate, which could diminish overall biological production of estuarine rocky shores (Hawkins et al., 2009).

Seagrass beds in Europe have dramatically declined over the past century, by 50–80% (Short and Wyllie-Echeverria, 1996). In the UK, there is no research-based evidence on how the area cover of seagrass beds will respond to climate change (Mieszkowska et al., 2013), although they are considered to be vulnerable to a plethora of changes associated with climate change (Jones et al., 2011). Current patterns of change are variable: there are indications of areal increases in Scotland and Wales (Smith et al., 2002 and Howsen, 2009), but retreat in the Isle of Wight (Mieszkowska et al., 2013). Seagrasses elsewhere have suffered from climate-related changes in temperature, storminess and prevailing winds (Bjork et al., 2008). Mediterranean seagrass beds are predicted to decline in response to temperature stress (Marba and Duarte, 2010). In the UK, forecasted temperatures are well within the tolerance levels for *Zostera noltii* (37 °C, Massa et al., 2009); however, *Zostera marina*, the other UK species, stops growing at >20 °C and starts dying at > 25 °C (Greve et al., 2003), and dieback during summer heat waves have been observed as far north as Denmark (Greve et al., 2003 and Winters et al., 2011). It is not unreasonable to expect similar patterns in the UK (Mieszkowska et al., 2013).

Sedimentary habitats account for 90–95% of estuarine intertidal habitats and are powerhouse production areas of invertebrates that feed higher trophic levels, including birds (McLusky and Elliott, 2004). Sediment habitats are already undergoing geomorphological change from climatic drivers including increasing wave height, surges and sea-level rise (Lowe et al., 2009); they are likely to respond to changes in river flow, although there are no detailed studies to support this prediction in the UK. A projected more energetic wave climate is likely to steepen benthic profiles (Mieszkowska et al., 2013) and diminish the abundance and diversity of sediment living animals through coarsening of sediments (Pethick and Crooks, 2000, Fujii and Raffaelli, 2008, Yates et al., 2010 and Fujii, 2012). Mudflats are thought to be particularly sensitive to climate change, through an assortment of drivers, including shifts in sediment supply and sediment coarsening (Gubbay and Earll, 2010 and Jones et al., 2011), and the risk of coastal squeeze as enhanced wave and tidal energy require mudflats to move up-estuary to less energetic areas (Pethick and Crooks, 2000). Modelling indicates an upstream shift in the position of the most productive intertidal flats of the Humber estuary could reduce estuarine production by 23%, because the area for flats is reduced by the narrowing of the estuary (Fujii and Raffaelli, 2008 and Fujii, 2012). Accurate predictions are difficult to make given the influences of factors such as inter-estuary variation in geomorphology, which regulates hydrological forcing and sediment supply, spatial variation in catchment processes, which regulate sediment supply and water chemistry, and the interactions of biological and physical variables on benthic organisms (Fujii, 2012). For instance, a decline in productivity with sediment coarsening might be countered by increased temperature and milder winters that boost benthic production, (Fujii, 2012) and/or by upper reaches of estuaries becoming more saline and accommodating of invertebrate diversity and biomass (Ysebaert et al., 2003).

The honeycomb worm *Sabellaria* sp. makes bulky biogenic reefs in intertidal and subtidal rocky areas. Reefs are colonised by an array of invertebrates and are a priority habitat for protection in the UK. *Sabellaria* is a 'warm-water' genus and the reefs have been extending northwards in the UK in recent decades, as a response to less severe winter temperatures (Frost et al., 2004); a pattern which is likely to continue into the future. Temperature rise is unlikely to markedly affect the distribution of UK mussel beds, as predicted conditions will well within their temperature tolerance; however, mussel condition might well be affected by eutrophication (Mieszkowska et al., 2013).

The potential impact of climate change on estuarine habitats is somewhat determined by their physical nature (biogenic, non-biogenic, hard, soft) and position (intertidal, subtidal, location in estuary). The persistence of biogenic habitats, such as mussel beds, is governed by the physiological tolerances of the habitat-forming species. Non-biogenic systems can continue to exist despite climate change, although some might become biologically unavailable. Thus, rocky shores will persist as a habitat, but are projected to decrease in area due to coastal squeeze (Jackson and McIlvenny, 2011). Intertidal habitats are naturally more influenced by short-term changes in weather than sub-tidal habitats. Ultimately, changes in community composition induced by climate change might affect the resilience of systems to other sources of environmental change, such as eutrophication and coastal development, although the effects of multi-stressor impacts on UK estuaries is limited. Major research gaps also exist on how biological feedbacks that affect physical conditions in estuaries will be influenced by sea-level rise. For instance, saltmarsh colonisation alters the forcing and distribution of hydrological energy, which facilitates further expansion of marshes and ultimately influences the tidal prism and sediment distribution. Whether or not such biological feedback mechanisms will be enhanced or reduced by change in climatic conditions is currently not known.

Larvae in Estuaries

Environmental cues synchronize the reproductive cycle of many marine invertebrates and fish (Kingsford et al., 2002). A broad diversity of taxa have the capacity to detect variations in current direction, magnetism, water pressure, water chemistry (e.g., amino acids, salinity), sound and vibration (e.g., waves breaking or fish assemblages), and light gradients (Forward et al., 2001). Some organisms can detect multiple cues (e.g., decapods and fishes) and are likely to have integrated sensory responses (Kingsford et al., 2002). The type of response may be endogenous and regulated by an internal clock (e.g., tidal stream transport), or a direct response to the environment (e.g., diel transport) (Queiroga and Blanton, 2005). Further, the phasing and extent of vertical migration both change throughout ontogeny (Queiroga and Blanton, 2005). These cyclic cues may affect a number of reproductive parameters, including sex determination, gametogenesis, spawning, transport, settlement, retention, connectivity, and metamorphosis (Lawrence and Soame, 2004).

For estuarine species, primary environmental cues are tidal currents and photoperiod. European shore crab larvae in estuaries within the North Sea use selective tidal stream transport to swim to the surface during the flood phase of the tide, which aids in-shore dispersal towards estuarine feeding grounds (Moksnes et al., 2014). Conversely, decapod crustacean larvae may swim to the surface during the ebb phase of the tide and migrate offshore (Queiroga and Blanton, 2005). In other regions, shore crab display nocturnal (diel) vertical migration to avoid predation (Moksnes et al., 2014). Estuarine species live in a highly dynamic environment where recruitment variability is a key determinant of population dynamics. Environmental requirements for successful recruitment may differ between co-occurring species, and therefore species may be advantaged or disadvantaged under climate change (Allen et al., 2008 and Jenkins et al., 2015). Climate change impacts on UK estuaries are likely to alter the environmental cues, which in turn will affect their reproductive parameters. For example, the phase relationship between temperature of the water column and photoperiod directly controls the metabolism and lifecycles of aquatic organisms, with most biological processes operating faster at higher temperatures, leading to shorter pelagic larval durations and time to metamorphosis (Whitehead et al., 2009 and González-Ortegón and Giménez, 2014).

Variability in salinity has been documented to both increase and decrease time to metamorphosis, depending on species (Forward et al., 2001). The estuarine copepod*Eurytemora affinis* (which exhibits a combination of tidal, salinity, and temperature cues) will change its migration in response to the location of the estuarine salinity maximum (Hough and Naylor, 1991). Indeed, evidence from Chaalali et al. (2013) shows climate-driven up-estuary migration of the species in the Gironde Estuary, France. The presence of aquatic vegetation and increased humic acids (a decomposition product of terrestrial plant material; Fox, 1981), both accelerate the time to metamorphosis of brachyuran crabs (Forward et al., 2001). Alternatively, cues that delay metamorphosis include hypoxia and chemical odours from potential predators (Forward et al., 2001). Thus, species that cue reproduction based on these signals are likely to be particularly vulnerable in the future (Lawrence and Soame, 2004). Consequently, climate change may cause local extirpations of populations in the extreme, or more likely impact fecundity, spawning success, and recruitment significantly.

However, change will ultimately depend on the relative speed of adaptation to climate change (Allen et al., 2008) and the degree of mixing between populations across the Metapopulation (Lawrence and Soame, 2004). Some larvae of estuarine species appear to be resilient to environmental variability and may be able to tolerate changes related with climate change. For instance, an important species, inhabiting brackish waters in the UK is *Palaemon varians*; this is a shrimp found in saltmarshes, coastal ponds and lagoons. Adults of *P. varians* can potentially extend the upper thermal range through acclimation (Ravaux et al., 2012). Larvae of this species appear to be exported and develop in estuarine-coastal areas (Fincham, 1979). Larvae show a wide range of salinity tolerance (5–42: Antonopoulou and Emson, 1988). In addition, larvae appear to tolerate high temperatures as well as food limitation. Metabolic rates increase with temperature (range: 5–25 °C), but total energy losses, calculated from oxygen consumption are lower than at low temperature. In this shrimp, energy reserves at the time of hatching appear to confer larvae a high capacity to tolerate food limitation (Oliphant and Thatje, 2014). Oliphant and Thatje (2013) and Oliphant et al. (2014) showed that there is sufficient variability among broods in per offspring investment to produce variability in development and starvation tolerance; if such variability is heritable, it may confer this species a sufficient capacity to adapt to variable conditions of food, temperature, and salinity.

Besides the hypothesised effect of environmental factors as cues, such factors operate as stressors, in particular in the early life history stages of marine invertebrates. Meteorological forcing, such as extreme rainfall events, promote salinity decreases in estuaries. Thus, low salinities can reduce growth rates especially during the larval stages, if these are weak osmoregulators or osmoconformer (e.g., decapod crustaceans:Torres et al., 2011). Low salinity, experienced by parents or at the egg/embryo stage, can have knock-on effects along the life cycle of estuarine organisms (Giménez, 2006,Jensen et al., 2014 and Chaparro et al., 2014). For instance, low salinity at the embryonic stage can affect the capacity of larvae to osmoregulate (Charmantier et al., 2002). Exposure of larvae to low salinity also affects juvenile size and growth rate (Giménez et al., 2004 and Rey et al., 2015).

It is likely that multiple stressors will operate on estuarine organisms in a synergistic way (Dolbeth et al., 2011 and Przeslawski et al., 2015). As a consequence of warming, temperature may modify the sensitivity to other environmental stressors, but also thermal sensitivity should be modulated by exposure to additional stressors (Sokolova and Portner, 2007). For instance, under low salinity, larvae may be more vulnerable to pollutants (González-Ortegón et al., 2013) or food limitation (González-Ortegón and Giménez, 2014). On the other hand, increased temperatures may increase the tolerance of low salinities (González-Ortegón and Giménez, 2014). Overall, an assessment of the impact of changes in estuaries will require an understanding of the correlated changes in environmental variables such as salinity and temperature.

Biota

Species living in estuaries experience a naturally highly variable physical environment that is often close to their physiological tolerances (Attrill and Rundle, 2002). Despite this, projected climate changes may have significant impacts on estuarine species and their ecosystems. Moreover, invasive species to estuaries have increased during the last 20 years, due to global temperature warming (Austin et al., 2010), leading to some dramatic effects on native assemblages (Thompson et al., 2002). Problems associated with invasive species, especially pathogens, are projected to increase over the next few decades (e.g., McDowell et al., 2014).

Projected sea-level rise and increased winter river discharge for the UK are thought to be the most important impacts of climate change on species, principally through associated habitat changes (Kimmerer, 2002 and IPCC, 2014). Habitat loss from sea-level rise, such as coastal squeeze in the absence of managed retreat, is likely to have important implications for estuarine food webs (Fujii, 2007, Fujii and Raffaelli,

2008 and Fujii, 2012). Although only from an individual study site in the Humber estuary, Fujii (2007)demonstrates that a 'squeezed' location can support fewer species and biomass of keystone benthic macroinvertebrates per unit area due to the narrower, steeper and less muddy beach profile, when compared with adjacent undefended shorelines. Yamanaka et al. (2013) have supported these findings with a study based on three estuaries with meiofauna included with macrofauna in the study. Evidence does appear to support that coastal squeeze will decrease benthos biomass; however Yamanaka et al. (2013) found that local conditions were important in describing macrobenthic biomass in intermediate-to-steep sloping estuary shores. This means that predicting the effect of coastal squeeze on macrobenthos without knowledge of complex local processes and conditions, and their interaction with the physical drivers affected by coastal squeeze, is likely complex if to be made reliably. Such observations do not incorporate how other feedbacks, such as changes in primary production, will interact with these trends.

Estuarine foodwebs may also be impacted by habitat modification from increased river discharge, changes in river-borne materials, and variations in the position and intensity of estuarine turbidity maxima and salinity stratification (Kirby, 2010 and Uncles et al., 2014). Changes in the salinity experienced by fauna in estuaries is likely to have an impact on species diversity and distribution: a projected increase in salinity range may reduce species diversity, as found in the Thames Estuary (Attrill, 2002), and move the focus of biomass maximum higher up estuaries (Fujii, 2007). Additionally, the occurrence of hypoxic events and the vulnerability of species in estuaries to these events is likely to increase (Rabalais et al., 2009 and Vaquer-Sunyer and Duarte, 2011). Key components such as benthic filter and deposit feeders have been observed to be, and are predicted to be, negatively affected through these changes (Fujii and Raffaelli, 2008). These functional groups are very important for nutrient cycling and supporting higher trophic levels such as predatory fish and birds (Austin and Rehfisch, 2003), and so can have far reaching implications for British estuarine ecology if affected beyond natural variation. Turbidity has been suggested to be important for survival and behaviour of mid-water organism such as juvenile fish through food supply and vulnerability to visual predators, which may present further implications to changes in river discharge and sediment suspension (Power et al., 2000).

In addition to these habitat modifications, climate changes to temperature may also indirectly affect populations and food webs (Walther et al., 2002). For example, projected milder UK winters may negatively impact infaunal bivalve reproductive output and recruitment directly and through affecting predator phenology (Philippart et al., 2003 and Beukema and Dekker, 2014). Predatory shore crabs *Carcinus maenas* and brown shrimp *Crangon crangon* recruit earlier after mild winters, which can result in a greater overlap with their post-larval bivalve prey and reduces bivalve and flatfish recruitment success, when compared with periods with proceeding colder winters (Power et al., 2000 and Beukema and Dekker, 2014). This again may reduce the biomass of benthic macroinvertebrate species that support higher trophic levels and the estuarine ecosystem. However, as acknowledged already, the complexities of local systems need considering before broad climate change predictions are applied to individual estuaries.

DISCUSSION

Global climate models, linked to impact models which predict biophysical and socio-economic impacts, propose that stabilizing global carbon dioxide levels at 550 ppm by 2150 could significantly reduce most stressors on the water cycle, compared with unmitigated emissions scenarios; for example, reducing flooding frequency by some 80–90% along the most vulnerable coastlines, which include estuaries in southeast England (Arnell et al., 2002). Even though these emissions targets will be difficult to achieve, the UK are attempting to show leadership, and are seeking international commitment to reduce carbon dioxide and other greenhouse gas emissions through the United Nations Framework Convention on Climate Change (UNFCCC). The UK government has committed to reducing the country's emissions by 60% from 1990 levels by 2050 – achieved by reducing the amount of energy we consume and by substantially increasing our use of renewable energy resources (King, 2004).

Even with these initiatives, in the short term (the 21st century) some consequences of climate change appear to be inevitable for the UK under any future emissions scenario (Arnell et al., 2002), such as sea-level rise. If we define the risk of climate-driven impact as the probability of the occurrence multiplied by the magnitude of the consequence, it is obvious that the level of risk is different for each affected party and for each region. Thus, the magnitude of potential climate-driven impacts on UK estuaries, described in this paper, and hence the level of risk, require further quantification in order to inform management strategies. Regarding flood risk and flood mitigation, Elliott et al. (2014)point out that the UK has a strategy to try to protect urban and industrial coasts but deem protection of farmland not to be economically justifiable. Therefore, rural areas require local adaptation strategies that rely on soft engineering and improved community awareness.

More effective coastal management in the UK requires reducing the potentially large uncertainties, associated with climate change projections based on limited simulation of climatic variability and complex impact predictions, so that levels of risk can be assigned accurately (Arnell et al., 2002). For example, unresolved temporal and spatial variability within climate models needs to be parameterised in the boundary forcing conditions of local-scale impact models (e.g., Lewis et al., 2011). In terms of predicting climate-driven impact, many researchers are using species distribution modelling to examine how ranges may shift under climate change. However, this approach has been criticised because most models have not included non-climate stressors, such as species competition and predation, habitat loss and land-use changes, eutrophication and invasive species, that could themselves influence distributions (McDowell et al., 2014). Additionally, state-of-the-art, process-based, morphological models are, in general, capable of predicting only the statistical behaviour defining channel geometry and meandering on time scales of decades or longer. Therefore, we must understand the relative importance of climatic and non-climatic variables in controlling present-day distributions to predict the impacts of climate change, and also understand the associated uncertainties so that we can measure the potential risk and improve management strategies.

CONCLUSIONS

In this review, we summarise published literature of historic observations and current anthropogenic climate change projections for the 21st century, and the anticipated processes and impacts to UK estuarine systems, including physical, biological, and ecological changes. Fig. 3 summarises the likely main, but not all, pathways from climatic drivers to processes to impacts on UK estuaries. There is also no indication from Fig. 3 of the scale of impact; for example, the consequences of flooding may be more significant than shifts in frontal positions, although this is difficult to quantify.

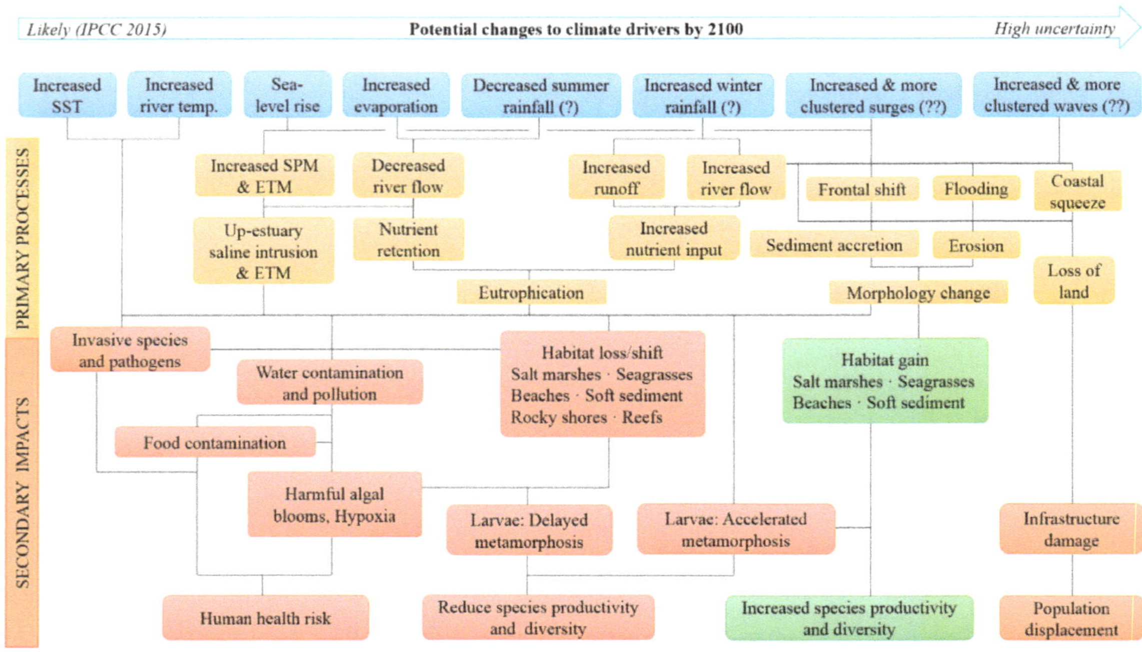

Figure 3: Flow diagram showing the main climatic drivers, and primary and secondary impacts of climate change to UK estuaries. Pathways (black lines) are directed *from* the bottom/side *to* the top of a 'driver/impact box'. Predominantly positive impacts are coloured green. (For interpretation of the references to colour in this figure legend, the reader is referred to the web version of this article.)

In terms of the climatic drivers shown in Fig. 3, sea surface temperatures around the UK have increased by 0.7 °C over the past 30 years, and are projected with high confidence to increase further by between 1.5 °C and 4 °C, by 2100, depending on our carbon emissions. Sea-levels have risen by 1–3 mm yr^{-1} during the 20th century, and are projected with high confidence to rise further by between 0.44 m and 0.74 m, by 2100, again depending on our carbon emissions but also with significant spatial variability. Precipitation and storminess have shown high variability over the past few decades, with a general pole-ward shift and increased intensity/decreased frequency of events. Although UK river flows have generally increased during autumn and winter over the past 50 years, flow rates have decreased during spring with no detectable trends during summer. Future UK projections, albeit with low confidence and high variability, are for more clustering of events, with increased winter precipitation, river flow, surge and wave intensity, and longer dry and calm periods in summer.

Potential changes to physical, *primary*, processes (see Fig. 3) include increased flooding and coastal squeeze, caused by increased sea-level, changing surge and wave climates, and changing river flow events. The capacity for estuaries to respond to flooding will depend in the short term on available 'floodable' space, and in the long term on morphodynamics relative to sea-level rise (e.g., sediment supply and accretion, and coastal realignment). Sea-level rise will likely cause a shift towards net sediment accretion in UK estuaries, but with reduced transport overall. Turbulent mixing – critical for water quality and coastal ecology – is primarily controlled by river flow variability. Hence, potential alterations to river flows will increase the temporal variability and length scales of estuarine fronts, stratification and mixing. For example, longer dry periods in summer, in combination with sea-level rise will push the salinity maximum further up-estuary and enhance estuarine nutrification, with negative impacts of eutrophication, harmful algal blooms and hypoxia. On the other hand, clustering of storm events will alter the temporal mixing variability in winter.

Some of the *secondary impacts* on the estuarine ecosystem (see Fig. 3) may ultimately result in a risk to human health, associated with ecological disruption and heightened water-borne and food-borne microbial pathogens concentrations, mainly due to projected increased temperatures and changing river flows. Changes in these climate drivers are also likely to influence the concentrations of toxic metals, organic chemicals, algal toxins, and human pathogen contaminants in seafood. The primary impact of coastal squeeze is the greatest threat to already declining UK salt marsh habitats, which in turn may reduce their species diversity and food webs. Some intertidal habitats and biogenic reefs – often priority habitats for protection in the UK – have gradually moved north in response to rising temperatures, bringing changes in biodiversity and ecological functioning – for example, algae in rocky shores will likely become less prevalent. Projected increases in mild winters may negatively impact infaunal bivalve reproduction. Invasive species to estuaries have increased, due to temperature warming, affecting native assemblages and introducing pathogens. However, temperature rise is unlikely to affect species that live within their temperature tolerances, such as mussels. Sedimentary habitats, particularly mudflat-living species, are sensitive to an assortment of climate change drivers and primary impacts including wave attack and sediment coarsening. Changing climate drivers are likely to alter environmental cues which synchronize the reproductive cycle of many marine invertebrates and fish. Estuarine species that cue reproduction based on salinity signals are likely to be particularly vulnerable to climate change. Eventually, climate change may have implications on the conservation status of UK estuaries.

We recognise there is variety and diversity of estuarine ecosystems in Britain, and variety in drivers of change that require more detailed quantification/prediction. We therefore recommend that future research should focus on improving our understandings of the primary and secondary impacts to UK estuaries, on a case-by-case basis, caused by potential changes to regional climatic drivers; for example, through improved modelling and model coupling, accompanied by long-term monitoring. Current UK research funding, such as the NERC Macronutrient Cycles Programme, aims to contribute to this via integrated climate-catchment-river-estuary process modelling combined with intensive sampling across several UK systems. Some emerging outcomes of this programme are that climate-driven, coupled, soil-catchment-river models are required to accurately predict future river flows and nutrient fluxes. Further, uncertainties associated with commonly-used daily-averaged river flows to infer impact downstream in the estuary appear to be significant, emphasising the need for catchment modelling.

ACKNOWLEDGEMENTS

Author contributions: Abstract and Introduction (all); Section 2 (P. Robins and M. Lewis); Section 3 (P. Robins, M. Lewis, S. Neill, A. Davies, S. Malham); Section 4 (P. Robins, J. MacDonald, M. Skov, L. Gimenez, T. Whitton). This review paper was commissioned by the Environment Agency of behalf of the Natural Environment Research Councils (NERC) Living with Environmental Change Programme (LWEC). P. Robins, A. Davies, S. Malham, T. Whitton, S. Neill and C. Jago wish to acknowledge the support of the SEACAMS (Sustainable Expansion of the Applied Coastal and Marine Sectors; www.seacams.ac.uk, Grant Number 80366) research project at Bangor University, the Welsh Government, the Higher Education Funding Council for Wales, the Welsh European Funding Office, and the European Regional Development Fund Convergence Programme. P. Robins, S. Malham, J. MacDonald, and C. Jago wish to acknowledge the NERC-funded Macronutrients Cycles Programme (Turf2surf project: www.turf2surf.org). S. Neill, P. Robins, M. Lewis and M. Skov wish to acknowledge the support the Sêr Cymru National Research Network for Low Carbon, Energy and the Environment (NRN-LCEE). We would also like to thank the reviews for their constructive comments on a previous version of this paper.

REFERENCES

1. Alexandersson, H., Tuomenvirta, H., Schmith, T., Iden, K., 2000. Trends of storms inNW Europe derived from an updated pressure data set. Clim. Res. 14, 71e73.
2. Allen, R.M., Buckley, Y.M., Marshall, D.J., 2008. Offspring size plasticity in response to intraspecific competition: an adaptive maternal effect across life-historystages. Am. Nat. 171, 225e237.
3. Allan, R., Tett, S., Alexander, L., 2009. Fluctuations in autumnewinter severe stormsover the British Isles: 1920 to present. Int. J. Climatol. 29, 357e371.
4. Andrews, J.E., Burgess, D., Cave, R.R., Coombes, E.G., Jickells, T.D., Parkes, D.J.,Turner, R.K., 2006. Biogeochemical value of managed realignment, Humberestuary, UK. Sci. Totl Environ. 371, 19e30.
5. Antonopoulou, E., Emson, R., 1988. The combined effects of temperature andsalinity on survival, moulting and metamorphosis of the larval stages of threespecies of palaemonid prawns. In: Ryland, J.S., Tyler, P.A. (Eds.), Reproduction,Genetics and Distributions of Marine Organisms: Proc. 23rd Eur. Mar. Biol.Symp., Swansea, 59 September 1988.
6. Arnell, N.W., Cannell, M.G.R., Hulme, M., Kovats, R.S., Mitchell, J.F.B., Nicholls, R.J.,Parry, M.L., Livermore, M.T.J., White, A., 2002. The consequences of CO2 stabilisationfor the impacts of climate change. Clim. Change 53, 413e446.
7. Arnell, N.W., 2003. Relative effects of multi-decadal climatic variability and changesin the mean and variability of climate due to global warming: future stream-flows in Britain. J. Hydrol. 270, 195e213.
8. Attrill, M.J., 1998. A Rehabilitated Estuarine Ecosystem: the Environment and Ecology of the Thames Estuary. Kluwer Academic Publisheres, London.
9. Attrill, M.J., 2002. A testable linear model for diversity trends in estuaries. J. Anim. Ecol. 71, 262e269.
10. Attrill, M.J., Rundle, S.D., 2002. Ecotone or ecocline: ecological boundaries in estuaries. Estuar. Coast. Shelf Sci. 55, 929e936.
11. Austin, G.E., Rehfisch, M.M., 2003. The likely impact of sea level rise on waders (Charadrii) wintering on estuaries. J. Nat. Conserv. 11, 43e58.
12. Austin, J., Zhang, L., Jones, R.N., Durack, P., Dawes, W., Hairsine, P., 2010. Climate change impact on water and salt balances: an assessment of the impact of climate change on catchment salt and water balances in the Murray-Darling Basin, Australia. Clim. change 100, 607e631.
13. Asmala, E., Bowers, D.G., Autio, R., Kaartokallio, H., Thomas, D.N., 2014. Qualitative changes of riverine dissolved organic matter at low salinities due to flocculation. J. Geophys. Res. Biogeosci. 119, 1919e1933.
14. Balvanera, P., Pfisterer, A.B., Buchmann, N., He, J.S., Nakashizuka, T., Raffaelli, D., Schmid, B., 2006. Quantifying the evidence for biodiversity effects on ecosystem functioning and services. Ecol. Lett. 9, 1146e1156.

15. Barbier, E.B., Hacker, S.D., Kennedy, C., Koch, E.W., Stier, A.C., Silliman, B.R., 2011. The value of estuarine and coastal ecosystem services. Ecol. Monogr. 81, 169e193.

16. Beck, M.W., Heck, K.L., Able, K.W., Childers, A.D., Eggleston, D.B., Gillanders, B.M., Halpern, B., Hays, C.G., Hoshino, K., Minello, T.J., Orth, R.J., Sheridan, P.F., Weinstein, M.P., 2001. The identification, conservation, and management of estuarine and marine nurseries for fish and invertebrates. Bioscience 51, 633e641.

17. Beukema, J., Dekker, R., 2014. Variability in predator abundance links winter temperaturesand bivalve recruitment: correlative evidence from long-term data in a tidal flat. Mar. Ecol. Prog. Ser. 513, 1e15.

18. Bijl, W., Flather, R., de Ronde, J.G., Schmith, T., 1999. Changing storminess? An analysis of long-term sea level data sets. Clim. Res. 11, 161e172.

19. Bjork, M., Short, F., McLeod, E., Beer, S., 2008. Managing Seagrasses for Resilience to Climate Change. IUCN, Gland, Switzerland.

20. Boorman, L.A., 2003. Saltmarsh Review. An Overview of Coastal Saltmarshes, Their Dynamic and Sensitivity Characteristics for Conservation and Management. JNCC Report No. 334, pp. 114. ISSN 0963-8091.

21. Bouma, T.J., van Belzen, J., Balke, T., Zhu, Z., Airoldi, L., Blight, A.J., Davies, A.J.,Galvan, C., Hawkins, S.J., Hoggart, S.P.G., Lara, J.L., Losada, I.J., Maza, M., Ondiviela, B., Skov, M.W., Strain, E.M., Thompson, R.C., Yang, S., Zanuttigh, B., Zhang, L., Herman, P.M.J., 2014. Identifying knowledge gaps hampering application

22. of intertidal habitats in coastal protection: opportunities & steps to take. Coast. Eng. 87, 147e157.

23. Brown, J., Turrell, W.R., Simpson, J.H., 1991. Aerial surveys of axial convergent frontsin UK estuaries and the implications for pollution. Mar. Pollut. Bull. 22,397e400.

24. Burt, T.P., Ferranti, E.J.S., 2012. Changing patterns of heavy rainfall in upland areas: acase study from northern England. Int. J. Climatol. 32, 518e532.

25. Cai, W.J., 2011. Estuarine and coastal ocean carbon paradox: CO2 sinks or sites of terrestrial carbon incineration? Annu. Rev. Mar. Sci. 3, 123e145.

26. Carini, R.J., Chickadel, C.C., Jessup, A.T., Thom on, J., 2015 May 1. Estimating wave energy dissipation in the surf zone using thermal infrared imagery. J. Geophys. Res. Oceans. Cazenave, A., Nerem, R.S., 2004. Present-day sea level change: observations and causes. Rev. Geophys. 42 http://dx.doi.org/10.1029/2003RG000139.

27. Chaalali, A., Chevillot, X., Beaugrand, G., David, V., Luczak, C., Boet, P., Sottolichio, A., €Sautour, B., 2013. Changes in the distribution of copepods in the Gironde estuary: a warming and marinisation consequence? Estuar. Coast. Shelf Sci. 134, 150e161.

28. Chaparro, O.R., Segura, C.J., Osores, S.J.A., Pechenik, J.A., Pardo, L.M., Cubillos, V.M., 2014. Consequences of maternal isolation from salinity stress for brooded embryos and future juveniles in the estuarine direct-developing gastropod Crepipatella dilatata. Mar. Biol. 161, 619e629.

29. Charmantier, G., Gimenez, L., Charmantier-Daures, M., Anger, 2002. Ontogeny of osmoregulation, physiological plastic-ity and larval export strategy in thegrapsid crab Chas-magnathus granulata (Crustacea Decapoda). Mar. Ecol. Prog. Ser. 229, 185e194.

30. Christierson, B.V., Vidal, J.-P., Wade, S.D., 2012. Using UKCP09 probablistic climate information for UK water resource planning. J. Hydrol. 424e425, 48e67.

31. Church, J.A., White, N.J., Coleman, R., Lambeck, K., Mitrovica, J.X., 2004. Estimates of the regional distribution of sea level rise over the 1950-2000 period. J. Clim. 17, 2609e2625.

32. Church, J.A., White, N.J., 2006. A 20th century acceleration in global sea level rise. Geophys. Res. Lett. 33 http://dx.doi.org/10.1029/2005GL024826.

33. Curriero, F.C., Patz, J.A., Rose, J.B., Lele, S., 2001. The association between extremeprecipitation and waterborne disease outbreaks in the United States, 1948-1994. Am. J. Public Health 91, 1194e1199.

34. Davidson, N.C., Laffoley, D.A., Doody, J.P., Way, L.S., Gordon, J., Key, R., Pienkowski, M.W., Mitchell, R., Duff, K., 1991. Nature Conservation and Estuaries in Great Britain. Nature Conservancy Council, Peterborough.

35. Davies, 1964. A morphogenic approach to w orld shorelines. Z. Geomorphol. 8,27e42.

36. Day, J.W., Pont, D., Hensel, P.F., Ibanez, C., 1995. Impacts of sea-level rise on deltas in ~ the Gulf of Mexico and the Mediterranean: the importance of pulsing events tosustainability. Estuaries 18, 636e647.

37. Debernard, J., Røed, L.P., 2008. Future wind, wave and storm surge climate in theNorthern Seas: a revisit. Tellus 60A, 427e438.

38. Dickson, M.E., Walkden, M.J., Hall, J.W., 2007. Systemic impacts of climate change onan eroding coastal region over the twenty-first century. Clim. Change 84,141e166.

39. Dissanayake, P., Brown, J., Karunarathna, H., 2015. Impacts of storm chronology onthe morphological changes of the Formby beach and dune system, UK. Nat. Hazards Earth Syst. Sci. Discuss. 3, 2565e2597.

40. Dolbeth, M., Cardoso, P.G., Grilo, T.F., Bordalo, M.D., Raffaelli, D., Pardal, M.A., 2011. Long-term changes in the production by estuarine macrobenthos affected by multiple stressors. Estuar. Coast. Shelf Sci. 92, 10e18.

41. Doody, P., 2008. Saltmarsh Conservation, Management and Restoration. Springer, London.

42. Dronkers, J., 1986. Tidal asymmetry and estuarine morphology. Neth. J. Sea Res. 20, 117e131.

43. Dyer, K.R., 1997. Estuaries: a Physical Introduction, second ed. John Wiley and Sons/ Wiley & Sons, Chichester. ISBN: 0-471-9741-4. 195 pp.

44. EA, 2011. The Extent of Saltmarsh in England and Wales: 2006e2009. EnvironmentAgency, Bristol, ISBN 978-1-84911-233-8.

45. Elliott, M., Cutts, N.D., Trono, A., 2014. A typology of marine and estuarine hazardsand risks as vectors of change: a review for vulnerable coasts and their

management. Ocean Coast. Manag. 93, 88e99.

46. Elliott, M., Borja, A., McQuatters-Gollop, A., Mazik, K., Birchenough, S., Andersen, J.H., Painting, S., Peck, M., 2015. Force majeure: will climate change affect our ability to attain good environmental status for marine biodiversity?Mar. Pollut. Bull. 95, 7e27.

47. Fagherazzi, S., Kirwan, M.L., Mudd, S.M., Guntenspergen, G.R., Temmerman, S.,D'Alpaos, A., Koppel, J., Rybczyk, J.M., Reyes, E., Craft, C., Clough, J., 2012. Numerical models of salt marsh evolution: ecological, geomorphic, and climatic factors. Rev. Geophys. 50 http://dx.doi.org/10.1029/2011RG000359.

48. Fagherazzi, S., Wiberg, P.L., Temmerman, S., Struyf, E., Zhao, Y., Raymond, P.E., 2013. Fluxes of water, sediments, and biogeochemical compounds in salt marshes.Ecol. Process 2, 3.

49. Forward, R.B., Tankersley, R.A., Rittschof, D., 2001. Cues for metamorphosis of Brachyurancrabs: an overview. Am. Zool. 41, 1108e1122.

50. Fincham, A.A., 1979. Larval development of British prawns and shrimps (Crustacea: Decapoda:Natantia). 2. Palaemonetes (Palaemonetes) varians (Leach, 1814) and morphological variation. Bull. Br. Mus. (Nat. Hist.) Zool. 35, 163e182.

51. Fowler, H.J., Kilsby, C.G., 2007. Using regional climate model data to simulate historical and future river flows in northwest England. Clim. Change 80, 337e367.

52. Fowler, H.J., Wilby, R.L., 2010. Detecting changes in seasonal precipitation extremesusing regional climate model projections: implications for managing fluvial flood risk. Water Resour. Res. 46 http://dx.doi.org/10.1029/2008WR007636.Fox, L.R., 1981. Defence and dynamics in plant-herbivore systems. Am. Zool. 21,853e886.

53. French, J., 2006. Tidal marsh sedimentation and resilience to environmentalchange: exploratory modelling of tidal, sea-level and sediment supply forcing in predominantly allochthonous systems. Mar. Geol. 235, 119e136.

54. Friedman-Huffman, D.E., Rose, J.B., 1999. The Continuing Threat of Waterborne Pathogens. In: Providing Safe Drinking Water in Small Systems: Technology, Operations, and Economics, p. 11.

55. Friedrichs, C.T., Aubrey, D.G., 1988. Non-linear tidal distortion in shallow well-mixed estuaries: a synthesis. Estuar. Coast. Shelf Sci. 27, 521e545.

56. Frost, M.T., Leaper, R., Mieszkowska, N., Moschella, P., Murua, J., Smyth, C.,Hawkins, S.J., 2004. Recovery of a Biodiversity Action Plan Species in Northwest England: Possible Role of Climate Change, Artificial Habitat and Water Quality Amelioration. Sabellaria alveolata. English Nature.

57. Fujii, T., 2007. Spatial patterns of benthic macrofauna in relation to environmental variables in an intertidal habitat in the Humber estuary, UK: developing a tool for estuarine shoreline management. Estuar. Coast. Shelf Sci. 75, 101e119.

58. Fujii, T., 2012. Climate change, sea-level rise and implications for coastal andestuarine shoreline management with particular reference to the ecology of intertidal benthic macrofauna in NW europe. Biology (Basel) 1, 597e616.

59. Fujii, T., Raffaelli, D., 2008. Sea-level rise, expected environmental changes, and responses of intertidal benthic macrofauna in the Humber estuary, UK. Mar. Ecol. Prog. Ser. 371, 23e35.

60. Gedan, K.B., Bertness, M.D., 2010. How will warming affect the salt marsh foundationspecies Spartina patens and its ecological role? Oecologia 164, 479e487.

61. Geng, Q., Sugi, M., 2003. Possible change of extratropical cyclone activity due to enhanced greenhouse gases and sulfate aerosols-study with a high-resolution AGCM. J. Clim. 16, 2262e2274.

62. Gimenez, L., 2006. Phenotypic links in complex life cycles: conclusions from studies with decapod crustaceans. Int. Comp. Biol. 46, 615e622.

63. Gimenez, L., Anger, K., Torres, G., 2004. Linking life history traits in successive phases of a complex life cycle: effects of larval biomass on early juvenile development in an estuarine crab Chasmagnathus granulata. Oikos 104,570e580.

64. Gonz alez-Ortegon, E., Blasco, J., Le Vay, L., Gim enez, L., 2013. A multiple stressor approach to study the toxicity and sub-lethal effects of pharmaceutical compounds on the larval development of a marine invertebrate. J. Hazard. Mater. 263, 233e238.

65. Gonz alez-Ortegon, E., Gim enez, L., 2014. Environmentally mediated phenotypic links and performance in larvae of a marine invertebrate. Mar. Ecol. Prog. Ser. 502, 185e195.

66. Greve, T.M., Borum, J., Pedersen, O., 2003. Meristematic oxygen variability ineelgrass (Zostera marina). Limnol. Oceanogr. 48, 210e216.

67. Gubbay, S., Earll, R., 2010. Assessing the Vulnerability of Marine Biodiversity in Wales to the Impacts of Climate Change: Workshop Report. CCW Contract Science Report No: 934, 21pp. CCW, Bangor.

68. Hannaford, J., 2015. Climate-driven changes in UK river flows A review of the evidence. Prog. Phys. Geogr. 39, 29e48.

69. Hannaford, J., Buys, G., 2012. Seasonal trends in UK river flow regimes and extremes in the UK. J. Hydrol. 475, 158e174.

70. Harland, B.J., Taylor, D., Wither, A., 2000. The distribution of mercury and other trace metals in the sediments of the Mersey Estuary over 25 years 1974e1998. Sci. Total Environ. 253, 45e62.

71. Hawkins, S.J., Moore, P.J., Burrows, M.T., Poloczanska, E., Mieszkowska, N., Herbert, R.J.H., Jenkins, S.R., Thompson, R.C., Genner, M.J., Southward, A.J., 2008.

72. Complex interactions in a rapidly changing world: responses of rocky shore communities to recent climate change. Clim. Res. 37, 123e133.

73. Hawkins, S.J., Sugden, H.E., Mieszkowska, N., Moore, P.J., Poloczanska, E., Leaper, R., Herbert, R.J.H., Genner, M.J., Moschella, P.S., Thompson, R.C., Jenkins, S.R., Southward, A.J., Burrows, M.T., 2009. Consequences

of climate-driven biodiversity changes for ecosystem functioning of North European rocky shores. Mar. Ecol. Prog. Ser. 396, 245e259.

74. Hanson, S., Nicholls, R., Ranger, N., Hallegatte, S., Corfee-Morlot, J., Herweijer, C., Chateau, J., 2011. A global ranking of port cities with high exposure to climate extremes. Clim. Change 104, 89e111.

75. Hiddink, J.G., Hofstede, R.T., 2008. Climate induced increases in species richness of marine fishes. Glob. Change Biol. 14, 453e460.

76. Hetland, R.D., Geyer, W.R., 2004. An idealized study of the structure of long,partially mixed estuaries. J. Phys. Oceanogr. 34, 2677e2691.

77. Hoitink, A.J.F., Hoekstra, P., Van Maren, D.S., 2003. Flow asymmetry associated withastronomical tides: implications for the residual transport of sediment. J. Geophys. Res. Oceans 108, C10.

78. Horrillo-Caraballo, J.M., Reeve, D.E., Simmonds, D., Fox, A., Thompson, R., Hoggart, S., Kwan, S.S.H., Greaves, D., 2013. Application of a source-pathwayreceptor-consequence (SPRC) methodology to the Teign Estuary, UK. J. Coast. Res.

79. Hough, A.R., Naylor, E., 1991. Field studies on retention of the planktonic copepod Eurytemora affinis in a mixed estuary. Mar. Ecol. Prog. Ser. 76, 115e122.

80. Howlett, E.R., Bowers, D.G., Malarkey, J., Jago, C.F., 2015. Stratification in the presenceof an axial convergent front: causes and implications. Estuar. Coast. ShelfSci. 161, 1e10.

81. Howard, T., Lowe, J., Horsburgh, K., 2010. Interpreting century-scale changes in North Sea storm surge climate derived from coupled model simulations. J. Clim.23, 6234e6247.

82. Howsen, C., 2009. Intertidal SAC Monitoring, Camarthen Bay and Estuaries SAC. CCW Marine Monitoring Report no 79. Countryside Council for Wales, Bangor.

83. Huang, G., Zhou, J., Lin, B., Chen, Q., Falconer, R., 2013. Distributed numerical hydrological and hydrodynamic modelling for large river catchment. In: Proceedings of the 35th IAHR World Congress, pp. 2420e2431.

84. IPCC, 2007. Synthesis Report: Contribution of Working Groups I, II and III to the Fourth Assessment Report of the Intergovernmental Panel on Climate Change. IPCC, Geneva, 104pp.

85. IPCC, 2013. Summary for policymakers. In: Stocker, T.F., Qin, D., Plattner, G.-K., Tignor, M., Allen, S.K., Boschung, J., Nauels, A., Xia, Y., Bex, V., Midgley, P.M. (Eds.), Climate Change 2013: the Physical Science Basis. Contribution of Working Group I to the Fifth Assessment Report of the Intergovernmental Panel on Climate Change. Cambridge University Press, United Kingdom and New York, NY, USA.

86. IPCC, 2014. In: Field, C.B., Barros, V.R., Dokken, D.J., Mach, K.J., Mastrandrea, M.D.,Bilir, T.E., Chatterjee, M., Ebi, K.L., Estrada, Y.O., Genova, R.C., Girma, B.,Kissel, E.S., Levy, A.N., MacCracken, S., Mastrandrea, P.R., White, L.L. (Eds.),Climate Change 2014: Impacts, Adaptation, and Vulnerability. Part A: Global and Sectoral Aspects. Contribution of Working Group II to the Fifth Assessment Report of the Intergovernmental Panel on Climate Change. Cambridge University Press, Cambridge, United Kingdom and New York, NY, USA, 1132 pp.

87. Ittekkot, V., 1988. Global trends in the nature of organic matter in river suspensions. Nature 332, 436e438.

88. Jackson, A.C., McIlvenny, J., 2011. Coastal squeeze on rocky shores in northern Scotland and some possible ecological impacts. J. Exp. Mar. Biol. Ecol. 400,314e321.

89. Jackson, C.R., Meister, R., Prudhomme, C., 2011. Modelling the effects of climate change and its uncertainty on UK Chalk groundwater resources from an ensemble of global climate model projections. J. Hydrol. 399, 12e28.

90. Jarvie, H.P., Jickells, T.D., Skeffington, R.A., Withers, P.J.A., 2012. Climate change and coupling of macronutrient cycles along the atmospheric, terrestrial, freshwater and estuarine continuum. Sci. Total Environ. 434, 252e258.

91. Jenkins, G.J., Perry, M.C., Prior, M.J.O., 2008. The Climate of the United Kingdom and Recent Trends. Met Office Hadley Centre, Exeter, UK. Jenkins, G., Murphy, J., Sexton, D., Lowe, J., 2009. UK Climate Projections: Briefing

92. Report. Met Office Hadley Centre, Exeter. http://ukclimateprojections.defra.gov. uk/.

93. Jenkins, G.P., Spooner, D., Conron, S., Morrongiello, J.R., 2015. Differing importance of salinity stratification and freshwater flow for the recruitment of apex species of estuarine fish. Mar. Ecol. Prog. Ser. 523, 125e144.

94. Jensen, N., Allen, R.M., Marshall, D.J., 2014. Adaptive maternal and paternal effects: gamete plasticity in response to parental stress. Funct. Ecol. 28, 724e733.

95. Jevrejeva, S., Grinsted, A., Moore, J.C., 2014. Upper limit for sea level projections by 2100. Environ. Res. Lett. 9, 104008.

96. Jickells, T., Andrews, J., Samways, G., Sanders, R., Malcolm, S., Sivyer, D., Parker, R., Nedwell, D., Trimmer, M., Ridgway, J., 2000. Nutrient fluxes through the Humber estuary-Past, present and future. AMBIO A J. Hum. Environ. 29, 130e135.

97. Jones, D., Bain, V., Dawson, S., Watt, T., 2011. Assessing the Vulnerability of Marine Biodiversity in Wales to the Impacts of Climate Change. Workshop Report. CCW contract Science Report no 934. Countryside Council for Wales, Bangor. Jones, M.R., Fowler, H.J., Kilsby, C.G., Blenkinsop, S., 2013. An assessment of changes in seasonal and annual extreme rainfall in the UK between 1961 and 2009. Int. J. Climatol. 33, 1178e1194.

98. Kaiser, M.J., Attrill, M.J., Jennings, S., Thomas, D.N., Barnes, D.K.A., Brierley, A.S., Hiddink, J.G., Kaartokallio, H., Polunin, N.V.C., Raffaelli, D.G., 2005. MarineEcology, Processes, Systems and Impacts. Oxford University Press, Oxford.

99. Kay, A.L., Davies, H.N., Bell, V.A., Jones, R.G., 2009. Comparison of uncertainty sources for climate change

impacts: flood frequency in England. Clim. Change 92, 41e63.

100. Kay, A.L., Bell, V.A., Blyth, E.M., Crooks, S.M., 2013a. Davies HN andReynard NS Ahydrological perspective on eva-poration: historical trends and future projections in Brit-ain. J. Water Clim Change 4 (3), 193e208.

101. Kay, A.L., Crooks, S.M., Davies, H.N., 2013b. Prudhomme C andReynard NS Probabilistic impacts of climatechange on flood frequency using response surfaces. I:England and Wales. Reg. Environ. Change 14 (3), 1215e1227.

102. Keef, C., Svensson, C., Tawn, J.A., 2009. Spatial dependence in extreme river flowsand precipitation for Great Britain. J. Hydrol. 378, 240e252.

103. Kennish, M.J., 2002. Environmental threats and environmental future of estuaries. Environ. Conserv. 29, 78e107.

104. Kimmerer, W.J., 2002. Effects of freshwater flow on abundance of estuarine organisms:physical effects or trophic linkages? Mar. Ecol. Prog. Ser. 243, 39e55.King, D.A., 2004. Climate change science: adapt, mitigate, or ignore? Science 303,176e177.

105. Kingsford, M.J., Leis, J.M., Shanks, A., Lindeman, K.C., Morgan, S.G., Pineda, J., 2002.Sensory environments, larval abilities and local self-recruitment. Bull. Mar. Sci. 70, 309e340.

106. Kirby, R., 2010. Distribution, transport and exchanges of fine sediment, with tidalpower implications: Severn Estuary, UK. Mar. Pollut. Bull. 61, 21e36.

107. Knight, P., Prime, T., Brown, J.M., Morrissey, K., Plater, A.J., 2015. Application of flood risk modelling in a web-based geospatial decision support tool for coastal adaptation to climate change. Nat. Hazards Earth Syst. Sci. Discuss. 3, 1615e1642.

108. Lanzoni, S., Seminara, G., 1998. On tide propagation in convergent estuaries. J. Geophys. Res. 103, 30793e30812.

109. Lawrence, A.J., Soame, J.M., 2004. The effects of climate change on the reproduction of coastal invertebrates. Ibis 146, 29e39.

110. Lee, S.V., Cundy, A.B., 2001. Heavy metal contamination and mixing processes in sediments from the Humber Estuary, Eastern England. Estuar. Coast. Shelf Sci.53, 619e636.

111. Lewis, M.J., Horsburgh, K., Bates, P., Smith, R., 2011. Quantifying the uncertainty in future coastal flood risk estimates for the UK. J. Coast. Res. 27, 870e881.

112. Lewis, M.J., Bates, P., Horsburgh, K., Neal, J., Schumann, G., 2013. A storm surge inundation model of the northern Bay of Bengal using publicly available data.Q. J. R. Meteorol. Soc. 139, 358e369.

113. Lewis, M.J., Neill, S.P., Hashemi, M.R., Reza, M., 2014. Realistic wave conditions andtheir influence on quantifying the tidal stream energy resource. Appl. Energy 136, 495e508.

114. Lipp, E.K., Kurz, R., Vincent, R., Rodriguez-Palacios, C., Farrah, S.R., Rose, J.B., 2001. The effects of seasonal variability and weather on microbial fecal pollution andenteric pathogens in a subtropical estuary. Estuaries 24, 266e276.

115. Lipp, E.K., Huq, A., Colwell, R., 2002. Effects of global climate on infectious disease: the cholera model. Clin. Microbiol. Rev. 15, 757e770.

116. Lowe, J.A., Howard, T.P., Padaens, A., Tinker, J., Holt, J., Wakelin, S., Milne, G., Leake, J., Wolf, J., Horsburgh, K., 2009. UK Climate Change Projections Science Report:Marine and Coastal Projections. Met Office, Hadley Centre, Exeter.

117. MacCready, P., 1998. Estuarine adjustment to changes in river flow and tidal mixing.J. Phys. Oceanogr. 29, 708e726.

118. Malham, S.K., Rajko-Nenow, P., Howlett, E., Tuson, K.E., Perkins, T.L., Pallett, D.W.,Wang, H., Jago, C.F., Jones, D.L., McDonald, J.E., 2014. The interaction of human microbial pathogens, particulate material and nutrients in estuarine environments and their impacts on recreational and shellfish waters. Environ. Sci.Process. Impacts 16, 2145.

119. Marani, M., D'Alpaos, A., Lanzoni, S., Carniello, L., Rinaldo, A., 2007. Biologicallycontrolledmultiple equilibria of tidal landforms and the fate of the Venice lagoon. Geophys. Res. Lett. 34, L1140.

120. Marba, N., Duarte, C.M., 2010. Mediterranean warming triggers seagrass (Posidonia oceanica) shoot mortality. Glob. Change Biol. 16, 2366e2375.

121. Marques, A., Nunes, M.L., Moore, S.K., Strom, M.S., 2010. Climate change and seafoodsafety: human health implications. Food Res. Int. 43, 1766e1779.

122. Maskell, J., Horsburgh, K., Lewis, M., Bates, P., 2013. Investigating river-surgeinteraction in idealised estuaries. J. Coast. Res. 30, 248e259.

123. Massa, S.I., Arnaud-Haond, S., Pearson, G.A., Serrao, E.A., 2009. Temperature toleranceand survival of intertidal populations of the seagrass Zostera noltii (Hornemann)in Southern Europe (Ria Formosa, Portugal). Hydrobiologia 619, 195e201.

124. McCabe, G.J., Clark, M.P., Serreze, M.C., 2001. Trends in Northern Hemisphere surface cyclone frequency and intensity. J. Clim. 14, 2763e2768.

125. McCann, D.L., Davies, A.G., Bennell, J.D., 2011. Bed roughness feedback in TELEMAC- 2D and SISYPHE. In: Proceedings of the XVIII Telemac and Mascaret User Club,pp. 99e104.

126. McDowell, W.G., Benson, A.J., Byers, J.E., 2014. Climate controls the distribution of a widespread invasive species: implications for future range expansion. Freshw. Biol. 59, 847e857.

127. Mcinnes, K.L., Walsh, K.J.E., Hubbert, G.D., Beer, T., 2003. Impact of sea-level rise and storm surges on a coastal community. Nat. Hazards 30, 187e207.

128. McLusky, D.S., Elliott, M., 2004. The Estuarine Ecosystem; Ecology, Threats and Management, third ed. OUP, Oxford. pp. 216.

129. McMichael, A.J., Woodruff, R.E., Hales, S., 2006. Climate change and human health: present and future

risks. Lancet 367, 859e869.

130. Menendez, M., Woodworth, P.L., 2010. Changes in extreme high water levels basedon a quasi-global tide-gauge dataset. J. Geophys. Res. 115, C10011.
131. Mieszkowska, N., Firth, L., Bentley, M., 2013. Impacts of climate change on intertidalhabitats. Mar. Clim. Change Impacts Partnersh. Sci. Rev. 2013, 180e190.
132. Moksnes, P.-O., Tryman, H.C.K., Hordoir, R., Jonsson, P.R., 2014. Larval behavior anddispersal mechanisms in shore crab larvae (Carcinus maenas): local adaptationsto different tidal environments? Limnol. Oceanogr. 59, 588e602.
133. Moore, R.D., Wolf, J., Souza, A.J., Flint, S.S., 2009. Morphological evolution of the DeeEstuary, Eastern Irish Sea, UK: a tidal asymmetry approach. Geomorphology 103, 588e596.
134. Morgan, B., Rate, A.W., Burton, E.D., 2012. Water chemistry and nutrient releaseduring the resuspension of FeS-rich sediments in a eutrophic estuarine system. Sci. Total Environ. 432, 47e56.
135. Najjar, R.G., Pyke, C.R., Adams, M.B., Breitburg, D., Hershner, C., Kemp, M.,Howarth, R., Mulholland, M.R., Paolisso, M., Secor, D., Sellner, K., Wardrop, D.,Wood, R., 2010. Potential climate-change impacts on the Chesapeake Bay.Estuar. Coast. Shelf Sci. 86, 1e20.
136. Neill, S.P., 2009. A numerical study of lateral grain size sorting by an estuarine front. Estuar. Coast. Shelf Sci. 81, 345e352.
137. Nedwell, D.B., Jickells, T.D., Trimmer, M., Sanders, R., 1999. Nutrients in Estuaries,Advances in Ecological Research, 08, pp. 60191e60199. http://dx.doi.org/10.1016/S0065-2504.
138. Nicholls, R.J., Hoozemans, F.M.J., Marchand, M., 1999. Increasing flood risk andwetland losses due to global sea-level rise: regional and global analyses. Glob. Environ. Change Human Policy Dimens. 9, S69eS87.
139. Nicholls, R.J., Marinova, N., Lowe, J.A., Brown, S., Vellinga, P., de Gusmao, D., ~ Hinkel, J., 2011. Sea-level rise and its possible impacts given a 'beyond 4 degrees C world' in the twenty-first century. Philos. Trans. R. Soc. A 369, 161e181.
140. Nicholls, R.J., Townend, I.H., Bradbury, A.P., Ramsbottom, D., Day, S.A., 2013. Planning for long-term coastal change: experiences from England and Wales. Ocean Eng. 71, 3e16.
141. O'Donnell, J., 1993. Surface fronts in estuaries: a review. Estuaries 16, 12e39.Oliphant, A., Thatje, S., 2014. Energetic adaptations to larval export within the brackish-water palaemonine shrimp, Palaemonetes varians. Mar. Ecol. Prog. Ser. 505, 177e191.
142. Oliphant, A., Thatje, S., 2013. Per offspring investment implications for crustaceanlarval development: evolutionary insights into endotrophy and abbreviateddevelopment. Mar. Ecol. Prog. Ser. 493, 207e217.
143. Oliphant, A., Ichino, M., Thatje, S., 2014. The influence of per offspring investment(POI) and starvation on larval developmental plasticity within the palaemonid shrimp, Palaemonetes varians. Mar. Biol. 161, 2069e2077.
144. Orford, J.D., Pethick, J.S., 2006. Challenging assumptions of coastal habitat formationin the 21st century. Earth Surf. Process. Landf. 31, 1625e1642.
145. Paciorek, C.J., Risbey, J.S., Ventura, V., Rosen, R.D., 2002. Multiple indices of NorthernHemisphere cyclone activity, winters 1949-99. J. Clim. 15, 1573e1590.
146. Perciasepe, R., 1998. EPA: ocean protection challenges e 'Daunting but manageable'.Sea Technol. 39, 25e29.
147. Pethick, J.S., 1994. Estuaries and wetlands: function and form. In: Falconer, R.A.,Goodwin, P. (Eds.), Wetland Management. Thomas Telford, London, pp. 75e87.
148. Pethick, J.S., Crooks, S., 2000. Development of a coastal vulnerability index: a geomorphological perspective. Environ. Conserv. 27, 359e367.
149. Philippart, C.J.M., van Aken, H.M., Beukema, J.J., Bos, O.G., Cadee, G.C., Dekker, R., 2003. Climate-related changes in recruitment of the bivalve Macoma balthica. Limnol. Oceanogr. 48, 2171e2185.
150. Power, M., Attrill, M.J., Thomas, R.M., 2000. Environmental factors and interactions affecting the temporal abundance of juvenile flatfish in the Thames Estuary.J. Sea Res. 43, 135e149.
151. Prandle, D., 2009. Estuaries: Dynamics, Mixing, Sedimentation, and Morphology.University Press, Cambridge.
152. Prandle, D., Lane, A., 2015. Sensitivity of estuaries to sea level rise: vulnerability indices. Estuar. Coast. Shelf Sci. 160, 60e68.
153. Prime, T., Brown, J.M., Plater, A.J., 2015. Physical and economic impacts of sea-level rise and low probability flooding events on coastal communities. PLoS One 10,e0117030.
154. Prudhomme, C., Davies, H., 2009a. Assessing uncertainties in climate change impactanalyses on the river flow regimes in the UK. Part 1: baseline climate. Clim. Change 93, 177e195.
155. Prudhomme, C., Davies, H., 2009b. Assessing uncertainties in climate change impact analyses on the river flow regimes in the UK. Part 2: future climate. Clim. Change 93, 197e222.
156. Prudhomme, C., Dadson, S., Morris, D., Williamson, J., Goodsell, G., Crooks, S., Boelee, L., Davies, H., Buys, G., Lafon, T., Watts, G., 2012. Future flows climate: an ensemble of 1-km climate change projections for hydrological application in Great Britain. Earth Syst. Sci. Data 4, 143e148.
157. Przeslawski, R., Byrne, M., Mellin, C., 2015. A review and meta-analysis of the effects of multiple abiotic stressors on marine embryos and larvae. Glob. Change Biol. 21, 2122e2140.
158. Pye, K., Blott, S.J., 2006. Coastal processes and morphological change in the Dunwich-Sizewell area, Suffolk, UK. J. Coast. Res. 22, 453e473.
159. Pye, K., Blott, S.J., 2014. The geomorphology of UK estuaries: the role of geologicalcontrols, antecedent

conditions and human activities. Estuar. Coast. Shelf Sci. 150, 196e214.

160. Quinn, N.D., Bates, P.D., Siddall, M., 2013. T he contribution to future flood risk in theSevern Estuary from extreme sea level rise due to ice sheet mass loss. J. Geophys. Res. 118, 5887e5898.

161. Quinn, N.D., Lewis, M., Wadey, M., Haigh, I., 2014. Assessing the variability in extreme storm-tide time-series for coastal flood risk assessment. J. Geophys. Res. 119, 4983e4998.

162. Queiroga, H., Blanton, J., 2005. Interactions between behaviour and physical forcing in the control of horizontal transport of decapod crustacean Larvae. Adv. Mar. Biol. 47, 107e214.

163. Rabalais, N.N., Turner, R.E., Diaz, R.J., Justic, D., 2009. Global change and eutrophication of coastal waters. ICES J. Mar. Sci. 66, 1528e1537.

164. Ravaux, J., Leger, N., Rabet, N., Morini, M., Zbinden, M., Thatje, S., Shillito, B., 2012. Adaptation to thermally variable environments: capacity for acclimation ofthermal limit and heat shock response in the shrimp Palaemonetes varians. J. Comp. Physiol. B 182, 899e907.

165. Rey, F., Neto, G.M.S., Rosa, R., Queiroga, H., Calado, R., 2015. Laboratory trials reveal that exposure to extreme raining events prior to metamorphosis affect thepost-settlement performance of an estuarine crab. Estuar. Coast. Shelf Sci. 154, 179e183.

166. Robins, P.E., Davies, A.G., 2010. Morphological controls in sandy estuaries: the in- fluence of tidal flats and bathymetry on sediment transport. Ocean Dyn. 60, 503e517.

167. Robins, P.E., Neill, S.P., Gimenez, L., 2012. A numerical study of marine larval dispersal in the presence of an axial convergent front. Estuar. Coast. Shelf Sci. 100, 172e185.

168. Robins, P.E., Lewis, M.J., Simpson, J.H., Howlett, E.R., Malham, S.K., 2014. Future variability of solute transport in a macrotidal estuary. Estuar. Coast. Shelf Sci. 151, 88e99.

169. Rodríguez, R.A., Gundy, P.M., Rijal, G.K., Gerba, C.P., 2012. The impact of combined sewage overflows on the viral contamination of receiving waters. Food Environ. Virol. 4, 34e40.

170. Rose, J.B., Epstein, P.R., Lipp, E.K., Sherman, B.H., Bernard, S.M., Patz, J.A., 2001.Climate variability and change in the United States: potential impacts on waterand foodborne diseases caused by microbiologic agents. Environ. Health Perspect. 109, 211e221.

171. Rose, J.B., 2005. Identification and characterization of biological risks for water. In: Management of Intentional and Accidental Water Pollution. Springer, Netherlands, 2006. 85e101. Rossington, S.K., Nicholls, R.J., Knaapen, M.A.F., Wang, Z.B., 2007. Morphological behaviour of UK estuaries under conditions of accelerating sea-level rise. In:

172. River, Coastal and Estuarine Morphodynamics. Enchede, Netherlands.Samuels, P.G., Burt, N., 2002. A new joint probability appraisal of flood risk. Proc. Int. Civ. Eng. Water Mar. Eng. 154, 109e115.

173. Senior, C.A., Jones, R.G., Lowe, J.A., Durman, C.F., Hudson, D., 2002. Predictions ofextreme precipitation and sea-level rise under climate change. Philos. Trans. R.Soc. 360, 1301e1311.

174. Sharps, E., Smart, J., Skov, M.W., Garbutt, A., Hiddink, J.G., 2015. Light grazing of saltmarshes is a direct and indirect cause of nest failure in common redshank Tringa totanus. IBIS 157, 239e249.

175. Shennan, I., Horton, B., 2002. Holocene land and sea level changes in Great Britain.J. Quat. Sci. 17, 511e526.

176. Shennan, I., Milne, G., Bradley, S., 2012. Late Holocene vertical land motion and relative sea-level changes: lessons from the British Isles. J. Quat. Sci. 27, 64e70.

177. Short, F.T., Wyllie-Echeverria, S., 1996. Natural and human-induced disturbance ofseagrasses. Environ. Conserv. 23, 17e27.

178. Simpson, J.H., Vennell, R., Souza, A.J., 2001. The salt fluxes in a tidally energeticestuary. Estuar. Coast. Shelf Sci. 52, 131e142.

179. Smith, P.M., Dixon, R.O.D., Cochrane, M.P., 2002. Plant Life of Edinburgh and the Lothians. Edinburg University Press, Edinburgh.

180. Smith, A., Bates, P., Freer, J., Wetterhall, F., 2013. Future flood projection: investigating the application of climate models across the UK. Hydrol. Process 28,

181. 2810e2823. Sokolova, I.M., Portner, H.O., 2007. The role of thermal environment in stress adaptation and tolerance: integration of multiple stressors. J. Therm. Biol. 32,117.

182. Stachel, B., Gotz, R., Herrmann, T., Krüger, F., Knoth, W., P € apke, O., Rauhut, U., € Reincke, H., Schwartz, R., Steeg, E., Uhlig, S., 2004. The Elbe flood in August 2002-occurrence of polychlorinated dibenzo-p-dioxins, polychlorinated dibenzofurans (PCDD/F) and dioxin-like PCB in suspended particulate matter (SPM), sediment and fish. Water Sci. Technol. 50, 309e316.

183. Statham, P.J., 2012. Nutrients in estuaries e an overview and the potential impacts of climate change. Sci. Total Environ. 434, 213e227.

184. Struyf, E., Damme, S.V., Meire, P., 2004. Possible effects of climate change on estuarine nutrient fluxes: a case study in the highly nitrified Schelde estuary (Belgium, The Netherlands). Estuar. Coast. Shelf Sci. 52, 131e142.

185. Svensson, C., Jones, D.A., 2002. Dependence between extreme sea surge, river flowand precipitation in eastern Britain. Int. J. Climatol. 22, 1149e1168.

186. Tappin, A.D., 2002. An examination of the fluxes of nitrogen and phosphorus in temperate and tropical estuaries: current estimates and uncertainties. Estuar. Coast. Shelf Sci. 55, 885e901.

187. Tessier, B., Billeaud, I., Sorrel, P., Delsinne, N., Lesueur, P., 2012. Infilling stratigraphy of macrotidal tide-dominated estuaries. Controlling mechanisms: sea-level fluctuations, bedrock morphology, sediment supply and climate changes (theexamples of the Seine estuary and the Mont-Saint-Michel Bay, English Channel,NW

France). Sediment. Geol. 279, 62e73.

188. Thompson, R.C., Crowe, T.P., Hawkins, S.J., 2002. Rocky intertidal communities: past environmental changes, present status and predictions for the next 25 years.Environ. Conserv. 29, 168e191.

189. Torres, G., Gimenez, L., Anger, K., 2011. Growth, tolerance to low salinity, and osmoregulation in decapod crustacean larvae. Aquat. Biol. 12, 249e260.

190. Uncles, R.J., 2003. From catchment to coastal zone: examples of the application of models to some long-term problems. Sci. Total Environ. 314, 567e588.

191. Uncles, R.J., Stephens, J.A., Harris, C., 2014. Freshwater, tidal and wave influences on a small estuary. Estuar. Coast. Shelf Sci. 150, 252e261.

192. Valdemarsen, T., Quintana, C.O., Kristensen, E., Flindt, M.R., 2014 Jan 1. Recovery of organic-enriched sediments through microbial degradation: implications for eutrophic estuaries. Mar. Ecol. Prog. Ser. 503, 41e58.

193. Vaquer-Sunyer, R., Duarte, C.M., 2011. Temperature effects on oxygen thresholds for hypoxia in marine benthic organisms. Glob. Change Biol. 17, 1788e1797.

194. Wadey, M.P., Haigh, I.D., Brown, J.M., 2014. A century of sea level data and the UK's 2013/14 storm surges: an assessment of extremes and clustering using the Newlyn tide gauge record. Ocean Sci. Discuss. 11, 1995e2028.

195. Wakelin, S.L., Woodworth, P.L., Flather, R.A., Williams, J.A., 2003. Sea level dependenceon the NAO over the NW European Continental Shelf. Geophys. Res. Lett. 30 http://dx.doi.org/10.1029/2003GL017041.

196. Walther, G.R., Post, E., Convey, P., Menzel, A., Parmesan, C., Beebee, T.J., Fromentin, J.M., Hoegh-Guldberg, O., Bairlein, F., 2002 Mar 28. Ecological responses to recent climate change. Nature 416 (6879), 389e395.

197. Watts, G., Battarbee, R., Bloomfield, J.P., Crossman, J., Daccache, A., Durance, I., Elliott, J.A., Garner, G., Hannaford, J., Hannah, D., Hess, T., Jackson, C., Kay, A., Kernan, M., Knox, J., Mackay, J., Monteith, D., Ormerod, S., Rance, J., Stuart, M., Wade, A., Wade, S., Weatherhead, K., Whitehead, P., Wilby, R., 2015. Climate change and water in the UK e past changes and future prospects. Prog. Phys. Geogr. 39, 6e28.

198. Whitehead, P.G., Wilby, R.L., Battarbee, R.W., Kernan, M., Wade, A.J., 2009. A review of the potential impacts of climate change on surface water quality. Hydrol. Sci. J. 54, 101e123.

199. Winters, G., Nelle, P., Fricke, B., Rauch, G., Reusch, T.B., 2011. Effects of a simulated heat wave on photophysiology and gene expression of high- and low-latitude populations of Zostera marina. Mar. Ecol. Prog. Ser. 435, 83e95.

200. Wood, R.A., Vellinga, M., Thorpe, R., 2003. Global warming and thermohaline circulationstability. Philos. Trans. R. Soc. Lond. A Math. Phys. Eng. Sci. 361, 1961e1975.

201. Woodworth, P.L., Blackman, D.L., 2004. Evidence for systematic changes in extreme high water since the mid-1970s. J. Clim. 17, 1190e1197.

202. Woodworth, P.L., Teferle, F.N., Bingley, R.M., Shennan, I., Williams, S.D., 2009. Trends in UK mean sea level revisited. Geophys. J. Int. 176, 19e30.

203. Worth, K., Weisse, R., von Storch, H.,2006. Climate change and North Sea storm surge extremes: an ensemble study of storm surge extremes expected in a changed climate projected by four different regional climate models. Ocean Dyn. 56, 3e15.

204. Yamanaka, T., Raffaelli, D., White, P.C., 2013. Non-linear interactions determine theimpact of sea-level rise on estuarine benthic biodiversity and ecosystem processes. PLoS One 8, e68160.

205. Yang, Z.Q., Wang, T.P., Voisin, N., Copping, A., 2015. Estuarine response to river flow and sea-level rise under future climate change and human development. Estuar. Coast. Shelf Sci. 156, 19e30.

206. Yates, C.J., Elith, J., Latimer, A.M., Le Maitre, D., Midgley, G.F., Schurr, F.M., West, A.G., 2010. Projecting climate change impacts on species distributions in megadiverse South African Cape and Southwest Australian Floristic Regions: opportunities and challenges. Austral Ecol. 35, 374e391.

207. Ysebaert, T., Herman, P.M.J., Meire, P., Craeymeersch, J., Verbeek, H., Heip, C.H.R., 2003. Large-scale spatial patterns in estuaries: estuarine macrobenthic communities in the Scheldt estuary, NW Europe. Estuar. Coast. Shelf Sci. 57, 335e355.

208. Zhang, K.Q., Douglas, B.C., Leatherman, S.P., 2004. Global warming and coastal erosion. Clim. Change 64, 41e58.

209. Zheng, F., Westra, S., Sisson, S.A., 2013. Quantifying the dependence between extreme rainfall and storm surge in the coastal zone. J. Hydrol. 505, 172e187.Zheng, F., Westra, S., Leonard, M., Sisson, S.A., 2014. Modeling dependence between extreme rainfall and storm surge to estimate coastal flooding risk. Water Resour. Res. 50, 2050e2071.

CHAPTER 15

APPLICATION OF A NEW METHODOLOGY FOR COASTAL MULTI-HAZARD-ASSESSMENT & MANAGEMENT ON THE STATE OF KARNATAKA, INDIA

Lars Rosendahl Appelquist[a], Thomas Balstrøm[b]

[a] UNEP DTU Partnership, Technical University of Denmark, UN City Copenhagen, Marmorvej 51, 2100 Copenhagen Ø, Denmark

[b] Department of Development and Planning, Aalborg University, A. C. Meyers Vænge 15, 2., 2450 Copenhagen SV, Denmark

ABSTRACT

This paper presents the application of a new methodology for coastal multi-hazard assessment & management under a changing global climate on the state of Karnataka, India. The recently published methodology termed the Coastal Hazard Wheel (CHW) is designed for local, regional and national hazard screening in areas with limited data availability, and covers the hazards of ecosystem disruption, gradual inundation, salt water intrusion, erosion and flooding. The application makes use of published geophysical data and remote sensing information and is showcasing how the CHW framework can be applied at a scale relevant for regional planning purposes. It uses a GIS approach to develop regional and sub-regional hazard maps as well as to produce relevant hazard risk data, and includes a discussion of uncertainties, limitations and management perspectives. The hazard assessment shows that 61 percent of Karnataka›s coastline has a high or very high inherent hazard of erosion, making erosion the most prevalent coastal hazard. The hazards of flooding and salt water intrusion are also relatively widespread as 39 percent of Karnataka›s coastline has a high or very high inherent hazard for both of these hazard types.

INTRODUCTION

The projected climate change will place significant stress on coastal regions worldwide and constitutes a particular challenge for developing countries where coastal development often happens rapidly and without prior investigation of natural dynamics. Improving the knowledge of the physical characteristics of coastal areas as well as their inherent natural hazards is therefore an important prerequisite for sustainable and safe coastal development. This paper tests the practical application of the CHW framework (Rosendahl Appelquist, 2012) through a multi-hazard assessment of the coastline of Karnataka, India, under a changing global climate (IPCC, 2007). The goal of the paper is both to showcase a practical procedure for applying the CHW framework for regional hazard assessments, and to develop hazard maps and hazard risk data for the hazards of ecosystem disruption, gradual inundation, salt water intrusion, erosion and flooding for the state of Karnataka, India.

As the CHW framework was published in late 2012, the hazard assessment for Karnataka is intended to test its practical applicability on a diverse and largely sedimentary coastline. Whereas most existing assessment systems are designed for areas with relatively good data availability (Thieler et al., 2000 and Ramieri et al., 2011), the CHW framework is

Citation: Lars Rosendahl Appelquist, Thomas Balstrøm, Application of a new methodology for coastal multi-hazard-assessment & management on the state of Karnataka, India, doi:10.1016/j.jenvman.2014.12.017

developed to be used for hazard screening and assessment in areas with limited geophysical data collection systems. The state of Karnataka is therefore considered a good test case as coastal data for this region is relatively sparse but not completely absent.

The CHW framework is designed to be applied in a stepwise manner, depending on the appropriate scale and resolution of the hazard assessment. At Step 1, the framework can be applied for regional and national hazard screening, and in most cases, this can be carried out based on publicly available geophysical data and remote sensing information. For areas that are of particular interest or are indicated as hazard hotspots in the hazard screening, a more detailed assessment can be carried out as Step 2. In this step, it is recommended to supplement the data obtained in Step 1 with representative field verification. If local hazard information is needed, Step 3 can be carried out by supplementing data from step one and two with detailed local data collection. The user of the CHW framework can choose only to carry out the step relevant for their specific needs, but should be aware of the appropriate data requirements for each assessment step. The step-wise approach means that data collection can be adjusted according to the scale and resolution of the assessment and should therefore lead to an appropriate balance between data requirements and assessment detail. As this paper focuses on regional hazard screening, the assessment relies solely on published geophysical data and remote sensing information.

The regional hazard screening for Karnataka can be carried out based on relatively simple means and should therefore be replicable without major difficulties in other locations worldwide. The paper is written so it can function as a guided example for coastal planners and developers who are interested in producing hazard maps and hazard data using the CHW framework. The data used for the assessment is available at low/no cost from the internet or regional institutions and it is expected that the same will be the case for most other world regions. For the assessment, it was decided to acquire some supplementary RapidEye satellite images to cover a few low-resolution gaps in ESRI›s ArcGIS image series from ArcGIS Online and this added some extra costs to the assessment. It is expected, however, that the quality of satellite images available in ArcGIS and Google Earth will continue to improve and supplementary satellite images should therefore not be necessary for most locations in the near future.

Since the CHW framework is based on geo-biophysical properties of natural coastal systems, it gives information on the inherent hazards of the different coastal environments. Where human activities have altered a coastal area, the inherent hazards for that generic coastal system are likely to be affected. The CHW framework is able to take most human alterations into account such as changes in sediment supply from river damming and changes in wave climate due to harbour construction. However, if the human activities alter a coastline to a level outside its natural occurrences such as by completely removing a mangrove forest in an otherwise natural mangrove area or constructing a large dike in a coastal plain, the framework is unable to take these changes into account. With the data sources used for this hazard screening, it can in many cases be quite difficult to capture smaller human alternations of the natural coastline, but as a Step 1 assessment, these alterations should not have a great impact on the general hazard profile of the coastline. For more detailed hazard assessments, however, human activities such as sand mining could have a significant impact at a local level and appropriate field verification is therefore recommended if Step 2 or 3 should be implemented.

THE COASTAL HAZARD WHEEL FRAMEWORK

The CHW framework is developed as a screening and assessment tool to assist coastal planners and decision-makers in determining the hazard profile of a particular coastal area under a changing global climate. This could be relevant for regional infrastructure planning, expansion of residential areas and protection of sensitive natural sites, as well as for determining hazard mitigation strategies for coastal stretches. The CHW framework is based on a specially designed coastal classification system that contains 113 generic coastal environments. The system incorporates the main geo-biophysical parameters determining the characteristics of coastal systems and aims to cover all coastal areas worldwide. It uses the coastal geological layout as a basis on which it adds the main dynamic parameters and processes acting in the coastal environment.

The framework provides information on the degree to which key climate-related hazards are inherent in a particular coastal environment, defined as the hazards being an integral part of the geo-biophysical properties of a coastal system when exposed to future climate change. The framework covers the inherent hazards of ecosystem disruption, gradual inundation, salt water intrusion, erosion and flooding, and a total of 565 generic hazard evaluations are included in the system, each graduated into four different hazard levels based on a scientific literature review. The framework is generally designed to be applied in locations with limited data availability and computing capacity.

The CHW framework is provided as a graphical tool – the Coastal Hazard Wheel – to facilitate its application for planning purposes. The user starts in the centre of the CHW and then moves outwards, ending with the inherent hazard evaluations in the outermost circles. Starting from the centre, the coastal classification parameters comes in the following order where each category is represented by a new circle: Geological layout, wave exposure, tidal range, flora/fauna, sediment balance and storm climate. The inherent hazards then come in the following order: Ecosystem disruption, gradual inundation, salt water intrusion, erosion and flooding. In the practical application of the assessment framework, the user should make a new assessment every time any of the classification parameters changes significantly. This can be done by visually assessing the coastal appearance either in the field or through remote sensing, combined with evaluating data for the individual dynamic parameters. When conducting the assessment, the user should be aware of human alterations of the natural environment and whether these alterations are of permanent character, as this would have an impact on the coastal classification and hazard levels. The CHW is shown in Fig. 1 and a detailed description of the assessment methodology, assumptions and limitations can be found in Rosendahl Appelquist (2012).

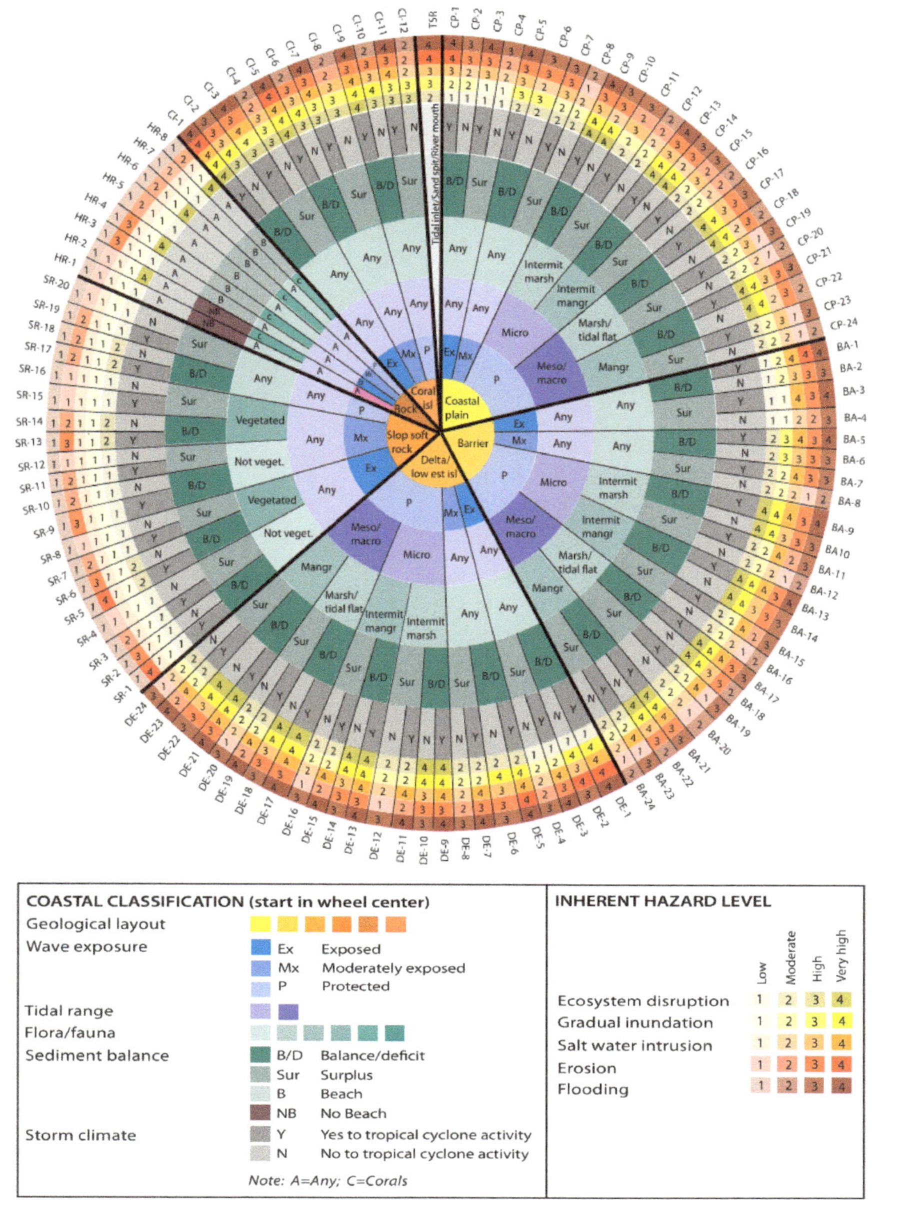

Figure1: The Coastal Hazard Wheel consisting of six geo-biophysical classification circles, five hazard circles and the coastal classification codes. In the classification code, CP stands for coastal plain, BA for barrier, DE for delta, SR for sloping soft rock, HR for sloping hard rock, CI for coral island and TSR for tidal inlet/sand spit/river mouth (Rosendahl Appelquist, 2012).

THE COASTLINE OF KARNATAKA

The state of Karnataka is bordering the Arabian Sea and has a tropical monsoon climate. The months from March to May constitute the hot season with the hottest temperatures occurring in May. The state receives heavy rainfall between June and September due to the SW monsoon and the average annual rainfall is close to 4000 mm of which about 80 percent is received during the SW monsoon season (Dwarakish et al., 2009 and Kumar et al., 2010). The heavy monsoon rainfall leads to increased river flows and sediment transport to the coastline (Jayappa et al., 2003). Winds are strong and mainly westerly or south-westerly during the SW monsoon months. In the remaining months, the wind generally blows from northern and eastern directions in the morning and from western and north-western directions in the evening (Jayappa et al., 2003). Deep-water waves approach the coast from south-western and north-western directions and the significant wave height, Hs, have been assessed to >3 m during the SW monsoon (Kumar et al., 2010). It has been observed that the long-shore currents are strongest and towards the south during the SW monsoon (Narayana et al., 2001).

The coastline of Karnataka can generally be divided into two main geomorphologic sections with somehow different characteristics. The northern part is composed of Precambrian crystalline gneiss, schist and granite rocks, fronted by a narrow coastal plain of alluvial or Tertiary deposits. In locations where the rock extends to the coastline, coastal cliffs and rocky shores are formed. The coastline displays characteristics of submergence with drowned river valleys, estuaries and many small inlets (Nayak and Hanamgond, 2010). The southern part of Karnataka has extensive straight beaches backed by estuaries with low estuarine islands and mangroves. Sand spits growing northwards often border the estuaries (Nayak and Hanamgond, 2010).

The northern part of Karnataka›s coastlines has a relatively low level of industrial development with small fishing villages located along the coast. However, due to a growing tourist industry, increased fishing intensity and industrial aquaculture, the coastal area is under growing pressure from human activities (Equations, 2000). The southern part of Karnataka›s coastline close to the city of Mangalore has been used for heavy industrial development for several decades. The transformation from traditional fishing and farming activities started with the construction of the New Mangalore Port in the 1970s and today, many large-scale industries including chemical and petroleum processing plants are located along this coastline. The port of Mangalore is India›s ninth largest harbour in terms of cargo handling and handles 75 percent of India›s coffee export (World Port Source, 2012). The entire coastline of Karnataka has been declared special tourism area for promotion of tourism (Equations, 2000).

The coastline of Karnataka generally faces severe erosion during the SW monsoon and accretion during the fair weather season (Jayappa et al., 2003). In southern Karnataka, research indicates that most of the sand lost during the SW monsoon is regained during the calmer months (Jayappa et al., 2003). However, some parts of Karnataka›s coastline show continuous and significant erosion (Dwarakish et al., 2009). In some locations, beach width has been reduced to zero due to reduction in sediment supply from human activities such as construction of breakwaters and seawalls and damming of rivers (Kumar and Jayappa, 2009).

Hard engineering structures including breakwaters, seawalls and revetments have been constructed along Karnataka›s coastline over the past decades with varying success. Soft measures such as beach nourishment have generally not been applied due to economic reasons (Jayappa et al., 2003) although nourishment has been carried out at Thannirbhavi in January 2000 (Kumar and Jayappa, 2009). Legal and illegal dredging and sand mining from beaches, estuaries and upstream rivers has resulted in sediment deficits in some locations (Jayappa et al., 2003) and a recent increase in sand mining has lead to accelerated erosion (Kumar and Jayappa, 2009).

DATA FOR THE HAZARD ASSESSMENT

The hazard assessment makes use of data that is available in the original CHW framework paper or that can be easily obtained from other sources. The only advanced tool used for the assessment is ESRSI's computer software, ArcGIS, which requires a license and some software-specific expertise. The complete list of data used for the assessment includes a geological map of Karnataka (Mundkur, 2010), the wave, tide and storm maps included in the original CHW framework paper and published by Masselink and Hughes (2003), supplementary information on local tidal range (Nayak and Hanamgond, 2010), the UNEP-WCMC World Atlas of Coral Reefs (Spalding et al., 2001), Google Earth satellite images with timeline and ground elevation functions (Google, 2012), Bing Maps available in ESRI's ArcGIS (ESRI, 2012 and Microsoft, 2012) and two sections of Rapideye satellite images covering some low resolution gaps in the Bing maps (GRAS, 2012). The following sections describe how each of the coastal classification circles of the CHW has been determined based on the available data, and a thorough description of the different CHW classification categories can be found inRosendahl Appelquist (2012).

Classification circle 1 – Geological Layout

The geological layout is determined based on an ordinary geological map and Google Earth's satellite images and ground elevation function. The geological layout type is found by combining information from these three data sources and a new evaluation is made every time any of the parameters i.e. geological base material, geomorphology and coastal slope changes significantly.

For Karnataka, the determination of geological base material is relatively straightforward as the coastline is mainly composed of laterites. However, as the assessment is carried out remotely, it is not possible to assess the compaction and cementation level of the laterites in the field. In the practical classification, sloping laterite coastlines have been grouped into the sloping soft rock coast category, while flat laterites have been grouped into one of the flat coastal categories. In cases where the sloping laterites are heavily cemented, this may lead to an overestimation of the hazard levels as the coast would otherwise fall into the sloping hard rock coast category. Additional field verification of the laterite cementation would therefore be appropriate for implementing Step 2 and 3. Another challenge to the categorisation of geological layout is that smaller hard rock headlands are not visible on the geological map of Karnataka although they are visible on Google Earth's satellite images. In most cases, however, it is sufficient to rely on Google Earth as these structures are relative easily identified.

The slope of the coastline is determined using Google Earth's ruler and ground elevation functions. The elevation function is based on a digital elevation model from NASA's Shuttle Radar Topography Mission and its altitude resolution varies by country. Large parts of USA currently have a resolution of 10 m but most other world regions including India have a lower resolution (Wikipedia, 2012). When the ground elevation is assessed, the smoothing of the contours by the elevation model can be easily noticed, which may lead to some errors in flat areas adjacent to elevated regions. However, as the coastal classification system only requires input on whether the coast is sloping more or less than 3–4%, 200 m inland of the MSL, the error is not expected to significantly affect the classification accuracy. Large sections of Karnataka's coastline are sloping to some degree and it is therefore necessary to be cautious when conducting the elevation assessment. The fact that several of the barriers along Karnataka's coastline have a slope of more than 3–4% also increases the need for a careful slope assessment.

The coastal morphology is determined based on a visual assessment of Google Earth's satellite images. Form elements presented as barriers, deltas, tidal inlets, sand spits and river mouths can be easily identified with a zoom level of 5–10 km. The remaining mainland coastline can be categorized based on geology and slope.

Classification circle 2 – the Wave Exposure

The wave climate is determined based on the wave maps in the original CHW paper (Rosendahl Appelquist, 2012 and Masselink and Hughes, 2003). Since Karnataka is located outside the areas with swell/monsoon wave climates, the level of wave exposure is dependent on the free fetch and wind speed. It was not possible to obtain detailed wind data for the region and it was therefore decided to rely solely on the free fetch to determine the exposure levels for this classification. As the wind is blowing from the open ocean during the SW monsoon season, the free fetch is likely to be an appropriate proxy for the possible wave heights. The assessment has used Google Earth to determine whether the free fetch for a given coastal stretch is less than 10 km, 10–100 km or above 100 km which are the defined boundaries for protected, moderately exposed and exposed coastlines in the CHW framework. Generally, the outer reaches of Karnataka›s coastline are directly exposed to the waves of the Arabian Sea and categorized as exposed while the coastlines of the inner estuaries are classified as protected. As the coastline varies between estuaries and open coast, the moderately exposed category has generally not been applied.

Classification circle 3 – the Tidal Range

The tidal range is determined based on the tidal range maps included in the original CHW paper (Rosendahl Appelquist, 2012 and Masselink and Hughes, 2003). However, as Karnataka is located close to the border between the micro- and meso-tidal types, supplementary data has been collected on the local tidal range. This data indicate that all of Karnataka generally stays within the micro-tidal category with tidal range increasing towards the northern part of the state (Nayak and Hanamgond, 2010). It was therefore decided to apply this category for the coastal classification. Meso-tidal conditions may be present inside some of the estuaries due to the local coastal configuration but because of the limited data availability it is difficult to verify. However, the tidal range in these locations is still expected to stay close to the border between micro- and meso-tide. The micro-tide category is therefore applied consistently to the full coastline of Karnataka.

Classification Circle 4 – the Flora/Fauna

The flora/fauna is determined based on a visual assessment of the coastline in Google Earth combined with information on its geographical location and global coral reef data. As Karnataka is situated in the tropical climate zone, flat protected coastlines such as coastal plains and barriers generally have some kind of mangrove vegetation in protected locations, but due to the relatively low tidal range, the mangrove areas are of intermittent character. Coral reefs are generally non-existent along Karnataka's coastline and it is uncertain whether past sporadic coral habitats still exist (Spalding et al., 2001). Therefore, the coral reef option has not been applied to any parts of the coastline.

Classification circle 5 – the Sediment Balance

The sediment balance evaluation uses remote sensing information from Google Earth's satellite images and timeline function to compare images of the coastline taken over the last decade. Generally, coastal stretches have been assumed to have a sediment balance/deficit unless it is very clear that they have a sediment surplus in order to avoid underestimating some of the hazard levels. For the exposed, littoral coastlines of Karnataka, it has to some degree been possible to get a reliable indication of the sediment balance using Google Earth's images from the last 5–10 years, as the changes in the vegetation line in most cases is clearly visible. For protected coastlines, however, it has been difficult to visually assess smaller temporal changes based on the satellite images and these coasts have therefore in many cases been placed in the balance/deficit category. In addition to the general challenge of estimating the sediment balance, Google Earth has some gaps in its timeline function meaning that some areas are only coved by one satellite image, making temporal assessments impossible. This is the case for the coastline at Kodi Bengare to Kemmannu; Kota; Marvanthe; and Ternamakki to Kasarkod and these coastlines have therefore been placed in the balance/deficit category. Sometimes only two images with a few years in between are available in Google Earth which also leads to uncertainty in the evaluations.

Classification circle 6 – the Storm Climate

The storm climate is determined based on the wave/storm maps included in the original CHW paper (Rosendahl Appelquist, 2012 and Masselink and Hughes, 2003). As Karnataka is indicated to be under tropical storm influence, the complete coastline is classified to be located in a tropical cyclone area.

The GIS Procedure

The coastal classification and hazard assessment procedure is carried out in ArcGIS based on a Hybrid Bing Map. As the resolution of the satellite images is generally better in Google Earth than in Bing Maps, is was considered to conduct the whole classification in Google Earth. However, due to the technical limitations of Google Earth, it was decided to conduct the classification in ArcGIS, using Google Earth as data source.

As a first stage a geodatabase is created in ArcGIS that will contain all coastal classification data as well as data on hazard levels. In order to have a relatively detailed and up-to-date digitized coastline of Karnataka which can be used for the coastal classification, a new line feature class is created in the geodatabase referencing the WGS1984 Web Mercator Auxiliary Sphere coordinate system. It should be noted that other coordinate systems may be more appropriate for other world locations. The line feature is then used for creating a digitized coastline of Karnataka by manually digitizing the coast at the approximate Mean Sea Level (MSL) with a zoom level of 2–4 km in the ArcGIS window. Because the satellite images are taken at different times during the tidal cycle, the line feature will most likely deviate from the actual MSL but this is considered of minor importance for the purpose of this assessment, as it only requires a relatively accurate and up-to-date coastline. The digitizing is carried out with an accuracy of about 5–10 m leaving gaps for river mouths and tidal inlets. Islands are digitized as separate units. This line feature then constitutes the foundation for the further coastal classification and the hazard maps.

To facilitate the assessment of the coastal slope and sediment balance, two supplementary line features are created in Google Earth. The line feature for facilitating the slope evaluation consists of a range of shore-parallel line sections that are drawn landwards of the coastline in all coastal areas with a slope greater than 3–4%. This enables the user to quickly determine whether a particular coastal area is sloping or not when carrying out the coastal classification. The slope of a particular coastal section is determined by manually placing the cursor over the first 200 m landwards of the coastline in Google Earth, taking note of elevation levels given in the button of the Google Earth window. This procedure is carried out for every approximately 100–200 m of coastline at a Google Earth zoom level of 2–4 km. The line feature for facilitating the sediment balance evaluation consists of a continuous line drawn on the approximate coastal vegetation line. When the coastal classification is carried out, the sediment balance can be assessed by comparing the satellite images taken at different times through Google Earth's timeline function, looking at how the coast has been developing compared to the digitized, most recent coastline. Since the satellite images are taken at different tide levels and time of the year, the beach width cannot be reliably used for determining the sediment balance, but the vegetation line is considered as a relatively good indicator for the general sediment balance.

The coastal classification based on the CHW is carried out on top of the digitized coastline by using a polygon feature created in the geodatabase with the same coordinate system as the line feature for the coastline. The polygons are used to split the original line feature into sections, each representing a different coastal environment defined in the CHW framework. The classification is done by manually drawing a separate polygon for each coastal classification category along the coastline, based on an evaluation of the classification parameters mentioned in the data section earlier. When drawing the polygons, it is important to enable a snapping environment to ensure that the polygons are snapped properly to each other. The name of the coastal environment in question is then typed into the attribute table for each polygon in the ID field. As the ID field in the attribute table only accepts numbers, the coastal environments in the CHW framework are assigned values between 1 and 113, with 1 given to the CHW type CP-1. Because the classification of each coastal stretch is carried out based on the CHW and the listed input data, the user has to decide on an appropriate coastal type and its extension before each polygon is completed. Sometimes a coastline can maintain the same properties for longer distances, meaning that the length of a polygon can range from less than fifty meters to several kilometres.

The polygons are subsequently used to divide the initial digitized coastline into sections, each representing a specific coastal category. The hazard levels given in the CHW and further described in the original CHW paper (Rosendahl Appelquist, 2012) are then typed into a separate attribute table that is joined to the attribute table of the coastal classification file. Based on this, five different hazard maps are created for the respective hazards types and the different hazard levels are assigned a colour code. Finally, a background land polygon and a text layer with city names are created to improve the readability of the hazard maps and the relevant hazard statistics is extracted from the GIS.

RESULTS

The results from the application of the CHW framework on the coastline of Karnataka are an overview table of the most common coastal types in Karnataka, an overview table of the prevalence of the different coastal hazards and a range of sub-regional and regional hazard maps. Table 1 shows the top 10 most common costal types in Karnataka in distance as well as in percentage of the total coastline. In this assessment, the total length of Karnataka's coastline has been calculated to 647 km which is significant more than many estimates given in the literature. This is the case as the coastline in the assessment includes the open ocean coastline as well as back-barriers, estuaries and islands. From the table, it can be seen that the 10 most common coastal types make up over 90 percent of Karnataka's coastline. The most common types are the sloping soft rock coasts, SR-5 and SR-17, followed by the sloping hard rock coast HR-1. Special coastal elements such as tidal inlets, sand spits and river mouths are also relatively common, making up 13 percent of the total coastline. The flat coastal environments, coastal plain CP-13, delta DE-13 and barrier BA-13 are also quite widespread making up 9 percent, 8 percent and 3 percent respectively.

Table 1: The top 10 most common coastal types in Karnataka.

Coastal type	Length (km)	Percent of coastline
Sloping soft rock 5 (SR-5)	146	23
Sloping soft rock 17 (SR-17)	118	18
Hard rock 1 (HR-1)	100	16
Tidal inlet/Sand spit/River mouth (TSR)	84	13
Coastal plain 13 (CP-13)	58	9
Delta 13 (DE-13)	49	8
Barrier 13 (BA-13)	16	3
Coastal plain 1 (CP-1)	14	2
Delta 15 (DE-15)	13	2
Barrier 1 (BA-1)	12	2
	609	94

The hazard profile of the coastline of Karnataka is shown in Table 2. The table shows the distribution of the different hazards and hazard levels as a percentage of the total coastline length. From the table it can be seen that erosion constitutes the most prevalent hazard type as 61 percent of Karnataka's coastline has a high or very high inherent hazard for erosion. The hazards of flooding and salt water intrusion are also relatively widespread as 39 percent of the coastline has a high or very high inherent hazard for both of these hazard types. 32 percent of the coastline has a high or very high inherent hazard of gradual inundation while 19 percent has a high or very high inherent hazard of ecosystem disruption.

Table 2: The distribution of hazard levels in percent for Karnataka's coastline

Hazards/Hazard level	Low	Moderate	High	Very high
Ecosystem disruption	24	56	0	19
Gradual inundation	61	6	13	19
Salt water intrusion	61	0	25	14
Erosion	16	24	21	40
Flooding	61	0	0	39

Fig. 2 shows the hazards of erosion and flooding for northern Karnataka and is an example of how the CHW framework can be used for sub-regional hazard mapping. The hazard class 1 is low inherent hazard, 2 is moderate inherent hazard, 3 is high inherent hazard and 4 is very high inherent hazard. The maps give a relatively good overview of areas that requires special attention and can provide a basis for sub-regional planning and management decisions.

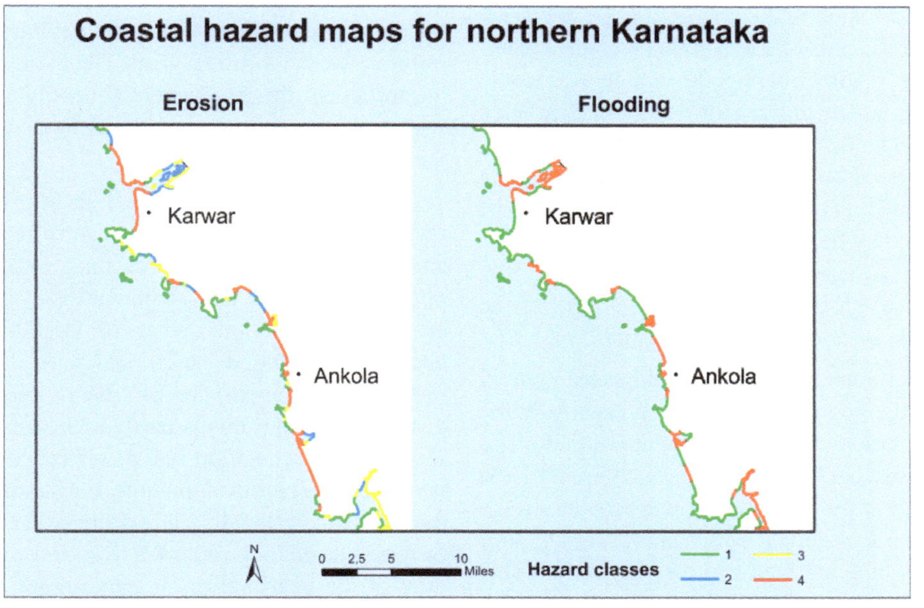

Figure 2: Coastal hazard maps for northern Karnataka.

Fig. 3 shows a range of overview hazard maps for the state of Karnataka and includes the hazards of ecosystem disruption, gradual inundation, salt water intrusion, erosion and flooding. The hazard classes are the same as for Fig. 2. Generally, the maps are not as applicable for planning and management purposes as the ones shown in Fig. 2 but gives a general overview of the hazard presence along the coastline of Karnataka and can be used for identifying hazard hotspots. For the inherent hazard of ecosystem disruption, it can be seen that the outer coastline of Karnataka generally has a low or moderate hazard level, while the very high hazard levels are found in relation to the estuaries. The same pattern can be seen for gradual inundation and salt water intrusion, while large sections of Karnataka's outer coastline have a high or very high hazard level for erosion. The high and very high flooding hazards can especially be found in association with the estuaries and some of the exposed coastal plains.

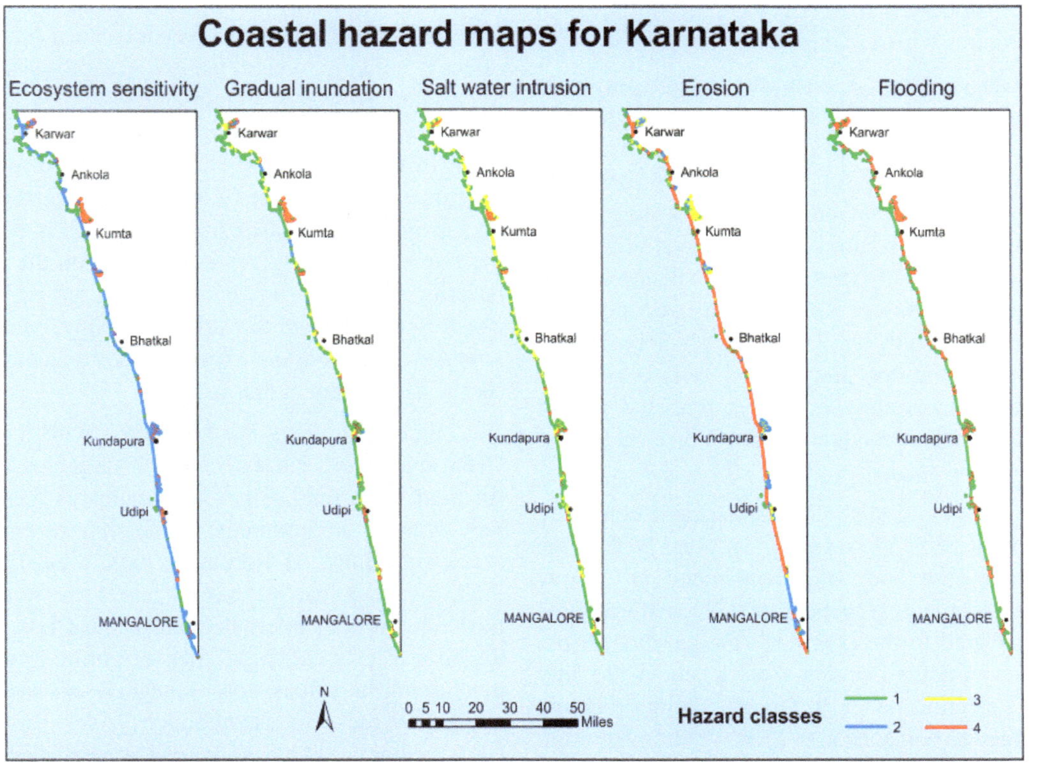

Figure 3: Overview maps of coastal hazards for Karnataka.

UNCERTAINTIES AND LIMITATIONS

The hazard assessment is carried out at sub-regional and regional scale, meaning that the hazard maps are not intended to guide local development activities but rather to assist regional planners and decision-makers in getting an overview of the hazard profile of the coastline and in indentifying hazard hotspots. Whereas the maps covering the whole Karnataka are good for providing an overall picture, the more detailed maps are more appropriate for sub-regional planning purposes. Since the assessment is based on published geophysical data and remote sensing information, several uncertainties exist that should be addressed by field verification if a more detailed assessment is needed. However, as a Step 1 assessment, it is considered to provide a reasonably reliable overview of the hazard presence and the location of hazard hotspots.

An important uncertainty that could be addressed by field verification relates to the geological layout and especially the compaction and cementation level of the coastal sediment. The coastal stretches composed of laterites could be compacted and cemented to various degrees and a particular coastal stretch could therefore fall into the sloping soft rock or sloping hard rock categories depending in their cementation level. This could change for different sections of the coastline and a random field assessment of the compaction/cementation levels of the sloping laterite coastlines could therefore provide an indication of the prevalence of the different conditions. The assumption that all sloping laterite coastlines fall into the sloping soft rock category is considered reasonable as most laterites becomes relativity soft if they are made wet. However, this may overestimate the hazard levels at locations where the laterites are heavily cemented. The relatively low resolution of the geological map of Karnataka also means that it is necessary to rely on the satellite images to identify smaller sloping hard rock features such as headlands. Additional field assessments could have been useful for verifying this, although the resolution of the Google Earth images is generally sufficient to identify these structures with a relatively high accuracy.

The flora/fauna category is also associated with some classification uncertainty that could be addressed with field verification as it is almost impossible to evaluate the percentage of vegetation cover on sloping soft rock coastlines based on the satellite images available in Google Earth and Bing Maps. Because of Karnataka's favourable climatic conditions for full year vegetation growth, it is assumed that all sloping soft rock coasts are vegetated unless clear counter-indications are present. As this parameter only has a minor effect on the hazard levels of ecosystem disruption and erosion, it is considered to be an acceptable uncertainty at this step in the hazard assessment, but for implementation of Step 2 or 3 additional investigations would be needed.

The satellite images used for the assessment constitute another source of possible uncertainty. In areas where the resolution of Google Earth and Bing Maps images are so low that it complicates detailed assessment of the coastline, some uncertainties are related to the coastline configuration. More problematic, however, is the fact that some locations are only coved by a single satellite image in Google Earth's timeline function or only have two images with a few years in between. In the first case, temporal assessment of the sediment balance is impossible while in the second, it is associated with significant uncertainties. This problem may be addressed for most world locations in the coming years as Google Earth continuously adds new satellite images, but for this test-assessment it constitutes a significant source of error. Furthermore, the sediment balance of protected coastal stretches is difficult to assess visually with the current resolution of the satellite images, but this may also improve in the coming years. To avoid underestimating the hazard levels, this assessment generally assumes that a coastal stretch has a sediment balance/deficit, unless it is very clear that it has a sediment surplus.

Since Google Earth and Bing Maps are comprised of a range of different images taken at different times of the day and year, one also compares images taken during different points in the tidal and sedimentary cycles. With the annual erosion/accumulation cycles of large parts of Karnataka's coastline mentioned earlier and a tidal range close to two meters, this comparison can be problematic. The possible error arising from this is partly addressed in the classification by using the vegetation line as reference when evaluating the temporal coastal developments, but it still adds some noise to the assessment. Ideally, the sediment balance should be based on satellite images captured over several years, at the same time of the year and at the same point in a tidal cycle. The current approach, however, is expected to provide acceptable results given the resolution and purpose of the assessment. If more detailed information is needed for planning purposes, aerial photos, field assessments and interviews could improve the reference data.

The human alteration of Karnataka's coastline constitutes another source of uncertainty. Coastal protection work has been carried out along Karnataka's coastline in the past decades, impacting the natural dynamics. At sub-regional and regional scale, however, these activities are not likely to have a major effect on the hazard profile, as they are relatively locally focused and mainly based on hard engineering approaches. Legal and illegal sand mining from the beaches, however, could have some impact on the sediment balance evaluations, but the effect is unlikely to significantly affect the hazard assessment at this step. However, an implementation of Step 2 or 3 would require further investigation of the scale and geographical focus of these activities. Heavily modified or artificial urban coastlines such as those of the city of Mangalore are likely to be surrounded by some errors in the CHW framework, since the framework only gives information on the natural inherent hazards of the coastline before it was turned into an artificial coast. But apart from this urban coastline, human alteration of Karnataka's coast is not expected to cause significant problems for the assessment at this step.

Some limitations are associated with the design of the CHW framework itself. The CHW defines a special category for tidal inlets/sand spits/river mouths as these generally are very dynamic environments with high hazard levels. However, a few tidal inlets in Karnataka have a headland next to the inlet, meaning that the hazard levels are significantly lower than for the tidal inlets defined in the CHW framework. As the hazard levels of these inlets are more in line with that of the sloping hard rock coast category, this category has been applied to these inlets, although it does not adhere to the CHW principles. Also, some of Karnataka's river mouths are so small that they could rather be considered a stream than a river mouth. The guidance given in the CHW framework to apply the river mouth category to the coastline 1 km on each side of the river mouth is therefore regarded as inappropriate. In this assessment, the river mouth category is therefore only extended 0.5 km on each side of the river if it is of stream-size.

The process of carrying out the practical classification process and drawing the polygons is also surrounded by some uncertainty as it based on a manual evaluation of the coastal data. Since the evaluation procedure for the different classification parameters are well defined in the original CHW paper (Rosendahl Appelquist, 2012) and in this paper, the assessment method is not expected to lead to significant greater uncertainty than an automated assessment, as that would still be based on some predefined evaluation procedures. However, the manual approach means that two parallel studies of the same area would be likely to come up with slightly different assessment results. As the CHW framework is designed as a screening tool that can be applied in developing countries and data-poor locations, it tries to strike a balance between simplicity, low-tech design, data requirements and accuracy. The magnitude of the uncertainty related to this manual procedure is therefore regarded as acceptable given the detail and purpose of the assessment, but it is important to keep this uncertainty and possible source of error in mind when using the CHW framework for practical assessments.

REGIONAL PLANNING AND MANAGEMENT PERSPECTIVES

The assessment process outlined in the previous sections is intended to showcase a procedure for applying the CHW framework for regional multi-hazard-assessments. The hazard maps developed for Karnataka can be used for identifying hazard hotspots, getting an overview of the hazard profile of the coastline and detecting areas where human activities may be at risk from future coastal dynamics. As broader coastal hazard assessments are generally non-existent for most developing countries, the methodology provides a possibility for planners and managers to increase their knowledge base in areas with limited data availability. Likewise, it offers a simple system for initial hazard screening in areas where data is readily available.

The hazards covered in the assessment framework are of very different character and hence have very different consequences for human activities. Ecosystem disruption, gradual inundation, salt water intrusion and to some degree erosion is likely to occur gradually and worsen with climate change. Flooding, on the other hand, is an abrupt and potentially disastrous event that will become more likely with rising sea level and increasing precipitation intensity and storm activity. The different hazards are to some degree related to each other, but only a few coastlines have high inherent hazard levels for all hazard types. Coastal planners and managers therefore need to address the specific hazard combination for each coastal stretch in question.

For the state of Karnataka, all hazard types are present but apply to different stretches of the coastline. The hazard of ecosystem disruption is especially related to the mangrove areas in the extensive protected estuary and back-barrier coasts and in the short term, it will probably not be possible to distinguish the climate change hazards to these ecosystems from the major current drivers of change such as overfishing and clearing of mangroves e.g. for aquaculture. In the longer term, however, climate change is likely to pose an additional risk to these systems due to especially sea level rise. Enhancing their resilience at this point should therefore be a priority and is likely to be economically viable as these environments provide valuable services such as flood protection and breeding ground for marine fisheries (Millennium Ecosystem Assessment, 2005).

The hazard of gradual inundation is mainly related to the low-lying protected estuary coasts and the coastal barriers of Karnataka. Barriers with a sediment deficit are already at significant risk and people in these areas may face losing their land permanently to the sea if no countermeasures are taken. Simple dikes could protect the areas to some degree, but on a longer term, a managed retreat or extensive dike systems may be necessary. If dikes are constructed, however, they should always be of a decent quality to avoid giving people a false sense of security of flood protection.

Salt water intrusion is especially a hazard to Karnataka's coastal plans and barriers. The magnitude of this hazard may increase due to human extraction of groundwater and hence it is essential to monitor the groundwater reservoirs and water extraction to avoid that salt water is replacing the current freshwater resources. Simple water balance calculations can be carried out for the barriers to see if the current water extraction practices are sustainable, but gradual inundation and flooding events can completely eliminate the groundwater reservoirs in these locations. In that case, other long-term options for freshwater supply should be investigated and a managed retreat from some of the barriers may be considered.

Erosion is a major general hazard to Karnataka's coastline and many areas are already suffering from the effects of this. Although the state only has limited experience with beach nourishment, possible nourishment schemes combined with groins or breakwaters may be a viable hazard mitigation option for densely populated sections of the coastline. The challenge in this regard is likely to be offshore sand availability, as large quantities of sediment may be needed. Since the cost of sand can vary tenfold depending on dredging conditions and sediment transport distance, it can be a costly management option if sediment is not readily available. A purely hard-engineering strategy may be less costly, but will destroy the natural dynamics of the coastline and the associated natural services, and for coastal stretches used for recreational activities this may not be a viable option. A managed retreat may be relevant for areas experiencing extensive erosion, but with a densely populated coastline, some kind of hold-the line strategy is likely to be necessary in most locations.

Flooding constitutes a serious hazard for the low estuary islands, barriers and coastal plains of Karnataka, and should be addressed properly due to its potential disastrous consequences. Flood warning systems and flood shelters could provide economically viable solutions in the short term, but as repeated floods can disrupt agricultural production, freshwater supply and infrastructure, some kind of dike system may be necessary as a long term solution. As most hazard mitigation options have effects on other hazards than the ones they are primarily designed to address, it is important to consider the possible effect of a given management option on all hazard types. Dikes and hard engineering measures are good at mitigating hazards of flooding and erosion, but often increase the hazards of ecosystem disruption as they disrupt the natural coastal dynamics. For each section of the coastline, it is therefore necessary to consider which hazards are the most important to mitigate and what consequences different mitigation strategies have on all hazards. Because flooding can have dramatic consequences on human activities and be potentially life threatening, mitigating this hazard may in many cases be given higher priority than other hazards such as ecosystem disruption. Hence coastal planner should not only look at which hazards are scoring highest in the CHW framework but also consider which hazards are most problematic to the human activities taking place in a particular coastal area.

A key parameter for deciding on appropriate mitigation strategies is therefore the human activities taking place in a coastal area. Measures of this could be population density, presence of important infrastructure, cultural heritage and various economic activities. As many countries have GIS data on economic activities and global population density data is publicly available (SEDAC, 2013) this information can be added to the GIS used for the coastal hazard assessment to identify areas with specific combinations of coastal hazards and human activities. In this way, the CHW framework can be used to identify areas with e.g. high flooding hazards and high population density. Combing the hazard maps with socioeconomic data could thereby provide a good base for supporting coastal management decisions.

CONCLUSION

The CHW framework has been very suitable for carrying out sub-regional and regional hazard assessments at the scale of the state of Karnataka. It has been possible to conduct the hazard assessment based on easily obtainable data and the assessment procedure outlined in this paper should be replicable in most other areas of the world yielding results of similar quality. The assessment is associated with some uncertainties as it relies solely on published geophysical data and interpretations of remote sensing information, but the uncertainties are considered acceptable given the resolution and goal of the assessment. For more detailed hazard assessments at Step 2 and 3, additional field verification is recommended to improve the assessment accuracy and reliability seen from a decision support perspective. Attempts have been made to keep the assessment procedure relatively simple, with a manual application of the coastal classification in the GIS. This makes the coastal classification process relatively straightforward but at the same time increases the possibility for human misjudgements due to the subjectivity of the procedure. Users should therefore be aware of these risks when using the CHW framework and the assessment procedure outlined in this paper. Supplementing the physical CHW assessment with socioeconomic data may in many cases be relevant for improving the information base for coastal planners and managers. This would provide CHW users with a combined picture of physical hazards and societal activities which could be relevant for supporting long-term planning decisions.

ACKNOWLEDGEMENTS

We would like to thank Kirsten Halsnæs, Per Skougaard Kaspersen, Amit Garg and Prakriti Naswa for fruitful discussions and advice.

REFERENCES

1. Dwarakish, G.S., Vinay, S.A., Natesan, Usha, Asano, Toshiyuki, Kakinuma, Taro,Venkataramana, Ketta, Jagadeesha Pai, B., Babita, M.K., 2009. Coastal vulnerabilityassessment of the future sea level rise in Udupi coastal zone of Karnatakastate, west coast of India. Ocean Coast. Manag. 52, 467e478.
2. Equations, 2000. Karnataka Coast: a Case for Better Protection. Equations,Bangalore.ESRI, 2012. ArcGIS, ArcMap 10.1. ESRI.Google, 2012. Google Earth. Available for download the 13-11-12 at: http://www.google.com/earth/index.html.
3. GRAS, 2012. Geo-referenced RapidEye Satellite Images. GRAS.IPCC, 2007. Climate change 2007: the physical science basis. In: Solomon, S., Qin, D.,Manning, M., Chen, Z., Marquis, M., Averyt, K.B., Tignor, M., Miller, H.L. (Eds.),Contribution of the Working Group I to the Fourth Assessment Report of theIntergovernmental Panel on Climate Change. Cambridge University Press,Cambridge UK.
4. Jayappa, K.S., Vijaya Kumar, G.T., Subrahmanya, K.R., 2003. Influence of coastalstructures on the beaches of Southern Karnataka, India. J. Coast. Res. 19 (2),389e408.
5. Kumar, A., Jayappa, K.S., 2009. Long and short-term shoreline changes alongMangalore coast, India. Int. J. Environ. Res. 3 (2), 177e188.
6. Kumar, A., Narayana, A.C., Jayappa, K.S., 2010. Shoreline changes and morphology ofspits along southern Karnataka, west coast of India: a remote sensing andstatistics-based approach. Geomorphology 120, 133e152.
7. Masselink, G., Hughes, M.G., 2003. Introduction to Coastal Processes and Geomorphology.Oxford University Press.
8. Microsoft, 2012. Bing Maps Hybrid. Available in ERSI ArcGIS.
9. Millennium Ecosystem Assessment, 2005. Ecosystems and Human Well-being:Synthesis. Island Press, Washington DC.
10. Mundkur, Ravi, 2010. Geological Map of Karnataka. Available online the 09-11-12at: http://geokarnataka.blogspot.dk/2010/01/geological-map-of-karnataka.html#!/2010/01/geological-map-of-karnataka.html.
11. Narayana, A.C., Manojkumar, P., Tatavarti, R., 2001. Beach dynamics related to themudbank along the southwest coast of India. In: McAnnally, W.H., Mehta, A.J.(Eds.), Coastal and Estuarine Fine Sediment Processes. Elsevier Science BV,pp. 495e507.
12. Nayak, G.N., Hanamgond, P.T., 2010. India. In: Bird ECF (Ed.), Encyclopaedia of theWorld's Coastal Landforms. Springer.
13. Ramieri, E., Hartley, A., Barbanti, A., Santos, F.D., Gomes, A., Hilden, M., Laihonen, P.,Marinova, N., Santini, M., 2011. Methods for Assessing Coastal Vulnerability toClimate Change. ETC CCA Technical Paper 1/2011. European Environment Agency.

14. Rosendahl Appelquist, L., 2012. Generic framework for meso-scale assessment of climate change hazards in coastal environments. J. Coast. Conserv. http://dx.doi.org/10.1007/s11852-012-0218-z.
15. SEDAC, 2013, NASA Socioeconomic Data and Applications Centre, Online at: http://sedac.ciesin.columbia.edu/ (accessed 25.02.13.).
16. Spalding, M.D., Ravilious, C., Green, E.P., 2001. World Atlas of Coral Reefs. UNEPWCMC.Thieler, E.R., Willams, J., Hammer-Klose, E., 2000. National Assessment of CoastalVulnerability to Future Sea-level Rise. Online at: http://woodshole.er.usgs.gov/project-pages/cvi/ (accessed 09.11.12.).
17. Wikipedia, 2012. Google Earth. Available online the 13-11-12 at: http://en.wikipedia.org/wiki/Google_Earth.
18. World Port Source, 2012. Port of Mangalore. Available online the 18-12-12 at:http://www.worldportsource.com/ports/review/IND_Port_of_Mangalore_1834.php.

CHAPTER 16

MULTI-SCALE CURVATURE FOR AUTOMATED IDENTIFICATION OF GLACIATED MOUNTAIN LAND-SCAPES

Günther Prasicek[a, b] Jan-Christoph Otto[b], David R. Montgomery[c], Lothar Schrott[d]

[a] Department of Geoinformatics – Z_GIS, University of Salzburg, Hellbrunnerstr. 34, 5020 Salzburg, Austria

[b] Department of Geography and Geology, University of Salzburg, 5020 Salzburg, Austria

[c] Department of Earth and Space Sciences, University of Washington, Seattle, WA 98195, USA

[d] Department of Geography, University of Bonn, 53115 Bonn, Germany

ABSTRACT

Erosion by glacial and fluvial processes shapes mountain landscapes in a long-recognized and characteristic way. Upland valleys incised by fluvial processes typically have a V-shaped cross-section with uniform and moderately steep slopes, whereas glacial valleys tend to have a U-shaped profile with a changing slope gradient. We present a novel regional approach to automatically differentiate between fluvial and glacial mountain landscapes based on the relation of multi-scale curvature and drainage area. Sample catchments are delineated and multiple moving window sizes are used to calculate per-cell curvature over a variety of scales ranging from the vicinity of the flow path at the valley bottom to catchment sections fully including valley sides. Single-scale curvature can take similar values for glaciated and non-glaciated catchments but a comparison of multi-scale curvature leads to different results according to the typical cross-sectional shapes. To adapt these differences for automated classification of mountain landscapes into areas with V- and U-shaped valleys, curvature values are correlated with drainage area and a new and simple morphometric parameter, the Difference of Minimum Curvature (*DMC*), is developed. At three study sites in the western United States the *DMC* thresholds determined from catchment analysis are used to automatically identify 5 × 5 km quadrats of glaciated and non-glaciated landscapes and the distinctions are validated by field-based geological and geomorphological maps. Our results demonstrate that *DMC* is a good predictor of glacial imprint, allowing automated delineation of glacially and fluvially incised mountain landscapes.

INTRODUCTION

The effect of glacial processes on the geometry of mountain landscapes has been studied since the 19th century and large scale features of alpine glaciation like cirques, hanging valleys, and U-shaped valley cross sections have been described and investigated by generations of geologists. The now-conventional interpretation of U-shaped glacial and V-shaped fluvial valleys probably originated in 1872, when Swiss geologist Franz Joseph Kaufmann concluded that round-bottomed valleys owe their form to glacial erosion (Kaufmann, 1872). In North America, Clarance King recognized the cross-sectional U-shape of the upper valleys in the glaciated district of the Uinta Mountains, Utah and the V-shaped profiles below, and attributed these differences to the effect of glacial erosion (King, 1878). William Morris Davis compiled a variety of morphologic attributes of glaciated mountain landscapes and presented hand-drawn illustrations of V-shaped and U-shaped valleys (Davis, 1906). For decades, consensus on this basic

Citation: Günther Prasicek, Jan-Christoph Otto, David R. Montgomery, Lothar Schrott, Multi-scale curvature for automated identification of glaciated mountain landscapes, doi:10.1016/j.geomorph.2013.11.026

distinction between fluvially and glacially carved valleys was primarily based on a plethora of similar qualitative reports, rather than on quantified and measurable attributes.

Quantitative descriptions of valley cross profiles can capture the essence of valley morphology and provide an effective tool to differentiate between valleys formed by different processes (Li et al., 2001). Two principal models are widely used to achieve mathematical approximation of glacial valley transects: a power law adopted bySvensson (1959) and a second-order polynomial first applied by Wheeler (1984). Both approximations show advantages and limitations in depicting valley cross profiles. Power laws have more potential for understanding cross-sectional shape, whereas quadratic equations offer a more robust description (Harbor and Wheeler, 1992 and Li et al., 2001).

In geomorphometry, referred to as quantitative land surface analysis based on digital terrain models (Hengl and Reuter, 2009), polynomials are fitted to a regular neighborhood of grid cells (e.g., a kernel of 3 × 3 cells) to calculate land surface parameters (LSPs) like slope and curvature. For curvature calculation, two approaches are widely used. Second order polynomials have been proposed by Evans (1972), and partial fourth order polynomials were adapted by Zevenbergen and Thorne (1987). The lower order approach is incorporated in the geographic information system Landserf for multi-scale LSP calculation (Wood, 1996). Although mathematical approximation of valley cross sections by power laws or polynomials is widely used (Graf, 1970,Doornkamp and King, 1971, Augustinus, 1992, James, 1996 and Schrott et al., 2003), to our knowledge, quantification of cross-sectional valley shape has never been done based on mathematical approximation of the three dimensional land surface instead of a two dimensional cross section. The advantages of a three dimensional approach would be automation, spatially continuous instead of discrete results, and the potential for automated mapping of glaciated valleys.

Identifying the location of recent and past glaciated areas has been an integrated part of glaciology since Agassiz (1840), and plays a crucial role in understanding climate variations and landscape evolution. Presence and extent of Pleistocene glaciation have been mapped throughout the globe, but knowledge is still incomplete in some regions (Ehlers and Gibbard, 2004 and Ehlers et al., 2011), and consensus remains elusive in others (Gualtieri et al., 2000, Grosswald and Hughes, 2002 and Owen et al., 2008). In addition, evidence for glacial remains on Mars is extensively investigated, and controversial, in planetary science (Head et al., 2003 and Head et al., 2010). Although the importance of glacial mapping is undoubted, implications of automated approaches are widely lacking and existing investigations reveal several drawbacks. d›Oleire-Oltmanns et al. (2013) developed a simple semantic model for automated delineation of drumlins and tested their approach in Bavaria. Agreements between mapped and reference landforms were satisfactory, but the study area covered only about 40 km² and did not include large variations in landform development. Sternai et al. (2011) introduced hypsokyrtomes, a specified derivative of hypsometric curves, to identify the regional glacial imprint of mountain ranges. While their results are promising, *a priori* knowledge about an important variable of glaciation, the mean long-term equilibrium line altitude (ELA), is a prerequisite to apply their approach.

Here we present and test a novel method to automatically identify glaciated mountain landscapes based on digital land surface analysis. We exploit the conventional wisdom of U-shaped and V-shaped valleys to gain simple geomorphometric semantics and identify glacial imprint in three mountain ranges across the western United States. Continuous DTMs are segmented into regular quadrangles of identical size, and finally those quadrangles are classified. We first investigate differences in multi-scale curvature of sample catchments revealing well-established fluvial and glacial morphology to define threshold values for differentiation. We then apply these thresholds to the study areas and validate our results using field mapping from prior studies. Our methodology is designed to identify glaciated valleys in a regional manner and to assign fluvially incised valleys and flat terrain to the general class *non-glaciated.*

STUDY AREAS

We test our approach in three study areas in the west of the United States: Sawtooth Mountains, southern Sierra Nevada and Olympic Mountains (Fig. 1). These mountain ranges were selected to test the performance of the approach presented below because of: 1) extensive Pleistocene glaciation; 2) no or very limited recent glaciation; 3) presence of proximal fluvially incised terrain not affected by glaciation; and 4) availability of field mapping of LGM extent or glacial remains for validation.

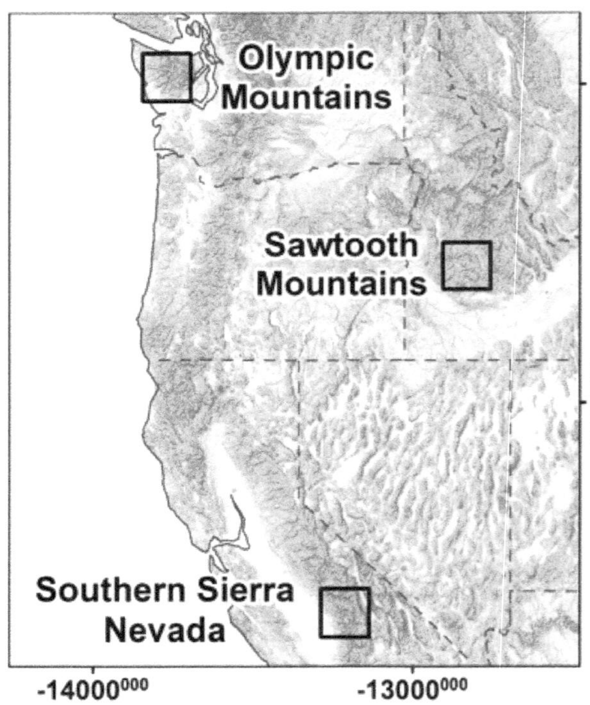

Figure 1: Location of study areas. Spatial reference: WGS84/World Mercator (EPSG 3395).

Sawtooth Mountains and southern Salmon River–Boise Mountains

The Sawtooth Mountains and their western drainages in the southern Salmon River–Boise Mountains area (Fig. 1) primarily consist of Cretaceous biotite granodiorite of the Idaho Batholith and Eocene biotite or hornblende-biotite granite of the Challis magmatic complex. A large block of metamorphic rocks of possible Precambrian age occurs near Stanley Basin (Reid, 1963). Northwest-striking faults of Miocene age and younger caused strong uplift of the rocks underlying the Sawtooth Range. Of these ruptures, only the Sawtooth Fault, an active, range-bounding normal fault on the eastern flank of the Sawtooth Mountains, is known to have had major movement within the last 130 ka (Breckenridge et al., 2003).

Extensive valley glaciers developed in the Sawtooth Range during the Pleistocene, fostered by moist Pacific air masses traversing central Idaho and encountering the mountain barrier (Thackray et al., 2004). Well-developed glacial landforms including deep valley troughs and high jagged peaks are abundant (Reid, 1963, Stanford, 1982 and Borgert et al., 1999). However, the western part of the study area has not been affected by glaciers, but shows extensive fluvial relief (Amerson et al., 2008) qualifying for an ideal study site to test our approach. Reconstructed late Pleistocene ELA fromMeyer et al. (2004) is used for validation of automated classification results. The ELA rises eastward across the study area from about 2250 to 2650 m. Maps of glacial deposits provide additional validation data (Stanford, 1982, Borgert et al., 1999, Kiilsgard et al., 2001, Kiilsgard et al., 2006 and Thackray et al., 2004).

Southern Sierra Nevada

The southern Sierra Nevada study area is located in California; about 150 km from the Nevada border (Fig. 1). It extends east–west from Great Basin to Central Valley and from Kings Canyon in the north to Kern Peak in the south. Large sections of the study area belong to Kings Canyon and Sequoia National Park. The bedrock is dominated by granite of Jurassic–late Cretaceous plutons of the Sierra Nevada Batholith (Moore, 1981 and Moore and Sisson, 1985). The physiographic history of the area now occupied by the Sierra Nevada remains controversial. Until recently, consensus was that uplift, mainly caused by westward block tilting of the entire range, occurred in several episodes over the last 10 Ma and produced the present elevation only in the Quaternary Period. Alternatively, recent studies argue that the Sierra Nevada was uplifted in the late Mesozoic and remained high or even subsided in the late Cenozoic (Henry, 2009).

The Sierra Nevada was repeatedly glaciated during the climatic fluctuations of the Pleistocene, and Wahrhaftig and Birman (1965) and Clark (1995) mapped the extent of late Pleistocene Tioga glaciers. Here we use subsequent mapping by Gillespie and Zehfuss (2004) for validation of automated identification of glacial mountain landscapes. As in the Sawtooth Mountains study area, glaciation in the southern Sierra Nevada was limited to the higher parts of the range, developing extensive valley glaciers and related morphology. The western drainages retain fluvially-incised morphology with deep canyons, whereas eastern drainages formerly occupied by glaciers descend abruptly to the Great Basin.

Southwestern Olympic Mountains

The study site covers the western, southern and central part of the Olympic Mountains (Fig. 1). We omitted the eastern and northern margins of the range to avoid complex interactions of continental and alpine glaciations. The central Olympic Mountains consist of marine sedimentary rocks of Oligocene to Eocene age, dominated by laminated and/or thin-bedded semischist and slate or phyllite. Miocene lithofeldspatic sandstone and siltstone form part of the western margin of the study area. Southernmost parts of the Olympics are built of the lower–middle Eocene Crescent Formation — tholeiitic basalt flows, basaltic flow breccia, and volcaniclastic conglomerate (Dragovich et al., 2002).

The Olympic Mountains form the first prominent barrier for moist Pacific air in Washington State, were repeatedly affected by alpine glaciations during the Pleistocene, and experience limited recent glaciation concentrated around Mount Olympus and Mount Anderson. Continental ice sheets only affected the northernmost sections of the range, which are not included in our analysis. In contrast to the other study areas, glaciation in the western Olympics was particularly heterogeneous, producing highly variable degrees of glacial imprint in neighboring valleys. Repeated presence of large valley glaciers alternates with no glaciers or smaller glaciers, generally restricted to headwaters (Montgomery, 2002). We used field mapping of Fraser and pre-Fraser glacial moraines (Thackray, 2001 and Dragovich et al., 2002) to validate our automated classification results.

METHODS

We developed a method to automatically map glaciated valleys in mountain landscapes based on valley cross-sectional shape. We assumed that valleys with prevailing fluvial imprint typically reveal a V-shaped cross-section with uniformly steep slopes, whereas glacial valleys tend to have a U-shaped profile with a changing slope gradient. Distinction between these two genetic types of valley transects can be performed by analysis of multi-scale curvature to depict valley shape. We adopted this automated land surface analysis approach for a three-dimensional environment and tested the methodology on real-world digital terrain data.

Curvature calculation using fitted polynomials

Curvature in general indicates to what extent an object is curved. In Fig. 2 the theoretical principles of the methodology are illustrated in 2D for cross-sectional profiles of artificial V-shaped (a) and U-shaped (b) valleys of similar dimensions. Horizontal bars represent different scales of investigation taking into account different parts of the cross sections. To calculate curvature, the following second order polynomial can be fitted to the profile for each scale of investigation (dotted curves in Fig. 2) and curvature is calculated as its second derivative:

$$y = ax^2 + bx + c \tag{1}$$

where x is horizontal distance, y is height, and a to c are constants.

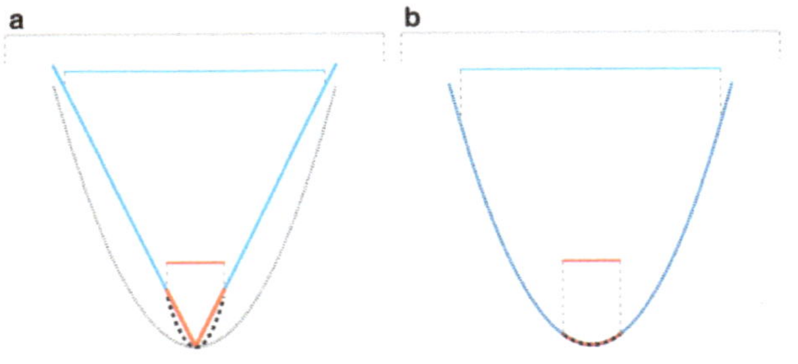

Figure 2: 2D scheme of multi-scale curvature analysis: Idealized cross sections of similar sized V-shaped (a) and U-shaped (b) valleys (bold blue) and thalweg subsections (bold red). Horizontal bars and thin, vertical dashed lines indicate valley parts investigated at different scales. Reference scale is marked by red bars, multi-scale valley analysis by blue bars and invalid analysis scale by light gray bars. Dotted lines indicate best-fit second order polynomials for valley cross sections (fine dots) and subsections (bold dots).

Note that the ideal glacial valley transect (Fig. 2b) is modeled by a parabola (James, 1996) and therefore is exactly matched by the fitted polynomial. For the ideal V-shaped case (Fig. 2a), the cross-sectional shape is a triangle and the shape of the fitted curve is identical at all scales. On the contrary, the shapes of the polynomials fitted to the U-shaped transect change with scale (Fig. 2b). In Fig. 2a, the curvature of the entire V-shaped cross section (blue V, dotted fitted parabola) is 3.71. The curvature depicting a subsection of the V-shaped graph (red V, boldly dotted fitted parabola) is 3.59. Both curvature values are similar because of the analogy in shape. In contrast, in Fig. 2b, the curvature of the entire U-shaped graph (3.71) is considerably different from the curvature of its subsection (0.74). It has to be emphasized that the fitted polynomials must be normalized according to their extent to achieve similar curvature values for objects of identical shape but different size, as presented in this investigation. Without normalization, curvature would be constant (scale-independent) for U-shaped valleys, whereas it would change with scale of investigation for V-shaped valleys. However, we think that producing similar curvature values for similarly shaped objects independent of size is more intuitive in landform analysis. Therefore, we follow Wood (1996): curvature is calculated in radians per 100 m. To account for differences in object size, the total change is given — a dimensionless ratio providing similar values for similar shapes independent of scale. For example, in Fig. 2 curvature for the entire cross section is obtained in radians per 5000 m, whereas for the subsections it is in radians per 1000 m. This procedure corresponds to resizing the cross sections according to a reference scale.

There are two ways to automatically differentiate between the artificial valley cross sections in Fig. 2: comparison of subsection curvature (red curves), or comparison of the difference of curvature values requiring curvature calculation for at least two scales (red curves and blue curves). The first solution requires high resolution data of a well-defined valley floor. In addition, it will suffer from ambiguity when applied to real world data because of variable size of cross sections and various slope angles resulting from variations in valley sizes and height–width ratios. Therefore, the second option is applied, yielding relative results, allowing for comparison of transects with variable size and height–width ratio, as well as for data of lower resolution.

The two-dimensional concept presented in Fig. 2 can be generalized to three dimensions to investigate a digital land surface, as intended in this work: based on a DTM, curvature is calculated for each cell over a multitude of scales using moving windows of variable width. In geomorphometry, curvature calculated for DTMs can highlight divergent and convergent parts of the landscape and has important implications for surface processes (Carson and Kirkby, 1972). To derive curvature on an irregular digital surface, a polynomial model is commonly interpolated and fitted to the original topography. In addition, a plane intersecting the surface has to be specified because curvature of a three-dimensional object varies with orientation. Curvature can then be calculated from the fully differentiable intersection graph of the fitted polynomial model and the specified plane. Concave features are depicted by negative curvature values in geomorphometry, which is the opposite of the general mathematical convention. Over the last several decades polynomials of varying order and different methods of fitting have been applied to DTMs built of regular grid cells. The most widely used approaches are those of Evans (1979) and Zevenbergen and Thorne (1987), which have been variously adapted by others (Mitášová and Hofierka, 1993, Moore et al., 1993 and Shary, 1995). Whatever polynomial model is used, it is fitted to a square neighborhood and centered on a cell in the grid. Evans (1979) suggests fitting the shape of the surface as an interpolated second-order polynomial based on a least squares fit over a 3 × 3 cell moving window:

$$z=ax^2+by^2+cxy+dx+ey+f \qquad (2)$$

where a to f are constants.

Zevenbergen and Thorne (1987) exactly fit a partial fourth-order polynomial with nine coefficients through the central cell and its eight neighbors on a rectangular grid:

$$z=ax^2y^2+bx^2y+cxy^2+dx^2+ey^2+fxy+gx+hy+I \qquad (3)$$

where a to i are constants.

The latter method is implemented in ArcGIS. Discussion about advantages and disadvantages of either approach is legion. Schmidt et al. (2003) provide abundant analysis and show that second-order polynomials are more robust, since they are adjusted to the original terrain by least squares fit and therefore perform a considerable amount of smoothing. Partial quartic models are highly sensitive to local variations in input data because they have to exactly meet all input cells. For our multi-scale approach we needed a sound method to interpolate and fit polynomials to varying sizes of neighborhoods. Therefore, we followed Wood (1996) and adopted the robust method of second-order polynomials (Eq. (2)), getting along with six coefficients independent of the number of incorporated grid cells.

Wood (1996) adapted the quadratic parameterization approach of Evans (1979) to perform computation of LSPs on varying neighborhoods for multi-scale analysis, and implemented it in the geographic information system LandSerf, which we used for computation. According to Evans (1979), the six coefficients needed for fitting of a quadratic surface to irregular topography can be derived from six simple equations due to the data arrangement in the regular 3 × 3 cell neighborhood. Conventional least squares fitting is therefore unnecessary. Wood (1996) replaced this simplification by a matrix solution to enable for a neighborhood of up to $n \times n$ cells, only limited by the smaller side of the DTM. In addition, Wood (2009) suggested that a characteristic scale is defined as the scale where an LSP calculated over multiple window widths (neighborhoods) becomes most extreme. We used this concept to perform multi-scale curvature calculation in a three-dimensional environment.

DTM Resolution and Scales of Investigation

The curvature of two scales represented by moving windows (neighborhoods) of regular shape has to be compared to apply this method: a scale depicting mainly the central part of the valley holding the valley floor and a scale including the entire cross section. The first has to be determined by the user and is termed reference concavity scale, because it is identical for all valley sizes. The second varies with cross-sectional valley extent and is termed maximum concavity scale. When analyzing a range of scales, starting with the reference scale, and assuming normalization of the fitted polynomials, maximum concavity scale can be defined by the particular scale leading to most extreme concavity according to Wood (2009). Assuming normalization according to scale, concavity increases with valley depth for a U-shaped valley, hence it shows most concave curvature when most of the valley transect is analyzed. V-shaped valleys can be expected to show similar concavity for all scales of investigation. In either case, the curvature value will become less concave as soon as the scale of investigation gets large enough to overlap with major ridges or neighboring valleys (Fig. 2, light grey bars), and those scales will therefore not be taken into account. This approach automatically adjusts to valley width, only limited by the defined minimum and maximum scales. The minimum scale is the aforementioned reference concavity scale and the maximum scale varies and no upper limit is imposed, up to the width of the study area. Finally two scales are emphasized: the reference concavity scale is user-defined and has the smallest moving window to only take into account central parts of the cross section; the maximum concavity scale is automatically defined by the most concave (most extreme) curvature value of all analyzed scales, following Wood (2009). In this publication, the moving window width (w) used to compute reference concavity is termed reference concavity window, and w used to compute maximum concavity is called maximum concavity window.

Comparison of the concavity of a valley subsection with the concavity calculated for the whole valley width for valleys of varying scale requires a DTM suitable to resolve narrow valleys and to accordingly define w to calculate curvature. Both DTM resolution and $w$can be used to increase or decrease the amount of detail taken into account. Therefore, these variables have to act in concert to provide accurate classification results. Finally, the minimum DTM resolution is determined by reference concavity scale and data points needed. The reference concavity window has to be small enough to not overlap ridges of the narrowest distinctive valleys (including ridges will cause the curvature to take a more convex value, not representing the actual valley) while being wide enough to not overemphasize minor fluvial channels on the floor of glacial valleys. Arbitrary manual measurements carried out across aerial images of all three study areas revealed a minimum cross-sectional width of well-developed valleys of approximately 250 m and a maximum width of upland glacio-fluvial incision of approximately 30 m. Following these findings, we defined a reference concavity scale of 225 m and a DTM resolution of 25 m (approximately 1 arc second on the average latitude of our study areas), allowing for considerable resolution of narrow valleys but smoothing out minor fluvial incisions on glacial valley floors. Consequently, we used the 1 arc second National Elevation Dataset (NED) of the United States.

Detection of the maximum concavity scale requires an adequately defined range of window widths to account for the large variations in valley sizes. For this, we defined a function that increases w by a minimum of 2% of the previous window width, but at least by two cells, starting with the reference concavity window. As mentioned above, in theory the maximum of w can be any scale large enough to cover the widest valleys. Here it is set to approximately 7 km.

The role of Drainage Area

In geomorphometry, drainage area is a regional LSP specifying the amount or the area of all grid cells located upstream of a specified cell (Gruber and Peckham, 2009). This parameter can be used to determine cells which carry the line of the lowest points along a valley as a streamline on a DTM, further referred to as *thalweg*. Mainly thalweg cells carry curvature information relevant to our approach, as depicted in Fig. 2. Cells forming ridges or hillslopes are therefore not to be taken into account but have to be ruled out by a drainage area cutoff. This causes only a subset of all grid cells to be left for classification. In order to perform a gapless landscape classification and to reduce the influence of local variations, we aimed to regionalize our results. For this, we followed Sternai et al. (2011)and subdivided the study areas into regular quadrats.

The drainage area cutoff directly influences the size of the quadrats. A high threshold value (large drainage area) would mainly include thalweg cells of well-developed valleys showing distinct fluvial or glacial morphology most suitable for our approach. But the higher the threshold, the fewer cells are left valid for landscape classification, leading to smaller samples or larger areas for regionalization. Therefore, a tradeoff has to be made between intended unambiguity of morphological differences and amount of grid cells available for classification. To specify the drainage area cutoff for our investigation, we assumed that distinct valley morphology does not occur at drainage areas smaller than the transition zone from divergent to convergent landscapes. This transition has been previously described based on DTMs with a resolution similar to our elevation datasets:Montgomery and Foufoula-Georgiou (1993) showed that transition from debris-flow dominated channels to alluvial channels occurred between 0.1 and 1 km^2 of the Tennessee Valley, California, indicated by an inflection on slope/area plots. Ijjasz-Vasquez and Bras (1995) recognized a transition zone closely related to those findings, and McNamara et al. (2006) found similar values in a slope/area plot of the Pang Khum Experimental Watershed in northern Thailand. Drainage areas of channel heads mapped in northern Italy by Tarolli and Dalla Fontana (2009) ranged between 0.001 and 0.1 km^2. All of these studies have been carried out in fluvial morphology, but Brardinoni and Hassan

(2006) proposed a threshold between colluvial and fluvial regimes at approximately 0.2 km^2 for formerly glaciated Coastal British Columbia. These findings indicate that convergent landscapes can be morphometrically identified for grid cells holding an upstream drainage area of ≥ 0.1 km^2 on DTMs with a resolution of about 1 arc second. Both drainage area cutoff and quadrat size control the amount of pixels per quadrat. To determine suitable values for both parameters we plotted them against each other (Fig. 3) and defined two criteria: 1) each quadrat has to contain grid cells with an upstream drainage area larger than the cutoff, and 2) quadrats should be as small as possible to allow a maximum of detail. Based on the results shown in Fig. 3, and mostly determined by the situation in the Olympic Mountains, the drainage area cutoff was set to 0.1 km^2 and the quadrat size to 5×5 km for this study.

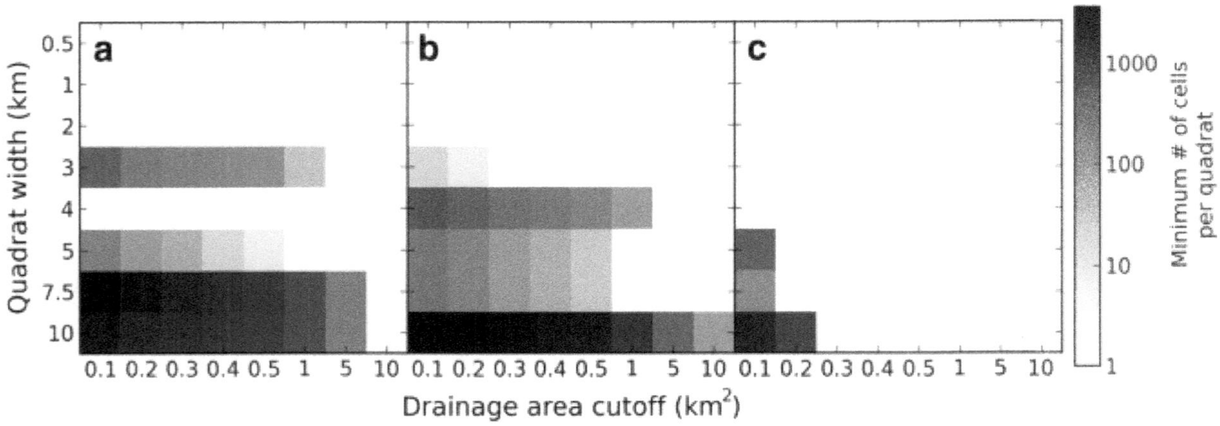

Figure 3: Drainage area cutoff plotted against quadrat width for the Sawtooth Mountains (a), the Sierra Nevada (b), and the Olympic Mountains (c). Minimum amount of valid grid cells per quadrat indicated by logarithmic color map (zero = blank). Partial quadrats are not included.

Curvature Type

A variety of curvatures, depicting different attributes of terrain shape for different purposes, has been defined in geomorphometry. Shary et al. (2002) state four major directions naturally marked on a surface (Fig. 4), and distinguish between two major types of curvatures: field-specific and field-invariant. The orientation of field-specific curvatures is defined by a vector field (solar irradiation, magnetic, electrical or gravity) and is exemplarily represented by gradient and a contour line in Fig. 4; field-invariant curvatures, also referred to as principal curvatures, are independent of any kind of vector field (line of minimum and maximum curvature in Fig. 4). Our investigation required a curvature type capable of depicting the cross-sectional valley shape for a point on the thalweg. Three major field-specific curvatures fulfill these requirements, dependent on gravity: contour curvature, tangential curvature, and cross-sectional curvature (Schmidt et al., 2003). Besides that, the cross-sectional valley shape is in most cases also represented by the most concave curvature. Therefore, minimum curvature provides relevant information as well. For a point *X* on a surface, minimum curvature is defined as the most concave curvature value given by any normal plane through *X* (Fig. 4). If *X* is located on the thalweg, the minimum curvature intersection is given by a plane perpendicular to the planar direction of flow through the valley. Exceptions to this rule may occur in two cases: poorly developed valleys, and valley parts with profile concavity exceeding cross-sectional concavity. The first exception may only apply to a limited amount of thalweg cells in specific kinds of mountainous areas, like plateaus. The second exception may occur below the drop section of hanging valleys. There, the longitudinal profile of the hanging valley thalweg represents the cross-sectional shape of the main valley, which can be interpreted to be *glacial* or *non-glacial*. Instead, a transect perpendicular to the hanging valley thalweg would follow the flank of the main valley in a more or less parallel direction, carrying no information particularly valuable for our investigation. In this case the behavior of minimum curvature is an advantage over the field-specific curvatures, as below hanging valley drop sections transects parallel or perpendicular to the thalweg can carry the desired information about glacial morphology. Therefore, calculation of minimum curvature is applied for its ability to represent both cross-sectional and profile curvature. Computation is based on the formula provided byEvans (1979) as implemented in the geographic information system LandSerf (Wood, 1996):

$$C_{min} = -a-b-\sqrt{(a-b)^2 + c^2} \tag{4}$$

where C_{min} is minimum curvature, and a, b, and c are the coefficients derived from fitting of the quadratic surface to irregular topography, common to Eq. (1).

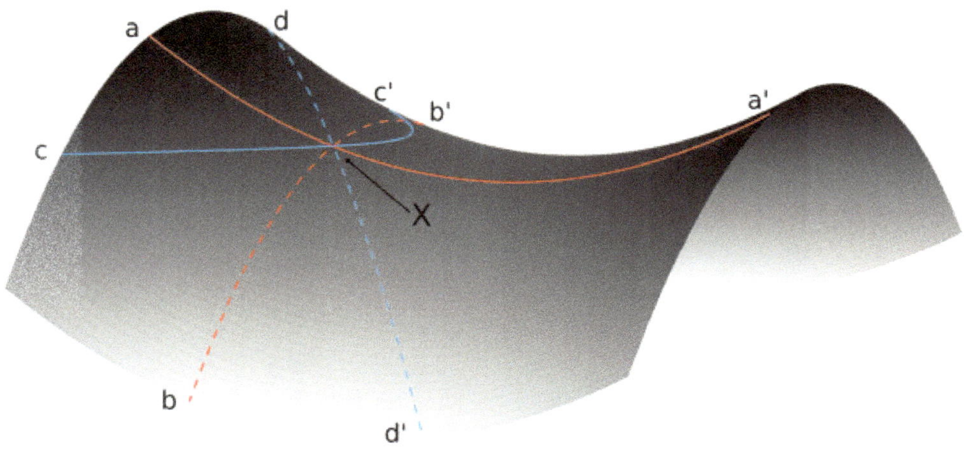

Figure 4: The four directions naturally marked on a surface at point *X*, (after Shary et al. (2002)): a–a′ and b–b′: main normal sections, line of minimum and of maximum curvature, respectively; c–c′: contour line; d–d′: gradient line.

Difference of Minimum Curvature

To reduce information content per cell from reference concavity and maximum concavity to a single quantity, a simple variable is defined: Difference of Minimum Curvature (*DMC*). It is calculated by subtracting reference concavity from maximum concavity:

$$DMC = C_{max} - C_{ref} \tag{5}$$

where C_{max} is maximum concavity being the most concave value over all investigated scales including reference concavity C_{ref}, which can only take negative values or zero for convergent terrain. In theory, V-shaped cross profiles would lead to a *DMC* value of zero. However, two factors will cause deviation when dealing with real fluvial cross sections: curvature is calculated for a second order polynomial fitted to a variable amount of surface points due to variable scales; and real fluvial cross sections are not perfectly V-shaped and their valley floors show some width. Therefore, an empirically determined *DMC* threshold is needed to identify glacial mountain landscapes. After setting the drainage area cutoff to rule out cells carrying divergent terrain, we adopted probability density functions of remaining valid cells (along drainage lines) in the fluvial and glacial sample areas to specify the *DMC* threshold. It is defined as the intersection point of the two probability density functions and determined separately for all three study areas. In general, a *DMC* value close to zero accounts for fluvial terrain, classified *non-glacial*, and lower (more negative) *DMC* values indicate glacial valleys. Flat areas reveal a similar reference and maximum concavity leading to a *DMC* value close to zero and therefore are classified *non-glacial* as well. Finally, *DMC* is calculated per grid cell and each valid cell (with an upstream drainage area larger than 0.1 km^2) is classified *glacial* or *non-glacial*. Classes are then assigned to quadrats according to the majority of grid cells. This approach minimizes the influence of extreme values as curvature is strongly dependent on the cross-sectional height–width ratio and a deeper incised valley would automatically lead to higher concavity and lower *DMC* values. To test whether the *DMC* differences of glacial and even deeply incised fluvial valleys are sufficiently large to prevent from ambiguity, we chose study areas comprising fluvial valleys with exceptional relief, including the Kings Canyon in the Sierra Nevada, one of the deepest canyons in North America. However, this dependency on the height–width ratio hinders the calculation of a 'degree of glaciatedness' from the continuous DMC values.

RESULTS

General Behavior of Multi-scale Curvature and DMC

We defined fluvial and glacial sample areas with well-documented fluvial or glacial imprint and analyzed the behavior of curvature, *w* and *DMC* over drainage area to determine the drainage area cutoff and the *DMC* threshold. Mean values of C_{ref}, C_{max}, and *w* are plotted against drainage area in Fig. 5. Three primary observations emphasizing morphological differences and supporting our hypotheses can be stated: i)C_{ref} and C_{max} decrease with drainage area for fluvial and glacial sample sites, ii) differences between the two concavity values vary significantly between glacial and fluvial sample catchments, and iii) *w* leading to maximum concavity is considerably larger for glacial than for fluvial sample catchments. This can also be observed per-cell in Fig. 6, where parts of the Sawtooth Mountains are displayed.

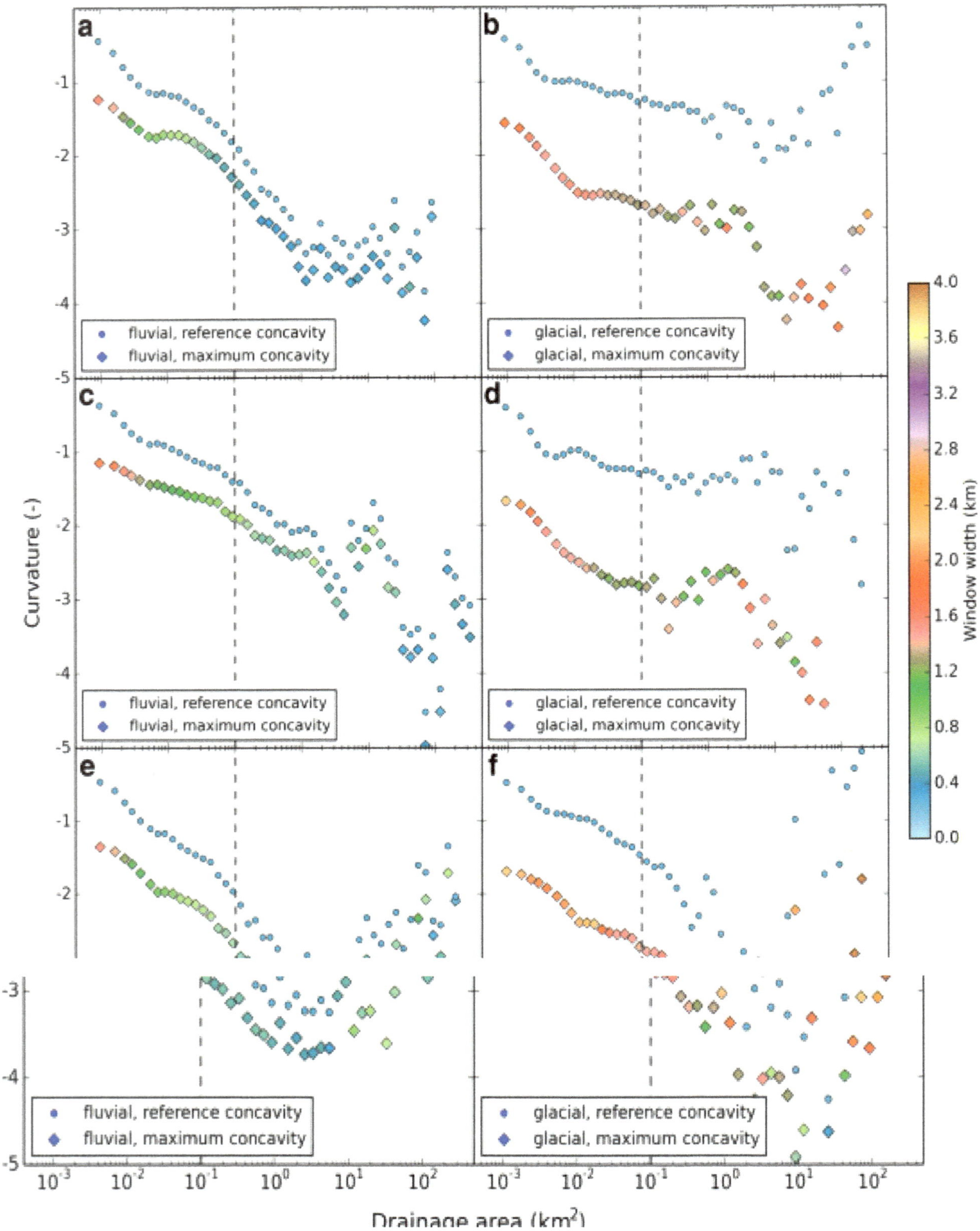

Figure 5: Reference concavity (C_{ref}), maximum concavity (C_{max}) and moving window width (w) plotted vs. drainage area for the fluvial and glacial sample catchments (displayed in Fig. 10, Fig. 11 and Fig. 12) in the Sawtooth Range (a, b), the Sierra Nevada (c, d), and the Olympic Mountains (e, f). Data points are average values calculated per drainage area bin. Bin sizes follow a logarithmic function to control amount of sample cells per bin. Dashed line indicates drainage area cutoff.

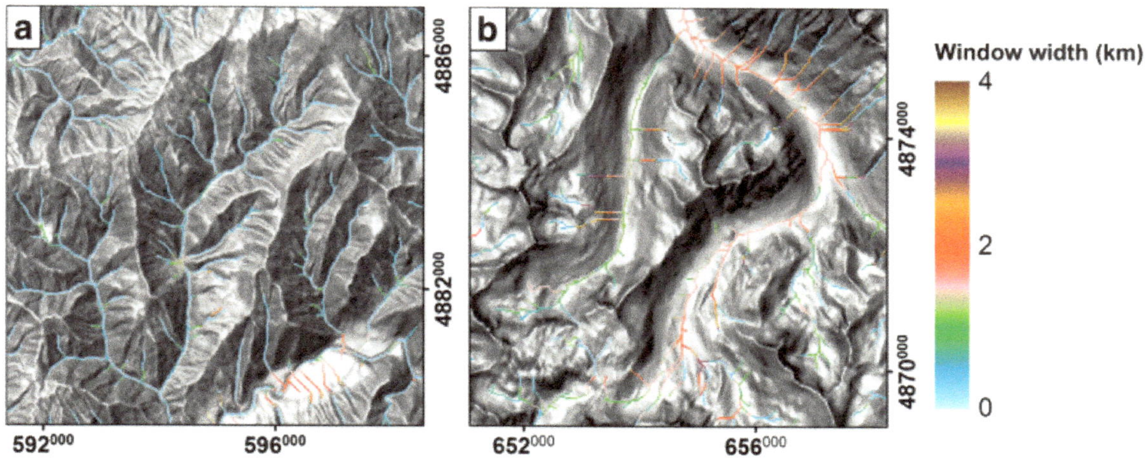

Figure 6: Spatial distribution of moving window width (w) for parts of fluvial (a) and glacial (b) sample catchments (displayed in Fig. 10) in the Sawtooth Mountains. Only thalweg cells with drainage area > 0.1 km^2 are displayed. Spatial reference: WGS84/UTM 11N (EPSG 32611).

The behavior of *DMC* for fluvial and glacial sample catchments of the three study areas is illustrated in Fig. 7, showing large glacial–fluvial differences of average *DMC* values. InFig. 8, *DMC* values are given per cell for cells holding an upstream drainage area > 0.1 km^2. Fluvial channels mostly reveal a *DMC* close to zero (Fig. 8a), but for glacial areas, *DMC* in general takes large values for valid cells (Fig. 8b). Fig. 9 shows the probability for a cell of being either *glacial* or *non-glacial* for a variety of drainage area cutoffs. Functions for fluvial and glacial sample catchments are considerably skewed. Fluvial *DMC* values reveal a narrow distribution and peak slightly below zero, whereas glacial density functions show a wide range of values. In addition, glacial values reveal a peak slightly below zero, a result of cells showing drainage areas larger than the cutoff combined with low relief. Nevertheless, at a specific *DMC* threshold probabilities switch. A cell holding a *DMC* value smaller than the threshold is more likely to indicate glacial terrain and a cell with a *DMC* value larger than the threshold probably belongs to an area dominated by fluvial incision or low relief. In Fig. 9, *DMC* thresholds for a drainage area cutoff of 0.1 km^2 are indicated by black diamonds and were determined separately for the three study areas. In addition, error margins are displayed as horizontal bars. They range from the *DMC* value where the probability of a cell being glacial is twice the probability being fluvial (lower margin) to the *DMC* value where the probability of a cell being fluvial is twice the probability being glacial (upper margin).

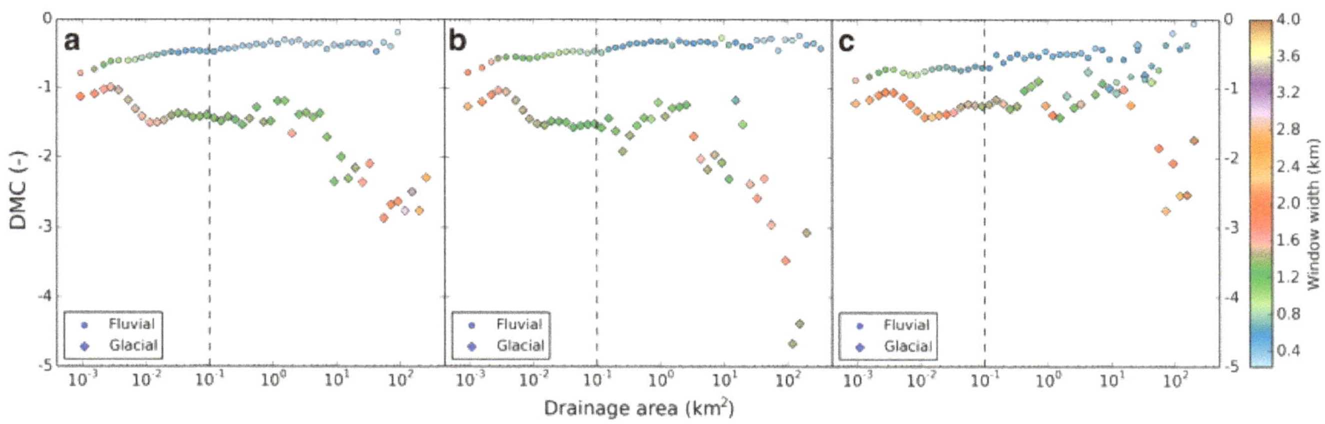

Figure 7: Difference of Minimum Curvature (*DMC*) plotted vs. drainage area for fluvial (circles) and glacial (diamonds) sample catchments (displayed in Fig. 10, Fig. 11 and Fig. 12) in the Sawtooth Range (a), the Sierra Nevada (b), and the Olympic Mountains (c). Data points are mean values calculated per drainage area bin. Bin sizes follow a logarithmic function to control amount of sample cells per bin. Dashed line indicates drainage area cutoff.

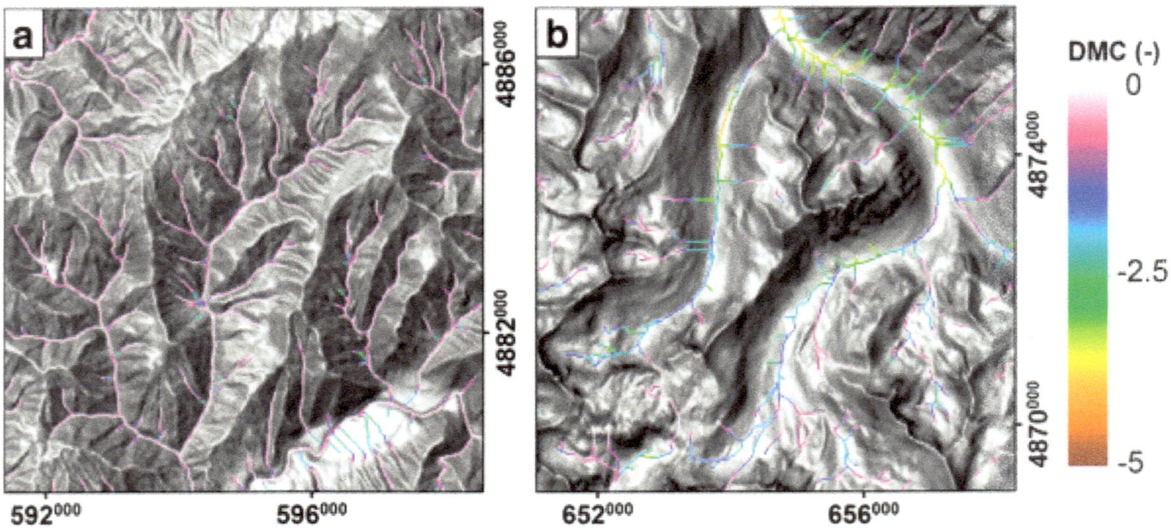

Figure 8: Spatial distribution of Difference of Minimum Curvature (*DMC*) for parts of fluvial (a) and glacial (b) sample catchments in the Sawtooth Mountains (displayed in Fig. 10). Only thalweg cells with drainage area > 0.1 km² are displayed. Spatial reference: WGS84/UTM 11N (EPSG 32611).

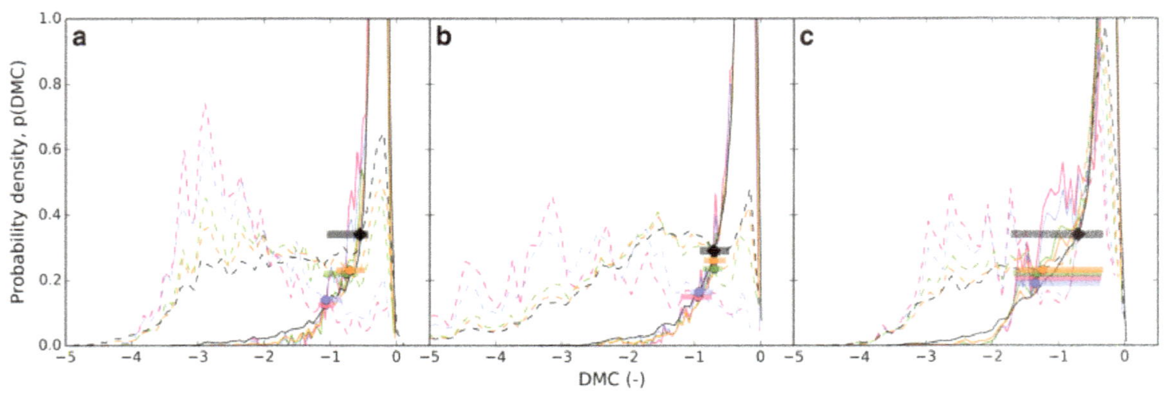

Figure 9: Probability density functions of Difference of Minimum Curvature (*DMC*) for fluvial (continuous) and glacial (dashed) sample catchments (displayed in Fig. 10, Fig. 11 and Fig. 12) in the Sawtooth Mountains (a), the Sierra Nevada (b), and the Olympic Mountains (c). Functions are given for drainage area cutoffs of 0.1 (black), 0.5 (orange), 1 (green), 5 (blue), and 10 (purple) km². *DMC* thresholds are marked by bold dots, the threshold applied for classification in this study is indicated with a black diamond.

Sawtooth Mountains and southern Salmon River–Boise Mountains

In the Sawtooth Mountains and the southern Salmon River–Bose Mountains, C_{ref}, C_{max}, and w plotted against drainage area reveal obvious differences between fluvial (Fig. 5a) and glacial (Fig. 5b) sample catchments. In contrast to the fluvial sample catchments, large differences occur between reference and maximum concavity of glacial areas. These variations are reflected in the resulting *DMC* values depicted in Fig. 7a. Probability density functions for fluvial and glacial terrain lead to the definition of a distinct *DMC*threshold of − 0.55 (Fig. 9a, black diamond), which was used to automatically identify glacial mountain landscapes. Results are shown in Fig. 10. The field mapped extents of glacial deposits and the ELA during the last glacial maximum are used to validate the output. The highest elevation portions of the Sawtooth Range with abundant glacial deposits and large areas elevated above the LGM ELA are consistently classified glacial with a sharp and homogenous transition to the Stanley Basin in the East. The western drainages feature several well-developed valleys carved by large valley glaciers, dominated by the South Fork Payette River catchment. In addition, several isolated zones of limited glaciation have been mapped in the study area, not all of them identified as such by the *DMC* threshold criterion. Lightning Ridge (1 in Fig. 10) and Steel Mountain (2) have been glaciated and are classified glacial, and testing based on manually taken cross sections shows distinct development of U-shaped valleys. Areas around Swanholm Peak (3) and Freeman Peak (4) have been glaciated as well, but are classified non-glacial. Manual testing of Swanholm Peak area exhibits four minor U-shaped valleys draining north of the main summit. Freeman Peak shows no well-developed glacial valleys. At the northern border of the study area a considerable fraction of formerly glaciated terrain is not classified accordingly by our algorithm. The area shows less relief than major glaciated parts of the study area and hardly any distinctive U-shaped valleys, a consequence of tectonic activities that also account for the development of the Stanley Basin (Anderson, 1947 and Reid, 1963). Garden Valley (5) and a short section of the South Fork Payette River drainage (6), both representing fluvial valleys with wide valley floors, are erroneously classified glacial.

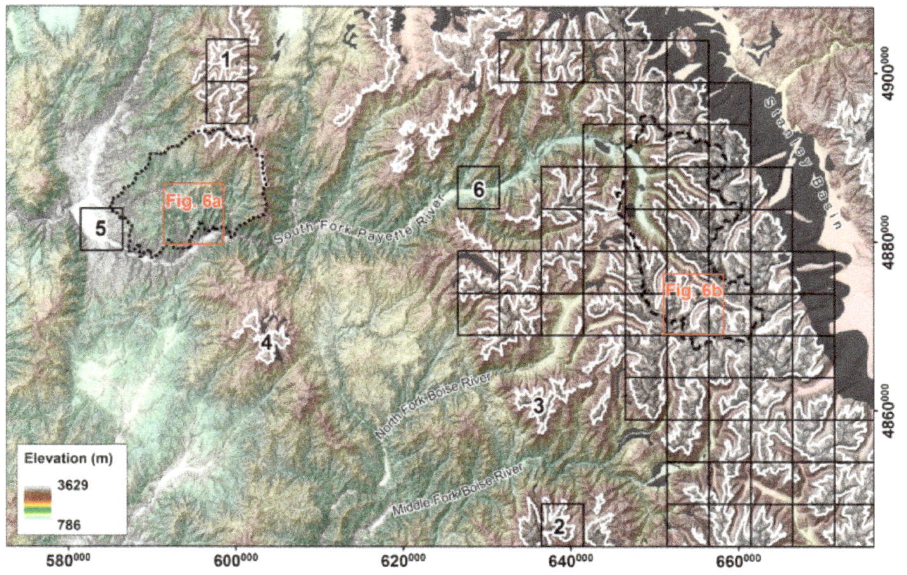

Figure 10: Identification of glacial (black quadrats) mountain regions of the Sawtooth Mountains, Idaho. Sample catchments are indicated by dotted outline (fluvial) and dashed outline (glacial). Red squares mark fluvial and glacial areas shown in Fig. 6 and Fig. 8. Field evidence for former glaciation: LGM ELA (white; Meyer et al. (2004) and glacial depositions (dark gray; (Stanford, 1982, Borgert et al., 1999, Kiilsgard et al., 2001,Kiilsgard et al., 2006 and Thackray et al., 2004). See text for detailed discussion of areas marked with bold letters. Spatial reference: WGS84/UTM 11N (EPSG 32611).

Southern Sierra Nevada

In the Sierra Nevada study area, behavior of C_{ref}, C_{max} and *DMC* for fluvial and glacial sample catchments is similar to the observations in the Sawtooth Range (Figs. 5c,d and7b). The *DMC* threshold used to identify formerly glaciated parts of the Southern Sierra Nevada is − 0.71 and differs slightly from the value applied to the Sawtooth Range (Fig. 9b).

The extent of the Sierra Nevada LGM used for validation of automated classification results shown in Fig. 11 was published by Gillespie and Zehfuss (2004). In contrast to the Sawtooth Mountains, the study area in the Sierra Nevada shows a very compact distribution of formerly glaciated areas, lacking isolated glaciers on single mountains. The border of this glaciated zone is well depicted by our algorithm and deeply incised fluvial canyons draining the western slopes of the mountain range are classified accordingly. Nevertheless, some formerly glaciated parts of the plateau lack distinct glacial valleys (1 in Fig. 11), leading to misclassification. In the upper drainage of the King›s Canyon (2) the maximum extent of the glacier tongue is not correctly represented by automated classification due to non-existent U-shaped valley cross sections. Two lakes in the westernmost fluvial part of the Sierra Nevada study area generate a geomorphometric fingerprint that is mistaken for a glacial valley by our algorithm, Flat Pine Lake (3) and Lake Kaweah (4). Where the Sierra Nevada declines to the Central Valley, another example of a flat floored fluvial valley is mistaken for glacial morphometry (5).

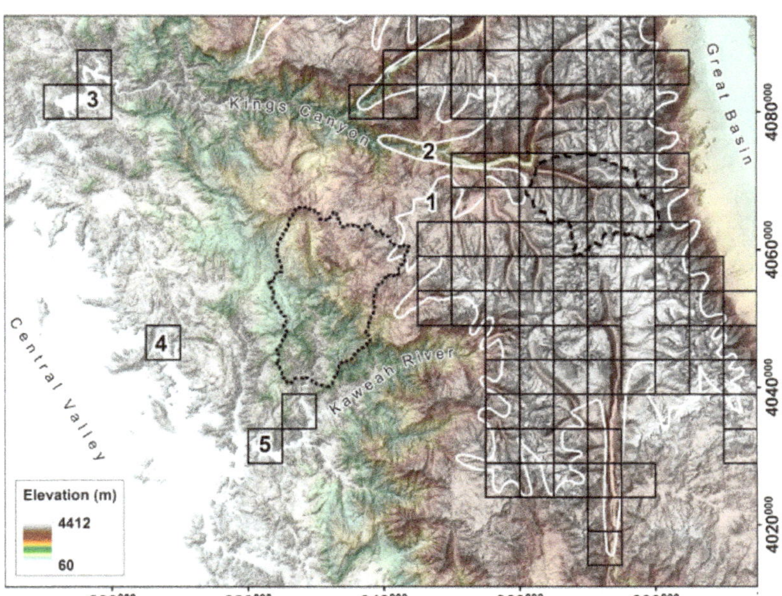

Figure 11: Classification of glacial (black quadrats) mountain regions of the Southern Sierra Nevada, California. Sample catchments are indicated by dotted outline (fluvial) and dashed outline (glacial). Field evidence for former glaciation: glacial extent during LGM (white; Gillespie and Zehfuss (2004). See Results for detailed discussion of areas marked with bold letters. Spatial reference: WGS84/UTM 11N (EPSG 32611).

Southwestern Olympic Mountains

Compared with the Sierra Nevada, the Olympic Mountains show a very heterogeneous pattern of glacial imprint, but unlike the Sawtooth Range, no isolated patches of limited past glaciation exist. Due to local climatic variations, valley morphology can change from one valley to the next (Thackray, 2001, Dragovich et al., 2002 and Montgomery, 2002), leading to a fringe transition zone between fluvial and glacial mountain areas. In addition, abundant sedimentation affects the morphology of fluvial catchments in the vicinities of former outlet glaciers, and highly-erosive rocks may cause features of former glaciation to be less persistent. This puts our automated classification algorithm to a particular test.

The behavior of C_{ref}, C_{max} and *DMC* differs considerably from the Sawtooth Mountains and the Sierra Nevada (Figs. 5e,f and 7c) causing *DMC* to be less distinctive than in the other study areas. This has implications on the specification of the *DMC* threshold as well (Fig. 9c). It cannot be determined as clearly because of a partially similar progression of the probability density functions. However, the probabilities finally switch at a *DMC* value of − 0.70, similar to the value of the Sierra Nevada.

For validation of the automated classification results (Fig. 12), manual mapping of Quaternary geology is used distinguishing depositions of Fraser and pre-Fraser ages (Dragovich et al., 2002). The formerly glaciated heart of the Olympics is classified accordingly including largest glaciated valleys of Hoh (1 in Fig. 12), Queets (2) and Quinault river (3). However, those valleys are not classified glacial down to the positions of terminal moraines, where relief is low and no distinct glacial valleys are developed. Minor western drainages are identified as *glacial* where Fraser age moraines have developed (4, 5, 6), but are identified *non-glacial* where only pre-Fraser deposits have been found (7, 8). Parts of the Upper Elwha River drainage are classified *non-glacial* as well (9) and manually drawn valley cross sections reveal distinct V-shapes. Detected glacial terrain also includes recent alpine ice (10).

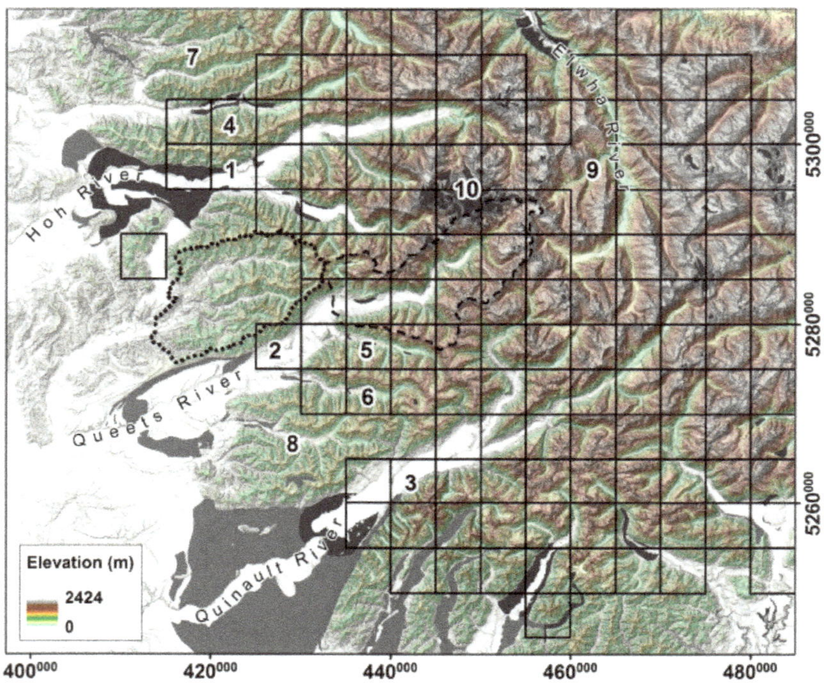

Figure 12: Classification of glacial (black quadrats) mountain regions of the Western Olympic Mountains, WA. Sample catchments are indicated by dotted outline (fluvial) and dashed outline (glacial). Field evidence for former glaciation: deposits of Fraser age (dark gray) and pre-Fraser age (light gray) alpine glaciations (Dragovich et al., 2002). See Results for detailed discussion of areas marked with bold letters. Spatial reference: WGS84/UTM 10N (EPSG 32610).

DISCUSSION

The extent of fluvial and glacial mountain landscapes automatically identified based on the acquired *DMC* thresholds conforms to the field mapping. While the big picture is depicted accurately, some systematic finer-scale misclassification illustrates limitations in our methodology. Fluvial drainages with a large amount of sediment fill are considered to be glacial by our algorithm. Such valleys can be found in all three study areas, and correspond to a limited number of misclassified quadrangles. For further application of this approach, an upper drainage area threshold might therefore be applied. Alternatively, a second indicator could be employed: sediment-filled fluvial valleys may have a *DMC* value similar to glacial ones, but the valley-sides are likely to be steeper for the latter, which can be identified by standard deviation of slope gradient (Evans, 1979). A major reason for abundant sedimentation in fluvial catchments is base level rise, which has been documented in the Sawtooth Mountains (Anderson, 1947 and Reid, 1963) with minor but visible impact on our classification. In the Olympic Mountains low-elevation fluvial catchments draining into valleys formerly occupied by large piedmont glaciers show abundant sedimentation as well. Glacial drainages can be considered non-glacial if the relief is very low, leading to a small *DMC* value because of the small curvature values. Such areas in the northern part of the Sawtooth Range study area illustrate the relief dependence of our automated classification approach. Clearly, distinct valleys are needed for differentiation, and both large plateau glaciation areas and broad piedmont moraine spreads cannot be identified.

w depicted in Fig. 5 and Fig. 7 indicates that cells with very small drainage area reveal maximum concavity at large values of *w*. These cells form ridges or hillslopes, where concavity increases for larger moving windows when parts of adjunct valleys are included in calculation. *w* finally takes values representing valley width at the location of the center cell for drainage areas larger than approximately 0.02–0.5 km^2 which we believe to mark onset of convergent terrain. The similarity of our assumptions with previous findings (Montgomery and Foufoula-Georgiou, 1993, Ijjasz-Vasquez and Bras, 1995, Brardinoni and Hassan, 2006, McNamara et al., 2006 and Tarolli and Dalla Fontana, 2009) suggests plots of characteristic curvature scales against drainage area to be suitable for identification of the hillslope/valley transition over multiple scales.

The drainage area cutoff of 0.1 km^2 was determined based on the transition zone from divergent to convergent terrain and on considerations about the amount of valid grid cells per quadrat (Section 3.3, Fig. 3). The chosen quadrat size of 5 × 5 km is coherent with the width of the widest valleys in all three study areas and the drainage area cutoff ensures valid grid cells in every quadrat. The probability density plots of *DMC* provided in Fig. 9show that especially the distribution of glacial *DMC* is greatly affected by the drainage area cutoff. *DMC* values close to zero are typical for fluvial valleys, but exist in glacial terrain as well. In the latter case, they are a result of flat terrain cells holding an upstream drainage area larger than the cutoff. They are less likely if the cutoff is raised, which leads to more negative glacial *DMC*, to larger differences in fluvial and glacial probability density functions, and to variations of the *DMC* thresholds. Although an increase in fluvial-glacial contrast would be favorable for distinction, we did not apply a higher drainage area cutoff because of the associated drop in the number of valid grid cells leading to empty quadrats for regionalization. Using only a slightly higher drainage area cutoff (e.g. 0.2 km^2), we would have to approximately quadruple quadrat area to avoid empty quadrats. Hence, the number of interpretable grid cells clearly is a governing factor in our methodology.

The *DMC* thresholds for differentiation between fluvial and glacial mountain landscapes have been calculated using separate sample catchments for each study area. Nonetheless, threshold values only differ slightly, by about 5% of the *DMC* range. In addition, error margins displayed in Fig. 9 indicate considerable overlap between the*DMC* thresholds determined for the three study areas. The error margins generally decrease with an increase in drainage area cutoff and they show a wide range in the Olympic Mountains — a result of sediment fill within the fluvial sample catchments causing less fluvial–glacial contrast. Full overlap of all three *DMC* thresholds with error margins exists for a drainage area cutoff of 0.1 km^2. For larger drainage area cutoffs the*DMC* thresholds of the Sawtooth Mountains are still situated within the error margins of the Southern Sierra Nevada and vice versa, but the thresholds of the Olympic Mountains are located outside the error margins of the other two study areas. This is caused by sediment-filled fluvial valleys in the Olympic Mountains and emphasizes the importance of typical fluvial and glacial sample catchments for determination of the *DMC* threshold. Nevertheless, because of minor differences for the 0.1 km^2 drainage area cutoff, the mean of the three thresholds − 0.65 can be applied to each of the three study areas leading to similar results (Fig. 13). These characteristics of *DMC* can be explained by the principles of the method: V-shaped cross sections theoretically reveal a *DMC* value of 0. However, in reality, fluvial valley sides are not straight and the valley floor has some width. Therefore, in most cases, a *DMC* value slightly below 0 can be expected, which fits our results. *DMC* values of glacial valleys tend to be more negative, causing the *DMC*threshold to mainly depend on the deviation of real fluvial cross sections from the ideal V-shape. Although this deviation may vary due to different influences, for example sediment fill in the fluvial sample catchments of the Olympic Mountains, our classification results account for the general applicability of the V-shape and U-shape concept. Hence, we assume *DMC* thresholds for different mountain ranges to vary in close vicinity of the values we found in the western United States and we expect the existence of a best-fit*DMC* threshold suitable to identify glacial and fluvial imprint in mountain ranges throughout the globe. However, this study presents only a first test of the approach at three study sites and further sampling will be necessary to define such a threshold.

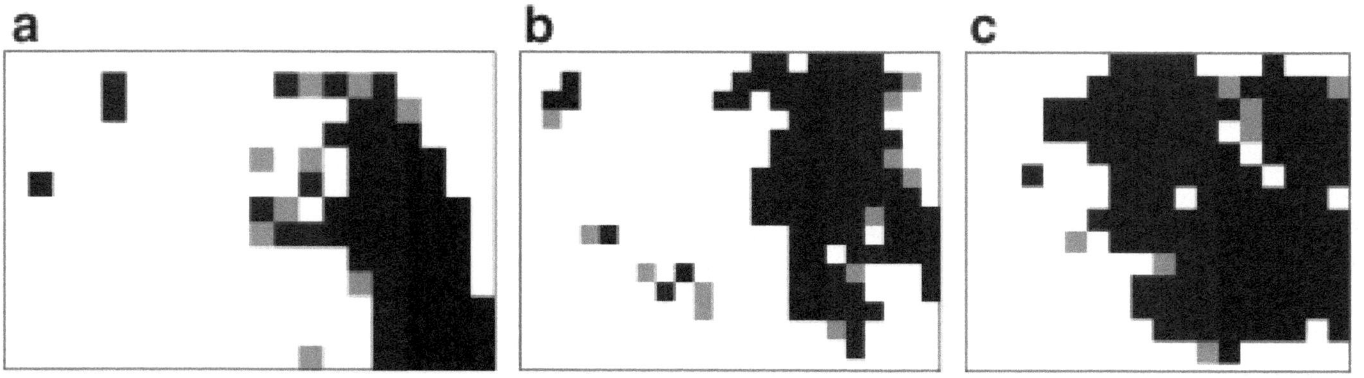

Figure 13: Classification of all three study areas based on a single DMC threshold, the mean (− 0.65) of the three separately defined thresholds. Gray quadrats indicate a change in classification. Sawtooth Mountains (a): change from glacial to fluvial; Sierra Nevada (b) and Olympic Mountains (c): change from fluvial to glacial.

Classification of mountain landscapes into glacial and non-glacial terrain does not account for gradual variation over time leading to intermediate stages and partial overprinting. The Olympic Mountains consist of rocks less resistant to erosion than the other study areas. Therefore we tried to explain our classification results with variations in time since deglaciation and related differences in partial readjustment by splitting validation data into Fraser age and pre-Fraser age remains. Whether these assumptions apply or not, it leads to the general question over which amount of time glacial or fluvial imprint may prevail and builds a bridge to the concept of paraglacial sedimentation (Church and Ryder, 1972). However, gradual changes of glacial landscapes cannot be investigated with the approach presented. As cross-sectional valley shape is strongly dependent on the height–width ratio, a deeper incised valley would inevitably result in higher glaciality. To tackle this problem and to account for ambiguity in the shape of glacial troughs and sediment-filled fluvial valleys, a multi-indicator approach for automated identification of glaciated terrain should be considered for future investigations.

CONCLUSION

The conventional geomorphological interpretation of genetic differences in cross sectional valley shapes related to glacial and fluvial erosion constitutes the basis of a novel regional approach quantitatively identifying formerly glaciated mountain landscapes. Concavity values calculated over multiple scales tend to be similar for V-shaped fluvial valleys, but vary over scale for U-shaped glacial valleys. These differences enable automated landscape classification. The approach was tested in three study areas in the western United States yielding promising results. Large areas modified by past glaciation are identified successfully, as are isolated patches of well-developed glacial valleys. Misclassification occurs mostly due to abundant sedimentation leading to wide-floored fluvial valleys and ambiguity in cross sectional valley shape. In general, relief plays an important role in the automated classification routine and distinct valley development is an essential prerequisite. Therefore, this automated approach is restricted to mountain landscapes.

Most mountain ranges around the globe have been investigated by glacial geomorphologists either directly in the field or by remote sensing, but a method like the one presented may be applied to terrain data of Mars and other planetary bodies to investigate the amount of glacial imprint. Further research is underway to combine multiple indicators of glacial terrain for automated detection of more subtle changes in cross-sectional valley morphology, and for calculation of a "glaciality index".

ACKNOWLEDGMENTS

We thank Glenn Thackray, Loudon Stanford, Byron Amerson, Grant Meyer and Alan Gillespie for sharing their abundant knowledge about the morphology of the study areas and for the provision of original data on surficial geology, LGM ELAs and glacial extents. Ian Evans and Alain Demoulin provided insightful reviews that substantially enhanced this work. The project was funded by the Austrian Science Fund (FWF) through the Doctoral College GIScience (Grant number: W1237-N23).

REFERENCE

1. Agassiz, L., 1840. Études sur les glaciers. Jent et Gassmann, Neuchâtel. Amerson, B.E., Montgomery, D.R., Meyer, G., 2008. Relative size of fluvial and glaciatedvalleys in central Idaho. Geomorphology 93, 537–547.
2. Anderson, A.L., 1947. Drainage diversion in the Northern Rocky Mountains of East-Central Idaho. J. Geol. 55, 61–75.
3. Augustinus, P.C., 1992. The influence of rock mass strength on glacial valley cross-profile morphometry: a case study from the Southern Alps, New Zealand. Earth Surf. Process. Land. 17, 39–51.
4. Borgert, J.A., Lundeen, K.A., Thackray, G.D., 1999. Glacial Geology of the Southeastern Sawtooth Mountains. In: Hughes, S.S., Thackray, G.D. (Eds.), Guidebook to the Geology of Eastern Idaho. Idaho Museum of Natural History, Pocatello, pp. 205–217.
5. Brardinoni, F., Hassan, M.A., 2006. Glacial erosion, evolution of river long profiles, and the organization of process domains in mountain drainage basins of coastal British Columbia. J. Geophys. Res. 111, F01013. http://dx.doi.org/10.1029/2005JF000358.
6. Breckenridge, R.M., Lewis, R.S., Adema, G.W., Weisz, D.W., 2003. Miocene and Younger Faults in Idaho. Idaho Geological Survey, Moscow.
7. Carson, M., Kirkby, M., 1972. Hillslope Form and Process. Cambridge University Press, Cambridge.
8. Church, M., Ryder, J.M., 1972. Paraglacial sedimentation: consideration of fluvial processes onditioned by glaciation. Geol. Soc. Am. Bull. 83, 3059–3072.
9. Clark, D.H., 1995. Extent, timing and climatic significance of latest Pleistocene and Holocene glaciation in the Sierra Nevada. Ph.D. University of Washington (193 pp.).
10. Davis, W.M., 1906. The sculpture of mountains by glaciers. Scott. Geogr. Mag. 22, 76–89.
11. d›Oleire-Oltmanns, S., Eisank, C., Dragut, L., Blaschke, T., 2013. An object-based workflow to extract landforms at multiple scales from two distinct data types. IEEE Geosci. Remote Sens. Lett. 10, 947–951.
12. Doornkamp, J.C., King, C.A.M., 1971. Numerical Analysis in Geomorphology. Arnold, London. Dragovich, J.D., Logan, R.L., Schasse, H.W., Walsh, T.J., Lingley, W.S., Norman, D.K., Gerstel, W.J., Lapen, T.J., Schuster, E.J., Meyers, K.D., 2002. Geologic Map of Washington - Northwest Quadrant. Washington State Department of Natural Resources, Olympia.
13. Ehlers, J., Gibbard, P.L. (Eds.), 2004. Quaternary Glaciations - Extent and Chronology, PartIII. Developments in Quaternary Science, 3. Elsevier.
14. Ehlers, J., Gibbard, P.L., Hughes, P.D., 2011. Quaternary Glaciations. Extent and Chronology: A Closer Look, 15. Elsevier (Access Online via).
15. Evans, I.S., 1972. General geomorphometry, derivatives of altitude, and descriptive statistics. In: Chorley, R.J. (Ed.), Spatial Analysis in Geomorphology. Methuen & Co. Ltd, London, pp. 17–90.

16. Evans, I.S., 1979. An an integrated system of terrain analysis and slope mapping. US Army European Research Office, London.
17. Gillespie, A.R., Zehfuss, P.H., 2004. Glaciations of the Sierra Nevada, California, USA. In:Ehlers, J., Gibbard, P.L. (Eds.), Quaternary Glaciations - Extent and Chronology, Part II. Elsevier, Amsterdam, pp. 51–62.
18. Graf, W.L., 1970. The geomorphology of the glacial valley cross section. Arct. Alp. Res. 2, 303–312.
19. Grosswald, M.G., Hughes, T.J., 2002. The Russian component of an Arctic Ice Sheet during the Last Glacial Maximum. Quat. Sci. Rev. 21, 121–146.
20. Gruber, S., Peckham, S., 2009. Chapter 7 Land-surface parameters and objects in hydrology. In: Hengl, T., Reuter, H.I. (Eds.), Developments in Soil Science. Elsevier, pp. 171–194.
21. Gualtieri, L., Glushkova, O., Brigham-Grette, J., 2000. Evidence for restricted ice extent during the last glacial maximum in the Koryak Mountains of Chukotka, far eastern Russia. Geol. Soc. Am. Bull. 112, 1106–1118.
22. Harbor, J.M., Wheeler, D.A., 1992. On the mathematical description of glaciated valleycross sections. Earth Surf. Process. Land. 17, 477–485.
23. Head, J.W., Mustard, J.F., Kreslavsky, M.A., Milliken, R.E., Marchant, D.R., 2003. Recent ice ages on Mars. Nature 426, 797–802.
24. Head, J.W., Marchant, D.R., Dickson, J.L., Kress, A.M., Baker, D.M., 2010. Northern midlatitude glaciation in the Late Amazonian period of Mars: criteria for the recognition of debris-covered glacier and valley glacier landsystem deposits. Earth Planet. Sci. Lett. 294, 306–320.
25. Hengl, T., Reuter, H.I. (Eds.), 2009. Geomorphometry - Concepts, Software, Applications. Developments in Soil Science, 33. Elsevier.
26. Henry, C.D., 2009. Uplift of the Sierra Nevada, California. Geology 37, 575–576.
27. Ijjasz-Vasquez, E.J., Bras, R.L., 1995. Scaling regimes of local slope versus contributing area in digital elevation models. Geomorphology 12, 299–311.
28. James, L.A., 1996. Polynomial and power functions for glacial valley cross-section morphology. Earth Surf. Process. Land. 21, 413–432.
29. Kaufmann, F.J., 1872. Gebiete der Kantone Bern, Luzern, Schwyz und Zug. Beitraege zur Geologischen Karte der Schweiz, 11. Geologische Commisiion der schweiz .naturforschenden Gesellschaft, Bern.
30. Kiilsgard, T.H., Stanford, L.R., Lewis, R.S., 2001. Geologic Map of the Idaho City 30 X 60 Minute Quadrangle, Idaho. Idaho Geological Survey, Moscow.
31. Kiilsgard, T.H., Stanford, L.R., Lewis, R.S., 2006. Geologic Map of the Deadwood River 30 X 60 Minute Quadrangle, Idaho. Idaho Geological Survey, Moscow.
32. King, C., 1878. Systematic Geology. US Geological Explorations of the Fortieth Parallel, 1. Department of the Interior, Washington.
33. Li, Y., Liu, G., Cui, Z., 2001. Glacial valley cross-profile morphology, Tian Shan Mountains, China. Geomorphology 38, 153–166.
34. McNamara, J.P., Ziegler, A.D., Wood, S.H., Vogler, J.B., 2006. Channel head locations with respect to geomorphologic thresholds derived from a digital elevation model: a case study in northern Thailand. For. Ecol. Manag. 224, 147–156.
35. Meyer, G.A., Fawcett, P.J., Locke,W.W., 2004. Late-Pleistocene equilibrium-line altitudes, atmospheric circulation, and timing of mountain glacier advances in the interior northwestern
36. United States. In: Haller, K.M., Wood, S.H. (Eds.), Geological Field Trips in Southern Idaho, Eastern Oregon, and Northern Nevada. U.S. Geological Survey, Reston, pp. 64–97.
37. Mitášová, H., Hofierka, J., 1993. Interpolation by regularized spline with tension: II. Application to terrain modeling and surface geometry analysis. Math. Geol. 25, 657–669.
38. Montgomery, D.R., 2002. Valley formation by fluvial and glacial erosion. Geology 30,1047–1050.
39. Montgomery, D.R., Foufoula-Georgiou, E., 1993. Channel network source representation using digital elevation models. Water Resour. Res. 29, 3925–3934.
40. Moore, J.G., 1981. Geologic Map of the Mount Whitney Quadranlge, Inyo and Tulare Counties, California. U.S. Geological Survey, Reston.
41. Moore, J.G., Sisson, T.W., 1985. Geologic Map of the Kern Peak Quadrangle, Tulare County, California. U.S. Geological Survey, Reston.
42. Moore, I.D., Gessler, P.E., Nielsen, G.A., Peterson, G.A., 1993. Soil attribute prediction using terrain analysis. Soil Sci. Soc. Am. J. 57, 443–452.
43. Owen, L.A., Caffee, M.W., Finkel, R.C., Seong, Y.B., 2008. Quaternary glaciation of the Himalayan–Tibetan orogen. J. Quat. Sci. 23, 513–531.
44. Reid, R.R., 1963. Reconnaissance Geology of the Sawtooth Range, 129. Idaho Bureau of Mines and Geology, Moscow.
45. Schmidt, J., Evans, I.S., Brinkmann, J., 2003. Comparison of polynomial models for land surface curvature calculation. Int. J. Geogr. Inf. Sci. 17, 797–814.
46. Schrott, L., Hufschmidt, G., Hankammer, M., Hoffmann, T., Dikau, R., 2003. Spatial distribution of sediment storage types and quantification of valley fill deposits in an alpine basin, Reintal, Bavarian Alps, Germany. Geomorphology 55, 45–63.
47. Shary, P.A., 1995. Land surface in gravity points classification by a complete system of curvatures. Math. Geol. 27, 373–390.
48. Shary, P.A., Sharaya, L.S., Mitusov, A.V., 2002. Fundamental quantitative methods of land surface analysis. Geoderma 107, 1–32.
49. Stanford, L., 1982. Surficial Geology of the Upper South Fork Payette River Area, South Central Idaho. Idaho Geological Survey, Moscow.

50. Sternai, P., Herman, F., Fox, M.R., Castelltort, S., 2011. Hypsometric analysis to identify spatially variable glacial erosion. J. Geophys. Res. 116, F03001. http://dx.doi.org/10.1029/ 2010JF001823.
51. Svensson, H., 1959. Is the cross-section of a glacial valley a parabola? J. Glaciol. 3, 362–363.
52. Tarolli, P., Dalla Fontana, G., 2009. Hillslope-to-valley transition morphology: new opportunities from high resolution DTMs. Geomorphology 113, 47–56.
53. Thackray, G.D., 2001. Extensive early and middle Wisconsin glaciation on the western Olympic Peninsula, Washington, and the variability of Pacific moisture delivery to the northwestern United States. Quat. Res. 55, 257–270.
54. Thackray, G.D., Lundeen, K.A., Borgert, J.A., 2004. Latest Pleistocene alpine glacier advances in the Sawtooth Mountains, Idaho, USA: reflections of midlatitude moisture transport at the close of the last glaciation. Geology 32, 225–228.
55. Wahrhaftig, C., Birman, J.H., 1965. The Quarternary of the Pacific mountain system in California. In: Wright, H.E., Frey, D.G. (Eds.), The Quaternary of the United States. Princeton University Press, Princeton, pp. 299–340.
56. Wheeler, D.A., 1984. Using parabolas to describe the cross-sections of glaciated valleys. Earth Surf. Process. Land. 9, 391–394.
57. Wood, J., 1996. The Geomorphological Characterisation of Digital Elevation Models. Doctoral Thesis University of Leicester.
58. Wood, J., 2009. Chapter 14 Geomorphometry in LandSerf. In: Hengl, T., Reuter, H.I. (Eds.), Developments in Soil Science. Elsevier, pp. 333–349.
59. Zevenbergen, L.W., Thorne, C.R., 1987. Quantitative analysis of land surface topography. Earth Surf. Process. Land. 12, 47–56.

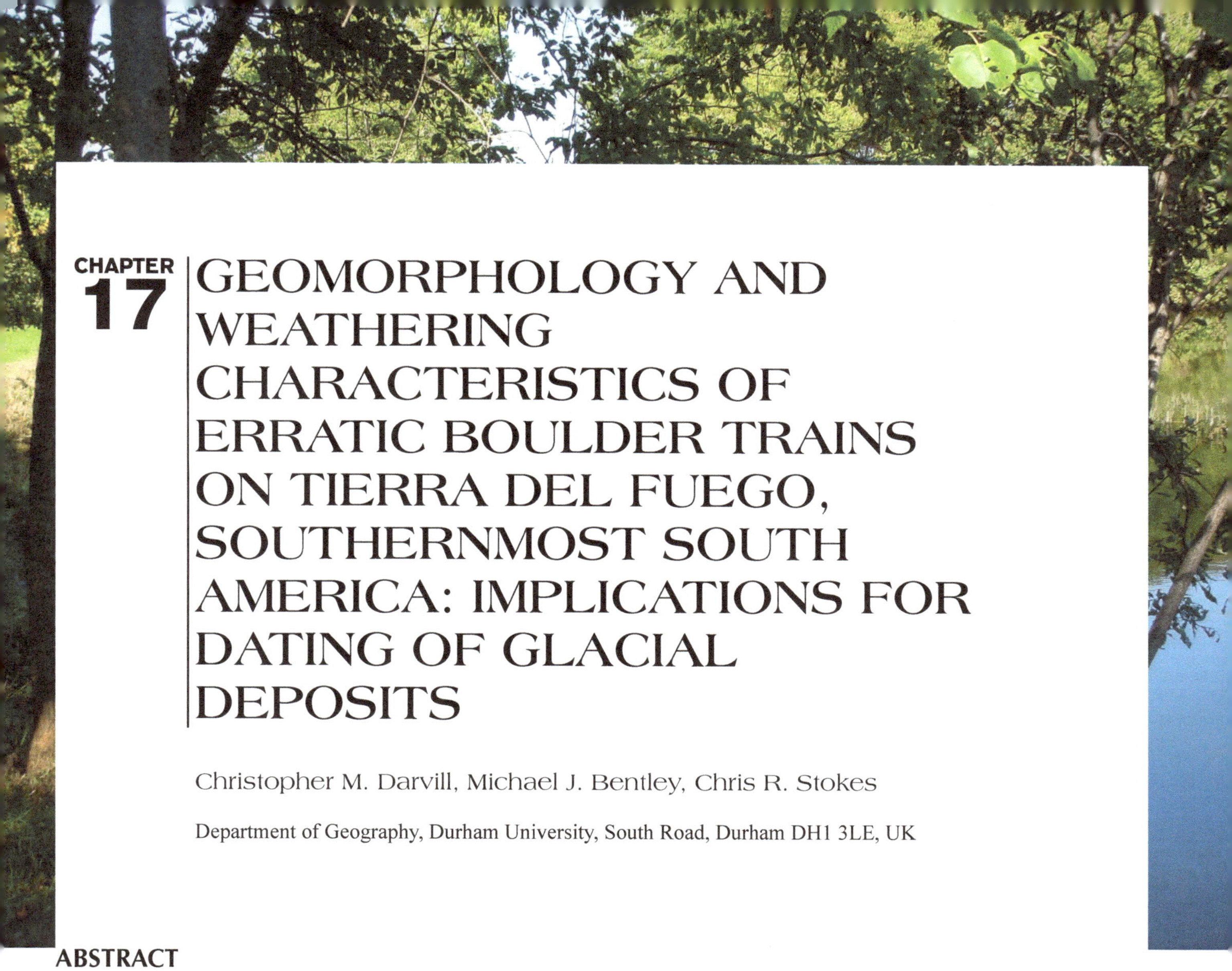

CHAPTER 17

GEOMORPHOLOGY AND WEATHERING CHARACTERISTICS OF ERRATIC BOULDER TRAINS ON TIERRA DEL FUEGO, SOUTHERNMOST SOUTH AMERICA: IMPLICATIONS FOR DATING OF GLACIAL DEPOSITS

Christopher M. Darvill, Michael J. Bentley, Chris R. Stokes

Department of Geography, Durham University, South Road, Durham DH1 3LE, UK

ABSTRACT

Erratic boulder trains (EBTs) are a useful glacial geomorphological feature because they reveal former ice flow trajectories and can be targeted for cosmogenic nuclide exposure dating. However, understanding how they are transported and deposited is important because this has implications for palaeoglaciological reconstructions and the pre-exposure and/or erosion of the boulders. In this study, we review previous work on EBTs, which indicates that they may form subglacially or supraglacially but that large angular boulders transported long distances generally reflect supraglacial transport. We then report detailed observations of EBTs from Tierra del Fuego, southernmost South America, where their characteristics provide a useful framework for the interpretation of previously published cosmogenic nuclide exposure dates. We present the first comprehensive map of the EBTs and analyse their spatial distribution, size, and physical appearance. Results suggest that they were produced by one or more supraglacial rock avalanches in the Cordillera Darwin and were then transported supraglacially for 100 s of kilometres before being deposited. Rock surface weathering analysis shows no significant difference in the weathering characteristics of a sequence of EBTs, previously hypothesized to be of significantly different age (i.e., different glacial cycles). We interpret this to indicate that the EBTs are much closer in age than previous work has implied. This emphasises the importance of understanding EBT formation when using them for cosmogenic nuclide exposure dating.

INTRODUCTION

Erratic boulder trains (EBTs) are a poorly understood glacial geomorphological feature. These linear clusters of erratic boulders record the flow lines of former glaciers by pinpointing the parent rock from which they have originated (Kujansuu and Saarnisto, 1990 and Evans, 2007) and have frequently been targeted for cosmogenic nuclide exposure dating (Jackson et al., 1997, Jackson et al., 1999, McCulloch et al., 2005,Kaplan et al., 2007, Kaplan et al., 2008, Ward et al., 2007, Evenson et al., 2009, Vincent et al., 2010 and Wilson et al., 2012). Consequently, they offer a valuable tool for reconstructing the nature and timing of former glacial advances.

Despite their importance to palaeoglaciology, these features are rarely reported in detail, and understanding their formation will help contextualise dating studies. This paper brings together previous literature on EBTs to assess how they form and presents detailed observations of examples from Tierra del Fuego, southernmost South America. The Tierra

Citation: Christopher M. Darvill, Michael J. Bentley, Chris R. Stokes, Geomorphology and weathering characteristics of erratic boulder trains on Tierra del Fuego, southernmost South America: Implications for dating of glacial deposits, doi:10.1016/j.geomorph.2014.09.017

del Fuego EBTs make an excellent case study because they are well preserved and easily distinguishable. They have also been investigated using cosmogenic nuclide exposure dating, but the resultant ages can be interpreted in two quite different ways (McCulloch et al., 2005, Kaplan et al., 2007, Kaplan et al., 2008 and Evenson et al., 2009). This study aims to test between these two opposing hypotheses by combining spatial and volumetric measurements with weathering proxies to gain a better understanding of EBT formation. In this way, we test the interpretation of cosmogenic nuclide exposure dates.

DEFINITION AND PREVIOUS WORK ON ERRATIC BOULDER TRAINS

The EBTs are a subset of dispersal trains, which includes any dispersal of a particular lithology by former ice flow (DiLabio, 1981, DiLabio, 1990, Dyke and Morris, 1988 and Evans, 2007). However, whilst EBTs are linear clusters of boulders, other dispersal trains are not necessarily linear or clustered and can include a wide range of grain sizes, surficial and within glacial deposits. Given the lack of any previous compilation in the literature, we begin by providing a brief review of the limited number of detailed studies of EBTs, summarised in Fig. 1 and Table 1, focusing on their formation and dating. Likely other EBTs exist, but they are rarely reported in the literature and are often only given cursory mention in wider studies.

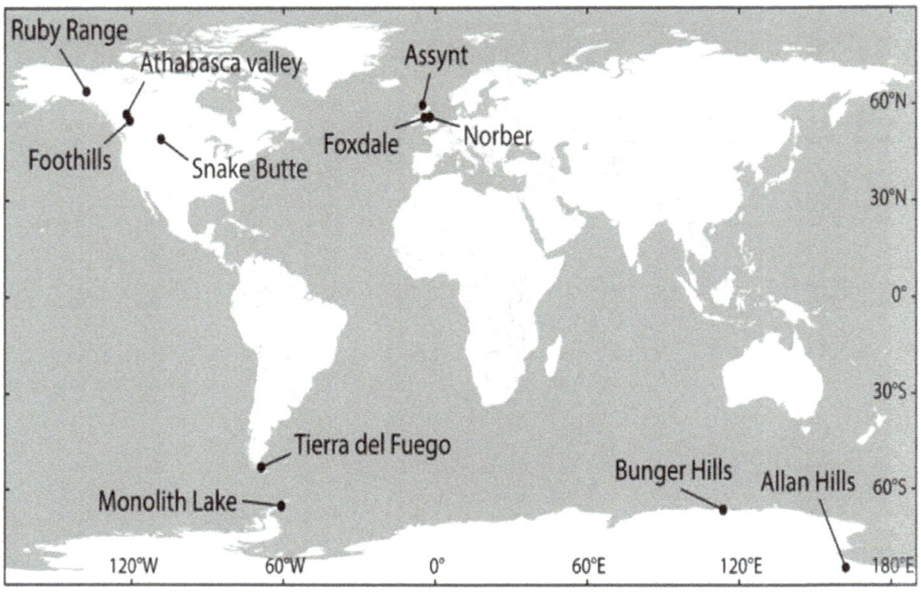

Figure 1: Map showing the locations of erratic boulder trains reviewed in this paper (see Section 2).

Table 1: Summary of the key characteristics of EBTs based on a review of the literature (NR = not reported)

EBT name	Location	Length of train(s)	Max. distance from source	Boulder diameter	Lithology	Suggested transport pathway	CNE dated?	Age	References
Foothills	Canada	>580 km	>580 km	1–41 m	Quartzite and pebbly quartzite	Supraglacial	^{36}Cl	18–12ka	Stalker (1956); Mountjoy (1958); Stalker (1976); Jackson et al. (1997); Jackson et al. (1999); Jackson and Little (2004)
Athabasca valley	Canada	ca. 70 km	ca. 120 km	Up to 1 m	Metamorphic schist	Supraglacial	-	-	Roed et al. (1967)
Ruby Range	Canada	ca. 5 km	ca. 5 km	Some >1.5 m	NR	NR	^{10}Be	54–51 ka	Ward et al. (2007)
Snake Butte	USA	ca. 79 km	ca. 80 km	Up to 23 m	Shonkinite	Subglacial?	-	-	Knechtel (1942)
Assynt	Scotland	9–14 km (4 trains)	>9 km	NR	Sandstone	NR	-	-	Lawson (1990); Lawson (1995)
Norber	England	>1 km	>1 km	Up to 4 m	Greywacke	Likely subglacial over a short distance	^{36}Cl	22–17 ka	Davis (1880); Goldie (2005); Huddart (2002); Vincent et al. (2010); Wilson et al. (2012)
Foxdale	Isle of Man	Up to 1 km	≤2 km	Up to 1 m	Granite	Subglacial	-	-	Roberts et al. (2007); Roberts (*pers. comm.*)
Bunger Hills	Antarctica	Up to 4 km?	≤4 km	NR	Dolerite	Subglacial but only a short distance	-	-	Adamson and Colhoun (1992); Augustinus et al. (1997)
Allan Hills	Antarctica	Up to 3 km?	≤3 km	Up to 3 m	Sandstone	Subglacial	-	-	Atkins et al. (2002)
Monolith Lake	Antarctica	ca. 9 km	ca. 12 km	Up to 5 m	Hyaloclastite	Likely supraglacial	-	-	Davies et al. (2013)
Tierra del Fuego	Chile/Argentina	4–15 km (4 trains) 95 km total	ca. 250 km	Up to 21 m	Granodiorite	Supraglacial	^{10}Be ^{26}Al ^{36}Cl	222–15 ka	Darwin (1841); Meglioli (1992); Coronato et al. (1999); Bentley et al. (2005); McCulloch et al. (2005); Kaplan et al. (2007); Kaplan et al. (2008); Evenson et al. (2009); This study

Formation: Subglacial Versus Supraglacial

No single model for the formation of EBTs exists, and it is possible that they can be formed in a variety of ways. This is not surprising given the reported variety in boulder size, train length, number of boulders, transport distance, and lithology (Table 1). Two hypotheses prevail: (i) subglacial entrainment and (ii) supraglacial debris.

The Norber EBT in England, Foxdale EBTs on the Isle of Man, Bunger Hills EBT and Allan Hills EBT in Antarctica, and Snake Butte EBT in the USA (see Fig. 1 and Table 1) are all interpreted to have formed subglacially. The Norber boulders have been transported laterally more than 1 km and 120 m vertically upward from their source lithology (Huddart, 2002 and Wilson et al., 2012). Given that the ice flowed over the source outcrop (Vincent et al., 2010), we suggest that subglacial transport of the boulders is most probable (though the formation mechanism has not been investigated further). The two Foxdale boulder trains were interpreted to have been initially transported and deposited subglacially by ice flowing southeastward, but with subsequent ice flowing southwestward and dispersing the larger train subglacially across a broader area of the southern part of the island (Roberts et al., 2007). In the Bunger Hills, Augustinus et al. (1997) suggested that a lack of glacial polish or facetting on the boulders implied subglacial transport over only a very short distance, thereby explaining the limited extent of the EBT. Likewise, Atkins et al. (2002) considered the boulders of the Allan Hills EBT to have been eroded by plucking of the Beacon sandstone bedrock prior to subglacial dragging and deposition on the stoss side of a bedrock ridge. Knechtel (1942) suggested that striations and polished surfaces of boulders of the Snake Butte EBT resulted from transport at the base of ice flowing southeastward and that they were then deposited with ground moraine.

In contrast, the Foothills EBT and Athabasca Valley EBT in Canada, Monolith Lake EBT in Antarctica, and Tierra del Fuego EBTs in Chile/Argentina are all suggested to have formed from material being deposited onto ice and then transported supraglacially. The Foothills boulder train formed in a medial ice position as two lobes converged around a nunatak and show no signs of subglacial transport (Jackson and Little, 2004). Similarly, the friable nature of the Monolith Lake boulders means that they are only likely to have survived if transported supraglacially (Davies et al., 2013). Meglioli (1992), Bentley et al. (2005), McCulloch et al. (2005), and Evenson et al. (2009) all noted the tight distribution, large size, angularity, and monolithology of the boulders on Tierra del Fuego, which are unlikely to have survived subglacial erosion and are instead indicative of supraglacial transport (Evenson et al., 2009).

The transport pathway has important implications for the likely exposure and depositional history of a boulder train. Too few detailed studies of EBTs exist to be able to clearly define their formation based on physical characteristics. However, our synthesis of previously published data suggests an apparent trend between transport distance, boulder size, and the proposed transport pathway (Fig. 2), with those moved greater distances (e.g. >~ 10 km) more likely to have been transported supraglacially. The relationship with boulder size is unsurprising given the association between transport pathway and boulder erosion (Boulton, 1978), but is important in the context of erratic dispersal more generally. For example, the principles of 'half-distance' transport (Salonen, 1986) and concentration peaks (DiLabio, 1981, DiLabio, 1990 and Boulton, 1996) may better relate to subglacial EBTs, whereas supraglacial EBTs are also controlled by the maximum transport distance and the preservation potential of the boulders.

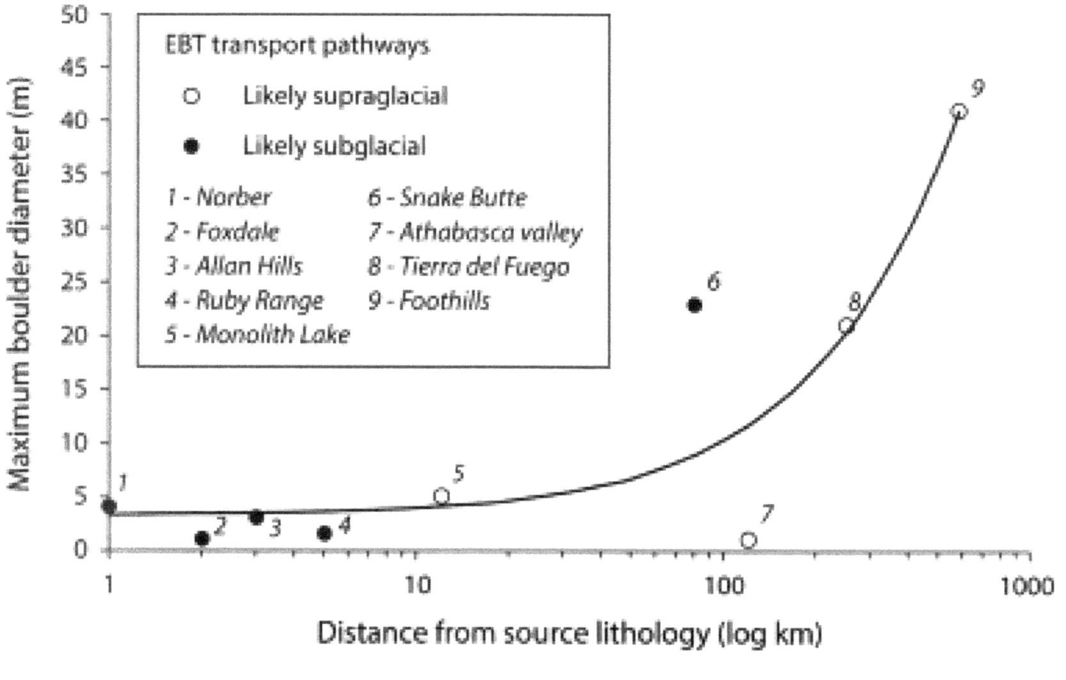

Figure 2: Maximum boulder diameter plotted against the maximum distance of boulders from their source lithology (note the log scale), as reported in studies of EBTs (labelled as per Table 1). Our review of the literature suggests that EBTs consisting of larger boulders (> 5 m) transported greater distances (> 10 km) are more likely to have been transported supraglacially. The Snake Butte EBT is an exception to this pattern;Knechtel (1942) inferred subglacial transport of these boulders. Of the reported studies, lithology does not appear to play a key role in determining the preservation of boulder trains through supraglacial or subglacial transport.

Cosmogenic Nuclide Exposure Dating

The EBTs are useful targets for cosmogenic nuclide exposure dating, with the added benefit of being able to trace the source and transport pathway of the samples. The Foothills boulder train in Canada was dated using ^{36}Cl by Jackson et al., 1997 and Jackson et al., 1999, yielding dates of 18–12 ka and demonstrating the limits of the Laurentide Ice Sheet during its Last Glacial Maximum (LGM). The dates implied a maximum transport time of ca. 3 ka (< 17% of the total exposure time; Jackson and Duk-Rodkin, 1996 and Jackson et al., 1997) and yielded one anomalously old age of ca. 53.3 ka, which Jackson et al. (1997) ascribed to pre-exposure.

The Norber boulder train in England was also dated to around 22–17 ka using ^{36}Cl, which helped establish the timing of deglaciation in the region (Vincent et al., 2010 and Wilson et al., 2012). The ^{10}Be dates from the Ruby Range boulder train in Canada were clustered and suggested that the train was deposited during the penultimate glacial episode, with dates of 54–51 ka (Ward et al., 2007). The authors suggested no significant influence of inheritance given that the same pre-exposure prior to deposition of all four samples is unlikely (Ward et al., 2007), and subglacial transport of the Norber and Ruby Range boulders may have removed any inheritance signal. Statistical modelling of supraglacial erosion and transport of boulders has also suggested that these processes can yield dateable inheritance signatures in erratic boulders (Applegate et al., 2010,Applegate et al., 2012 and Heyman et al., 2011).

As noted above, the Tierra del Fuego EBTs have been targeted for cosmogenic nuclide exposure dating (McCulloch et al., 2005, Kaplan et al., 2007, Kaplan et al., 2008 and Evenson et al., 2009). The established regional age model for the timing of glaciations implies that a series of four glacial limits are successively less extensive (i.e., 'nested') and correspond to different glacial cycles: MIS 12, 10, 6, and 2 (Fig. 3). This is based on palaeomagnetism, uranium-series dating, correlation between marine terraces, and relative weathering indices and augmented by radiocarbon, amino acid racemization, and tephra dating of the younger limits (Meglioli, 1992, Coronato et al., 2004, Rabassa, 2008 and Rabassa et al., 2011; and references therein). Published cosmogenic nuclide exposure dates from two EBTs on the MIS 2 (LGM) limit close to Bahía Inútil cluster around 20 ka (Fig. 4 and Fig. 5) and agree well with radiocarbon dates of deglaciation in the region (Heusser, 2003, McCulloch et al., 2005, Kaplan et al., 2008 and Hall et al., 2013). However, two EBTs on the outer two limits yielded dates significantly younger than expected, dominantly between 30 and 15 ka for the putative MIS 12 limit ($n = 7$) and two dates of 24 and 222 ka for the putative MIS 10 limit (Kaplan et al., 2007 and Evenson et al., 2009).

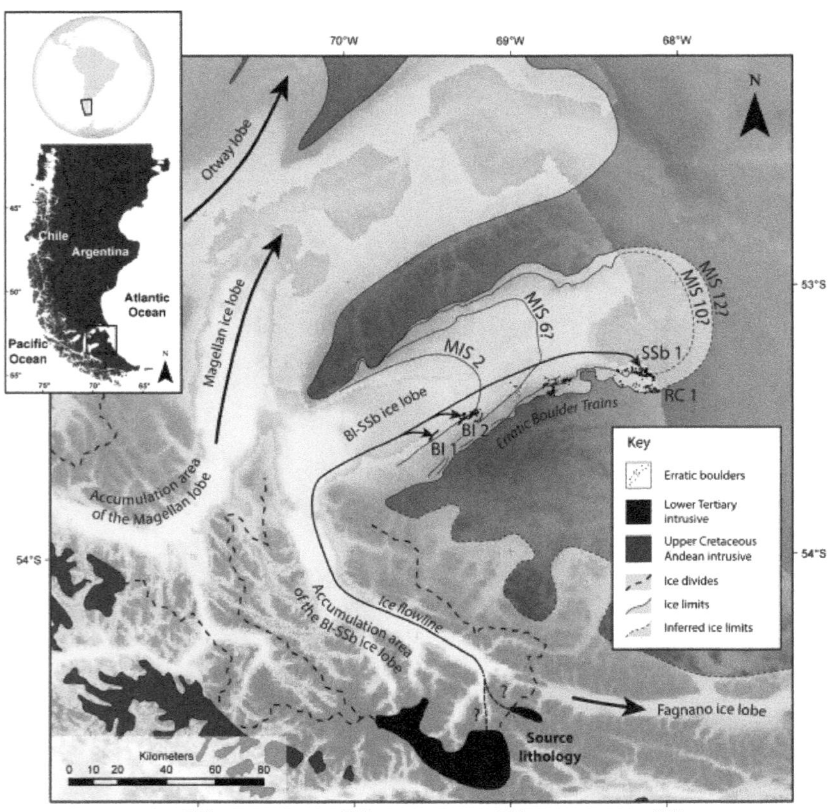

Figure 3: The location of the study area, showing the extent of former ice lobes in southernmost South America. Thick dashed lines indicate inferred ice divides, and thin lines show the outermost limit of ice lobes based on geomorphology mapped by Darvill et al. (2014) or inferred (dashed) from the literature. For the Bahía Inútil–San Sebastián (BI–SSb) ice lobe, the hypothesised limits and their ages are shown to illustrate how the EBTs relate to them. All other ice lobes are shown at maximum extent, though this does not imply that they advanced or retreated to these limits at the same time. A former flow line from the source lithology in the Darwin Cordillera to the EBTs on Tierra del Fuego is illustrated. The two possible origins of the granodiorite lithology are shown (from Natland et al., 1974), but only the lower Tertiary intrusive unit outcrops in the former accumulation area of the BI–SSb ice lobe.

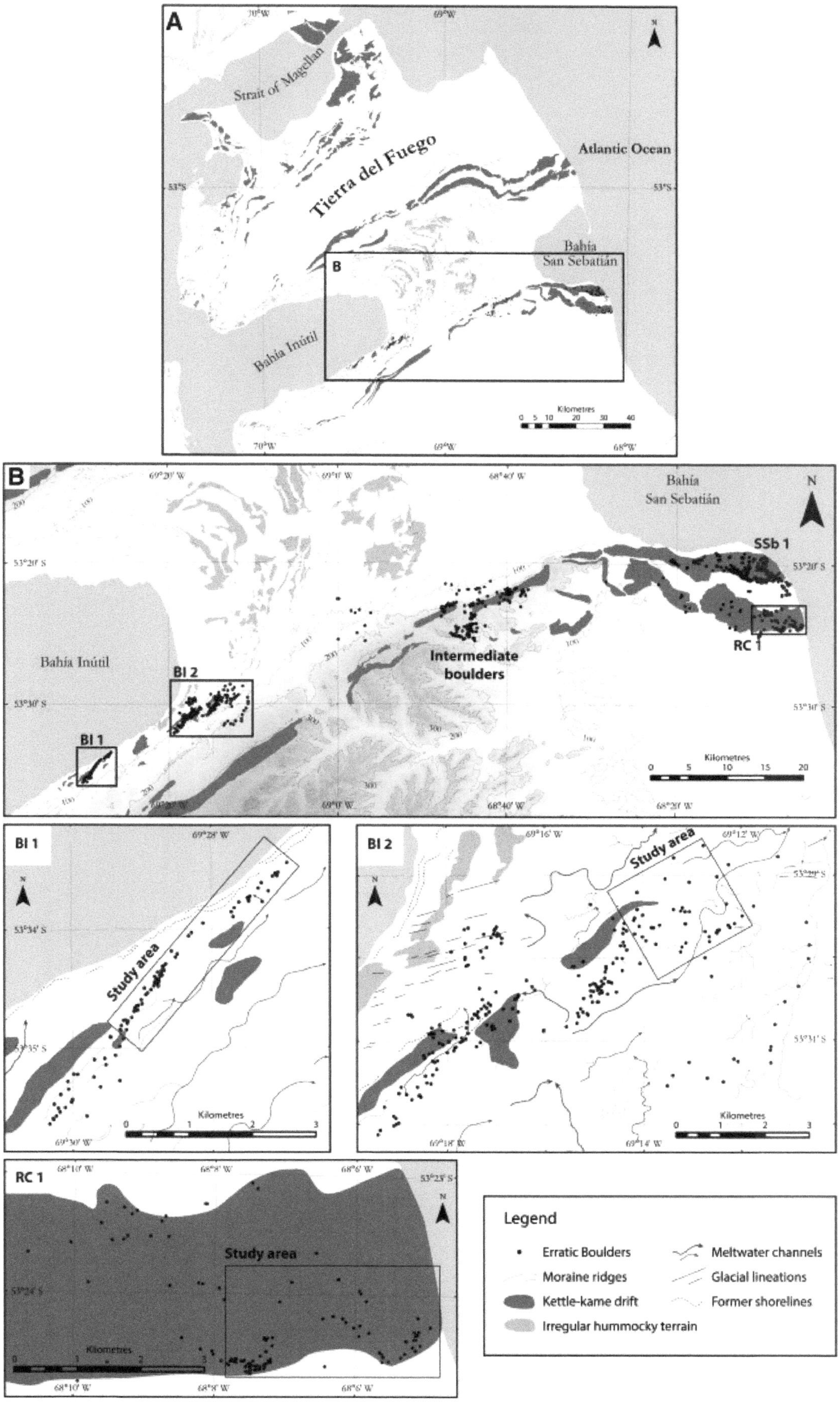

Figure 4: (A) Simplified overview of the glacial geomorphology from Darvill et al. (2014) for Tierra del Fuego. (B) Detailed glacial geomorphological maps showing the boulders mapped in this study on Tierra del Fuego. From our mapping, the boulders clearly form EBTs (rather than simply being boulders on moraines): the trains cut across the moraine morphology, form isolated discrete clusters, and are not found on equivalent moraines to the north. Boxes in the overview map are shown enlarged below — these three boulder trains are examined in this study. The sampled boulders are highlighted within each respective EBT.

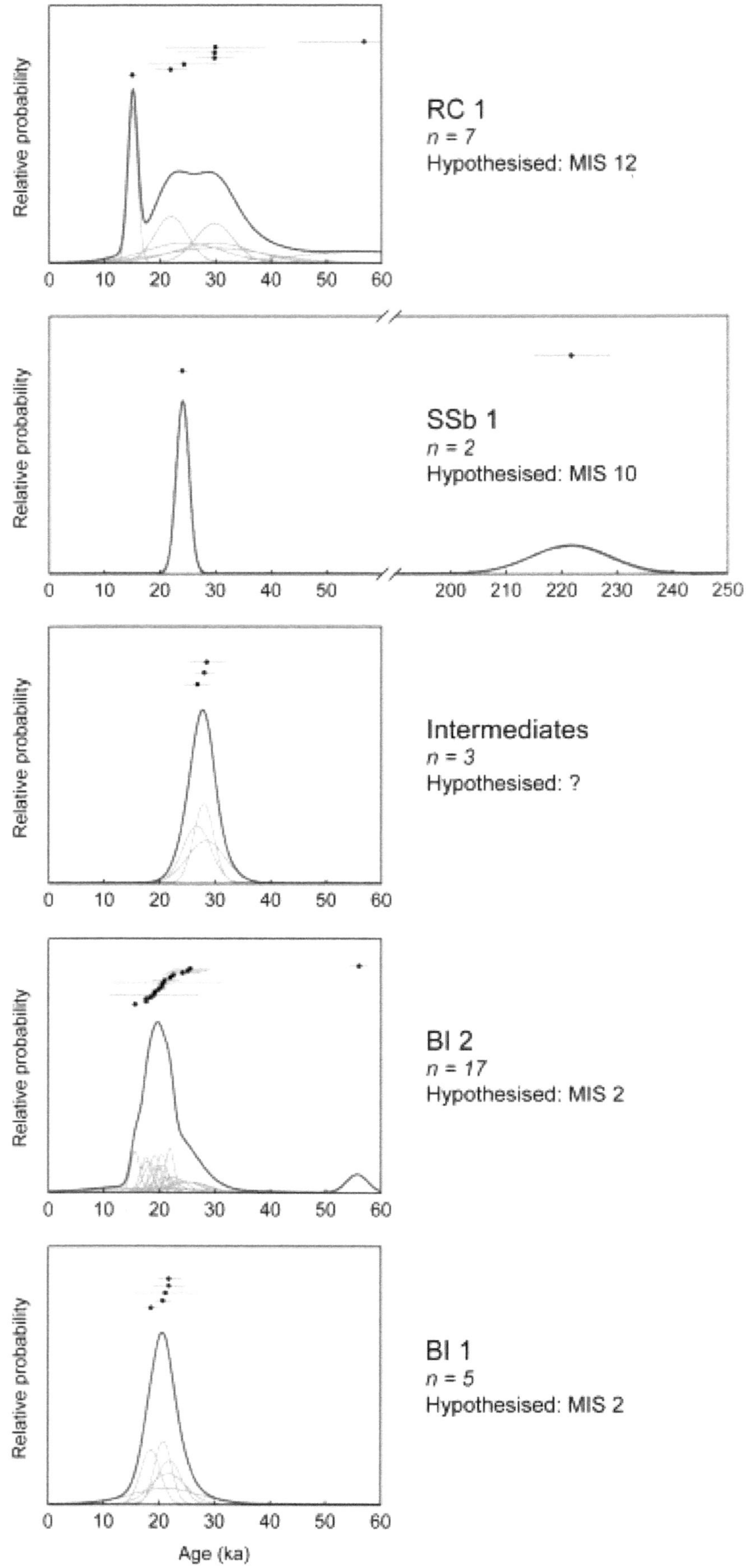

Figure 5: Previous ^{10}Be cosmogenic nuclide exposure dates from the boulder trains, as mean ages with standard errors (plotted in age order) and as probability plots; recalculated from the original data of McCulloch et al. (2005), Kaplan et al., 2007 and Kaplan et al., 2008, and Evenson et al. (2009); and using the production rate of Putnam et al. (2010) and the scaling model of Lal (1991) and Stone (2000). Note the substantial difference between the cosmogenic nuclide exposure dates and the hypothesised ages for RC 1 and SSb 1.

Our review of the literature highlights that EBTs can form in different ways, with different erosion, transport, and depositional histories. For supraglacial EBTs, this can result in incomplete erosion of inherited nuclide concentrations, which is important to understand if using the features for cosmogenic nuclide exposure dating.

This Study

Kaplan et al. (2007) suggested that the occasional older dates for the EBTs on Tierra del Fuego were closer to the true age of the glacial limits and proposed that intense, episodic exhumation and/or erosion of the EBTs resulted in the samples yielding anomalously young ages (i.e., they are 'old' boulders that were exhumed; Fig. 6). However, an alternative hypothesis, not previously considered, is that all of the EBTs in Tierra del Fuego were deposited during the last glacial cycle (i.e., they are 'young' boulders; Fig. 6), and occasional samples are anomalously old owing to inheritance. The 'old' hypothesis fits with the established age model but requires intense physical exhumation and erosion to explain the dates (Kaplan et al., 2007). The 'young' hypothesis does not require such extreme processes but questions the established age model and implies that ice was much more extensive during the last glacial cycle. This study tests between these two opposing hypotheses, and a summary of the expected weathering characteristics that might be found under these two different scenarios is given in Table 2.

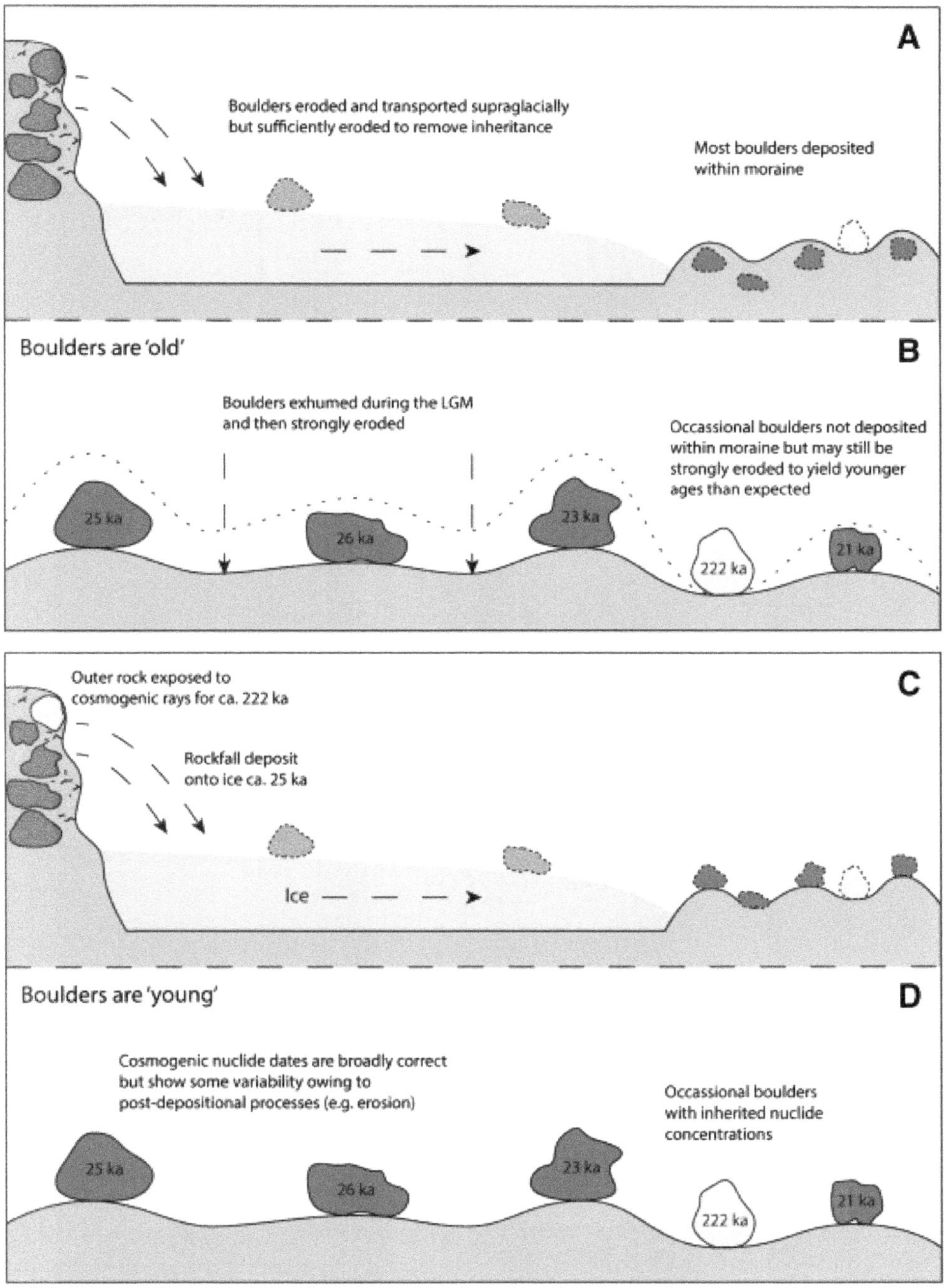

Figure 6: The different cosmogenic nuclide exposure dates from the RC 1 and SSb 1 boulders on Tierra del Fuego can be explained by two different hypotheses. (A) and (B) Illustrate the first hypothesis, proposed byKaplan et al. (2007), which suggests that the majority of the boulders were exhumed and eroded subsequent to deposition, yielding anomalously young ages. (C) and (D) Illustrate the second hypothesis, proposed in this study, which suggests that occasional boulders may be anomalously old owing to inheritance because they were not sufficiently eroded by supraglacial transport. In this case, the majority of the cosmogenic nuclide exposure dates are a better representative of the age of the EBTs.

Table 2: Age and likely weathering characteristics for the Tierra del Fuego EBTs

	Boulder train		
Properties	BI 1/BI 2	RC 1 if 'old'[a]	RC 1 if 'young'[b]
Hypothesised age	20 ka[c]	ca. 450 ka[c]	30–20 ka[c]
Cosmogenic ages	ca. 20 ka[d]	ca. 30–15 ka[e]	ca. 30–15 ka[e]
Nature of erosion	Possibly some erosion	Intense exhumation and erosion[f]	Possibly some erosion
Likely agents of erosion	Wind erosion, frost action, dissolution, mild salt-spray weathering	Wind erosion, frost action, dissolution, salt-spray weathering	Wind erosion, frost action, dissolution, salt-spray weathering
Possible weathering rates	0.5–12 mm ka^{-1}[g]	> 25 mm ka^{-1}[h]	0.5–12 mm ka^{-1}[g]
Likely surface erosion	≤ 240 mm[i]	500–> 11,250 mm[j]	≤ 360 mm[k]
Likely erosional difference compared to BI 1/BI 2	0 mm	260–> 11,010 mm	≤ 120 mm
Roughness compared to BI 1/BI 2	N/A	Significantly rougher	Possibly slightly rougher or the same
Hardness compared to BI 1/BI 2	N/A	Significantly weaker	Possibly slightly weaker or the same

[a]As hypothesised by Kaplan et al. (2007).

[b]As hypothesised in this study.

[c]Approximate ages, based on Meglioli (1992); Kaplan et al., 2007 and Kaplan et al., 2008.

[d]McCulloch et al. (2005); Kaplan et al. (2008); Evenson et al. (2009).

[e]Kaplan et al. (2007); Evenson et al. (2009).

[f]Kaplan et al (2007).

[g]Kaplan et al. (2005) calculated apparent erosion rates of roughly 0.5 to 2.5 mm ka^{-1} over > 760 ka for boulders deposited by the Lago Buenos Aires lobe in northern Patagonia. Kaplan et al. (2007)calculated apparent erosion rates of roughly 5 to 12 mm ka^{-1} over ca. 50–120 ka for boulders deposited by the Río Gallegos lobe in southern Patagonia.

[h]Apparent erosion rate estimated by Kaplan et al. (2007).

[i]Assuming up to 12 mm ka^{-1} erosion over 20 ka.

[j]Assuming > 25 mm ka^{-1} erosion following rapid exhumation at 20 ka or continuous erosion at > 25 mm ka^{-1} over 450 ka. Kaplan et al. (2007) suggested that continuous erosion is unlikely given preservation of original surface geomorphology, so the amount of erosion is probably between these two end members.

[k]Assuming up to 12 mm ka^{-1} erosion over up to 30 ka.

METHODS

Mapping and Sampling

We began by mapping all erratic boulders in the study area and then selected a sample of 150 boulders from three of the EBTs to compare trains hypothesised to be of similar and differing ages (Fig. 3). Many of the boulders are sufficiently large and clear against the surrounding landscape for them to be mapped from remote imagery, so a map was produced to show their distribution using a combination of remote sensing analysis and field-checking. Aerial photographs from the Servicio Aerofotométrico de la Fuerza Aérea de Chile were used where possible as well as Google Earth™ imagery (version 7). Field-checking verified the broad spread of mapped boulders, and the locations of the 150 sampled boulders were recorded using a handheld Magellan eXplorist 610 GPS device. All boulders within a given area were sampled provided they were > 2 m in height. This was to avoid the effects of snow/vegetation cover and variable erosion on smaller boulders and to avoid the chances of sampling fragments that may have broken off subsequent to deposition. Boulders were described in terms of basic lithology, surface appearance, and setting.

Size approximation, angularity, and appearance

Boulders varied in size and accessibility, so boulder dimensions and volume were estimated using eight photographs taken around each boulder at roughly equal bearings. All photographs included a 1-m measuring-staff that was later used to gauge boulder height, width, and depth. Volume was then calculated in three ways to give approximate upper, middle, and lower values (supplementary material: SM1.2). Boulder angularity was recorded using a visual scale (Benn, 2004) and a relative measurement technique (Kirkbride, 2005; SM1.1). The surficial characteristics and lithology of the boulders were also noted.

Schmidt Hammer

The Schmidt hammer has been used to measure rock-surface hardness as an indicator of the amount of time that rock has been exposed to subaerial processes (Matthews and Owen, 2010; see SM1.4). Rebound values (*R*-values) produced by the impact of the hammer decline with increased rock-surface weathering, and so can be used as a relative measure of exposure history (McCarroll, 1991, McCarroll and Nesje, 1993,Nesje et al., 1994, Goudie, 2006 and Shakesby et al., 2011). We used an N-type Schmidt hammer to analyse 50 boulders from two of the boulder trains thought to be from different glacial cycles (BI 1 and RC 1). Fifty Schmidt hammer blows were recorded per boulder, with a total of 2500 blows per boulder train (5000 blows in total). On each boulder, 10 blows were recorded per face (approximately: north, east, south, west, and top) to explore whether aspect is a control on rock-surface weathering.

Profile Gauge

Rock surface roughness can be used to compare the effects of weathering where it is assumed that initial surface texture was roughly the same (McCarroll and Nesje, 1993 and McCarroll and Nesje, 1996). This can then be used as an indicator of the relative time that a rock surface has been exposed to weathering processes. A profile gauge is a quick and easy tool for measuring rock surface roughness (McCarroll and Nesje, 1996), and we used a 25-cm gauge consisting of 250 × 1 mm independent pins to sample planar rock surfaces (SM1.3). The gauge was pressed firmly against the surface and then traced onto graph paper in the field and later digitised. Pin positions were recorded every 8, 16, 24 and 32 mm to evaluate different roughness wavelengths. We measured roughness profiles for 50 boulders in three boulder trains (BI 1, BI 2, and RC 2). Five profiles were recorded per boulder, making a total of 250 profiles per boulder train (750 profiles in total). On each boulder, one profile was recorded per face (approximately: north, east, south, west, and top) to show whether roughness varied with aspect.

RESULTS

Distribution

We mapped a total of 1248 boulders and distinguished four EBTs. From innermost (ice-proximal) to outermost (ice-distal), these are: BI 1, BI 2, SSb 1, and RC 1 (Fig. 4). In addition, a large spread of boulders lies between Bahía Inútil and Bahía San Sebastián (intermediates in Fig. 4B). These are dispersed rather than tightly clustered and so are not classed as an EBT for the purposes of our study. However, we highlight that these intermediate boulders bridge the gap between BI 1/ BI 2 and SSb 1/RC 1. It was not possible to access the SSb 1 EBT or some of the intermediate boulders for field-checking or analysis, but both were sampled for cosmogenic nuclide exposure dating by Kaplan et al. (2007). A summary of the characteristics of the boulder trains is given in Table 3 and example photos are given in Fig. 7.

Table 3: The spatial characteristics and geomorphological context of the EBTs on Tierra del Fuego.

EBT	No. boulders mapped	Approx. train length (km)	Approx. train width (km)	Approx. train height (m asl)	Spatial distribution	Geomorphological context	Hypothesised ages
BI 1	98	5.5	0.5	50–70	A highly linear train extending to the coast of Bahía Inútil.	On a moraine belt, parallel to meltwater channels and with one meltwater channel dissecting it.	MIS 2
BI 2	238	11.0	7.5	60–220	Elements of linearity, but spread over a wider area than BI 1. May consist of more than one EBT, but not possible to distinguish in the field.	Deposited across numerous moraines, with several meltwater channels cutting through.	MIS 2
Intermediates	267	26.0	8.0	30–80	Boulder density is much less than the other EBTs, so the boulders were not studied further.	Lies across several moraines.	Possibly relates to more than one glacial limit
SSb 1	476	15.0	4.0	0–80	Boulder density varies, delimited to the north and south by the extent of kettle and kame drift.	Sits on a broad band of kettle and kame drift and continues eastward into the Atlantic Ocean.	MIS 10
RC 1	162	20.0	4.0	0–40	Boulder density varies, delimited to the north and south by the extent of kettle and kame drift. There are two tight, linear clusters of boulders roughly 1.7 km long and 1.5 km long.	Sits on a broad band of kettle and kame drift and continues eastward into the Atlantic Ocean.	MIS 12

Figure 7: Example photographs of the EBTs on Tierra del Fuego. (A) and (B) are from BI 1; (C) and (D) are from BI 2; and (E) and (F) are from RC 1. Note the large size, angularity, and clustering of the boulders. Additional colour photographs can be found in SM2.

Volume Characteristics

The total volume of rock within the measured boulders in BI 1, BI 2, and RC 1 combined is > 22,000 m^3, within which the total volume of BI 1 is > 5000 m^3, BI 2 is > 14,000 m^3, and RC 1 is > 2000 m^3 (Fig. 8). The largest boulders were in BI 2, with three boulders exceeding 1000 m^3. Of all boulders sampled, only one was found to be taller than it was wide, and we interpret this to indicate that almost all boulders are stable and 'at rest'. Several examples of 4–6 near-consecutive boulders were found to increase in volume down-ice (Fig. 8). This is masked in the field by scattered smaller boulders, and the pattern is least distinct in BI 2 where boulders are more dispersed. However, we suggest that these trends are real because they cannot be explained by measurement uncertainty or by changing the point from which EBT distance is measured.

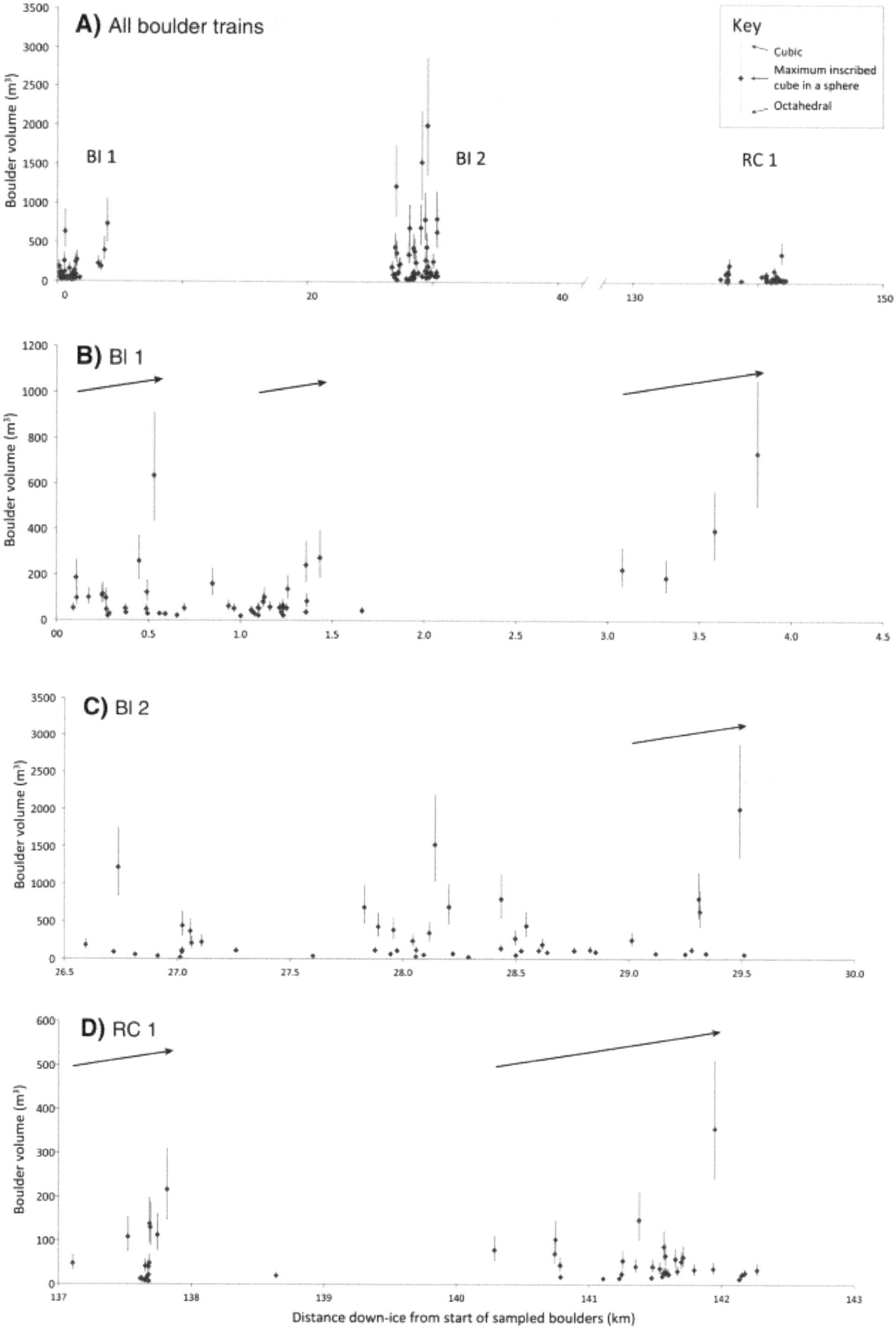

Figure 8: The results of boulder volume measurements. (A) Shows the results for all three boulder trains, which are shown individually beneath: (B) BI 1; (C) BI 2; (D) RC 1. Arrows highlight an apparent repeated pattern of increasing down-ice volume across several near-consecutive boulders in all three EBTs.

Angularity

Little difference in angularity was recorded between boulder trains, and all showed a dominance of angular or subangular boulders, supported by RA values of 64% for BI 1/BI 2 and 54% for RC 1 (see Fig. 9).

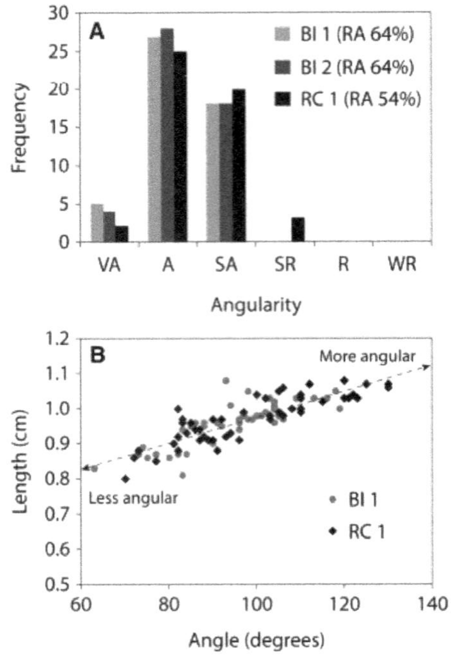

Figure 9: The results of angularity measurements. (A) Is a frequency histogram of observed boulder roundness based on criteria in SM1.1, showing a dominance of angular boulders supported by high RA values (percentage of angular and very angular boulders). (B) Shows the similarity in boulder edge roundness between BI 1 and RC 1.

Rock Surface Hardness

The Schmidt hammer *R*-values were corrected for the angle at which the Schmidt hammer was held and for instrumental drift (SM1.4). The total mean *R*-values for the two boulder trains are 43.6 ± 19.9 for BI 1 and 37.7 ± 19.7 for RC 1 (Fig. 10), and the means of different aspect faces also show consistently lower *R*-values for RC 1 than BI 1, in the range of 2.8–8.4. A Mann–Whitney U test reveals a statistical difference between the median values of the two EBTs at the 0.95 significance level ($p = < 0.05$), when comparing total values and for each of the aspect faces.

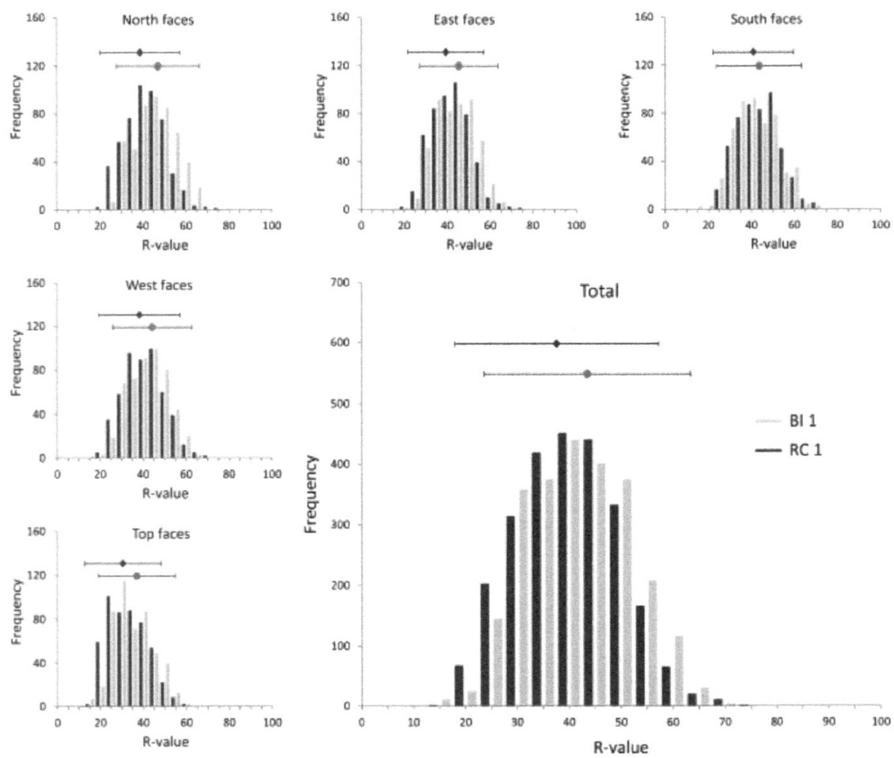

Figure 10: Results of Schmidt hammer rock hardness analysis on the BI 1 and RC 1 EBTs, as total data and split into directional faces. Mean data are shown with 2σ errors, in addition to frequency histograms that illustrate the overlap between the two boulder trains.

Rock Surface Roughness

Following McCarroll and Nesje (1996), we present our roughness data in two ways. Index A uses the standard deviation of measurements and removes the influence of rock surface slope, whereas root mean square (RMS) represents roughness more accurately as deviation from the mean but will be affected by surface curvature (SM1.3). The profile gauge was used on all three boulder trains, and a strong correlation was found between Index A and RMS values (Fig. 11), indicating that surface curvature did not affect the roughness results. Consequently, RMS values are used to assess differences in surface roughness (Fig. 11). Reducing the wavelength reduces the mean roughness and spread of values, but McCarroll and Nesje (1996) suggested that lower wavelengths were unlikely to capture true surface roughness. We found that the greatest rock crystal size was generally > 20 mm; hence we focus on the 32-mm wavelength data. The different number and spacing of data points along the profiles when the wavelength was varied did not affect the mean results, suggesting that the results indicate actual changes in roughness and are not caused by sampling strategy.

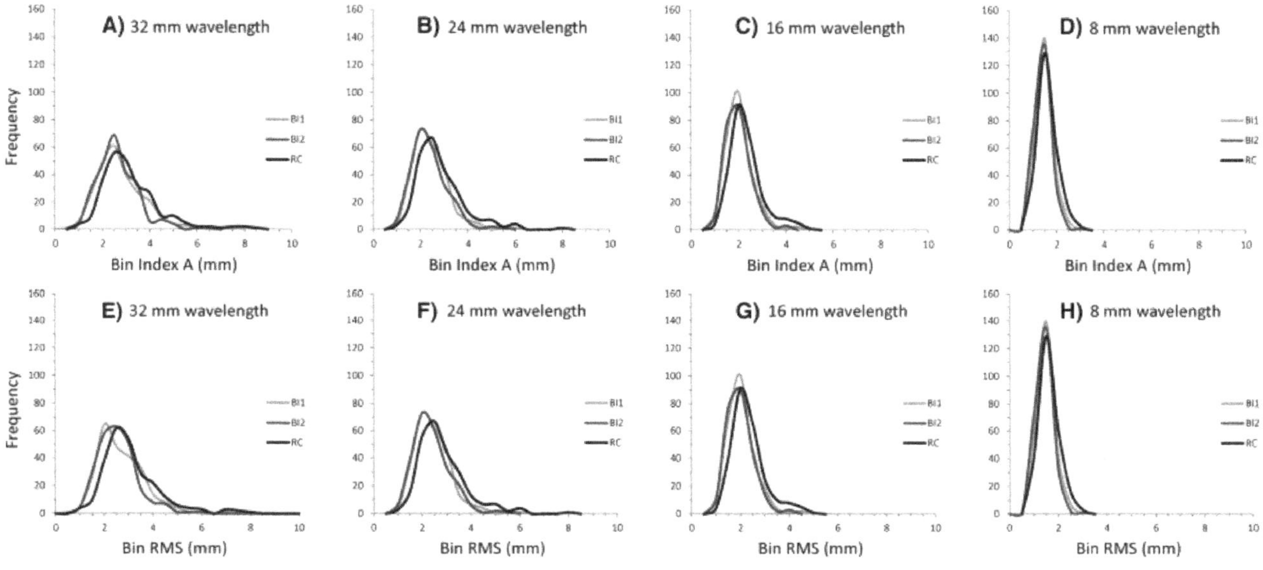

Figure 11: Results of profile gauge rock surface roughness analysis on the BI 1, BI 2, and RC 1 EBTs, shown as frequency distributions. (A–D) are Index A results and (E–H) are RMS results. Note the similarity between the results of the two methods. RMS data is used in this study, focussing on the 32-mm wavelength.

The total mean values for the three boulder trains are 2.4 ± 1.8 mm for BI 1, 2.3 ± 1.6 mm for BI 2, and 2.8 ± 2.3 mm for RC 1 at the 32-mm wavelength (Fig. 11). Like the Schmidt hammer data, patterns exists within the roughness results, with RC 1 consistently showing greater roughness – mean values 0.19–0.64 mm greater than the lowest mean value from BI 1 or BI 2 – for the total and different aspect faces. A Mann–Whitney U test shows a statistical difference between the total median values of all three EBTs at the 0.95 significance level ($p = < 0.05$). Likewise, some, but not all, of the aspect faces also show a statistical difference when comparing all three of the EBTs, and this applies when comparing BI 1 and BI 2 as well as BI 1/BI 2 and RC 1.

DISCUSSION

Source

Kaplan et al. (2007) and Evenson et al. (2009) described the boulders from BI 1, BI 2, and RC 1 as hornblende granodiorites. The only source for this lithology is a small area of the Cordillera Darwin mountain range (Fig. 3), where lower Tertiary intrusive units outcrop within the former glacial accumulation area (Natland et al., 1974, Nelson et al., 1980 and Evenson et al., 2009). This source suggests that the Bahía Inútil–San Sebastián (BI–SSb) ice lobe originated from the central Cordillera Darwin to the south. It also helps to position the ice divide between the BI–SSb and Fagnano ice lobes at peak glaciation; and the presence of supraglacial rock debris on Tierra del Fuego implies that granodiorite nunataks existed within the central Cordillera Darwin during peak glaciation. Thus, peak ice thickness could not have exceeded around 2500 m, which acts as a constraint on ice cap reconstructions.

The large boulders in Tierra del Fuego (up to 1000 m^3) are unlikely to have been transported subglacially given that erosion during the distance from their source (ca. 250 km) should have considerably reduced their size (Fig. 2). Furthermore, our data show that the boulders are dominantly angular or subangular. So many angular boulders up to 21 m in diameter are unlikely to have been transported beneath the ice over such a long distance. We found no evidence of subglacial abrasion (e.g. polishing or striae), and the angularity of the boulders suggests that such signs have not been removed by weathering. A similar conclusion was reached by Evenson et al. (2009). As such, the most likely formation mechanism is supraglacial debris at the source location in the Cordillera Darwin.

Meglioli (1992), Bentley et al. (2005), McCulloch et al. (2005), and Evenson et al. (2009)all noted the tight distribution, large size, angularity, and monolithology of the EBTs on Tierra del Fuego; and we suggest that these features match the characteristics expected of a supraglacial rock avalanche deposit

(Shulmeister et al., 2009). Studies routinely estimate total rock avalanche deposit volumes of > 1,000,000 m^3 (Shugar and Clague, 2011, Sosio et al., 2012 and Delaney and Evans, 2014) and even > 10,000,000 m^3(Jibson et al., 2006, McColl and Davies, 2011, Shugar and Clague, 2011 and Sosio et al., 2012), far in excess of our minimum estimates of > 22,000 m^3 (or > 150,000 m^3 assuming that the coarse fraction accounts for around 15% of the total volume; Delaney and Evans, 2014). This suggests that the size of the deposit is not unreasonable for a supraglacial rock avalanche. The clustering of the EBTs suggests that the erosion of the boulders occurred in discrete events and implies that gradual erosion at the source is improbable. Debuttressing during glacial retreat could have resulted in episodic deposition of debris caused by joint failures. However, we note that the boulders cut across a range of glacial geomorphology – both small moraine ridges and kettle and kame topography (Fig. 4) – and that the RC 1 EBT was deposited when the ice lobe was still fully extended. Additionally, whilst the EBTs might have been deposited during recession, they represent flowline features, not ice-marginal deposits.

The production of boulder trains as a result of supraglacial rock avalanches has not been explicitly examined in previous work. Shulmeister et al. (2009) noted that for large supraglacial rock avalanches, a greater proportion of thicker debris remains on the glacier surface rather than being incorporated into the subglacial system, and a characteristic 'carapace' of coarser debris generally caps deposits (Reznichenko et al., 2011). Coarse distal rims are found in numerous deposits (Hewitt, 1999, Hewitt, 2009,Chevalier et al., 2009, McColl and Davies, 2011 and Shugar and Clague, 2011), and theoretically, these could result in boulder trains, especially if divergent flowlines drew the boulders into a train prior to deposition (Evenson et al., 2009). However, large boulders may also be found across the debris sheet, with the rim simply representing a greater density of large boulders caused by bulldozing (Shugar and Clague, 2011). It thus remains unclear exactly how a rock avalanche debris sheet is transformed into EBTs, but we now discuss a conceptual model for their transport and deposition.

Transport and Deposition

The lateral position of the EBTs on Tierra del Fuego probably resulted from ice to the north (the Magellan lobe) deflecting the BI–SSb lobe to the east (Fig. 3), with divergent flow lines focusing the boulders in a lateral position (Evenson et al., 2009). Thus we envisage that the boulders were transported supraglacially from the central Cordillera Darwin, east of Isla Dawson, before turning sharply east into Bahía Inútil to be deposited on the south side of the former ice lobe (Fig. 3). Given that the Cordillera Darwin has experienced seismic activity in the past (Cunningham, 1993, Klepeis, 1994 and Bentley and McCulloch, 2005), the four distinct boulder trains may logically have resulted from separate rock avalanche pulses (Larsen et al., 2005 and Jibson et al., 2006). Ice crevassing, subaerial weathering, and meltwater may have acted to remove much of the finer material (Shulmeister et al., 2009), and ice flow and the divergence of flow lines then dragged the distal rim and boulder carapace into trains (Evenson et al., 2009).

Our mapping shows that the EBTs are significantly more extensive than previously thought (Evenson et al., 2009) and occur at several places along the southern limit of the former ice lobe. There are also numerous boulders that do not cluster as boulder trains (intermediates) but act as a continuum between BI 1/BI 2 and SSb 1/RC 1. Whilst the EBTs may represent distinct spatial and temporal events (as proposed by Kaplan et al., 2007, and Evenson et al., 2009), they could also represent rock avalanche pulses during the same period, possibly linked to seismic activity (Larsen et al., 2005 and Chevalier et al., 2009). Studies in New Zealand have suggested that supraglacial rock avalanches may result in the deposition of nonclimatic moraines (Anderson and Mackintosh, 2006,Tovar et al., 2008, Shulmeister et al., 2009 and Reznichenko et al., 2011). However, given the small area and isolated nature of the EBTs compared to the area covered by the BI–SSb ice lobe, a similar debris-induced, nonclimatic model is improbable.

The patterns of increasing down-ice boulder volume observed within each of the EBTs (Fig. 8) are unlikely to have been preserved in any scenario other than deposition onto – and transport on top of – the former ice. However, the formation of these patterns is unclear. Little evidence suggests that gravitational sorting of avalanche debris occurs (Marangunic and Bull, 1968 and Shugar and Clague, 2011), but Hewitt (2009), Shugar and Clague (2011), and Delaney and Evans (2014) described trains of debris several metres wide and parallel to the direction of debris flow. These likely resulted from snow ploughing of a large boulder, with finer material following behind on exposed glacial ice (Delaney and Evans, 2014), and currently offer the best analogue for the patterns of boulder size trends seen in Tierra del Fuego.

Rock Surface Weathering

The Schmidt hammer data show a statistical difference between the RC 1 and BI 1 EBTs, indicating that the RC 1 boulders have been subjected to a greater degree of weathering than the BI 1 boulders. However, the differences are not as great as might be expected if intense, episodic erosion has taken place (Table 2). The *R*-values showed a total average difference of 5.9, whereas McCarroll and Nesje (1993) demonstrated differences of 25–40 between Little Ice Age and Late Glacial sites in western Norway; and Matthews and Owen (2010) highlighted differences of 21–35 between Little Ice Age and Preboreal sites in southern Norway. Rock weathering (and therefore *R*-values) could have reached saturation in Tierra del Fuego. However, despite agreement that *R*-values will progress toward a dynamic equilibrium over time (White et al., 1998, Engel, 2007,Sánchez et al., 2009, Černá and Engel, 2011 and Stahl et al., 2013), numerous studies have effectively distinguished LGM and pre-LGM deposits using the technique on timescales of 10s to 100 s ka (Ballantyne et al., 1997, Rae et al., 2004, Černá and Engel, 2011 and Stahl et al., 2013). Very high rates of weathering are required to reduce the ages of the RC 1 boulders (Table 2), yet no obvious jump in *R*-values occurs between these and the BI 1/BI 2 boulder trains.

The roughness data show a statistical difference between the total values RC 1 and BI 1/BI 2, but also between the BI 1 and BI 2 total values. Because BI 1 and BI 2 are assumed to be roughly the same age (lying within the same ice marginal deposits;Fig. 3), this implies that the difference between RC 1 and BI 1/BI 2 is not necessarily related to a significant difference in age. Differences between BI 1 and BI 2 could be driven by the highly variable values measured for the top aspect faces of the boulders, but statistical similarities also exist between the BI 1/BI 2 and RC 1 aspect faces, suggesting that any differences are unlikely to be the result of a great difference in age between these EBTs. Most importantly, and like the Schmidt hammer data, we do not find a jump in values between RC 1 and BI 1/BI 2 that might be expected if intense, episodic erosion has taken place (Table 2). The average total difference in roughness between RC 1 and BI 2 is 0.5 mm at the 32-mm wavelength. Given that the large crystal size is > 20 mm and that the averages for all three boulder trains are between 2.3 and 2.8 mm, this difference is negligible. McCarroll and Nesje (1996) recorded much greater differences in roughness values when studying salt spray and chemical weathering of boulders. We cannot directly compare studies of different lithologies in different environments, but we would expect a clear difference in roughness values between the RC 1 and BI 1/BI 2 boulder trains if intense erosion has occurred.

Kaplan et al. (2007) suggested that proximity to the coastline may have caused higher rates of salt-spray weathering of the RC 1 boulders. Similarly, the dominant effect of the westerly winds in the region could have caused increased aeolian abrasion. Fig. 12shows the results for averaged rock surface hardness and rock surface roughness from each boulder, with total boulder values and east/west faces shown. Correlations between these results and the distance from the start of the BI 1 and RC 1 boulder trains are negligible. Given the variability in the data, there is no indication that salt-spray weathering or aeolian abrasion has had a variable effect on boulders within RC 1 or a markedly greater effect on RC 1 than BI 1.

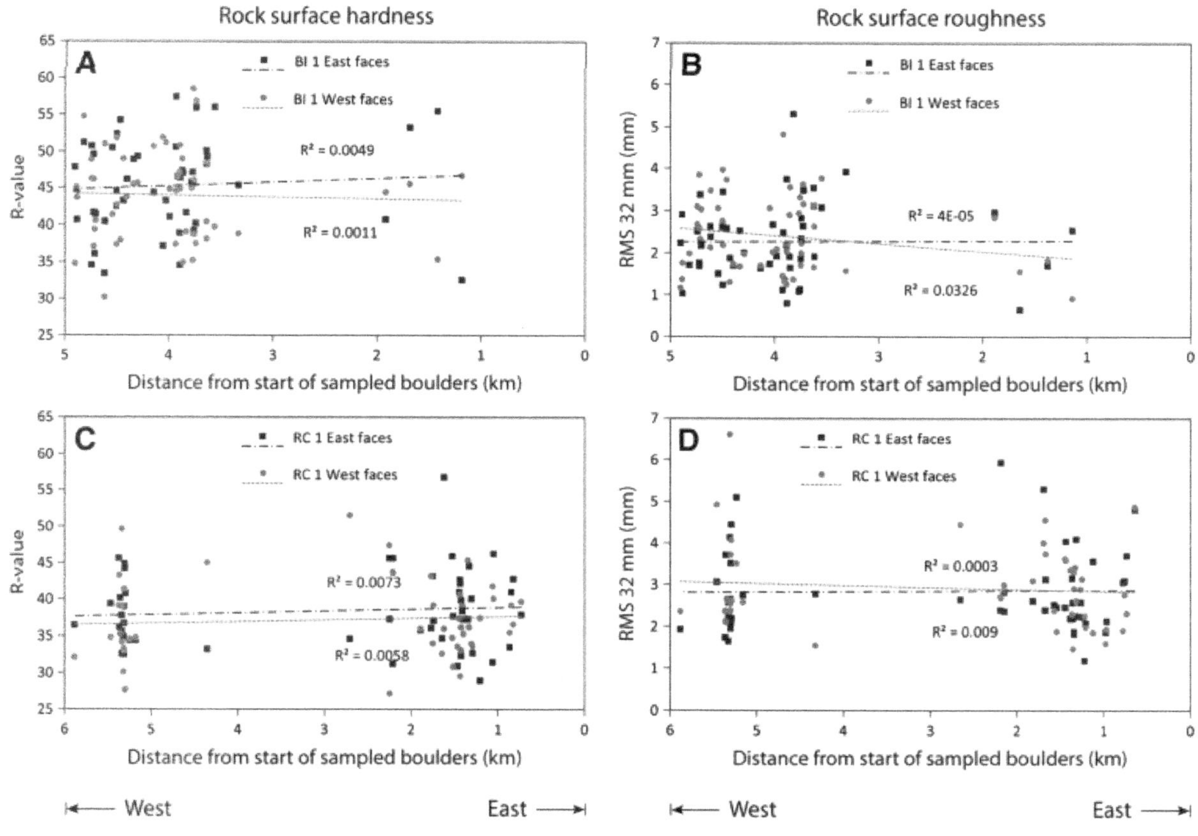

Figure 12: The rock surface hardness and roughness results from Fig. 11 and Fig. 12 plotted to illustrate variations in aspect with distance along the boulder trains (data points relate to individual boulders and are only shown for BI 1 and RC 1). (A) Shows hardness results for the east and west faces of BI 1, and (B) shows roughness results for the east and west faces of BI 1. (C) Shows hardness results for the east and west faces of RC 1, and (D) shows roughness results for the east and west faces of RC 1. In all graphs, boulder values toward the right are located farther to the east and those toward the left are located farther to the west. No discernible trends are found in either of the boulder trains, as illustrated by very low R^2 values, suggesting that salt weathering has not affected RC 1 significantly more than BI 1, as hypothesised byKaplan et al. (2007), despite closer proximity to the coast.

Weathering could have been operating on a scale greater than that recorded by our proxies, but whilst examples of microgullies and potholes are abundant on the top surfaces of the RC 1 boulders (Kaplan et al., 2007), similar features exist on the BI 1 and BI 2 boulders, we found little difference in top surface weathering between the boulder trains (see SM2). The similarity in weathering results between the EBTs suggests that they have probably experienced similar local climatic conditions. This is important given that the intense weathering needed to reduce the ages of the RC 1 boulder train during, or since, the LGM should have also reduced the ages of the BI 1 and BI 2 boulder trains. The cosmogenic exposure dates for the BI 1 and BI 2 boulder trains are believed to be good estimates of the time of deposition (around the LGM), agreeing with radiocarbon dates in the area (Heusser, 2003, McCulloch et al., 2005, Kaplan et al., 2008 and Hall et al., 2013), so factors that reduced the RC 1 boulder dates by hundreds of thousands of years are highly unlikely to have also reduced the BI 1 and BI 2 dates.

Alternative (post-)Depositional Model

The EBTs on Tierra del Fuego were thought to have been deposited during different glacial episodes (MIS 12, 10, and 2), but cosmogenic nuclide exposure dating from all of the EBTs yielded dates predominantly around 21 ka (Kaplan et al., 2007 and Evenson et al., 2009). One hypothesis (the boulders are 'old') explains this anomaly by invoking intensive post-depositional exhumation and erosion of the RC 1 boulders (Kaplan et al., 2007), but a second (the boulders are 'young') suggests that supraglacial debris may have an inheritance signature. Our study demonstrates that the distribution, volume, and monolithology of the boulder trains are indicative of supraglacial transport of rock avalanche material, and our rock surface weathering data does not support intensive post-depositional exhumation and erosion of the RC 1 boulders. Given this information, our data support the second, 'young' hypothesis, whereby the EBTs were transported and deposited at roughly the same time (i.e. within a few thousands of years, rather than separated by 100 s of thousands of years). Under this scenario, one would anticipate mostly young ages from the boulders, but with some anomalously old dates resulting from an inheritance signal owing to pre-exposure of boulders in the cliff face, which were not then sufficiently eroded during supraglacial transport (Applegate et al., 2010,Applegate et al., 2012 and Heyman et al., 2011).

This scenario is similar to that envisaged for the Foothills boulder train in Alberta, whereJackson et al. (1997) suggested that an erratic boulder that yielded a date of ca. 53 ka (four times older than the next oldest date) was improbable and most likely caused by pre-exposure prior to glacial transport. This implies that, for a supraglacial EBT deposit, a fraction of boulders should be expected to yield anomalously old dates. Assuming that the majority of ^{10}Be cosmogenic exposure dates are roughly correct, three anomalously old dates are apparent from the Tierra del Fuego boulder trains (Kaplan et al., 2007 and Evenson et al., 2009). For RC 1, one date (ca. 57 ka) out of 7 and for BI 2, one date (ca. 56 ka) out of 17 may have been pre-exposed. The similarity between these ages could indicate a consistent (pre-)exposure pattern, but more samples would be needed to support this idea. The outlier from SSb 1 is significantly older at ca. 222 ka, although this is only one of two samples from that boulder train.

Leaving aside the anomalously old dates, substantial variability remains in the cosmogenic nuclide exposure dates of the RC 1 EBT, a fact that Kaplan et al. (2007)highlighted. Variability in the dates could have been caused by variability in the residence time on the ice during transport. Hubbard et al. (2005) modelled ice flow velocities of up to 2000 m a^{-1} for the northern Patagonian Ice Sheet. Assuming similar velocities and a transport distance of up to 250 km for the BI–SSb lobe, transport time may have been as little as 0.125 ka: insignificant given dating uncertainty. Slower velocities of 100 m a^{-1} or even 50 m a^{-1} would have resulted in residence times of 2.5 or 5 ka, respectively; but given that the clustering of the EBTs suggests that they were deposited as discrete events, this could only have resulted in age differences between, not within, boulder trains.

Rather, we suggest three mechanisms that could have produced the variability in RC 1. Firstly, RC 1 has probably experienced greater erosion than BI 1/BI 2 as it was deposited first, causing variable reduction in cosmogenic nuclide exposure dates. Second, we agree with Kaplan et al. (2007) that many boulders in the RC 1 train show signs of surface weathering, though we find similar signs in the BI 1/ BI 2 EBTs. Sampling of the RC 1 boulder train may have been more affected by weathered surfaces than BI 1/BI 2 because fewer larger boulders are found in RC 1 and this could have resulted in sampling of boulders that were 'less ideal' than those in the other boulder trains.

Third, Jackson et al. (1997) suggested that boulder orientation can affect the spread of dates in a boulder train. We have demonstrated that all boulders in Tierra del Fuego are likely 'at rest', but do not know how long they were sitting on melting ice before reaching this state. The RC 1 and SSb 1 boulder trains are located on large bands of kettle kame drift, a characteristic deposit of 'dead ice' terrain (Dyke and Evans, 2005 and Schomacker, 2008), whereas BI 1 and BI 2 are located on small, sharp moraines interspersed by meltwater channels. Ground ice within the kettle and kame topography could have taken a long time to fully melt and settle, and occasional examples of boulders exposed at the coastline within the kettle and kame deposits may be a result of this slow process of settling (see SM2). Thus, the RC 1 boulder train may have taken ca. 10 ka to come to rest and implies that ground ice melting may have prevailed long after the recession of the ice lobe.

To summarise, the main implication behind this alternative history of the EBTs on Tierra del Fuego is that the established age model for the timing of glaciations in the region may need re-examining. Whilst this study provides a new approach to interpreting the cosmogenic nuclide exposure dates, further independent age controls for the glacial limits are needed to investigate the timing of glacial advances.

CONCLUSION

Erratic boulder trains make valuable targets for flow-line indicators and cosmogenic nuclide exposure dating, but few have been reported in detail and further studies are needed to understand their formation, transport, and deposition. Based on the limited number of EBTs that have been reported in the literature, we identify an apparent trend between boulder size, transport distance, and likely mode of transport. The EBTs containing boulders greater than ca. 5 m in diameter and/or demonstrating transport from the source lithology of greater than ca. 10 km are likely to have been transported supraglacially.

In Tierra del Fuego, the distribution and volume of three boulder trains suggest that they were formed by a rock avalanche and transported and deposited supraglacially. This highlights a former ice flow line from the Cordillera Darwin and helps to constrain the position of ice divides and maximum ice surface elevation. Using a variety of techniques and measurements, we do not identify any major changes in rock surface weathering characteristics across EBTs that have been previously interpreted to be from different glacial cycles. Thus we suggest that they are much closer in age (probably within a few thousand years), which is also consistent with previous cosmogenic nuclide exposure dates from the boulders. Occasional, anomalously old samples should be expected from supraglacial boulder trains caused by pre-exposure prior to and during glacial transport, and an understanding of the likely history of an EBT should be an integral part of using the features for dating.

ACKNOWLEDGEMENTS

We are very grateful to William Christiansen, Harold Lovell, Mark Hulbert, and Paul Lincoln for their fantastic assistance during field work. The research was primarily funded by a UK NERC PhD studentship awarded to CMD at Durham University, but was also supported by an Explorers Club Exploration Grant, Quaternary Research Association New Research Workers' Award, Santander Mobility Grant, the Durham University Faculty of Social Sciences and Health, and Durham University Geography Department. Thanks to Jorge Rabassa and Andrea Coronato at CADIC-CONICET in Ushuaia, Juan Carlos Aravena at CEQUA in Punta Arenas, the Fernandez family of Estancia San Clemente, the Vilamera family of Estancia Sara, and Juan Robertson of Estancia Tres Hermanos for their assistance. This paper has benefitted greatly from the insightful comments of three anonymous reviewers and the editor, Richard Marston.

REFERENCE

1. Adamson, D., Colhoun, E., 1992. Late Quaternary glaciation and deglaciation of the Bunger Hills, Antarctica. Antarct. Sci. 4, 435–446.
2. Anderson, B., Mackintosh, A., 2006. Temperature change is the major driver of late-glacialand Holocene glacier fluctuations in New Zealand. Geology 34, 121–124. http://dx.doi.org/10.1130/g22151.1.
3. Applegate, P.J., Urban, N.M., Laabs, B.J.C., Keller, K., Alley, R.B., 2010. Modeling the statistical distributions of cosmogenic exposure dates from moraines. Geosci. Model Dev. 3, 293–307.
4. Applegate, P.J., Urban, N.M., Keller, K., Lowell, T.V., Laabs, B.J.C., Kelly, M.A., Alley, R.B., 2012. Improved moraine age interpretations through explicit matching of geomorphic process models to cosmogenic nuclide measurements from single landforms. Quat. Res. 77, 293–304. http://dx.doi.org/10.1016/j.yqres.2011.12.002.
5. Atkins, C.B., Barrett, P.J., Hicock, S.R., 2002. Cold glaciers erode and deposit: evidence from Allan Hills, Antarctica. Geology 30, 659–662. http://dx.doi.org/10.1130/0091-7613 ((2002) 030 b 0659:cgeade N 2.0.co;2).
6. Augustinus, P.C., Gore, D.B., Leishman, M.R., Zwartz, D., Colhoun, E.A., 1997. Reconstruction of ice flow across the Bunger Hills, East Antarctica. Antarct. Sci. 9, 347–354.
7. Ballantyne, C.K., McCarroll, D., Nesje, A., Dahl, S.O., 1997. Periglacial trimlines, former nunataks and the altitude of the last ice sheet in Wester Ross, northwest Scotland. J. Quat. Sci. 12, 225–238.
8. Benn, D.I., 2004. Clast morphology. In: Evans, D.J.A., Benn, D.I. (Eds.), A Practical Guide to the Study of Glacial Sediments. Arnold, London, pp. 78–92.
9. Bentley, M.J., McCulloch, R.D., 2005. Impact of neotectonics on the record of glacier and sea level fluctuations, Strait of Magellan, southern Chile. Phys. Geogr. 87A, 393–402. http://dx.doi.org/10.1111/j.0435-3676.2005.00265.x.
10. Bentley, M.J., Sugden, D.E., Hulton, N.R.J., McCulloch, R.D., 2005. The landforms and pattern of deglaciation in the Strait of Magellan and Bahia Inútil, southernmost South America. Phys. Geogr. 87A, 313–333. http://dx.doi.org/10.1111/j.0435-3676.2005.00261.x.
11. Boulton, G.S., 1978. Boulder shapes and grain-size distributions of debris as indicators of transport paths through a glacier and till genesis. Sedimentology 25, 773–799. http:// dx.doi.org/10.1111/j.1365-3091.1978.tb00329.x.
12. Boulton, G.S., 1996. Theory of glacial erosion, transport and deposition as a consequence of subglacial sediment deformation. J. Glaciol. 42, 43–62.
13. Černá, B., Engel, Z., 2011. Surface and sub-surface Schmidt hammer rebound value variation for a granite outcrop. Earth Surf. Process. Landf. 36, 170–179. http://dx.doi.org/ 10.1002/esp.2029.
14. Chevalier, G., Davies, T., McSaveney, M., 2009. The prehistoric Mt Wilberg rock avalanche, Westland, New Zealand. Landslides 6, 253–262. http://dx.doi.org/10.1007/s10346- 009-0156-5.
15. Coronato, A., Roig, C., Rabassa, J., Meglioli, A., 1999. Erratic boulder field of pre-Illinoian age at Punta Sinaí, Tierra del Fuego, southernmost South America. XV INQUA International Congress, Durham, South Africa, pp. 47–48.
16. Coronato, A., Martínez, O., Rabassa, J., 2004. Glaciations in Argentine Patagonia, southern South America. In: Ehlers, J., Gibbard, P.L. (Eds.), Developments in Quaternary Sciences. Elsevier, pp. 49–67 http://dx.doi.org/10.1016/s1571-0866(04)80111-8.
17. Cunningham, W.D., 1993. Strike-slip faults in the southernmost andes and the development of the Patagonian orocline. Tectonics 12, 169–186. http://dx.doi.org/10.1029/ 92tc01790.
18. Darvill, C.M., Stokes, C.R., Bentley, M.J., Lovell, H., 2014. A glacial geomorphological map of the southernmost ice lobes of Patagonia: the Bahía Inútil–San Sebastián, Magellan, Otway, Skyring and Río Gallegos lobes. J. Map. 1–21 http://dx.doi.org/10.1080/17445647.2014.890134.
19. Darwin, C., 1841. On the distribution of the erratic boulders and on the contemporaneous unstratified deposits of South America. Trans. Geol. Soc. London 2–6, 415–431. http://dx.doi.org/10.1144/transgslb.6.2.415.
20. Davies, B.J., Glasser, N.F., Carrivick, J.L., Hambrey, M.J., Smellie, J.L., Nývlt, D., 2013. Landscape evolution and ice-sheet behaviour in a semi-arid polar environment: James Ross Island, NE Antarctic Peninsula. Geol. Soc. Lond., Spec. Publ. 381, 353–395. http://dx.doi.org/10.1144/sp381.1.
21. Davis, J.W., 1880. On a group of erratic boulders at Norber, near Clapham, in Yorkshire. Proc. Yorks. Geol. Polytech. Soc. 7, 266–273. http://dx.doi.org/10.1144/pygs.7.3.266.

22. Delaney, K., Evans, S., 2014. The 1997 Mount Munday landslide (British Columbia) and the behaviour of rock avalanches on glacier surfaces. Landslides 1–18 http://dx.doi. org/10.1007/s10346-013-0456-7.

23. DiLabio, R.N.W., 1981. Glacial dispersal of rocks and minerals at the south end of Lac Mistassini, Quebec, with special reference to the Icon dispersal train. Geol. Surv. Can. Bull. 323, 46.

24. DiLabio, R.N.W., 1990. Glacial dispersal trains. In: Kujansuu, R., Saarnisto, M. (Eds.), Glacial Indicator Tracing. Balkema Publishers, Rotterdam, Netherlands, pp. 109–122.

25. Dyke, A.S., Evans, D.J.A., 2005. Ice-marginal terrestrial landsystems: Northern Laurentideand Innuitian ice sheet margins. In: Evans, D.J.A. (Ed.), Glacial Landsystems. Hodder Arnold, London.

26. Dyke, A.S., Morris, T.F., 1988. Drumlin fields, dispersal trains, and ice streams in Arctic Canada. Can. Geogr. 32, 86–90.

27. Engel, Z., 2007. Measurement and age assignment of intact rock strength in the Krkonoše Mountains, Czech Republic. Z. Geomorphol. 51, 69–80. http://dx.doi. org/10.1127/ 0372-8854/2007/0051s-0069.

28. Evans, D.J.A., 2007. Glacial erratics and till dispersal indicators. In: Elias, S.A. (Ed.), Encyclopedia of Quaternary Science. Elsevier, Oxford, pp. 975–978 http://dx.doi.org/10. 1016/B0-44-452747-8/00089-2.

29. Evenson, E.B., Burkhart, P.A., Gosse, J.C., Baker, G.S., Jackofsky, D., Meglioli, A., Dalziel, I., Kraus, S., Alley, R.B., Berti, C., 2009. Enigmatic boulder trains, supraglacial rock avalanches, and the origin of "Darwin›s boulders," Tierra del Fuego. GSA Today 19, 4–10. http://dx.doi.org/10.1130/GSATG72A.1.

30. Goldie, H.S., 2005. Erratic judgements: re-evaluating solutional erosion rates of limestonesusing erratic-pedestal sites, including Norber, Yorkshire. Area 37, 433–442. http://dx.doi.org/10.1111/j.1475-4762.2005.00653.x.

31. Goudie, A.S., 2006. The Schmidt Hammer in geomorphological research. Prog. Phys. Geogr. 30, 703–718. http://dx.doi.org/10.1177/0309133306071954.

32. Hall, B.L., Porter, C.T., Denton, G.H., Lowell, T.V., Bromley, G.R.M., 2013. Extensive recession of Cordillera Darwin glaciers in southernmost South America during Heinrich Stadial 1. Quat. Sci. Rev. 62, 49–55. http://dx.doi.org/10.1016/j.quascirev.2012.11.026.

33. Heusser, C.J., 2003. Ice Age Southern Andes: A Chronicle of Paleoecological Events.Elsevier http://dx.doi.org/10.1016/S1571-0866(03)80001-5.

34. Hewitt, K., 1999. Quaternary moraines vs catastrophic rock avalanches in the KarakoramHimalaya, Northern Pakistan. Quat. Res. 51, 220–237. http://dx.doi. org/10.1006/qres. 1999.2033.

35. Hewitt, K., 2009. Rock avalanches that travel onto glaciers and related developments, Karakoram Himalaya, Inner Asia. Geomorphology 103, 66–79. http://dx.doi. org/10. 1016/j.geomorph.2007.10.017.

36. Heyman, J., Stroeven, A.P., Harbor, J.M., Caffee, M.W., 2011. Too young or too old: evaluating cosmogenic exposure dating based on an analysis of compiled boulder exposure ages. Earth Planet. Sci. Lett. 302, 71–80. http://dx.doi.org/10.1016/j.epsl.2010.11.040.

37. Hubbard, A., Hein, A.S., Kaplan, M.R., Hulton, N.R.J., Glasser, N., 2005. A modelling reconstruction of the last glacial maximum ice sheet and its deglaciation in the vicinity of the Northern Patagonian Icefield, South America. Phys. Geogr. 87A, 375–391. http://dx.doi. org/10.1111/j.0435-3676.2005.00264.x.

38. Huddart, D., 2002. Norber erratics. In: Huddart, D., Glasser, N. (Eds.), Quaternary of Northern England. JNCC, Peterborough, pp. 200–203.

39. Jackson, L.E., Duk-Rodkin, A., 1996. Quaternary geology of the ice-free corridor: glacial controls on the peopling of the New World. In: Akazawa, T., Szathmary, E. (Eds.), Prehistoric Mongoloid Dispersals. Oxford University Press, Oxford, pp. 214–227.

40. Jackson, L.E., Little, E.C., 2004. A single continental glaciation of Rocky Mountain Foothills,south-western Alberta, Canada. In: Ehlers, J., Gibbard, P.L. (Eds.), Developments inQuaternary Sciences. Elsevier, pp. 29–38 http://dx.doi.org/10.1016/S1571-0866(04) 80183-0.

41. Jackson, L.E., Phillips, F.M., Shimamura, K., Little, E.C., 1997. Cosmogenic 36Cl dating of the Foothills erratics train, Alberta, Canada. Geology 25, 195–198. http://dx.doi.org/10. 1130/0091-7613 ((1997)025b0195 :ccdotfN2.3.co;2).

42. Jackson, L.E., Phillips, F.M., Little, E.C., 1999. Cosmogenic 36Cl dating of the maximum limitof the Laurentide Ice Sheet in southwestern Alberta. Can. J. Earth Sci. 36, 1347–1356. http://dx.doi.org/10.1139/e99-038.

43. Jibson, R.W., Harp, E.L., Schulz, W., Keefer, D.K., 2006. Large rock avalanches triggered by the M 7.9 Denali Fault, Alaska, earthquake of 3 November 2002. Eng. Geol. 83, 144–160. http://dx.doi.org/10.1016/j. enggeo.2005.06.029.

44. Kaplan, M.R., Douglass, D.C., Singer, B.S., Caffee, M.W., 2005. Cosmogenic nuclide chronologyof pre-last glacial maximum moraines at Lago Buenos Aires, 46 degrees S,Argentina. Quat. Res. 63, 301–315. http://dx.doi.org/10.1016/j.yqres.2004.12.003.

45. Kaplan, M.R., Coronato, A., Hulton, N.R.J., Rabassa, J.O., Kubik, P.W., Freeman, S.P.H.T.,2007. Cosmogenic nuclide measurements in southernmost South America and implications for landscape change. Geomorphology 87, 284–301. http://dx.doi.org/10. 1016/j.geomorph.2006.10.005.

46. Kaplan, M.R., Fogwill, C.J., Sugden, D.E., Hulton, N., Kubik, P.W., Freeman, S.P.H.T., 2008. Southern Patagonian glacial chronology for the Last Glacial period and implications for Southern Ocean climate. Quat. Sci. Rev. 27, 284–294. http://dx.doi.org/10.1016/j. quascirev.2007.09.013.

47. Kirkbride, M.P., 2005. Boulder edge-roundness as an indicator of relative age: a lochnagar case study. Scott.

Geogr. J. 121, 219–236. http://dx.doi.org/10.1080/00369220518737232.

48. Klepeis, K.A., 1994. Relationship between uplift of the metamorphic core of the southernmostAndes and shortening in the Magallanes foreland fold and thrust belt, Tierra delFuego, Chile. Tectonics 13, 882–904. http://dx.doi.org/10.1029/94tc00628.

49. Knechtel, M.M., 1942. Snake Butte boulder train and related glacial phenomena, northcentralMontana. Geol. Soc. Am. Bull. 53, 917–936. http://dx.doi.org/10.1130/gsab-53-917.

50. Kujansuu, R., Saarnisto, M., 1990. Glacial Indicator Tracing. Balkema Publishers, Rotterdam,Netherlands.

51. Lal, D., 1991. Cosmic ray labeling of erosion surfaces: in situ nuclide production rates anderosion models. Earth Planet. Sci. Lett. 104, 424–439. http://dx.doi.org/10.1016/0012-821X(91)90220-C.

52. Larsen, S.H., Davies, T.R.H., McSaveney, M.J., 2005. A possible coseismic landslide origin oflate Holocene moraines of the Southern Alps, New Zealand. N. Z. J. Geol. Geophys. 48,311–314. http://dx.doi.org/10.1080/00288306.2005.9515117.

53. Lawson, T.J., 1990. Former ice movement in Assynt, Sutherland, as shown by the distribution of glacial erratics. Scott. J. Geol. 26, 25–32. http://dx.doi.org/10.1144/sjg26010025.

54. Lawson, T.J., 1995. Boulder trains as indicators of former ice flow in Assynt, N.W. Scotland. Quat. Newsl. 75, 15–21.

55. Marangunic, C., Bull, C., 1968. The Landslide on the Sherman Glacier, The Great Alaska Earthquake of 1964–Hydrology. Pt. A. National Academy of Sciences, Washington, DC, pp. 383–394.

56. Matthews, J.A., Owen, G., 2010. Schmidt hammer exposure-age dating: developing linear age-calibration curves using Holocene bedrock surfaces from the Jotunheimen–Jostedalsbreen regions of southern Norway. Boreas 39, 105–115. http://dx.doi.org/10.1111/j.1502-3885.2009.00107.x.

57. McCarroll, D., 1991. The age and origin of Neoglacial moraines in Jotunheimen, southern Norway: new evidence from weathering-based data. Boreas 20, 283–295. http://dx. doi.org/10.1111/j.1502-3885.1991.tb00156.x.

58. McCarroll, D., Nesje, A., 1993. The vertical extent of ice sheets in Nordfjord, western Norway: measuring degree of rock surface weathering. Boreas 22, 255–265. http://dx.doi.org/10.1111/j.1502-3885.1993.tb00185.x.

59. McCarroll, D., Nesje, A., 1996. Rock surface roughness as an indicator of degree of rock surface weathering. Earth Surf. Process. Landf. 21, 963–977. http://dx.doi.org/10.1002/ (sici)1096-9837 ((199610)21:10b963::aid-esp643N3.0.co;2-j).

60. McColl, S.T., Davies, T.R., 2011. Evidence for a rock-avalanche origin for 'The Hillocks' "moraine", Otago, New Zealand. Geomorphology 127, 216–224. http://dx.doi.org/10. 1016/j.geomorph.2010.12.017.

61. McCulloch, R.D., Fogwill, C.J., Sugden, D.E., Bentley, M.J., Kubik, P.W., 2005. Chronology of the last glaciation in central Strait of Magellan and Bahia Inutil, southernmost South America. Phys. Geogr. 87A, 289–312. http://dx.doi.org/10.1111/j.0435-3676.2005. 00260.

62. Meglioli, A., 1992. Glacial Geology and Chronology of southernmost Patagonia and Tierradel Fuego, Argentina and Chile(Ph.D. Thesis) Geology. Lehigh University.

63. Mountjoy, E.W., 1958. Jasper area Alberta, a source for the Foothills erratics train. Bull. Can. Petrol. Geol. 6, 218–226.

64. Natland, M., Gonzalez, P., Canon, A., Ernst, M., 1974. Geology and paleontology of Magellanes Basin: a system of stages for correlation of Magellanes Basin sediments. Mem. Geol. Soc. Am. 139, 3–57.

65. Nelson, E.P., Dalziel, I.W.D., Milnes, A.G., 1980. Structural geology of the Cordillera Darwin; collisional-style orogenesis in the southernmost Chilean Andes. Eclogae Geol. Helv. 73, 727–751.

66. Nesje, A., McCarroll, D., Dahl, S.O., 1994. Degree of rock surface weathering as an indicator of ice-sheet thickness along an east–west transect across southern Norway. J. Quat.Sci. 9, 337–347. http://dx.doi.org/10.1002/jqs.3390090404.

67. Putnam, A.E., Schaefer, J.M., Barrell, D.J.A., Vandergoes, M., Denton, G.H., Kaplan, M.R., Finkel, R.C., Schwartz, R., Goehring, B.M., Kelley, S.E., 2010. In situ cosmogenic 10Be production-rate calibration from the Southern Alps, New Zealand. Quat. Geochronol.5, 392–409. http://dx.doi.org/10.1016/j.quageo.2009.12.001.

68. Rabassa, J., 2008. Late Cenozoic glaciations in Patagonia and Tierra del Fuego. In: Rabassa, J. (Ed.), Developments in Quaternary Sciences. Elsevier, pp. 151–204 http://dx.doi.org/10.1016/s1571-0866(07)10008-7.

69. Rabassa, J., Coronato, A., MartÍNez, O., 2011. Late Cenozoic glaciations in Patagonia and Tierra del Fuego: an updated review. Biol. J. Linn. Soc. 103, 316–335. http://dx.doi. org/10.1111/j.1095-8312.2011.01681.x.

70. Rae, A.C., Harrison, S., Mighall, T., Dawson, A.G., 2004. Periglacial trimlines and nunataks of the Last Glacial Maximum: the Gap of Dunloe, southwest Ireland. J. Quat. Sci. 19, 87–97. http://dx.doi.org/10.1002/jqs.807.

71. Reznichenko, N.V., Davies, T.R.H., Alexander, D.J., 2011. Effects of rock avalanches on glacier behaviour and moraine formation. Geomorphology 132, 327–338. http://dx.doi. org/10.1016/j.geomorph.2011.05.019.

72. Roberts, D.H., Dackombe, R.V., Thomas, G.S.P., 2007. Palaeo-ice streaming in the central sector of the British–Irish Ice Sheet during the Last Glacial Maximum: evidence from the northern Irish Sea Basin. Boreas 36, 115–129. http://dx.doi.org/10.1111/j.1502-3885.2007.tb01186.x.

73. Roed, M.A., Mountjoy, E.W., Rutter, N.W., 1967. The Athabasca Valley Erratics Train, Alberta and Pleistocene ice movements across the continental divide. Can. J. Earth Sci. 4, 625–632. http://dx.doi.org/10.1139/e67-039.

74. Salonen, V.-P., 1986. Glacial Transport Distance Distributions of Surface Boulders in Finland. Geologian tutkimuskeskus.

75. Sánchez, J.S., Mosquera, D.F., Romaní, J.R.V., 2009. Assessing the age-weathering correspondence of cosmogenic 21Ne dated Pleistocene surfaces by the Schmidt Hammer.Earth Surf. Process. Landf. 34, 1121–1125. http://dx.doi.org/10.1002/esp.1802.

76. Schomacker, A., 2008. What controls dead-ice melting under different climate conditions? A discussion. Earth Sci. Rev. 90, 103–113. http://dx.doi.org/10.1016/j.earscirev.2008.08.003.

77. Shakesby, R.A., Matthews, J.A., Karlén, W., Los, S.O., 2011. The Schmidt hammer as a Holocene calibrated-age dating technique: testing the form of the R-value-age relationship and defining the predicted-age errors. The Holocene 21, 615–628. http:// dx.doi.org/10.1177/0959683610391322.

78. Shugar, D.H., Clague, J.J., 2011. The sedimentology and geomorphology of rock avalanche deposits on glaciers. Sedimentology 58, 1762–1783. http://dx.doi.org/10.1111/j.1365-3091.2011.01238.x.

79. Shulmeister, J., Davies, T.R., Evans, D.J.A., Hyatt, O.M., Tovar, D.S., 2009. Catastrophic landslides, glacier behaviour and moraine formation — a view from an active plate margin. Quat. Sci. Rev. 28, 1085–1096. http://dx.doi.org/10.1016/j.quascirev.2008.11.015.

80. Sosio, R., Crosta, G.B., Chen, J.H., Hungr, O., 2012. Modelling rock avalanche propagation onto glaciers. Quat. Sci. Rev. 47, 23–40. http://dx.doi.org/10.1016/j.quascirev.2012. 05.010.

81. Stahl, T., Winkler, S., Quigley, M., Bebbington, M., Duffy, B., Duke, D., 2013. Schmidt hammer exposure-age dating (SHD) of late Quaternary fluvial terraces in New Zealand. Earth Surf. Process. Landf. 38, 1838–1850. http://dx.doi.org/10.1002/esp.3427.

82. Stalker, A.M., 1956. The erratics train, foothills of Alberta. Geol. Surv. Can. Bull. 37, 31. Stalker, A.M., 1976. Megablocks, or the enormous erratics of the Albertan Prairies. Geol. Surv. Can. Pap. 76-1C, 185–188.

83. Stone, J.O., 2000. Air pressure and cosmogenic isotope production. J. Geophys. Res. 105, 23753–23759. http://dx.doi.org/10.1029/2000jb900181.

84. Tovar, S.D., Shulmeister, J., Davies, T.R., Geosci, Nature, 2008. Evidence for a landslide origin of New Zealand/'s Waiho Loop moraine. Nat. Geosci. 1, 524–526. http://dx.doi.org/10.1038/ngeo249.

85. Vincent, P.J., Wilson, P., Lord, T.C., Schnabel, C., Wilcken, K.M., 2010. Cosmogenic isotope (36Cl) surface exposure dating of the Norber erratics, Yorkshire Dales: furtherconstraints on the timing of the LGM deglaciation in Britain. Proc. Geol. Assoc. 121,24–31. http://dx.doi.org/10.1016/j.pgeola.2009.12.009.

86. Ward, B.C., Bond, J.D., Gosse, J.C., 2007. Evidence for a 55–50 ka (early Wisconsin) glaciationof the Cordilleran ice sheet, Yukon Territory, Canada. Quat. Res. 68, 141–150. http://dx.doi.org/10.1016/j.yqres.2007.04.002.

87. White, K., Bryant, R., Drake, N., 1998. Techniques for measuring rock weathering: application to a dated fan segment sequence in southern Tunisia. Earth Surf. Process. Landf. 23, 1031–1043.

88. Wilson, P., Barrows, T.T., Lord, T.C., Vincent, P.J., 2012. Surface lowering of limestone pavement as determined by osmogenic (36Cl) analysis. Earth Surf. Process. Landf. 37, 1518–1526. http://dx.doi.org/10.1002/esp.3260.

CHAPTER 18

THE MANAGE DRAIN LOAD DATABASE: REVIEW AND COMPILATION OF MORE THAN FIFTY YEARS OF NORTH AMERICAN DRAINAGE NUTRIENT

L.E. Christianson[a], R.D. Harmel[b]

[a] The Conservation Fund, 1098 Turner Road, Shepherdstown, WV 25443, USA

[b] Grassland Soil and Water Research Laboratory, USDA-ARS, 808 East Blackland Road, Temple, TX 76502, USA

ABSTRACT

As agriculture in the 21st century is faced with increasing pressure to reduce negative environmental impacts while continuing to efficiently produce food, fiber, and fuel, it becomes ever more important to reflect upon more than half a century of drainage water quality research to identify paths forward. This work provided a quantitative review of the water quality and crop yield impacts of artificially drained agronomic systems across North America by compiling data from drainage nutrient studies in the "Measured Annual Nutrient loads from AGricultural Environments" (MANAGE) database. Of the nearly 400 studies reviewed, 91 individual journal publications and 1279 site-years were included in the new MANAGE Drain Load table with data spanning 1961–2012. Across site-years, the mean and median percent of precipitation occurring as drainage were 25 and 20%, respectively, with wet years resulting in significantly greater drainage discharge and nutrient loads. Water quality and crop yield impacts due to management factors such as cropping system, tillage, and drainage design were investigated. This work provided an important opportunity to evaluate gaps in drainage nutrient research. In addition to the current analyses, the resulting MANAGE drainage database will facilitate further analyses and improved understanding of the agronomic and environmental impacts of artificial drainage.

INTRODUCTION

A strong "development ethos" in North America first demonstrated through early settlement water development projects set the context for the 21st century's widespread use of artificial agricultural drainage (Skaggs and van Schilfgaarde, 1999). Drainage legislations and works in the United States are reported as early as the mid-1700s even prior to the signing of the Declaration of Independence (Shirmohammadi et al., 1995). The first documented use of tile drainage in the US occurred in 1835 by a New York farmer who imported "horseshoe-type drain tile" patterns from Scotland (Ritter et al., 1995). Since these early beginnings, artificial agricultural drainage has been a source of scientific study and policy debate (Madramootoo et al., 2007) and will likely continue to be so for years to come.

The economic benefit of improved artificial drainage is ultimately the most important driver for such installations. Improved drainage enhances crop growth and yield (Portch et al., 1968, Stout and Schnabel, 1994 and Tan et al., 1993), and also reduces on-farm risk by increasing the number of days available for field activities (Fausey et al., 1995 and Skaggs and van Schilfgaarde, 1999). Installation of subsurface drainage systems costs from $740–$1480 per ha but can boost

Citation: L.E. Christianson, R.D. Harmel, The MANAGE Drain Load database: Review and compilation of more than fifty years of North American drainage nutrient studies, doi:10.1016/j.agwat.2015.06.021

yields by 5–25% (Blann et al., 2009). Pavelis (1987) reported the average US replacement value of drainage was $914 per ha, and the total net value of drainage capital in the US was nearly $25 billion in 1985 dollars. With high crop and land prices in the late 2000's (Nickerson et al., 2012), the value of said infrastructure is doubtless much higher.

Despite the agronomic and rural economy benefits associated with artificial drainage, it does change natural hydrology and is a major conduit for nutrient transport (David et al., 2010). Foundational work by Randall and Goss (2008) identified controllable and uncontrollable factors that impact drainage water quality. Uncontrollable factors include the amount and temporal distribution of precipitation, climate during the non-growing season, soil type and organic matter, the latter of which can be influenced by management practices which are controllable. Controllable factors pertain to human-induced choices such as cropping system, tillage practices, and nutrient management (Randall and Mulla, 2001). Drainage system design is also an important controllable factor not only for drainage efficiency but for water quality (Randall and Goss, 2008,Sands et al., 2008 and Skaggs et al., 1994).

Installation of tile drainage can reduce surface runoff volume and peak outflow rates by providing storage capacity in the soil above the tiles (Ball Coelho et al., 2012a, Blann et al., 2009 and Robinson and Rycroft, 1999). In areas where prioritization between pollutants is required, subsurface drainage may indeed be a strategy for reducing surface runoff-associated sediment and phosphorus (P) transport (Ball Coelho et al., 2012b, Bottcher et al., 1981, Fausey et al., 1995 and Gold and Loudon, 1989). However, additional mitigation strategies for soluble pollutants, particularly nitrate-nitrogen, will be necessary if drainage is implemented to reduce sediment and particulate P loads. Conveniently, installation of subsurface systems allows opportunity to treat some dissolved pollutants through diversion of outflow and water table control (Fausey et al., 1995). Through good design, drainage systems can optimize both agronomic and environmental goals (Skaggs et al., 1994).

As policy debate and regulatory interest related to water quality continue to grow, it becomes important to reflect upon decades of drainage research in North America to create a future vision for drained agricultural lands. With increased computing power and more sophisticated hydrologic and biogeophysical modeling efforts, there is clearly a need for the large number of drainage nutrient studies to be compiled and analyzed to enhance understanding of the state of drainage science and to develop improved drainage models. Fortunately, an existing framework is available for such a drainage-oriented compilation. The "Measured Annual Nutrient loads from AGricultural Environments" (MANAGE) database aims to "compil[e] measured annual nitrogen and phosphorus load data representing field-scale transport from agricultural land uses in the USA into a readily accessible, easily queried format" (Harmel et al., 2008). This free and publically available water quality database was developed in Microsoft Access by the United States Department of Agriculture, Agricultural Research Service, Grassland, Soil, and Water Research Laboratory in Temple, Texas (www.ars.usda.gov/spa/manage-nutrient). The agricultural runoff and forest-focused tables within MANAGE include over 1800 watershed-years from 300 nutrient load records (i.e., sites or plots) with database fields pertaining to study location, tillage type, conservation practice, soil type/group, fertilizer application, and nutrient loss (Harmel et al., 2006). This well-established database was the ideal platform for this work aimed at integrating and compiling water quality and crop yield information from drained landscapes in North America. The specific objectives were to further develop the MANAGE database through addition of drainage studies and to analyze the resulting pooled information to investigate drainage trends and impacts during the past fifty years. This work was a part of broader efforts to evaluate the nutrient loading and economic impacts associated with the 4R nutrient management strategies.

MATERIALS AND METHODS

Literature was reviewed between April and October 2014 for nitrogen (N) and phosphorus (P) drainage loads and crop yields, and site-years deemed acceptable were entered into a new "Drain Load" database table in MANAGE (Microsoft Access). Information sources were identified through web and journal searches and by tracing citations in relevant papers and review articles. In total, 394 individual publications were reviewed. Data in MANAGE are based on a robust, previously peer-reviewed selection process (Harmel et al., 2006 and Harmel et al., 2008). Suitable studies met the following criteria: peer-reviewed, from study areas of at least 0.009 ha with a homogenous land use within North America, not a rainfall simulation or lysimeter study, and include load data from at least one year. For the new Drain Load table, irrigation-drainage systems common in the western US and controlled drainage treatments were not included (i.e., only nutrient loads from free, unrestricted outlets were included). The most prevalent unsuitability reasons for the MANAGE Drain Load database were that (1) a given study did not contain an annual nutrient load (e.g., the study reported only nutrient concentrations or hydrology information, the study reported event-based sampling rather than annual values), (2) the study was not, in fact, a drainage study (e.g., a soil leaching or groundwater seepage study using porous cup samplers or lysimeters), (3) the study was from outside North America, or (4) the paper was a review with no original data. There were a few notable drainage studies that were necessarily excluded from the Drain Load table. For example, work at the Waseca, MN research station used drainage plots of only 0.0055 ha (6.1 m × 9.1 m), smaller than the 0.009 ha threshold. Nevertheless, these studies and others were used to inform the text-based literature review (Randall and Vetsch, 2005 and Randall et al., 2003). In at least 25 studies, it was necessary to extract data from published graphs and figures using Data Thief® software (Johnson and Curtis, 2001 and Tonitto et al., 2006). In one case, it was necessary to contact the lead author to clarify the number of site-years and study details (Evans et al., 1995).

Data on dissolved, particulate, and total N and P loads were sought, and existing database fields in MANAGE's Ag Load table served as the template for the new Drain Load table. Recently, several new fields were added across all of MANAGE's tables to enhance ability to make 4R-related comparisons. These fields included both N and P crop uptake, yield, and fertilizer timing ("At Planting, within 1 week of plant", "Out of Season, >2 months before plant", "Pre-Plant, 2 months-1 week before plant", and "Side/Top Dress, >1 week after plant"). Specifically for the Drain Load table, new drainage-related fields included: drain type ("surface", "subsurface (with inlets)", or "subsurface (no inlet specified)"), drain spacing, and drain depth. In the case of surface drainage, drain spacing referred to spacing between ditches and drain depth was the ditch depth. The largest deviation from the existing MANAGE format was that in the new Drain Load table, each record represented an individual site-year, whereas previously, each record represented a site with data pooled. The site-year approach was taken here as it was easier to quantify trends when each record was weighted equally across time.

Drain Load table data were analyzed with counts (e.g., histograms), box plots, and regression analyses. Due to the large dataset and high variability, it was necessary to "bin" similar groups within certain categorical fields. For example, for hydrologic analysis, "wet" and "dry" years were separated. The approximate mean (846 ± 219 mm) and median (828 mm) across all precipitation records (n = 889) were used as separation points, with precipitation values less than 820 mm or greater than 850 mm considered "dry" or "wet" site-years, respectively. Site-years with precipitation values falling between these separation points (n = 28 or 3%) were excluded to provide a distinct break point. The data were non-normally distributed, thus were analyzed using Kruskal–Wallis one-way analysis of variance tests based on rank which uses median values. Mann–Whitney Sum t-tests based on rank and median values were used when comparing only two treatments (Sigma Plot 12.5).

RESULTS AND DISCUSSION

Site-years Over Time and Space

A total of 91 individual journal publications and 1279 site-years were included in the MANAGE Drain Load table. While this was based on a comprehensive search, MANAGE is intended to be dynamic with periodic additions (e.g., Harmel et al., 2006 and Harmel et al., 2008).

Geography

The majority of nutrient load site-years were from Midwestern states which was unsurprising considering the prevalence of drainage, primarily subsurface drainage, in this region (Fig. 1, Pavelis, 1987, Sugg, 2007 and Zucker and Brown, 1998). Iowa and Illinois alone accounted for 50% of the site-years. Canadian drainage studies were also clearly important with Ontario the second most predominant state/province (255 site-years). There were relatively fewer site-years from the eastern Midwest (Indiana, Michigan, and Ohio). There were also few nutrient load studies from the southeast and Mid-Atlantic states, despite the long history and prevalence of drainage in this region (Madramootoo et al., 2007 and Thomas et al., 1995). South Carolina and Delaware had no studies represented in spite of a high prevalence of drained cropland in each state (both approximately 25% in Pavelis, 1987). Any spatial estimate of drained lands is necessarily viewed with some uncertainty as there is no comprehensive assemblage of drainage records kept by any central authority (Blann et al., 2009).

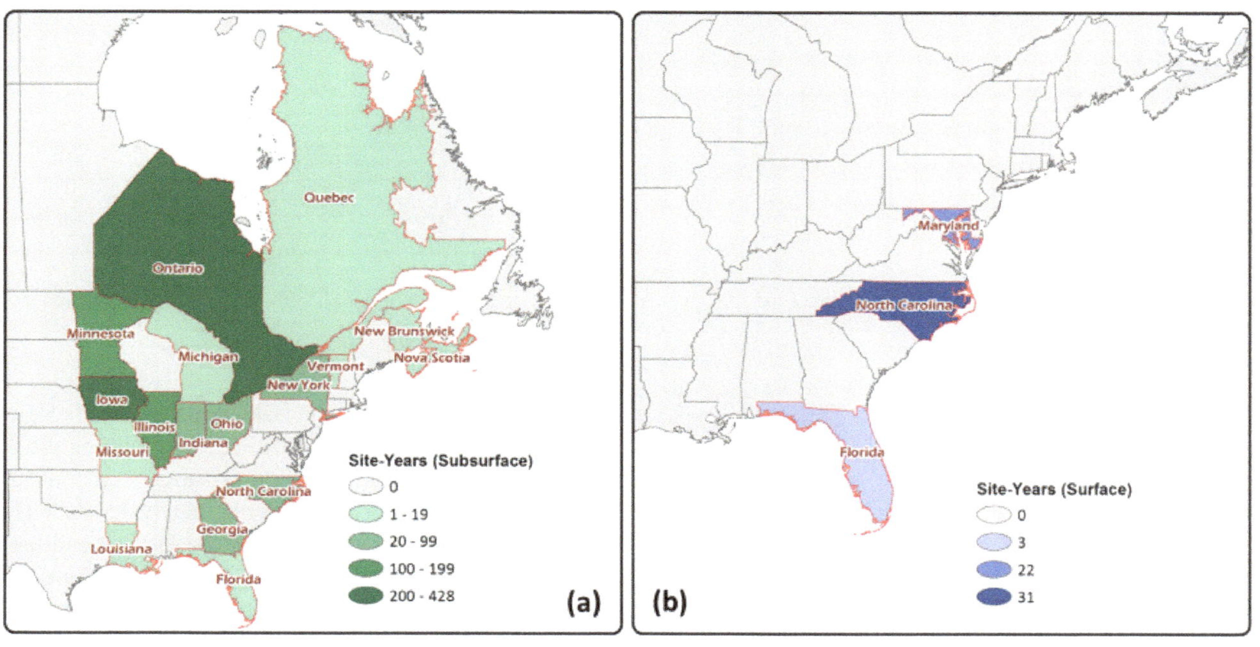

Figure 1: MANAGE Drain Load subsurface (a) vs. surface drainage site-years (b).

Much more nutrient load information was available for subsurface drainage compared to surface drainage (1177 vs. 56 site-years, respectively, Fig. 1). This may be complicated by the fact that, in canal or ditch drainage, it is sometimes difficult to differentiate between surface and subsurface discharge as flows are generally combined groundwater and surface runoff (Evans et al., 1995). Many reviewed surface drainage ditch studies were not included in the Drain Load table as they represented more than a single land use or did not report annual nutrient loads (e.g., presented storm event data). Nevertheless, the practice of surface drainage is widespread across North America (Strock et al., 2007). Because ditch systems are unique in terms of geomorphology and nutrient cycling (Needelman et al., 2007), future ditch nutrient transport studies may an important contribution to the MANAGE Drain Load table.

Surface intakes or inlets are an important component of many subsurface drainage systems, but mentions of these were curiously lacking across the literature review. Of the 1177 subsurface drainage site-years, only 22 (2%) were from studies that specified occurrence of surface intakes. Ball Coelho et al. (2010) established terminology of an "open" versus "closed" system referring to subsurface drainage systems with or without surface intakes, respectively. While several studies reported that surface inlets had a fairly small contribution to flow and nutrient loading (Ball Coelho et al., 2012a, Ball Coelho et al., 2012b and Ginting et al., 2000), pollution reduction approaches for intakes is an area of active research. Oolman and Wilson (2003) recommended use of slotted standpipes to control sediment and Smith and Livingston (2013) recommended blind inlets compared to tile risers.

Timeline

Although drainage nutrient studies date back to at least the late 19th century (Lawes et al., 1882), site-years in the Drain Load database ranged from 1961 to 2012 (Fig. 2). There was a notable increase in the number of dissolved N site-years in the 1990's, potentially in response to nutrient concerns in the Mississippi River basin (Turner and Rabalais, 2003). More emphasis has been placed upon the study of these loads compared to total N/P and dissolved P (Fig. 2; note y-axes scales). The majority of total N and total P site-years stemmed from the 1960's.

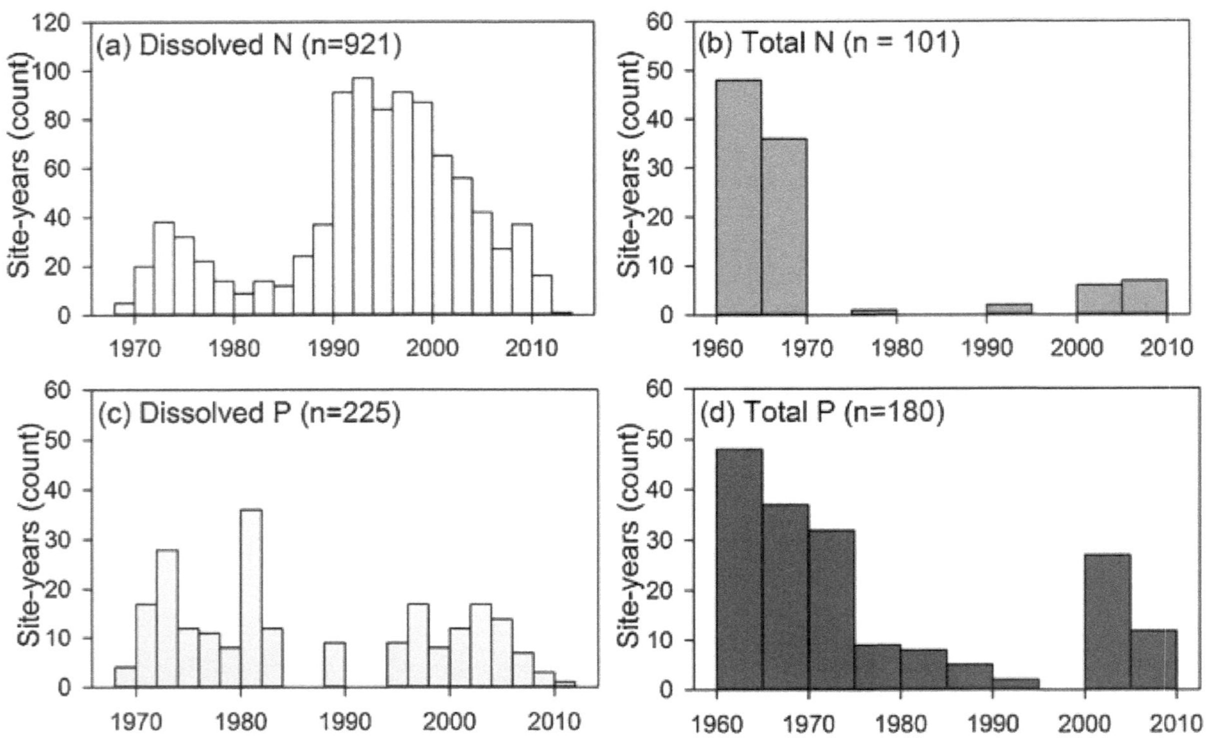

Figure 2: Histograms of Drain Load database dissolved nitrogen (a), total nitrogen (b), dissolved phosphorus (c), and total phosphorus (d) site-year timing.

Uncontrollable Factors: Variable Climate and Hydrology

Precipitation and drainage discharge are strongly correlated with wetter years generating more drainage (Randall and Iragavarapu, 1995 and Randall and Mulla, 2001). While total annual precipitation volume affects drainage discharge (Fig. 3a), it is now also thought that in some locations, seasonal rainfall totals may be more relevant than annual (e.g., March–June in the upper Midwest; Bakhsh et al., 2007, Nguyen et al., 2013 and Sands et al., 2008). Studies reporting the percent of annual precipitation occurring as subsurface drainage tend to give values ranging from approximately 15–40% (Fig. 4). Where this could be calculated for the Drain Load table, the mean and median were 25 and 20%, respectively, ($n = 827$; Fig. 4). Some variability can be explained by separating wet and dry years, with wet years resulting in a significantly higher percentage (Fig. 3b).

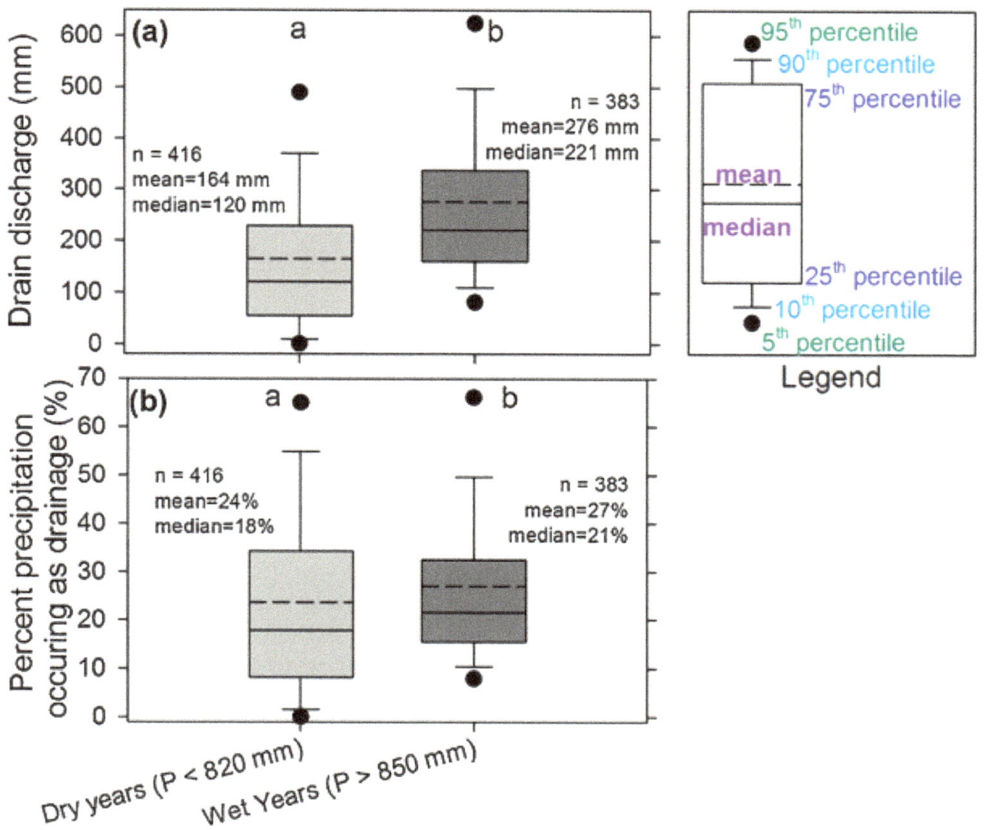

Figure 3: Drainage discharge (a) and percentage of precipitation occurring as drainage (b) for the MANAGE Drain Load table; medians with the same letters are not statistically significantly different based on a Mann–Whitney Rank Sum *t*-test.

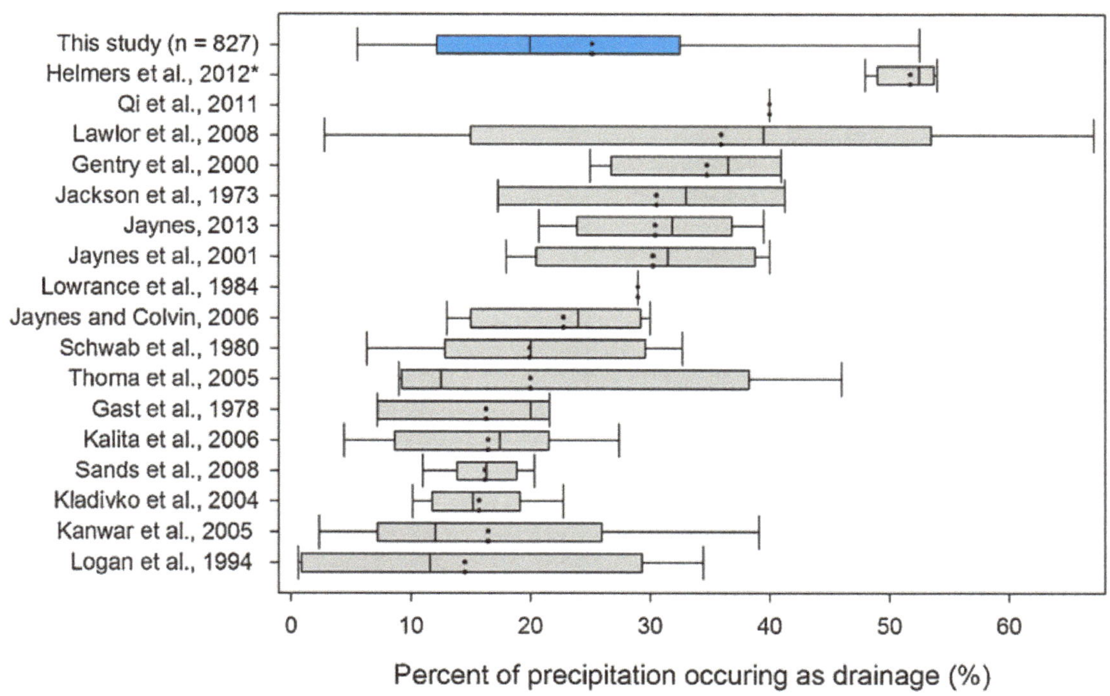

Figure 4: Percentage of precipitation occurring as subsurface drainage as reported by 17 studies (199 site-years) and over the Drain Load table; *Helmers et al. (2012b) considered the drainage season March–November only.

Because drainage nutrient loads are highly dependent upon drainage volume (Bakhsh et al., 2002, Bolton et al., 1970 and Nguyen et al., 2013), which is clearly related to precipitation trends, it follows that wetter years will have greater nutrient loads (Fig. 5). The difference between wet and dry years was highly significant for dissolved N ($p < 0.001$), dissolved P ($p < 0.001$) and total P ($p = 0.002$) loads; total N loads were also significantly different between binned wet and dry years but at a lower level potentially due to the low site-year count ($p = 0.072$).

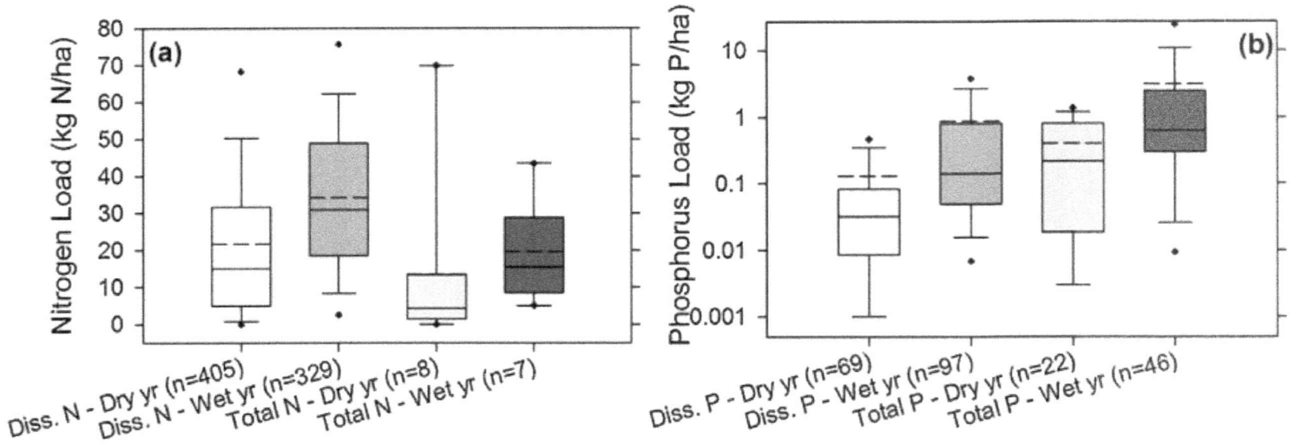

Figure 5: Dissolved and total nitrogen (a) and phosphorus (b; log axis) load ranges shown by wet or dry year from the MANAGE Drain Load database.

There is now significant interest in understanding the impacts of increasingly variable climates on agriculture. In this context, "wet" and "dry" years may occur with higher frequency which has special implications for drainage discharge and nutrient loadings. Events such as hail storms or droughts that reduce crop growth may result in additional residual N in the soil available for drainage losses in subsequent years (Bakhsh et al., 2002, Bakhsh et al., 2005, Gentry et al., 2000 and Gentry et al., 1998). Dry or drought years are especially associated with low nitrate losses, but high losses and/or concentrations the following year, particularly if the following year is wetter than average (Drury et al., 1993, Kladivko et al., 2004, Logan et al., 1994, Mitsch et al., 2001, Randall and Iragavarapu, 1995, Randall and Mulla, 2001, Sands et al., 2008 and Tan et al., 2002b). For example, Bjorneberg et al. (1996) reported nitrate losses and annual flow weighted nitrate concentrations greater than 100 kg N/ha and 65 mg N/L during the wet year in a dry/wet cycle. To put this in context of production, Randall and Iragavarapu (1995) estimated N loss during a dry year was less than 3% of the applied N, but in the wet year, this increased to 25–70%.

Table 1: Median (count) for precipitation, drainage discharge, dissolved nitrogen load and crop yields from eight studies that reported a drought followed by a wet year (Gentry et al., 2000, Gentry et al., 1998, Kladivko et al., 2004, Logan et al., 1994, Randall et al., 1997, Randall and Iragavarapu, 1995 and Sands et al., 2008) compared against the remainder of the Drain Load dataset grouped by above- and below-average precipitation; medians with the same letters are not statistically significantly different based on a Kruskal–Wallis one way analysis of variance on ranks

	Precipitation	Drainage discharge	Precipitation lost as drainage	Dissolved nitrogen load	Yield	
					Corn	Soybean
	mm		%	kg N/ha	Mg/ha	
Selected dry/drought years	562(46)d	42(46)c	8.3(46)b	3.1(46)c	5.9(21)c	2.4(7)bc
Selected wet-following-drought years	933(40)b	180(40)b	18(40)a	28.3(40)ab	6.6(17) bc	2.6(6)ab
Remaining Drain Load table pooled dry years [a]	689(381)c	132(357)b	20(357)a	18.1(346)b	8.5(173)b	3.3(59)ab
Remaining Drain Load table pooled wet years[a]	1001(394)a	230(356)a	22(356)a	30.2(302)a	9.2(125)a	3.4(47)a

[a]Selected drought years and wet-following-drought years removed from the pooled data.

Aggregated data from studies where a reportedly dry year was followed by a wet year showed statistically significant differences between years in terms of precipitation, drainage discharge, and dissolved N loads, but not crop yield (Table 1: "selected" site-years). Compared against the entire dataset (i.e., Fig. 3 and Fig. 5a but with the "selected" site-years removed), the drought years from the selected studies were indeed significantly drier than the pooled dry years. Drainage from the selected drought years was a significantly lower percentage of the precipitation compared to the other years (42 mm discharge; 8.3% of drought year precipitation). It is plausible that a year following a drought would experience a relatively lower percentage of the precipitation eluted as drainage due to a lingering soil moisture deficit. This was observed in terms of median values (18 versus 22% for selected wet and pooled wet years, respectively), although this result was not significant. While literature indicates these wet-following-drought years pose an elevated concern for extreme N loss, statistically, these selected wet years did not result in greater dissolved N loads than the pooled wet years (medians: 28.3 vs 30.2 kg N/ha; Table 1). Nevertheless, the mean dissolved N load from the selected wet years was slightly higher than the mean from the pooled wet years (35.8 and 33.9 kg N/ha, respectively, data not shown) indicating the impacts of climate variability on drainage nutrient loads is a potential topic meriting further investigation.

These data lend evidence to the premise that surplus residual soil N following a poor crop yielding-year may be available for leaching in the future. Corn and soybean yields were lower in the selected drought years and wet-following-drought years compared to their respective pooled datasets. These differences in corn yields were significant (Table 1: 5.9 versus 8.5 Mg/ha and 6.6 versus 9.2 Mg/ha for dry and wet years, respectively).

Controllable Factors

Crop Selection and Rotation

The most prevalent cropping systems within the MANAGE Drain Load database were a corn and soybean rotation and continuous corn (43 and 23% of site-years, respectively;Fig. 6). Corn additionally featured prominently as part of other rotations (e.g., Corn–Oat–Alfalfa, Corn–Soybean–Oats, Corn–Wheat–Soybean). Thirty-five individual crops were represented, with over half the site-years planted to corn (including seed, silage, and white corn; Fig. 6). Soybeans were also a major contributor at 27% of site-years, followed by alfalfa (6%), grasses ("grass", prairie, miscanthus, switchgrass; 5%), and oats (2%). "Other" crops included barley, cabbage, carrots, citrus, cotton, onions, peas, peanuts, potato, rye, snap beans, sugarcane, and wheat (5%).

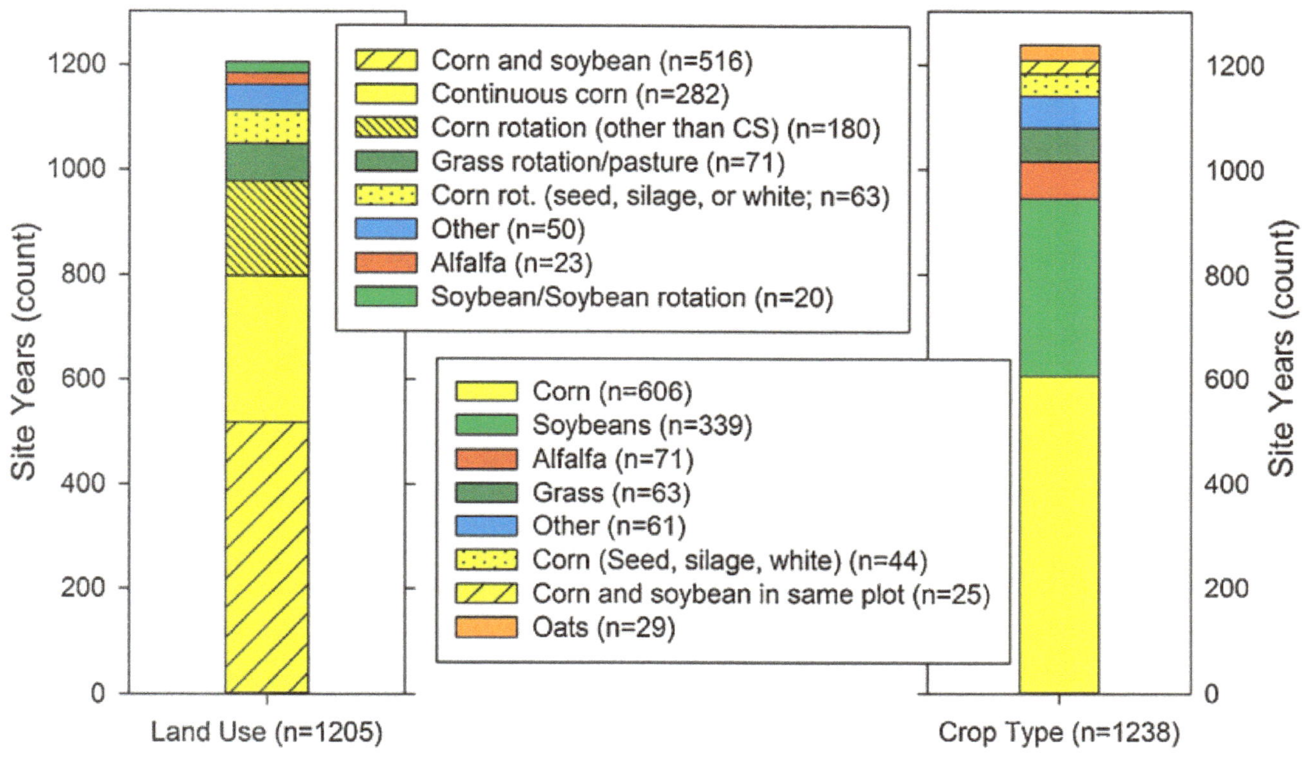

Figure 6: Land use (left) and crop type (right) across the MANAGE Drain Load database.

There was no significant difference in drainage discharge or dissolved N load between the two most common cropping rotations, continuous corn and a corn-soybean rotation, in either wet or dry years (Fig. 7a–d). Higher flow weighted nitrate concentrations have been observed from continuous corn systems compared corn-soybean rotations (Kanwar et al., 1997 and Randall et al., 1997), with several studies indicating continuous corn will also result in greater nitrate loads (Kanwar et al., 1997, Owens et al., 1995 and Weed and Kanwar, 1996). However, Klocke et al. (1999) noted higher N leachate loads from a corn-soybean rotation. Helmers et al., 2012b and Kanwar et al. (1997) both reported higher corn yields from corn rotated with soybeans versus continuously grown corn (both at recommended N application rates); this difference was shown to be significant here in both wet and dry years (Fig. 7e and f). The lack of significant difference in discharge and N load between the two systems is consistent with the variability in literature (Bakhsh et al., 2005). When only the corn rotations were included in the statistical analyses (i.e., Alfalfa and Grass treatments not included), continuous corn resulted in significantly greater discharge than the corn phase of the corn-soybean rotation ($p = 0.020$) and significantly greater dissolved N loads than the soybean phase ($p = 0.031$) but both only in wet years. Dry years still showed no difference between the three treatments in discharge or N load ($p = 0.333$ and $p = 0.272$; i.e., consistent with the broader analysis in Fig. 7).

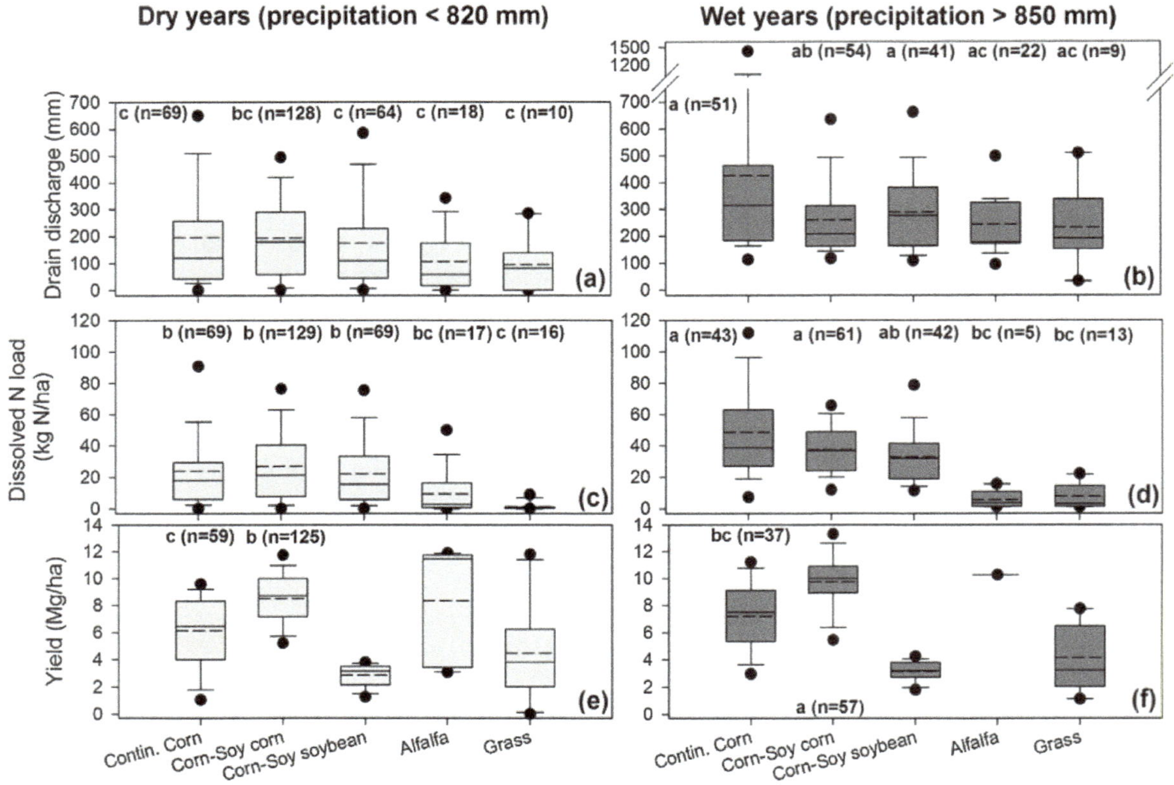

Figure 7: Drainage discharge (a, b), dissolved nitrogen load (c, d), and crop yield (e, f) between crops grouped by wet and dry years; medians with the same letters are not statistically significantly different based on a Kruskal–Wallis one way analysis of variance on ranks.

There was no drainage difference between the two crop phases of a corn-soybean rotation (Fig. 7a–b), which was consistent with some findings (Lawlor et al., 2008 and Logan et al., 1994), but not with other reports that soybeans produce greater drainage discharge than corn (Bakhsh et al., 2007 and Drury et al., 2014). Investigating further into only studies where corn and soybeans were both grown in the same year on separate plots or fields, resulted in no statistically significant difference between the drain discharge from the two crop phases (Fig. 8a; n = 31 from 6 studies; Bakhsh et al., 2009,Bakhsh et al., 2002, Bakhsh et al., 2005, Bakhsh et al., 2007, Bjorneberg et al., 1996 and Randall et al., 1997). The slope indicated that every 1.0 mm increase in discharge from corn would result in a 1.0 mm increase in the discharge from soybeans; however this relationship is offset by 20 mm (i.e., the y-intercept) indicating that soybeans result in a relatively minimal 20 mm greater discharge than corn within the rotation.

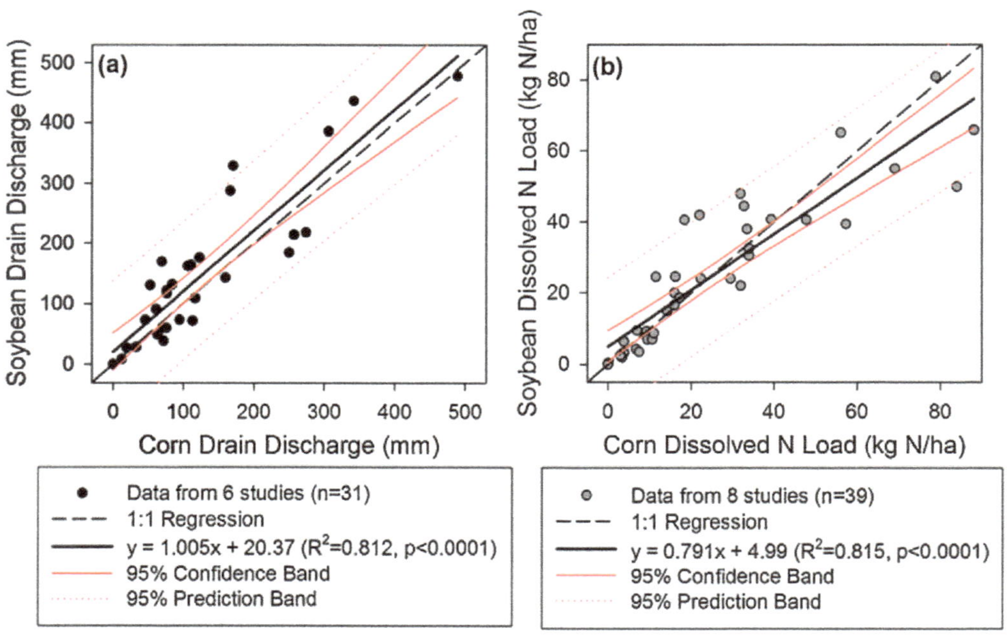

Figure 8: Corn versus soybean drainage discharge (a) and dissolved N load (b) from selected studies where corn and soybeans were both grown in the same year on separate plots or fields.

In terms of corn versus soybean phase N loss, Jaynes and Colvin (2006) and Qi et al. (2011) both observed the soybean year had lower annual flow weighted nitrate concentrations (not significant), although Logan et al. (1994) reported the soybean phase nitrate concentrations were as high or higher than the maize phase. Strock et al. (2004) reported soybeans had higher N loss and lower residual soil N compared to corn.Zhu and Fox (2003) found that soybeans may leach more N than corn at low corn N application rates, but the two phases were not different at an application rate of 200 kg N/ha to corn. Randall and Vetsch (2005) documented that 54 versus 46% of the N load occurred in the corn versus soybean phases (i.e., very similar), and Bakhsh et al. (2005)saw no significant N load difference between a corn-soybean and soybean-corn rotation over a six year study. Here, based on the selected corn-soybean studies, the slope of the dissolved N load regression between the two phases showed that for every additional kilogram N per ha lost in drainage from the corn phase, only 0.8 kg N/ha would be lost from the soybean phase (Fig. 8b; n = 39 from 8 studies; studies in Fig. 8a plus Lawlor et al., 2008 and Logan et al., 1994). This lends support to findings of greater N loads from the corn phase. Nevertheless, the 95% confidence bands of this regression overlapped the 1:1 line for the most commonly reported N load range, likely indicating no practical difference in dissolved N loads exists between the two phases. From a field-scale perspective, this lends credence to the notion that corn–soybean rotations will result in a net negative N balance (Gentry et al., 2009 and Jaynes and Karlen, 2008), as both crop phases leach similar N loads, soybeans typically do not receive N fertilization, and they symbiotically fix less N than is exported in grain harvest. This finding also supports the investigation of corn-soybean rotations as a combined system (i.e., a given research plot planted to both corn and soybeans in a given year; e.g., Lawlor et al., 2008, Lawlor et al., 2011 and Nguyen et al., 2013).

In the upper Midwest and Canada, the majority of the annual drainage volume and nitrate loss occurs during the spring when conventional row crops are not yet growing or during the very early growing season (Bakhsh et al., 2007, Ball Coelho et al., 2012a, Bjorneberg et al., 1996, Bryant et al., 1987, Gangbazo et al., 1997, Gentry et al., 2000, Kladivko et al., 2004, Lawlor et al., 2008, Randall and Vetsch, 2005 and Randall et al., 2003). To address this crop-soil-water imbalance and to reduce this "asynchronous production and uptake of nitrate in the soil" (Cambardella et al., 1999 and Sands et al., 2008), the use of perennials and diverse cropping rotations is an important area of research interest (Dinnes et al., 2002). Perennial crops (e.g., alfalfa, miscanthus, switchgrass, perennial forage) are widely thought to reduce both subsurface drainage discharge and nitrate loads compared to annual crops (Benoit, 1973, Burwell et al., 1976, Qi et al., 2011 and Smith et al., 2013). Randall et al. (1997) reported that drainage nitrate losses were more than thirty-fold higher for row crops than perennials due to more prolonged uptake of water and N of the latter. Here, there was no significant drainage-reduction benefit of perennials compared to conventional crops in either dry or wet years, although alfalfa and grasses consistently had lower median/mean discharge volumes (Fig. 7a and b). This lack of significance for perennial-induced reduction in drain flow may have been due to the relatively low site-year count for alfalfa and grass crops. In terms of dissolved N, the alfalfa and grass median loads were always less than 4 kg N/ha whereas the other treatments were greater than 15 and 30 kg N/ha in dry and wet years, respectively.

Wetter years produced more drainage discharge regardless of the cropping option, though not always significantly (Fig. 7a and b). It is plausible perennial crops provide buffering of drainage volumes and N loads in wet years, as both alfalfa and grass showed no significant increase in discharge or load between wet and dry years, whereas continuous corn and the soybean phase of the corn–soybean rotation produced significantly more drainage in wet years, and continuous corn and the corn phase produce significantly greater N loads (Fig. 7c and d). Some caution is issued, however, when considering a perennial legume such as alfalfa, as this crop can result in drainage N losses especially after plow-down or overwintering (Fleming, 1990 and McCracken et al., 1994). Additionally, it is worth noting that a drainage water quality benefit of perennials has not exclusively been observed across all years in all studies where multiple cropping strategies were employed (Kanwar et al., 2005 and Tan et al., 2002b). While choice of cropping system is one of the most "controllable" factors for having a major impact on drainage N loads (Randall and Goss, 2008), adequate economic returns are often challenging for the production of perennials (Randall and Mulla, 2001) and such a major paradigm shift in farming practice may be difficult to overcome (Qi et al., 2011).

Tillage Management

Conventional tillage (moldboard plow, "conventional tillage"), conservation tillage (ridge till, chisel plow, "conservation tillage"), no till, and pasture each accounted for 430, 202, 170, and 34 site-years, respectively, of the total 836 site-years where a tillage practice was reported. Drain discharge differences between tillage types were most apparent during the dry years with the practice of conventional tillage yielding significantly greater discharge than conservation and pasture tillage practices (Table 2). Across literature, drainage discharge is reported to be greater with no-tillage compared to conventional tillage (or other forms of conservation tillage) due, in part, to increased macropore flow under no-till (Bakhsh et al., 2007, Bjorneberg et al., 1996, Blann et al., 2009, Patni et al., 1996, Randall and Iragavarapu, 1995 and Tan et al., 2002a). This drainage volume difference between no-till and conventional tillage may be as high as 2 to 3 times (Endale et al., 2010). However, the increased presence of macropores under no-tillage systems may cause infiltrating water to have decreased soil contact and relatively lower drainage nitrate concentrations compared with intensive tillage (Angle et al., 1993, Bjorneberg et al., 1996, Kanwar et al., 1988 and Kanwar et al., 1997). Thus, this potential for greater flow volume from no-till systems is confounded by this practice's reduced drainage nitrate concentrations; this combination may mask any significant differences in N loading due to no-till. No-till sometimes results in greater N loads than more conventional tillage practices (Bakhsh et al., 2002, Kanwar et al., 1997 and Patni et al., 1996), but not always (Francesconi et al., 2014 and Randall and Iragavarapu, 1995). Here, the practice of no-till had lower dissolved N load means and medians compared to the conventional and conservation treatments in both wet and dry years, though this difference was not statistically significant.

Table 2: Median (count) for drain discharge, percentage of precipitation resulting as drainage, and dissolved nitrogen loads in dry and wet years for four tillage practices; medians with the same letters are not statistically significantly different based on a Kruskal–Wallis one way analysis of variance on ranks

	Drainage discharge		Percent of precipitation occurring as drainage		Dissolved N load	
	mm		%		kg N/ha	
	Dry[a]	Wet[a]	Dry	Wet	Dry	Wet
Conventional	165 (221)b	259 (137) a	23 (222)ab	25 (138)a	23.0 (217)b	37.0 (141)ab
Conservation	92 (74) cd	221 (49) ab	13 (74)bc	20 (49)ab	13.6 (72)b	37.3 (47)a
No till	120 (66) bc	215 (52) ab	18 (67)b	20 (52)ab	9.1 (63)bc	22.0 (51)ab
Pasture	0 (11)d	200 (8)ab	0 (11)c	22 (7)ab	0.0 (11)c	3.0 (3)bc

[a]Dry years: precipitation <820 mm; wet years: precipitation >850 mm.

Weed and Kanwar (1996) reported higher drainage N losses for chisel plow compared to ridge plow, moldboard plow, and no-till. However, Karlen et al. (1998) reported their two lowest 15-yr N loads were from chisel and ridge tillage systems (467, 369, 352, and 466 kg N/ha for moldboard plow, chisel plow, ridge tillage, and no tillage treatments, respectively). Drury et al. (1993) similarly reported lower N losses and drainage discharge from a conservation tillage treatment compared to conventional tillage, which was generally consistent with this analysis, but not significantly (Table 2). In the end, tillage management may play a fairly minor role in predicting drainage N loads (Kanwar et al., 1997, Randall and Goss, 2008 and Randall and Mulla, 2001).

Not only did the pasture treatment experience increased drainage volumes in wet versus dry years, but this was also the only treatment where the percentage of precipitation experienced as drainage statistically increased in the wet year (Table 2). This may mean during a wet year, increases in drainage flow may result even with a "conservation-oriented" land cover and tillage approach. Nevertheless, importantly for water quality goals, the pasture treatment consistently leached the lowest dissolved N loads (Table 2; note the low site-year count).

Drainage System Design

Tile drainage spacing in the Drain Load database ranged from 2.5 to 43 m (Fig. 9a; one 100 m spacing outlier was removed from analysis; 27 and 24% used 7.6 and 28.5 m spacing, respectively). Drain depths ranged from 0.5 to 1.5 m with a third of studies using 1.2 m which is consistent with conventional tile drain depth in the US Midwest (Fig. 9b). Due to the low number of site-years and lack of information reported for surface drainage studies, this drainage design section focused on subsurface design trends.

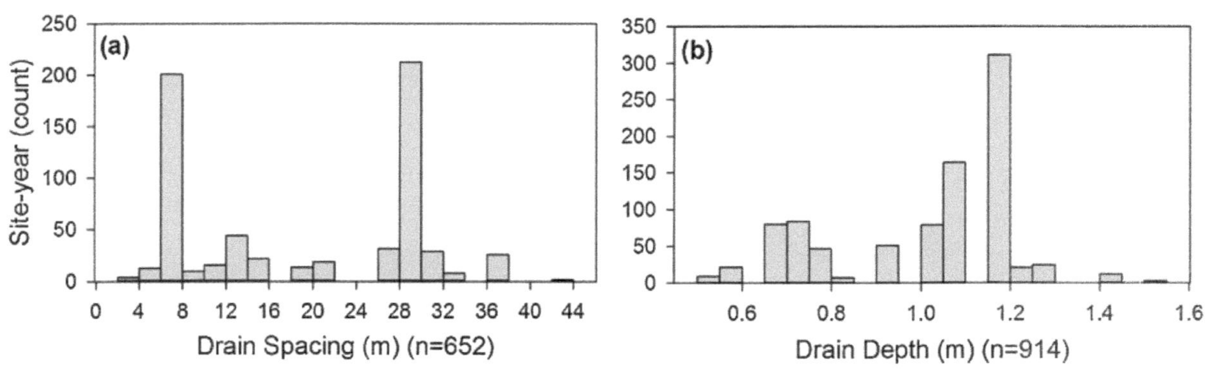

Figure 9: istograms of Drain Load database site-years by tile drain spacing (a) or tile drain installation depth (b).

Wider drain spacing reduces drainage discharge and N loading (Davis et al., 2000,Hoover and Schwab, 1969, Kladivko et al., 2004 and Sands et al., 2008), but may decrease crop yield if trafficability is reduced (Bolton et al., 1980 and Skaggs et al., 2005). The increase in nitrate loading with narrow drain spacing is a factor of increased flow rather than differences in nitrate concentrations (Kladivko et al., 2004). Binning the Drain Load tile spacing data into discrete groups confirmed narrow drain spacing tended to elute greater dissolved N loads, although differences in drainage discharge between spacings were less clear (Fig. 10a and b). Here, a smaller bubble size indicated a larger data population and thus reduced uncertainty surrounding the median of a group of binned drainage spacings.

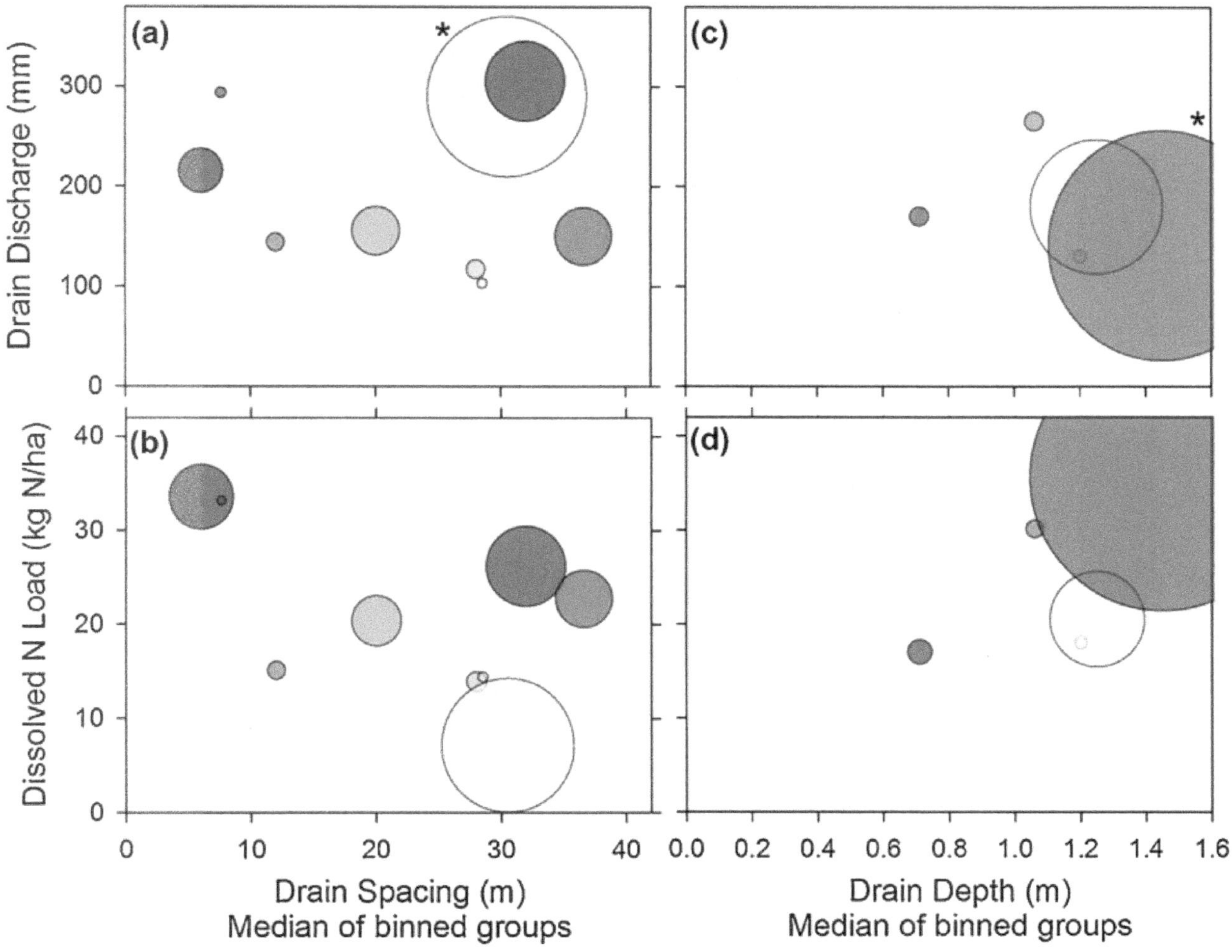

Figure 10: Median drain spacing (a and b) and drain depth (c and d) of binned spacing- or depth-groups graphed against the median drain discharge (a, c) and median dissolved nitrogen load (b, d); bubble size determined by the scaled inverse of the population for each binned group (spacing bubble size = 10/n; depth bubble size = 25/n), thus smaller bubbles indicate larger populations and greater certainty; the starred spacing and depth bubbles were scaled by 1 and 10, respectively, due to small populations (n = 1 and 7, respectively).

Shallow placement of drainage pipes decreases drainage discharge and associated nitrate loads, the latter of which was corroborated here (Fig. 10c and d; Burchell et al., 2005, Davis et al., 2000 and Gordon et al., 2000a; Sands et al., 2008; Schwab et al., 1980; Smith and Kellman, 2011). However, shallow drain placement must allow sufficient depth to avoid structural failures (ASABE, 2014 and Gordon et al., 2000), and take into consideration potential yield impacts (Gordon et al., 2000, Kalita and Kanwar, 1993 and Smith and Kellman, 2011), though these may be minimal (Helmers et al., 2012a).

The reduction in dissolved N load due to shallow drain placement is largely thought to be due to reduction in drainage volume rather than changes in nitrate concentration (Sands et al., 2008). The impact of shallow drainage on nitrate concentrations is variable; Sands et al. (2008) found no significant difference between shallow and deep drainage flow weighted nitrate concentration (though the shallow treatment concentrations tended to be lower), and Helmers et al. (2012a) reported shallow drainage had higher concentrations than conventional drainage. Regressing drainage discharge by dissolved N load for five binned groups of drain depths revealed an interesting trend in the binned groups' regressions (Fig. 11). As the drain installation depth increased, the regression slope increased and y-intercept generally decreased. The slopes indicated at a given drainage discharge, a deeper drain depth would result in a disproportionately higher dissolved N load. For example, application of these slopes yields that for 100 mm additional drainage discharge between two hypothetical years (e.g., a normal vs a wet year), a drain placed within 1.0 m of the soil surface would generate 8.4 kg N/ha of additional N load in the wet year, whereas a drain placed 1.2 m below the soil surface would generate an additional 11 kg N/ha. The difference in y-intercepts is also interesting, as these seem to indicate that in dry years (i.e., when drain discharge is around 0 mm at the y-intercept), deeper drains have lower N loads. This may be rooted in the concept that deeper placement of drains provides greater volume of soil storage for precipitation. In dry years, this increased storage capacity and the theoretical potential for increased denitrification over this volume may act to reduce N loads. While there is wide scatter between these data, the impact of drainage spacing, particularly in wet versus dry years, is an important topic meriting further investigation.

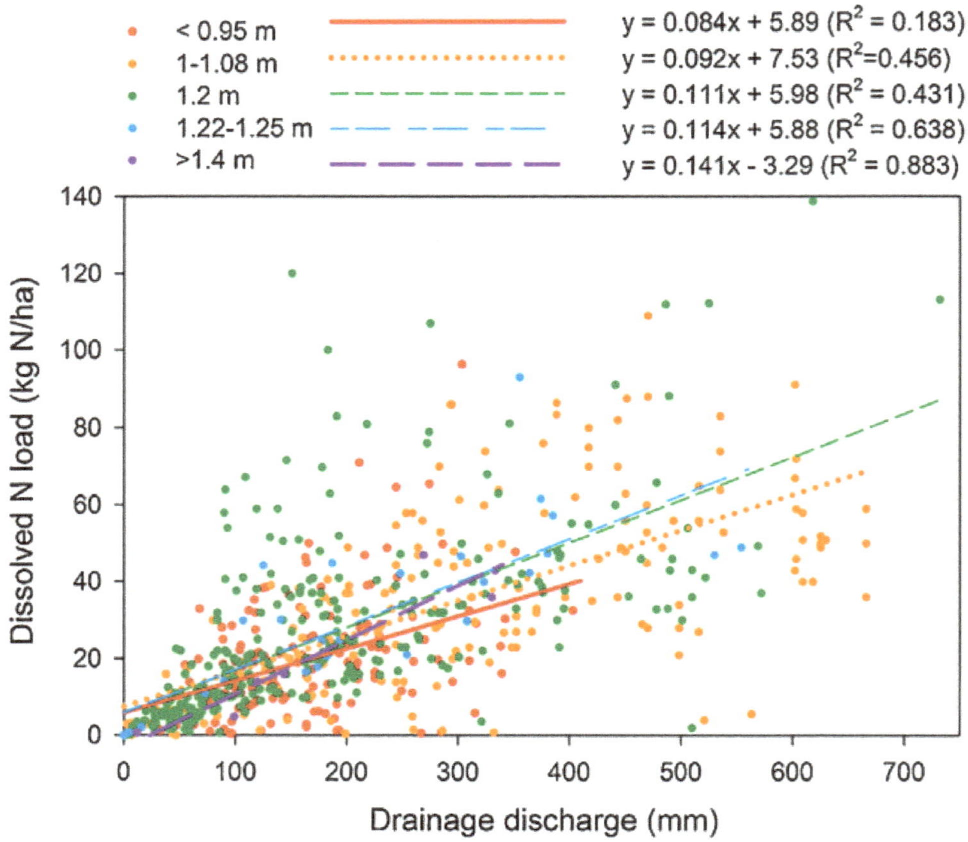

Figure 11: Drainage discharge and dissolved nitrogen load relationships for five groups of binned drain tile installation depths.

Knowledge Gaps

An important outcome of any review is the identification of potential gaps in scientific understanding. Drain Load database development underscored several areas for future drainage research including more intensive year-round monitoring (in some locations) and improved monitoring in newly drained areas, ditch drained areas, and of surface intakes. Perhaps most importantly, more long-term studies with coordinated controls across multiple sites and years would improve knowledge of drainage-associated nutrient loads. Randall and Vetsch (2005) noted the common lack of statistically significant differences in N loss between treatments is not surprising considering this metric compounds variability associated with both drainage flow and nitrate concentration. For more robust statistical comparisons, long-term studies are necessary to increase statistical power to overcome both the strong effect and high variability of precipitation (Randall et al., 2003).

In the northern Midwest, winter drainage usually ceases due to low precipitation or frozen soil (Kalita et al., 2006). Challenging field conditions, restricted site access, and limitations of monitoring equipment mean that drainage potentially occurring over this period, particularly snowmelt drainage, will not be captured (Ball Coelho et al., 2010, Ball Coelho et al., 2012a, Fleming, 1990 and Milburn et al., 1990). Drainage hydrology during the late winter/early spring is complicated by diurnal freeze/thaw cycles, rain falling on snow, and the breaking of the 'freeze line' which may result in a very rapid transition from no drainage to pipe-full flow (Ball Coelho et al., 2012b and Bottcher et al., 1981). Studies that have been able to avoid retiring monitoring equipment over the winter have noted the importance of monitoring year-round, and suggest that better and more intensive monitoring strategies are needed to capture these critical missed periods (Ball Coelho et al., 2012a and Milburn and MacLeod, 1991). Snowmelt drainage can contribute significantly to total annual drainage volume and nutrient/sediment loading (Ball Coelho et al., 2012a, Ball Coelho et al., 2012b, Gangbazo et al., 1997, Jamieson et al., 2003,Klatt et al., 2003 and Milburn et al., 1990). In a study specifically intended to investigate snowmelt contributions, Ginting et al. (2000) found that snowmelt drainage mobilized dissolved pollutants, whereas storm event drainage may be responsible for more of the particulate and sediment-bound nutrient loads.

By noting the Drain Load studies that retired monitoring equipment over the winter, it was possible to separate potential effects between studies that monitored all year versus studies that represented early spring through late fall as the annual period (Table 3). The site-years covering a full year had significantly higher precipitation and drainage discharge compared to the site-years where the monitoring equipment was retired for the winter. Confoundingly, the studies where it was necessary to winterize monitoring equipment yielded significantly higher dissolved N loads than full year-studies, which may be an indication the former were not underestimating loads. There is likely a spatial/geographic aspect for this discussion, however, as the majority of the "not full year" studies were from Iowa. It may be that certain locations will not significantly underestimate nutrient losses by excluding winter months (e.g., Iowa), whereas full years of monitoring will be critical for other locations (e.g., Ontario, New York, Indiana).

Table 3: Median (count) for precipitation, drainage discharge, and dissolved nitrogen loads from studies reporting a full year versus those reporting early spring though late fall; medians with the same letters are not statistically significantly different based on a Mann–Whitney rank sum *t*-test

	Precipitation	Drainage discharge	Dissolved N load
	mm		kg N/ha
Not a full year	790 (331)b	170 (393)b	24.0 (401)a
Full year	889 (489)a	191 (565)a	20.0 (517)b

There are clearly gaps in peer-reviewed records of annual drainage nutrient loads from certain geographic regions and from certain types of drainage systems (Fig. 2). The lack of site-years from areas widely practicing ditch drainage (e.g., Maryland, Delaware) complicates efforts to develop drainage best management practice guidance specific for these locations (Shirmohammadi et al., 1995). Fortunately, recent efforts are aiming to address this (Bryant et al., 2012 and Penn et al., 2007). Improved knowledge about the water quality impacts of surface intakes could also contribute to water quality improvement efforts (Ginting et al., 2000 and Schilling and Helmers, 2008). There were only four studies included in the Drain Load database that specifically mentioned surface inlets (Ball Coelho et al., 2012a, Ball Coelho et al., 2012b, Bottcher et al., 1981 and Hanway and Laflen, 1974), despite their widespread implementation.

It is critical for the Dakotas and other newly drained areas (e.g., Jia et al., 2012 and Nash et al., 2014) to continue to build their library of region-specific drainage water quality data especially considering new drainage systems, and their associated soil disturbance, pose a special concern for water quality (Fausey, 1983, Ritter et al., 1995 and Roberts et al., 1986). The long history of drainage in many locations across North America means that aging infrastructure is now starting to be replaced and upgraded. This is a pivotal opportunity to upgrade systems not only for improved crop growth and yield but also to advance water quality goals (Strock et al., 2010). The redesign of existing drainage systems means understanding drainage design impacts becomes increasingly important.

CONCLUSIONS

Artificial drainage will remain a vital component of many agricultural systems across North America, and improved understanding of management impacts, especially under variable climate conditions, can help point the way to a more sustainable future. This compilation of nearly 1300 drainage nutrient load site-years in the new MANAGE Drain Load table facilitated quantitative analyses of the history of drainage water quality research in North America.

Across site-years, the mean and median percent of precipitation occurring as drainage were 25 and 20%, respectively, with wet years resulting in significantly greater drainage discharge, percentages of precipitation eluted as drainage, and nutrient loads. In terms of controllable factors, no significant difference was observed in drainage discharge or N loads between continuous corn and corn-soybean cropping systems, although corn in rotation showed significantly greater yields. The evidence also supported investigation of corn-soybean rotations as a single system as there was no practical difference in discharge or N load between the two phases. Alfalfa and grasses provided notable N loading benefits compared to row crops and small grains, but these comparisons were limited by low site-year counts. Consistent with literature, tillage management resulted in no clear best practice to reduce drainage discharge or N loads. Wider drain spacing and shallower drain placement tended to decrease N loss in subsurface drainage, but the aggregated effects on drainage discharge were less apparent.

As drainage water quality research continues into the 21st century, MANAGE's Drain Load table would benefit from annual nutrient loading data in newly drained areas, ditch drained areas, and areas where surface intakes are specified. Most importantly, more long-term drainage nutrient transport studies with coordinated controls across multiple sites and years would increase statistical power for more robust comparisons in the future.

ACKNOWLEDGEMENTS

The authors gratefully acknowledge the 4R Research Fund for support of this work. The USDA is an equal opportunity provider and employer.

REFERENCES

1. Angle, J., Gross, C., Hill, R., McIntosh, M., 1993. Soil nitrate concentrations under corn as affected by tillage, manure, and fertilizer applications. J. Environ. Qual. 22, 141–147.
2. ASABE, 2014. EP 260 Design and Construction of Subsurface Drains on Agricultural Lands in Humid Areas. American Society of Agricultural and BiologicalEngineers, St. Joseph, Michigan.
3. Bakhsh, A., Kanwar, R., Baker, J., Sawyer, J., Malarino, A., 2009. Annual swine manure applications to soybean under corn-soybean rotation. Trans. ASABE 52, 751–757.
4. Bakhsh, A., Kanwar, R.S., Bailey, T.B., Cambardella, C.A., Karlen, D.L., Colvin, T.S., 2002. Cropping system effects on NO3-N loss with subsurface drainage water. Trans ASAE 45, 9.
5. Bakhsh, A., Kanwar, R.S., Karlen, D.L., 2005. Effects of liquid swine manure applications on NO3–N leaching losses to subsurface drainage water fromloamy soils in Iowa. Agric. Ecosyst. Environ. 109, 118–128.
6. Bakhsh, A., Kanwar, R.S., Pederson, C., Bailey, T.B., 2007. N-Source effects ontemporal distribution of NO3–N leaching losses to subsurface drainage water. Water Air Soil Pollut. 181, 35–50.

7. Ball Coelho, B., Bruin, A.J., Staton, S., Hayman, D., 2010. Sediment and nutrient contributions from subsurface drains and point sources to an agricultural watershed. Air Soil Water Res. 3, 1–21.

8. Ball Coelho, B., Lapen, D., Murray, R., Topp, E., Bruin, A., Khan, B., 2012a. Nitrogen loading to offsite waters from liquid swine manure application under different drainage and tillage practices. Agric. Water Manage. 104, 40–50.

9. Ball Coelho, B., Murray, R., Lapen, D., Topp, E., Bruin, A., 2012b. Phosphorus and sediment loading to surface waters from liquid swine manure application under different drainage and tillage practices. Agric. Water Manage. 104,51–61.

10. Benoit, G.R., 1973. Effect of agricultural management of wet sloping soil on nitrate and phosphorus in surface and subsurface water. Water Res. Res. 9, 1296–1303.

11. Bjorneberg, D., Kanwar, R., Melvin, S., 1996. Seasonal changes in flow and nitrate-N loss from subsurface drains. Trans. ASAE 39, 961–976.

12. Blann, K.L., Anderson, J.L., Sands, G.R., Vondracek, B., 2009. Effects of agricultural drainage on aquatic ecosystems: a review. Crit. Rev. Environ. Sci. Technol. 39,909–1001.

13. Bolton, E., Aylesworth, J., Hore, F., 1970. Nutrient losses through tile drains under three cropping systems and two fertility levels ona Brookston clay soil. Can. J. Soil Sci. 50, 275–279.

14. Bolton, E., Dirks, V., Hore, F., 1980. Corn, soybean and wheat yields on Brookston clay drained by plastic tubing installed by two methods at seven spacings and two depths. Can. Agric. Eng. 22, 145–148.

15. Bottcher, A., Monke, E., Huggins, L., 1981. Nutrient and sediment loadings from asubsurface drainage system. Trans. ASAE 24, 1221–1226.

16. Bryant, G., Irwin, R., Stone, J., 1987. Tile drain discharge under different crops. Can.J. Agric. Eng. 29, 117–122.

17. Bryant, R.B., Buda, A.R., Kleinman, P.J.A., Church, C.D., Saporito, L.S., Folmar, G.J., Bose, S., Allen, A.L., 2012. Using flue gas desulfurization gypsum to removedissolved phosphorus from agricultural drainage waters. J. Environ. Qual. 41, 664–671.

18. Burchell, M., Skaggs, R., Chescheir, G., Gilliam, J., Arnold, L., 2005. Shallowsubsurface drains to reduce nitrate losses from drained agricultural lands.Trans. ASAE 48, 1079–1089.

19. Burwell, R., Schuman, G., Saxton, K., Heinemann, H., 1976. Nitrogen in subsurface discharge from agricultural watersheds. J. Environ. Qual. 5, 325–329.

20. Cambardella, C.A., Moorman, T.B., Jaynes, D.B., Hatfield, J.L., Parkin, T.B., Simpkins,W.W., Karlen, D.L., 1999. Water quality in Walnut Creek watershed:Nitrate-Nitrogen in soils, subsurface drainage water, and shallow groundwater. J. Environ. Qual. 28, 25–34.

21. David, M.B., Drinkwater, L.E., McIsaac, G.F., 2010. Sources of nitrate yields in theMississippi River Basin. J. Environ. Qual. 39, 1657–1667.

22. Davis, D.M., Gowda, P.H., Mulla, D.J., Randall, G.W., 2000. Modeling nitrate nitrogenleaching in response to nitrogen fertilizer rate and tile drain depth or spacingfor Southern Minnesota, USA. J. Environ. Qual. 29, 1568–1581.

23. Dinnes, D.L., Karlen, D.L., Jaynes, D.B., Kaspar, T.C., Hatfield, J.L., Colvin, T.S.,Cambardella, C.A., 2002. Nitrogen management strategies to reduce nitrateleaching in tile-drained midwestern soils. Agron. J. 94, 153–171.

24. Drury, C., Tan, C., Welacky, T., Reynolds, W., Zhang, T., Oloya, T., McLaughlin, N.,Gaynor, J., 2014. Reducing nitrate loss in tile drainage water with cover crops and water-table management systems. J. Environ. Qual. 43, 587–598.

25. Drury, C.F., Findlay, W.I., Gaynor, J.D., McKenney, D.J., 1993. Influence of tillage onnitrate loss in surface runoff and tile drainage. Soil Sci. Soc. Am. J. 57, 797–802.

26. Endale, D.M., Schomberg, H.H., Jenkins, M.B., Franklin, D.H., Fisher, D.S., 2010. Management implications of conservation tillage and poultry litter use for Southern Piedmont USA cropping systems. Nutr. Cycl. Agroecosyst. 88,299–313.

27. Evans, R.O., Wayne Skaggs, R., Wendell Gilliam, J., 1995. Controlled versus conventional drainage effects on water quality. J. Irrig. Drain. Eng. 121, 271–276.

28. Fausey, N., 1983. Shallow subsurface drain performance in Clermont soil. Trans.ASAE 26, 782–784.

29. Fausey, N.R., Brown, L.C., Belcher, H.W., Kanwar, R.S., 1995. Drainage and water quality in Great Lakes and cornbelt states. J. Irrig. Drain. Eng. 121, 283–288.

30. Fleming, R.J., 1990. Impact of agricultural practices on tile water quality. Proc. Of the Int'l Meeting of the ASAE Paper No. 902,028. In: American Society of Agricultural and Biological Engineers. St. Joseph, Michigan.

31. Francesconi, W., Smith, D., Heathman, G., Williams, C., Wang, X., 2014. Monitoringand APEX modeling of no-till and reduced-till in tile drained agricultural landscapes for water quality. Trans. ASABE 57, 777–789.

32. Gangbazo, G., Pesant, A., Coté, D., Barnett, G., Cluis, D., 1997. Spring runoff anddrainage n and p losses from hog-manured corn. J. Am. Water Res. Assoc. 33,405–411.

33. Gentry, L.E., David, M.B., Below, F.E., Royer, T.V., McIsaac, G.F., 2009. Nitrogen massbalance of a tile-drained agricultural watershed in East-Central Illinois. J.Environ. Qual. 38, 1841–1847.

34. Gentry, L.E., David, M.B., Smith-Starks, K.M., Kovacic, D.A., 2000. Nitrogen fertilizerand herbicide transport from tile drained fields. J. Environ. Qual. 29,232–240.

35. Gentry, L.E., David, M.B., Smith, K.M., Kovacic, D.A., 1998. Nitrogen cycling and tile drainage nitrate loss in a corn/soybean watershed. Agric. Ecosyst. Environ. 68, 85.

36. Ginting, D., Moncrief, J., Gupta, S., 2000. Runoff, solids, and contaminant losses into surface tile inlets draining lacustrine depressions. J. Environ. Qual. 29, 551–560.

37. Gold, A., Loudon, T., 1989. Tillage effects on surface runoff water quality from artificially drained cropland. Trans. ASAE 32, 1329–1334.

38. Gordon, R., Madani, A., Boyd, N., Astatkie, T., Jamieson, R., Caldwell, K., 2000. Subsurface Nitrate–N leaching loss as affected by drainage size and depth in a shallow slowly–permeable soil. Can. Water Res. J. 25, 331–341.

39. Hanway, J.J., Laflen, J.M., 1974. Plant nutrient losses from tile-outlet terraces. J. Environ. Qual. 3, 351–356.

40. Harmel, D., Potter, S., Casebolt, P., Reckhow, K., Green, C., Haney, R., 2006.Compilation of measured nutrient load data for agricultural land uses in the United States. J. Am. Water Res. Assoc. 42, 1163–1178.

41. Harmel, D., Qian, S., Reckhow, K., Casebolt, P., 2008. The MANAGE database: nutrient load and site characteristic updates and runoff concentration data. J.Environ. Qual. 37, 2403–2406.

42. Helmers, M., Christianson, R., Brenneman, G., Lockett, D., Pederson, C., 2012a. Water table, drainage, and yield response to drainage water management insoutheast Iowa. J. Soil Water Conserv. 67, 495–501.

43. Helmers, M.J., Zhou, X., Baker, J.L., Melvin, S.W., Lemke, D.W., 2012b. Nitrogen loss on tile-drained Mollisols as affected by nitrogen application rate under continuous corn and corn-soybean rotation systems. Can. J. Soil Sci. 92,493–499.

44. Hoover, J.R., Schwab, G.O., 1969. Effect of tile depth, spacing, and croppingpractices on drain discharge. Trans. ASAE 12, 150–152.

45. Jamieson, A., Madramootoo, C., Enright, P., 2003. Phosphorus losses in surface andsubsurface runoff from a snowmelt event on an agricultural field in Quebec.Can. Biosyst. Eng. 1, 1–17.

46. Jaynes, D.B., Colvin, T.S., 2006. Corn yield and nitrate loss in subsurface drainage from midseason nitrogen fertilizer application. Agron. J. 98, 1479–1487.

47. Jaynes, D.B., Karlen, D., 2008. Chapter 11: Sustaining soil resources while managing nutrients. In: Final Report: Gulf Hypoxia and Local Water Quality Concern Workshop. ASABE, Michigan, pp. 149–158.

48. Jia, X., DeSutter, T., Lin, Z., Schuh, W., Steele, D., 2012. Subsurface drainage andsubirrigation effects on water quality in southeast North Dakota. Trans. ASABE55, 1757–1769.

49. Johnson, D.W., Curtis, P.S., 2001. Effects of forest management on soil C and Nstorage: meta analysis. Forest Ecol. Manage. 140, 227–238.

50. Kalita, P.K., Algoazany, A.S., Mitchell, J.K., Cooke, R.A.C., Hirschi, M.C., 2006. Subsurface water quality from a flat tile-drained watershed in Illinois, U.S.A.Agric. Ecosyst. Environ. 115, 183–193.

51. Kalita, P.K., Kanwar, R.S., 1993. Effect of water-table management practices on thetransport of nitrate-n to shallow groundwater. Trans. ASABE 36, 413–422.

52. Kanwar, R.S., Baker, J.L., Baker, D.G., 1988. Tillage and split n-fertilization effects onsubsurface drainage water quality and crop yields. Trans. ASABE 31, 453–461.

53. Kanwar, R.S., Colvin, T.S., Karlen, D.L., 1997. Ridge, moldboard, chisel, and no-tilleffects on tile water quality beneath two cropping systems. J. Prod. Agric. 10, 227.

54. Kanwar, R.S., Cruse, R.M., Ghaffarzadeh, M., Bakhsh, A., Karlen, D.L., Bailey, T.B.,2005. Corn-soybean and alternative cropping systems effects on NO3-Nleaching losses in subsurface drainage water. App. Eng. Agric. 21, 181–188.

55. Karlen, D.L., Kumar, A., Kanwar, R.S., Cambardella, C.A., Colvin, T.S., 1998. Tillagesystem effects on 15-year carbon-based and simulated N budgets in a tile-drained Iowa field. Soil Tillage Res. 48, 155–165.

56. Kladivko, E.J., Frankenberger, J.R., Jaynes, D.B., Meek, D.W., Jenkinson, B.J., Fausey,N.R., 2004. Nitrate leaching to subsurface drains as affected by drain spacing and changes in crop production system. J. Environ. Qual. 33, 1803–1813.

57. Klatt, J., Mallarino, A., Downing, J., Kopaska, J., Wittry, D., 2003. Soil phosphorus,management practices, and their relationship to phosphorus delivery in the Iowa Clear Lake agricultural watershed. J. Environ. Qual. 32, 2140–2149.

58. Klocke, N., Watts, D.G., Schneekloth, J., Davison, D.R., Todd, R., Parkhurst, A.M., 1999. Nitrate leaching in irrigated corn and soybean in a semi-arid climate. Trans. ASAE 42, 1621–1630.

59. Lawes, J.B., Gilbert, J.H., Warington, R., 1882. On the amount and composition of the rain and drainage-waters collected at Rothamsted. J. R. Agric. Soc. Engl. XVIII, 1–150.

60. Lawlor, P.A., Helmers, M.J., Baker, J.L., Melvin, S.W., Lemke, D.W., 2008. Nitrogen application rate effect on nitrate-nitrogen concentration and loss in subsurface drainage for a corn-soybean rotation. Trans. ASABE 51, 83–94.

61. Lawlor, P.A., Helmers, M.J., Baker, J.L., Melvin, S.W., Lemke, D.W., 2011. Comparison of liquid swine manure and ammonia nitrogen application timing on subsurface drainage water quality in Iowa. Trans. ASABE 54, 973–981.

62. Logan, T.J., Eckert, D.J., Beak, D.G., 1994. Tillage, crop and climatic effects of runoff and tile drainage losses of nitrate and four herbicides. Soil Tillage Res. 30, 75–103.

63. Madramootoo, C.A., Johnston, W.R., Ayars, J.E., Evans, R.O., Fausey, N.R., 2007. Agricultural drainage management, quality and disposal issues in North America. Irrig. Drain. 56, S35–S45.

64. McCracken, D.V., Smith, M.S., Grove, J.H., Blevins, R.L., MacKown, C.T., 1994. Nitrate leaching as influenced by cover cropping and nitrogen source. J. Soil Sci. Soc. Am. 58, 1476–1483.

65. Milburn, P., MacLeod, J., 1991. Considerations for tile drainage-water quality studies in temperature regions. Appl. Eng. Agric. 7, 209–215.

66. Milburn, P., Richards, J., Gartley, C., Pollock, T., O'Neill, H., Bailey, H., 1990. Nitrate leaching from systematically tiled potato fields in New Brunswick, Canada. J. Environ. Qual. 19, 448–454.

67. Mitsch, W.J., Day, J.W., Gilliam, J.W., Groffman, P.M., Hey, D.L., Randall, G.W., Wang, N., 2001. Reducing nitrogen loading to the Gulf of Mexico from the Mississippi River Basin. Bioscience 51, 373–388.

68. Nash, P.R., Nelson, K.A., Motavalli, P.P., 2014. Reducing nitrogen loss in subsurface tile drainage water with managed drainage and polymer-coated urea in a riverbottom soil. J. Water Res. Prot. 6, 988–997.

69. Needelman, B.A., Kleinman, P.J.A., Strock, J.S., Allen, A.L., 2007. Drainage ditches: Improved management of agricultural drainage ditches for water quality protection: An overview. J. Soil Water Constr. 62, 171–178.

70. Nguyen, H.Q., Kanwar, R.S., Hoover, N.L., Dixon, P.M., Hobbs, J.M., Soupir, M.L., 2013. Long-Term effects of poultry manure application on nitrate leaching in tile drain water. Trans. ASABE 56, 91.

71. Nickerson, C., Morehart, M., Kuethe, T., Beckman, J., Ifft, J., Williams, R., 2012.Trends in U.S. Farmland Values and Ownership EIB-92. United States Department of Agriculture Economic Research Service, Washington DC.

72. Oolman, E., Wilson, B., 2003. Sediment control practices for surface tile inlets. Appl. Eng. Agric. 19, 161–170.

73. Owens, L., Edwards, W., Shipitalo, M., 1995. Nitrate leaching through lysimeters in a corn-soybean rotation. J. Soil Sci. Soc. Am. 59, 902–907.

74. Patni, N.K., Masse, L., Jui, P.Y., 1996. Tile effluent quality and chemical losses underconventional and no tillage – Part 1: flow and nitrate. Trans. ASAE 39, 1665–1672.

75. Pavelis, G.A., 1987. Farm Drainage in the United States: History, Status, and Prospects Publication No. 1455. United States Department of Agriculture Economic Research Service, Washington DC.

76. Penn, C.J., Bryant, R.B., Kleinman, P.J.A., Allen, A.L., 2007. Removing dissolved phosphorus from drainage ditch water with phosphorus sorbing materials. J. Soil Water Constr. 62, 269–276.

77. Portch, S., MacKenzie, A.F., Steppler, H.A., 1968. Effect of fertilizers, soil drainage class and year upon protein yield and content of oats. Agron. J. 60, 672–674.

78. Qi, Z., Helmers, M.J., Christianson, R.D., Pederson, C.H., 2011. Nitrate-nitrogen losses through subsurface drainage under various agricultural land covers. J. Environ. Qual. 40, 1578–1585.

79. Randall, G.W., Goss, M.J., 2008. Chapter 6: Nitrate losses to surface water through subsurface, tile drainage. In: Follett, R.F., Hatfield, J.L. (Eds.), Nitrogen in the Environment: Sources, Problems, and Management. , 2nd ed. Elsevier Science, Amsterdam, pp. 145–176.

80. Randall, G.W., Huggins, D.R., Russelle, M.P., Fuchs, D.J., Nelson, W.W., Anderson, J.L., 1997. Nitrate losses through subsurface tile drainage in conservation reserve program, alfalfa, and row crop systems. J. Environ. Qual. 26, 1240–1247.

81. Randall, G.W., Iragavarapu, T.K., 1995. Impact of long-term tillage systems for continuous corn on nitrate leaching to tile drainage. J. Environ. Qual. 24, 360–366.

82. Randall, G.W., Mulla, D.J., 2001. Nitrate nitrogen in surface waters as influenced by climatic conditions and agricultural practices. J. Environ. Qual. 30, 337–344.

83. Randall, G.W., Vetsch, J.A., 2005. Nitrate losses in subsurface drainage from a corn–soybean rotation as affected by fall and spring application of nitrogen and Nitrapyrin. J. Environ. Qual. 34, 590–597.

84. Randall, G.W., Vetsch, J.A., Huffman, J.R., 2003. Nitrate losses in subsurface drainage from a corn–soybean rotation as affected by time of nitrogen application and use of Nitrapyrin. J. Environ. Qual. 32, 1764–1772.

85. Ritter, W., Rudra, R., Milburn, P., Prasher, S., 1995. Drainage and water quality in Northern United States and Eastern Canada. J. Irrig. Drain. Eng 121, 296–301.

86. Roberts, G., Hudson, J., Blackie, J., 1986. Effect of upland pasture improvement on nutrient release in flow from a 'natural'lysimeter and a field drain. Agric. Water Manage. 11, 231–245.

87. Robinson, M., Rycroft, D.W., 1999. Chapter 23: The impact of drainage on streamflow. In: Skaggs, R., van Schilfgaarde, J. (Eds.), Agricultural Drainage Agronomy Monograph No. 38. ASA CSSA SSSA, Madison, Wisconsin, pp.767–800.

88. Sands, G.R., Song, I., Busman, L.M., Hansen, B.J., 2008. The effects of subsurface drainage depth and intensity on nitrate loads in the Northern Cornbelt. Trans.ASABE 51, 937–946.

89. Schilling, K.E., Helmers, M., 2008. Tile drainage as karst: conduit flow and diffuse flow in a tile-drained watershed. J. Hydrol. 349, 291–301.

90. Schwab, G., Fausey, N., Kopcak, D., 1980. Sediment and chemical content ofagricultural drainage water. Trans. ASAE 23, 1446–1449.

91. Shirmohammadi, A., Wenberg, R., Ritter, W., Wright, F., 1995. Effect of agricultural drainage on water quality in Mid-Atlantic states. J. Irrig. Drain. Eng. 121, 302–306.

92. Skaggs, R., Breve, M., Gilliam, J., 1994. Hydrologic and water quality impacts of agricultural drainage. Crit. Rev. Environ. Sci. Technol. 24, 1–32.

93. Skaggs, R., Youssef, M., Chescheir, G., Gilliam, J., 2005. Effect of drainage intensityon nitrogen losses from drained lands. Trans. ASAE 48, 2169–2177.

94. Skaggs, R.W., van Schilfgaarde, J., 1999. Agricultural Drainage Agronomy Monograph No. 38. ASA CSSA SSSA, Wisconsin.

95. Smith, C.M., David, M.B., Mitchell, C.A., Masters, M.D., Anderson-Teixeira, K.J., Bernacchi, C.J., DeLucia, E.H., 2013. Reduced nitrogen losses after conversion ofrow crop agriculture to perennial biofuel crops. J. Environ. Qual. 42, 219–228.

96. Smith, D.R., Livingston, S.J., 2013. Managing farmed closed depressional areas using blind inlets to minimize phosphorus and nitrogen losses. Soil UseManage. 29, 94–102.

97. Smith, E.L., Kellman, L.M., 2011. Nitrate loading and isotopic signatures in subsurface agricultural drainage systems. J. Environ. Qual. 40, 1257–1265.

98. Stout, W.L., Schnabel, R.R., 1994. Soil drainage influence on biomass and nitrogenaccumulation by ryegrass. Agron. J. 86, 111–116.

99. Strock, J.S., Dell, C.J., Schmidt, J.P., 2007. Managing natural processes in drainage ditches for nonpoint source nitrogen control. J. Soil Water Constr. 62, 188–196.

100. Strock, J.S., Kleinman, P.J.A., King, K.W., Delgado, J.A., 2010. Drainage water management for water quality protection. J. Soil Water Constr. 65, 131A–136A.Strock, J.S., Porter, P.M., Russelle, M.P., 2004. Cover cropping to reduce nitrate loss through subsurface drainage in the Northern U.S. S. Corn Belt. J. Environ. Qual. 33, 1010–1016.

101. Sugg, Z., 2007. Assessing U.S. Farm Drainage: Can GIS Lead to Better Estimates ofSubsurface Drainage Extent? World Resources Institute, Washington DC. Tan, C.S., Drury, C.F., Gaynor, J.D., Welacky, T.W., 1993. Integrated soil, crop and water management system to abate herbicide and nitrate contamination of the Great Lakes. Water Sci. Technol. 28, 497–507.

102. Tan, C.S., Drury, C.F., Reynolds, W.D., Gaynor, J.D., Zhang, T.Q., Ng, H.Y., 2002a.Effect of long-term conventional tillage and no-tillage systems on soil and water quality at the field scale. Water Sci. Technol. 46, 183–190.

103. Tan, C.S., Drury, C.F., Reynolds, W.D., Groenevelt, P.H., Dadfar, H., 2002b. Water and nitrate loss through tiles under a clay loam soil in Ontario after 42 years of consistent fertilization and crop rotation. Agric. Ecosyst. Environ. 93, 121–130.

104. Thomas, D., Perry, C., Evans, R., Izuno, F., Stone, K., Wendell Gilliam, J., 1995. Agricultural drainage effects on water quality in Southeastern U.S. J. Irrig. Drain. Eng. 121, 277–282.

105. Tonitto, C., David, M.B., Drinkwater, L.E., 2006. Replacing bare fallows with cover crops in fertilizer-intensive cropping systems: A meta-analysis of crop yield and N dynamics. Agric. Ecosyst. Environ. 112, 58–72.

106. Turner, R.E., Rabalais, N.N., 2003. Linking land scape and water quality in the Mississippi River Basin for 200 years. Bioscience 53, 563–572.

107. Weed, D.A.J., Kanwar, R.S., 1996. Nitrate and water present in and flowing from root-zone soil. J. Environ. Qual. 25, 709–719.

108. Zhu, Y., Fox, R.H., 2003. Corn–soybean rotation effects on nitrate leaching. Agron. J. 95, 1028–1033.

109. Zucker, L.A., Brown, L.C., 1998. Agricultural drainage: Water quality impacts and subsurface drainage studies in the Midwest. Ohio State University Extension Bulletin 871. The Ohio State University, Ohio.

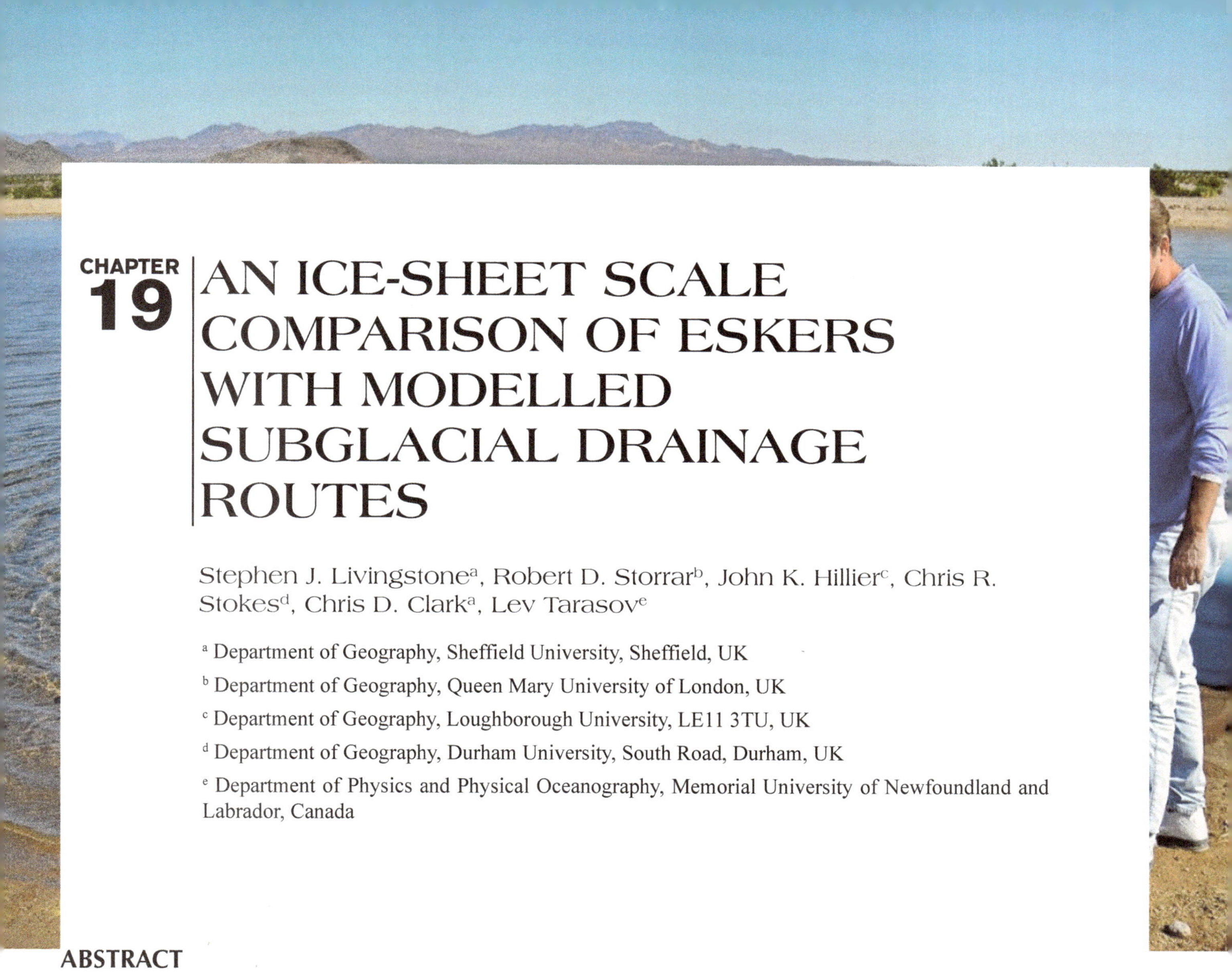

CHAPTER 19

AN ICE-SHEET SCALE COMPARISON OF ESKERS WITH MODELLED SUBGLACIAL DRAINAGE ROUTES

Stephen J. Livingstone[a], Robert D. Storrar[b], John K. Hillier[c], Chris R. Stokes[d], Chris D. Clark[a], Lev Tarasov[e]

[a] Department of Geography, Sheffield University, Sheffield, UK

[b] Department of Geography, Queen Mary University of London, UK

[c] Department of Geography, Loughborough University, LE11 3TU, UK

[d] Department of Geography, Durham University, South Road, Durham, UK

[e] Department of Physics and Physical Oceanography, Memorial University of Newfoundland and Labrador, Canada

ABSTRACT

Eskers record the signature of channelised meltwater drainage during deglaciation providing vital information on the nature and evolution of subglacial drainage. In this paper, we compare the spatial pattern of eskers beneath the former Laurentide Ice Sheet with subglacial drainage routes diagnosed at discrete time intervals from the results of a numerical ice-sheet model. Perhaps surprisingly, we show that eskers predominantly occur in regions where modelled subglacial water flow is low. Eskers and modelled subglacial drainage routes were found to typically match over distances of < 10 km, and most eskers show a better agreement with the routes close to the ice margin just prior to deglaciation. This supports a time-transgressive esker pattern, with formation in short (< 10 km) segments of conduit close behind a retreating ice margin, and probably associated with thin, stagnant or sluggish ice. Esker-forming conduits were probably dominated by supraglacially fed meltwater inputs. We also show that modelled subglacial drainage routes containing the largest concentrations of meltwater show a close correlation with palaeo-ice stream locations. The paucity of eskers along the terrestrial portion of these palaeo-ice streams and meltwater routes is probably because of the prevalence of distributed drainage and the high erosion potential of fast-flowing ice.

INTRODUCTION

Eskers are slightly sinuous ridges composed of glaciofluvial sand and gravel that are deposited in subglacial, englacial, or supraglacial drainage channels (e.g., Banerjee and McDonald, 1975, Brennand, 2000 and Storrar et al., 2014a). They can extend for tens to hundreds of kilometres (taking into account small gaps), reach in excess of 50 m in height, and are typically arranged roughly parallel to each other (e.g., Prest et al., 1968,Banerjee and McDonald, 1975, Shilts, 1984, Shreve, 1985a, Shreve, 1985b, Aylsworth and Shilts, 1989, Clark and Walder, 1994, Boulton et al., 2009, Storrar et al., 2014a and Storrar et al., 2014b). Eskers may therefore provide vital information about channelised drainage. This is significant because observations from modern ice sheets reveal the important role of water in lubricating the bed and facilitating rapid ice flow (Bartholomew et al., 2011 and Sundal et al., 2011). In particular, the configuration of the subglacial drainage network and how it evolves to accommodate water inputs is critical (e.g., Budd et al., 1979, Alley et al., 1986 and Iken and Bindschadler, 1986). Two end-member drainage configurations are typically envisaged (e.g., Walder and Fowler, 1994): (i) an efficient channelised system commonly associated with lower water pressures, lower ice velocities, and higher

Citation: Stephen J. Livingstone, Robert D. Storrar, John K. Hillier, Chris R. Stokes, Chris D. Clark, Lev Tarasov, An ice-sheet scale comparison of eskers with modelled subglacial drainage routes, doi:10.1016/j.geomorph.2015.06.016

water discharges; and (ii) an inefficient distributed system (e.g., linked cavities, braided canals, and porous till layer) commonly associated with higher water pressures, higher ice velocities, and lower water discharges. However, because of the difficulties in directly observing the drainage of water at the bed of ice masses, we have a limited understanding of the distribution and geometry of the subglacial drainage network and a lack of data at the spatial and temporal scales necessary to constrain or test subglacial hydrological models (e.g., Hewitt, 2011 and Werder et al., 2013).

Presently, investigating eskers located under contemporary ice sheets is physically difficult. Thus the imprint of eskers recorded on the bed of former ice sheets has a clear advantage over data from contemporary ice sheets because we can directly observe the expression of meltwater drainage over large spatial scales. However, despite the use of eskers to reconstruct and constrain ice-retreat histories (e.g., Dyke and Prest, 1987 and Margold et al., 2013), very few studies have investigated their pattern at the ice-sheet scale (Aylsworth and Shilts, 1989, Clark and Walder, 1994, Storrar et al., 2014a and Storrar et al., 2014b). This is because it is not known whether eskers form synchronously in long conduits (cf. Brennand, 1994) or if they represent a time-integrated signature of drainage deposition throughout deglaciation (e.g., Banerjee and McDonald, 1975, Shilts, 1984, Dyke and Dredge, 1989, Kleman et al., 1997 and Hooke and Fastook, 2007). Consequently, the extraction and interpretation of information about where they form in relation to subglacial water generation and ice dynamics is difficult.

A new approach to understanding the pattern of eskers is to compare their distribution and orientation with numerical models of subglacial meltwater drainage at discrete intervals throughout deglaciation. In this study our aim is to compare the expression of eskers on the bed of the former Laurentide Ice Sheet (LIS) with subglacial drainage routes predicted from the results of a numerical ice sheet model. This builds on previous work that has compared eskers and hydraulic gradient routes at glaciers (e.g., Syverson et al., 1994), but represents the first attempt to compare modelled drainage with eskers at the ice-sheet scale and at discrete time intervals.

METHODS

Mapping Esker Networks

This paper uses 3749 interpolated eskers (mostly > 2 km long) mapped by Storrar et al. (2013). The crestlines of esker ridges were mapped from Landsat 7 Enhanced Thematic Mapper (ETM) + imagery of Canada, which has a resolution of ~ 30 m and ~ 15 m in the panchromatic band. Eskers were typically mapped at a scale of 1:40,000 and were identified based on the criteria set out by Margold and Jansson (2012). Shorter eskers (< 2 km long) were more difficult to identify in Landsat imagery. Comparison with mapping from aerial photographs suggests that ~ 75% of eskers were identified and that 81% of those missed are < 2 km long (Storrar et al., 2013).

To enable the effective comparison of eskers with modelled subglacial drainage routes, we used the interpolated esker data set produced by Storrar et al. (2014a). This data set was derived by interpolating a straight line (over short distances in the majority of cases) between aligned esker ridges that appear genetically related (i.e., formed in the same conduit) and merging with the mapped ridges to produce a single esker. It was produced to fill gaps that may have resulted from fragmentary deposition, post-depositional erosion or, submergence beneath lakes and is therefore thought to give a better indication of where the esker-forming conduits were located (Storrar et al., 2014a). We refer to this data set throughout the paper simply as 'eskers'.

Modelling Subglacial Meltwater Drainage

Subglacial meltwater drainage was modelled using the method outlined in Livingstone et al., 2013a and Livingstone et al., 2013b. Hydraulic potential surfaces (ϕ) of the LIS were calculated from the Shreve equation (Shreve, 1972):

$$\phi = \rho_w gh + \rho_i gH \quad (1)$$

where ρ_w is the density of water; ρ_i is the density of ice; g is the acceleration of gravity;h is the bed elevation; and H is the ice thickness. We calculated the subglacial drainage routes every 500 years for the period between 12 and 7 ka BP, which encompassed the largest retreat distance (hundreds of kilometres) during deglaciation and was over the predominantly hard crystalline bedrock on the Canadian Shield (seeDyke, 2004). The bed elevation data (h) were constructed at 5-km resolution from Gebco_08 digital elevation model (DEM), and the palaeo-ice surfaces and palaeo-bed topographies (corrected for isostasy) were derived from ice-sheet model output from one of the higher probability runs (LT9927) from the ensemble-based analyses of the LIS using the three-dimensional (3D) Glacial Systems Model (GSM) (Tarasov et al., 2012). The GSM includes a 3D thermomechanically coupled shallow ice-sheet model, bed-thermal model, visco-elastic bedrock response, and coupled surface drainage and pro-glacial lake solver. The GSM is calibrated against a large set of observational constraints, including geological and geomorphological evidence and is able to reproduce ice stream locations and ice-margin positions (Stokes and Tarasov, 2010 and Tarasov et al., 2012). The LT9927 is from the subensemble of runs used byLivingstone et al., 2013a and Livingstone et al., 2013b. Their analysis showed that the modelled distribution of subglacial lakes and major drainage routes is a robust result achieved irrespective of the model run used from this subensemble (see Fig. 8 fromLivingstone et al., 2013a and Livingstone et al., 2013b). Given this and the time required for analysing each run, we base our analysis just on LT9927 for this study. The 1° longitude by 0.5° latitude resolution model output was regridded at 5-km cell size. Subglacial drainage routes (i.e., the direction that water flows) were constructed from the hydraulic potential surfaces, using simple GIS routing tools as per Livingstone et al., 2013a and Livingstone et al., 2013b. Basal meltwater production (cm/y) generated from the GSM was used to weight flow accumulation down subglacial drainage routes. Each cell was given an accumulative basal meltwater value of all the cells that flow into it, hereafter referred to as the 'modelled subglacial flow concentration'. Output cells with a high flow accumulation represent a drainage route along which subglacial meltwater is concentrated. To allow for basal meltwater production caused by likely subgrid topographic variation, we set a minimum basal meltwater output (0.1 cm/y) in regions of the bed where the temperature was 0 to − 2 °C below pressure melting point. Meltwater may also enter the subglacial system from supraglacial sources, although these are not reproduced here because of the difficulty of modelling this process. Thus, we use basal meltwater production simply to indicate where meltwater is likely to concentrate rather than suggesting that all meltwater is necessarily produced at the bed.

Comparison of Eskers and Modelled Subglacial Drainage Routes

To our knowledge, a comparison between the pattern of eskers and modelled subglacial drainage routes has not been previously undertaken at the ice-sheet scale. Thus, in this analysis, we explore first-order relationships between the location and orientation of eskers and modelled drainage routes (Fig. 1). Approaches for these two comparisons are described below.

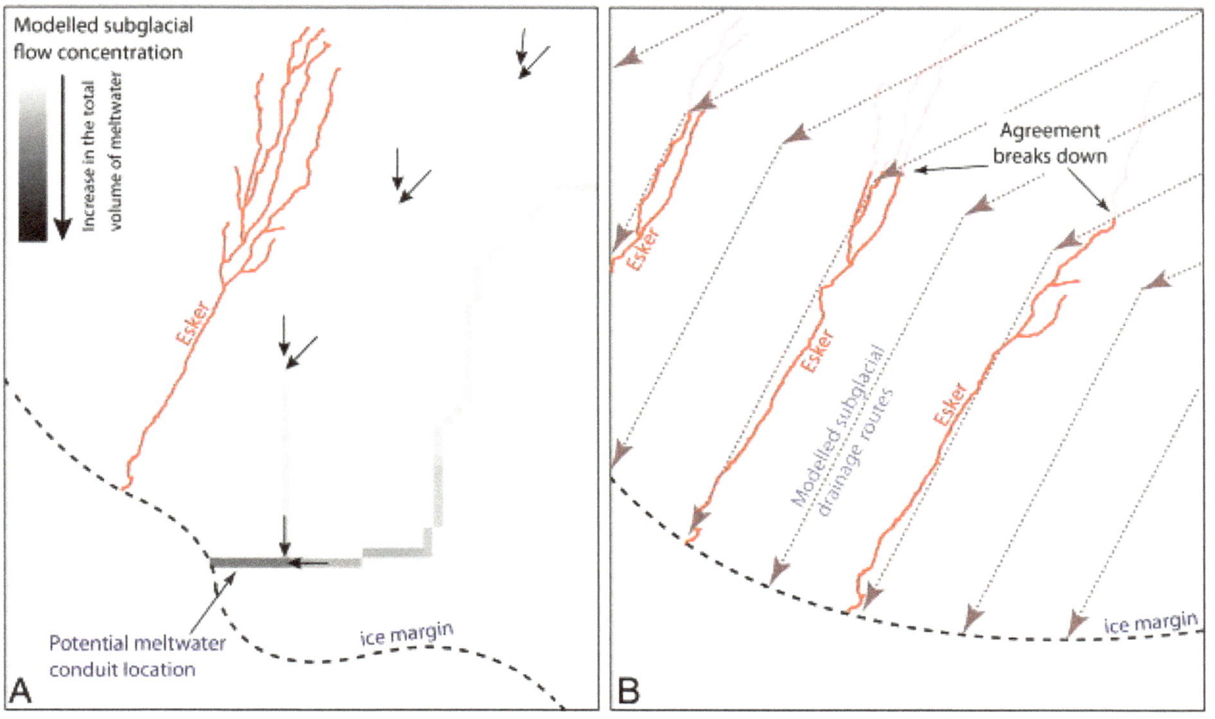

Figure 1: Cartoons illustrating the methods used to investigate the relationship between eskers and modelled subglacial drainage routes. (A) Spatial conformity of eskers and modelled subglacial flow concentration. The drainage routes are the total concentrated flow being transferred into each downstream cell, with each cell weighted by the basal meltwater production (cm/y) output from GSM. In this example, the darker colours indicate greater meltwater flow concentration, which pick out a dendritic drainage network. (B) Directional conformity of eskers and modelled subglacial drainage routes. Eskers are traced from the ice-margin up ice until the agreement with the modelled drainage directions breaks down. Directional conformity between modelled drainage routes and eskers represents the length along which an esker-forming meltwater conduit could have occurred.

Spatial Conformity of Eskers and Modelled Subglacial Flow Concentration

Output cells with a high subglacial flow concentration indicate regions where large volumes of meltwater are routed, and these represent potential meltwater conduit locations (Fig. 1A). These should correspond to esker locations, as this is where Röthlisberger channels are theorised to form (e.g., Röthlisberger, 1972 and Shreve, 1972).

To investigate how the spatial pattern of eskers relates to the routing of subglacial meltwater beneath the LIS, we compared modelled subglacial flow concentration with the esker pattern at 500-year time slices from 12 to 7 ka BP (Fig. 1A). At each time slice we extracted flow concentration values of all cells that contain eskers and all cells covered by ice in the model domain. We also identified cells that match with the spatial extent of palaeo-ice stream locations recently compiled in Margold et al. (2015) and cells that match the terrestrial portion of the palaeo-ice sheet bed (i.e., where the mapping was carried out) (Storrar et al., 2013). The probability density function of modelled subglacial flow concentration was calculated for each of the variables extracted and results displayed as a ratio between each probability density function and the probability density function of all the cells covered by ice in the model domain. Statistical significance was evaluated with a binomial test.

To further identify any spatial match between eskers and major subglacial drainage routes we used a flow concentration of > 20 cm/y to identify potential meltwater conduit locations (Fig. 1A). A value of 1 was assigned to cells where the flow concentration exceeded 20 cm/y and a value of 0 to those that did not (see also Livingstone et al., 2013b). This was done for every time slice and the values (0 s and 1 s) then summed together to produce a composite map of potential meltwater conduits and their persistence over time.

Directional Conformity of Eskers and Modelled Subglacial Drainage Routes

Eskers provide a composite history of the former drainage and direction of meltwater flow beneath an ice sheet throughout deglaciation. To try and disentangle what is assumed to be a time-integrated pattern, the match between the orientations of eskers and modelled subglacial drainage routes beneath the LIS were compared at 500-year intervals between 12 and 7 ka BP (Fig. 1B) (Livingstone et al., 2013a). This comparison was used to investigate (i) how far up-ice from the margin eskers align with the modelled drainage directions; (ii) whether eskers show a better alignment closer to or farther away from the ice margin; and (iii) whether any particular time slice is matched to the entire esker pattern.

A tracking algorithm was developed to compare the directional conformity of eskers and modelled drainage routes (see supplementary information). The tracking proceeds up-ice, moving from the margin of the LIS toward its centre. Tracking was initiated for all eskers at each time slice, with esker-model agreement starting the first time an esker's orientation matched the modelled subglacial drainage direction of an overlying cell and continuing until the agreement first breaks down (Fig. 1B). The length (L) that each esker was tracked and the percentage (f) of overall length tracked $((\Sigma L_{tracked}/\Sigma L\text{total}) \times 100)$ was used to investigate the upstream conformity of eskers with modelled drainage directions. We also compared (i) the timing and (ii) the distance from the ice margin between the best and average fits of the modelled subglacial drainage and esker directions at each point along an esker that was tracked in ≥ 2 time periods. This was used as a measure of when (timing) and where (distance from the ice margin) eskers best fit the modelled drainage directions.

The algorithm has three main parameters that could be varied: (i) the maximum angle of divergence (θ) allowed between the orientations of the esker and modelled drainage direction; (ii) the length scale at which the eskers are smoothed (E); and (iii) the spatial scale (W) used to smooth the modelled subglacial drainage direction grid. Preferred values for the parameters are 45° for θ and a smoothing scale of 5 km for E and W, which is the same resolution at which the model output was regridded and drainage routes constructed. The supplementary section details a sensitivity analysis of θ, E, and W.

Limitations

Our preliminary approach has three major limitations. First, the results use the predictions of a numerical model, which is not reality. However, the GSM includes basic glaciological physics and is calibrated against a large set of observational constraints (see Tarasov et al., 2012), including relative sea-level data, present-day rates of surface uplift, and an ice-margin chronology (± 250- to 1000-year uncertainty) derived from geological and geomorphological evidence (Dyke, 2004). The LT9927 is therefore glaciologically self-consistent and is a sample from the most well-constrained distribution of possible deglaciation chronologies to date. Future extension of this study to a subensemble from the calibration will enable uncertainty quantification of our initial results herein.

Secondly, the difference between the initial resolution of the numerical ice model used to derive the ice surface and bed topographies (~ 50-km resolution) and the (tens of metre) scale at which eskers were mapped is large. Even the smoothed (5-km resolution) ice surfaces used to generate the hydraulic potential surfaces are relatively coarse, which might artificially create drainage routes that are farther apart than they would be in reality. However, away from the ice margin and ice streams/outlet glaciers, ice sheets are relatively smooth and a 50-km resolution is adequate.

Thirdly, this paper uses a simple representation of subglacial hydrology, which provides a first approximation of where water should seek to flow and pond relative to the ice-sheet model selected. Crucially, the subglacial drainage system and overlying ice are not dynamically coupled, and it does not allow drainage configurations to evolve. We therefore assume that water pressure is equal to ice-overburden pressure uniformly across the bed. This is a reasonable assumption when averaged over coarse ice-sheet scales where geometry is the key control (Livingstone et al., 2013a and Livingstone et al., 2013b), but unlikely to be valid at smaller scales dominated by local processes (e.g.,Gulley et al., 2012).

RESULTS

Spatial Conformity of Eskers and Modelled Subglacial Flow Concentration

Fig. 2 reveals that the probability of palaeo-ice streams occurring in regions with high subglacial flow concentrations is significantly higher than chance, even if the resolution is reduced ten-fold (binomial test for flow concentrations > 20 cm/y, $p < 0.001$). In contrast, cells that match esker locations show a lower probability for subglacial flow concentration values > 100 cm/y. At values < 100 cm/y, eskers occur significantly more often than chance (binomial test, $p < 0.001$ and $p < 0.01$ if the resolution is reduced ten-fold) and, importantly, display a stronger correlation than the mapping domain. Eskers exhibit a higher probability than ice streams for subglacial flow concentration values < 20 cm/y.

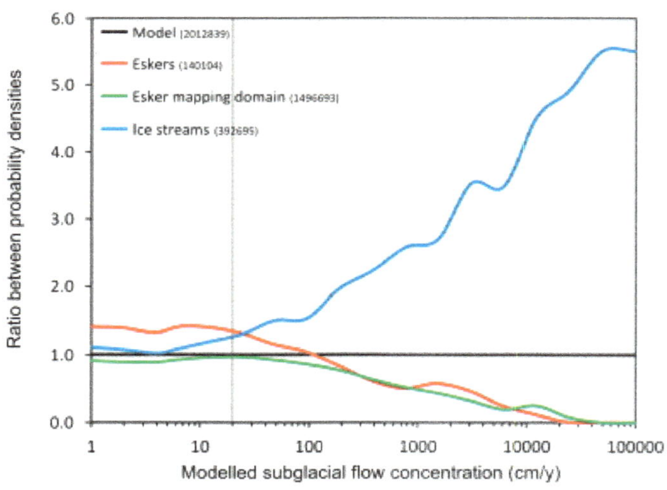

Figure 2: Spatial relationship of eskers, the domain used to map the eskers (i.e. terrestrial portion of the former LIS bed), and ice streams with varying subglacial flow concentration (ranging from 1 to 100,000 cm/y). Results are displayed as a ratio between each probability density function and the probability density function of all the cells covered by ice in the model domain (i.e., the population). The 20-cm/y drainage cutoff we use to produce Fig. 3 is shown by the vertical line. Values above 1 indicate that subglacial flow concentration happens more often than you would expect by a random distribution. Numbers in brackets are the number of cells used to produce each probability density function. Note how eskers are associated with lower subglacial flow concentrations and ice streams with higher subglacial flow concentrations.

The spatial pattern between eskers and potential subglacial meltwater conduits (subglacial flow concentration of > 20 cm/y) is shown in Fig. 3. While localised agreement can be identified, for example in the district of Keewatin, the general pattern is of eskers occurring in the gaps between potential meltwater conduits. This arrangement is particularly noticeable in the Northwest Territories and Quebec (Figs. 3B,C). Moreover, the correlation between mapped eskers and the persistence of potential meltwater conduits throughout deglaciation is not clear (Fig. 3). This is demonstrated in the Ungava Bay region (Fig. 3C), where conduits occur repeatedly and in similar locations throughout deglaciation, but eskers are rare.

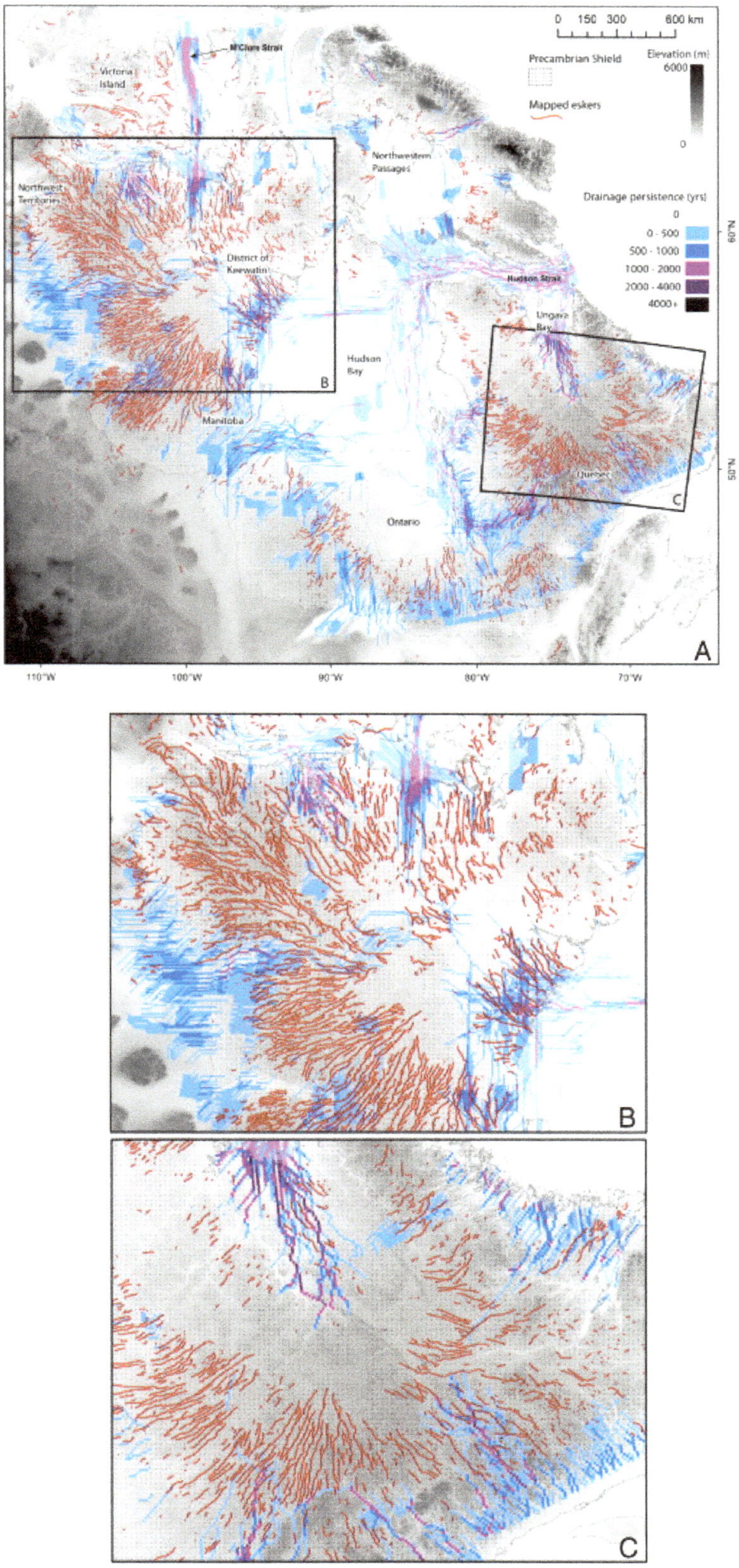

Figure 3: Spatial association between eskers and potential meltwater conduit locations. (A) Composite of all possible modelled subglacial drainage routes from 12 to 7 ka BP (Fig. 1A, see text for details). The purple lines represent potential meltwater conduits that were relatively persistent over time, and the light blue lines are those that only formed during one modelled time slice. Panels (B) and (C) are closeups of Keewatin and Quebec, respectively. (For interpretation of the references to colour in this figure legend, the reader is referred to the web version of this article.)

Directional Conformity of Eskers and Modelled Subglacial Drainage Routes

The length of directional conformity, L, between esker ridges and modelled subglacial drainage routes ranges from < 1 to > 100 km (Fig. 4). The tracking algorithm matches with 65% of the total length of all eskers. The median length of L is 5.4 km, the 5th and 95th percentiles are 0.5 and 42 km, respectively, and the tail of outliers includes matches of > 100 km in all except the 7.5- and 7-ka time slices. Length distributions for L are similar at each modelled time slice, but the tail of outliers is shorter compared to the eskers (Fig. 4). Indeed, longer eskers conform less well to the direction of modelled subglacial drainage routes, with the fraction of > ~ 50-km-long eskers tracked < 20% (Fig. 5).

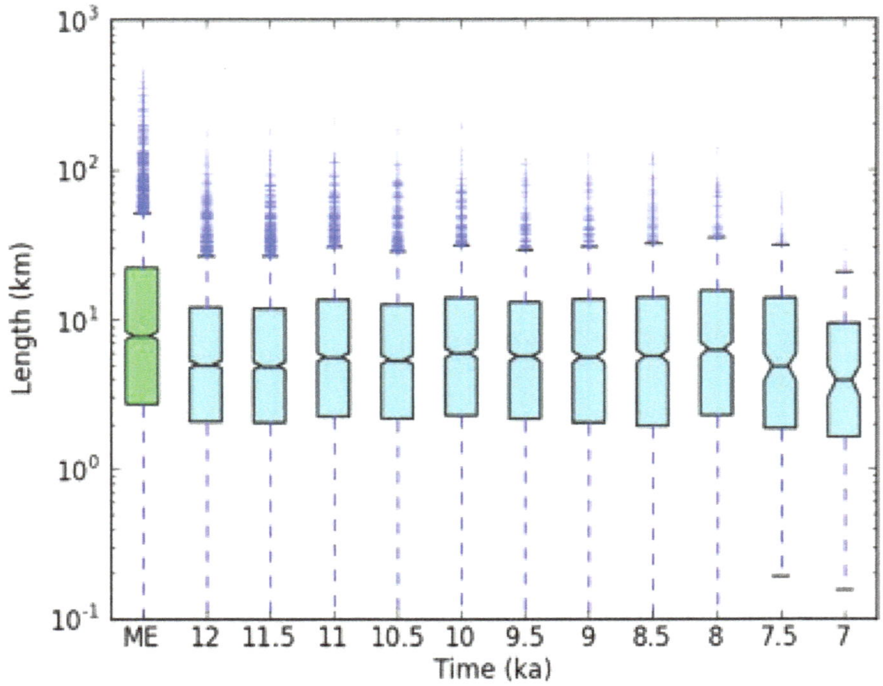

Figure 4: Box and whisker plots showing the length distribution of the mapped eskers (ME) and the agreement between eskers and modelled subglacial drainage directions at each time slice. The black indented line is the median, the edges of the box are the 25th and 75th percentiles, the whiskers are 1.5 times the interquartile range, and the crosses are outliers. Note the median length tracked is < 10 km, although with some matches of > 100 km.

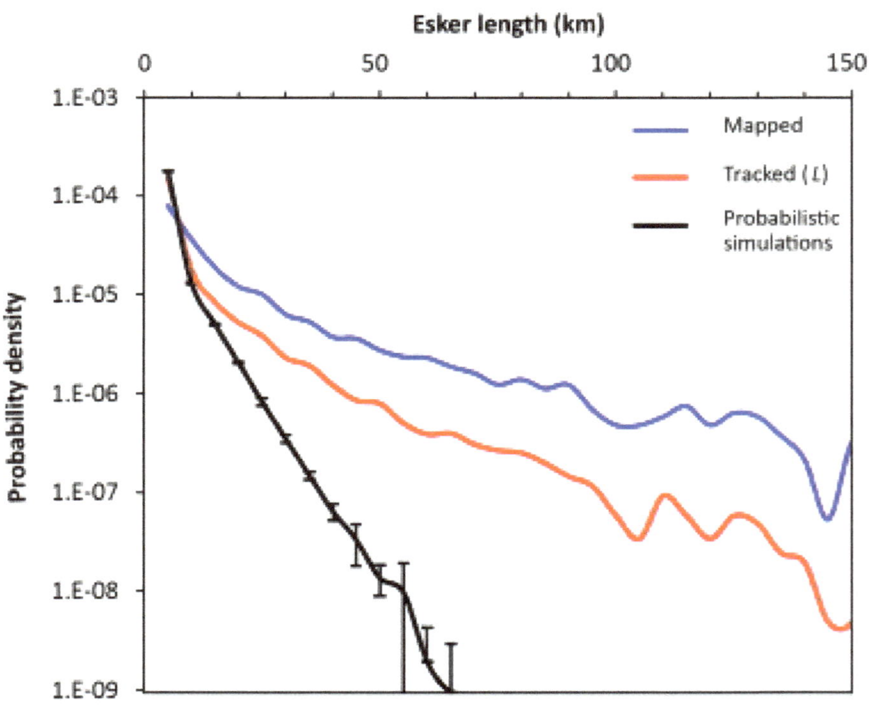

Figure 5: Probability density plot showing the distribution of eskers (blue line), tracked eskers (L) (red line), and the average and standard deviation of five stochastic simulations that look at the probability of tracking continuing along a randomly wandering esker (see text for details) (black line). The simulations use the length distribution of the eskers. As θ is $\pm$ 45° or, 90° of a possible 180° forward arc, the probability p of a 5-km step randomly orientated with respect to the previous one also being tracked is 0.5.

The difference in distance from the ice margin between the best and average fits of the modelled subglacial drainage and esker directions (Fig. 6A) reveals a positively skewed distribution (skew: + 3.09), with the peak offset toward negative values (i.e., the fit is better than average when closer to the ice margin). Fig. 6B illustrates the difference in time between the best and average fits of the modelled subglacial drainage and esker directions. The result is a positively skewed distribution (skew: + 2.2), with the peak offset toward negative values, i.e., the fit is better than average when eskers are formed during later time slices.

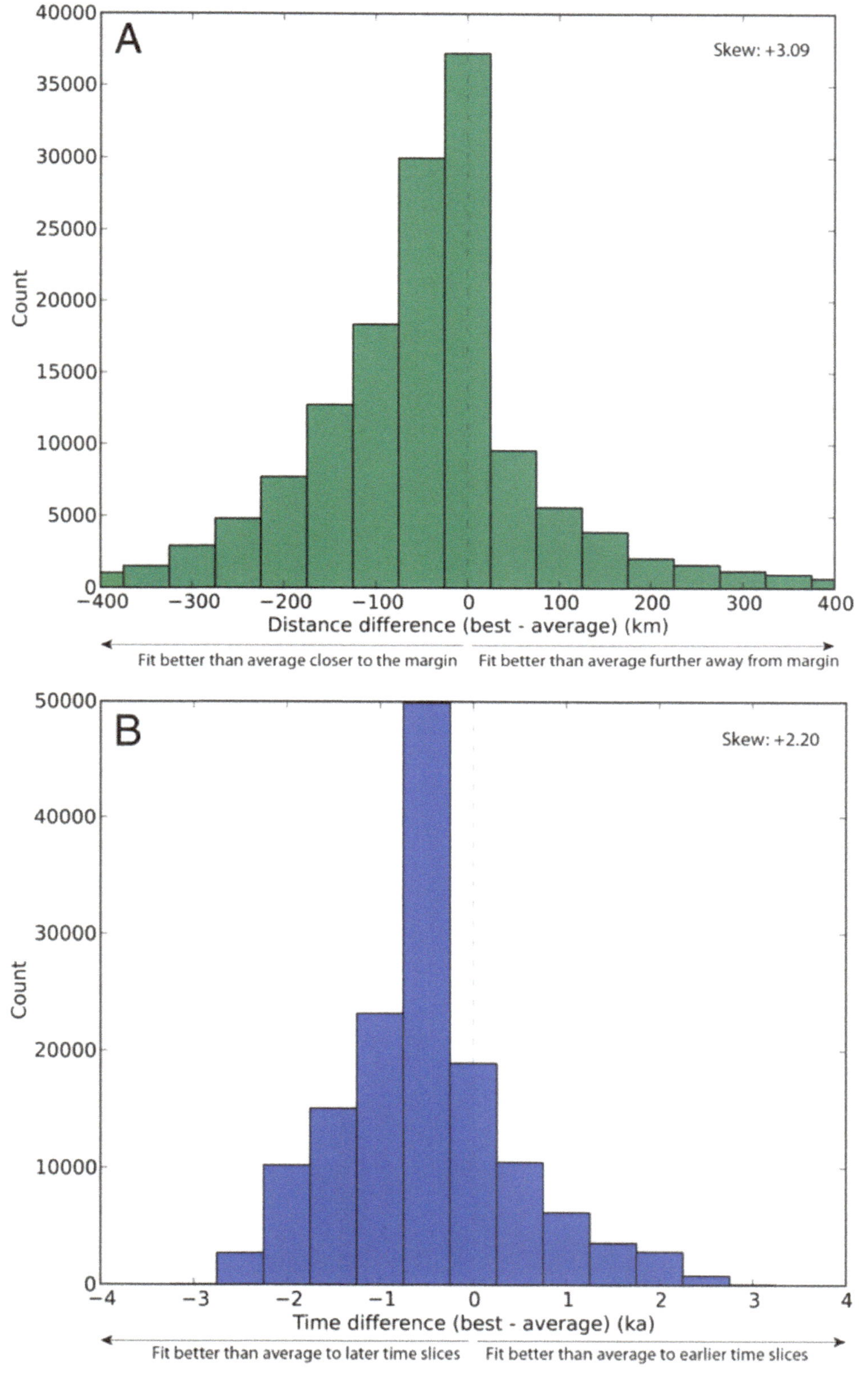

Figure 6: Histogram showing, for any point that is tracked in ≥ 2 time periods: (A) the difference in distance from the ice margin between the best and average fits of the modelled subglacial drainage and esker directions. Negative values indicate that the ice margin is closer when the modelled drainage direction best fits the esker direction. Bins are in 50-km intervals. (B) The difference in time between the best and average fits of the modelled subglacial drainage and esker directions. If the best fit between the modelled subglacial drainage and esker directions occurred later than the average fit (i.e., when the ice sheet was smaller) the result is negative. Bins are in 500-year intervals.

To evaluate whether or not the distribution of L may have been produced by chance, we compared the results against the probability of tracking continuously along a randomly wandering esker (Fig. 5). As θ is ± 45° or, 90° of a possible 180° forward arc, the probability p of a 5-km step (i.e., matching E) randomly orientated with respect to the previous one also being tracked is 0.5. The chance of proceeding to the nth step is p^n. We ran five stochastic simulations using the length distribution of the eskers. Fig. 5shows the probability density of the eskers (blue line), the tracked eskers (L) (red line), and an average of the five stochastic simulations (black line). In initial tracking acquisition, and progressing from the first to the second 5-km step, L does little better than random. This may be because of resolution issues, measurement 'noise', or local factors dominating ice flow at this scale or near the margin. Further tracking (> 10 km) occurs much more frequently than chance, which implies an association between eskers and the modelled subglacial drainage direction output.

DISCUSSION

Spatial Association Between eskers, Palaeo-ice Stream Locations, and Modelled Subglacial flow Concentration

Our results show that eskers do not coincide with the dominant drainage routes where high concentrations of meltwater (> 100 cm/y) are modelled and routed subglacially (Fig. 2 and Fig. 3). Indeed, there is a highly significant statistical tendency for eskers to occur in regions of low subglacial flow concentration (< 100 cm/y) in the gaps between major drainage routes (i.e., left of the vertical line in Fig. 2). The highest modelled subglacial flow concentrations are associated with palaeo-ice stream locations (Fig. 2), where eskers are rare (Fig. 7). In particular, major marine-terminating palaeo-ice stream troughs with large drainage basins, such as Hudson and M›Clure straits, are predicted to have had significant volumes of subglacial meltwater routed down them (Fig. 3). This is unsurprising given that ice streams require warm-based ice, produce large volumes of subglacial water by enhanced frictional melting (Engelhardt and Kamb, 1997 and Kamb, 2001), and typically occur in topographic troughs (cf. Winsborrow et al., 2010) where meltwater is focussed. The pattern of meltwater drainage is therefore strongly controlled by the evolution of fast ice flow in the GSM. Significantly, fast ice flow in the model evolves freely as basal ice approaches the pressure melting point, and it is able to reproduce most major palaeo-ice streams (Stokes and Tarasov, 2010) though the shallow ice approximation employed and model grid resolution would caution against overinterpretation of small-scale flow features.

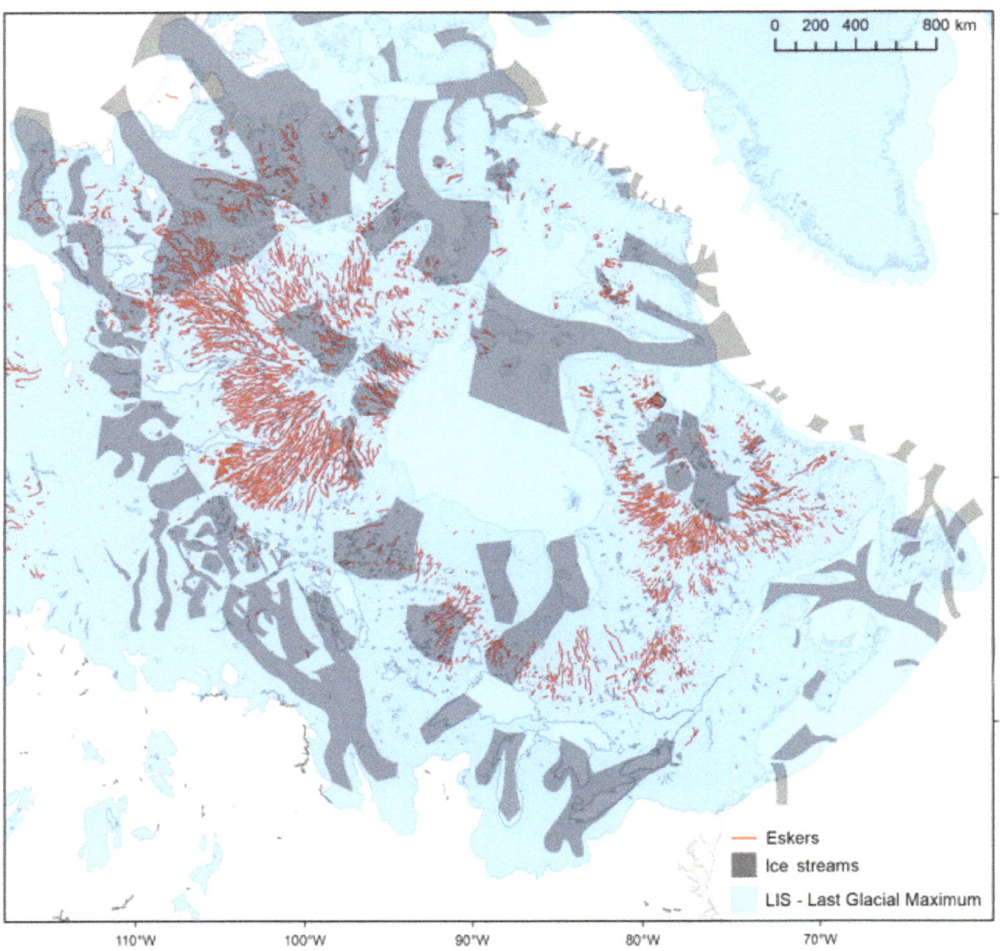

Figure 7: The distribution of inferred palaeo-ice streams (from Margold et al., 2015) and mapped eskers (Storrar et al., 2013) beneath the former Laurentide Ice Sheet. In general, eskers are rarely found on the bed of palaeo-ice streams.

We suggest a number of explanations for the relative absence of eskers on the terrestrial portions of Laurentide palaeo-ice stream beds where subglacial water was concentrated (Fig. 7). First, eskers tend to form on hard beds in Röthlisberger channels, while ice streams are more typically floored by soft sediments that are predicted to drain water through a distributed network (Clark and Walder, 1994). Secondly, increased ice creep associated with high basal velocities will promote the formation of high pressure distributed drainage networks (e.g., Kamb, 1987 and Bell, 2008). Thirdly, any eskers that are deposited are unlikely to be preserved because of the enhanced erosional potential of fast-flowing ice (Boulton, 1996).

The formation of eskers in inter-ice-stream regions, characterised by low subglacial flow concentration, may seem counterintuitive because conduit location and growth are thought to be associated with high water discharges (e.g., Shreve, 1972, Schoof, 2010,Hewitt, 2011 and Werder et al., 2013). Indeed, it implies that large volumes of subglacial meltwater were not significant in the formation of conduits and eskers beneath the LIS, a point further demonstrated by the absence of suitable palaeo-subglacial lake locations on the relatively flat Canadian Shield (Livingstone et al., 2013a). An alternative source of water is therefore needed, and the most likely option is the penetration of surface meltwater to the bed (e.g., Zwally et al., 2002). Critically, our model does not allow moulin formation or supraglacial inputs, instead assuming that all meltwater was generated subglacially. However, eskers on the Canadian Shield become more frequent later during deglaciation (e.g., Prest et al., 1968 and Aylsworth and Shilts, 1989), and this coincides with increased rates of atmospheric warming and ice-margin recession (Storrar et al., 2014b). This suggests a genetic link between supraglacial meltwater inputs and esker formation (see also Shilts, 1984, St-Onge, 1984, Aylsworth and Shilts, 1989, Hooke and Fastook, 2007 and Burke et al., 2012).

Basal conditions in inter-ice-stream regions may also have been conducive to subglacial conduit and esker formation. For instance, slower ice flow, where creep closure and glacial erosion are reduced, is more suitable for Röthlisberger channel formation and is consistent with observations of eskers in other deglaciated and modern landscapes (e.g.,Price, 1966). Moreover, conduits could have been relatively small compared to the size of the esker produced, with growth occurring slowly by the meltout of dirty ice rather than by large discharges through a large channel (e.g., Clark and Walder, 1994, Warren and Ashley, 1994, Hooke, 2005 and Hooke and Fastook, 2007). However, sedimentological investigations of eskers comprising horizontally bedded sand and gravel deposits extending all the way across their width suggests conduits can also be the same size as the esker (e.g., Brennand, 1994).

Directional Conformity of Eskers and Modelled Subglacial Drainage Routes

The length of agreement between modelled subglacial drainage directions and eskers is typically < 10 km (Fig. 4) and the ability to track eskers is improved when the ice-margin is closer (Fig. 6A) just prior to deglaciation (Fig. 6B). In addition, no single modelled time slice is able to reproduce the entire esker pattern. This suggests that the majority of eskers were deposited time-transgressively in short segments of conduit close to the retreating ice margin (as frequently inferred; e.g., Banerjee and McDonald, 1975, Shilts, 1984, St-Onge, 1984, Bolduc et al., 1987, Dyke and Dredge, 1989 and Hooke and Fastook, 2007). This interpretation is consistent with a supraglacially fed drainage system, where more meltwater is produced and able to penetrate to the bed at lower elevations and in thin ice, near the margin (cf. Hooke and Fastook, 2007 and Storrar et al., 2014b).

Agreement between modelled subglacial drainage directions and eskers can also be tracked for hundreds of kilometres up ice in rare cases (Fig. 4 and Fig. 5). These might be explained in two ways: (i) synchronous formation in subglacial conduits that penetrated deep into the ice sheet interior (e.g., Brennand, 2000); or (ii) incremental formation in a stable conduit that migrated up ice during uniform ice retreat. Where the pattern of retreat was uniform, we cannot distinguish between options (i) and (ii). However, we suggest that synchronous formation of very long (> 100 km) eskers is unlikely given the poor correlation with subglacial meltwater concentration (Fig. 2 and Fig. 3) and the high elevation and great thickness of ice in the interior of the LIS (e.g., Peltier, 2004 and Tarasov et al., 2012), which would have suppressed the generation and penetration of surface melt to the bed and increased conduit closure by ice creep, respectively.

Further Work

This paper uses a simple model to compare the spatial pattern of esker networks with modelled subglacial drainage routes, and as such, a certain degree of fallibility is unavoidable. The next step is to rerun the analysis using a higher model resolution or to take numerical models that can account for dynamic changes in subglacial water pressure and drainage configuration (e.g., Schoof, 2010, Hewitt, 2011 and Werder et al., 2013) and apply them to real world situations where the results can be compared to geomorphological data. In particular, sediment erosion, transport, and deposition need to be included in these models to produce the landform imprint that results from subglacial drainage routing and evolution. Our hope is that this preliminary effort offers a useful template for future work.

CONCLUSIONS

In this paper, we explore a new approach that compares the expression of eskers beneath the former LIS with subglacial drainage routes predicted from the results of a numerical ice-sheet model at specific time steps during deglaciation.

Eskers are more abundant in non-ice-stream regions where modelled subglacial flow concentration is low. Moreover, we find that the length of agreement between modelled subglacial drainage directions and eskers is typically < 10 km, and the agreement is improved when eskers lie close to the ice margin just prior to deglaciation. This suggests the majority of eskers were deposited time-transgressively in < 10-km segments of conduits close to a retreating ice margin and in thin, sluggish or stagnant ice where conduit closure by ice creep is less significant. We suggest that supraglacial meltwater inputs were important for supporting Röthlisberger channel and esker-forming processes.

We find that modelled subglacial drainage routes dominated by high flow concentration are associated with palaeo-ice stream locations. The paucity of eskers along the terrestrial portion of palaeo-ice streams is probably a result of the high velocities and soft-sediment beds that promote distributed rather than channelized drainage, while any eskers that do form are unlikely to be preserved beneath fast-flowing ice.

ACKNOWLEDGEMENTS

This work was supported by an NERC Early Career Research Fellowship awarded to SJL (NE/H015256/1) and an NERC PhD studentship awarded to RDS. It is a contribution to the "Meltwater routing and Ocean–Cryosphere–Atmosphere response" (MOCA) network. We thank Jakob Heyman and an anonymous reviewer for their constructive reviews.

REFERENCES

1. Alley, R.B., Blankenship, D.D., Bentley, C.R., Rooney, S.T., 1986. Deformation of till beneath Ice Stream B, West Antarctica. Nature 322, 57–59.
2. Aylsworth, J.M., Shilts, W.W., 1989. Glacial features around the Keewatin Divide, Districtsof Mackenzie and Keewatin. Geological Survey of Canada Paperspp. 21–88.
3. Banerjee, I., McDonald, B.C., 1975. Nature of esker sedimentation. In: Jopling, A.V.,McDonald, B.C. (Eds.), Glaciofluvial and Glaciolacustrine Sedimentation. Soc. Econ.Paleontol. Mineral., Special Publication 23, pp. 304–320.
4. Bartholomew, I., Nienow, P., Sole, A., Palmer, S., Wadham, J., 2011. Supraglacial forcing ofsubglacial drainage in the ablation zone of the Greenland ice sheet. Geophys. Res.Lett. 38, L08502.
5. Bell, R.E., 2008. The role of subglacial water in ice-sheet mass balance. Nat. Geosci. 1,297–304.
6. Bolduc, A.M., Klassen, R.A., Evenson, E.B., 1987. Cobble lithologies in eskers of central Labrador. Geol. Surv. Can. Pap. 87A, 43–51.
7. Boulton, G.S., 1996. Theory of glacial erosion, transport and deposition as a consequence of subglacial sediment deformation. J. Glaciol. 42 (140), 43–62.
8. Boulton, G.S., Hagdorn, M., Maillot, P.B., Zatsepin, S., 2009. Drainage beneath ice sheets: groundwater-channel coupling and the origin of esker systems from former ice sheets. Quat. Sci. Rev. 28, 621–638.
9. Brennand, T.A., 1994. Macroforms, large bedforms and rhythmic sedimentary sequences in subglacial eskers, south-central Ontario: implications for esker genesis and meltwater regime. Sediment. Geol. 91, 9–55.
10. Brennand, T.A., 2000. Deglacial meltwater drainage and glaciodynamics: inferences from Laurentide eskers, Canada. Geomorphology 32, 263–293.
11. Budd, W.F., Keage, P.L., Blundy, N.A., 1979. Empirical studies of ice sliding. J. Glaciol. 23,157–170.
12. Burke, M.J., Brennand, T.A., Perkins, A.J., 2012. Evolution of the subglacial hydrologic system beneath the rapidly decaying Cordilleran Ice Sheet caused by ice-dammed lake drainage: implications for meltwater induced ice acceleration. Quat. Sci. Rev50, 125–140.
13. Clark, P.U., Walder, J.S., 1994. Subglacial drainage, eskers and deforming beds beneath theLaurentide and Eurasian ice sheets. Geol. Soc. Am. Bull. 106, 304–314.
14. Dyke, A.S., 2004. An outline of North American deglaciation with emphasis on central and northern Canada. In: Ehlers, J., Gibbard, P.L. (Eds.), Quaternary Glaciations—Extent and Chronology, Part II. Vol. 2b. Elsevier, pp. 373–424.
15. Dyke, A.S., Dredge, L.A., 1989. Quaternary geology of the northwestern Canadian Shield. In: Fulton, R.J. (Ed.), Quaternary Geology of Canada and Greenland. Geological Survey of Canada. Geol. Can 1, pp. 189–214.
16. Dyke, A.S., Prest, V., 1987. Late Wisconsin and Holocene history of the Laurentide IceSheet. Géog. Phys. Quatern. 41, 237–263.
17. Engelhardt, H., Kamb, B., 1997. Basal hydraulic system of a West Antarctic ice stream: constraints from borehole observations. J. Glaciol. 43, 207–230.
18. Gulley, J.D., Grabiec, M., Martin, J.B., Jania, J., Catania, G., Glowacki, P., 2012. The effect of discrete recharge by moulins and heterogeneity in flow-path efficiency at glacier beds on subglacial hydrology. J. Glaciol. 58, 926–940.
19. Hewitt, I.J., 2011. Modelling distributed and channelized subglacial drainage: the spacing of channels. J. Glaciol. 57, 302–314.
20. Hooke, R.L., 2005. Principles of Glacier Mechanics. Cambridge University Press. Hooke, R.L., Fastook, J., 2007. Thermal conditions at the bed of the Laurentide Ice Sheet in Maine during deglaciation: implication for esker formation. J. Glaciol. 53, 646–658.
21. Iken, A., Bindschadler, R.A., 1986. Combined measurements of subglacial water pressure and surface velocity of Findelengletscher, Switzerland: conclusions about drainage system and sliding mechanism. J. Glaciol. 32, 101–119.
22. Kamb, B., 1987. Glacier surge mechanism based on linked cavity configuration of the basal water conduit system. J. Geophys. Res. 92, 9083–9100.
23. Kamb, B., 2001. Basal zone of the West Antarctic ice streams and its role in lubrication of their rapid motion. The West Antarctic Ice Sheet: Behaviour and Environment pp. 157–199.
24. Kleman, J.I., Hättestrand, C., Borgström, I., Stroeven, A., 1997. Fennoscandian palaeoglaciologyreconstructed using a glacial geological inversion model. J. Glaciol. 43, 283–299.
25. Livingstone, S.J., Clark, C.D., Tarasov, L., 2013a. North American palaeo-subglacial lakes and their meltwater drainage pathways: predictions and geomorphological clues to their origin. Earth Planet. Sci. Lett. 375, 13–33.
26. Livingstone, S.J., Clark, C.D., Woodward, J., Kingslake, J., 201 3b. Potential subglacial lake locations and meltwater drainage pathways beneath the Antarctic and Greenland ice sheets. Cryosphere 7, 1721–1740.

27. Margold, M., Jansson, K.N., 2012. Evaluation of data sou rces for mapping glacial meltwater features. Int. J. Remote Sens. 33, 2355–2377.

28. Margold, M., Jansson, K.N., Kleman, J., Stroeven, A.P., Clague, J.J., 2013. Retreat pattern of the Cordilleran Ice Sheet in central British Columbia at the end of the last glaciation reconstructed from glacial meltwater landforms. Boreas 42,830–847.

29. Margold, M., Stokes, C.R., Clark, C.D., Kleman, J., 2015. Ice streams in the Laurentide Ice Sheet: a new mapping inventory. Journal of Maps 11 (3), 380–395.

30. Peltier, W.R., 2004. Global glacial isostasy and the surface of the ice-age Earth: the ICE-5G (VM2) model and GRACE. Ann. Rev. Earth Planet. Sci. 32, 111–149.

31. Prest VK, Grant DR, Rampton N (1968) Glacial map of Canada. Geological Survey of Canada, Map 1253A, scale 1:5,000,000, 1 sheet. Price, R.J., 1966. Eskers near the casement glacier, Alaska. Geogr. Ann. 48, 111–125.

32. Röthlisberger, H., 1972. Water pressure in intra- and subglacial channels. J. Glaciol. 11,177–203.

33. Schoof, C., 2010. Ice-sheet acceleration driven by melt supply variability. Nature 468,803–806.

34. Shilts, W.W., 1984. Esker sedimentation models, deep Rose Lake Map area, district of Keewatin. Current Research, Part B, Geological Survey of Canada, Paper. 84-1B, pp. 217–222.

35. Shreve, R.L., 1972. Movement of water in glaciers. J. Glaciol. 11, 205–214.

36. Shreve, R.L., 1985a. Esker characteristics in terms of glacier physics, Katahdin eskersystem, Maine. Geol. Soc. Am. Bull. 96, 639–646.

37. Shreve, R.L., 1985b. Late Wisconsin ice-surface profile calculated from esker paths andtypes, Katahdin esker system, Maine. Quat. Res. 23, 27–37.

38. Stokes, C.R., Tarasov, L., 2010. Ice streaming in the Laurentide Ice Sheet: a first comparisonbetween data-calibrated numerical model output and geological evidence. Geophys. Res. Lett. 37, L01501. http://dx.doi.org/10.1029/2009GL040990.

39. St-Onge, D.A., 1984. Surficial deposits of the Redrock Lake area, District of Mackenzie. Geological Survey of Canada Paper 84-1A, pp. 271–278.

40. Storrar, R.D., Stokes, C.R., Evans, D.J.A., 2013. A map of large Canadian eskers from Landsat satellite imagery. J. Maps 9, 456–473.

41. Storrar, R.D., Stokes, C.R., Evans, D.J.A., 2014a. Morphometry and pattern of a large samples (N20,000) of Canadian eskers and implications for subglacial drainage beneath ice sheets. Quat. Sci. Rev. 105, 1–25.

42. Storrar, R.D., Stokes, C.R., Evans, D.J.A., 2014b. Increased channelization of subglacial drainage during deglaciation of the Laurentide Ice Sheet. Geology 42, 239–242.

43. Sundal, A.V., Shepherd, A., Nienow, P., Hanna, E., Palmer, S., Huybrechts, P., 2011. Meltinduced speed-up of Greenland ice sheet offset by efficient subglacial drainage.Nature 469, 521–524.

44. Syverson, K.M., Gaffield, S.J., Mickelson, D.M., 1994. Comparison of esker morphology and sedimentology with former ice-surface topography, Burroughs Glacier, Alaska. Geol. Soc. Am. Bull. 106, 1130–1142.

45. Tarasov, L., Dyke, A.S., Neal, R.M., Peltier, W.R., 2012. A data-calibrated distribution of deglacial chronologies for the North American ice complex from glaciological modelling. Earth Planet. Sci. Lett. 315–316, 30–40.

46. Walder, J.S., Fowler, A., 1994. Channelised subglacial drainage over a deformable bed. J. Glaciol. 40, 3–15.

47. Warren, W.P., Ashley, G.M., 1994. Origins of the ice-contact stratified ridges (eskers) o f Ireland. J. Sediment. Res. 64, 433–449.

48. Werder, M.A., Hewitt, I.J., Schoof, C.G., Flowers, G.E., 2013. Modelling channelized and distributed subglacial drainage in two dimensions. J. Geophys. Res. 118, 2140–2158.

49. Winsborrow, M.C.M., Clark, C.D., Stokes, C.R., 2010. What controls the location of ice streams? Earth Sci. Rev. 103, 45–59.

50. Zwally, H.J., Abdalati, W., Herring, T., Larsen, K., Saba, J., Steffen, K., 2002. Surface meltinduced acceleration of Greenland ice-sheet flow. Science 297, 218–222.

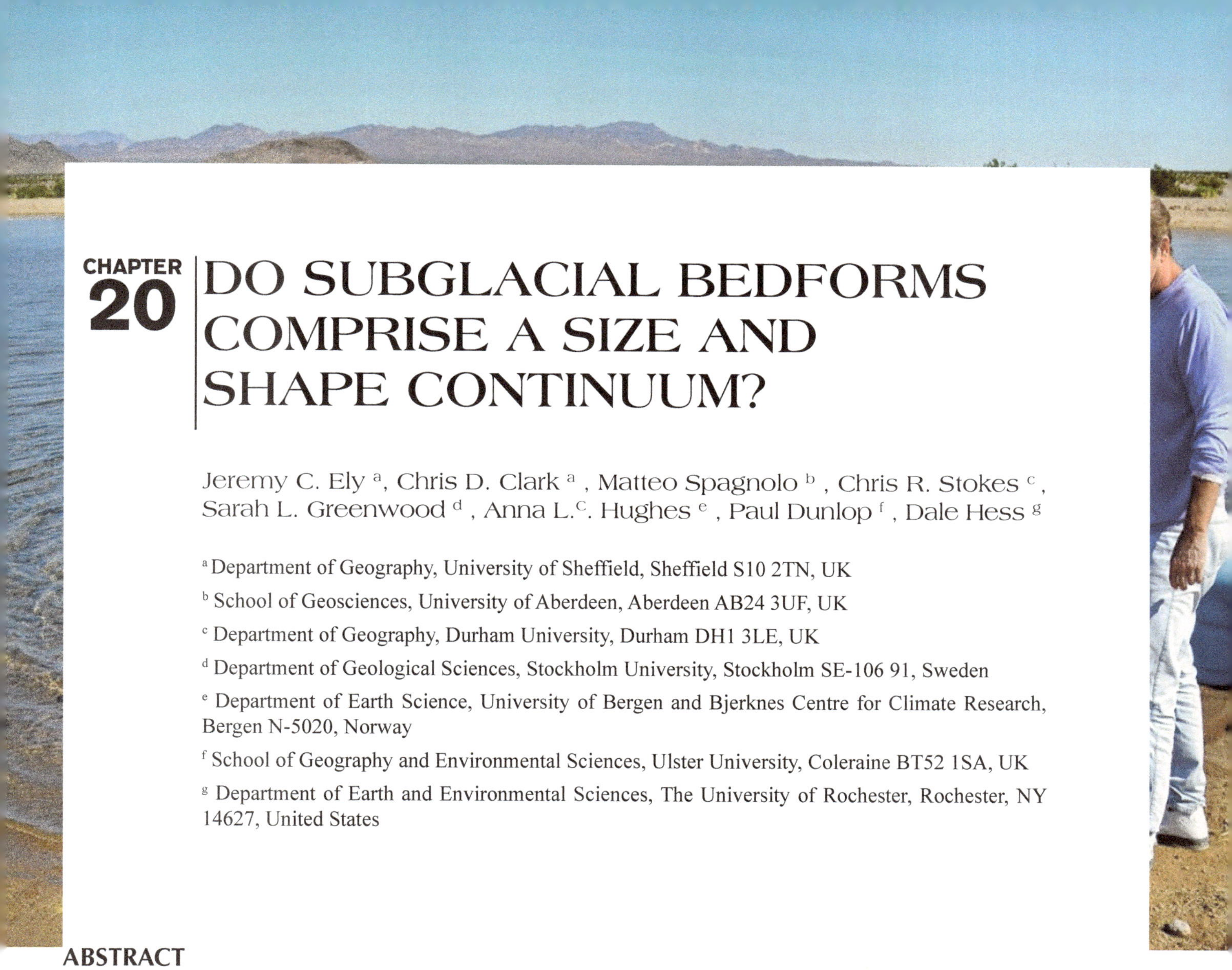

CHAPTER 20

DO SUBGLACIAL BEDFORMS COMPRISE A SIZE AND SHAPE CONTINUUM?

Jeremy C. Ely [a], Chris D. Clark [a], Matteo Spagnolo [b], Chris R. Stokes [c], Sarah L. Greenwood [d], Anna L.C. Hughes [e], Paul Dunlop [f], Dale Hess [g]

[a] Department of Geography, University of Sheffield, Sheffield S10 2TN, UK

[b] School of Geosciences, University of Aberdeen, Aberdeen AB24 3UF, UK

[c] Department of Geography, Durham University, Durham DH1 3LE, UK

[d] Department of Geological Sciences, Stockholm University, Stockholm SE-106 91, Sweden

[e] Department of Earth Science, University of Bergen and Bjerknes Centre for Climate Research, Bergen N-5020, Norway

[f] School of Geography and Environmental Sciences, Ulster University, Coleraine BT52 1SA, UK

[g] Department of Earth and Environmental Sciences, The University of Rochester, Rochester, NY 14627, United States

ABSTRACT

Understanding the evolution of the ice-bed interface is fundamentally important for gaining insight into the dynamics of ice masses and how subglacial landforms are created. However, the formation of the suite of landforms generated at this boundary — subglacial bedforms — is a contentious issue that is yet to be fully resolved. Bedforms formed in aeolian, fluvial, and marine environments either belong to separate morphological populations or are thought to represent a continuum of forms generated by the same governing processes. For subglacial bedforms, a size and shape continuum has been hypothesised, yet it has not been fully tested. Here we analyse the largest data set of subglacial bedform size and shape measurements ever collated (96,900 bedforms). Our results show that flutes form a distinct population of narrow bedforms. However, no clear distinction was found between drumlins and megascale glacial lineations (MSGLs), which form a continuum of subglacial lineations. A continuum of subglacial ribs also exists, with no clear size or shape distinctions indicating separate populations. Furthermore, an underreported class of bedform with no clear orientation to ice flow (quasi-circular bedforms) overlaps with the ribbed and lineation continua and typically occurs in spatial transition zones between the two, potentially merging these three bedform types into a larger continuum.

INTRODUCTION

The interface between moving water, air, or ice and unconsolidated sediment is often populated by undulating landforms, collectively referred to as bedforms (e.g., Allen, 1968, Wilson, 1973, Aario, 1977 and Rose and Letzer, 1977). Rather than being individuals, bedforms commonly occur in swathes or fields: configurations which cover large portions of Earth›s deserts, river beds, and ocean floors (e.g., Costello and Southard, 1981, Amos and King, 1984 and Carling, 1999), as well as the surfaces of extraterrestrial bodies (Cutts and Smith, 1973, Kargel and Strom, 1992 and Radebaugh et al., 2008). The abundance of subglacial bedforms on deglaciated terrain (e.g.,Aylsworth and Shilts, 1989, Ottesen et al., 2005, Greenwood and Clark, 2008, Larter et al., 2009, Trommelen and Ross, 2010 and McHenry and Dunlop, 2015), their emergence from receding ice margins (Johnson et al., 2010), and their repeated detection by geophysical surveys of contemporary ice-sheet beds (Rooney et al., 1987, Smith et al., 2007 and King et al., 2009) indicates that they are a key component of the subglacial environment, where processes that regulate ice flow occur (Alley et al., 1986, Engelhardt and Kamb, 1998 and Kleman and Glasser, 2007). Elucidating

Citation: Jeremy C. Ely, Chris D. Clark, Matteo Spagnolo, Chris R. Stokes, Sarah L. Greenwood, Anna L.C. Hughes, Paul Dunlop, Dale Hess, Do subglacial bedforms comprise a size and shape continuum?, Geomorphology, Volume 257, 15 March 2016, Pages 108-119, ISSN 0169-555X, http://dx.doi.org/10.1016/j.geomorph.2016.01.001.

the genesis of subglacial bedforms is an important goal in geomorphology and for understanding ice dynamics.

In addition to their composition, the morphological properties of bedforms provide constraints for hypotheses and models of their formation (e.g., Jackson, 1975, Clark et al., 2009 and Worman et al., 2013). Fluvial, aeolian, and marine bedforms can form distinct, often hierarchical, populations (Allen, 1968). In deserts, for example, discrete populations of bedforms are found, increasing in size from ripples to dunes then draas (Wilson, 1973 and Lancaster, 1988). Conversely, bedforms can also belong to populations within which there are no clear size distinctions, forming size and shape continua [e.g., aeolian ripples (Ellwood et al., 1975), aeolian dunes (Lancaster, 2013, p. 159), and subaqueous dunes (Ashley, 1990)]. For flutes, drumlins, megascale glacial lineations (MSGLs), and ribbed moraines, whether they form separate distinct morphological populations or form a single continuous size and shape population with no natural breaks (i.e., a size and shape continuum) is unclear.

Subglacial bedforms are often subdivided and named on the basis of perceived distinctions in scale and morphology. Commonly, subglacial bedforms that form aligned with ice-flow direction are divided into drumlins, MSGLs, and flutes (Fig. 1); whilst ribbed moraines and megaribs form transverse to flow direction (Fig. 1; Greenwood and Kleman, 2010 and Klages et al., 2013). Quasi-circular bedforms, which have no clear orientation to ice-flow direction, are less frequently studied but have been previously noted (Hill, 1973, Markgren and Lassila, 1980, Bouchard, 1989, Smith and Wise, 2007 and Greenwood and Clark, 2008). Beyond this, a wealth of further nomenclature exists (e.g., megaflutes, fluting, megadrumlins, and minor ribbed, Blatnick, Rogen, and Niemsel moraines).

Figure. 1: Different types of subglacial bedform and their orientation to ice-flow direction (denoted by black arrow). Scales are approximate. (A) A drumlin in NE England. These streamlined hills are typically 250–1000 m along-flow and 120–300 m across-flow (Clark et al., 2009). (B) Transverse ridges, termed ribbed moraine, in Nunavut, Canada, typically 300–1200 m across-flow and 150–300 m along-flow (Hättestrand and Kleman, 1999). (C) A flute formed parallel to ice-flow direction, Svalbard. Note the accumulation of sediment in the lee of a boulder. (D) A MSGL in Nunavut, Canada. MSGLs typically are 100–200 m across-flow and 1–9 km along-flow (Spagnolo et al., 2014a) but have been reported to be much longer (e.g., 180 km, Andreassen et al., 2008). All photographs from www.shef.ac.uk/drumlins.

Despite the array of terms applied to the varieties comprising subglacial bedforms, a long-standing hypothesis is that they actually belong to a continuum of size and shape (Aario, 1977, Rose and Letzer, 1977 and Rose, 1987). This 'continuum hypothesis' was originally based upon observations of spatial transitions between subglacial bedform types and orientations along the direction of ice flow, suggesting a single continuum of form stretching from ribbed moraine, through quasi-circular forms, to drumlins (Aario, 1977, Markgren and Lassila, 1980 and Punkari, 1984). However, subsequent work on the continuum hypothesis has focussed solely upon flow-aligned subglacial bedforms (e.g., Rose, 1987 and Stokes et al., 2013b). Furthermore, the relatively recent discovery of 'megascale' subglacial bedforms (Clark, 1993 and Greenwood and Kleman, 2010) might imply a distinct variety of bedform rather than end-members of a continuum. Morphological studies often focus upon previously labelled categories of subglacial bedforms [e.g. drumlins in Clark et al. (2009) and Maclachlan and Eyles (2013); ribbed moraines in Hättestrand (1997) and Dunlop and Clark (2006b); MSGL in Spagnolo et al. (2014a)]. A comprehensive study that looks at all landform types together to compare and contrast their size and shape is missing. Here, we present and analyse a data set of 96,900 size and shape measurements of subglacial bedforms from numerous locations, spanning the range of types in the literature, in order to investigate whether there is a continuum of subglacial bedforms or whether separate size and shape populations exist. Note that we do not consider bedforms composed purely of bedrock (e.g., whalebacks, roche moutonnées, megagrooves), nor do we consider hybrid forms such as crag and tails, as they are often regarded as different from subglacial bedforms created in sediment (e.g., Dionne, 1987, Stokes et al., 2011 and Lane et al., 2015).

DATA AND METHODS

Data

A database of 96,900 mapped subglacial bedforms was compiled from previous studies and additional mapping, which was conducted using standard remote sensing techniques (Table 1). A variety of reported bedform morphologies from a wide range of sites were chosen from the literature (Fig. 2; Table 1). Mapped bedforms were grouped by locality and reported type (i.e., as reported in previous studies) so that further analysis could ascertain the similarities or differences between these types. At each location, the highest resolution remote sensing data available were used to map and derive landforms metrics (Table 2). Only data with resolution higher than 30 m were used, as measurements of bedform length and width derived from coarser resolution data have been shown to misrepresent subglacial bedform size and shape (Napieralski and Nalepa, 2010). Each bedform is represented in our data set as a smooth polygon, manually digitised directly into a geographic information system (GIS) around the break of slope on hill-shaded digital elevation models illuminated from multiple directions (Smith and Clark, 2005 and Hughes et al., 2010). On satellite imagery, landforms are detectable as a change in vegetation and/or soil moisture (e.g., Spagnolo et al., 2014a). Automated mapping techniques (e.g., Saha et al., 2011 and Maclachlan and Eyles, 2013) were avoided as they require predefinition of parameters such as shape and scale and thus would introduce bias into our results.

Table 1: Location of mapped subglacial bedforms; all bedforms have previously been described in the literature. Where mapping was not available from the original study a '#' denotes that further mapping was conducted at a previously described site

Reported bedform type (number of bedforms)	Location	Central coordinates (decimal degrees)	Number of bedforms	Data set	Study describing site (# denotes mapping has not previously been analysed)	Location number
Drumlins (n = 42,495)	Britain	54.069, 2.258	30,304	NEXTMap DEM	Hughes et al. (2010)	1
	New York State, USA	44.830, − 75.268	5650	USGS NED	Hess and Briner (2009)	2
	Alta, Norway	69.476, 23.139	1638	Landsat ETM +	Spagnolo et al. (2010)	3
	Ungava Bay, Quebec, Canada	57.176, − 67.299	5903	Landsat ETM +	Spagnolo et al. (2010)	4
MSGL (n = 31,668)	Cameron Hills, Alberta, Canada	59.950, − 117.826	581	SPOT	Brown et al. (2011) #	5
	Dubawnt Lake, Nunavut, Canada	64.171, − 100.415	17,038	Landsat ETM +	Stokes et al. (2013b)	6
	Eskimo Bay, Nunavut, Canada	61.387, − 94.829	4499	SPOT	Clark (1993)#	7
	Great Bear Lake, NWT, Canada	64.454, − 122.069	1260	SPOT	Winsborrow et al. (2004)#	8
	Haldane Ice Stream, NWT, Canada	67.008, − 121.299	489	SPOT	Winsborrow et al. (2004)#	9
	Liard Ice Stream, NWT, Canada	61.217, − 121.701	340	SPOT	Brown et al. (2011) #	10
	Great Slave Lake, NWT, Canada	61.703, − 116.576	784	SPOT	Brown et al. (2011) #	11
	Payne Bay, Quebec, Canada	59.618, − 70.430	531	SPOT	This study #	12
	M'Clintock Channel Ice Stream, Canada	72.743, − 105.753	2796	Landsat ETM +	Storrar and Stokes (2007) #	13
	West James Bay Ice Stream	54.478, − 87.280	3350	SPOT	Clark (1993)#	14
Quasi-circular bedforms (n = 1955)	Ireland	53.723, − 7.803	1955	Landmap DEM, SRTM, Landsat ETM +	Greenwood and Clark (2008)	15

Flutes (n = 664)	Skeiðarajökull, Iceland	63.977, − 17.219	101	NERC ARSF aerial photography	Waller et al. (2008) #	16
	Breiðamerkurjökull, Iceland	64.071, − 16.327	131	NERC ARSF Aerial Photography	Evans and Twigg (2002) #	17
	Conwaybreen, Svalbard	78.994, 12.486	432	NERC ARSF aerial photography	This study #	18
Ribbed moraine (n = 18,384)	Ireland	53.723, − 7.803	5464	Landmap Dem, SRTM, Landsat ETM +	Greenwood and Clark (2008)	19
	Lac Naococane, Quebec, Canada	52.972, − 70.921	501	Landsat	Dunlop and Clark (2006a) #	20
	St. Lawrence Valley, New York	44.831, − 75.268	921	USGS NED	Carl (1978)#	21
	Lake Rogen, Sweden	62.328, 12.394	3,357	Landsat ETM +	Dunlop and Clark (2006b) #	22
	Ungava Bay, Quebec, Canada	59.618, − 70.430	7582	Landsat ETM +	Dunlop and Clark (2006b) #	23
	Great Slave Lake, NWT, Canada	61.703, − 116.576	559	SPOT	Brown et al. (2011) #	24
Mega subglacial ribs (n = 733)	Keewatin, Canada	64.171, − 100.415	733	Landsat ETM + and SPOT	Greenwood and Kleman (2010)	25

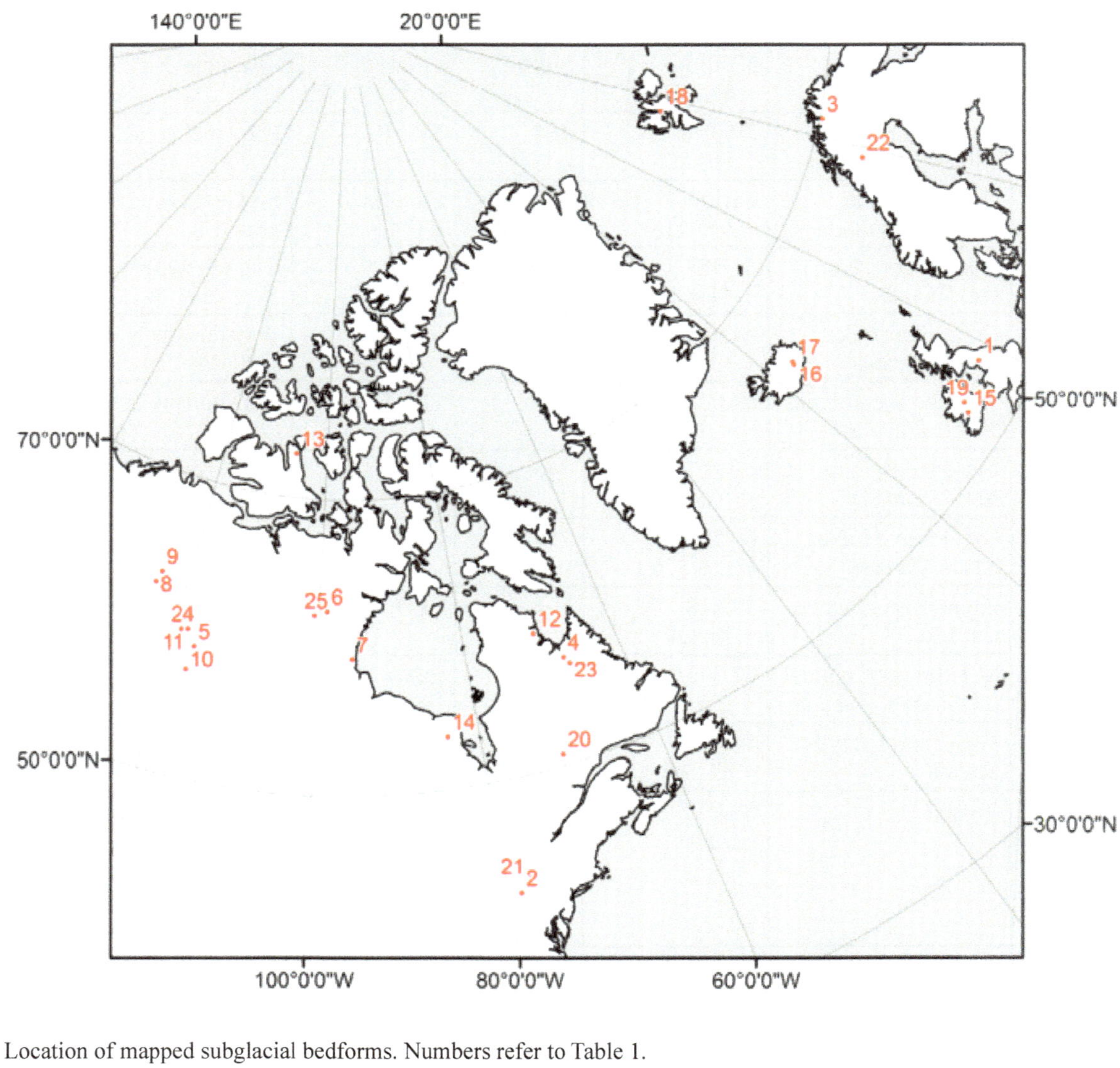

Figure. 2: Location of mapped subglacial bedforms. Numbers refer to Table 1.

Table 2: Data type and source for the mapping described in Table 1

Data set	Type of data	Horizontal resolution (m)	Source
NEXTMap Great Britain™	DEM	5	http://arsf.nerc.ac.uk/
USGS NED	DEM	10 m	http://ned.usgs.gov/
Landsat ETM +	Imagery	15 pan-chromatic, 30 colour	http://earthexplorer.usgs.gov/
SPOT	Imagery	10 pan-chromatic, 20 colour	http://geobase.ca/
Landmap DEM	DEM	25	http://landmap.ac.uk/
NERC ARSF aerial photography	Imagery	0.15	http://arsf.nerc.ac.uk/
SRTM	DEM	30	http://earthexplorer.usgs.gov/

Methods

The length (L) and width (W) of each mapped polygon was estimated via Euler's approximation for an ellipse (Clark et al., 2009). A limitation of this approximation is that it slightly underestimates the length (and overestimates the width) of highly elongate or irregular polygons, but this error is insignificant compared to the size of the bedforms (Clark et al., 2009) and typically is less than 3%. These polygon measurements were then converted into a-axis (distance down-ice) and b-axis (distance across-ice) measurements according to bedform orientation to ice flow (Fig. 3). For near-circular bedforms, bedform a-axis and b-axis were manually measured within a GIS using regional ice-flow patterns as an indicator of flow alignment (Fig. 3). Elongation ratio, often used as a proxy for shape (e.g., Rose, 1987, Clark et al., 2009 and Dowling et al., 2015), was simply measured as a-axis divided by b-axis.

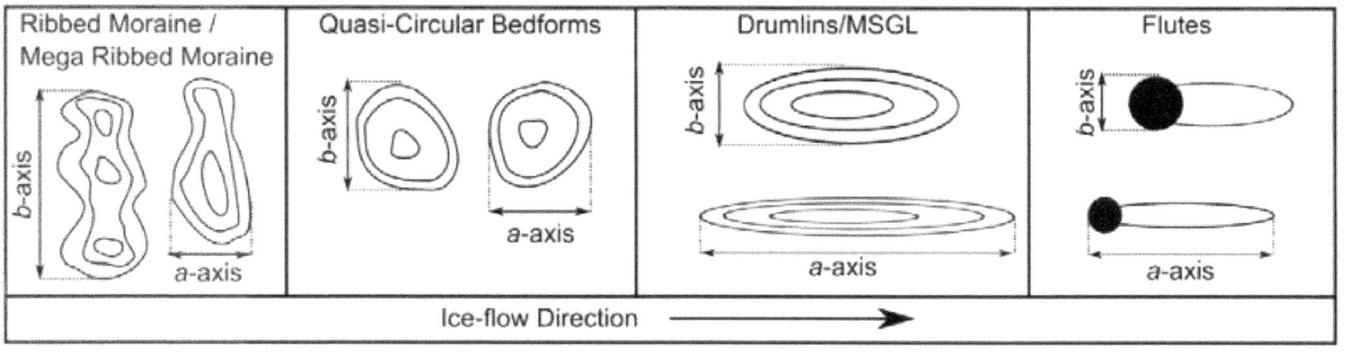

Figure. 3: Schematic of derived bed form axes.

Testing the subglacial bed form continuum hypothesis requires the detection and definition of populations within our data set. If the variously-named types of bed forms are found to vary continuously in size and shape such that, for example, a large drumlin is the same as a small MSGL, then a size and shape continuum between them exists (Fig. 4A and B). When plotted on a scatter graph, this would show a single continuous data cloud, or cluster (Fig. 4B). On the contrary, if the derived metrics reveal gaps or jumps in scale or shape then they are better interpreted as discrete phenomena (Fig. 4C and D), leading to the rejection of the continuum hypothesis. In this case, a scatter graph of bedform metrics would show several distinct clusters (Fig. 4D). Several approaches are taken here to assess the degree to which separate populations, or clusters, can be detected within our data set.

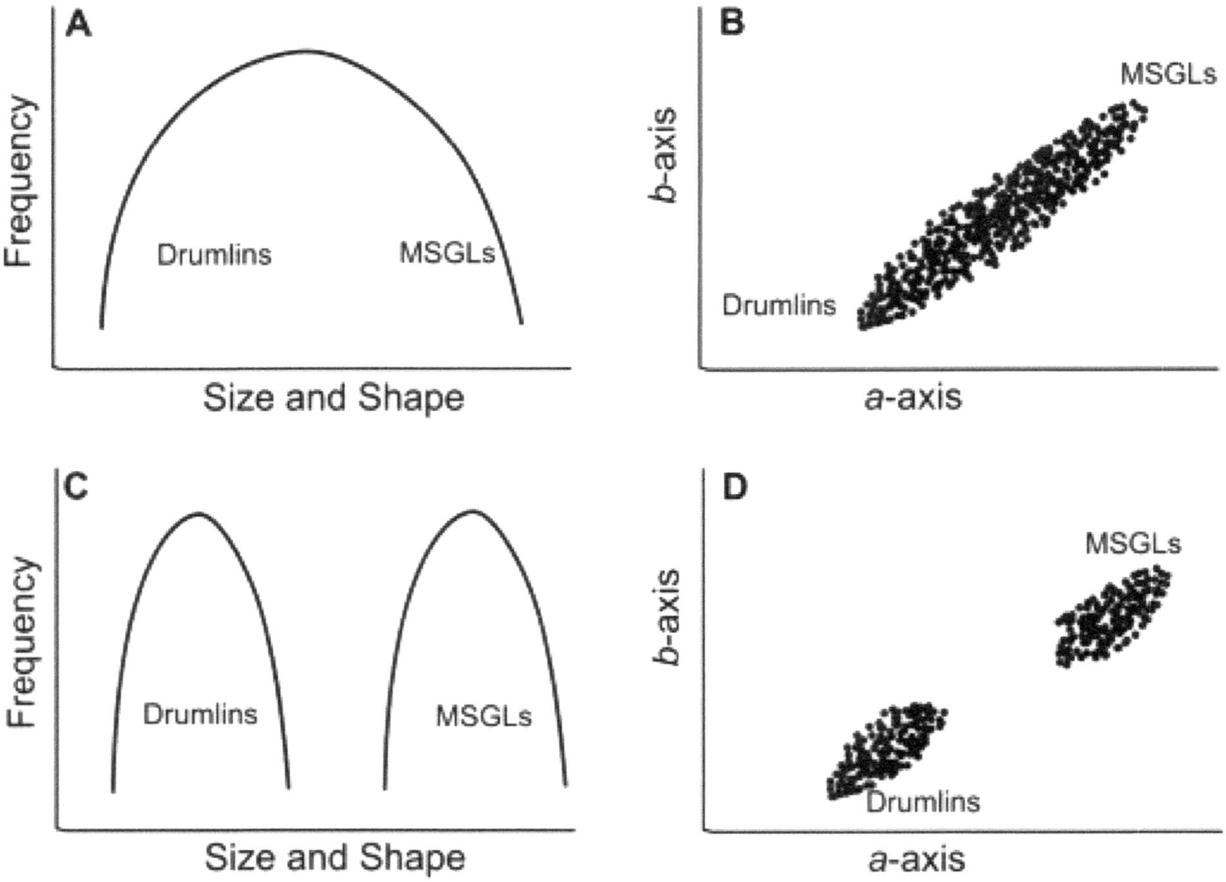

Figure. 4: Schematic of how a continuum (A and B) or different populations (C and D) may be detected in our data set. If a continuous sequence in the size and shape of bedforms exists, e.g., between drumlins and MSGLs (A), then a scatter plot of their metrics would show a single cluster. If separate size and shape populations occur (C), then separate clusters would be shown on a scatter plot (D).

The human eye is an excellent tool for detecting clusters (Jain, 2010). As such, our first attempt at detecting clusters in our data set was to plot the data and visually assess clustering qualitatively. That said, different interpreters may see different clusters. Thus, in an effort to make our analysis more objective and quantitative we employed the density-based clustering algorithm DBSCAN (Ester et al., 1996) using the package 'fpc' in R statistical software. No clustering algorithm or technique is perfect because no objective definition of a cluster exists (Estivill-Castro, 2002). However, DBSCAN was chosen as it requires no predefinition of the number of expected clusters within a data set; hence it requires no prior categorisation of bedforms into separate populations. The algorithm requires two input parameters: the window size (ε) and the minimum number of points within a cluster (MinPts). In order to define dense regions (clusters) within a point cloud, DBSCAN creates a window at each point and determines which points are sufficiently reachable from each other in order to warrant definition as a cluster (Ester et al., 1996). Core points are deemed reachable if there are more than MinPts within the search radius of ε from that point, and are included within the cluster owing to point density. Additionally, points are also deemed cohesive with the cluster if a core point is within the radius of ε. If neither of these criteria are met, the point is deemed to be outside of the cluster or is left unclustered. Sensitivity analysis of cluster-definition to window size (ε) was run from $\varepsilon = 1.0$ to $\varepsilon = 0.01$ with steps of $\varepsilon = 0.01$. The sensitivity of minimum points per cluster (MinPts) was also tested at values of 10, 50, and 100. The DBSCAN procedure was only conducted on the independent variables of length and width, elongation being derived from the two.

A second approach based on a direct, visual assessment of the degree of overlap between assigned bedform category metrics was also applied (Table 1). In covariance plots of the measured variables (e.g., length vs. width), the density of observations per category were calculated and contoured. Here, the highest density contour contained 75% of the observations, the middle 95%, and the lowest 99%. The degree of overlap between adjacent categories was then evaluated in order to test the validity of frequently used bedform nomenclature.

RESULTS

The size and shape of all 96,900 subglacial bedforms in our data set are displayed in Fig. 5. The most striking aspect is that data are concentrated into a narrow range of values. When plotted on linear axes (Fig. 5A), there appears to be two clouds of data which merge toward the origin of the plot. Larger bedforms (> 10,000 m) are less frequent in our data set, but plot as extensions of the same data clouds rather than forming separate clusters. Due to the skew imposed by these large values, data was also plotted on a logarithmic scale (Fig. 5B–D). At first order, two clusters of data are visually discernible: a small cloud of narrow and elongate bedforms aligned with flow direction, and a much larger cloud comprising the remaining bedforms (Fig. 5D). Within this larger cloud, density variations occur, perhaps recording three populations that (in places) merge and overlap, or maybe this is a single cluster that has been incompletely sampled. Within the larger cloud, these subclusters exhibit ellipsoid scatters with apparent trends (red ellipses in Fig. 5D), with bedforms with a larger a-axis than b-axis following a separate trajectory to transverse bedforms. Bedforms with no clear orientation (i.e., circular) fall between these two groups.

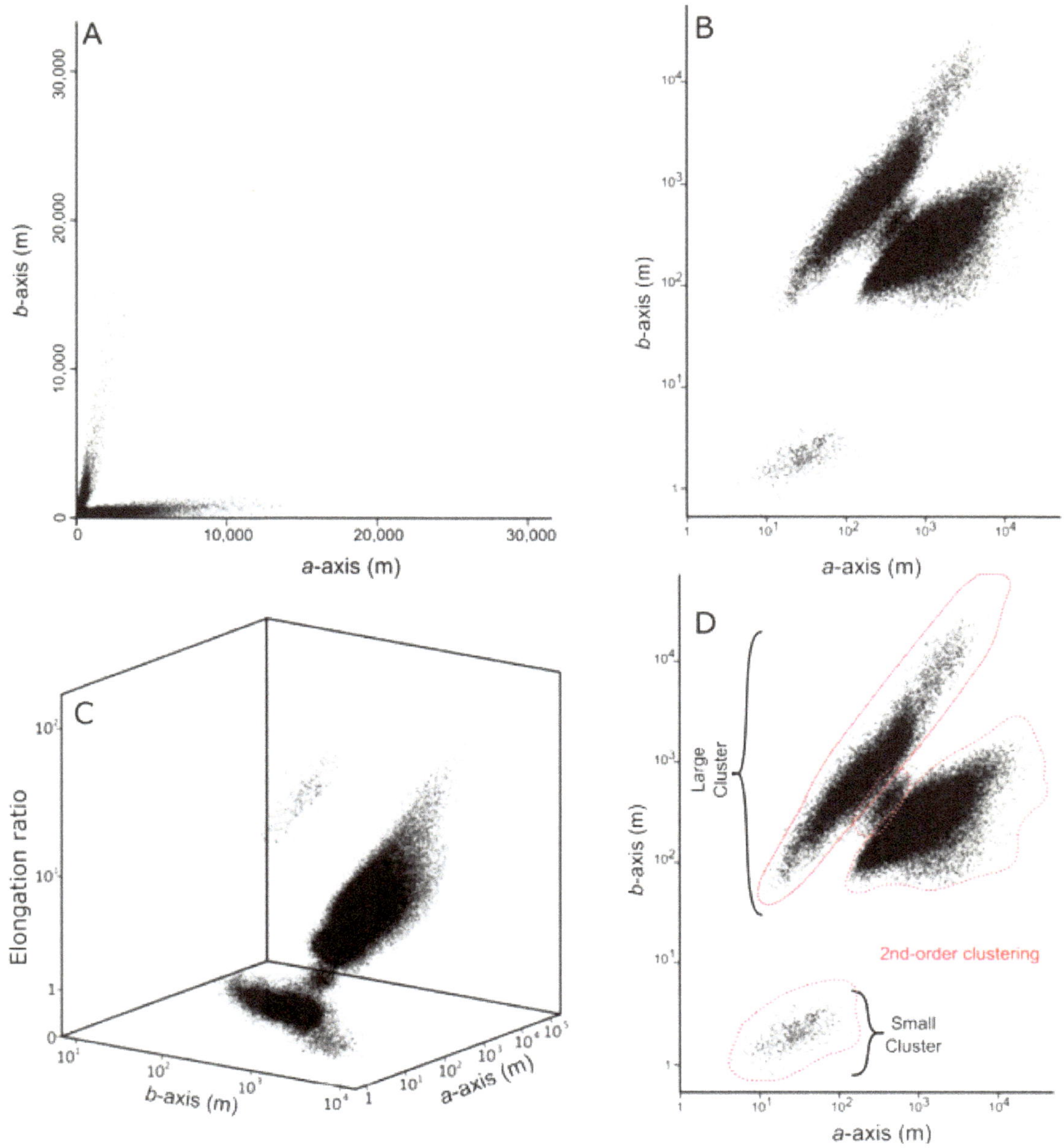

Figure. 5: The size and shape of all 96,900 subglacial bedforms, paying no regard to their nomenclature. (A) Plot ofa-axis (down-flow) and b-axis (across flow) dimensions. (B) The same dimensions, plotted on log–log axes. (C) Combined plot of a-axis, b-axis and elongation ratio (a/b). Although elongation ratio is dependent upon a and b axes, it is plotted for visualisation purposes. (D) Two possible qualitative (visual) interpretations of clusters. Perhaps the data reveals just two clusters, or on closer inspection it might be possible to distinguish four. See text for further details.

To assess the degree to which separate populations are quantitatively distinctive, the density-based clustering algorithm (DBSCAN) was used with an extensive sensitivity analysis (results are summarised in Fig. 6). At the extremes of the window size parameter (Ɛ), the algorithm groups the data into inappropriately sized clusters. When Ɛis large, the window size is such that the whole data set is seen as a single cluster (Fig. 5A). When Ɛ is small, either numerous small clusters or no clusters at all are detected because of an insufficient search radius (Fig. 6O). The most common result is that two clusters are detected (Fig. 6B–F, K, L, and N). These two clusters occur in the same positions as the first-order visual clustering (Fig. 5D). The larger cluster is not separated when the search window size parameter is adjusted from 0.5 to 0.06 (Fig. 6H–J). Beyond this, the larger cloud is separated into two clusters, one with bedforms aligned with flow direction and another with bedforms transverse to flow (Fig. 6K–N). For only a very small parameter space, DBSCAN distinguishes subglacial bedforms with no clear orientation with flow direction from flow-aligned and flow-transverse bedforms (Fig. 6M). Our results were found to be insensitive to the minimum number of points parameter (MinPts).

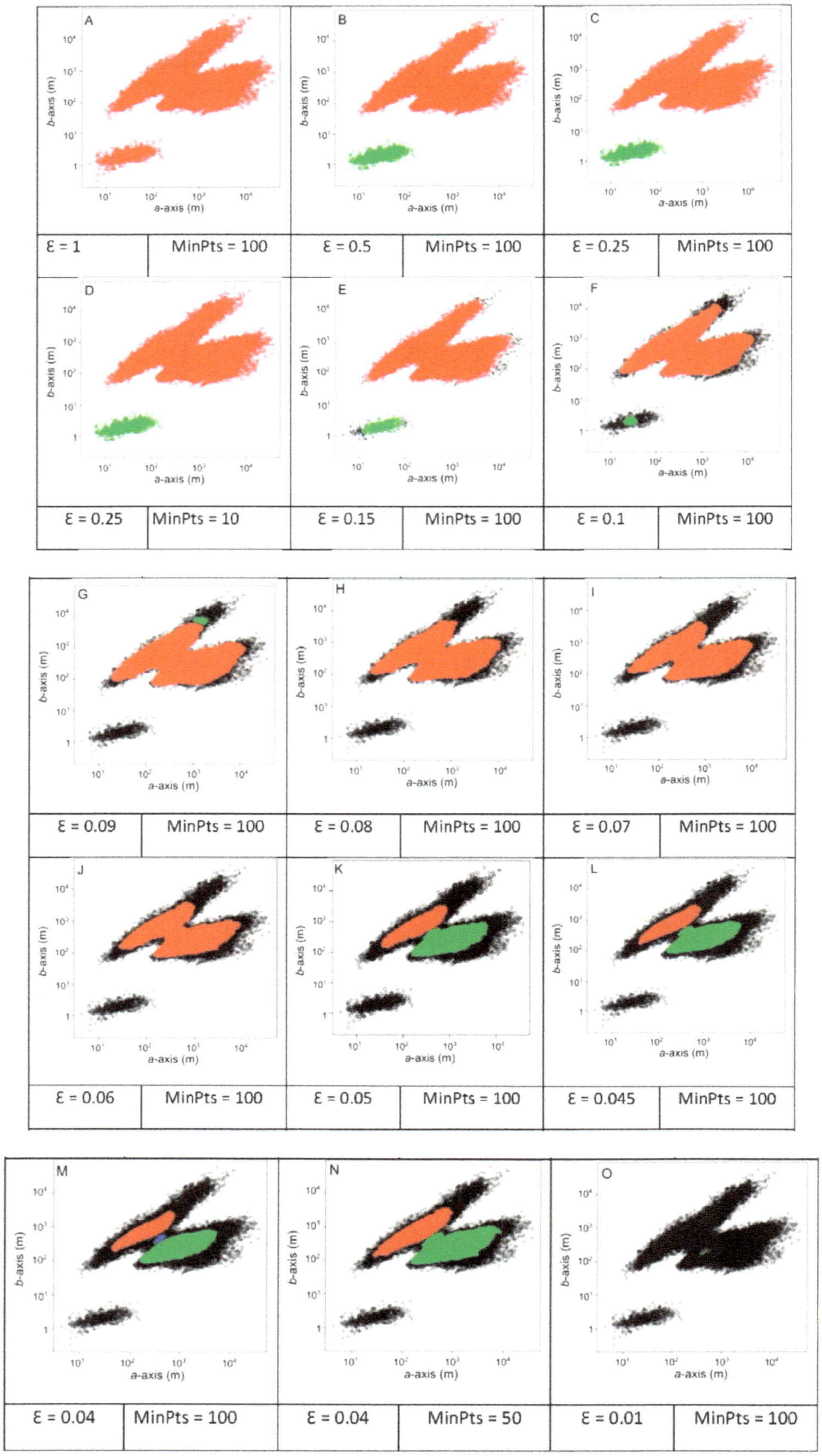

Figure. 6: DBSCAN sensitivity analysis to assess how many clusters exist using different parameter settings. Red, green, and blue indicate different clusters. Black points are disregarded by the clustering algorithm as being outside of any cluster caused by a lack of density and cohesiveness with other points. The most common result is that two clusters are detected (B, C, D, E, and F): narrow elongate bedforms and a larger cluster comprising of the remaining data. This occurs when Ɛ = 0.5–0.1, and MinPts = 10–100 (B–F). ForƐ = 0.09–0.06 (G–J), the smaller cluster is not detected, but the large cluster remains. When Ɛ = 0.05–0.045, clusters distinguish between flow alignment (K). At Ɛ = 0.04 and MinPts = 100, a further small cluster between flow alignments is detected (blue in 'M'). Yet, a slight change in parameter values alters this result (N). When Ɛ is too large, the whole data set is considered a cluster (A), or when Ɛ is too small, no clusters or clusters of an inappropriate size are detected (O).

When each bedform›s previously determined category was considered, the group of small bedforms that were distinguished visually (Fig. 5) and by DBSCAN (Fig. 6B–F), are revealed to be those that are commonly known as flutes (Fig. 7 and Fig. 8). The distinguishing factor of flutes is their b-axis, which is much smaller than for all other subglacial bedforms (Fig. 7), and has a narrow range and small measures of variance (Table 3). Overlap occurs between all other categories of bedform, as shown by the intersection of density contours belonging to adjacent categories (Fig. 7) and the frequency distribution of their measured metrics (Fig. 8). The strongest overlap occurs between drumlins and MSGLs, which for all parameter combinations occurs between 75% contours. The median b-axis, and the 10th and 90th percentiles of the drumlin and MSGL categories are remarkably similar (Table 3), indicating a consistency between the two groups. This contributes to the clustering of drumlins and MSGL together, visually and in DBSCAN (Fig. 5 and Fig. 6). The elongation ratio of ribbed moraines and megaribs is also similar, with low coefficients of variation (Table 3). This indicates a similar shape. The size range of quasi-circular bedforms is much narrower compared to all other subglacial bedforms, except for flutes. Additionally, the coefficient of variation for their size metrics is small, hence their selective positioning on Fig. 5. The category of quasi-circular bedforms may contain bedforms some might categorise as either drumlins or ribbed moraines. However, the mean elongation ratio of 1.07 (Table 3) and the high density of data on Fig. 5, Fig. 6, Fig. 7 and Fig. 8 where the elongation ratio is 1 clearly indicates the existence of near-circular forms.

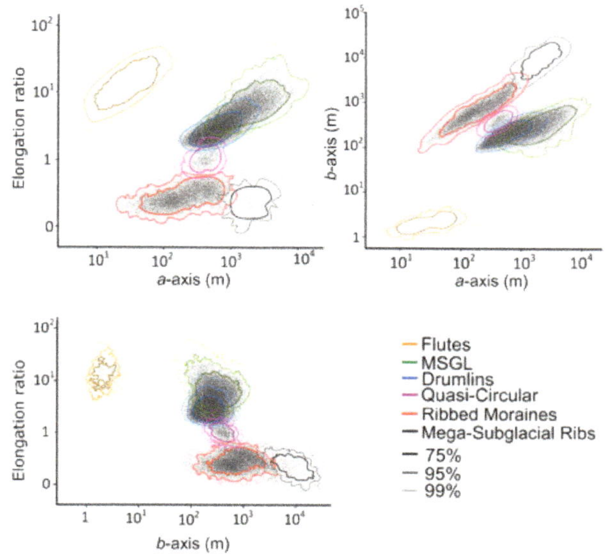

Figure. 7: Density contour plots to assess the degree of separation and overlap between named bedform types. Flutes consistently plot separately to all other bedform types, whilst overlap (a-axis, b-axis plot) occurs between other subglacial bedform types. Most notably, this overlap occurs between drumlins and MSGL (overlapping 75% contours) and between ribbed moraine and megasubglacial ribs (overlapping 90% contours). Quasi-circular bedforms overlap with ribbed moraine (90% contours) and drumlin (75%) contours. Note that elongation ratio is derived from length and width. However, it is plotted against its components here for illustrative purposes.

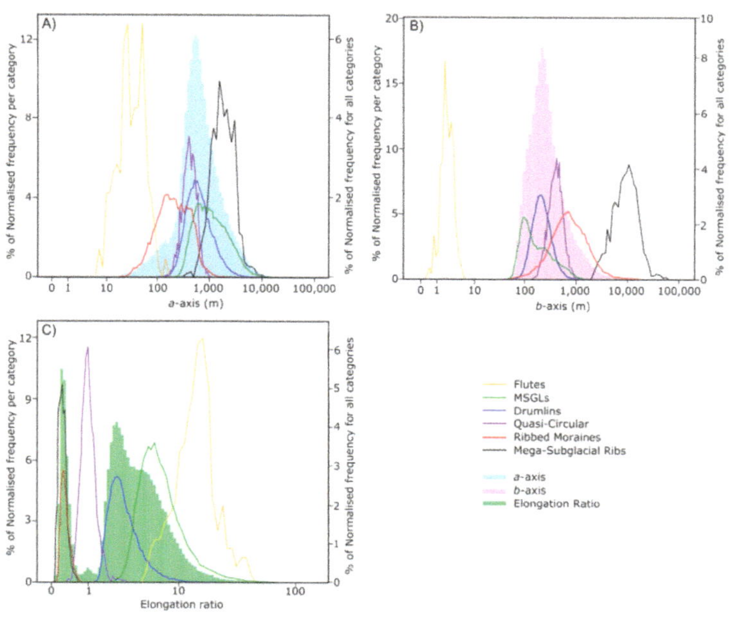

Figure. 8: Frequency distributions for (A) a-axis, (B) b-axis, and (C) elongation ratio. In each case, the probability distribution function of each bedform is plotted, normalised to the total population of their respective category in order to permit comparisons. This is plotted on the primary y-axis. The histograms (in solid colours) show the distribution of the whole population, plotted against the secondary y-axis.

Table 3: Summary statistics of subglacial bedform categories

		Range (m)	Median (m)	10th percentile (m)	90th percentile (m)	Coefficient of variation
Flutes	a-Axis	138.1	30.5	12.9	68.4	0.6
	b-Axis	5.6	2.1	1.3	3.4	0.4
	Elongation	41.0	15.3	7.9	25.2	0.4
Megasubglacial ribs	a-Axis	11765.4	1980.4	1094.6	3706.4	0.6
	b-Axis	53940.1	9286.3	4283.4	17692.7	0.6
	Elongation	0.55	0.2	0.1	0.4	0.4
Ribbed Moraines	a-Axis	3683.6	196.9	61.2	606.5	0.9
	b-Axis	16749.8	701.1	260.1	1941.8	1.0
	Elongation	0.75	0.27	0.2	0.4	0.4
Circular Bedform	a-Axis	876.4	416.9	274.6	615.0	0.3
	b-Axis	1002.7	409.2	264.3	596.6	0.3
	Elongation	2.64	1.0	0.7	1.4	0.3
Drumlins	a-Axis	10602.1	603.5	323.4	1372.6	0.7
	b-Axis	1509.7	213.6	131.2	371.7	0.5
	Elongation	36.20	2.8	1.9	4.9	0.5
MSGLs	a-Axis	37268.2	1102.1	427.4	3912.9	1.2
	b-Axis	2725.8	150.1	81.7	479.2	0.9
	Elongation	131.62	6.4	4.0	12.9	0.8

To assess if there are any systematic controls on bedform metrics from their geographic location, Fig. 9 plots the outlines of the data at each locality shown in Table 1. No single location plots separately to the others, with all locations showing multiple overlap with other areas.

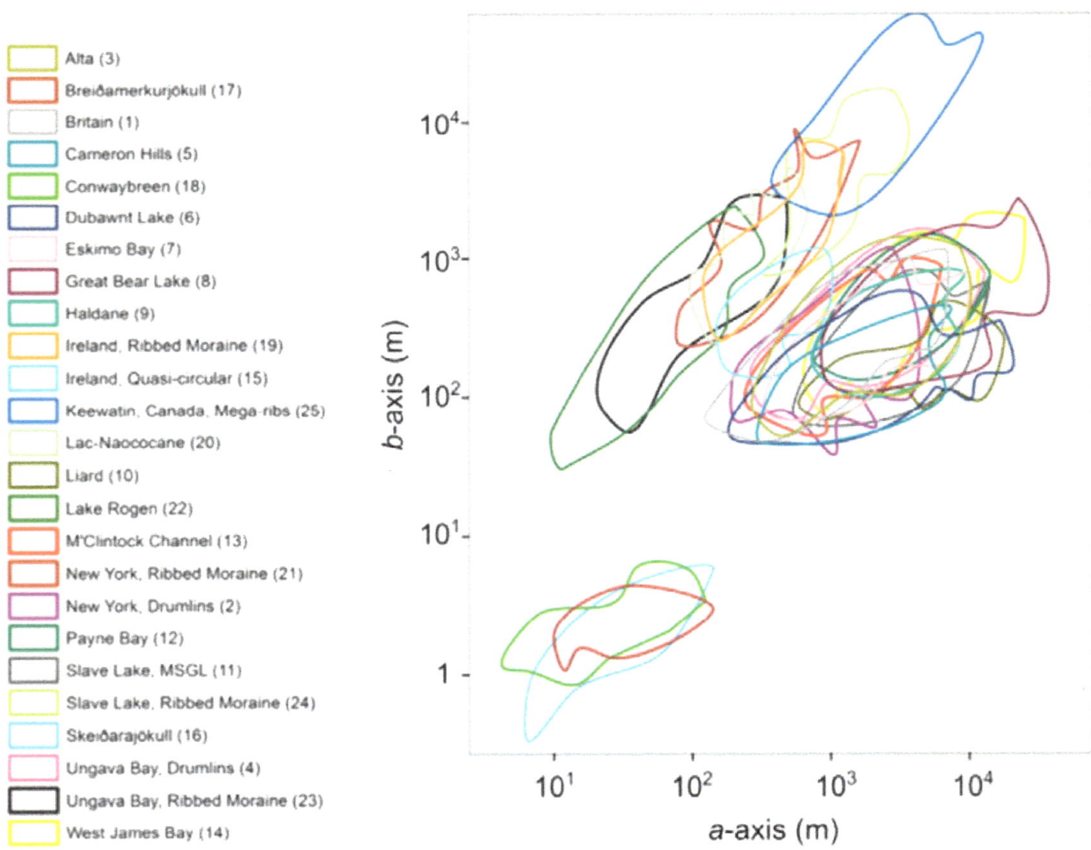

Figure. 9: Outlines of the length and width of bedform scatter plots grouped by location. Locations are listed in Table 1. Note the overlap between locations, such that data from no single location is found to be unique.

DISCUSSION

Is There a Subglacial Bedform Size and Shape Continuum, Several Continua, Or are they Separate Phenomena?

The number of populations or clusters, and therefore the number of continua, within our data set depends, to some extent, upon the interpretation of Figs. 5, Fig. 7 and Fig. 8, as well as the parameters chosen for DBSCAN (Fig. 6). However, a consistent result is that flutes plot separately to all other subglacial bedforms (Fig. 5, Fig. 6, Fig. 7, Fig. 8 and Fig. 9), forming a distinct population. Given their smaller size, flutes could only be mapped from aerial photography (Table 1 and Table 2), rather than digital terrain models or satellite images, but we do not interpret their separation to be a consequence of higher resolution imagery being used for mapping. This interpretation is supported by observations of flutes that reach over 1.5 km long (exceeding the length of many drumlins) whilst maintaining their narrow width (Kjær et al., 2006). Flutes are often found superimposed on top of drumlins (e.g., Boulton, 1987, Hart, 1995 and Waller et al., 2008), emphasising the scale difference between the two. We therefore interpret flutes to be morphologically distinct to all other subglacial bedforms, forming a group of narrow and elongate subglacial bedforms. Thus, a clear difference exists in scale between flutes and other subglacial bedforms. This is consistent with previous suppositions and interpretations (Boulton, 1976, Rose, 1987 and Clark, 1993).

Aside from flutes, it is difficult to distinguish separate populations of bedforms that are elongated with ice-flow direction. A consistency in b-axis lengths and a lack of any stepwise jump in a-axis lengths between drumlins and MSGLs means that these two categories of flow-aligned bedforms were never separated by any combination of parameters in our quantitative analysis (Fig. 6). They display the highest level of overlap on Fig. 7 and Fig. 8. We therefore concur with previous work using a smaller data set (Stokes et al., 2013b) that drumlins and MSGL form a size and shape continuum of subglacial lineations.

Ribbed moraines and megascale ribs also possess overlapping size and shape metrics (Fig. 7 and Fig. 8), and were only partially separated by one combination of DBSCAN parameters (Fig. 6G), which may be a consequence of a smaller sample of mega-ribbed moraines. Both populations have a consistent elongation ratio of ≈ 0.3 (Table 3). We find no grounds to suggest different morphological populations of ribbed moraines, as suggested elsewhere (e.g., Hättestrand, 1997), as there is no clear break or stepwise relationship in the metrics to indicate separate size and shape categories. Furthermore, putatively different types and scales of subglacial ribs often display close spatial relationships (Greenwood and Kleman, 2010), and have been observed to occur in potentially evolutionary sequences (Markgren and Lassila, 1980). Therefore, we propose that ribbed moraines and mega-ribs form a size and shape continuum of transverse subglacial bedforms that can be called subglacial ribs.

The existence of quasi-circular bedforms that overlap with drumlins and ribbed moraines (Figs. 5, Fig. 7 and Fig. 8) raises the possibility of a single bedform continuum comprising ribs to quasi-circular forms to lineations (e.g., Aario, 1977). All three are often categorised as one cluster by DBSCAN (Fig. 6B–J). When they are split into separate clusters by the quantitative analyses (Fig. 6K–N) it may be a consequence of a genuine difference in the shape and scale of subglacial ribs and lineations (Fig. 5D) and/or a small sample size of quasi-circular forms (Table 1). The bridging between ribbed moraines and drumlins provided by quasi-circular bedforms only occurs within a narrow range of a- andb-axis lengths, centred around 450 m (Fig. 5 and Fig. 8).

Quasi-circular bedforms — such as the Blattnick moraine of Markgren and Lassila (1980), the mammillary hills of Aario (1977), the circular forms noted by Knight et al. (1999), and the ovoid forms noted by Smith and Wise (2007) — are often reported in transition zones between ribbed moraines and drumlins. Quasi-circular forms were most notable in several locations across the bed of the Irish-ice sheet (Greenwood and Clark, 2008). Given their potential importance, further examples of quasi-circular bedforms were sought and are shown in Fig. 10. These examples illustrate how they occur in gradual down-ice transitions between subglacial ribs and lineations and sometimes superimposed upon subglacial ribs. Since subglacial bedforms are often assigned a label based upon flow orientation, perhaps quasi-circular forms have been confusing to identify and classify and are more common than is reported in the literature. As they may form an important link between subglacial ribs and lineations, further work on quasi-circular bedforms is required.

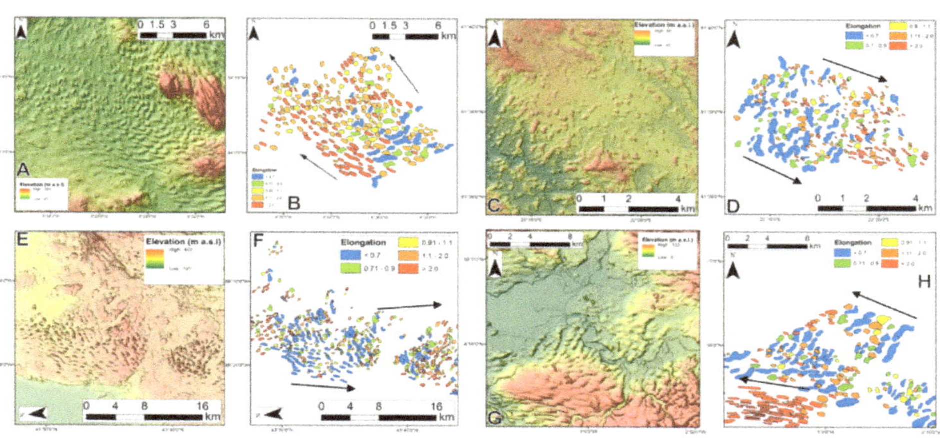

Figure. 10: Examples of spatial transitions between ribbed moraines and drumlins. Note how in each case, bedforms with no clear alignment to flow (elongation ratio of 0.9 to 1.1; yellow) were noted. (A) Hill-shaded SRTM DEM (30 m) and mapping (B) of bedforms in County Roscom-

mon, Ireland. (C) 2 m LiDAR-derived DEM and mapping (D) of bedforms near Harjavalta, Finland. (E) Hill-shaded 2 m LiDAR DEM and mapping (F) of bedforms in Fon Du Lac County, Wisconsin, USA. (G) Hill-shaded NEXTMAP (5 m) DEM and mapping of bedforms (H) in Cumbria, UK.

Controls on the Size and Shape of Subglacial Bedforms

A correspondence between bedform size and shape and the flow characteristics of the geomorphic agent is often invoked for aeolian, fluvial, and marine bedforms (Allen, 1968,Rubin and Ikeda, 1990 and Reffet et al., 2010). Similar links have been sought for subglacial bedforms (e.g., Rose and Letzer, 1977). Several lines of evidence that suggest that subglacial bedform size and shape is primarily determined by properties of the overlying ice mass. Firstly, bedforms within a flowset usually have a similar size, shape, and alignment to flow direction (Clark, 1999). Thus, different flow events, with different glaciological properties, produce different bedform morphologies. Secondly, bedforms are often observed to evolve gradually along former ice flow trajectories. Such evolutionary sequences can take the form of increases in the elongation of lineations (e.g., Ó Cofaigh et al., 2002, Stokes and Clark, 2002, Briner, 2007 and Stokes et al., 2013b) or a switch from ribs to drumlins (e.g., Fig. 8; Aario, 1977, Aylsworth and Shilts, 1989, Dyke et al., 1992, Knight et al., 1999 and Dunlop and Clark, 2006b). Both of these situations have been interpreted to indicate ice acceleration along flow, particularly in ice-stream settings (Aario, 1977, Hart, 1999, Cofaigh et al., 2002, Stokes and Clark, 2002,Briner, 2007 and Stokes et al., 2013a). Support for this is found from modern ice streams where drumlins have been observed beneath the onset zone of Rutford Ice Stream (King et al., 2007) and MSGLs farther down-flow where ice velocity is higher (King et al., 2009). Subglacial ribs can also be found in evolutionary sequences (Markgren and Lassila, 1980 and Greenwood and Kleman, 2010), suggesting a gradual change in boundary conditions, potentially induced by ice-flow properties. Thirdly, bedform morphology conforms strongly to glaciological variables at the ice-sheet scale (Greenwood and Clark, 2010). At the local scale, lithological, sedimentological, and topographical differences across a bedform field may correspond to morphological differences in the bedforms/landforms (e.g., Raukas and Tavast, 1994, Rattas and Piotrowski, 2003 and Greenwood and Clark, 2010). As deformable sediment can accommodate accelerated ice flow (Alley et al., 1986), the glaciological and sedimentological properties of an ice mass can be interlinked. For example, where there is a boundary between a hard bedrock and soft deformable sediment, there is often a corresponding increase in bedform elongation, interpreted to correspond to an acceleration in ice flow (e.g., Wellner et al., 2001 and Cofaigh et al., 2002). Therefore, given the complex interaction between ice flow and the nature of its underlying substrate, the exact influence of each of these two components can be rarely discerned.

The most commonly invoked glaciological control on subglacial bedform morphology is ice velocity. Ribbed moraines typically are found near ice divides or cold-based regions (Hättestrand, 1997 and Trommelen et al., 2014) or on ice-stream beds where they have been inferred to record deceleration immediately prior to ice stream shutdown (Stokes et al., 2008). This suggests slower ice flow is associated with their formation, even if the precise mechanisms are unknown. Drumlins are found farther away from ice divides and in the onset zones of palaeo (Stokes and Clark, 1999 and Anderson and Fretwell, 2008) and contemporary ice streams (King et al., 2007), whereas more elongate subglacial lineations (MSGLs) have been associated with ice streams (Clark, 1993, Clark, 1994,Stokes and Clark, 1999, King et al., 2009 and Spagnolo et al., 2014a). Ice velocity seems to be a primary influence upon bedform orientation and elongation within and between ribs and lineations. Flutes have a separate morphology to other subglacial bedforms (Fig. 5, Fig. 6, Fig. 7, Fig. 8 and Fig. 9). Thus, it is likely that they form under a separate set of boundary conditions or by a process different to the larger bedforms.

The size and shape variations of subglacial bedforms appear to be predominately explicable in terms of position within the ice sheet and by variations in glaciological properties. In contrast, bedform sedimentary composition is found to be incredibly varied (overview of ribbed moraine in Hättestrand and Kleman, 1999; drumlins reviewed inStokes et al., 2011, see references therein) and no simple match has been found between sediment properties and the variously named types. Sediment within bedforms can contain evidence of deposition (e.g., Dardis et al., 1984, Fisher and Shaw, 1992,Newman and Mickelson, 1994, Möller, 2010, Spagnolo et al., 2014b and Hopkins et al., 2016) and erosion (e.g., Boyce and Eyles, 1991, Kerr and Eyles, 2007 and Ó Cofaigh et al., 2013), as well as several phases of development (Newman et al., 1990 and Zelčs and Dreimanis, 1997). Thus, competing erosional and depositional processes are likely to occur at the ice-bed interface during bedform production (Hart, 1997). The bedform conundrum is that the order and simplicity in size and shape that we observe, and which tends to lead to presumptions of a unifying process to create them, clashes with the complexity and variation found in the sedimentary properties, which tends to lead to suggestions of a whole range of different processes. We do not solve this problem here, but suggest that our data could be used as a test for any hypothesis or model of bedform formation, noting that the size and shape of bedforms as described here (Fig. 5, Fig. 6,Fig. 7, Fig. 8 and Fig. 9) require explanation, as does their varied internal structure.

CONCLUSIONS

To test the hypothesis that subglacial bedforms comprise a size and shape continuum across the variously-named types, we analysed 96,900 measurements of subglacial bedform size and shape. The approach was to assess if bedforms vary continuously in size and shape, and thereby comprise a continuum, or whether gaps or jumps exist between the variously named types, indicating that they should be interpreted as discrete phenomena and therefore leading to the continuum hypothesis being rejected. Qualitative and quantitative (cluster) analyses of the data were employed to assess the degree to which separate populations exist. Although there is inherent subjectivity in any type of cluster analysis, be it visual or quantitative, the convergence of results from our analyses leads us to the following conclusions:

Subglacial flutes form a distinct cluster of narrower subglacial bedforms, clearly separable from other bedforms.

- Drumlins and MSGL are end members of a size and shape continuum of flow-aligned subglacial bedforms (subglacial lineations).
- Transverse subglacial bedforms belong to a subglacial rib continuum that spans ribbed moraine through to mega-ribbed moraine, with no justification found here for separate entities existing within this class.
- An underreported class of bedforms that are quasi-circular potentially bridge the more obvious continua of lineations and ribs suggesting that a single subglacial continuum spanning ribs, quasi-circular forms, and lineations might exist. To test this possibility, further work on quasi-circular forms is required to increase the sample size and examine the range of scales at which they exist.

ACKNOWLEDGEMENTS

The authors thank R.D. Larter, M. Ross, 2 anonymous reviewers, and the editor for their useful comments which improved this manuscript. J.C.E. thanks Kathy and Chris Denison for funding his PhD. This work was initiated and supported by a NERC grant (NE/D011175/1) to C.D.C. M.S. was supported by a NERC new investigator grant (NE/J004766/1) and C.R.S., a Philip Leverhulme Prize. S.L.G. acknowledges the University of Sheffield, the Swedish Research Council, and Linnaeus grants to Johan Kleman and the Bolin Centre for Climate Research. A.L.C.H acknowledges BGS NERC PhD studentship (NE/S/A/2004/12102). We thank Andrew Fowler and Richard Hindmarsh for fruitful discussions.

REFERENCES

1. Aario, R., 1977. Classification and terminology of morainic landforms in Finland. Boreas 6 (2), 87–100.
2. Allen, J.R.L., 1968. The nature and origin of bed-form hierarchies. Sedimentology 10 (3), 161–182.
3. Alley, R.B., Blankenship, D.D., Bentley, C.R., Rooney, S., 1986. Deformation of till beneath ice stream B, West Antarctica. Nature 322, 57–59.
4. Amos, C.L., King, E.L., 1984. Bedforms of the Canadian eastern seaboard: a comparisonwith global occurrences. Mar. Geol. 57 (1), 167–208.
5. Anderson, J.B., Fretwell, L.O., 2008. Geomorphology of the onset area of a paleo-ice stream, Marguerite Bay, Antarctic Peninsula. Earth Surf. Process. Landf. 33 (4), 503–512.
6. Andreassen, K., Laberg, J.S., Vorren, T.O., 2008. Seafloor geomorphology of the SW Barents Sea and its glaci-dynamic implications. Geomorphology 97 (1), 157–177.
7. Ashley, G.M., 1990. Classification of large-scale subaqueous bedforms: a new look at an old problem—SEPM bedforms and bedding structures. J. Sediment. Res. 60 (1), 160–172.
8. Aylsworth, J.M., Shilts, W.W., 1989. Bedforms of the Keewatin ice sheet, Canada. Sediment. Geol. 62 (2), 407–428.
9. Bouchard, M.A., 1989. Subglacial landforms and deposits in central and northern Quebec, Canada, with emphasis on Rogen moraines. Sediment. Geol. 62 (2), 293–308.
10. Boulton, G.S., 1976. The origin of glacially fluted surfaces—observations and theory. J. Glaciol. 17, 287–309.
11. Boulton, G.S., 1987. A theory of drumlin formation by subglacial sediment deformation. In: Menzies, J., Rose, J. (Eds.), Drumlin Symposium. Balkema, Rotterdam, pp. 25–80.
12. Boyce, J.I., Eyles, N., 1991. Drumlins carved by deforming till streams below the Laurentide ice sheet. Geology 19 (8), 787–790.
13. Briner, J.P., 2007. Supporting evidence from the New York drumlin field that elongate subglacial bedforms indicate fast ice flow. Boreas 36 (2), 143–147.
14. Brown, V.H., Stokes, C.R., O'Cofaigh, C., 2011. The glacial geomorphology of the North– West sector of the Laurentide Ice Sheet. J. Maps 7 (1), 409–428.
15. Carl, J.D., 1978. Ribbed moraine–drumlin transition belt, St. Lawrence Valley, New York. Geology 6 (9), 562–566.
16. Carling, P.A., 1999. Subaqueous gravel dunes. J. Sediment. Res. 69 (3).
17. Clark, C.D., 1993. Mega-scale glacial lineations and cross-cutting ice-flow landforms. Earth Surf. Process. Landf. 18 (1), 1–29.
18. Clark, C.D., 1994. Large-scale ice-moulding: a discussion of genesis and glaciological significance. Sediment. Geol. 91 (1), 253–268.
19. Clark, C.D., 1999. Glaciodynamic context of subglacial bedform generation and preservation. Ann. Glaciol. 28 (1), 23–32.
20. Clark, C.D., Hughes, A.L., Greenwood, S.L., Spagnolo, M., Ng, F.S., 2009. Size and shape characteristics of drumlins, derived from a large sample, and associated scaling laws. Quat. Sci. Rev. 28 (7), 677–692.
21. Costello, W.R., Southard, J.B., 1981. Flume experiments on lower-flow-regime bed forms in coarse sand. J. Sediment. Res. 51 (3).
22. Cutts, J.A., Smith, R.S.U., 1973. Eolian deposits and dunes on Mars. J. Geophys. Res. 78 (20), 4139–4154.
23. Dardis, G.F., McCabe, A.M., Mitchell, W.I., 1984. Characteristics and origins of lee-side stratification sequences in late Pleistocene drumlins, Northern Ireland. Earth Surf. Process. Landf. 9 (5), 409–424.
24. Dionne, J.C., 1987. Tadpole rock (rock drumlin): a glacial stream moulded form. In: Menzies, J., Rose, J. (Eds.), Drumlin Symposium. Balkema, Rotterdam, pp. 149–159.
25. Dowling, T.P.F., Spagnolo, M., Möller, P., 2015. Morphometry and core type of streamlined bedforms in southern Sweden from high resolution LiDAR. Geomorphology 236, 54–63.
26. Dunlop, P., Clark, C.D., 2006a. Distribution of ribbed moraine in the Lac Naocokane Region, Central Québec, Canada. J. Maps 2 (1), 59–70.

27. Dunlop, P., Clark, C.D., 2006b. The morphological characteristics of ribbed moraine. Quat. Sci. Rev. 25 (13), 1668–1691.
28. Dyke, A.S., Morris, T.F., Green, D.E.C., England, J., 1992. Quaternary geology of Prince of Wales Island, Arctic Canada. Geol. Surv. Can. Mem. 433, 142.
29. Ellwood, J.M., Evans, P.D., Wilson, I.G., 1975. Small scale aeolian bedforms. J. Sediment. Res. 45 (2), 554–561.
30. Engelhardt, H., Kamb, B., 1998. Basal sliding of ice stream B, West Antarctica. J. Glaciol. 44 (147), 223–230.
31. Ester, M., Kriegel, H.P., Sander, J., Xu, X., 1996. A density-based algorithm for discovering clusters in large spatial databases with noise. Kdd 96 (34), 226–231.
32. Estivill-Castro, V., 2002. Why so many clustering algorithms: a position paper. ACM SIGKDD Explorations Newsletter 4 (1), 65–75.
33. Evans, D.J., Twigg, D.R., 2002. The active temperate glacial landsystem: a model based on Breiðamerkurjökull and Fjallsjökull, Iceland. Quat. Sci. Rev. 21 (20), 2143–2177.
34. Fisher, T.G., Shaw, J., 1992. A depositional model for Rogen moraine, with examples from the Avalon Peninsula, Newfoundland. Can. J. Earth Sci. 29 (4), 669–686.
35. Greenwood, S.L., Clark, C.D., 2008. Subglacial bedforms of the Irish ice sheet. J. Maps 4 (1), 332–357.
36. Greenwood, S.L., Clark, C.D., 2010. The sensitivity of subglacial bedform size and distribution to substrate lithological control. Sediment. Geol. 232 (3), 130–144.
37. Greenwood, S.L., Kleman, J., 2010. Glacial landforms of extreme size in the Keewatin sector of the Laurentide Ice Sheet. Quat. Sci. Rev. 29 (15), 1894–1910.
38. Hart, J.K., 1995. Drumlin formation in southern Anglesey and Arvon, northwest Wales. J. Quat. Sci. 10 (1), 3–14.
39. Hart, J.K., 1997. The relationship between drumlins and other forms of subglacial glaciotectonic deformation. Quat. Sci. Rev. 16 (1), 93–107.
40. Hart, J.K., 1999. Identifying fast ice flow from landform assemblages in the geological record: a discussion. Ann. Glaciol. 28 (1), 59–66.
41. Hättestrand, C., 1997. Ribbed moraines in Sweden—distribution pattern and palaeoglaciological implications. Sediment. Geol. 111 (1), 41–56.
42. Hättestrand, C., Kleman, J., 1999. Ribbed moraine formation. Quat. Sci. Rev. 18 (1), 43–61.
43. Hess, D.P., Briner, J.P., 2009. Geospatial analysis of controls on subglacial bedform morphometry in the New York Drumlin Field—implications for Laurentide Ice Sheet dynamics. Earth Surf. Process. Landf. 34 (8), 1126–1135.
44. Hill, A.R., 1973. The distribution of drumlins in County Down, Ireland. Ann. Assoc. Am. Geogr. 63 (2), 226–240.
45. Hopkins, N.R., Evenson, E.B., Kodama, K.P., Kozlowski, A., 2016. An anisotropy of magnetic susceptibility (AMS) investigation of the till fabric of drumlins: support for an accretionary origin. Boreas 45 (1), 100–108.
46. Hughes, A.L., Clark, C.D., Jordan, C.J., 2010. Subglacial bedforms of the last British Ice Sheet. J. Maps 6 (1), 543–563.
47. Jackson, R.G., 1975. Hierarchical attributes and a unifying model of bed forms composed of cohesionless material and produced by shearing flow. Geol. Soc. Am. Bull. 86 (11), 1523–1533.
48. Jain, A.K., 2010. Data clustering: 50 years beyond K-means. Pattern Recogn. Lett. 31 (8), 651–666.
49. Johnson, M.D., Schomacker, A., Benediktsson, Í.Ö., Geiger, A.J., Ferguson, A., Ingólfsson, Ó., 2010. Active drumlin field revealed at the margin of Múlajökull, Iceland: a surge-type glacier. Geology 38 (10), 943–946.
50. Kargel, J.S., Strom, R.G., 1992. Ancient glaciation on Mars. Geology 20 (1), 3–7.
51. Kerr, M., Eyles, N., 2007. Origin of drumlins on the floor of Lake Ontario and in upper New York State. Sediment. Geol. 193 (1), 7–20.
52. King, E.C., Woodward, J., Smith, A.M., 2007. Seismic and radar observations of subglacial bed forms beneath the onset zone of Rutford Ice Stream, Antarctica. J. Glaciol. 53 (183), 665–672.
53. King, E.C., Hindmarsh, R.C., Stokes, C.R., 2009. Formation of mega-scale glacial lineations observed beneath a West Antarctic ice stream. Nat. Geosci. 2 (8), 585–588.
54. Kjær, K.H., Larsen, E., van der Meer, J., Ingólfsson, Ó., Krüger, J., Benediktsson, Í.Ö., Knudsen, C.G., Schomacker, A., 2006. Subglacial decoupling at the sediment/bedrock interface: a new mechanism for rapid flowing ice. Quat. Sci. Rev. 25 (21), 2704–2712.
55. Klages, J.P., Kuhn, G., Hillenbrand, C.D., Graham, A.G.C., Smith, J.A., Larter, R.D., Gohl, K., 2013. First geomorphological record and glacial history of an inter-ice stream ridge on the West Antarctic continental shelf. Quat. Sci. Rev. 61, 47–61.
56. Kleman, J., Glasser, N.F., 2007. The subglacial thermal organisation (STO) of ice sheets. Quat. Sci. Rev. 26 (5), 585–597.
57. Knight, J., McCarron, S.G., McCabe, A.M., 1999. Landform modification by palaeo-ice streams in east-central Ireland. Ann. Glaciol. 28 (1), 161–167.
58. Lancaster, N., 1988. Controls of eolian dune size and spacing. Geology 16 (11), 972–975. Lancaster, N., 2013. Geomorphology of Desert Dunes. Routledge, p. 159.
59. Lane, T.P., Roberts, D.H., Rea, B.R., Ó Cofaigh, C., Vieli, A., 2015. Controls on bedrock bedform development beneath the Uummannaq Ice Stream onset zone, West Greenland. Geomorphology 231, 301–313.
60. Larter, R.D., Graham, A.G., Gohl, K., Kuhn, G., Hillenbrand, C.D., Smith, J.A., Deen, T.J., Livermore, R.A., Schenke, H.W., 2009. Subglacial bedforms reveal complex basal regime in a zone of paleo-ice stream convergence, Amundsen Sea embayment, West Antarctica. Geology 37 (5), 411–414.

61. Maclachlan, J.C., Eyles, C.H., 2013. Quantitative geomorphological analysis of drumlins in the Peterborough drumlin field, Ontario, Canada. Geografiska Annaler: Series A, Physical Geography 95 (2), 125–144.

62. Markgren, M., Lassila, M., 1980. Problems of moraine morphology: Rogen moraine and Blattnick moraine. Boreas 9 (4), 271–274.

63. McHenry, M., Dunlop, P., 2015. The subglacial imprint of the last Newfoundland Ice Sheet, Canada. J. Maps 1–22 (ahead-of-print).

64. Möller, P., 2010. Melt-out till and ribbed moraine formation, a case study from south Sweden. Sediment. Geol. 232 (3), 161–180.

65. Napieralski, J., Nalepa, N., 2010. The application of control charts to determine the effect of grid cell size on landform morphometry. Comput. Geosci. 36 (2), 222–230.

66. Newman, W.A., Mickelson, D.M., 1994. Genesis of Boston Harbor drumlins, Massachusetts. Sediment. Geol. 91 (1), 333–343.

67. Newman, W.A., Berg, R.C., Rosen, P.S., Glass, H.D., 1990. Pleistocene stratigraphy of the Boston Harbor drumlins, Massachusetts. Quat. Res. 34 (2), 148–159.

68. Ó Cofaigh, C., Pudsey, C.J., Dowdeswell, J.A., Morris, P., 2002. Evolution of subglacial bedforms along a paleo-ice stream, Antarctic Peninsula continental shelf. Geophys. Res. Lett. 29 (8), 41-1.

69. Ó Cofaigh, C., Stokes, C.R., Lian, O.B., Clark, C.D., Tulacyzk, S., 2013. Formation of megascale glacial lineations on the Dubawnt Lake Ice Stream bed: 2. Sedimentology and stratigraphy. Quat. Sci. Rev. 77, 210–227.

70. Ottesen, D., Dowdeswell, J.A., Rise, L., 2005. Submarine landforms and the reconstruction of fast-flowing ice streams within a large Quaternary ice sheet: the 2500-km-long Norwegian–Svalbard margin (57–80 N). Geol. Soc. Am. Bull. 117 (7–8), 1033–1050.

71. Punkari, M., 1984. The relations between glacial dynamics and the tills in the eastern part of the Baltic shield. In: Königsson, L.-K. (Ed.), Ten Years of Nordic Till Research, Striae. 20, pp. 49–54.

72. Radebaugh, J., Lorenz, R.D., Lunine, J.I., Wall, S.D., Boubin, G., Reffet, E., Kirk, R.L., Lopes, R.M., Stofan, E.R., Soderblom, L., Allison, M., Janssen, M., Paillou, P., Callahan, P.,

73. Spencer, C., The Cassini Radar Team, 2008. Dunes on Titan observed by Cassini RADAR. Icarus 194 (2), 690–703.

74. Rattas, M., Piotrowski, J.A., 2003. Influence of bedrock permeability and till grain size on the formation of the Saadjärve drumlin field, Estonia, under an east-Baltic Weichselian ice stream. Boreas 32 (1), 167–177.

75. Raukas, A., Tavast, E., 1994. Drumlin location as a response to bedrock topography on the southeastern slope of the Fennoscandian Shield. Sediment. Geol. 91 (1), 373–382.

76. Reffet, E., du Pont, S.C., Hersen, P., Douady, S., 2010. Formation and stability of transverse and longitudinal sand dunes. Geology 38 (6), 491–494.

77. Rooney, S.T., Blankenship, D.D., Alley, R.B., Bentley, C.R., 1987. Till beneath ice stream B: 2. Structure and continuity. J. Geophys. Res. Solid Earth (1978–2012) 92 (B9), 8913–8920.

78. Rose, J., 1987. Drumlins as part of glacier bedform continuum. In: Menzies, J., Rose, J. (Eds.), Drumlin Symposium. Balkema, Rotterdamn, pp. 103–118.

79. Rose, J., Letzer, J.M., 1977. Superimposed drumlins. J. Glaciol. 18 (80), 471–480.

80. Rubin, D.M., Ikeda, H., 1990. Flume experiments on the alignment of transverse, oblique, and longitudinal dunes in directionally varying flows. Sedimentology 37 (4), 673–684.

81. Saha, K., Wells, N.A., Munro-Stasiuk, M., 2011. An object-oriented approach to automated landform mapping: a case study of drumlins. Comput. Geosci. 37 (9), 1324–1336.

82. Smith, M.J., Clark, C.D., 2005. Methods for the visualization of digital elevation models for landform mapping. Earth Surf. Process. Landf. 30 (7), 885–900.

83. Smith, M.J., Wise, S.M., 2007. Problems of bias in mapping linear landforms from satellite imagery. Int. J. Appl. Earth Obs. Geoinf. 9 (1), 65–78.

84. Smith, A.M., Murray, T., Nicholls, K.W., Makinson, K., Ađalgeirsdóttir, G., Behar, A.E., Vaughan, D.G., 2007. Rapid erosion, drumlin formation, and changing hydrology beneath an Antarctic ice stream. Geology 35 (2), 127–130.

85. Spagnolo, M., Clark, C.D., Hughes, A.L., Dunlop, P., Stokes, C.R., 2010. The planar shape of drumlins. Sediment. Geol. 232 (3), 119–129.

86. Spagnolo, M., Clark, C.D., Ely, J.C., Stokes, C.R., Anderson, J.B., Andreassen, K., Graham, A.G.C., King, E.C., 2014a. Size, shape and spatial arrangement of mega-scale glacial lineations from a large and diverse dataset. Earth Surf. Process. Landf. 39 (11), 1432–1448.

87. Spagnolo, M., King, E.C., Ashmore, D.W., Rea, B.R., Ely, J.C., Clark, C.D., 2014b. Looking through drumlins: testing the application of ground-penetrating radar. J. Glaciol. 60 (224), 1126–1134.

88. Stokes, C.R., Clark, C.D., 1999. Geomorphological criteria for identifying Pleistocene ice streams. Ann. Glaciol. 28 (1), 67–74.

89. Stokes, C.R., Clark, C.D., 2002. Are long subglacial bedforms indicative of fast ice flow? Boreas 31 (3), 239–249.

90. Stokes, C.R., Lian, O.B., Tulaczyk, S., Clark, C.D., 2008. Superimposition of ribbed moraines on a palaeo-ice-stream bed: implications for ice stream dynamics and shutdown. Earth Surf. Process. Landf. 33 (4), 593–609.

91. Stokes, C.R., Spagnolo, M., Clark, C.D., 2011. The composition and internal structure of drumlins: complexity, commonality, and implications for a unifying theory of their formation. Earth Sci. Rev. 107 (3), 398–422.

92. Stokes, C.R., Fowler, A.C., Clark, C.D., Hindmarsh, R.C., Spagnolo, M., 2013a. The instability theory of drumlin formation and its explanation of their varied composition and internal structure. Quat. Sci. Rev. 62, 77–96.
93. Stokes, C.R., Spagnolo, M., Clark, C.D., Ó Cofaigh, C., Lian, O.B., Dunstone, R.B., 2013b. Formation of mega-scale glacial lineations on the Dubawnt Lake Ice Stream bed: 1. Size, shape and spacing from a large remote sensing dataset. Quat. Sci. Rev. 77, 190–209.
94. Storrar, R., Stokes, C.R., 2007. A glacial geomorphological map of Victoria Island, Canadian Arctic. J. Maps 3 (1), 191–210.
95. Trommelen, M., Ross, M., 2010. Subglacial landforms in northern Manitoba, Canada, based on remote sensing data. J. Maps 6 (1), 618–638.
96. Trommelen, M.S., Ross, M., Ismail, A., 2014. Ribbed moraines in northern Manitoba, Canada: characteristics and preservation as part of a subglacial bed mosaic near the core regions of ice sheets. Quat. Sci. Rev. 87, 135–155.
97. Waller, R.I., Van, D.A.T., Knudsen, Ó., 2008. Subglacial bedforms and conditions associated with the 1991 surge of Skeiðarárjökull, Iceland. Boreas 37 (2), 179–194.
98. Wellner, J.S., Lowe, A.L., Shipp, S.S., Anderson, J.B., 2001. Distribution of glacial geomorphic features on the Antarctic continental shelf and correlation with substrate: implications
99. for ice behavior. J. Glaciol. 47 (158), 397–411. Wilson, I.G., 1973. Ergs. Sediment. Geol. 10 (2), 77–106.
100. Winsborrow, M., Clark, C.D., Stokes, C.R., 2004. Ice streams of the Laurentide ice sheet. Géog. Phys. Quatern. 58 (2–3), 269–280.
101. Worman, S.L., Murray, A.B., Littlewood, R., Andreotti, B., Claudin, P., 2013. Modeling emergent large-scale structures of barchan dune fields. Geology 41 (10), 1059–1062.
102. Zelčs, V., Dreimanis, A., 1997. Morphology, internal structure and genesis of the Burtnieks drumlin field, Northern Vidzeme, Latvia. Sediment. Geol. 111 (1), 73–90.

INDEX

S

T

U

W

X